physica status solidi (a)

ISSN 0031-8965 * VOL. 49 · NO. 1 · SEPTEMBER 1978

AKADEMIE-VERLAG · BERLIN

Classification Scheme

1. Structure of Crystalline Solids
   1.1 Perfectly Periodic Structure
   1.2 Solid-State Phase Transformations
   1.3 Alloys. Metallurgy
   1.4 Microstructure (Magnetic Domains See 18; Ferroelectric Domains See 14.4.1)
   1.5 Films
   1.6 Surfaces

2. Non-Crystalline State

3. Crystal Growth

4. Bonding Properties

5. Mössbauer Spectroscopy

6. Lattice Dynamics. Phonons

7. Acoustic Properties

8. Thermal Properties

9. Diffusion

10. Defect Properties (Irradiation Defects See 11)
    10.1 Metals
    10.2 Non-Metals

11. Irradiation Effects (X-Ray Diffraction Investigations See 1 and 10)

12. Mechanical Properties (Plastic Deformations See 10)
    12.1 Metals
    12.2 Non-Metals

13. Electron States
    13.1 Band Structure
    13.2 Fermi Surfaces
    13.3 Surface and Interface States
    13.4 Impurity and Defect States
    13.5 Elementary Excitations (Phonons See 6)
        13.5.1 Excitons
        13.5.2 Plasmons
        13.5.3 Polarons
        13.5.4 Magnons

14. Electrical Properties. Transport Phenomena
    14.1 Metals. Semi-Metals
    14.2 Superconductivity. Superconducting Materials and Devices
    14.3 Semiconductors
        14.3.1 Films
        14.3.2 Surfaces and Interfaces
        14.3.3 Device. Junctions (Contact Problems See 14.3.4)
        14.3.4 High-Field Phenomena, Space-Charge Effects, Inhomogeneities, Injected Carriers (Electroluminescense See 20.3; Junctions See 14.3.3)
    14.4 Dielectrics
        14.4.1 Ferroelectrics

15. Thermoelectric and Thermomagnetic Properties

16. Photoconductivity. Photovoltaic Effects

17. Emission of Electrons and Ions
    17.1 Field Emission Microscope Investigations

18. Magnetic Properties
    18.1 Paramagnetic Properties
    18.2 Ferromagnetic Properties
        18.2.1 Ferromagnetic Films
    18.3 Ferrimagnetic Properties
    18.4 Antiferromagnetic Properties

(Continued on cover three)

19. Magnetic Resonance

20. Optical Properties
    20.1 Spectra. Optical Constants (X-Ray Spectra See 20)
    20.2 Lasers
    20.3 Luminescence (X-Ray Spectra See 20)

21. Metals. Alloys
    21.1 Transition Metals (Group Sc to Zn) and Their Alloys
        21.1.1 Fe and Fe Alloys
    21.2 Alkali Metals
    21.3 Alkali-Earth Metals
    21.4 Rare-Earth Metals
    21.5 Actinides
    21.6 Noble Metals
    21.7 Semi-Metals (Semiconducting Alloys 22.7)

22. Semiconductors and Ionic Crystals
    22.1 Elements
        22.1.1 Germanium
        22.1.2 Silicon
        22.1.3 Group VI Elements
    22.2 III–V Compounds
        22.2.1 Arsenides
        22.2.2 Phosphides
        22.2.3 Antimonides
        22.2.4 Mixed Crystals
    22.3 IV–IV, II–V, and III–VI Compounds
        22.3.1 Mixed Crystals
    22.4 II–VI Compounds
        22.4.1 Sulphides
        22.4.2 Selenides
        22.4.3 Tellurides
        22.4.4 Mixed Crystals
    22.5 Halides
        22.5.1 Silver Halides
        22.5.2 Alkali Halides
        22.5.3 Alkali-Earth Halides
        22.5.4 Mixed Crystals
    22.6 Simple Oxides (CdO, ZnO See 22.4)
        22.6.1 Mixed Crystals
    22.7 Semiconducting Intermetallic Compounds
    22.8 Tertiary and Higher Compounds
        22.8.1 Salts of Oxy-Acids (Carbonates, Phosphates, Silicates Including Mica etc.)
        22.8.2 Spinels, Garnets, Ferrites
    22.9 Organic Semiconductors

23. Solidified Gases

Attention

The preceding classification scheme is closely akin to the scheme used by Physics Abstracts to classify the various fields of solid state physics. A paper will either be classified under one or more of the subsections or, if none of these are relevant, it will just appear under the general section heading (i. e. sections and subsections are equivalent in this sense).

The *subject classification* given at the head of each paper begins with the main subject matter of the paper, and this is followed, if necessary, by figures referring to subsidiary subject matter and the type of solids investigated.

The papers published in this issue are listed in the order of their subject classification at the end of the table of contents.

Author Index

physica status solidi (a)

applied research

Volume 49 · Number 1 · Pages 1 to 420, K1 to K100, and A1 to A8

September 16, 1978

PSSA 49(1) 1—420, K1—K100, A1—A8 (1978)

ISSN 0031-8965

AKADEMIE-VERLAG · BERLIN

Subscriptions and orders for single copies should be directed

in the DDR to the Postzeitungsvertrieb, to a book-shop or to the Akademie-Verlag, DDR-108 Berlin, Leipziger Straße 3 – 4;

in the other socialist countries to a book-shop for foreign-language literature or to the competent news-distributing agency;

in the BRD and BERLIN (WEST) to a book-shop or to the wholesale distributing agency Kunst und Wissen, Erich Bieber, 7 Stuttgart 1, Wilhelmstraße 4 – 6;

in ÖSTERREICH to the Globus-Buchvertrieb, 1201 Wien, Höchstädtplatz 3;

in the British Commonwealth and USA to Academic Press Inc. (London) Limited, 24 – 28 Oval Road, London N. W. 1 or to Academic Press Inc., 111, Fifth Avenue, New York 10003, USA;

in other countries to the international book- and journal-selling trade, to Buchexport, Volkseigener Außenhandelsbetrieb der Deutschen Demokratischen Republik, DDR-701 Leipzig, Postfach 160, or to the Akademie-Verlag, DDR-108 Berlin, Leipziger Straße 3 – 4.

Editorial Note:

"physica status solidi (a)" undertakes that an original paper *accepted for publication* before the 23rd of any month will be published within 50 days of this date unless the author requests a postponement. In special cases there may be some delay between *receipt* and *acceptance* of a paper due to the review and, if necessary, revision of the paper.

Schriftleiter und verantwortlich für den Inhalt: Professor Dr. Dr. h. c. P. Görlich, DDR 102-Berlin, Neue Schönhauser Str. 20 bzw. DDR-69 Jena, Schillbachstr. 24. Verlag: Akademie-Verlag, DDR-108 Berlin, Leipziger Straße 3 – 4; Fernruf: 2236221 und 2236229; Telex-Nr.: 114420; Bank: Staatsbank der DDR, Berlin, Kto.-Nr.: 6836-26-20712. Chefredakteur: Dr. H.-J. Hänsch. Redaktionskollegium: Prof. Dr. E. Gutsche, Dr. H.-J. Hänsch, Dr. H. Lange, Dr. S. Oberländer. Anschrift der Redaktion: DDR 102-Berlin, Neue Schönhauser Str. 20; Fernruf: 2823380. Veröffentlicht unter der Lizenznummer 1620 des Presseamtes beim Vorsitzenden des Ministerrates der Deutschen Demokratischen Republik. Gesamtherstellung: VEB Druckerei „Thomas Müntzer", DDR-582 Bad Langensalza. Erscheinungsweise: Die Zeitschrift „physica status solidi (a)" erscheint jeweils am 16. eines jeden Monats. Jährlich erscheinen 6 Bände zu je 2 Heften. Bezugspreis eines Bandes 160,– M zuzüglich Versandspesen (Preis für die DDR 120,– M). Bestellnummer dieses Bandes: 1085/49.

AN (EDV) 20735

PROFESSOR Dr. rer. nat. FRITZ STÖCKMANN

IN HONOUR OF

HIS 60th BIRTHDAY

On the 19th of October Professor Fritz Stöckmann celebrates his 60th birthday. He is Professor at the Institute of Applied Physics at the University of Karlsruhe. Professor Stöckmann served as Head of the Physics Department and Dean of the Faculty of Science. In the scientific community outside the university he was for several years chairman of the Physics Commitee of the German Science Foundation (Deutsche Forschungsgemeinschaft). In 1960 Professor Stöckmann has essentially contributed to the preparation and foundation of physica status solidi. He is member of the Board of Editors of physica status solidi (b) since its foundation and has permanently encouraged the further development of the journal during these 20 years. His 60th birthday is an occasion of thanking him for his unresting activity and of wishing him health and creativity for the future.

Professor Stöckmann attended the Universities of Göttingen, Freiburg, and München, obtaining his Dr. rer. nat. 1942 in the Physical Institute in Göttingen. After becoming Privatdozent in 1950, he moved to Darmstadt and, after a stay of some months at the Purdue University in Lafayette, Indiana, he was appointed full Professor of Applied Physics at the University of Karlsruhe in 1959.

Professor Stöckmann began his research with investigations of the photoconductivity in various semiconductors, such as germanium and zinc oxide. In a long and successful career, Stöckmann has made essential and basic contributions to many aspects of photoeffects in solids. The role of injection of majority and minority charge carriers from contacts and its influence on the performance of a photoconductor has thoroughly been followed up by him. Furthermore, he elucidated the importance of the many time constants, such as life time, trapping time, and response time. He has remained ever faithful to his field, and during recent years he has achieved a complete synopsis of the relation of photoconduction and the presence of crystal imperfections, such as impurities or defects.

Photoconductors are his field of research. They are, however, but one of his many areas of interest in physics. He is particularly inclined towards optics and astronomy. Due to his profound knowledge of physics and his gift of presentation he has given many of his students and colleagues insight into problems outside his realm of research with a clarity they could hardly hope to obtain from some specialists in these fields.

The combination of his modest working style with his excellent achievements is respected and admired by friends and colleagues. His sympathy in human and public affairs, his advice in academic matters, and his competent and experienced judgement, together with his promptness for help and support in scientific problems, might for many more years enrich the lives of all those who have the pleasure and the privilege of living and working with him.

PROFESSOR Dr. rer. nat. FRITZ STÖCKMANN

IN HONOUR OF

HIS 60th BIRTHDAY

On the 19th of October Professor Fritz Stöckmann celebrates his 60th birthday. He is Professor at the Institute of Applied Physics at the University of Karlsruhe. Professor Stöckmann served as Head of the Physics Department and Dean of the Faculty of Science. In the scientific community outside the university he was for several years chairman of the Physics Commitee of the German Science Foundation (Deutsche Forschungsgemeinschaft). In 1960 Professor Stöckmann has essentially contributed to the preparation and foundation of physica status solidi. He is member of the Board of Editors of physica status solidi (b) since its foundation and has permanently encouraged the further development of the journal during these 20 years. His 60th birthday is an occasion of thanking him for his unresting activity and of wishing him health and creativity for the future.

Professor Stöckmann attended the Universities of Göttingen, Freiburg, and München, obtaining his Dr. rer. nat. 1942 in the Physical Institute in Göttingen. After becoming Privatdozent in 1950, he moved to Darmstadt and, after a stay of some months at the Purdue University in Lafayette, Indiana, he was appointed full Professor of Applied Physics at the University of Karlsruhe in 1959.

Professor Stöckmann began his research with investigations of the photoconductivity in various semiconductors, such as germanium and zinc oxide. In a long and successful career, Stöckmann has made essential and basic contributions to many aspects of photoeffects in solids. The role of injection of majority and minority charge carriers from contacts and its influence on the performance of a photoconductor has thoroughly been followed up by him. Furthermore, he elucidated the importance of the many time constants, such as life time, trapping time, and response time. He has remained ever faithful to his field, and during recent years he has achieved a complete synopsis of the relation of photoconduction and the presence of crystal imperfections, such as impurities or defects.

Photoconductors are his field of research. They are, however, but one of his many areas of interest in physics. He is particularly inclined towards optics and astronomy. Due to his profound knowledge of physics and his gift of presentation he has given many of his students and colleagues insight into problems outside his realm of research with a clarity they could hardly hope to obtain from some specialists in these fields.

The combination of his modest working style with his excellent achievements is respected and admired by friends and colleagues. His sympathy in human and public affairs, his advice in academic matters, and his competent and experienced judgement, together with his promptness for help and support in scientific problems, might for many more years enrich the lives of all those who have the pleasure and the privilege of living and working with him.

Contents

Short Notes

Systematic List

Contents of Volume 49 Continued on Page 423

phys. stat. sol. (a) 49, 13 (1978)

Subject classification: 16; 14.3.3; 22; 22.4.1

University of Delaware and SES, Inc., Newark[1])

The Physics of the Photovoltaic Effect

By

K. W. Böer

In honour of Prof. Dr. F. Stöckmann's 60th birthday

The main parts of a photovoltaic cell and their function are described. Photovoltaic cells are then classified in respect to their operation. The operation of typical cells is analyzed with the goal to obtain current–voltage characteristics in a self-consistent physical model. This is achieved by connecting the emitter diffusion current with the voltage drop in the junction by a doubly acting boundary condition, the electron density at the emitter–junction interface. The consequently obtained characteristics have near the open circuit voltage (Boltzmann range), the form of the commonly used shifted diode characteristic, however, with parameters in substantially improved agreement with the experiment. Outside the Boltzmann range, integration of transport and Poisson equation yields the shape of the characteristics. This theory is then extended to include photovoltaic cells with dominant interface recombination. Such interface recombination causes mostly a lowering of the open circuit voltage. The agreement between theory and experiment is surprisingly good for CdS/Cu_2S solar cells.

Die Hauptteile einer photovoltaischen Zelle und ihre Funktion werden beschrieben. Die Zellen werden dann hinsichtlich ihrer Wirkungsweise klassifiziert. Die Wirkungsweise typischer Zellen wird analysiert, um Strom–Spannungs-Kennlinien in einem selbstkonsistenten physikalischen Modell zu erhalten. Das wird erreicht, indem der Emitter-Diffusionsstrom mit dem Spannungsabfall im Übergang durch eine doppelt agierende Randbedingung, nämlich die Elektronendichte an der Grenzschicht zwischen beiden Gebieten, verbunden wird. Die Charakteristiken haben im Bereich der Leerlaufspannung die übliche Form einer um den Kurzschlußstrom verschobenen Diodencharakteristik, jedoch mit Parametern, die erheblich besser mit dem Experiment übereinstimmen. Diese Theorie wird dann für Zellen mit dominierender Zwischenschichtrekombination erweitert. Solche Rekombination verursacht im wesentlichen eine Verringerung der Leerlaufspannung. Die erzielte Übereinstimmung zwischen Theorie und Experiment für CdS/Cu_2S-Solarzellen ist erstaunlich gut.

1. Introduction

It is generally accepted that minority carriers, generated by light and diffusing to the junction, are responsible for the current in a photovoltaic cell, and that the voltage drop across the junction accounts for the desired fraction of the voltage drop across the cell.

However, the complexity of the physical model, involving the transport equations of both carriers, and the Poisson equation with explicit spatial coordinate has caused most researchers in the past to shy away from a systematic discussion of the physics of the photovoltaic cell, except for a few specific discussions of isolated problems. Such discussions were mainly related to the minority carrier diffusion, collection efficiency, quasi-Fermi level analysis with the aim to estimate the open-circuit voltage, and certain recombination phenomena in the junction region which seem to influence the current–voltage characteristics [1 to 4]. Only recently more effort was devoted to discuss the spatial distribution of space charge, field, and potential in the junction region [5, 6].

---

[1]) Newark, Delaware 19711, USA.

None of the early discussions was used to quantitatively develop a current–voltage characteristic. For this, usually an electrotechnical model was used, describing the solar cell as a diode in parallel with a current generator, having the effect of a parallel shift of the diode characteristic by the short-circuit current. However, mathematically such parallel shift is not permissible since the solutions of a nonlinear system of differential equation do not transform linearly [6].

Moreover, the mostly used adjustment parameters (to obtain improved agreement with the experiment), when explained as shunt and series resistance, and as exponential correction factor, indicating the type of junction recombination, may be, in fact, misleading since similar curve shapes may be obtained for entirely different reasons [7].

Therefore, an attempt was made to develop a self-consistent physical theory of the current–voltage characteristics [5, 8]. Necessary simplification (without distorting any of the prominent features of the characteristics) are most easily introduced for an *abrupt frontwall* heterojunction cell. However, the following discussion is not restricted to such a frontwall heterojunction and may be applied (with proper caution) to homojunctions.

In this paper we will first identify the main parts of a photovoltaic cell and define their operational properties. This will lead to a general classification of different types of cells. We will then describe quantitatively the operation of each part and finally develop the current–voltage characteristics by joining the different parts of the cell with major attention to the pertinent boundary conditions between these parts.

2. Main Parts of a Photovoltaic Cell

A photovoltaic cell has two active parts, the emitter and the junction, and three classes of passive parts, the semiconductive assist-layers (as defined below), the electrodes, and the window/encapsulant. Dependent on the type of solar cell (front, backwall, Schottky junction, etc.), one or more of these layers may be combined with others or may be omitted. The following listing gives a general overview of these layers and their main purposes in the operation of the cell.

2.1 Active parts

The *emitter* absorbs most of the light, generates minority carriers, and permits diffusion[2]) of these carriers to the junction. The magnitude and sign of the diffusion current is determined by the minority carrier density at the emitter-to-junction interface. The majority current in the emitter is carried by drift.

The *junction* provides almost all the voltage drop[3]) of the cell. In the junction, the minority carriers from the emitter are effectively separated from the emitter and, with sufficient reverse bias, can only return to the emitter through the external circuit.

An *interface layer* may exist between emitter and junction (or may be a small part thereof) in which the defect density is substantially larger[4]) than in the adjacent layers, and major carrier recombination occurs. This recombination influences potential and current distribution.

2.2 Semiconductive assist-layers

A semiconductive *collector* layer may extend from the junction and permit the junction to widen with reverse bias. This layer must be wide enough to prevent junction interference with the electrode (important for polycrystalline layers).

[2]) Diffusion rather than drift is the main cause of the photo-e.m.f.

[3]) The voltage drop which occurs in other parts of the cell contributes to the undesired series resistance.

[4]) Such an interface layer may be caused, e.g., by a lattice mismatch in heterojunction or by doping compensation causing substantial donor–acceptor pair formation.

A *surface passivation layer* on the emitter, opposite to the junction, is occasionally applied to reduce surface recombination. The passivation layer may be created by heavy doping (e.g. p^+ on p in Si homojunction cells) or, preferably, by heterojunction windows (GaAlAs on GaAs or Cu_2O on Cu_2S).

A *current-assist* conductive and transparent *top coating* in order to reduce lateral current densities, hence to permit lesser doping of the front (active) semiconductive layer.

2.3 Electrodes

The *grid* (front) *electrode* permits sufficient light penetration and allows current collection without major series resistance limitation. Proper gridline spacing and geometry and intimate gridline contact is required for avoiding excessive lateral current densities.

The *back electrode* is usually a massive metal layer with "ohmic" contact properties, often acting as part of the encapsulation. Multi-metal layers may be used to fulfil mechanical, chemical, and electrical needs.

2.4 Window encapsulant

Single or multi-layer *optical coatings* may fulfil a multitude of needs, such as mechanical support of grids, antireflecting coating, and hermetic seal to block penetration of active gases to the semiconductor and/or metal surface, hence preventing photochemical surface reactions and corrosion.

3. Types of Photovoltaic Cells

Dependent whether the emitter faces the light or is covered by the junction (and collector) one distinguishes *front* or *backwall cells.*

3.1 Frontwall cells

The top layer of a frontwall solar cell absorbs most of the light and acts as emitter. This has the advantage that the light is absorbed directly after penetrating through the window and is not partially lost in other semiconducting layers (collector and junction) as is the case in backwall cells.

The disadvantage of frontwall cells is the need for a higher majority carrier density to carry the *lateral* current to the grid; this tends to increase the recombination and to narrow the band gap, hence to lower short-circuit current and open-circuit voltage. It also creates most of the minority carriers close to the outer surface, hence frontwall cells are more sensitive to surface recombination.

3.2 Backwall cell

The light penetrates through collector and junction before it is absorbed in the emitter. This requires a very thin collector layer if made of the same basic material (homojunction) or, preferably, a wider band-gap collector, acting as window for most of the spectrum of the impinging light.

Collector-doping (to provide sufficient conductivity) and thickness need to be optimized to minimize optical and series resistance losses.

3.3 Schottky cell

In a Schottky metal–semiconductor junction cell, the emitter and the junction are of the same material.[5]) In frontwall operation, a metallic current-assist conductive and

[5]) Usually with homogeneous doping.

transparent top coating (see Section 2.2) is used causing some transmission losses and requiring diligence in producing a sufficiently thin continuous layer. This problem can be eliminated for a backwell Schottky cell, however, at the expenses of increased influence of surface recombination and requiring a longer diffusion length.

There is evidence that Schottky cells are in fact heterojunction cells with exceedingly thin junction layers (often oxides).

3.4 Bipolar generation cells

In bipolar generation cells, there is significant minority-carrier generation throughout the emitter, the junction, and at least in part of the collector, which causes the quasi-Fermi levels to remain spread throughout the junction. This yields a slightly larger open circuit voltage. It is believed that all better solar cells are of this type. However, since the collector and the junction are not ideally suited for photovoltaic power generation[6]), only a minor fraction of the light should be absorbed here.

3.5 Cell with dominant interface recombination

Due to a substantial density of interface recombination centers in such cells, major interface recombination takes place, causing a reduction in open-circuit voltage. In reverse bias, nevertheless, a substantial amount of minority carriers can leak out into the collector and result in an almost unimpeded short-circuit current. However, when the interface recombination is caused by a double layer resulting in a potential spike, tunneling through this spike will reduce also the short-circuit current [8].

We will return to this subject of interface recombination as one of the main topics of this paper in Section 4.3 for a quantitative analysis.

3.6 Cells with current limiting junction

If the junction region lies in a material with negative differential conductivity, the current through this region may saturate if the field extends to values for domain formation to occur. Current limitation through the cell is then caused by electron density limitation at the junction interface (see Section 4.1.1 and [5]).

The field limitation observed in this region permits the use of relatively impure material which, without negative differential conductivity, would develop too high fields in the junction, causing excessive tunneling.

4. Cell Operation

4.1 Emitter

An effective emitter absorbs most of the light, generating a substantial increase in minority carrier density (for definiteness here assumed to be electrons):

$$n_{10} = g\tau_n \tag{1}$$

(index 1 refers to the emitter, index 0 to the bulk, neglecting surface effects), g is the (assumed homogeneous[7])) generation rate and τ_n the minority carrier lifetime.

Electrons generated in the emitter may recombine in the bulk (characterized by τ_n), diffuse to the surface and recombine here (characterized by the surface recombination velocity s), or diffuse to the junction (see below).

[6]) The field in the junction contributes to drift current losses and the doping of the collector is usually not designed for high collection efficiency (excessive recombination).

[7]) This assumption is used to obtain a simple solution of the diffusion equation and causes a lesser error than the often used exponential dependence of g on x [9].

4.1.1 Diffusion current

Key to the understanding of the operation of the photovoltaic cell is the fact that the diffusion of the electrons to (or from) the junction is determined by the electron density, n_j at the emitter–junction interface (boundary condition in the mathematical sense). If n_j is lower than n_{10}, electrons diffuse from the emitter bulk to the junction with a diffusion current which is the larger, the lower n_j is (reverse bias)[8]). If n_j is larger than n_{10}, electrons diffuse from the junction into the emitter (forward bias).

The diffusion current

$$j_n = \mu kT \frac{dn}{dx} \tag{2}$$

can be calculated for

$$n(x) = \tau_n \left(g + \frac{1}{e} \frac{\partial j_n}{\partial x}\right) \tag{3}$$

yielding[9])

$$j_n(x = 0) = e \frac{L_n}{\tau_n} (n_j - n_{10}) \tanh\left(\frac{x_m}{L_n}\right) \tag{4}$$

($x = 0$ and $x = -d_1$ are the position of the junction interface and the emitter surface, respectively), L_n is the diffusion length of electrons and x_m is the position of the electron density maximum in the emitter in reverse bias. x_m depends on surface recombination and bias, but is typically of the order of $0.5d_1$ (see Fig. 1).

For better photovoltaic cells, L_n is substantially larger than d_1. Neglecting detail (for such see [9]) equation (4) can be simplified to

$$j_n(x = 0) = e \frac{x_m}{\tau_n} (n_j - n_{10}) \tag{5}$$

with $x_m \approx d_1$ for negligible surface recombination.

The relation of this current to typical current–voltage characteristics becomes obvious when j_n of (4) is plotted as function of $\ln n_j$ (see Fig. 2): A current–voltage characteristic could be derived if $\ln n_j$ can be related to the applied voltage. This is indeed possible as shown in the next section.

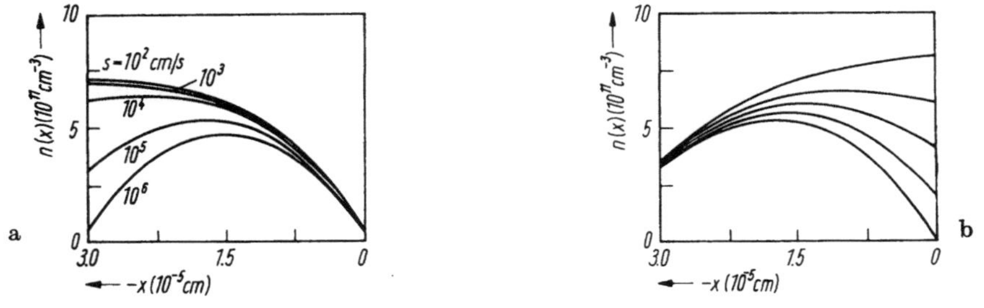

Fig. 1. Electron density profile in the emitter for $L_n = 10^{-5}$ cm, $g\tau_n = 7.5 \times 10^{11}$ cm^{-3} and s or j_n as family parameters in a) or b), respectively. $s = 10^5$ cm/s in b)

[8]) However, one should remember that n_j is limited by the Richardson-Dushman emission [3].

[9]) In the given form, using x_m, the lengthy expression is hidden in the tanh-term and shows in a rather transparent way the major dependence of j_n on n_j.

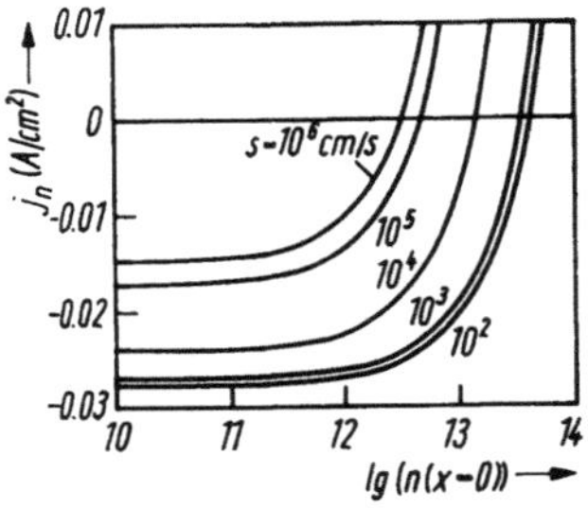

Fig. 1

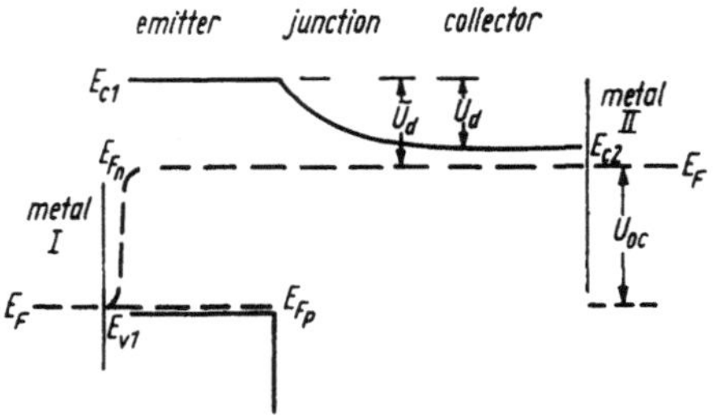

Fig. 3

Fig. 2. Electron current density at $x = 0$ as function of $\ln n_j$ for s as family parameter, $L_n = 7.5 \times 10^{-5}$ cm and $g = 6 \times 10^{21}$ cm^{-3} s^{-1}

Fig. 3. Simple band model of a heterojunction with minority carrier generation in the emitter, however, neglecting interface recombination at the emitter–junction interface or marked recombination influence on E_{Fp} at the emitter–metal interface

4.1.2 Open-circuit voltage

In order to obtain the pertinent relation to the applied voltage, one needs to evaluate the various contributions to the voltage drop across the cell. The main[10]) voltage drop across the cell is best divided into two parts, the open-circuit voltage and the additional voltage drop in the junction when current flows. Neglecting interface recombination[11]), the open-circuit voltage is identical to the separation of the two quasi-Fermi levels in the emitter (see Fig. 3):

$$U_{oc} = \frac{1}{e}\left[E_{g1} - (E_{c1} - E_{Fn}) - (E_{Fp} - E_{v1}) - \delta E_{g1}\right] \tag{6}$$

with E_{g1} the band gap of the emitter, and δE_{g1} the (possible) narrowing of this gap at high doping levels. With

$$E_{c1} - E_{Fn} = \frac{kT}{e} \ln\left(\frac{N_{c1}}{g\tau_n}\right) \tag{7}$$

the well-known logarithmic dependence of U_{oc} on the generation rate is explained ($E_{Fp} - E_{v1}$ and δE_{g1} are essentially independent of g).

4.2 Junction

The voltage drop in the junction is best related to open-circuit conditions. At open-circuit voltage, it is the diffusion voltage

$$U_d = \frac{kT}{e} \ln\left(\frac{n_{20}}{n_{10}}\right). \tag{8}$$

4.2.1 Applied voltage

When applying a voltage to the cell which is different from the open-circuit voltage, a current will flow, and a voltage different from the diffusion voltage will drop across

[10]) Neglecting any voltage drop caused by series resistances.

[11]) Obviously neglection of interface recombination usually results in a larger than observed open-circuit voltage, and sometimes different slope of U_{oc} with g and/or with T.

the junction:

$$U_j = U_d + (U_{appl} - U_{oc}) . \tag{9}$$

In contrast to common practice, we will label the applied voltage in reference to the open-circuit voltage, and refer to a forward bias when $U_{appl} - U_{oc} > 0$ and to a reverse bias when $U_{appl} - U_{oc} < 0$. This provides a current relationship in the junction similar to an unilluminated diode.

The voltage drop across the junction is consequently calculated from the solution of the transport equation[12])

$$j_n = e\mu nF - \mu kT \frac{dn}{dx} \tag{10a}$$

and the Poisson equation

$$\frac{dF}{dx} = \frac{e}{\varepsilon\varepsilon_0} \varrho(n, x) \tag{10b}$$

with

$$F = -\frac{dU}{dx} \tag{10c}$$

and requires in addition to the boundary condition n_j, as mentioned in Section 4.1, two others for F and U. We will return to these in Section 4.4.

4.2.2 Solution near U_{oc}

For j_n much smaller than drift or diffusion current (Boltzmann range), the solution of (10) is easily given:

$$n = n_{20} \exp\left(-\frac{eU(x)}{kT}\right) \tag{11}$$

yielding

$$n_j(x = 0, j_n \approx 0) = n_{20} \exp\left(-\frac{eU_j}{kT}\right) \tag{11a}$$

or

$$U_{appl} = U_{oc} + \frac{kT}{e} \ln\left(\frac{n_j}{n_{10}}\right). \tag{12}$$

Equation (12) indeed relates $\ln n_j$ to the applied voltage, as suggested from the discussion of Fig. 2 in Section 4.1, and yields, when combined with (4a),

$$j_n = e\frac{L_n n_{10}}{\tau_n} \left\{\exp\left(-\frac{e[U_{appl} - U_{oc}]}{kT}\right) - 1\right\} \tanh\left(\frac{x_m}{L_n}\right), \tag{13}$$

which may be rewritten to the well-known current–voltage relation for ideal photovoltaic cells:

$$j_n = j_0 \exp\left(-\frac{e\varphi}{kT}\right) \exp\left(\frac{eU_{appl}}{kT}\right) - j_n , \tag{13a}$$

however, with somewhat modified parameters:

$$j_0 = e\frac{L_n}{\tau_n} N_{c2} \tanh\left(\frac{x_m}{L_n}\right), \tag{13b}$$

---

[12]) For simplicity we have neglected the hole transport equation here. This is justified in frontwall heterojunctions.

$$\varphi = U_{\rm d} + U_{\rm oc}\,, \tag{13c}$$

$$U_{\rm d} = \frac{kT}{e} \ln \frac{N_{\rm c2}}{N_{10}}, \tag{13d}$$

$$j_{\rm n} = e \frac{L_{\rm n}}{\tau_{\rm n}} n_{10} \tanh\left(\frac{x_{\rm m}}{L_{\rm n}}\right) = eL_{\rm n}g \tanh\left(\frac{x_{\rm m}}{L_{\rm n}}\right). \tag{13e}$$

Equations (13) show a much improved agreement with the experiment, since j_0 is several orders of magnitude smaller than in previous theories[13]), and the measured φ is indeed equal to the sum of the measured $U_{\rm oc}$ and $U_{\rm n}$.

It should also be recognized that φ in the given theory has no other meaning than shown in (13c) and is not to be confused with a barrier height for electron transport in reverse direction (such a barrier height is indeed $U_{\rm d}$ and *not* φ).

4.2.3 Current–voltage characteristics

It should, however, be remembered that (13) holds only for the Boltzmann range, i.e. near the open-circuit voltage, and not for the entire current–voltage characteristic. For values of the applied voltage more than $\approx kT/e$ from $U_{\rm oc}$ one needs to integrate (10) properly, hence one needs to know more about the defect distribution in the junction and its influence on the space charge development with widening (reverse bias) or shrinking (forward bias) of the junction region. For each point of the characteristic, $j_{\rm n}$ enters as parameter in the transport equation. Numerical integration can be accomplished for any given model. The shape of the obtained current–voltage characteristic depends on the junction model chosen and is not the subject of this paper.

4.3 Interface layer

For most photovoltaic devices, the agreement between the given theory and experiment is not yet satisfactory for reasons that a major factor in these devices was neglected, namely interface recombination. Let us now assume that such interface recombination plays a *dominant* role. In contrast to some initial belief this will *not* render such

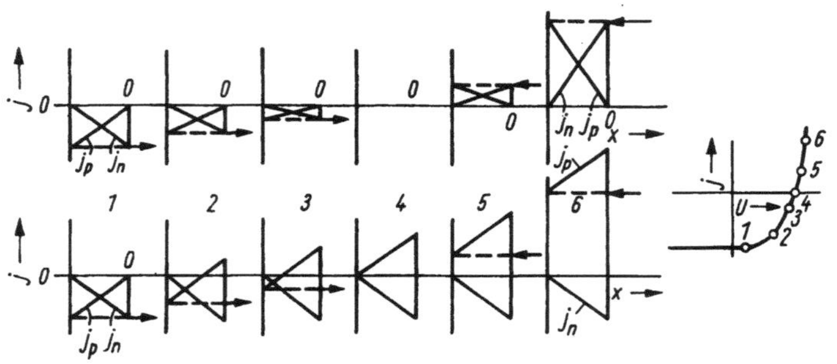

Fig. 4. Current distribution $j_{\rm n}(x)$ and $j_{\rm p}(x)$ in the collector at six different points of the current–voltage characteristics as shown in the insert. Upper row: zero interface recombination; lower row: dominant interface recombination. Arrow indicates direction of electron current

[13]) $j_0 = ev_{\rm n}N_{\rm c}$ with $v_{\rm n}$, the r.m.s. electron velocity, for the classical diffusion theory or $j_0 = e\mu F_{\rm c}N_{\rm c}$ with $F_{\rm c}$, the field at the junction-emitter interface, for the drift theory of the junction [10].

a photovoltaic device ineffective, but usually only reduce the open-circuit voltage to a moderate degree without major reduction of the short-circuit current.

This somewhat unexpected result is based on the fact that interface recombination and minority carrier "leakage" through the junction interface into junction and collector are competing processes, and the latter can be made highly efficient when n_j at the junction interface is sufficiently lowered.

On the other hand, the minority carrier density at this interface in open-circuit voltage condition can only be lowered to a degree which is given by the ratio of interface recombination velocity (a value which cannot exceed the thermal velocity of electrons) to the diffusion velocity L_n/τ_n, causing a lowering of the quasi-Fermi level, hence U_{oc}, by usually not more than 0.2 eV (see Section 4.4).

4.3.1 Negligible interface recombination

In order to understand this behaviour, let us first analyze the current distribution in the emitter for different values of the total current *without* interface recombination. Such distribution is shown for six different cases in the upper row of Fig. 4. In each case, the intersect with the ordinate indicates the outer surface, the position 0 at the x-axis indicates the junction interface of the emitter. Junction and collector ($x > 0$) experience no carrier generation. (Outer-) surface recombination is neglected here.

For open-circuit condition (case 4) the minority carrier density in the collector does not change with x, there is no diffusion current, hence $j_n = 0$ and $j_p = 0$ throughout the entire collector.

At current saturation (case 1), the electron density profile is similar to the upper curve in Fig. 1a. Consequently, the electron current increases (linearly, as shown in [9]) with decreasing distance from the junction interface. For reasons of total current continuity ($j_n + j_p = j =$ const) the hole current must decrease accordingly. Since the hole density is much larger than the electron density in the collector, such hole current profile can be achieved by a very small field (see Fig. 5) which easily self-adjusts by a small space charge of $\approx 10^8$ cm^{-3} in the example of Fig. 5 (for $d_1 \approx \approx 2 \times 10^{-5}$ cm and $\varepsilon \approx 10$). Hence, at the outer surface (assumed to be covered with a transparent electrode), the entire current is carried by hole drift which changes via generation into electron (diffusion) current with increasing distance from the outer surface (upper graph of Fig. 5). At the junction interface finally the hole current has vanished and the entire current is carried by electron diffusion. Case 1 of Fig. 4 reproduces the second part of Fig. 5.

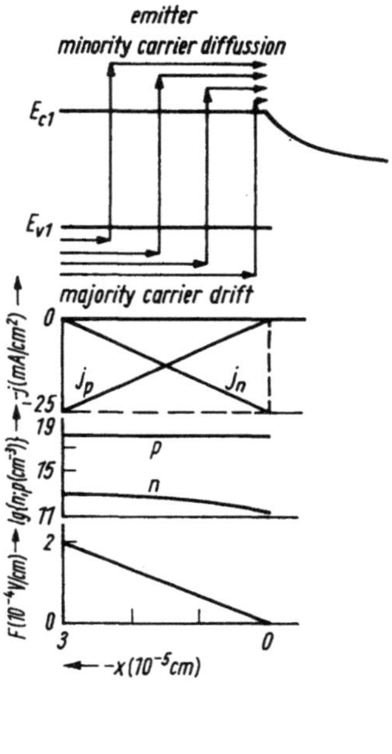

Fig. 5. Current, carrier density (logarithmic plot), and field profile in the collector at saturation current (reverse bias), no surface recombination. Upper graph: simplified band model with current schematics

With increased applied voltage (lower reverse bias), (cases 2 and 3), the distributions remain similar, however, the slopes are reduced.

In forward bias (cases 5 and 6), the sign of both j_n and j_p are reversed. However, again the electron current is largest at the junction interface and zero at the emitter surface. The current reverts from electron diffusion to hole drift by bulk recombination.[14]

4.3.2 *Dominant interface recombination*

With dominant interface recombination (lower row of Fig. 4) let us again start discussing the open-circuit condition first (case 4). Here zero current is achieved by an electron and hole current flowing in the same direction (i.e. opposite in sign), hence cancelling each other. This current is caused by major interface recombination.

In reverse bias, the electron current profile, in the emitter remains unchanged: Dominant interface recombination was assumed, hence the maximum possible electron (diffusion) current was already (at open-circuit condition) drawn towards the junction interface. However, lowering the electron density at the junction side of the interface (at $x = 0^+$) causes electrons to leak out from the emitter and produces a net current into the junction. This net current can (at best) be (almost) equal to the electron current at $x = 0^-$ and requires a shift of the triangular hole current downward with essentially no remaining recombination (hole) current at the interface as shown in case 1. Reverse currents before saturation are shown as cases 2 and 3.

In forward bias, excess electrons at the junction interface recombine here, hence shift the hole current upward (cases 5 and 6) while again leaving the electron current in the emitter unchanged.

It therefore is indicated that, in contrast to the case of negligible interface recombination, where the *change in electron current* at the interface produced the net current, with dominant interface recombination, the *change in hole current* causes the change in net current. As seen from the lower row of Fig. 4, such change in hole current is accomplished by a shift of the hole current at the interface. Since holes can esaily be supplied from the electrode but not from the junction through the interface (in the given example of a p-type emitter) the sequence from case 6 to case 1 must stop at case 1 $\big($decreasing $j_p(x = 0)\big)$ with $j_p(x = 0) = 0$. This again causes current saturation.

4.4 *Interface boundary conditions*

As shown in Section 4.1 and 4.2 for *zero interface recombination*, the interface boundary condition n_j connects the current flow from the emitter and the voltage drop in the junction: n_j was given by (4) as a function of the electron current at the junction interface. n_j is also the boundary condition for integration of (10a). The other boundary conditions are F_j, the field and $-U_j$, the potential at the junction interface ($U = 0$ at the collector electrode).[15]

The electron density distribution for a photovoltaic cell with negligible interface recombination is schematically illustrated in Fig. 6 for a variety of bias conditions;

---

[14]) Even though the slope of j_n and j_p must add up to 180° at each x, the behaviour shown in cases 5 and 6 is one example of recombination extending over the entire emitter evenly. Other recombination distributions are possible.

[15]) In order to obtain the yet unknown boundary conditions F_j and U_j one may first start integration of (10) from the collector bulk towards the junction interface. Since the solution approaches a singular point ($\mathrm{d}E/\mathrm{d}x = \mathrm{d}n/\mathrm{d}x = 0$) in the bulk, a boundary condition close to $j = j_n = e\mu n_{20}F_{20}$, such as $n_{20} - \delta n$ and $F_{20} + \delta F$ with $\delta n \ll n_{20}$ and $\delta F \ll F_{20}$ must be chosen with the numerical integration proceeding towards the junction. F_j and U_j can then be obtained from the position (x) at with $n(x) = n_j$.

the numbers indicate conditions similar to the ones given in Fig. 4, clearly illustrating the double role of n_j as boundary condition for emitter and junction. In the lower part of Fig. 6, conduction band and quasi-Fermi level are shown for open-circuit condition.

For *dominant interface recombination*, this condition must be re-evaluated. As shown in the lower row of Fig. 4, the electron density distribution in the emitter remains essentially the same independent of bias; hence n_j is *not* a boundary condition for the emitter. However, since $n(x)$ was fixed in the emitter by interface recombination, which is an effect of importance for photogenerated carriers but not for thermally generated carriers, one expects that, by introducing such interface recombination, little changes occur in junction and collector.

As a consequence for all but current saturation, n_j shows a jump[16]) with $n(x = 0^-) < < n(x = 0^+)$ (see Fig. 7). The band diagram shown for open-circuit condition in the lower part of Fig. 7 illustrates the reason for this jump: The quasi-Fermi level for electrons is lowered in the emitter near the junction; however, since the electron distribution in junction and collector remains the same, E_{Fn2} and $E_{c2}(x)$ slides down by the amount $E_{Fn10} - E_{Fn1}(x = 0^-)$, hence E_c experiences a jump at the interface of the same amount:

$$\Delta E_c = E_{c1}(x = 0^-) - E_{c2}(x = 0^+) = kT \ln \left(\frac{n_{2jR0}}{n_{1jR0}}\right) \quad (14)$$

with $n_{1jR0} = n_j(x = 0^-, j = 0)$ and $n_{2jR0} = n_j(x = 0^+, j = 0)$ for dominant interface recombination (R).

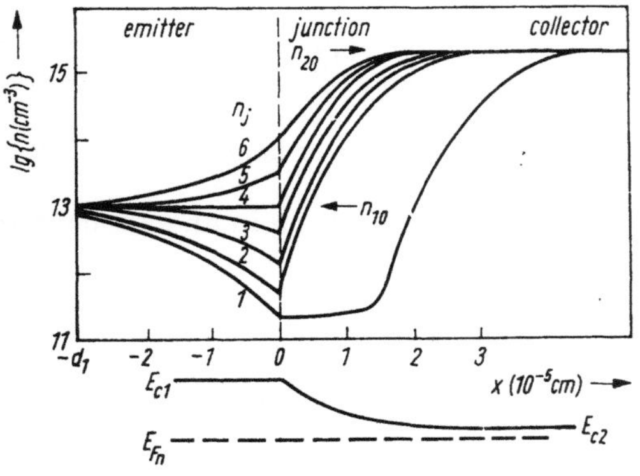

Fig. 6

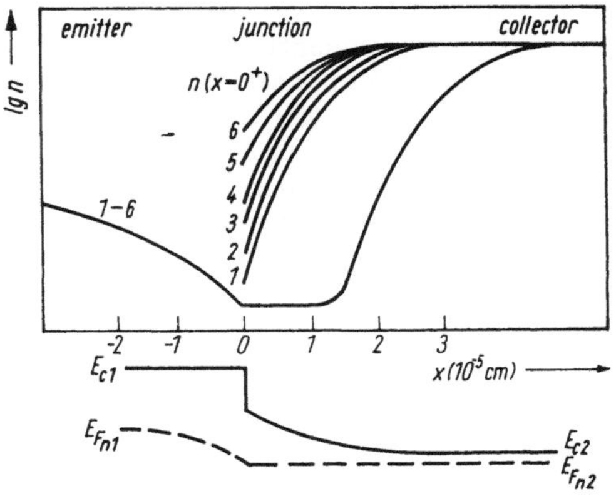

Fig. 7

Fig. 6. Electron density profile (at the edge of the conduction bands) in emitter, junction, and collector for different bias (currents). The family numbering refers to the same set of numbers in Fig. 4. The lower diagram depicts conduction band and electron quasi-Fermi level of such a junction in open-circuit condition

Fig. 7. Electron density profile and band scheme similar to Fig. 6, however, for dominant interface recombination.

16) It should be recognized that in Fig. 7 the electron densities at the lower edge of the conduction band is plotted, hence the jump. When drawn for any $n_j(E)$ at the same energy, at open-circuit condition most of the jump disappears expect for the difference in density of states in emitter and junction. In reverse bias, however, for any given energy, the electron density at $x = 0^+$ is smaller than at $x = 0^-$, providing the driving force for the current (the inverse holds for forward bias).

ΔE_c can be calculated from $j = j_n + j_p = 0$ with

$$j_n \approx e \frac{x_m}{\tau_n} (n_{1jR0} - n_{10}) , \tag{15a}$$

$$j_p = e s_j n_{1jR0} \tag{15b}$$

yielding (with $n_{2jR0} \approx n_{10}$)

$$\Delta E_c = kT \ln \left[\frac{\tau_n}{x_m} \left(s_j + \frac{\tau_m}{x_n} \right) \right]. \tag{14a}$$

This causes a reduction in the open-circuit voltage by the same amount, or, with (5) and (6), yields

$$U_{oc} \approx \frac{1}{e} E_{g1} - \frac{kT}{e} \ln \left[\frac{N_{c1}}{g x_m} \left(s_j + \frac{x_m}{\tau_n} \right) \right]. \tag{16}$$

(We neglect here $E_{Fp} - E_v + \delta E_{g1}$ and assume a simple current relation for the electron transport through the junction.[17])

5. Experimental Evidence

The given theory provides improved understanding and yields better agreement with the experiment:

a) The theory is self-consistent and, for description of the current–voltage characteristic, does not require the mathematically inacceptable parallel shift of the dark-diode characteristic.

b) The pre-exponential factor, j_0, in the photovoltaic diode equation is several orders of magnitude lower than $\approx 10^7$ A/cm² required by previous theories. Experimental values lie indeed in the 10^3 to 10^6 A/cm² range.

c) φ is not the barrier height but the sum of U_{oc} and the barrier height U_p, in agreement with the experiment.

d) j_0 and φ are connected via τ_n (see (6), (7), and (13)) resulting in an exponential relationship $j_0(\varphi)$:

$$j_0 = e N_{c1} s_j \exp \left(- \frac{E_{g1} - e\varphi}{kT} \right). \tag{17}$$

In CdS/Cu_2S solar cells such behaviour is indeed measured [11] in good agreement with the experimental slope (Fig. 8), however, shifted by 168 mV towards lower

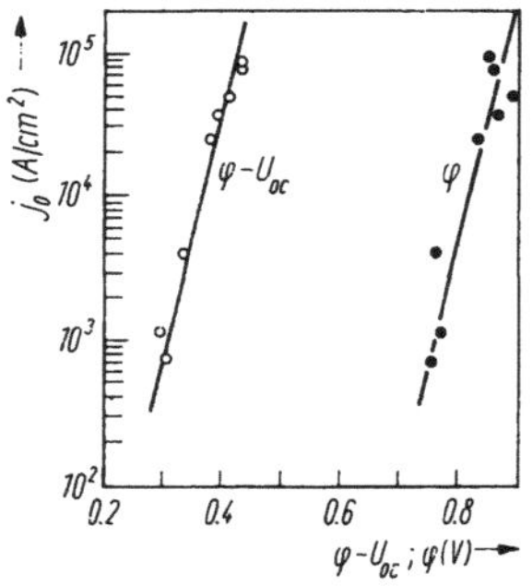

Fig. 8. Pre-exponential factor j_0 (equation (13b)) as function of φ or $\varphi - U_{oc}$ (equation (17)) as observed [11] for a set of CdS/Cu_2S solar cells with $U_{oc} \approx 450$ mV

[17]) In fact, the net electron current through the junction is given as the difference of two current components for electron transfer from the emitter to the junction and in reverse direction and each one of these components is the integral of all electron transfers over all energies.

φ-values ($N_{c1} \approx 10^{19}$ cm^{-3}, $s_j \approx 5 \times 10^6$ cm/s, and $E_{g1} \approx 1.18$ eV). This is nearly the same difference as measured in a set of these cells from the theoretical open-circuit voltage, and calculated from (16) ($U_{oc}^{(th)} = 614$ mV, $U_{oc}^{(exp)} = 450$ mV). Hence, when j_0 is plotted versus $\varphi - U_{oc}$ (Fig. 8) excellent agreement between theory and experiment is observed:

$$j_0 = egx_m \exp\left[-\frac{e(\varphi - U_{oc})}{kT}\right] \tag{17a}$$

yielding $gx_m = 4.5 \times 10^{16}$ cm^{-2} s^{-1}, a value within the experimental error expected from excitation (AM1: $g \approx 5 \times 10^{21}$ cm^{-3} s^{-1}) and thickness of Cu_2S ($x_m \approx 10^{-5}$ cm).

This result emphasizes the importance of (13c). The reason for the shift of U_{oc} is probably related to the rough surface of the CdS/Cu_2S layer [8], accounting for approximately half of the shift, the rest being probably caused by gap narrowing in the degenerate Cu_yS with $1.99 < y < 1.996$ for typical cells. With a density of copper vacancies (acting as shallow acceptors) of

$$[V_{Cu}] = (2 - y)\, 4.3 \times 10^{22}\ \text{cm}^{-3} \tag{18}$$

one estimates 2×10^{19} cm$^{2} < [V_{Cu}] < 5 \times 10^{20}$ cm^{-3}. Having no other information about band-gap narrowing in Cu_2S one may use a highly simplified band-tailing approach

$$\delta E_{g1} \approx \frac{kT}{e} \ln\left(\frac{[V_{Cu}]}{N_{v1}}\right) \tag{19}$$

yielding $0 < \delta E_{g1} < 80$ mV for the above given range of y. Hence one expects for multicrystalline CdS/Cu_2S solar cells with rough surfaces a typical range of

$$440\ \text{mV} < V_{oc} < 520\ \text{mV}\,, \tag{20}$$

again in good agreement with the experiment.

6. Summary

The commonly used parallel shift of a diode characteristic by the short-circuit current is mathematically only permissible in a small range near the open-circuit voltage [6]. This can be shown by connecting the diffusion current of minority carriers in the emitter with the voltage drop calculated by integrating the transport equation of the same carriers in the junction in the Boltzmann range, and using the electron density, n_j, at the interface between emitter and junction as connecting boundary condition. This means that n_j determines j_n in the emitter (equation (4)) *and* it determines the voltage drop in the junction (equation (11a)).

This remains the case also outside the Boltzmann range, however, it necessitates proper integration of transport *and* Poisson equation. The solutions obviously then become space-charge dependent, determining the *shape* of the current–voltage characteristics.

The analysis near the open-circuit condition provides insight into the transport mechanism through the junction interface. A moderate reduction of the open-circuit voltage is observed if major interface recombination takes place. This causes a band-edge jump of the same magnitude (as the reduction of U_{oc}). Such a band edge jump replaces the jump often assumed in heterojunction and believed to be caused by differences in electron affinities; it removes the difficulties encountered in previous models when discussing vanishing currents for open-circuit conditions.

References

[1] P. A. Crossley, G. T. Noel, and M. Wolf, Final Report NASA AED-R3346, Washington D. C. (1968).
[2] A. L. Milnes and D. L. Feucht, Heterojunctions and Metal–Semiconductor Junction, Academic Press, New York 1972.
[3] A. Rothwarf and K. W. Böer, Progr. Solid. State Chem. **10**, 71 (1975).
[4] R. Hill, Thin-Film Solar Cells in Active and Passive Thin-Film Devices, Ed. T. J. Coutts, Academic Press, London 1978.
[5] K. W. Böer, Phys. Rev. B **13**, 5373 (1976).
[6] K. W. Böer, Proc. 12th IEEE Photovolt. Spec. Conf., Baton Rouge 1976 (p. 475).
[7] K. W. Böer, Proc. Electro-Chem. Soc., Atlanta, Ga. 1977.
[8] K. W. Böer, Proc. Internat. Solar Energy Soc., New Delhi 1978.
[9] K. W. Böer, phys. stat. sol. (a) **40**, 355 (1977).
[10] E. Spenke, Electronic Semiconductors, Springer-Verlag, Berlin/Heidelberg/New York 1965.
[11] H. Hadley, Semi-Annual Report, NSF/RANN/AER72-0347804PR75/4, National Science Foundation, 1975(p.70).

(Received May 5, 1978)

phys. stat. sol. (a) 49, 27 (1978)

Subject classification: 14.3.2; 13.3; 13.4; 14.3.4; 16; 22.4

II. Physikalisches Institut der Rheinisch-Westfälischen Technischen Hochschule Aachen

Interpretation of Surface Phenomena on ZnO by the Compensation Model

By

G. HEILAND and D. KOHL

In honour of Prof. Dr. F. STÖCKMANN's 60th birthday

Field effect mobility and surface photoconductivity on ZnO crystals increase for several orders of magnitude with dark surface conductivity (electron accumulation). An explanation was given previously: Only a small increasing fraction of the injected electrons appears in the conduction band whereas the main part is trapped by surface states. In the new compensation model the trapping action of the surface states is replaced by that of bulk defects changing their condition of charge within an accumulation layer during variations of band bending: From the temperature dependence of con ductivity and Hall effect a set of three bulk defect levels is evaluated. The same set is used for a newcalculation of the accumulation layer and for a quantitative interpretation of photoconductivity and field effect measurements. Also surface photovoltage and photolysis data are discussed. All these experimental results are obtained on single crystals bearing an accumulation layer. Properties of thin polycrystalline ZnO films are also treated. The concept of a drift mobility is of central importance for the compensation model. As a result of compensation space charge layers are much thinner than previously assumed.

Feldeffektbeweglichkeit und Oberflächen-Photoleitfähigkeit von ZnO-Kristallen wachsen mit der Oberflächen-Dunkelleitfähigkeit (Elektronen-Anreicherungsschicht) über mehrere Größenordnungen. Eine Deutung wurde früher gegeben: Nur ein kleiner mit der Oberflächenleitfähigkeit zunehmender Bruchteil der injizierten Elektronen erscheint im Leitungsband, der größere Teil wird in Oberflächenzuständen festgelegt. In dem neuen Kompensationsmodell wird die bindende Wirkung der Oberflächenzustände durch die von Volumen-Störstellen ersetzt. Diese wechseln ihren Ladungszustand in einer Anreicherungsschicht bei Änderungen der Bandverbiegung. Aus dem Temperaturgang von Leitfähigkeit und Halleffekt wird ein Satz von drei Volumen-Störniveaus hergeleitet. Der gleiche Satz dient zu einer neuen Berechnung der Anreicherungsschicht und zur quantitativen Interpretation von Photoleitungs- und Feldeffektmessungen. Auch Oberflächenphotospannung und Photolyse werden diskutiert. Alle diese experimentellen Ergebnisse wurden an Einkristallen mit Anreicherungsschicht erhalten. Auch die Eigenschaften dünner polykristalliner ZnO-Schichten werden behandelt. Das Konzept der Driftbeweglichkeit besitzt für das Kompensationsmodell zentrale Bedeutung. Infolge der Kompensation sind die Raumladungsschichten erheblich dünner als bisher angenommen.

1. Introduction

In ZnO crystals without intentional admixtures always a considerable n-type conductivity has been found in spite of the wide band gap of about 3 eV [1]. The donors are attributed to an excess of Zn. By doping with Cu or Li the conductivity is diminished, but no p-type conductivity has been observed so far. Gases can produce additional donors or acceptors at the surface by adsorption and reaction. The resulting space charge layers have been calculated previously from Poissons equation and occupation statistics under the assumption of only one bulk donor level [1].

In the present paper only accumulation layers are discussed: The concentration of conduction electrons is increased compared with the bulk values and causes a positive surface conductivity $\Delta\sigma$ (see Section 4, equation (9)). The photoconductivity $\delta\sigma_{ph}$ of

ZnO has been found to be a surface phenomenon in many cases: It rises over several orders of magnitude with increasing dark surface conductivity $\Delta\sigma$ as

$$\delta\sigma_{\text{ph}} \sim (\Delta\sigma)^m \; ; \qquad m \text{ between 1 and } \tfrac{1}{2} \, . \tag{1}$$

This holds for single crystals at low temperature with various irradiations: Light from the UV [1] (intrinsic excitation, example in Fig. 10) through the visible [2] to the near IR spectral region [3, 4] (excitation out of bulk defects or surface states into the conduction band), furthermore electron bombardment (1 to 5 keV) [1].

The same relation (1) has also been found for the change of dark surface conductivity due to induced electrons (field effect, Fig. 9) [6]. It has been demonstrated not only for single crystals at low temperature, but also and at first for thin polycrystalline layers [1] (room temperature) irradiated with UV light or with electrons.

At first an explanation for the results on layers was given under the assumption of interstitial Zn atoms giving off a first electron in the dark and a second one by the action of light [5]. The single crystal results and especially the field effect data stimulated a different explanation of (1) in terms of a continuous distribution of electronic surface states in energy [6, 7]: The density of states near the Fermi level is independent of energy for low surface conductivity but grows exponentially towards the conduction band. This model allowed a quantitative description of the data. However, the stability of these surface states against various kinds of pretreatment was surprising: For a certain crystal the derived effective density of states at the Fermi level depended only on the surface conductivity $\Delta\sigma$, i.e. the band bending, and this relation was not modified by the way of pretreatment as exposure to gases or moderate annealing in vacuum.

The stability of the trapping states becomes understandable if they are not surface states but bulk states within the space charge layer. This steadiness represents an important argument in favour of the present compensation model besides the agreement with bulk Hall effect results. As a consequence, in this model surface states are only required for production of the initial band bending: donors above the conduction band edge (e.g., produced by adsorbed atomic hydrogen) or acceptors several kT below the Fermi level (e.g., produced by adsorbed oxygen).

2. Bulk Hall Effect and Bulk Defects

In the present paper the term "defect" includes donors and acceptors as well as intrinsic and impurity centers. The proposed compensation model is based on measurements of bulk conductivity and Hall effect between 90 and 600 K. From the electron concentration in the conduction band the position of the Fermi level is derived as a function of temperature, $E_F(T)$. Positions of the Fermi level at various temperatures are shown in Fig. 4a for ZnO crystal 11.

The temperature dependence $E_F(T)$ has to be attributed to the influence of bulk defects. For the choice of levels and concentrations of defects the following rules have been applied:

The number of levels should be as small as possible. One level is sufficient above $(E_F(90\text{ K}) + 2kT)$ or below $(E_F(600\text{ K}) - 2kT)$, since levels outside this energy range do not change their condition of charge between 90 and 600 K. For levels inside this energy range Fermi statistics is used to calculate the occupation.

The set of levels defined in this way need not be unique. Up to now we have examined the data for six crystals. For each crystal about fifty various sets of three levels within the energy range as defined above have been chosen. In no case a convergence to more than one set of levels was found.

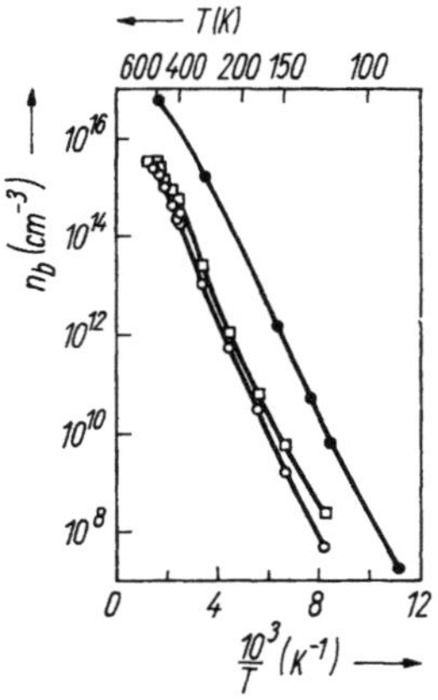

Fig. 1. Bulk electron concentration n_b as a function of temperature. ZnO crystal 11 (●, weakly Cu-doped, $\sigma_b(293\ K) = 6.8 \times 10^{-2}\ \Omega^{-1}\ cm^{-1}$); n_b derived from conductivity measurements [6] and Hall mobility data [8, 9]. ZnO crystals 391 (○) and 392 (□) (Cu-doped, $\sigma_b(293\ K) = 1.1 \times 10^{-3}$ and $5.2 \times 10^{-4}\ \Omega^{-1}\ cm^{-1}$, respectively); n_b derived from Hall effect measurements [10]

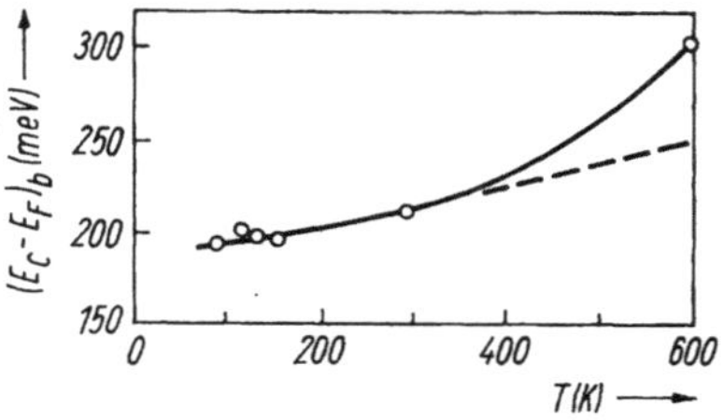

Fig. 2. Position of the Fermi level E_F below the conduction band edge E_C in the bulk as a function of temperature. Crystal 11. The points are derived from data in Fig. 1. The dashed line shows the theoretical interpretation by a model with one donor level. The solid line represents the result of a three-level model. For parameter values for both models see caption of Fig. 4

Fig. 1 shows the concentration of conduction electrons for three crystals as determined from conductivity and Hall effect measurements. The derived position of the Fermi level is given in Fig. 2 for crystal 11 (points), because on this crystal also detailed measurements of field effect and photoconductivity (Fig. 9) have been performed. As may be seen from Fig. 2 the slope of the curve cannot be explained by a single donor level. Compensation is now introduced by a second defect level (acceptor). Hausmann and Teuerle [8] have shown that in this case also a linear dependence $E_F(T)$ appears if one rules out very low acceptor densities. At least a third level is needed to describe the slope in Fig. 2 correctly. The three-level-term scheme of Fig. 4b was used to calculate the theoretical curve of Fig. 2. It should be emphasized that more than three levels are not required for an adequate fit.

Also results of more recent Hall measurements are given in Fig. 1 and 3. Again a set of three levels as shown in Fig. 3 gives an interpretation for the temperature dependence of the Fermi level position and of the electron concentration.

These examples demonstrate that a compensation model with three defect levels allows a description of the bulk Fermi level position in a wide range of temperatures. Now the same model is applied to the explanation of surface properties.

3. Calculation of the Accumulation Layer

The set of three bulk defect levels as shown in Fig. 4b is used with limitation to the case of a non-degenerate conduction band ($(E_C - E_F) > 0$ at the surface of the crystal). The surface donors causing the accumulation are situated above the conduction band edge. A schematic energy diagram is given in Fig. 5. Fermi statistics control the occupation of conduction band and defects. The concentration of space charge is given by

$$\varrho(z) = -e[n(z) - N^+_{D1}(z) - N^+_{D2}(z) + N^-_A(z)] . \tag{2}$$

The Poisson equation

$$\frac{d^2 V}{dz^2} = - \frac{\varrho(z)}{\varepsilon \varepsilon_0} \tag{3}$$

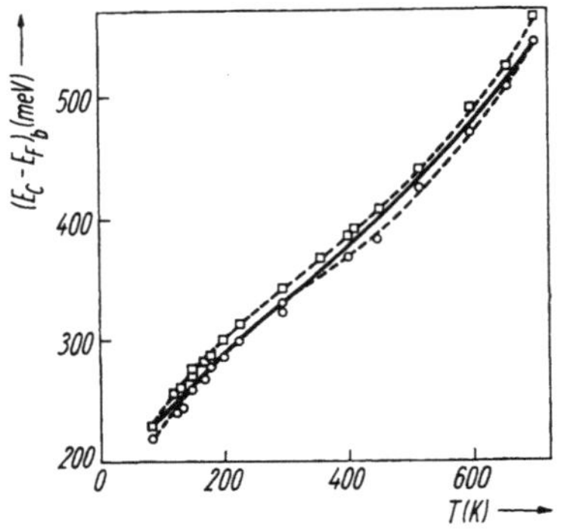

Fig. 3

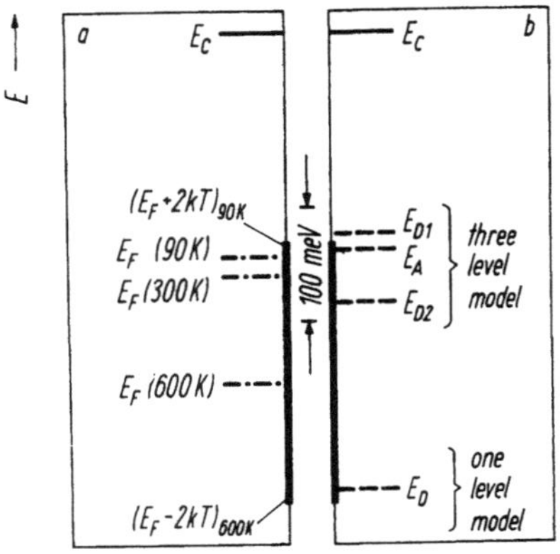

Fig. 4

Fig. 3. Position of the Fermi level E_F below the conduction band edge E_C in the bulk as a function of temperature. Points and dashed lines are derived from data in Fig. 1 (○ crystal 391, □ 392). The solid line shows the result for a three-level model with $E_C - E_{D1} = 50$ meV; $N_{D1} = 1.0 \times 10^{16}$ cm^{-3}; $E_C - E_{D2} = 300$ meV; $N_{D2} = 1.0 \times 10^{16}$ cm^{-3}; $E_C - E_A = 180$ meV; $N_A = 6.0 \times 10^{18}$ cm^{-3}

Fig. 4. Energy scheme for the bulk of crystal 11. a) Position of the Fermi level from bulk Hall effect data for various temperatures. b) Positions of the defect levels for a model with one donor level and for the compensation model with three defect levels. One donor level: $E_C - E_D = 393$ meV; $N_D = 2.5 \times 10^{18}$ cm^{-3}. Three defect levels: $E_C - E_{D1} = 170$ meV; $N_{D1} = 4.6 \times 10^{16}$ cm^{-3}; $E_C - E_A = 184$ meV; $N_A = 2.0 \times 10^{17}$ cm^{-3}; $E_C - E_{D2} = 227$ meV; $N_{D2} = 4.6 \times 10^{16}$ cm^{-3}

yields after integration

$$\frac{\mathrm{d}v}{\mathrm{d}z} = \frac{1}{L_{\mathrm{Deb}}} \sqrt{2} \left(\frac{N_C}{n_b} u_{E_C} + \frac{N_{D1}}{n_b} u_{E_{D1}} + \frac{N_{D2}}{n_b} u_{E_{D2}} + \frac{N_A}{n_b} u_{E_A} \right)^{1/2}, \tag{4}$$

where $V(z)$ electrostatic potential, $v(z) = eV(z)/kT$, z space coordinate normal to the surface ($z = 0$) (Fig. 5), ε dielectric constant, L_{Deb} Debye length (see (6)),

$$u_i = \ln\left[1 + \frac{e^v - 1}{1 + \exp e_{tF_i}}\right] - \frac{v}{1 + \exp e_{tF_i}} \; [11],$$

where

$$e_{tF_i} = \frac{E_i - E_F}{kT}.$$

The result of a second integration, $V(z)$, is used in a plot of the conduction band edge near the surface (Fig. 6a). $\mathrm{d}V/\mathrm{d}z$ depends in a direct way only on $V(z)$ but not on z. Therefore, it is possible to describe all kinds of non-degenerate accumulation ayers for one set of bulk defect levels by one single curve. The position of the origin $z = 0$ has to be chosen properly for each bending of the bands. The corresponding distribution of space charge in the conduction band and in the ionized defects is shown in Fig. 6b.

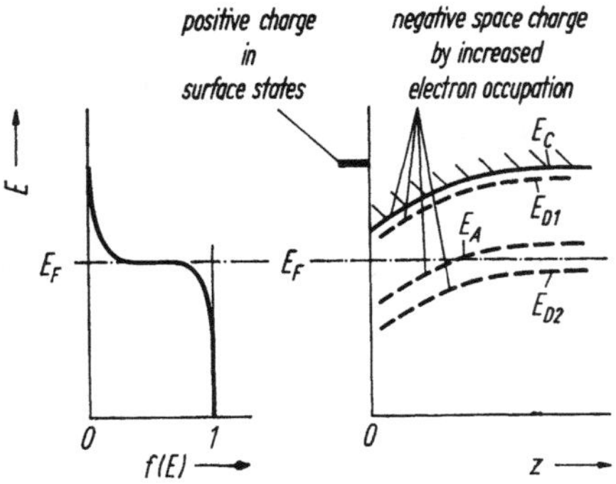

Fig. 5

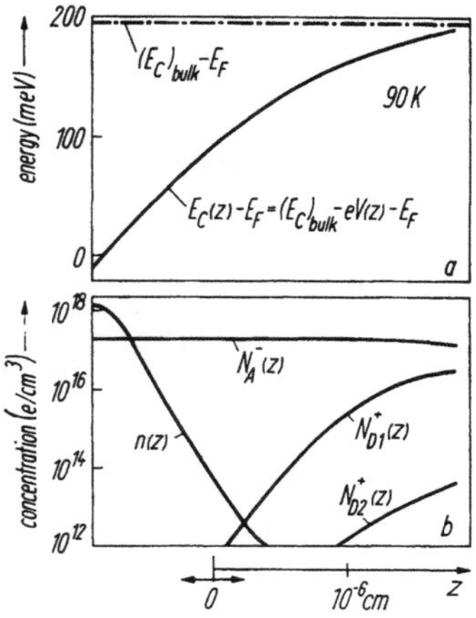

Fig. 6

Fig. 5. Energy scheme for the crystal surface with an accumulation layer according to the compensation model. $f(E)$ occupation probability, z distance from the surface

Fig. 6. Compensation model with three levels for crystal 11 (see Fig. 4 and 5). The origin of the abscissa, $z = 0$, shifts to the left with increasing total charge surface density Q_{sc}. a) Position $(E_C(z) - E_F)$ of the conduction band edge as a function of distance z from the surface in an accumulation layer. b) Corresponding concentration of charges: n for conduction electrons, N_{D1}^+, N_{D2}^+, N_A^- for ionized defects

Centers of gravity are defined for the mobile electrons and for the total charge of defects and mobile electrons by

$$L_{el} = \frac{\int_0^\infty z n(z)\,dz}{\int_0^\infty n(z)\,dz}\,; \qquad L_{sc} = \frac{\int_0^\infty z \varrho(z)\,dz}{\int_0^\infty \varrho(z)\,dz}\,. \tag{5}$$

The first quantity is important for the surface Hall effect and scattering of conduction electrons, the second quantity can be investigated by surface capacity measurements [11]. Fig. 7 shows the distance of both centers from the surface for various temperatures. In the case of L_{sc} two ranges can be distinguished: For weak accumulation the fixed charge in defects predominates. The extension of the space charge layer grows with band bending like in Schottky depletion layers. For strong accumulation the electrons become more important. Since the electron concentration $n(z)$ can attain higher values than the net concentration of ionized defects, L_{sc} finally decreases in this regime. The mean distance L_{el} of electrons from the surface reaches a maximum when the acceptor level of relative high concentration crosses the Fermi level near the surface. At low temperatures the mobile electrons are kept very close to the surface with L_{el} in the order of 10^{-7} cm, also for weak accumulation. Without compensation the extension of the space charge layer is given by the Debye length, e.g. for $T = 90$ K and $n_b = 2 \times 10^7$ cm^{-3}:

$$L_{Deb} = \left(\frac{\varepsilon\varepsilon_0 kT}{e^2 n_b}\right)^{1/2} = 4 \times 10^{-2}\ \text{cm}\,. \tag{6}$$

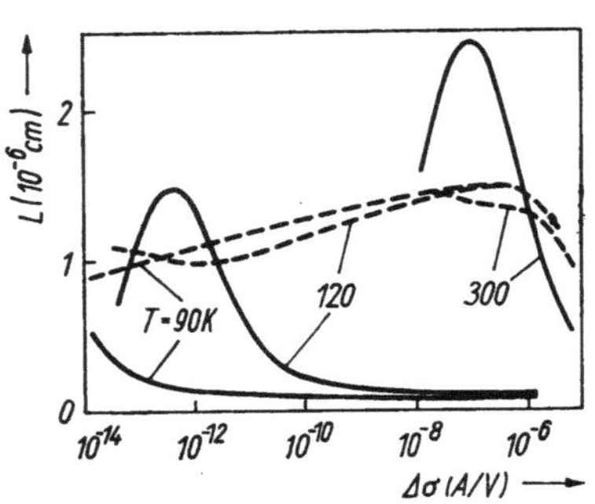

Fig. 7

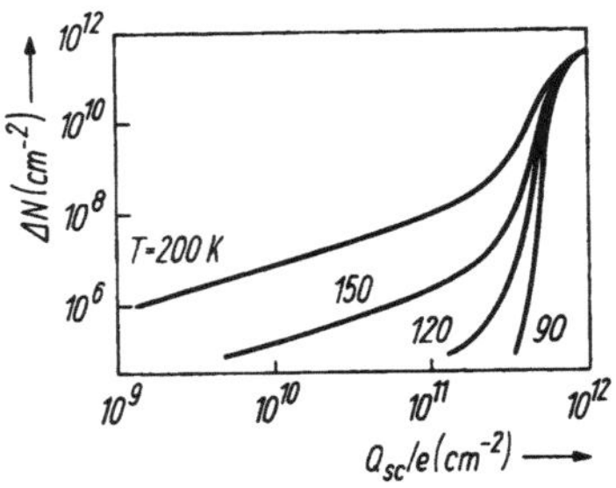

Fig. 8

Fig. 7. Distance of the two centers of gravity from the surface, L_{el} for the mobile electronic charge (solid curves) and L_{sc} for the total space charge (dashed curves), respectively, at several temperatures, calculated with data of crystal 11 in Fig. 4. For definitions see equation (5). $\mu_s = 100$ cm²/Vs

Fig. 8. Surface density ΔN of mobile electrons as a function of surface density Q_{sc}/e of the total space charge, for several temperatures, calculated with data of crystal 11. See Fig. 4 and equation (9)

The surface density of the total space charge is defined as (see (2))

$$\frac{Q_{sc}}{e} = \frac{\int_0^\infty \varrho(z)\,dz}{e}. \tag{7}$$

From Fig. 8 it can be seen that a considerable total space charge of about 5×10^{11} e/cm² is required before the electron density rises to values above 10^8 cm^{-2}. Especially at low temperature the measurable range of surface electron density ΔN below degeneracy includes five orders of magnitude, whereas the total space charge density Q_{sc} varies only by a factor of two.

We have not taken notice of the *image potential* φ_i. This additional potential for electrons near the surface is due to the abrupt change of the dielectric constant ε at the surface,

$$\varphi_i = \frac{e}{4\pi\, 2a\varepsilon\varepsilon_0}\, \frac{\varepsilon_1 - \varepsilon_2}{\varepsilon_1 + \varepsilon_2} = 67 \text{ meV} \tag{8}$$

with $\varepsilon_1 = 8.5$ (ZnO), $\varepsilon_2 = 1$ (vacuum), $a = 10^{-8}$ cm (distance from surface). The potential φ_i repels the electrons from the surface.

4. Drift Mobility

The surface density of mobile electrons is given as

$$\Delta N = \int_0^\infty [n(z) - n_0]\,dz \tag{9}$$

and the surface conductivity by $\Delta\sigma = e\mu_s\,\Delta N$, with n_b concentration of mobile electrons in the bulk, μ_s electron mobility at the surface.

In the field effect experiment a charge Q_{in} is induced into the surface and distributed between surface states (δQ_{ss}) and space charge layer (δQ_{sc}). The latter change includes variations in surface density of mobile electrons $\delta N \equiv \delta(\Delta N)$ and in the net charge of bulk defects near the surface $\delta(Q_{\text{def}})$:

$$Q_{\text{in}} = \delta Q_{\text{ss}} + \delta Q_{\text{sc}} = \delta Q_{\text{ss}} + e\,\delta N + \delta Q_{\text{def}}\,. \tag{10}$$

The change $\delta\sigma$ of surface conductivity $\Delta\sigma$ is given as

$$\delta\sigma = \mu_{\text{s}} e\,\delta N = \mu_{\text{fd}} Q_{\text{in}}\,. \tag{11}$$

The field effect mobility is defined as

$$\mu_{\text{fd}} = \mu_{\text{s}} e\,\frac{\delta N}{Q_{\text{in}}}\,. \tag{12}$$

If neither bulk defects nor surface states change their conditions of charge by application of the external field, all the induced charge appears as a change of surface electron density and μ_{fd} equals μ_{s}. The trapping of a fraction of the induced charge by surface states or bulk defects diminishes the field effect mobility.

As a *drift mobility* μ_{d} the field effect mobility relates to the total change of electron density in the conduction band and in bound states. This concept is not limited to the field effect but can be applied also to other experiments where a change of electron density is induced and distributed in a quasi-equilibrium between conduction band and bound states. In our model δQ_{ss} apparently can be neglected for the interpretation of experimental data (see Section 5). The drift mobility then reduces from (12) and (10) to

$$\mu_{\text{d}} = \mu_{\text{s}} e\,\frac{\delta N}{\delta Q_{\text{sc}}} = \mu_{\text{s}} e\,\frac{\delta N}{e\,\delta N + \delta Q_{\text{def}}}\,. \tag{13}$$

5. Field Effect and Photoconductivity

Fig. 9 shows typical results of *field effect measurements* on a ZnO prism face. The surface conductivity $\Delta\sigma$ is varied by pretreatments like adsorption of oxygen or atomic hydrogen and also by moderate annealing in vacuum. The field effect mobility μ_{fd} (see Section 4) rises with the surface conductivity over five orders of magnitude. With (11) this corresponds to (1).

The compensation model allows an interpretation of Fig. 9. The trapping of the induced charge takes place in the set of three bulk defect levels changing there occupation with a variation of band bending (Fig. 5). The data of Fig. 4 are used for the calculation of ΔN, Q_{sc}, and then μ_{d} (equation (13)). The result is given in Fig. 9 by the solid straight line which ends with beginning degeneracy ($E_{\text{C}} - E_{\text{F}} = 0$) at the surface. The surface mobility μ_{s} was taken to 30 cm²/Vs from surface Hall effect [10]. It is contained in $\Delta\sigma$ and in μ_{d} as a constant factor. Therefore, the choice of μ_{s} in-

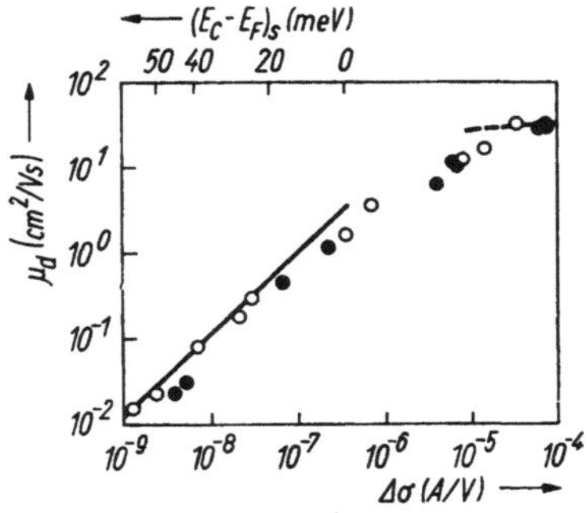

Fig. 9. Prism face. Drift mobility μ_{d} derived from field effect (○) and also from photoconductivity measurements (●) for crystal 11 as a function of surface conductivity $\Delta\sigma$. Vacuum, chopped light $10^4\ \text{s}^{-1}$, $\hbar\omega = 3.4$ eV. After Heiland [6]. The solid line shows the result of a calculation using the three defect levels as derived from bulk conductivity and bulk Hall effect (Fig. 4), without any further adjustment of parameters ($\mu_{\text{s}} = 30\ \text{cm}^2\ \text{V}^{-1}\text{s}^{-1}$). The measured surface Hall mobility μ_{s} (crystal 37) is given for the degenerate case as a dashed line, after Veuhoff and Kohl [10]. The distance $(E_{\text{C}} - E_{\text{F}})_{\text{s}}$ between conduction band and Fermi level at the surface is indicatedbove, $T = 90$ K

fluences the curve only slightly. The scale at the upper edge of Fig. 9 gives the distance between the Fermi level and the conduction band at the surface, whereas in the bulk $(E_C - E_F)_b = 194$ meV: A measurable surface conductivity requires a band bending $eV_s > 100$ meV and a total space charge density $Q_{sc} > 5 \times 10^{11}$ e/cm² (Fig. 8).

From the agreement of the calculated curve with the experimental data (points) we conclude that an interpretation of the field effect is possible without any surface states near the Fermi level on the ZnO prism face.

A drift mobility can also be derived from measurements of *surface photoconductivity with* chopped *band gap light* under certain conditions [6]. During each cycle of irradiation the same surface density of electrons is freed and distributed between conduction band and defects. Results are shown in Fig. 9 and can be explained by the same set of bulk defect levels as used for calculation of the field effect. The dependence of drift mobility on surface conductivity $\Delta\sigma$ appears again in the photoconductivity with steady irradiation $\delta\sigma_{ph}$ (Fig. 10). This can be understood if the lifetime of excited states does not change very much with surface conductivity $\Delta\sigma$ [7].

The concept of compensation by bulk defects also facilitates the understanding of the *photoconductivity* excited by *photon energies below the band gap* ($\hbar\omega < 3$ eV). Because of the weak absorption single crystals with a diameter of about 1 mm are homogeneously excited. However, the weak photoconductivity depends also on surface conductivity $\Delta\sigma$ according to (1) and in a similar way like in Fig. 10 for $\hbar\omega = 3.4$ eV [12]. According to our model the drift mobility in the bulk is very low since the Fermi level is pinned by a high concentration of defects, similar to the pinning of the Fermi level at semiconductor surfaces by a high density of surface states. In contrast the drift mobility at the surface increases with accumulation and a growing fraction of light-induced electrons appears in the conduction band. This explains the fact that also with homogeneous excitation throughout the crystal the photoconductivity of ZnO is observed as a surface phenomenon.

6. Surface Photovoltage

Results of more recent measurements are given in Fig. 10 [12]. Again a continuous rise of surface photoconductivity $\delta\sigma_{ph}$ with dark surface conductivity $\Delta\sigma$ is found (steady irradiation). Bulk data are not available for the two crystals. At the same surfaces measurements of surface photovoltage δV_{ph} (change of surface potential by irradiation) have been performed and are plotted in the lower part of Fig. 10. Under the assumption of a quasi-equilibrium during steady irradiation a relation between

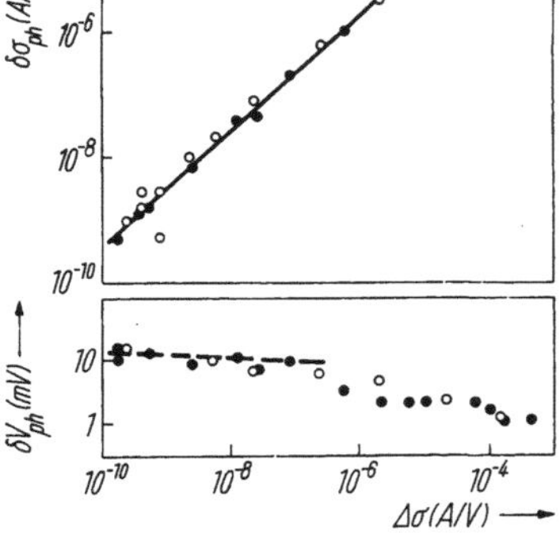

Fig. 10. Prism face. Surface photoconductivity $\delta\sigma_{ph}$ and surface photovoltage δV_{ph} measured as a function of surface conductivity $\Delta\sigma$ in the dark. After Ulrich and Heiland [12] (● crystal 20, ○ crystal 21, $T = 100$ K, steady irradiation $\hbar\omega = 3.4$ eV, absorbed photon flux 4×10^{14} cm^{-2}s^{-1}). The dashed line gives the result of a calculation deriving the photovoltage from the measured photoconductivity by means of the compensation model with defect levels according to Fig. 4, see equation (14). The model is again limited to the non-degenerate range. $\mu_s = 30$ cm²/Vs

$\delta\sigma_{ph}$ and δV_{ph} can be derived:

$$\delta V_{ph} = \frac{\delta N}{\delta N/\delta V_s} = \frac{\delta\sigma_{ph}/e\mu_s}{\delta N/\delta V_s}. \tag{14}$$

The function $\Delta N(V_s)$ and its derivative (see Chapters 3 and 4) can be calculated by means of the compensation model using bulk data of crystal 11 (Fig. 4). The dashed line in Fig. 10 represents the result in the non-degenerate range. Since the transition probability does not appear in the calculation, only the surface mobility μ_s remains as a parameter which played a negligible role in the field effect. A mobility of 30 cm^2/Vs reconciles the theoretical curve with the experimental points (cf. Fig. 3). However, without direct information about the carrier density ΔN the interpretation remains preliminary.

7. Photolysis

UV irradiation causes not only photoconductivity but also, especially in vacuum, a slow and continuous rise of surface conductivity. After termination of the irradiation the increased surface conductivity persists in vacuum. Experiments have been carried out on clean cleaved surfaces excluding photodesorption of adsorbed gases and demonstrating that photolysis takes place [13]. Oxygen is removed and the remaining excess zinc forms donors. In this way the surface conductivity can be increased by several orders of magnitude. The slope of the conductivity versus time curve depends on the particular crystal and its pretreatment. If oxygen is admitted after the first photolysis run the surface conductivity decreases again. A subsequent run is accelerated by a factor of 100. However, an exponential increase with time is found again.

The exponential rise follows straightforwardly from the compensation model: New donors are produced at constant rate during irradiation. Their levels must be above the Fermi level. Complete ionization then means a constant rate of electron injection into the space charge layer:

$$\frac{dQ_{sc}}{dt} = \text{const}.$$

This corresponds to a field effect experiment with the applied external field increasing at a constant rate. The slope of the observed surface conductivity follows with equation (11):

$$\frac{d(\Delta\sigma)}{dt} = \mu_{fd}\frac{dQ_{sc}}{dt} = \mu_{fd}\,\text{const}. \tag{15}$$

Fig. 9 shows that the field effect mobility μ_{fd} is roughly proportional to $\Delta\sigma$.

The slope results to

$$\frac{d(\Delta\sigma)}{dt} \sim \Delta\sigma. \tag{16}$$

An integration demonstrates the exponential rise of surface conductivity with time.

8. Thin Layers

Relation (1) was first observed on polycrystalline layers prepared by vapor deposition of zinc on a quartz or glass substrate and by subsequent heating and oxidation in air [1]. The thickness amounts to about 10^{-5} cm. From the results of various authors a drift mobility at room temperature has been derived and is shown in Fig. 11 of [14]. Like in equation (13) it is related to a change of the total electron density in the conduction band and in bound states. A linear increase of drift mobility with surface conductivity has been found not only for the additional conductivity which is induced by irradiation with UV light or with electrons (1 to 5 keV), but also for the decrease of conductivity by adsorption of a known number of oxygen molecules binding electrons [15, 1]. Furthermore a superlinear rise of surface conductivity with time has been observed during electron bombardment (5 keV) corresponding to the exponential rise on single crystal surfaces discussed as photolysis in the previous Section [1].

The relation between drift mobility and surface conductivity can be explained again by the compensation model. The concentration of defects has to be assumed one or two orders of magnitude higher than for the crystals discussed above. For the decrease of conductivity by oxygen adsorption the slope follows also from the compensation model. The chemisorbed oxygen binds electrons not only from the conduction band but also out of defects near the surface. The low absolute value of the drift mobility can be attributed to simultaneous physisorption of neutral oxygen molecules. In consideration of the polycrystalline structure and a possibly inhomogeneous distribution of bulk defects no quantitative evaluation is given.

9. Conclusion

The linear rise of drift mobility and photoconductivity with dark surface conductivity (Fig. 9 and 10), and thus also equation (1), follow in a simple way from the fact that the derivative of an exponential function (Boltzmann) is proportional to the function itself. A necessary condition is fulfilled by means of the trapping process: The change of surface potential upon an induced charge, $\delta V_s/\delta Q_{sc}$, varies only slowly during a variation of surface conductivity, $\Delta\sigma$, over orders of magnitude. This holds already for the surface state model and equally well for the present compensation model. For the interpretation of a whole sequence of surface phenomena electronic surface states are now replaced by bulk defects. However, there are proofs from the band bending for the existence of other types of surface states: the well-known results after treatment of the surface with reducing or oxidizing agents [1] and also the new data on clean cleaved surfaces [18].

For the six crystals analysed so far, two of the three bulk levels were chosen as donors. The concentrations N_{D1} and N_{D2} were found equal for each of five crystals (see Fig. 3 and 4). No explanation in terms of defect center models can be given at present. Three levels are sufficient for an interpretation of the experiments mentioned above. However, other kinds of measurements may require the introduction of more levels, especially if an extended fraction of the energy gap is involved.

High-temperature treatments in vacuum and under oxygen [16] or hydrogen [1] can change the concentration of bulk defects near the surface. The present experiments are explained without any change of bulk defect concentrations during variation of surface conductivity by various pretreatments.

The idea of a very thin surface layer at 90 K for weak and for medium accumulation (Fig. 7) raises some questions. Quantization of the space charge layer [17] might take place also for low surface electron densities. Furthermore the penetration depth $1/K$ of the strongly absorbed UV light exceeds at 90 K the thickness of the space charge layer by about a factor of ten and that of the conduction electron layer by about a factor of 100. However, the compensation model holds also for photoconductivity in the visible spectral range with homogeneous bulk excitation.

It was not intended here to discuss the complete mechanism of photoconductivity and photovoltage. Our point is only that a variety of experiments, where a change of surface electron density is involved, can be explained by one single concept of compensation.

Acknowledgement

This work was supported by the Deutsche Forschungsgemeinschaft in the frame of the Sonderforschungsbereich SFB 56.

References

[1] G. Heiland, E. Mollwo, and F. Stöckmann, Solid State Phys. **8**, 193 (1959).
[2] P. Ulrich and G. Heiland, to be published.
[3] H. Lüth and G. Heiland, phys. stat. sol. (a) **14**, 573 (1972).
[4] H. Lüth, Surface Sci. **37**, 90 (1973).
[5] E. Mollwo and F. Stöckmann, Ann. Phys. (Leipzig) (6. Folge) **3**, 240 (1948).
[6] G. Heiland, J. Phys. Chem. Solids **6**, 155 (1958).
[7] G. Heiland, J. Phys. Chem. Solids **22**, 227 (1961).
[8] A. Hausmann and W. Teuerle, Z. Phys. **259**, 189 (1973).
[9] G. Müller and R. Helbig, J. Phys. Chem. Solids **32**, 1971 (1971).
[10] E. Veuhoff and D. Kohl, to be published.
[11] A. Many, Y. Goldstein, and N. B. Grover, Semiconductor Surfaces, North-Holland Publ. Co., Amsterdam 1965.
[12] P. Ulrich and G. Heiland, to be published.
[13] D. Kohl and G. Heiland, Surface Sci. **63**, 96 (1977).
[14] G. Heiland, Disc. Faraday Soc. **28**, 168 (1959).
[15] E. Mollwo, Forschungen und Fortschr. **26**, Suppl. 3, 13 (1950).
[16] W. Göpel, Surface Sci. **62**, 165 (1977).
[17] G. Dorda, Festkörperprobleme **13**, 215 (1973).
[18] H. Moormann, D. Kohl, and G. Heiland, I. European Conf. Surface Science, Amsterdam 1978; Surface Sci. **73**, 160 (1978).

(Received June 2, 1978)

phys. stat. sol. (a) 49, 39 (1978)

Subject classification: 1.6; 21; 21.6; 22.1.2

Institut für Angewandte Physik, Lehrstuhl für Festkörperphysik, der Universität Erlangen/Nürnberg, Erlangen[1])

Reconstructed Surfaces

By

K. MÜLLER and K. HEINZ

In honour of Prof. Dr. F. STÖCKMANN's 60th birthday

Crystalline solids respond differently to the creation of a surface. In the case of most metals the surface layer saves its structure from the bulk while many semiconductors show reconstructed surfaces. As a representative the surface of Si(111) with its large (7 × 7) coincidence unit mesh is discussed. However, also a few metal surfaces are reconstructed. It is shown for Ir(100) that the surface layer can reorganize into a structure which still fits partially to the bulk. Moreover the case of Pt(100) shows that also free reorganization can occur. It is demonstrated that in both cases the surface and bulk layer structures are incoherent and that the total electron diffraction pattern arises by multiple diffraction. It is further shown how the different reconstructed surfaces are affected by adsorbing atoms or molecules.

Kristalline Festkörper reagieren unterschiedlich auf die Bildung einer Oberfläche. Im Falle der meisten Metalle bewahrt die Oberflächenlage die Struktur des Volumens, während viele Halbleiter rekonstruierte Oberflächen zeigen. Als repräsentatives Beispiel wird die Oberfläche von Si(111) mit ihrer großen (7 × 7) Koinzidenzeinheitsmasche diskutiert. Jedoch rekonstruieren auch einige Metalloberflächen. Es wird für Ir(100) gezeigt, daß die Oberflächenschicht sich in eine Struktur reorganisieren kann, die sich noch teilweise an das Volumen anpaßt. Darüberhinaus zeigt der Fall von Pt(100), daß auch freie Reorganisation auftreten kann. Es wird gezeigt, daß in beiden Fällen die Oberflächen- und Volumenlagenstrukturen inkohärent sind, und daß das vollständige Elektronenbeugungsbild durch Mehrfachbeugung erzeugt wird. Es wird weiter gezeigt, wie die unterschiedlich rekonstruierten Oberflächen durch adsorbierte Atome oder Moleküle beeinflußt werden.

1. Surfaces – A Simple Cut through the Bulk of a Solid?

The physics of crystal surfaces is in many respects different from that of the corresponding bulk. So, for example, the symmetry break at the surface involves additional features in the density of states (surface states) and forces the electron Bloch waves to obey non-periodic boundary conditions. In general, nearly all physical properties of the bulk are quantitatively modified at the surface.

In most cases the lateral periodicity of a surface coincides with that of the bulk and the only consequence of the establishment of the surface is a slight change in the distances between the first few atomic layers. However, it is known from several crystalline substances that the lateral structure of the surface layer can be different from that of the bulk layers. These examples show that the establishment of a surface can also induce qualitative modifications. In these cases the geometric arrangement of atoms within a bulk layer is not compatible with the minimum of the total free energy when the layer becomes a surface layer. The surface atoms reorganize resulting in a new geometric structure which again corresponds to a minimum of the free energy. The remaining interaction between the first layers decides, in which way the new structure fits to the bulk substrate.

[1]) Erwin-Rommel-Str. 1, D-8520 Erlangen, BRD.

Mostly surface reconstruction happens to semiconductors and only in a few cases to metals. This is a consequence of the fact that the truncation of polar bonds disturbs the solid more seriously than that of metallic bonds. So, with respect to the adsorption of adatoms, reconstructed semiconductor surfaces can be expected to be more stable than those of metals. Indeed adsorption strongly influences reconstructed metal surfaces and can be used to investigate their reorganization behaviour. It is by this argument that this paper, which intends to give a small survey concerning reconstructed surfaces, emphasizes the field of metal surfaces more than that of semiconductors.

2. Reconstructed Surfaces and LEED

The most suitable methods to investigate solids with respect to their structure are by diffraction experiments. However, X-ray diffraction is insensitive to modifications of a few layers only. So low-energy electron diffraction (LEED) has become the mostly used method to investigate the structure of surfaces. Incident electrons with an energy of less than 1000 eV interact very strongly with matter and therefore penetrate only some few atomic layers into the solid. Simultaneously their wavelength corresponds to the interatomic distances and a diffraction pattern of the surface arises. Both, the geometry of the pattern as well as the intensities of the different diffraction spots present the whole information of the surface structure. However, the interpretation of intensities is seriously complicated by multiple diffraction processes. This complication holds also for LEED patterns when reconstructed surfaces are concerned.

2.1 LEED patterns

The periodic arrangement of atoms within a layer causes the incident electrons to exchange momentum only by discrete amounts, i.e. by vectors of its reciprocal lattice (Fig. 1a). Energy conservation means that incident and diffracted wave vectors lie on a common sphere (Ewald sphere) which determines the diffraction angles by the intersection of the lattice rods and the sphere. Diffraction contributions from several identical layers keep the arrangements of the rods, i.e. the diffraction pattern is unchanged (Fig. 1b). However, because of mutual interference the lattice rods become structured and their thickness at the Ewald sphere corresponds to the diffraction intensity in kinematic approximation. Yet, as the latter is really not valid because of multiple diffraction, the rods are only important to construct the diffraction angles. They are therefore deleted in Fig. 1c, d.

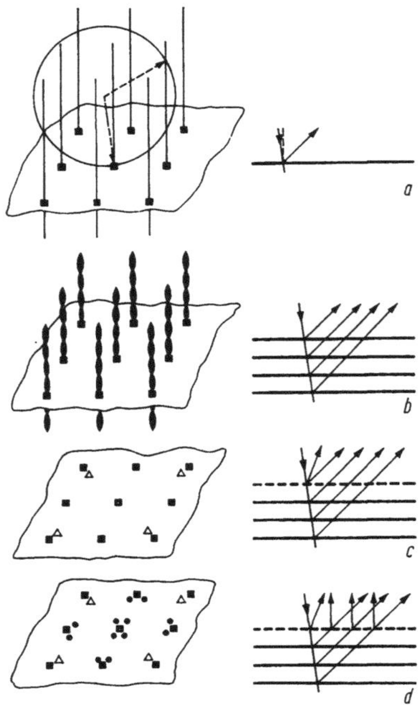

Fig. 1. Reciprocal lattices and diffraction processes for a) one single layer, b) several identical layers, c) identical bulk layers and one incoherent surface layer. In d) the "effective" reciprocal lattice is shown which works in LEED by multiple diffraction

Multiple diffraction in the case of identical layers cannot change the diffraction pattern, because the momentum exchanges at the different layers cannot be distinguished, i.e. their reciprocal lattices coincide. For differently structured layers, however, two other cases can appear. In the first one the layers coincide with a certain periodicity. So the reciprocal lattice of one of them must contain that of the other and therefore already produces the whole diffraction pattern. Multiple diffraction then causes no extra spots. This case is realized in the great majority of reconstructed semiconductor surfaces and adsorption layers on non-reconstructed surfaces.

The second case, which is verified for reconstructed metal surfaces, is totally different. The reciprocal lattices of the surface and the bulk layers do not coincide, that means, their contributions by single diffraction cannot interfere. As the total reciprocal lattice corresponds to the Fourier transform of the total real space it is the incoherent sum of the original lattices of the different layers (Fig. 1c). However, multiple diffraction mixes the lattices by subsequent momentum exchange. Therefore, the "effective" total reciprocal lattice which works in LEED must be understood as the coherent sum of the original lattices. This is equivalent to their mutual linear combination which can be easily performed by shifting the origin of one lattice to all points of the other. This has been done in Fig. 1d for some points near the origin only, because the exchange of larger momentum is less probable. The "effective" reciprocal lattice can be constructed by Fourier transformation, too. However, then the incoherent structures of different layers must be assembled in real space into one single layer in order to account for multiple diffraction by single diffraction only [1].

2.2 LEED intensities

In order to extract the total information from LEED experiments, the measurement and evaluation of diffraction intensities is necessary. Even in the case of "simple" surfaces, where multiple diffraction does not lead to extra spots, the intensities are strongly affected by these processes. It is only a few years that intensity calculation including multiple diffraction can be done with a certain degree of reliability. A rather complete survey of the theory was given by Pendry [2].

Concerning the measurement, the most serious point of the problem arises by the fact that a diffraction spot moves in real space when the electron energy is varied to yield an intensity spectrum. The classical methods which use a Faraday cup or a spot photometer are highly uncomfortable and need hours for measurement. At least in the case of surfaces which are sensitive to residual gas adsorption, these methods are less reliable.

Therefore other methods have been developed recently. One of them photographs the whole LEED pattern at various energies and later on measures the intensities by a densitometer [3]. The most rapid and elegant method, however, makes use of a video camera and a processing computer [4]. As shown schematically in Fig. 2 the camera takes the diffraction picture which is displayed on the monitor. Simultaneously the video signal is put to the computer via an electronic window (sample/hold amplifier) which is operated in a way to perform the intensity distribution along a vertical cut through the pattern. The cut appears on the monitor for control and its intensity profile is shown on the left-hand side of the computer display. The profile is integrated between two cursors (arrows in Fig. 2) which are set by the computer to select the beam wanted. Then the cut is shifted over the whole spot in digital steps each of which containing the integration of the corresponding profile. The procedure stops when the signal decreases below background. Then the total sum of all integrals is background subtracted and stored into the computer which simultaneously displays the result as an intensity–energy point at the right-hand side of the display. A signal

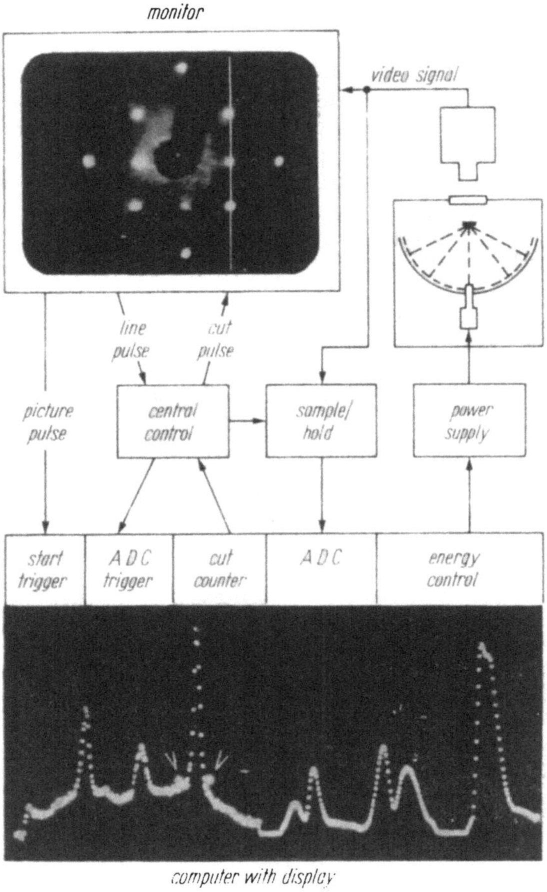

Fig. 2. Principle of the intensity measurement with a computer controlled television camera

is now sent to the power supply in order to vary the electron energy. The software determines the new position of the spot maximum and the whole procedure is repeated. In this way the processor follows the moving spot electronically and finally the whole spectrum appears on the display. It comes out that a total of 5 min is necessary for a spectrum of 600 intensity–energy points. However, within this time only one diffraction spot is measured. Therefore, recently the method was further developed using a video recorder in addition [5]. The intensities of all spots are now simultaneously stored with the rate of the half picture frequency and are evaluated later on by the method described. So a total measuring time of about 12 s is necessary for the total of all spots giving 600 points spectra.

The method described is extremely well suited to record intensities of reconstructed surfaces. As it is clear from Fig. 1d the diffraction spots of these surfaces can be very close to each other in general. This circumstance additionally complicates the measurements of intensities by the classical methods because neighboured spots have to be strictly distinguished. Electronic equipment, however, accounts to this problem easily.

Besides the intensity measurement, also intensity calculation is severely complicated in the case of reconstructed surfaces. This occurs by the enormous increase of the number of plane waves which have to be taken into account. It is proportional to the number of spots, which is n^2 when n is the number of spots of a single plane. Most approximation methods for dynamic intensity calculation require an amount of computing time and storage which varies with the square of the total number of plane waves. So in the case of reconstructed surfaces n^4 instead of n^2 determines the computing amount. For a realistic case, $n = 20$, the increase is by a factor of 400 which usually cannot be handled. However, it sometimes turns out, that more rigorous approximations can avoid the full dynamical calculation and then at least part of information can be extracted from the measured intensities [6 to 8].

3. Semiconductor Surfaces

In contrast to metals the majority of polar semiconductor surfaces are reconstructed and therefore show extra spots in the diffraction pattern [9]. In many cases the kind of reconstruction depends on the kind of preparation and irreversible conversion from one to the other can be caused, e.g. by temperature treatment. All superstructures correspond to coincidence structures, i.e. the periodicities of the surface and bulk layer are integer multiples of each other. So these semiconductors belong to the first of the two groups of reconstructed surfaces we discussed above.

3.1 The clean Si(111) surface

When a clean Si (111) is established by a crystal cleavage in an UHV equipment a (2 × 1) superstructure results which transforms in a stable (7 × 7) superstructure by annealing [10]. A typical LEED pattern of this surface is given in Fig. 3a, where the energy was chosen to be 122 eV in order to show an array of several unit meshes.

Several microscopic models have been developed to account for the large (7 × 7) unit mesh in real space. They assume surface vacancies [10] or extra atoms [11] within the first layer, or propose periodic raising and lowering of surface atoms [12] or pairing of them [13]. So far none of them could be definitely proved which is

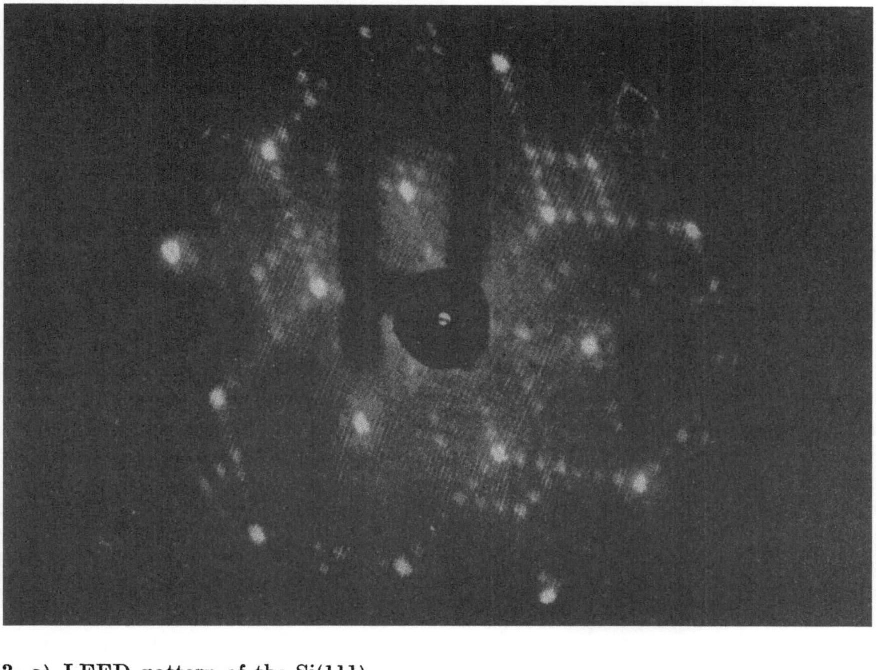

a

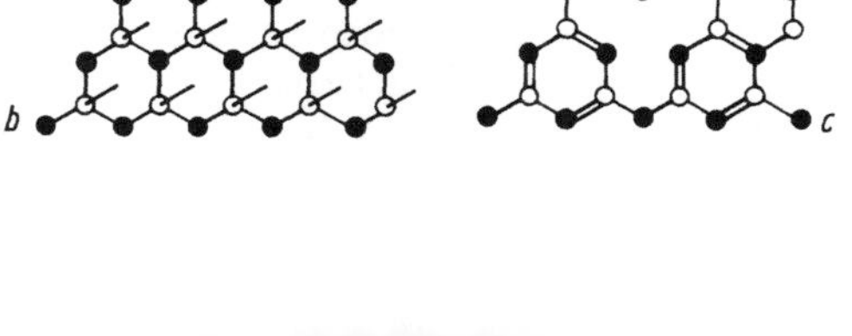

Fig. 3. a) LEED pattern of the Si(111) surface at $E = 122$ eV, b) arrangement of first- and secondlayer atoms (○ and ●, respectively) with dangling bonds in an undisturbed surface double layer, c) surface double layer after reconstruction with double bonds and vacancies

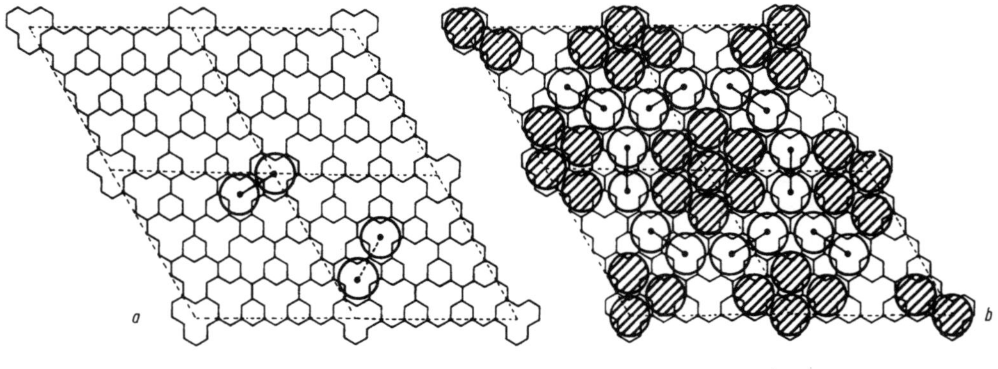

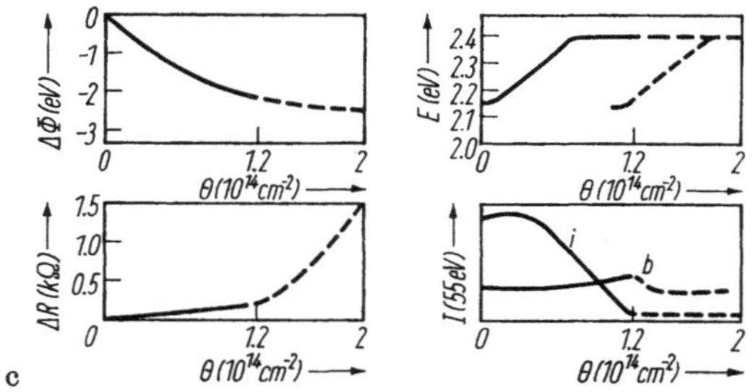

Fig. 4. a) Arrangement of four (7 × 7) unit meshes of the Si(111) surface with cesium atoms adsorbed at different pairs of sites (●——● 6.3, ●- - -● 7.2 Å), b) complete coverage with equal pairs of cesium atoms (●——● 6.3 Å), c) variation of various physical quantities with coverage. For details see the text

a consequence of the failure of dynamical calculation methods for LEED intensities in the case of large superstructures. It is only recently that kinematical calculations have been done for the intensities of the non-integer spots and they seem to favour a further model which requires a special deformation of the first four layers [6]. However, kinematical calculations are of very limited value and there are also experimental arguments for the other models. Perhaps a combination of some of them might resemble the real situation at best. So it is for instance possible that a coexistence of the vacancy model and the four-layer puckering model could fit to the majority of experimental data. Especially adsorption experiments are well suited to test the vacancy model because vacancies should stand out for special adsorption sites. A typical experiment of this kind is discussed in Section 3.2 and it is therefore appropriate to introduce the vacancy model in some more detail.

In the undisturbed, i.e. not reconstructed case the surface atoms would be arranged corresponding to the ideal diamond structure. Then those of the first and second layer which are often said to form a double layer, form a periodic array of hexagons. This is shown in Fig. 3b where open and closed circles represent the upper and lower atoms of the double layer, respectively. However, this undisturbed model is affected by dangling sp^3 bonds which also show up in Fig. 3b. They have to be saturated after the establishment of the surface and this is managed by the appearance of double bonds between first- and second-layer atoms (Fig. 3c). As a consequence some of the surface atoms are no longer bonded. They leave the surface and corresponding vacancies appear. This model is also favoured by theoretical considerations [14, 15] and leads to a (7 × 7) superstructure. Fig. 4a shows an arrangement of four superstructure unit meshes separated by broken lines. Within each mesh 13 atoms are missing.

3.2 The cesium covered Si(111) surface

If we take surface vacancies as a reality, they should show up significantly in adsorption experiments as long as the superstructure is saved during adsorption. Though all vacancy sites seem to be equivalent with respect to their nearest neighbours of Si atoms they are different with respect to their mutual distance. This is shown in Fig. 4a where distances of 6.3 and 7.2 Å appear. For adsorbing atoms with a diameter considerable below these values, this should not affect the adsorption. However, when atoms with a comparable diameter are chosen, mutual interaction between adatoms is possible.

Therefore an adsorption experiment using cesium atoms with a diameter of about 6 Å was performed [16]. A first apparent result is that the (7×7) superstructure is saved. The change of electronic work function $\Delta\Phi$ and of surface resistivity ΔR as well as the binding energy E of cesium atoms and the intensities I of an integer (i) and broken (b) diffraction spot were measured as a function of cesium coverage. Fig. 4c collects the results. In the beginning of the adsorption there is a rapid decrease of the electronic work function and a similar strong increase of the binding energy. There is no considerable change of surface resistivity. This can be interpreted by the assumption that the cesium atoms begin to adsorb at sites with the highest mutual coordination number and the nearest distance from each other. By mutual interaction they so increase their binding energy. This behaviour is opposite to the behaviour of metal surfaces where the binding energy usually decreases with increasing coverage. However, there are also other examples for this unusual dependence of metal atoms on semiconductors, e.g. Zn/GaAs [17].

So in the first stage of adsorption all sites at the corners of the unit meshes are filled leading to an intermediate state which is shown in Fig. 4b by shadowed adatoms. Small islands of Cs atoms have grown giving a partial coverage responsible for the decrease of the work function. The islands are well seperated and consequently the surface resistivity cannot change. Similarly, as there is no closed layer the diffraction intensities should not be affected too much in the very beginning. Then, as observed, the intensity of the bulk spot (i) should decrease relative to the broken surface layer spot (b) because the superstructure features are emphasized by adsorption into vacancy sites. With further increasing coverage the remaining sites with again low and then increasing binding energy must be filled until the coverage saturates. The corresponding results of this adsorption stage are given by broken lines in Fig. 4c. As the coverage is only completed there is no more considerable change of Φ in contrast to that of R which is due to the fact that the islands now become bridged resulting in a metal-like layer. This free and easy interpretation of the measurements within the vacancy model may be taken as a support of this model though other models should not be rigorously excluded.

4. Metal Surfaces

4.1 LEED patterns and structure models for Pt(100) and Ir(100)

The existence of reconstructed surfaces clearly demonstrates the sensitivity of crystals to severe disturbances such as the creation of a surface. In contrast to semiconductors metal surfaces only exceptionally reorganize with respect to their lateral structure. So most of them belong to the first of the following cases, which are in principle possible and occur as a result of the strict cut of all bonds along a net plane [18]:

(i) Dilatation or contraction of the surface layer distance with the atomic order unchanged with respect to the bulk,

(ii) reorganization of the first layer's atomic order but still influenced by the bulk substrate,

(iii) free reorganization of the surface layer.

Iridium and platinum (100) surfaces are examples for the cases (ii) and (iii), respectively, and we are to report about them in the following.

The (100) surface of iridium displays a (5 × 1) superstructure LEED pattern. This simply indicates bulk periodicity of the sample along one of the bulk unit mesh vectors and five times this value along the other. However, the surface layer unit mesh is not simply (5 × 1) of the bulk one. According to a model which is original to Ignatiev et al. [19] and favoured by calculations (Section 4.2) as well as by experimental data (Section 4.3) the surface layer atoms are assumed to coalesce from the original quadratic bulk structure to a nearly close-packed order. In this way they respond to the lack of nearest neighbours at the vacuum side. They increase their lateral binding strength along with an decrease of their corresponding distance. This implies that most surface atoms leave their bulk sites in order to occupy new positions still partly influenced by the layer below. It is this interaction between the two outermost layers that prevents the surface from becoming ideally close-packed. In this way it turns out that with respect to the orthogonal axis of the substrate unit mesh the surface can only exchange momentum which is a multiple of $\frac{6}{5}$ of the corresponding

a c

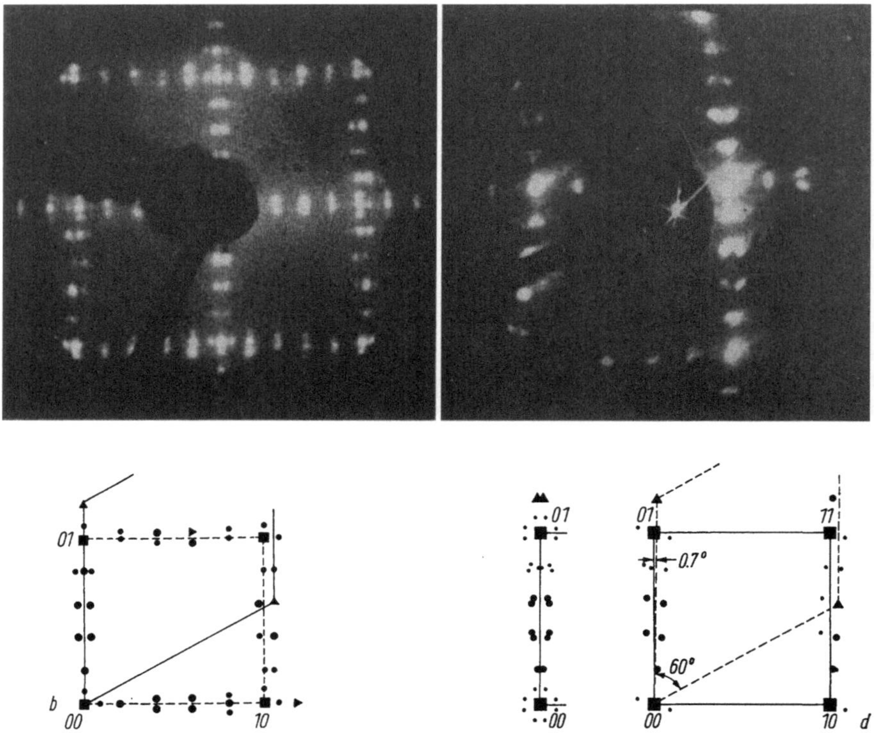

Fig. 5. LEED patterns of a) normal and c) rotated reconstructed surfaces of Pt(100). The parts b) and d) give the reciprocal vectors of the quadratic bulk layer (■) and of the hexagonal surface layer (▲), respectively, from which the double diffraction spots (●) can be constructed by linear combination

substrate momentum. So, for example the spot $\frac{1}{5}0$ can only occur by double diffraction at the surface and bulk layer exchanging momentum corresponding to $\frac{6}{5}0$ and $\bar{1}0$, respectively.

The case of an undisturbed hexagonal surface layer of atoms on top of a bulk (100) plane is realized in the Pt(100) surface. This situation represents the case (iii) mentioned above which might apply to surfaces other than Pt(100) as well. Its fairly complicated LEED pattern (Fig. 5a) contains the (5×1) features of that of Ir, however, with most of the spots split up. As demonstrated in Fig. 5b a thorough analysis of the described surface model associates all spots with either one surface layer (■ or ▲) or with double diffractions at both gratings (•). According to this model interpretation the Pt surface atoms form a layer of ideal hexagonal order regardless of atomic sites below [7, 8, 20]. By symmetry arguments two orthogonal and equivalent domains occur. Apparently the interaction of atoms within the surface highly dominates their interaction with the second layer. The structure phase indicated by the pattern in Fig. 5a is stable only in a moderate temperature range [18]. Elevated temperatures above 1100 K cause the surface to irreversibly take a slightly different position with a modified LEED pattern (Fig. 5c). While the ideal hexagonal symmetry of the surface layer remains, its unit mesh vectors appear to have rotated by a small angle of 0.7° with respect to their former position. This is demonstrated at the right-hand side of Fig. 5d. The left-hand side gives the total of all spots along one axis which arise by the two equivalent and orthogonal arrangements of the hexagons on the substrate each of which is rotated by $\pm 0.7°$. The new phase is stable at all temperatures indicating that a new equilibrium has been reached although the two surface layers are still geometrically incoherent.

4.2 Support of the model by simplified structure calculations

It turns out that this equilibrium is characterized by the maximum mutual overlap of the surface and bulk layer atom cross-sections. This can be interpreted as a maximum of an effective coordination number. Alternatively the overlap of the appropriate binding orbitals can be taken instead. It is surprising indeed, that a simple calculation of this overlap on the basis of the surface model reveals a maximum at $\pm 0.7°$ rotation between the top layers [18]. The quantitative agreement with the experimental result indicates that the described overlap is a simple measure of the interaction between surface layers.

It is interesting to note that a similar calculation for the case of Ir(100) reveals its maximum at the angle of 0°, and therefore no rotation is expected which is in agreement with the measurement.

The (100) surface of gold shows a behaviour similar to that of platinum: there is an ideal hexagonal surface layer [20, 21]. Other metals will probably exhibit reconstructed surfaces, too, perhaps in a less pronounced manner. The W(100) surface, for example, is most likely to be reconstructed [7, 22], but the corresponding $c(2 \times 2)$ diffraction pattern has so far mostly been attributed to impurities.

4.3 Support of the model by LEED intensities

As was pointed out in Section 2.2 some progress has been made in order to use also measured intensities for structure determination. The complexity of the full dynamical theory, however, defines the state of art: a surface model structure deduced from the pattern is used as input for the theory. Calculated intensity-versus-energy dependences can be compared to the measured curves. It is the degree of agreement that is decisive and often favours one out of several alternative models. Although such procedures have been very successful in the determination of simple adsorption structures they

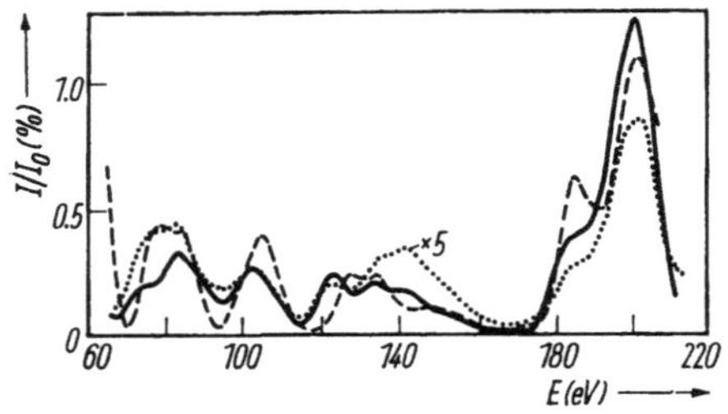

Fig. 6. Normalized intensity spectrum of the 10 beam of both, the "normal" and reconstructed surfaces of Ir(100). The scale at the intensity axis refers to a calculation for the (1 × 1) surface (dashed line) which has been done in RFS approximation. Solid and dotted curves show the measured intensity spectra at the (1 × 1) and (5 × 1) surfaces, respectively

have not yet been applicable to reconstructed surfaces such as Pt(100) or Si(111). On a lower level, however, some possibilities for support of the models by intensity measurements remain. If one develops the full dynamical theory in terms of multiple diffraction processes distinct momentum exchanges associated with single diffraction processes at the different layers can be deduced from intensity maxima. In the case of Pt(100) only reciprocal lattice vectors called for by the model have been found in the intensity spectra [8, 23].

Even better support was possible for the iridium surface model, for which not only the reconstructed but also the undisturbed (1 × 1) surface can be prepared and examined. As shown in Fig. 6 the intensity spectra of a certain bulk layer diffraction beam, e.g. the 10 beam, agree rather well for both cases with respect to their structure [24]. This behaviour is expected by the model since the surface layer cannot create a bulk beam because of its hexagonal structure. So the surface layer can only contribute by third- or higher-order diffraction processes to the intensity of a bulk spot and therefore no serious modifications of the structure of the spectra should be expected. However, the absolute level of intensities is clearly lowered in the case of the reconstructed surface. This again favours the model according to which the surface layer contributes only slightly to diffraction intensities but may considerably attenuate the incident electron beam by inelastic processes. Fortunately the absence of reconstruction at the simple (1 × 1) surface makes dynamical intensity calculations possible. The state of art is demonstrated in Fig. 6 by the broken line [25] using a perturbation scheme which works by renormalization of forward scattering (RFS) and a development in terms of backward scattering events [2].

5. Adsorption Structures and Phase Transitions

Examples for the behaviour of surfaces free of contaminants have been discussed. Adsorbed species give also rise to various surface effects, and much attention has been paid in the literature to the wealth of structures of adsorption layers. In the context of this paper, however, adsorption experiments attain even more importance by the fact that reconstructed surfaces are often unstable under the influence of an adsorption layer. This is a consequence of the fact that at least part of the bonds which had been cut during the creation of the surface are reestablished. Several cases are kown where adsorbed gas molecules [26, 27] or alkali atoms [7, 28, 29] induce a phase transition of the reconstructed substrate surface back to the bulk structure. These investigations are particularly complicated since they involve two mobile atom layers interacting with each other. Apart from the apparent importance of these experiments for catalytic applications they essentially represent very interesting two-dimensional phase transitions under variable boundary conditions. LEED intensities are sensitive to the degree of order and therefore should be applicable to the problem. In the course of adsorption experiments of Cs on Pt(100), which induce a transition of the Pt surface, intensities of certain spots in the diffraction pattern undergo drastic changes as a function of Cs coverage [30]. More detailed investigations of this kind should reveal information even about the dynamics of two-dimensional phase transitions. However, they should be preceded by a considerable improvement of intensity calculations.

Acknowledgements

The authors appreciate the contributions of several coworkers to this paper whose results have been partly used prior to full publication. Support by Deutsche Forschungsgemeinschaft is greatly acknowledged.

References

[1] K. Heinz, Z. Naturf. **29a**, 1514 (1974).
[2] J. B. Pendry, Low Energy Electron Diffraction, Academic Press, London/New York 1974.
[3] P. C. Stair, T. J. Kaminska, L. L. Kesmodel, and G. A. Somorjai, Phys. Rev. B **11**, 623 (1975).
[4] P. Heilmann, E. Lang, K. Heinz, and K. Müller, Appl. Phys. **9**, 247 (1976).
[5] P. Heilmann, E. Lang, G. Hanke, K. Heinz, and K. Müller, to be published.
[6] J. D. Levine, S. H. McFarlane, and P. Mark, Phys. Rev. B **16**, 5415 (1977).
[7] K. Müller, P. Heilmann, K. Heinz, and G. G. Waldecker, Vakuum Tech. 8, 227 (1976).
[8] K. Heinz, P. Heilmann, and K. Müller, Z. Naturf. **32a**, 28 (1977).
[9] For a review see: E. Tosatti, Festkörperprobleme **XV**, 113 (1975).
[10] J. J. Lander and J. J. Morrison, J. appl. Phys. **34**, 1403 (1963).
[11] W. A. Harrison, Surface Sci. **55**, 1 (1976).
[12] D. Haneman, Phys. Rev. **170**, 705 (1968); **121**, 1093 (1961).
[13] J. J. Lander, G. W. Gobeli, and J. J. Morrison, J. appl. Phys. **34**, 2298 (1963).
[14] J. C. Phillips, Surface Sci. **40**, 459 (1973).
[15] K. C. Pandey and J. C. Phillips, Phys. Rev. Letters **34**, 1450 (1975).
[16] P. Wagner, K. Müller, and K. Heinz, Third Interdisciplinary Surface Science Conf., York (England), March 1977; Surface Sci. **68**, 189 (1977).
[17] G. Laurence, B. A. Joyce, C. T. Foxon, A. P. Janssen, G. S. Samuel, and J. A. Venables, Surface Sci. **68**, 190 (1977).
[18] P. Heilmann, K. Heinz, and K. Müller, to be published.
[19] A. Ignatiev, A. V. Jones, and T. N. Rhodin, Surface Sci. **30**, 573 (1972).
[20] D. G. Fedak and N. A. Gjostein, Surface Sci. 8, 77 (1967).
[21] P. W. Palmberg and T. N. Rhodin, J. chem. Phys. **49**, 134 (1968).
[22] M. K. Debe and D. A. King, Phys. Rev. Letters **39**, 708 (1977).
[23] K. Heinz, N. Lieske, and K. Müller, Z. Naturf. **31a**, 1520 (1976).
[24] E. Lang, K. Heinz, and K. Müller, to be published.
[25] H. Köhler and K. Heinz, to be published.
[26] J. Küppers and H. Michel, Verh. DPG (IV) **13**, 580 (1978).
[27] C. R. Helms, H. P. Bonzel, and S. Kelemen, J. chem. Phys. **65**, 1773 (1976).
[28] H. Steininger, E. Lang, K. Heinz, and K. Müller, to be published.
[29] T. Grandke and K. Heinz, Z. Naturf. **32a**, 1049 (1977).
[30] N. Lieske, Diplomarbeit, Erlangen 1975.

(Received June 23, 1978)

phys. stat. sol. (a) **49**, 51 (1978)

Subject classification: 1.2 and 14.4.1; 22.8.1

Fachrichtung 11.3 Technische Physik der Universität des Saarlandes, Saarbrücken[1])

Optical Investigation of the Cubic–Tetragonal Phase Transition in $BaTiO_3$ Single Crystals

By

H. E. MÜSER, W. KUHN, and J. ALBERS

In honour of Prof. Dr. F. STÖCKMANN's 60th birthday

In the paraelectric phase of pure $BaTiO_3$ single crystals birefringence patterns can be observed, which in literature are explained by the existence of Schottky surface layers. It is shown, that the patterns are solely caused by the mechanical preparation of the crystals. Optical and electron microscopic investigations reveal plastically distorted surfaces layers, which result from the polishing process and have a thickness of about 1 μm. These surface layers exhibit an increased transition temperature and create a local variation of the Curie point in the whole crystal. This fact can be explained by energy considerations.

Die in der paraelektrischen Phase des $BaTiO_3$ beobachtbare Doppelbrechung ist nicht, wie in der Literatur angegeben, durch eine Schottky-Randschicht, sondern allein durch die mechanische Bearbeitung der Kristalle bedingt. Durch optische und elektronenmikroskopische Untersuchungen wird nachgewiesen, daß durch das Polieren der Kristalle eine plastisch deformierte Randschicht mit einer Dicke von etwa 1 μm erzeugt wird. Die Randschicht besitzt eine erhöhte Umwandlungstemperatur und ruft im gesamten Kristall eine ortsabhängige Curiepunktanhebung hervor, die anhand energetischer Betrachtungen begründet werden kann.

1. Introduction

The behaviour of ferroelectric crystals is considerably influenced by deviations from the ideal crystal structure, e.g. by impurities and surface layers. Most investigations on $BaTiO_3$ single crystals were done on thin plates grown from melts with fluxes like $BaCl_2$ or KF. Such crystals contain impurities of about 1 mol% and exhibit distorted surface layers [1]. Layer thicknesses were found between 10^{-3} and 10^{-6} cm, the results depend on the method of measurement [2]. On the other hand, if barium titanate single crystals have been grown from TiO_2-rich melts, very pure crystals can be produced differing in many properties from flux grown crystals [3, 4].

From optical investigations above the cubic-tetragonal phase transition temperature T_c it was reported, that such pure single crystals exhibit surface layers up to 1 mm thickness, thus near T_c even relatively large crystals should consist of surface layers only [5]. Furthermore from the assumption, that the surface layers are of the Schottky type, it was concluded that near T_c the undistorted bulk material should have a dielectric constant larger than 10^6 [6]. However in dielectric measurements, done for examining this statement, a dielectric constant of the usual order of only 10^4 was found [7]. Thus, the interpretation mentioned above seems to be dubious. The following optical investigations should contribute to an elucidation of these problems.

2. Experimental

All investigated material was grown by top-seeded solution method from TiO_2-rich melt [8]. Each grown crystal had a mass of about 30 g. By means of a wire saw the

[1]) D-6600 Saarbrücken, BRD.

crystals were cut into rectangular blocks. The edges of the samples have lengths of 3 to 5 mm and are parallel to the cubic crystal axes. The surfaces were polished on lead plates by diamond powder of 9, 6, 3, 1, and 0.25 μm grain size (Metadi aerosol spray) successively to optical quality.

A polarizing microscope Leitz Orthoplan-Pol was used for the optical investigations. The temperature was controlled by a Leitz microscope heater type 1300, and it was measured by NiCr–Ni thermocouples, which gave a resolution of about 0.05 K. By appropriate thermal isolation the temperature differences within the sample were kept in the same range. The thermocouples were absolutely calibrated against a quartz thermometer. All investigated samples showed nearly the same behaviour, but the corresponding temperatures differ from sample to sample by some tenths of a degree. The electron microscopic investigations were performed by a JOEL JEM 200-A electron microscope at 200 kV.

3. Structure of the Surface Layers

The birefringence pattern of the first (100) crystal face (F_1) at 403.6 K is shown in Fig. 1a. The crystal is oriented in diagonal position between crossed polarizers. In the middle of the crystal the birefingence is nearly zero. Parallel to the edges stripes can be seen having a distance period d near the dark region and a period $d/2$ near the edges. Monochromatic light was used for illumination ($\lambda = 589$ nm). If the temperature is lowered, the fringe area grows into the dark region in the middle, just as it was reported by Höchli and Pohl [5]. At higher magnification, however, a sawtooth-like structure of the stripes can be observed (Fig. 1b). This structure hardly can be understood by the explanation on the basis of Schottky layers [5, 6, 9].

In Fig. 2 and 3 corresponding photographs of both other (100) faces (F_2 and F_3, respectively) of the same sample are shown. F_1, F_2, and F_3 are orthogonal to each other and are a priori equivalent. Thus, Fig. 2 and 3 give an additional argument against the theory, that the electro-optical effect of a Schottky layer would be the reason for the appearance of the pattern in Fig. 1a. But all phenomena from the Fig. 1 to 3 quantitatively can be explained by the assumption, that in the paraelectric temperature range triangular prismatic regions near the sample edges are already ferroelectric and have

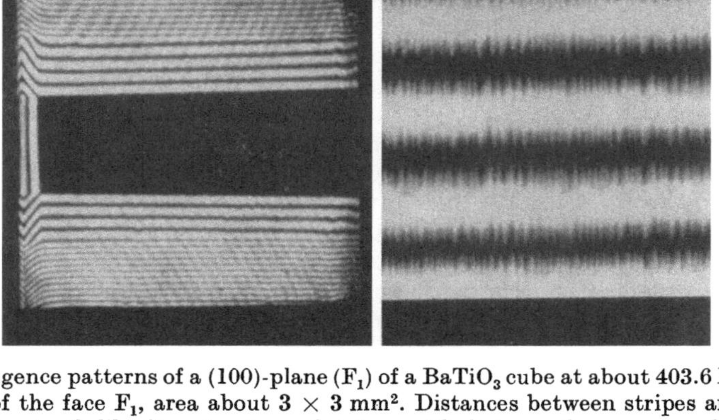

a b

Fig. 1. Birefringence patterns of a (100)-plane (F_1) of a $BaTiO_3$ cube at about 403.6 K ($\lambda = 589$ nm). a) Total view of the face F_1, area about 3×3 mm². Distances between stripes are d or $d/2$, $d = 107$ μm. b) Sawtooth-like fine structure of stripes with distance d near a sample edge. Temperature is a few 10^{-2} K higher than in Fig. 1a

Fig. 2

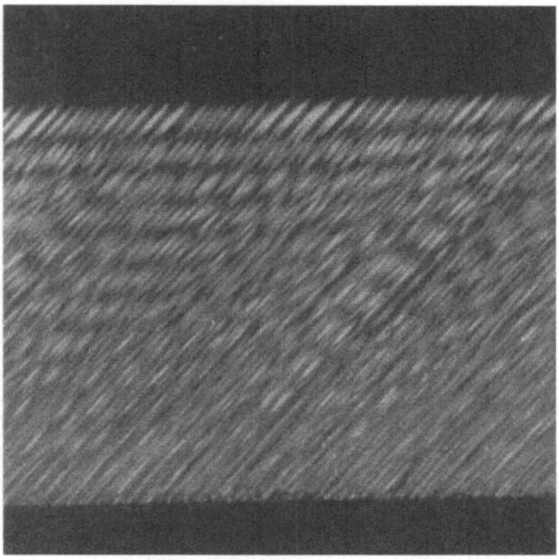

Fig. 3

Fig. 2. Total view of the birefringence of a second (100)-plane (F_2) of the same sample as in Fig. 1; $T = 403.6$ K

Fig. 3. Birefringence of the third (100)-plane (F_3) of the same sample as in Fig. 1 and 2. The distance from the crystal edge at the lower part of the figure to the beginning of the upper dark region (towards the middle of the sample) is about 0.24 mm; $T \approx 403.6$ K

a domain structure with 90°-walls as schematically shown in Fig. 4. The boundaries between the ferroelectric regions and the paraelectric region in the middle of the sample are plane phase transition fronts (PTF). If temperature is lowered the PTF's move into the middle of the sample.

The PTF orientation with respect to the crystal axes can be calculated by means of the Wechsler-Liebermann-Read theory of martensitic phase transitions [10], if the lattice parameters of both phases are known (see e.g. [11]). Using the lattice constants given by Clarke [12] the following angles (definitions in Fig. 4) are found for a temperature of 403 K: $\alpha = (90{,}1 \pm 0{,}1)°$, $\beta = (38 \pm 3)°$, $\gamma = (52 \pm 3)°$. Within the accuracy of measurement of about 1° calculated and observed angles agree. Furthermore, the ratio of thicknesses of both types of 90° domains also agrees with the theory [10].

In F_1 (Fig. 1 and 4) the spontaneous polarization P_s of the domains is alternately parallel and orthogonal to the surface of the sample. Due to the optical symmetry only those domains, in which P_s is oriented parallel to the surface, contribute to double refraction. The sawtooth structure (Fig. 1b) occurs, because the domain walls include an angle of

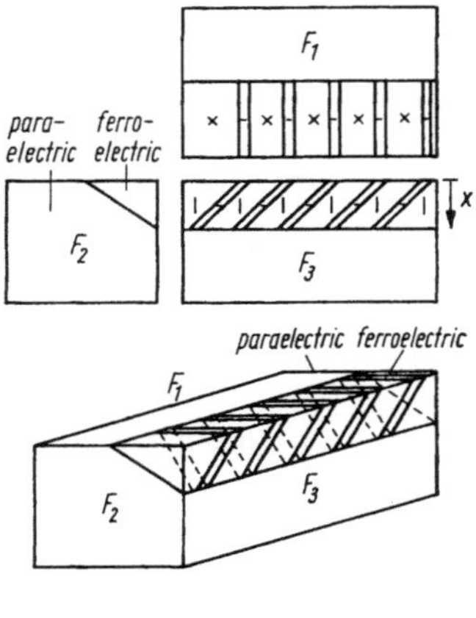

Fig. 4. Schematic representation of one prismatic ferroelectric region. P_s is parallel to F_3 and parallel to the dashed lines plotted in F_3

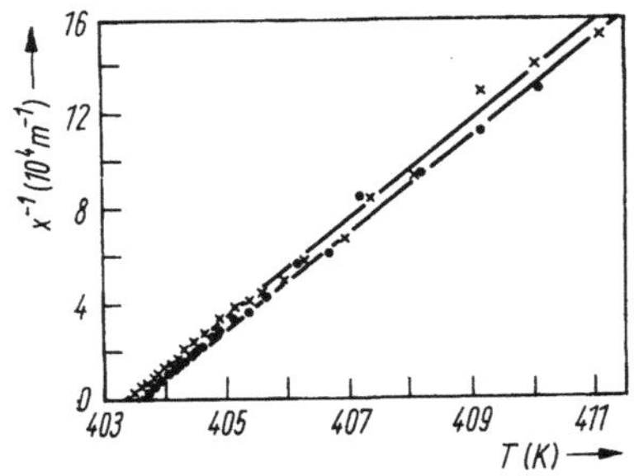

Fig. 5. Reciprocal of the width x of the ferroelectric region as a function of temperature. × decreasing and ● increasing temperature

45° with the surface. The distance between the stripes in Fig. 1 is constant, for the thickness of the ferroelectric region decreases linearly with the distance from the edge of the sample. If the optical phase difference results from one prismatic region only, so the stripes have a distance d as in Fig. 1a from the inside. If the light has passed through two ferroelectric regions, the distance of the stripes is only $d/2$ as in the outer part of the pattern in Fig. 1a.

The situation in F_2 (Fig. 2) easily can be understood from Fig. 4. The direction of domain walls (under 45° to the sample edges) can be seen in Fig. 3. Furthermore, there is a superposition of birefringence stripes parallel to the edges, similar to those in Fig. 1 but with smaller distances of about $d/5$ or $d/10$, respectively. These stripes are rather irregular, because the light is travelling not exactly parallel to the 90°-domain walls.

4. Temperature Dependence of the PTF Position

In Fig. 5 the reciprocal of the width x of the ferroelectric region (Fig. 4) is plotted versus temperature T. It holds the linear relation

$$\frac{1}{x} = g(T - T_1) . \qquad (1)$$

If the temperature is decreased, T_1 is slightly lower as if the temperature is increased. At about $T_c = 403.5$ K, very near to T_1, the whole crystal becomes ferroelectric. Above T_c the position of the PTF varies reversibly with temperature. Small birefringent ferroelectric regions can be observed near the sample edges up to temperatures of more than 10 K above T_c. Thus, the width of the ferroelectric region shows a similar behaviour as the width of the interference region reported by Höchli and Pohl [5]. So we believe that both phenomena are caused by the same reason. In both cases the width increases critically, if the Curie point is approached. If the explanation by means of a Schottky layer whould be correct, the width should critically depend on the dielectric constant [5, 6]. Thus, the divergence should occur at the Curie-Weiß temperature only, which is about 15 K below T_c.

5. Deformation of the Surface

If the crystal is cooled down into the ferroelectric range, just that domain structure occurs which had been present during polishing. This holds for as many heating and cooling cycles as one likes. But the domain structure is changed completely, if a layer of about 1 μm thickness is etched from all faces of the sample (for instance by H_3PO_4 at 430 K). Then all the phenomena describe aboved no longer can be observed. The phase transition occurs abruptly for cooling at about $T_c^- = 402$ K and for heating at about $T_c^+ = 404$ K. It should be remarked, that all transition temperatures both

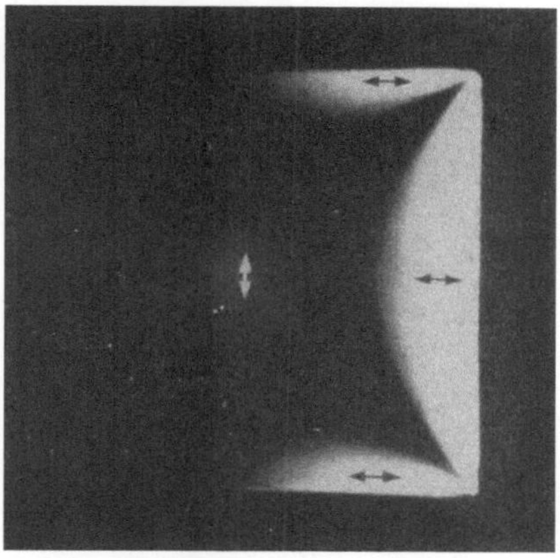

Fig. 6. Stress birefringence of a $BaTiO_3$ cube. The right-hand face was polished; $T = 443$ K. The arrows indicate n_γ

for polished and etched samples depend on the thermal prehistory of the sample. They are the higher the longer the sample was kept in the ferroelectric phase. Differences are up to 3 K. The transition temperatures reported here are the lowest ones which were observed; they can be reached only if the crystal was kept in the paraelectric phase for many hours.

If all faces of an etched sample are polished again, all the phenomena described above occur again. If only one face is polished at room temperature, after the next heating and cooling cycle through the transition temperature the spontaneous polarization in all domains of the whole crystal is parallel to the polished face. This fact holds for all following heating and cooling cycles until an additional polishing or etching process is done. Etching and polishing can be repeated again and again without any change of the facts given above.

In the cubic phase of polished crystals there is an additional type of double refraction, which is nearly independent of temperature. It is weaker than the double refraction due to P_s and is shown in Fig. 6. It increases from the middle of the sample towards the edges and becomes largest near the corners (order of magnitude: 10^{-4}). A similar behaviour was reported by Aso for polished crystals of $SrTiO_3$ [13]. This type of double refraction can also be eliminated by etching. To all appearance it is caused by mechanical stress which is generated by the polishing process.

After polishing a thin surface layer exhibits a *plastic* deformation, which causes an *elastic* strain in the whole crystal. Because in the cubic phase $BaTiO_3$ becomes optically positive by uniaxial pressure [14], from the orientation of the largest refraction index n_γ one can distinguish between dilatation and compression. Corresponding measurements with a view to completing the observations of Fig. 6 had the result, that the surface layer is dilated. If the crystal has been polarized before the measurement, the dilatation is larger parallel to the original direction of P_s than orthogonal to it (Fig. 7).

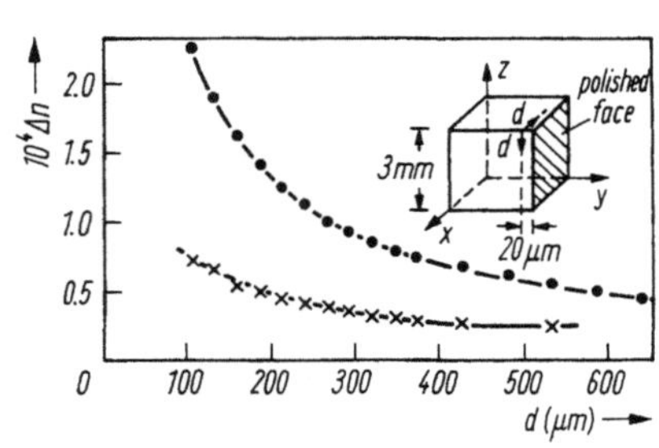

Fig. 7. Stress birefringence depending on the distance from the crystal edge. During polishing P_s was in z-direction. × $d = d(x)$, light parallel z; ○ $d = d(z)$, light parallel x. $T = 443$ K

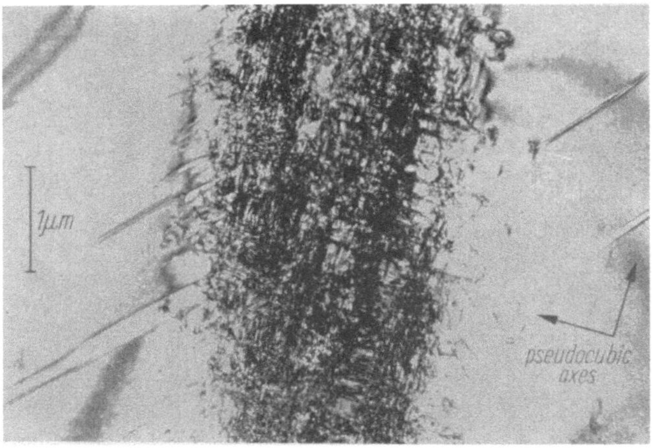

Fig. 8. Electron microscope photograph showing dislocations caused by a relatively large grain of the polishing powder. About 0.5 µm was etched from the polished side

A few surface layers were investigated by the electron microscope, too. Thin samples were etched by ion beam down to get regions with thicknesses below 1 µm. Thus, transmission photographs could be taken. A dense net of dislocations near the polished surface was found. At some places the density of the disolcations was so large, that the dislocations could not be resolved. Counting out the dislocations by the method of Ham [15], we found a total length of dislocations of more than 6×10^7 m/m^2 within a surface layer of less than 1 µm thickness. If about 0.5 µm was etched from the polished side of the sample, dislocations could be observed only near some rather straight lines as shown in Fig. 8. Obviously we see here the middle of the trace of a relatively large grain of the polishing powder. If about 1 µm has been etched from the polished side, all dislocations have disappeared. Because in general the volume grows with increasing dislocation density, the dilatation of the surface layer and thus the deformation of the whole volume of the sample can be understood. Furthermore, the interaction of dislocations and spontaneous ferroelectric deformation causes the memory effect with regard to the domain structure as mentioned at the beginning of this section.

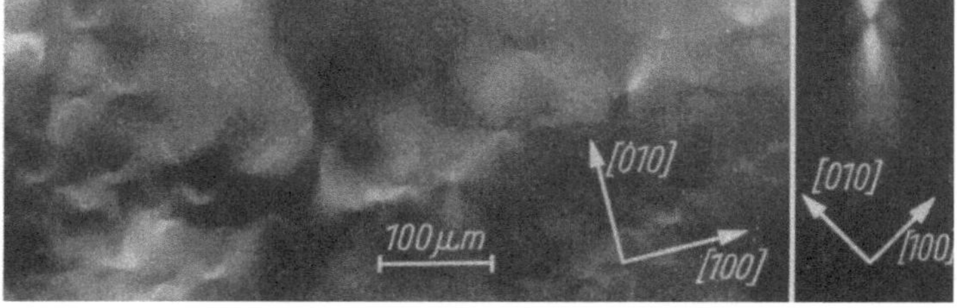

Fig. 9. Birefringence figures of dislocations created during crystal growth. $T = 443$ K. a) General view with high density of dislocations. b) Single dislocation

If step by step 100 nm at a time are etched from the surface, and then the investigations of Fig. 6 and 7 are repeated, it can be found, that all effects rapidly decrease. If about 200 nm are etched, the double refraction is less than half the original value. But only if about 1 μm has been etched, the effects can no longer be observed.

Finally, a third type of double refraction can be seen in Fig. 9. It is restricted to the near surrounding of special spots. Because of its typical form [16] it looks likely, that this type of double refraction is caused by dislocations which are originated during crystal growth. Most of them are parallel to the crystal pulling direction. For our investigations here we used samples with no or only a few optically visible dislocations.

6. Discussion

From our investigations it follows that the distortion of polished surfaces leads to an increase of the transition temperature of barium titanate. So we attribute a cubic-tetragonal phase transition temperature T_c^s to the surface layer and assume that T_c^s is considerably higher than the transition temperature T_c of the undistorted material. The increase may result from a difference between the elastic energy of the dislocations in the cubic phase and in the tetragonal phase.

In the simplifying model of Fig. 10 the density of the Gibbs function is defined G and G^s in the undistorted bulk material and in the surface layer, respectively. The cubic paraelectric phase and the tetragonal ferroelectric phase are marked by a subscript p and f, respectively. Then the equilibrium position x (Fig. 10) of the PTF can be found from the condition, that the Gibbs function of the whole crystal is independent of a small variation δx of the PTF position. Assuming, that the PTF is strictly in accordance with the WLR theory [10], and neglecting the energy of the PTF and of the domain walls, the condition is

$$\delta \int G \, dV = [(G_f^s - G_p^s)\, t + (G_f - G_p)\,(x \tan \gamma - t)]\, l\, \delta x = 0\,. \tag{2}$$

l is the length of the crystal edge orthogonal to the plane of Fig. 10. Furthermore, we assume that in first approximation the differences between the Gibbs functions of both phases are proportional to the difference between actual temperature and corresponding transition temperature:

$$G_f - G_p = a(T - T_c)\,, \tag{3}$$

$$G_f^s - G_p^s = a(T - T_c^s)\,. \tag{4}$$

From (2) to (4) it follows:

$$\frac{1}{x} = \frac{\tan \gamma}{d(T_c^s - T_c)}\,(T - T_c)\,. \tag{5}$$

This is just the experimental equation (1). If the factor in (3) is slightly different from

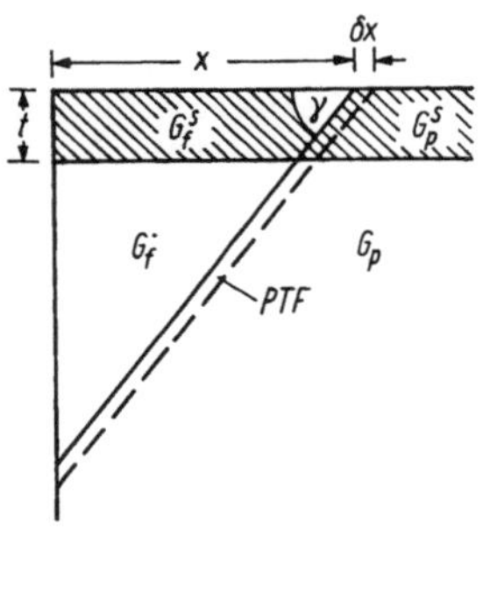

Fig. 10. Schematic representation of the ferroelectric region with the phase transition front PTF and the distorted surface layer of thickness t

that one in (4), the result is only a little changed. Equating (5) with (1), we find

$$T_c^s - T_c = \frac{\tan \gamma}{gt}. \tag{6}$$

The experimental values are $g = 2 \times 10^4\ m^{-1}\ K^{-1}$ (Fig. 5), $t = 1\ \mu m$, and $\gamma = 52°$. Thus, we find $T_c^s - T_c = 64$ K. Because of the diverse simplifications this result is a rough approximation only, but the order of magnitude seems to be correct. In the very thin crystal foils which we used for the investigations by the electron microscope, up to 28 K above T_c birefringence could be observed. Obviously this effect was caused by ferroelectric regions.

Because the ferroelectric behaviour is strictly connected with the mechanical treatment of the samples it can be assumed, that a number of published results was influenced by distorted surface layers. For instance, the dielectric constant of barium titanate depends on thickness [7], if polished crystals are investigated; this dependence vanishes, if the crystals are etched. Furthermore, in small $BaTiO_3$ particles deformations were found [17]; we assume that this deformation is no special effect of small particles but of the mechanical treatment during the crushing process. Also the flexure resonances, which were observed in the paraelectric phases of $BaTiO_3$, KH_2PO_4, and other ferroelectrics [9, 18], perhaps are caused by the piezoelectric effect of ferroelectric surface regions. Recently, in the case of $SrTiO_3$, Darlington and O'Connor [19] discussed the possibility that a central mode could be caused by dislocations in a surface layer.

Acknowledgements

This work has been supported by the Deutsche Forschungsgemeinschaft within the scope of the Sonderforschungsbereich 130 "Ferroelektrika". We are also indebted to Prof. H. Gleiter and M. Schuler from the Fachrichtung Werkstoffwissenschaften of our university for helpful assistance concerning the investigations by electron microscope.

References

[1] H. Arend and J. Novák, Kristall und Technik **1**, 93 (1966).
[2] F. Jona and G. Shirane, Ferroelectric Crystals, Pergamon Press, London 1962.
[3] A. Linz, V. Belruss, and C. S. Naiman, J. Electrochem. Soc. **112**, 60C (1965).
[4] H. Sasaki, Japan. J. appl. Phys. **4**, 24 (1965).
[5] U. T. Höchli and D. W. Pohl, Ferroelectrics **13**, 403 (1976).
[6] U. T. Höchli and H. Mueller, Ferroelectrics **13**, 399 (1976).
[7] R. Dudler, J. Albers, and H. E. Müser, contribution 2 E-4 to IMF4, Leningrad 1977, to be published in Ferroelectrics.
[8] J. Albers, Top Seeded Solution Growth of $BaTiO_3$ Single Crystals from the TiO_2-rich Melt, presented at ECCG1, Zürich 1976 (abstract D7, p. 150).
[9] U. T. Höchli, J. Phys. C. **8**, 3896 (1975).
[10] M. S. Wechsler, D. S. Liebermann, and T. A. Read, Trans. AIME **197**, 1503 (1953).
[11] M. DiDomenico, Jr. and S. H. Wemple, Phys. Rev. **155**, 539 (1967).
[12] R. Clarke, J. appl. Cryst. **9**, 335 (1976).
[13] K. Aso, Japan. J. appl. Phys. **15**, 1243 (1976).
[14] M. G. Cohen, M. DiDomenico, Jr., and S. H. Wemple, Phys. Rev. B **1**, 4334 (1970).
[15] R. K. Ham, Phil. Mag. **6**, 1183 (1961).
[16] D. A. Jenkins and J. J. Hren, Phil. Mag. **33**, 173 (1976).
[17] M. Anliker, H. R. Brugger, and W. Känzig, Helv. Phys. Acta **27**, 99 (1954).
[18] U. T. Höchli and K. Fossheim, J. Phys. C. **8**, 1928 (1975).
[19] C. N. W. Darlington and D. A. O'Connor, J. Phys. C **9**, 3561 (1976).

(Received June 15, 1978)

phys. stat. sol. (a) 49, 59 (1978)

Subject classification: 8; 22.1.3

Institut für Festkörperphysik der Technischen Universität Berlin, Berlin (West)

Heat Capacities and Thermodynamic Properties of α-Monoclinic Selenium in the Range 3 to 300 K

By

M. MEISSNER, A. TAUSEND, and D. WOBIG

In honour of Prof. Dr. F. STÖCKMANN's 60th birthday

The molar heat capacities at constant pressure of small single crystals of α-monoclinic selenium are measured by applying a heat pulse technique in the temperature range 3 to 300 K. Below 5 K the molar heat as a function of temperature obeys Debye's T^3-law yielding a Debye temperature $\theta_D = 128.3$ K. The molar heat at constant volume can be described over the entire temperature range by applying a combination of Debye- and Einstein-terms for the frequency distribution function, with respect to the ring molecular structure of monoclinic selenium. Furthermore, values of thermodynamic functions are calculated. By combining the results of this investigation and data on trigonal selenium given by Chang and Bestful the Gibbs free energy of conversion is obtained. This calculation shows that α-monoclinic selenium is unstable in the temperature range investigated.

Die molaren Wärmekapazitäten bei konstantem Druck von kleinen monoklinen α-Seleneinkristallen werden mittels einer Heizimpulstechnik im Temperaturbereich von 3 bis 300 K untersucht. Unterhalb 5 K befolgt die Molwärme als Funktion der Temperatur das Debyesche T^3-Gesetz und ergibt eine Debyetemperature von $\theta_D = 128{,}3$ K. Die Molwärme bei konstantem Volumen läßt sich im gesamten Temperaturbereich mittels einer Kombination von Debye- und Einsteintermen für die Frequenzverteilungsfunktion bezüglich der molekularen Ringstruktur von monoklinem Selen beschreiben. Darüberhinaus werden die Werte der thermodynamischen Funktionen berechnet. Die Ergebnisse dieser Untersuchung ergeben zusammen mit den von Chang und Bestful gegebenen Werten für trigonales Selen die Gibbssche freie Konversionsenergie. Die Berechnung zeigt, das monoklines α-Selen im untersuchten Temperaturbereich instabil ist.

1. Introduction

The thermodynamic properties of trigonal selenium have been thoroughly explored in the investigations by Anderson [1], DeSorbo [2], Lasjaunias [3], Chang and Bestful [4], and Grønvold [5]. No such satisfactory situation exists for monoclinic selenium. The heat capacity values of monoclinic selenium in the temperature ranges from 103 to 383 K and 288 to 348 K reported by Gattow and Heinrich [6] and Mondain-Monval [7], respectively, differ considerably from each other. A recent investigation in the low-temperature region from 2 to 20 K by Wu and Luo [8] exhibits a cubic T-law below 3 K. Above 3 K an excess specific heat has been found in this work which is correlated with the relative stabilities of amorphous, monoclinic, and trigonal selenium. However, it is not quite clear whether the growth of monoclinic selenium of a saturated solution of methylene iodide instead of CS_2 results in crystals of comparable quality. For trigonal selenium large differences in the low-temperature heat capacity data are found when trigonal crystals were investigated grown by a high-pressure technique [9] and from doped melts [10].

In addition to the low-temperature heat capacity of monoclinic selenium it is important to get data up to room temperature. These data give information on the enthalpy

Table 1

Smoothed values of the molar heat capacity C_p, enthalpy $\{H(T) - H(0)\}$ and entropy $S(T)$ of α-monoclinic selenium

T (K)	C_p ($J\ mol^{-1}\ K^{-1}$)	$H(T) - H(0)$ ($J\ mol^{-1}$)	$S(T)$ ($J\ mol^{-1}\ K^{-1}$)
3	0.025	0.0186	0.0083
4	0.059	0.050	0.019
5	0.115	0.089	0.027
7	0.318	0.409	0.084
10	0.887	1.88	0.264
15	2.28	9.06	0.849
20	3.78	23.6	1.69
30	6.54	74.2	3.75
40	8.94	151	5.97
50	11.0	250	8.19
60	12.8	368	10.4
70	14.4	504	12.5
80	15.8	655	14.5
90	17.1	818	16.4
100	18.1	994	18.3
120	19.9	1374	21.8
140	21.2	1784	24.9
160	22.2	2217	27.8
180	22.9	2668	30.5
200	23.6	3132	32.9
220	24.1	3609	35.2
240	24.5	4094	37.3
260	24.9	4588	39.3
280	25.2	5088	41.2
298.2	25.5	5549	42.8
300	25.5	5595	42.9

$\{H(T) - H(0)\}$ and the entropy $S(T)$. Starting with a study on the thermodynamic stability of α-monoclinic selenium from 170 K up to the melting point [11] we present now specific heat measurements of small single crystals in the temperature range from 3 to 300 K.

2. Experimental

High purity selenium (99.999%) was obtained from American Smelting and Refining Co. Single crystals of α-monoclinic selenium were grown from a saturated solution of CS_2. The growth rate was very small. After some weeks the largest crystals grown were of size $2 \times 2 \times 2\ mm^3$. Further details of growing monoclinic crystals are given elsewhere [11].

Heat capacity studies reported here were performed by applying a heat pulse technique. The theory and the experimental details of this method are reported in previous papers [12, 13]. For measurements two α-monoclinic crystals exhibiting parallel planes were chosen. The crystals were of size as given above. The front surface of one crystal was blackened with soot in order to absorb the energy from an extern flash light source.

The back side of the lower crystal is mounted to a copper sample holder by GE 7031 low-temperature varnish. The temperature rising between the two samples is measured with a 0.1 mm diameter gold–iron/chromel thermocouple. One junction of the thermocouple wires was fixed onto the surface of the sample holder in good thermal contact in order to obtain the equilibrium temperature. The other junction of the thermocouple consists of a thin indium plate (thickness ≈ 0.08 mm) connecting the two thermocouple wires. This junction was fixed between the two crystals by a GE 7031 varnish layer. In our experiments we have found that the above procedure gives good thermal and mechanical contact in the whole temperature range. The influence of thermal contact resistance which arises in the intermediary layer of indium and GE 7031 varnish could be neglected. However, after some cooling and heating cycles between liquid helium and room temperature the crystals were destroyed by this thermal procedure, probable due to internal stresses arising from imperfect sections of the crystals. As mentioned previously [13], only relative values of molar heat capacities are obtained by the heat pulse technique. Therefore a fitting of the C_p versus T curve to one absolute value at least have to be performed. This fitting condition is given by values obtained from measurements carried out by differential scanning calorimetry (DSC) at about 300 K [11].

3. Results

The experimental results of the investigation are shown in Fig. 1. The results are reproducible with an accuracy of 5%. From Fig. 1 it follows that the molar heat capacity obeys the Debye T^3-law below 5 K. The Debye temperature θ_D is found to be 128.3 K. This value agrees with θ_D given by Wu and Luo [8]. As shown in Fig. 2, however, an excess heat capacity was not observed. This result suggests that crystals applied in this work may provide a better quality, which means a more ideal crystalline structure. Similar results are obtained when trigonal crystals where investigated.

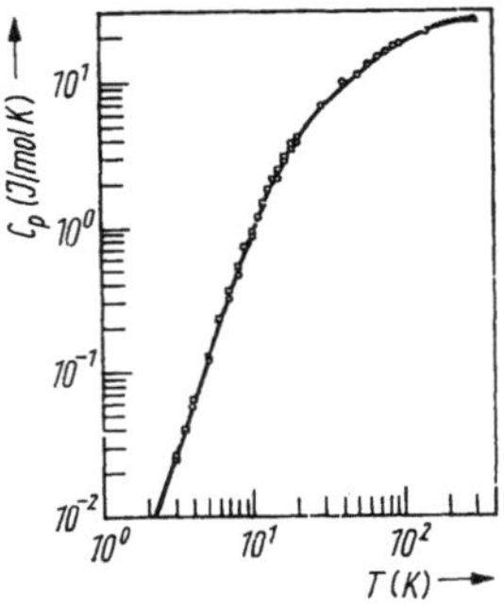

Fig. 1

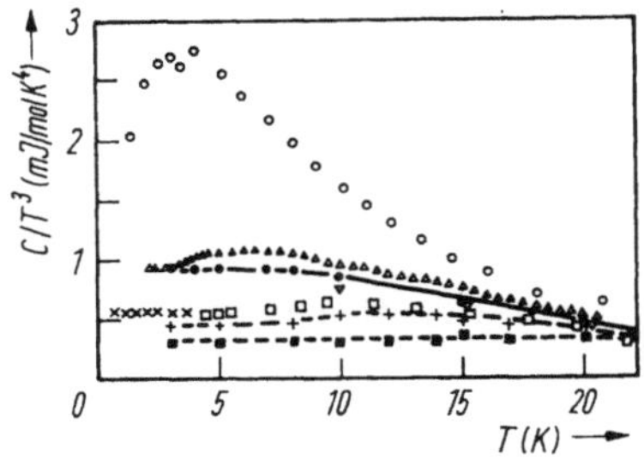

Fig. 2

Fig. 1. Molar heat capacity of α-monoclinic selenium at constant pressure as a function of temperature. □ Wu and Luo [8], ○ heat pulse method, △ differential scanning calorimetry, ——— best fit according to (5) and (6), constants are given in Table 2

Fig. 2. Curves of C_V/T^3 vs. T for amorphous, monoclinic, and trigonal selenium. ○ amorphous selenium (Lasjaunias [3]), △ α-monoclinic selenium crystals grown from a saturated solution of methylene iodide (Wu and Luo [8]), ● α-monoclinic selenium crystals grown from a saturated solution of CS_2 (present work), ▽ polycrystalline trigonal selenium (Chang and Bestful [4]), × trigonal selenium crystal prepared by a high-pressure method (Lasjaunias [3]), □ polycrystalline trigonal selenium (Berton and Lasjaunias [27]), + trigonal selenium crystal prepared by a high-pressure method (Große et al. [28] and Meißner and Wobig [15]), ■ trigonal selenium crystal grown from a thallium-doped melt (Große et al. [28] and Meißner and Wobig [15]), ——— calculated curve from (5) and (6), – – – – fit, given in [15]

According to Fig. 2 polycrystalline and single crystals of trigonal selenium grown by a high-pressure method reveal an excess specific heat. In contrary to this result crystals grown by the Bridgman technique from thallium-doped melts reveal no deviation from Debye's T^3-law. As it could be confirmed by measurements of ultrasonic attenuation and Laue back reflection photographs [14] the latter crystals are of better quality. A more detailed discussion on the results concerning the low-temperature specific heat data on amorphous, monoclinic, and trigonal selenium with respect to the degree of crystallinity will be presented in a succeeding paper [15].

4. Discussion

In order to compare the experimentally determined heat capacity with a simple lattice theory model for a single crystal of monoclinic selenium, we started with results from X-ray and infrared measurements on the ring molecular structure of this crystal. Our model was based on the idea that the mean features of the low-temperature heat capacity arises from a phonon density of states which depends strongly on the frequency modes of the eight-membered selenium ring molecule. Like the Tarassov model resulting in a good approximation of the one-dimensional chain-like behaviour of trigonal selenium, we propose a model resuming the ring-like structure of monoclinic selenium.

A general expression for the molar heat capacity at constant volume C_v of a solid is given by

$$C_v = k \int_0^\infty \frac{\exp\left(\frac{\hbar\omega}{kT}\right)}{\left(\exp\left(\frac{\hbar\omega}{kT}\right) - 1\right)^2} \left(\frac{\hbar\omega}{kT}\right)^2 g(\omega)\, d\omega\,, \tag{1}$$

where ω is the vibrational frequency, T the absolute temperature, k the Boltzmann constant, $h = 2\pi\hbar$ the Planck constant, and $g(\omega)$ the vibrational distribution function. Integrating of $g(\omega)$ gives the number of frequencies

$$\int_0^\infty g(\omega)\, d\omega = 3N\,, \tag{2}$$

where N is the Avogadro number.

The evaluation of the frequency distribution function $g(\omega)$ is thus imperative for correct evaluation of the heat capacity. Providing molecular crystals, a combination of Debye and Einstein terms are often sufficient for a first approximation of $g(\omega)$ [16]. Obviously such a situation exists in our case. Thus, X-ray measurements reported by Burbank [17], Marsh et al. [18], and Unger and Cherin [19] reveal that α-monoclinic selenium is a molecular crystal composed of eight-membered puckered ring molecules. The binding forces between the atoms in a given ring are covalent, whereas the forces holding the rings in a monoclinic lattice are of the weaker Van der Waals type. Lucovsky [20] therefore expected that the infrared and Raman activity of this crystal reflects the symmetry characteristics of the molecular unit. The Se_8-molecule has symmetry D_4-$\bar{8}$m2. For this group there are at most eleven discrete vibrational frequencies [21]. Therefore, in a first rough estimate, the Se_8 rings were assumed to vibrate almost unaffected by the weak Van der Waals forces. In this case the heat capacity contribution of the molecules is represented by Einstein functions. The corresponding frequencies were taken from infrared and Raman spectra given by Lucovsky [20], Mooradian and Wright [22]. As usual a Debye function for the acoustic portion of the vibrational distribution function is taken into account [16]. Thus, following the procedure given by Leibfried and Brenig [23] one obtains for $g(\omega)$ as represented in Fig. 3

$$g(\omega) = \frac{9N}{\omega_D^3}\,\omega^2 + \sum_{j=1}^{11} \frac{3N_E}{11}\,\delta(\omega - \omega_{Fj})\,, \tag{3}$$

where N_E is the sum of Einstein oscillators and N the sum of all oscillators. Integrating of (3) with respect to equation (2), regarding that the Debye term is different from zero only up to the cut-off frequency ω_G yields

$$\frac{N_E}{N} = 1 - \frac{\omega_G^3}{\omega_D^3} = 1 - \frac{\theta_G^3}{\theta_D^3}, \tag{4}$$

where $\theta = \hbar\omega/k$.

Inserting (3) into (1) with respect to (4) one obtains

$$C_v = 3R\left\{\left(\frac{\theta_G}{\theta_D}\right)^3 D_3\left(\frac{\theta_G}{T}\right) + \left(1 - \left(\frac{\theta_G}{\theta_D}\right)^3\right)\frac{1}{11}\sum_{j=1}^{11} P\left(\frac{\theta_{Ej}}{T}\right)\right\}, \tag{5}$$

where

$$D_3(x) = \frac{3}{x^3}\int_0^x \frac{t^4\,e^t}{(e^t - 1)^2}\,dt$$

and

$$P(x) = \frac{x^2\,e^x}{(e^x - 1)^2}$$

are the Debye and Einstein functions, respectively.

In order to compare our measured C_p-values with those to be calculated from (5) a conversion from C_p to C_v is necessary. This was done by applying C_p-values of monoclinic selenium at about 360 K [11]. At these temperatures ($T \approx 2.8\theta_D$) C_v was assumed to be approximately equal to $3R$. Providing this assumption to be valid a constant

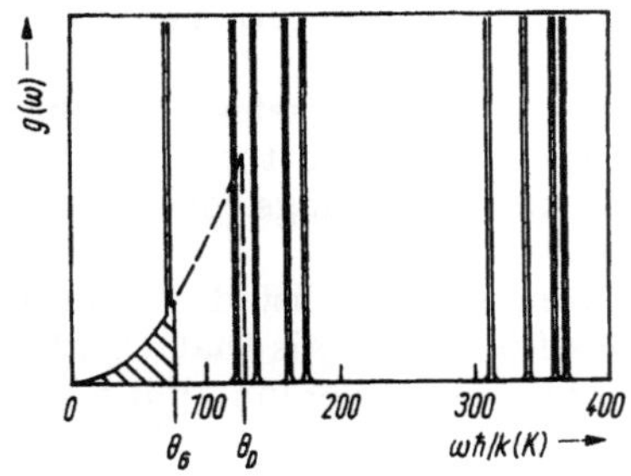

Fig. 3

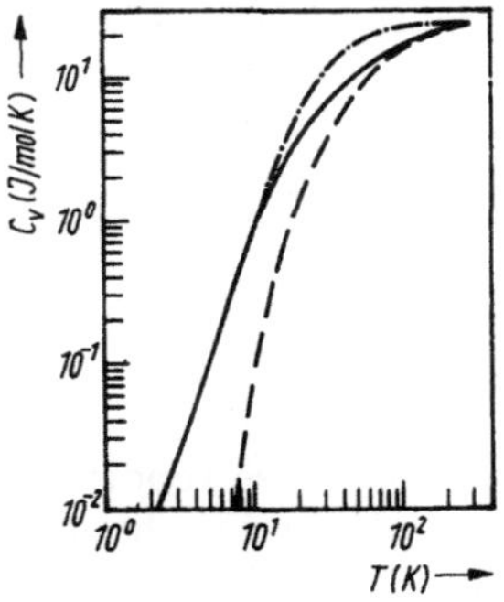

Fig. 4

Fig. 3. Estimated frequency distribution function $g(\omega)$ of α-monoclinic selenium, see text

Fig. 4. Molar heat capacities at constant volume providing different equations. ——— calculated curve from (5), - · - · - calculated curve from $C_v = 3RD_3(\theta_D/T)$ with $\theta_D = 128.3$ K; - - - - calculated curve from $C_v = 3R\,\frac{1}{11}\sum_{j=1}^{11} P(\theta_{Ej}/T)$ providing only Einstein functions $P(\theta_{Ej}/T)$ with θ_{Ej} as given in Table 2

A has been calculated solving the following thermodynamic relationship:

$$C_p - C_v = A C_v T \,, \tag{6}$$

where A is proportional to the product of the Grüneisen parameter and the linear expansion coefficient. According to our DSC measurements A is 2.4×10^{-4} K^{-1} at 363 K.

The last unknown quantity θ_G given in (5) was obtained by the method of least squares

$$S(\theta_G) = \sum_{i=1}^{n} \left(C_p^{ex}(T_i) - C_p^{th}(\theta_G, \theta_D, \theta_{E_j}, T_i)\right)^2 = \text{minimum} \,, \tag{7}$$

where $C_p^{ex}(T_i)$ are the measured values at temperatures T_i, $C_p^{th}(\theta_G, \theta_D, \theta_{E_j}, T_i)$ the theoretical values calculated from (5) and (6), and n designates the whole number of measuring points. The numerical values of the constants used in (7) are given in Table 2. The function $S(\theta_G)$ was found to be minimum at $\theta_G = 76$ K.

Table 2

Characteristic temperatures applied in equations (5) and (7); Debye temperature θ_D taken from our low-temperature measurements; Einstein-temperatures θ_{E_j} ($j = 1$ to 11) calculated from Se_8 mode frequencies as found by infrared and Raman measurements [20 to 22]; values given in K

θ_D	θ_{E_1}	θ_{E_2}	θ_{E_3}	θ_{E_4}	θ_{E_5}	θ_{E_6}	θ_{E_7}	θ_{E_8}	θ_{E_9}	$\theta_{E_{10}}$	$\theta_{E_{11}}$
128.3	72	121	137	161	173	184	311	338	358	366	366

As shown in Fig. 1 the C_p versus T curve calculated by (5) and (6) applying the numerical values reported in Table 2 gives a good fit to the measured C_p data. It may be mentioned in this connection that the C_p versus T curve also fits the C_p data measured by DSC in the temperature range 180 to 363 K [11].

With respect to equation (4) one obtains the fraction N_E/N of Einstein oscillators which is found to be 79%. Fig. 4 finally shows C_v as a function of temperature calculated from (5). In order to demonstrate the influence of Debye and Einstein terms also C_v versus T curves are presented where only Debye and Einstein terms are taken into account.

From Fig. 4 it is evident that at low temperatures monoclinic selenium behaves as a solid with rigid Se_8-ring molecules vibrating in the lattice. On the other hand, at comparatively high temperatures mainly the vibrations of Se_8-rings contribute to the molar heat.

In Table 1 smoothed values of thermodynamic functions are listed. A comparison of our data with those reported by Chang and Bestful [4] reveals that C_p (298.2 K) of monoclinic selenium is about 2% larger than C_p of trigonal selenium, but 6 and 30% less than C_p of monoclinic selenium reported by Mondain-Monval [7] and Gattow and Heinrich [6], respectively. By combining our results with those on trigonal selenium given by Chang and Bestful [4] and Grønvold [5] the Gibbs free energy $\Delta G(T)$ of the conversion from monoclinic to trigonal selenium can becalculated. This is performed with regard to the following thermodynamic relations

$$\Delta G(T) = \Delta H(T) - T\,\Delta S(T) \,, \tag{8}$$

$$\Delta H(T) = H(T)^{tr} - H(T)^{mon} = \int_0^T C_p^{tr}\,dT - \int_0^T C_p^{mon}\,dT + \{H(0\text{ K})^{tr} - H(0\text{ K})^{mon}\} \,, \tag{9}$$

$$\Delta S(T) = S(T)^{tr} - S(T)^{mon} = \int_0^T (C_p^{tr}/T)\,dT - \int_0^T (C_p^{mon}/T)\,dT\,, \tag{10}$$

where ΔH and ΔS is the enthalpy and entropy of conversion, respectively, "tr" and "mon" designates the trigonal and monoclinic phases, and $\Delta H(0\text{ K})$ is the enthalpy at absolute zero.

The unknown quantity $\{H(0\text{ K})^{tr} - H(0\text{ K})^{mon}\}$ is to be obtained from equation (9)

$$\Delta H(0\text{ K}) = \{H(0\text{ K})^{tr} - H(0\text{ K})^{mon}\} = \Delta H(T_c) - \int_0^T C_p^{tr}\,dT + \int_0^T C_p^{mon}\,dT\,, \tag{11}$$

where $\Delta H(T_c)$ is the enthalpy of conversion measured at the temperature T_c. However, $\Delta H(T_c)$ values are rather uncertain. Gattow [24] found by measurements of the enthalpy of combustion that ΔH (298.2 K) is −6.7 kJ/mol. Mondain-Monval [7] measured a much smaller value. He obtained H (417 K) = −0.7 kJ/mol. The latter value could be confirmed by DSC measurements [25]. Furthermore, this investigation suggests that $H(T_c)$ depends on crystal quality. Thus Murphy et al. [26] using powdered samples found ΔH (413 K) = −2.1 kJ/mol, a value which could be confirmed in a recent investigation [11] applying such mechanically pretreated samples. In addition to this result the investigation of good-quality crystals yielded $\Delta H(T_c)$ values between −1.3 and −1.6 kJ/mol. T_c was found to be about 400 K (at lower heating rates). Therefore $\Delta H(T_c) = -1.3$ kJ/mol was applied for calculation. Such a value is assumed to be representative for a nearly perfect crystal. However, this assumption has no substantial influence on results concerning the thermodynamic stability of monoclinic selenium.

According to (11) one obtains the difference of enthalpies of trigonal and monoclinic selenium H (0 K) = −1.2 kJ/mol where $\{H(400\text{ K})^{mon} - H(0\text{ K})^{mon}\}$ is calculated from (5) and (6) and $\{H(400\text{ K})^{tr} - H(0\text{ K})^{tr}\}$ is taken from tables given by Chang and Bestful [4] and Grønvold [5]. Further values are tabulated in Table 3.

Table 3

Enthalpy ΔH, entropy ΔS, and Gibbs free energy ΔG of conversion from monoclinic to trigonal selenium

T (K)	ΔH (kJ/mol)	ΔS (J/mol K)	ΔG (kJ/mol)
0	−1.2	0	−1.2
100	−1.2	−0.24	−1.2
200	−1.2	−0.33	−1.1
300	−1.2	−0.48	−1.1

From Table 3 it follows that α-monoclinic selenium is the thermodynamic unstable phase over the entire temperature range investigated in this work.

5. Conclusion

The molar heat capacity measurements of monoclinic selenium crystals yield Debye's cube law at temperatures below 5 K. The corresponding Debye temperature is $\theta_D = 128.3$ K. These results are in quite good agreement with those found by Wu and Luo [8]. However, in contrast to our measurements the former authors observed a small bump in the C_p/T^3 versus T curve at about 7 K. This fact may be due to crystal imperfections. Such a correlation between the low-temperature excess specific heat and the quality of crystals is found in the case of trigonal selenium [15].

In this study it is shown that a simple one-parameter model is suitable to give a good representation of our C_p data. This model suggests that the assumption of the frequency distribution function expressed by a combination of Debye and Einstein terms resulting from the molecular ring structure provides a certain reliability. On the other hand, it is well known that thermal properties are not very sensitive to the actual frequency distribution. Therefore, the frequency distribution function $g(\omega)$ shall be determined with the help of inelastic neutron scattering.

Apart from uncertainties arising from enthalpy of conversion $\{H^{tr} - H^{mon}\}$ this study shows that monoclinic selenium is unstable over the entire temperature range. This result is in accordance with an investigation performed most recently [11].

Acknowledgement

This work has been supported by the Deutsche Forschungsgemeinschaft.

References

[1] C. T. Anderson, J. Amer. Chem. Soc. **59**, 1036 (1937).
[2] W. DeSorbo, J. Chem. Phys. **21**, 1144 (1953).
[3] J. C. Lasjaunias, C. R. Acad. Sci. (France) **269B**, 763 (1969).
[4] S. S. Chang and A. B. Bestful, J. Chem. Thermodynamics **6**, 325 (1974).
[5] F. Grønvold, J. Chem. Thermodynamics **5**, 525 (1973).
[6] G. Gattow and G. Heinrich, Z. anorg. allg. Chem. **331**, 256 (1964).
[7] P. Mondain-Monval, C. R. Acad. Sci. (France) **182**, 1468 (1926).
[8] C. T. Wu and H. L. Luo, J. low-Temp. Phys. **23**, 487 (1976).
[9] D. E. Harrison, Recent Adv. Selenium Phys. European Selium–Tellurium Comitee, Pergamon Press, 1965 (p. 67).
[10] R. C. Keezer, The Physics of Selenium and Tellurium, Edited by W. Ch. Cooper, Pergamon Press, 1969 (p. 103).
[11] H. Gobrecht, P. Franz, M. Scheiba, H. Kunze, and D. Wobig, to be published.
[12] H. Gobrecht, H. Nelkowski, W. Seifert, and R. Krüger, Wärme- und Stoffübertragung **3**, 96 (1970).
[13] R. Krüger, M. Meissner, J. Mimkes, and A. Tausend, phys. stat. sol. (a) **17**, 471 (1973).
[14] H. Gobrecht and P. Kersten, Solid State Commun. **16**, 773 (1975).
[15] M. Meissner and D. Wobig, to be published.
[16] S. S. Mitra, Solid State Physics **13**, 41 (1962).
[17] R. D. Burbank, Acta cryst. **4**, 140 (1951).
[18] R. E. Marsh, L. Pauling, and J. D. Mc Cullough, Acta cryst. **6**, 71 (1953).
[19] P. Unger and P. Cherin, The Physics of Selenium and Tellurium, Ed. by W. Ch. Cooper, Pergamon Press, 1969 (p. 223).
[20] G. Lucovsky, The Physics of Selenium and Tellurium, Ed. by W. Ch. Cooper, Pergamon Pess, 1969 (p. 255).
[21] D. W. Scott, J. T. McCullough, and F. H. Kruse, J. Mol. Spectroscopy **13**, 313 (1964).
[22] A. Mooradian and G.B. Wright, The Physics of Selenium and Tellurium, Ed. by W. Ch. Cooper, Pergamon Press, 1969 (p. 269).
[23] G. Leibfried and G. Brenig, Z. Phys. **134**, 451 (1953).
[24] G. Gattow, Z. anorg. allge. Chem. **277**, 37 (1954).
[25] H. Gobrecht, G. Willers, and D. Wobig, Z. phys. Chem. NF **77**, 197 (1972).
[26] K. E. Murphy, M. B. Altmann, and B. Wunderlich, J. appl. Phys. **48**, 4122 (1977).
[27] A. Berton and J. C. Lasjaunias, phys. stat. sol. (b) **55**, K 79 (1973).
[28] R. Grosse, P. Krause, M. Meissner, and A. Tausend, J. Phys. C **11**, 45 (1978).

(Received June 9, 1978)

phys. stat. sol. (a) 49, 67 (1978)

Subject classification: 14.3.1; 22.4

Institut für Angewandte Physik der Technischen Hochschule Darmstadt

Temperature Dependence of the Magnetoresistance in CdO

By

M. ALTWEIN, H. FINKENRATH, and D. GERHARDT

In honour of Prof. Dr. F. STÖCKMANN's 60th birthday

The transverse magnetoresistance of very narrow and very broad samples of highly degenerate CdO (carrier density a few 10^{19} cm^{-3}) is measured up to 0.8 T in the temperature range 3 to 293 K. At room temperature the resistance change of almost all samples is positive and proportional to B^2. With decreasing temperature $\Delta\varrho/\varrho_0$ becomes increasingly negative. The field dependence of the negative component agrees well with a semiempirical relation derived by Khosla and Fischer. The influence of the sample geometry upon the magnetoresistance increases considerably at lower temperatures.

Der transversale Magnetowiderstand sehr schmaler und sehr breiter Proben von hochentartetem CdO (Trägerdichte einige 10^{19} cm^{-3}) wird bis 0,8 T im Temperaturbereich 3 bis 293 K gemessen. Bei Raumtemperatur ist die Widerstandsänderung fast aller Proben positiv und proportional B^2. Mit abnehmender Temperatur wird $\Delta\varrho/\varrho_0$ zunehmend negativ. Die Feldabhängigkeit der negativen Komponente stimmt mit einer von Khosla und Fischer angegebenen halbempirischen Beziehung überein. Der Einfluß der Probenform auf den Magnetowiderstand nimmt bei tiefen Temperaturen beträchtlich zu.

1. Introduction

Due to a high degree of nonstoichiometry cadmium oxide is a highly degenerate n-type semiconductor. The conductivity is related to an excess of cadmium in the lattice.

In a number of heavily doped semiconductors a negative magnetoresistance was observed in a region of nearly metallic conductivity, mostly at very low temperatures. Obviously, the magnetoresistance of CdO has never been investigated up to now. Only in [1] a negative mangetoresistance at 4.2 K is briefly mentioned. Because of the high carrier degeneracy the positive magnetoresistance component should be extremely weak.

Several negative magnetoresistance measurements could be explained by a theory of Toyozawa [2]. According to this theory, the negative magnetoresistance is caused by field-dependent carrier scattering on localized magnetic moments. However, in many cases only qualitative features of the experimental results were in agreement with the Toyozawa model. Especially the measurements on degenerate CdS [3] disagreed quantitatively with calculations based on the second-order perturbation expansion in the above model. Applying a scattering theory for the tunneling in magnetic fields to the considered problem, Khosla and Fischer [3] derived a semi-empirical relation for the negative magnetoresistance which did fit their own data for CdS as well as the results for other degenerate semiconductors showing negative magnetoresistance.

Also in our measurements on CdO the negative component of the magnetoresistance did not agree quantitatively with the Toyozawa theory. However, the results fit the equation of Khosla and Fischer very well. The positive magnetoresistance, observed at room temperature, can — for the most part — be explained by the conventional transport theory.

2. Experimental

The samples — in the form of thin films — were produced by reactive cathode sputtering of cadmium in an O_2-Ar gas mixture. In order to investigate the influence of the sample shape on the magnetoresistance, we prepared very narrow samples (length to width ratio 5:1) and also very broad ones (length to width ratio inverse). Low-field Hall-voltage measurements yielded electron concentrations between 3.2 and $5.3 \times 10^{19}\ cm^{-3}$.

Due to the high electron degeneracy an extremely weak positive contribution to the transverse magnetoresistance was expected. To measure it we used a method whose main characteristics are described as follows.

The static magnetic field was produced inside a helium gas-flow cryostat by means of a small built-in electromagnet originally designed for low-temperature Hall measurements [4]. The static field $B_= \leqq 0.8$ T was superimposed by a very weak a.c. field $\hat{B} \sin \omega t\ (\hat{B} \ll B_=)$. Compared to the zero-field resistance R_0, the field-dependent variation $\Delta R(B)$ of the sample resistance was very small in each case. The sample with the resistance $R_0 + \Delta R(B)$ was series-connected with a load resistance R_l to a d.c. voltage U_a. Due to the current modulation in the sample, caused by the a.c. field, the voltage drop U_l across the load resistance also contained an a.c. component: $U_l = U_{l=} + \hat{U}_l \sin \omega t$. For the amplitude $\hat{U}_l$ the expansion of ΔR yields

$$\hat{U}_l = -\frac{R_l U_a}{(R_0 + R_l)^2} \frac{d\Delta R}{dB} \hat{B} .$$

This voltage, filtered out and measured as a function of $B_=$, allowed calculation of the resistance variation under the influence of the magnetic field by means of a simple numerical integration:

$$\Delta R(B) = -\frac{(R_0 + R_l)^2}{R_l U_a} \int_0^B \frac{\hat{U}_l(B_=)}{\hat{B}} dB_= .$$

3. Results and Discussion

Fig. 1 shows a number of results yielded by room temperature measurements. A direct comparison with the results of the transport theory [5] is possible by plotting the magnetoresistance $\Delta\varrho/\varrho_0$ versus $(\mu B)^2$ (μ mobility, B magnetic induction). Except for one case the magnetoresistance is positive and exhibits a square-law dependence.

The resistance increase is strongly dependent on the sample shape. For very broad samples (subscript b) with unrestrained action of the deflecting Lorentz force upon the carriers, the transport theory yields in case of low magnetic fields ($\mu B \ll 1$):

$$\left(\frac{\Delta\varrho}{\varrho_0}\right)_b = \frac{\langle\tau^3\rangle}{\langle\tau\rangle^3} (\mu B)^2 \tag{1a}$$

($\langle\tau^n\rangle$ denotes the average value of the n-th power of the scattering relaxation time).

In very narrow samples (subscript n) the Lorentz force is, to a great extent, compensated by the Hall field influence. Nevertheless, because of drift velocity deviations from the mean value a very small resistance increase remains here, too:

$$\left(\frac{\Delta\varrho}{\varrho_0}\right)_n = \left[\frac{\langle\tau^3\rangle}{\langle\tau\rangle^3} - \frac{\langle\tau^2\rangle^2}{\langle\tau\rangle^4}\right] (\mu B)^2 . \tag{1b}$$

In case of high carrier degeneracy, i.e. Fermi energy (related to the band edge) $E_F \gg kT$, and of predominant carrier scattering at ionized impurities — which holds to a great extent for CdO — application of (1a) and (1b) results in

$$\left(\frac{\Delta\varrho}{\varrho_0}\right)_b = \left[1 + \frac{9\pi^2}{4}\left(\frac{kT}{E_F}\right)^2\right](\mu B)^2\,, \tag{2a}$$

$$\left(\frac{\Delta\varrho}{\varrho_0}\right)_n = \frac{3\pi^2}{4}\left(\frac{kT}{E_F}\right)^2 (\mu B)^2\,. \tag{2b}$$

Equations (2a) and (2b) — with a mean value $E_F = 13kT$ at room temperature for all samples — are plotted by the dash-dotted lines in Fig. 1. The experimental curves have, in principle, the same shape, but the slopes are too little for the broad samples and too large for the narrow ones. These deviations are most likely caused by the non-ideal sample shape and the neglect of other carrier scattering processes besides the impurity scattering. One narrow sample (3.2×10^{19} cm^{-3}) exhibits a striking particularity: a distinct superposition of a negative magnetoresistance, probably at room temperature observed for the first time.

At lower temperatures the extremely small positive magnetoresistance was completely superimposed by the considerably larger negative contribution. The results of measurements at a narrow sample (electron concentration 5.3×10^{19} cm^{-3}) are shown in Fig. 2. The measurement curves of all other narrow samples show about the same shape. The negative magnetoresistance is most pronounced at very low temperatures, decreases with increasing temperature and, at room temperature, changes to the small positive values.

A quantitative agreement between our results and the localized magnetic moment model of Toyozawa [2] or the preceding similar model of Yosida [6], respectively, could not be achieved. According to Yosida's theory for the resistance in dilute magnetic alloys, the negative magnetoresistance should be proportional to the square of

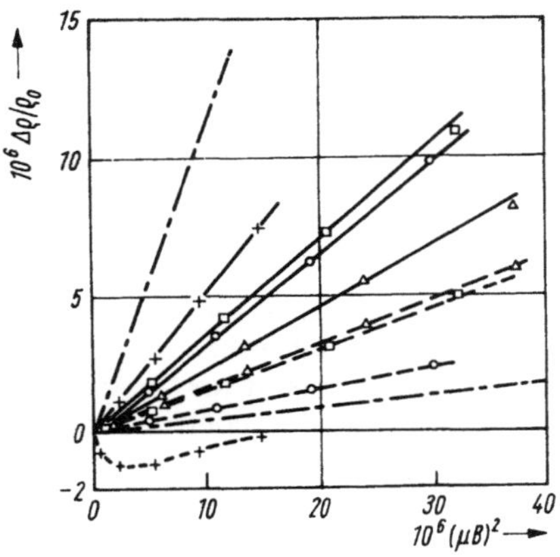

Fig. 1

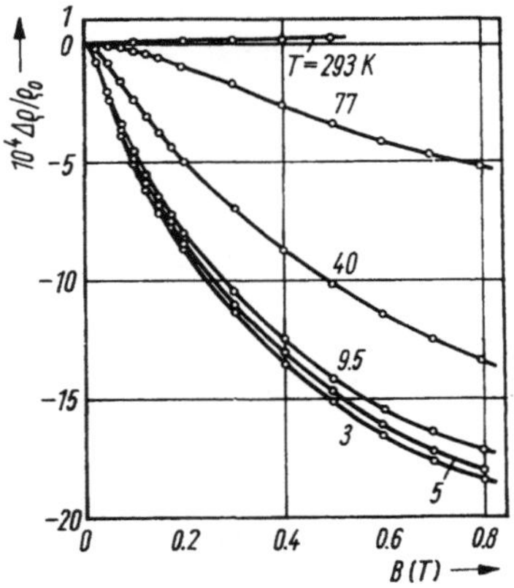

Fig. 2

Fig. 1. Room-temperature magnetoresistance versus the squared product of mobility and magnetic induction for various broad (———) and narrow (– – – –) samples. The dash-dotted lines represent the limiting cases for the sample geometry after (2a) and (2b). □ electron concentration 5.3×10^{19}, ○ 4.2×10^{19}, △ 4.0×10^{19}, + 3.2×10^{19} cm^{-3}

Fig. 2. Magnetoresistance of a narrow sample versus magnetic field for various temperatures (electron concentration 5.3×10^{19} cm^{-3})

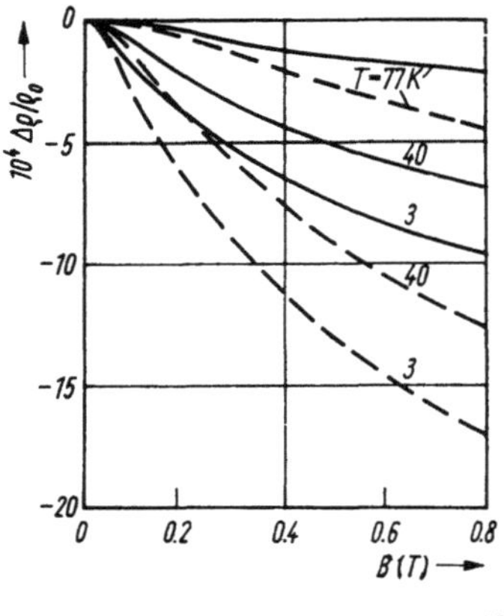

Fig. 3. Magnetoresistance of a broad (———) and a narrow (- - - -) sample of equal electron concentration, indicating the strong influence of the sample shape (electron concentration 3.2×10^{19} cm^{-3})

the average magnetization M of the localized spins:

$$-\frac{\Delta\varrho}{\varrho_0} \sim M^2 .$$

Especially at low fields should be $-\Delta\varrho/\varrho_0 \sim B^2$. However, our data could not be fitted to this square law dependence. In all cases, the power law exponent at low fields was distinctly smaller than 2. Also it was not possible to plot the measurement curves for different temperatures as unified function of $B/(T + \theta)$, as for instance in [7] and [8].

There exist a few theoretical models which try to explain the negative magnetoresistance, disregarding the existence of localized magnetic moments. The two-band model of Hedgcock and Raudorf [9], further developed by Giovannini and Hedgcock [10], cannot be used to explain our measurements. The very special assumptions of the model certainly are not applicable to the energy band structure, modified by high disorder, in the highly degenerate CdO. Moreover, one consequence of the model — the proportionality of the zero-field resistance to T^{-1} —is contrary to our experimental results.

A theory of Didora and Nitsovich [11], based on the assumption of mutual scattering of electrons with opposite spin in a narrow energy band, also offers no explanation, since the presumed narrow conduction band does not exist in CdO.

Analyzing their measurements on CdS Khosla and Fischer [3] concluded that in the calculation of the s–d exchange interaction the second-order perturbation theory is not sufficient to interpret the negative magnetoresistance. They suggested that higher-order terms should be included and derived a semi-empirical expression of the form

$$-\frac{\Delta\varrho}{\varrho_0} = C_1 \ln\left(1 + C_2 B^2\right) . \tag{3}$$

The coefficients C_1 and C_2 are functions of various physical parameters related to the exchange interaction.

Neglecting the weak positive component we were able to fit all low-temperature CdO measurements ($T \leqq 77$ K) remarkably good to (3). The solid lines of Fig. 2 (except the 293 K curve) represent the results of the least-squares fit for the given sample. We do not list the thereby determined numerous values of the coefficients C_1 and C_2 since they depend on mostly unknown parameters and, thus, are difficult to interpret. We only refer to the temperature dependence of the coefficient C_2 for which in [3], in accordance with the theoretical expectation, an approximate proportionality to T^{-2} was found. In all CdO samples we also observed that C_2 decreased monotonously if the temperature was increased from 3 to 77 K. Yet below 20 K the decrease was about proportional to $T^{-1/2}$, above 40 K, however, with about $T^{-2.5}$ it became quite strongly temperature dependent. Because of the small concentration range of the investigated samples, a dependence of parameters C_1 and C_2 upon the electron concentration could not be established.

We finally point out a still unexplained problem: a strong geometry dependence of the negative magnetoresistance. One would expect the negative component of the magnetoresistance — commonly interpreted as a carrier scattering phenomenon — to be independent of the sample shape. Yet in all our measurements, as the example in Fig. 3 shows, did the differences between the values determined from broad (solid lines) and narrow (dashed lines) samples increase considerably with decreasing temperature. At room temperature the maximum difference amounts to approximately 10^{-5}, at 3 K it reaches almost 10^{-3}.

This result cannot be explained with the assumption of simple additive superposition of the positive and negative component. Then at each temperature the difference $(\Delta\varrho/\varrho_0)_b - (\Delta\varrho/\varrho_0)_n$ should be equal to the very small difference of the positive components for 293 K.

In order to understand the experimental results, disregarding the additive superposition we tried to solve the transport equations for the current density components in the plane perpendicular to the magnetic field direction [5], taking into account a field dependence of the carrier scattering. This was done tentatively and quite qualitatively assuming the negative magnetoresistance to be caused by a field-dependent mobility $\mu = \mu_0 + \Delta\mu(B)$. In fact, we concluded from the simple calculation that with increasing $\Delta\mu(B)$, corresponding to an increase of the negative resistance component at lower temperatures, $(\Delta\varrho/\varrho_0)_b - (\Delta\varrho/\varrho_0)_n$ increases, too. Compared to the experimental result, however, the increase was considerably smaller. Therefore, this result remains still unexplained.

References

[1] R. P. Benedict and D. C. Look, Phys. Rev. B **2**, 4949 (1970).
[2] Y. Toyozawa, J. Phys. Soc. Japan **17**, 986 (1962).
[3] R. P. Khosla and J. R. Fischer, Phys. Rev. B **2**, 4084 (1970).
[4] M. Altwein, H. Finkenrath, and T. Stöckel, J. Phys. E **6**, 623 (1973).
[5] O. Madelung, in: HdB Phys., Ed. S. Flügge. Vol. XX, Springer Verlag, Berlin 1957 (p. 73).
[6] K. Yosida, Phys. Rev. **107**, 396 (1957).
[7] Y. Katayama and S. Tanaka, Phys. Rev. **153**, 873 (1967).
[8] L. Halbo and R. J. Sladek, Phys. Rev. **173**, 794 (1968).
[9] F. T. Hedgcock and T. W. Raudorf, Solid State Commun. **8**, 1819 (1970).
[10] B. Giovannini and F. T. Hedgcock, Solid State Commun. **11**, 367 (1972).
[11] I. D. Didora and V. M. Nitsovich, Soviet Phys. — Solid State **18**, 1214 (1976).

(Received June 2, 1978)

phys. stat. sol. (a) 49, 73 (1978)

Subject classification: 2; 21.1; 22.1.1

Physikalisches Institut der Universität Karlsruhe[1])

Zur Struktur amorpher metallischer Schichten des Systems Germanium–Kupfer

Von

H.-J. Nowak, H. Leitz und W. Buckel

Herrn Prof. Dr. F. Stöckmann zum 60. Geburtstag gewidmet

Amorphe Filme von Ge, Ge + 23 At% Cu und Ge + 40 At% Cu werden durch abschreckende Kondensation hergestellt. Elektronenbeugungsdiagramme dieser Filme werden in situ unmittelbar nach der abschreckenden Kondensation und nach dem Tempern aufgenommen. Die daraus erhaltenen Interferenzfunktionen können zumindest qualitativ mit Hilfe von Mikrokristallmodellen verstanden werden. Dabei müssen Mikrokristalle des Diamantgitters, einer primitiv-kubischen Nahordnung und einer hexagonalen Struktur zugrunde gelegt werden. Der mit wachsendem Cu-Gehalt zunehmende metallische Charakter der Filme kann aus diesen Modellen mit der Abnahme der kovalenten Bindungsanteile erklärt werden.

Amorphous films of Ge, Ge + 23 at% Cu, and Ge + 40 at% Cu are produced by quench condensation. Electron diffraction intensities are recorded on the films in situ immediately after the quench condensation and during annealing. The interference functions can be interpreted, at least qualitatively, by microcrystal models using crystallites of the diamond, the primitive-cubic, and the hexagonal structure. The increasing metallic character of the films with increasing Cu-content can be understood by the occurrence of structures with less covalent bindings.

1. Einleitung

Der amorphe Zustand von Halbleitern, mit Diamantstruktur in der kristallinen Modifikation, wurde in den zurückliegenden Jahren eingehend untersucht [1, 2]. Man weiß heute, daß die Nahordnung im amorphen Zustand des Ge und Si die gleiche ist wie im Diamantgitter. Jedes Atom hat vier nächste Nachbarn mit rein kovalenten Bindungen [2]. Durch eine etwas geänderte Vernetzung geht die Fernordnung verloren.

In diesem Zustand bleiben die charakteristischen Halbleitereigenschaften erhalten. Bei geeigneter Präparation ist die Energielücke zwischen Valenz- und Leitungsband so gut ausgebildet, daß durch entsprechende Dotierung p- und n-leitende Proben erzeugt werden können [3]. Dieser amorphe Zustand mit der Nahordnung des Kristalls ist recht stabil. Man muß die Proben auf einige hundert Grad Celsius erwärmen, um den Übergang in das Diamantgitter zu aktivieren.

Röntgen-amorphe Zustände mit wesentlich anderen Eigenschaften kann man erzeugen, indem man den Ge- oder Si-Dampf zusammen mit einem Edelmetall, z. B. Cu oder Au, auf eine sehr kalte Unterlage abschreckend kondensiert. Für Edelmetallkonzentrationen größer ≈ 20 At% entstehen dabei rein metallische Proben, die zudem supraleitend werden [4]. Im Gegensatz zum Verhalten des „gitterähnlich" amorphen Zustands der reinen Ge- bzw. Si-Proben ist der amorphe Zustand der Legierungen wesentlich instabiler. Weit unterhalb von Zimmertemperatur werden bereits starke irreversible Veränderungen des elektrischen Widerstandes beobachtet. Erste Struktur-

[1]) Engesserstraße 7, D-7500 Karlsruhe 1, BRD.

untersuchungen mit Hilfe der Elektronenbeugung [5] haben ergeben, daß auch diese Legierungsproben röntgen-amorph aufwachsen. Die beim Tempern beobachteten Veränderungen sind verknüpft mit Veränderungen im Elektronenbeugungsbild, wobei jedoch keine Umwandlung in einen kristallinen Zustand auftritt. Die Kristallisation erfolgt erst bei höheren Temperaturen.[2]) Daraus muß man schließen, daß die irreversiblen Veränderungen, die beim Tempern beobachtet werden, Veränderungen in der Nahordnung entsprechen.

Solche Veränderungen der Nahordnung in abschreckend kondensierten Schichten sind wohlbekannt [6]. So entstehen z. B. Bi- und Ga-Schichten bei der Kondensation aus dem Dampf auf eine Unterlage bei He-Temperaturen mit Nahordnungen, die vergleichbar sind mit denjenigen in Hochdruckphasen oder in der Flüssigkeit [7, 8]. Für amorphe Sn-Schichten[3]) konnte die Ähnlichkeit der Nahordnung zu der einer Hochdruckphase mit Hilfe des Mößbauer-Effektes nachgewiesen werden [9].

Da auch Ge und Si metallische Hochdruckphasen besitzen, die zudem supraleitend werden [10], liegt es nahe zu vermuten, daß in den abschreckend kondensierten Ge–Cu-Schichten eine veränderte Nahordnung eingefroren wird, die weniger kovalente Bindungen besitzt und damit mehr Ähnlichkeit zu den Nahordnungen in flüssigem Ge oder in der tetragonalen Hochdruckphase hat.

In der vorliegenden Arbeit werden Elektronenbeugungsdiagramme mit Hilfe eines Kristallzählers unmittelbar nach der abschreckenden Kondensation und im Verlauf des Temperns aufgenommen. Anhand der Intensitätskurven wird versucht, eine quantitative Strukturanalyse dieser Schichten durchzuführen. Da die Nahordnung des Ge studiert werden soll, werden nur Legierungen auf der Ge-reichen Seite des Systems, nämlich Ge + 23 At% Cu und Ge + 40 At% Cu untersucht. Die Ergebnisse an Legierungen auf der Cu-reichen Seite, bei denen es gelingt, mit $\approx$ 15 At% Ge-Zusatz das Cu in einer amorphen Phase zu stabilisieren, gehören in den Problemkreis der amorphen Metalle und werden an anderer Stelle diskutiert.

2. Experimentelles

Die Legierungen werden in Quarzampullen erschmolzen und durch Eintauchen in Wasser abgeschreckt. Aus diesen Legierungen wurden durch Zerkleinern Körner hergestellt, die aus einem Drehrohrofen [11] einzeln verdampft wurden. Jedes Körnchen ist so klein, daß es auf der Unterlage nur ca. eine Atomlage zur gesamten Schicht beiträgt. Auf diese Weise kann der Einfluß der Fraktionierung vermieden werden.

Die Schichten wurden auf eine sehr dünne Movitalfolie kondensiert, die über eine Kreisblende mit 20 µm Durchmesser gespannt und auf He-Temperaturen abgekühlt war.[4]) Gleichzeitig mit dieser für die Elektronenbeugung geeigneten Schicht wurde auf eine kristalline Quarzplatte eine Schicht kondensiert, an der das Widerstandsverhalten untersucht werden konnte.[5]) Die verwendeten Schichtdicken lagen bei $\approx$ 20 nm. Um den Einfluß der Restgase möglichst gering zu halten, war der Probenraum im Kryostaten durch mehrere Strahlungsschilde bei der Temperatur der flüssigen Luft und des flüssigen Heliums von den warmen Teilen der Apparatur völlig abgeschirmt. Lediglich zur Kondensation und während der Aufnahme der Beugungsdiagramme wurden möglichst kleine Blenden geöffnet.

[2]) Die Kristallisationstemperatur wird niedriger mit wachsendem Gehalt an Edelmetall.

[3]) Der amorphe Zustand kann beim Zinn nur durch die gleichzeitige Kondensation einer zweiten Komponenten, z. B. Cu mit Konzentrationen über $\approx$ 10 At% stabilisiert werden.

[4]) Von Untersuchungen an anderen Substanzen, z. B. Bi und Ga, die sehr tiefe Kristallisationstemperaturen besitzen, ist bekannt, daß die Erwärmung bei der Kondensation auch im Zentrum der Blende $\approx$ 15 K nicht überschreitet.

[5]) Ein Einfluß der Unterlage liegt nicht vor. Dies wurde dadurch getestet, daß die Quarzplatte mit einer Movitalschicht überzogen wurde.

Für die Elektronenbeugungsdiagramme wurden monochromatische Elektronen mit einer Beschleunigungsspannung von 68 kV verwendet.[6]) Der Elektronenstrahl wird durch eine Eintrittsblende mit einem Durchmesser von 50 μm und durch die Blende auf dem Präparathalter (Durchmesser 20 μm) kollimiert. Dies liefert am Kristallzähler einen Strahldurchmesser von ≈ 50 μm. Unmittelbar vor dem Zähler ist eine Schlitzblende mit einer Breite von 50 μm angebracht. Bei dieser Anordnung betrug das Winkelauflösungsvermögen $2{,}5 \times 10^{-4}$ rad. Zur Überprüfung der Apparatekonstante $L\lambda$ (L Abstand Präparat zu Zähler) und der durch die Apparatur bedingten Halbwertsbreite wurden Pb-Schichten als Eichsubstanz vermessen.

Für die Registrierung wurde der Zähler mit konstanter Geschwindigkeit bewegt. Eine Registrierung des gesamten Beugungsdiagramms erfolgte in ≈ 30 min.[7]) Dabei wurde bis zu s-Werten von 15 nm^{-1} registriert ($s = 2 \sin \theta/\lambda$). Alle getemperten Schichten wurden zur Aufnahme der zugehörigen Elektronenbeugungsdiagramme auf 80 K abgekühlt.

Um für die Analyse der Beugungsdiagramme den Untergrund zu kennen, der von inelastischen Streuprozessen herrührt, wurden die Schichten stets auch im kristallinen Zustand gemessen. Die Kristallisation kann bei Zimmertemperatur mit Hilfe des Elektronenstrahls erzwungen werden. Die Interferenzfunktion $i(s)$ und die zugehörige radiale Verteilungsfunktion wurden mit Standardmethoden ermittelt [12, 13]. Es wird

$$i(s) = \frac{I_a^{\exp}(s) - a\langle f^2(s)\rangle - b\big(I_{\mathrm{Uk}}(s) + c_k\big)}{a\langle f(s)\rangle^2}.$$

$I_a^{\exp}(s)$ gemessene Elektronenintensität, $I_{\mathrm{Uk}}(s)$ Untergrundintensität des kristallinen Films, $\langle f^2(s)\rangle = c_{\mathrm{Ge}} f_{\mathrm{Ge}}^2(s) + c_{\mathrm{Cu}} f_{\mathrm{Cu}}^2(s)$. Da $f_{\mathrm{Ge}}(s) \approx f_{\mathrm{Cu}}(s)$, kann man $\langle f(s)\rangle^2 \approx \langle f^2(s)\rangle$ annehmen.

3. Meßergebnisse und Diskussion

Fig. 1 zeigt die Beugungsintensitäten für einen reinen Ge-Film, einen Ge + 23 At% Cu-Film und für einen Ge + 40 At% Cu-Film. Dabei sind jeweils in den Teilfiguren auf der linken Seite die Intensitäten unmittelbar nach der Kondensation bei 4 K (durchgezogene Kurve) und nach dem Erwärmen auf Temperaturen etwas über Zimmertemperatur (gestrichelte Kurve) aufgetragen. Es sei hier noch einmal erwähnt, daß die Beugungsbilder der getemperten Filme bei 80 K aufgenommen wurden. Die Filme sind nach dieser Erwärmung noch immer röntgen-amorph. Die Veränderungen in den Intensitätskurven, besonders bei den Legierungsfilmen, zeigen jedoch, daß strukturelle Veränderungen in der Nahordnung ablaufen.

Die Kristallisation kann nun durch eine genügend hohe Elektronenintensität aktiviert werden. In den Teilfiguren auf der rechten Seite sind die Intensitätskurven dargestellt, wie sie nach der Kristallisation erhalten werden. Für den reinen Ge-Film werden, wie zu erwarten, die Linien des Diamantgitters gefunden. Bei den Legierungsfilmen treten neben den Linien des Ge-Gitters die Reflexe einer hexagonal dichtesten Packung mit einem c/a-Verhältnis von 1,6 und dem Wert $a = 0{,}264$ nm auf. Bei diesen Strukturanteilen handelt es sich wahrscheinlich um die intermetallische Verbindung Cu_5Ge, die sogenannte ζ-Phase. Für diese Phase wird bei Ge-Konzentrationen von 19,3 bis 19,4 At% $c/a = 1{,}61$ und $a = 0{,}2658$ nm angegeben [14].

[6]) Zur Erzeugung des Elektronenstrahl wurde eine EO, ÜM 1120f-Kathode der Firma Siemens verwendet. Die Beschleunigungsspannung wird über eine Widerstandskette auf 2‰ genau bestimmt. Um den Einfluß von Spannungsschwankungen auf die Elektronenintensität zu unterdrücken, wurde die Vorspannung am Wehneltzylinder mit der Hochspannung gekoppelt.

[7]) Beim Durchgang durch das zentrale Maximum wurde der Zähler abgedeckt.

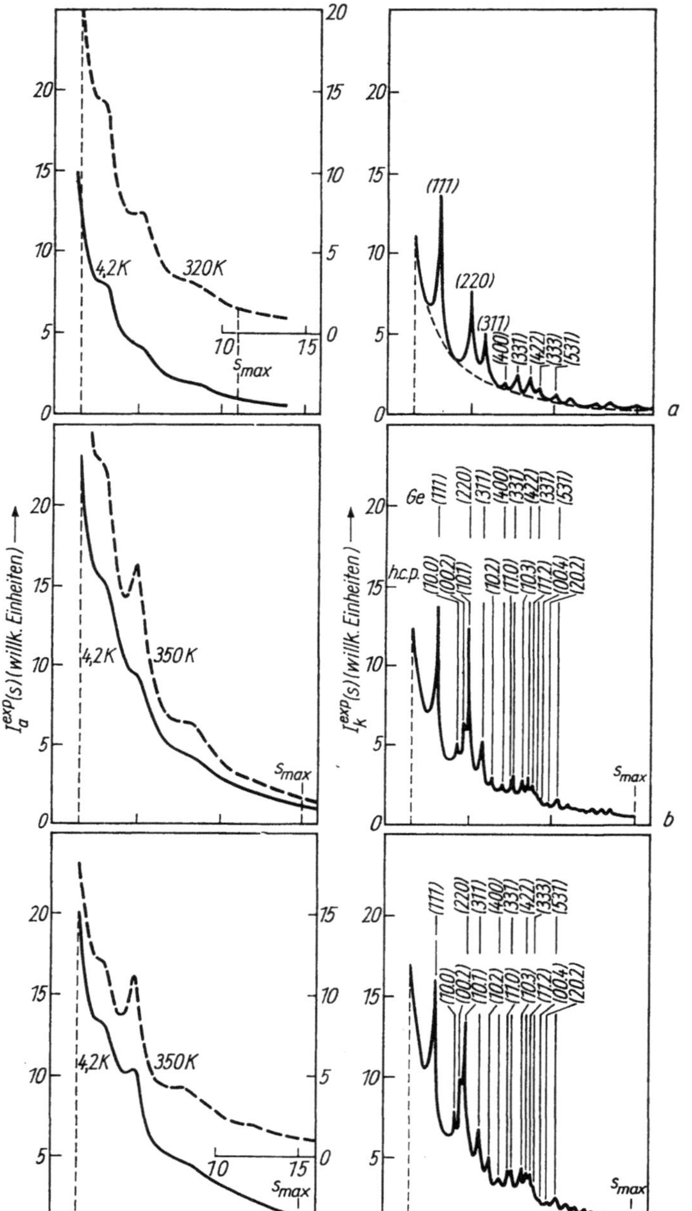

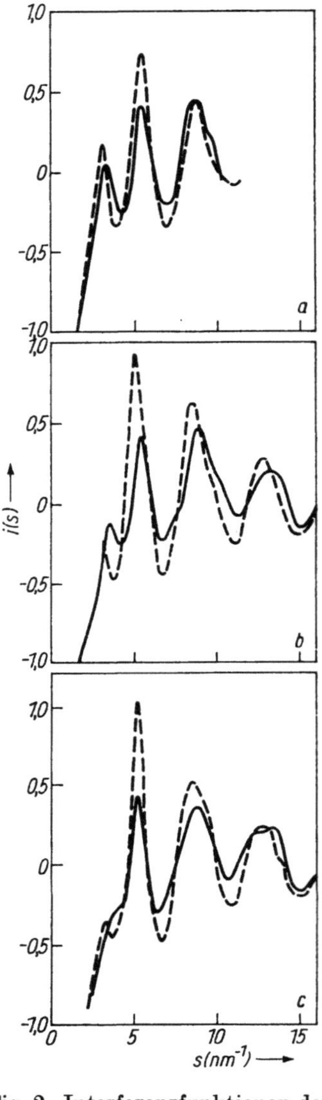

Fig. 2. Interferenzfunktionen der amorphen Filme aus den linken Teilbildern der Fig. 1 bestimmt. Ausgezogene bzw. gestrichelte Kurven gehören wieder zu den Filmen unmittelbar nach der Kondensation und nach dem Tempern

Fig. 1. Elektronenbeugungsdiagramme von abschreckend kondensierten Ge–Cu-Filmen. a) Reines Ge; b) Ge + 23 At% Cu; c) Ge + 40 At% Cu. In den Teilbildern auf der linken Seite sind die Diagramme für die amorphen Filme unmittelbar nach der Kondensation bei ≈ 10 K (ausgezogene Kurven) und nach dem Tempern auf die angegebenen Temperaturen (gestrichelte Kurven) dargestellt. In den Teilbildern auf der rechten Seite sind die Diagramme der kristallisierten Filme dargestellt

Aus den Intensitätskurven wurden nun die Interferenzfunktionen bestimmt.[8]) In der Fig. 2 sind diese Interferenzfunktionen wiedergegeben. Die ausgezogenen bzw. gestrichelten Kurven gehören wieder wie in der Fig. 1 zu den Filmen unmittelbar nach der Kondensation bzw. nach dem Erwärmen auf die in der Fig. 1 angegebenen Temperaturen.

Diese Interferenzkurven enthalten alle Informationen über die geometrische Anordnung der Atome, soweit sie uns aus diesen Experimenten zugänglich ist. Die Veränderungen, die innerhalb des amorphen Zustands beim Tempern auftreten, werden in den Interferenzfunktionen sehr deutlich sichtbar. Aus den Interferenzfunktionen lassen sich durch eine Fouriertransformation die radialen Verteilungsfunktionen bestimmen. Wir verzichten hier auf die Wiedergabe der Verteilungsfunktionen, da wir im folgenden die Analyse der Beugungsdiagramme anhand der Interferenzfunktionen durchführen wollen. Dabei betrachten wir Modelle mit statistisch verteilten Mikrokristallen, wobei über die Nahordnung in den Mikrokristallen plausible Annahmen (z. B. Anordnung der Atome wie im Diamantgitter oder in einer für dieses System bekannten anderen Struktur) gemacht werden. Für diese Modelle werden die Interferenzfunktionen berechnet und mit den experimentell bestimmten verglichen. Wenn es gelingt, auf diese Weise Modelle zu finden, deren Interferenzfunktionen mit den gemessenen übereinstimmen, kann man ~~hoffen, daß~~ die Modelle ein gewisses Bild der Atomanordnungen in unseren Proben vermitteln.

3.1 Reine Ge-Filme

Auch die reinen abschreckend kondensierten Ge-Filme zeigen beim Tempern irreversible Widerstandsänderungen innerhalb des röntgen-amorphen Zustands [5]. In Übereinstimmung damit stehen die irreversiblen Veränderungen der Interferenzfunktion beim Tempern. Dabei entwickelt sich der Film zu *der* gitteramorphen Phase hin, die man von amorphen, bei Zimmertemperatur kondensierten Ge-Filmen kennt. Die Atomverteilungsfunktion liefert nach dem Tempern auf 350 K nahezu die Koordinationszahl 4 für die nächsten Nachbarn.

Das wesentliche Problem ist nun, die Interferenzfunktion des abschreckend kondensierten Films zu analysieren. Dazu wird zunächst ein Modell verwendet, das aus statistisch verteilten Mikrokristallen des Diamantgitters besteht, wobei einmal zehn Atome und zum anderen 26 Atome pro Mikrokristall zugrunde gelegt werden. Dabei werden die Mikrokristalle so aus dem Diamantgitter „geschnitten", daß die Atome möglichst wenig freie Bindungen besitzen, d. h. eine möglichst große Koordinationszahl haben. Bei zehn Atomen kann man eine mittlere erste Koordination von 2,4 und bei 26 Atomen von 2,77 (gegenüber 4 im Diamantgitter) erhalten. Als interatomarer Abstand wird 0,245 nm angenommen, das ist der Abstand nächster Nachbarn im Diamantgitter. In Fig. 3 sind die für diese Modelle berechneten Interferenzkurven zusammen mit der an dem frisch kondensierten Film gemessenen dargestellt. Man sieht, daß beträchtliche Abweichungen, insbesondere bezüglich der relativen Intensitäten des 2. und 3. Maximums vorhanden sind.

Es wurde deshalb nach anderen Strukturen gesucht. Weder die für eine supraleitende Hochdruckphase bekannte tetragonale Struktur noch die ebenfalls bei hohem Druck auftretende GeIII-Phase mit Wurtzitstruktur [15] ergaben bessere Übereinstimmung der Interferenzfunktionen. Von Haug et al. [5] wurde vermutet, daß die abschreckend kondensierten Filme des Ge–Cu-Systems mit zunehmendem Cu-Gehalt in wachsendem Maße flüssigkeitsähnlichen Charakter erhalten. Es wurde deshalb hier auch die Interferenzfunktion von flüssigem Ge [16] zum Vergleich herangezogen. Die

[8]) In Fig. 1 ist als Beispiel der für die Bestimmung der Interferenzfunktion wichtige inelastische Untergrund eingezeichnete (Teilbild a, recht Seite).

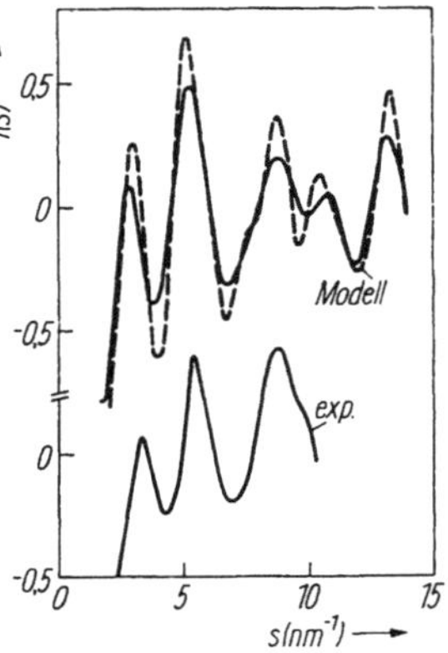

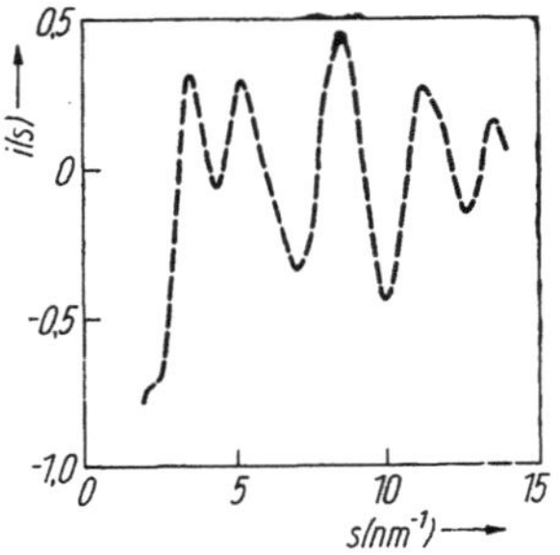

Fig. 4. Interferenzfunktion statistisch verteilter Mikrokristalle (19 Atome pro Kristallit) mit primitiv-kubischer Nahordnung

Fig. 3. Vergleich der Interferenzfunktion des reinen Ge-Films unmittelbar nach der Kondensation (untere Kurve) mit den Interferenzfunktionen für Mikrokristallmodelle. 10 Atome/Mikrokristall: ausgezogene Kurve, 26 Atome/Mikrokristall: gestrichelte Kurve

Übereinstimmung war noch unbefriedigender als für die verschiedenen mikrokristallinen Modelle. Es zeigte sich stets, daß das Verhältnis der Intensitäten vom 2. und 3. Interferenzmaximum Schwierigkeiten machte.

Es wurde deshalb nach einer Modellstruktur gesucht, bei der die Intensität des 3. Maximums vergleichbar sein sollte mit der des 2. Maximums. Hierfür bot sich als einfache Struktur die primitiv kubische an. Die Fig. 4 zeigt die Interferenzfunktion einer Modellsubstanz aus statistisch verteilten Mikrokristallen mit jeweils 19 Atomen in primitiv kubischer Struktur (Zentralatom mit 6 nächsten und 12 übernächsten Nachbarn). Als Abstand nächster Nachbarn wurde 0,27 nm gewählt. Das ist der Wert, wie er für flüssiges Ge gefunden wurde. Diese Interferenzfunktion weist wenig Ähnlichkeit mit der gemessenen auf. Sie hat aber als wesentliches Merkmal eine sogar etwas höhere Intensität des 3. Maximums gegenüber der Intensität des 2. Maximums.

Man könnte nun vermuten, daß die beobachtete Interferenzfunktion aus einer Überlagerung von zwei Interferenzfunktionen zustandekommt. Es wäre auch vom Kondensationsprozeß her durchaus verständlich, daß nur ein Teil der Ge-Atome in einer neuen Nahordnung eingefroren wird, während ein anderer Teil in der Lage ist, die kovalenten Bindungen des Diamantgitters aufzubauen. Ursache für das dazu notwendige unterschiedliche Wachstum könnte der Einbau von Restgasen sein.

Es wurde deshalb mit Hilfe der Debye-Formel die Interferenzfunktion einer Mischung statistisch verteilter Mikrokristalle mit der Struktur des Diamantgitters einerseits und mit einer primitiv kubischen Struktur andererseits berechnet. Als Parameter gehen die Größen der Mikrokristalle, die relative Häufigkeit der beiden Strukturen (angegeben in Prozent der in ihnen enthaltenen Atome) sowie die Abstände der nächsten Nachbarn in beiden Strukturen ein.

In Fig. 5 ist die Interferenzfunktion einer solchen Mischung gestrichelt dargestellt. Um die in diesem Modell bestmögliche Anpassung an die gemessene Interferenzfunktion (durchgezogene Kurve) zu erhalten, wurde der folgende Parametersatz verwendet. 40% der Atome liegen in Mikrokristallen der Diamantstruktur vor, für die eine Zahl von 26 Atomen pro Kristallit und ein Abstand nächster Nachbarn von 0,245 nm angenommen wurden. Die restlichen 60% der Atome liegen in Mikrokristallen mit primitiv kubischer Struktur vor, für die 19 Atome pro Kristallit und ein Abstand von 0,26 nm zwischen nächsten Nachbarn angenommen wurde.

Die erzielte Übereinstimmung ist qualitativ befriedigend. Die große Breite des beobachteten 3. Maximums gegenüber dem des gerechneten Modells kann leicht mit

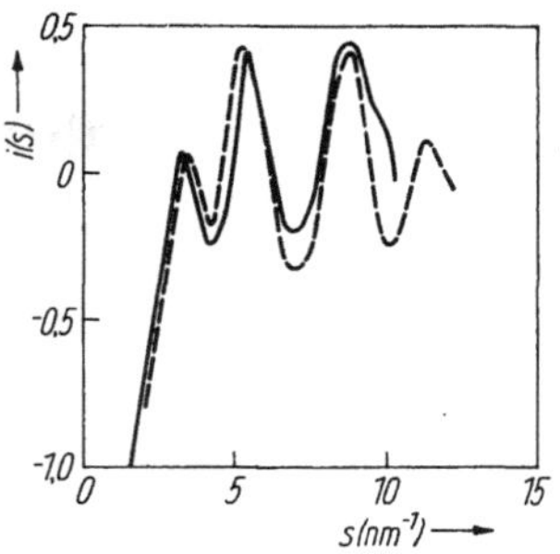

Fig. 5. Vergleich der Interferenzfunktion des frisch kondensierten reinen Ge-Films (ausgezogene Kurve) mit der eines Mikrokristallmodells (gestrichelte Kurve), bei dem eine Mischung von Mikrokristallen des Diamantgitters und solchen mit primitiv-kubischer Nahordnung angenommen wurde

einer geringfügigen Lagerverschmierung der Atome in den Nahordnungsbereichen erklärt werden.

Der hohe Prozentsatz primitiv kubischer Struktur macht es nun verständlich, daß die frisch kondensierten Filme einen um drei bis vier Zehnerpotenzen kleineren elektrischen Widerstand haben als die bei Zimmertemperatur aufgedampften Filme. Natürlich stellt dieses Modell nur einen Versuch dar, die beobachtete Interferenzfunktion zu deuten.

3.2 Ge–Cu-Filme

Bei der Analyse der an Legierungsfilmen gewonnenen Interferenzfunktionen gehen wir ähnlich vor wie bei den reinen Ge-Filmen. Auch hier ist die Interpretation der Ergebnisse für die getemperten Filme einfacher als für die frisch kondensierten. Es liegt nahe anzunehmen, daß sich beim Tempern auf 350 K Nahordnungsbereiche der hexagonalen Phase bilden, die bei der Kristallisation als neue Phase mit eigener Struktur sichtbar wird. Daneben werden die restlichen Ge-Atome, ähnlich wie bei den reinen getemperten Ge-Filmen in der Nahordnung des Diamantgitters vorliegen.

Es wurde deshalb die Interferenzfunktion für ein Modell berechnet, bei dem statistisch verteilte Mikrokristalle des Diamantgitters neben solchen einer hexagonal dichtesten Packung angenommen wurden. Die erforderlichen Parameter wurden wie folgt festgelegt: Das Cu bildet mit einem kleinen Anteil Ge die hexagonale Struktur (Cu_5Ge), wobei pro Kristall 21 Atome und ein Abstand nächster Nachbarn von 0,264 nm, wie er für die hexagonale ζ-Phase bekannt ist, zugrunde gelegt wurden. Für die Mikrokristalle des Diamantgitters wurden wie bei den reinen Ge-Filmen 26 Atome pro Kristallit und ein Nächstnachbarnabstand von 0,245 nm verwendet.

Die so erhaltene Interferenzfunktion ist für die Legierung Ge + 40 At% Cu in Fig. 6 zusammen mit der experimentell gewonnenen Interferenzfunktion dargestellt. Die Übereinstimmung ist recht befriedigend. Auch für die Legierung mit 23 At% Cu wird eine ähnlich gute Anpassung der Modellrechnung an die experimentelle Interferenzfunktion erreicht. Man wird deshalb mit einigem Recht feststellen können, daß die Veränderungen in den Nahordnungsbereichen, die beim Tempern ablaufen, darin bestehen, in Mikrokristallen des reinen Ge die kovalenten Bindungen des Diamantgitters besser auszubilden und das Cu in Mikrobereichen der hexagonalen ζ-Phase festzulegen.

Schwieriger ist auch hier die Analyse der Interferenzfunktion der frisch kondensierten Filme, die das wesentliche Problem der vorliegenden Arbeit darstellen. Um den Einfluß des Cu-Zusatzes auf die Interferenzfunktion deutlich zu machen, sind in Fig. 7 die Interferenzfunktionen für die drei untersuchten Cu-Konzentrationen (0, 23 und 40 At% Cu) zusammengestellt. Besonders auffällig ist der Abbau des 1. Maximums mit wachsender Cu-Konzentration. Auch die vergleichsweise große Intensität des 3. Maximums, zu deren Deutung wir im Modell für das reine Ge Mikrobereiche mit

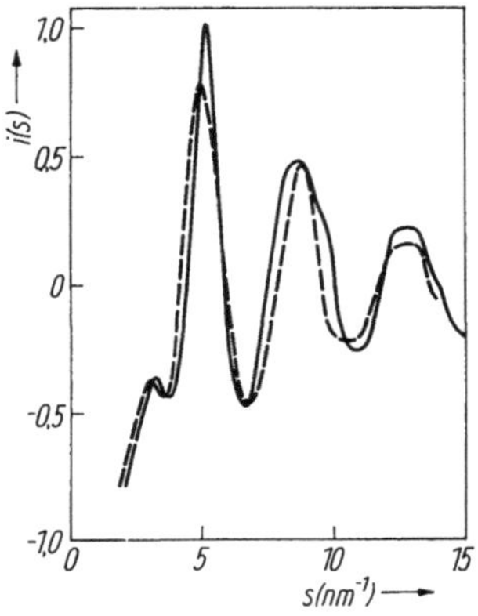

Fig. 6

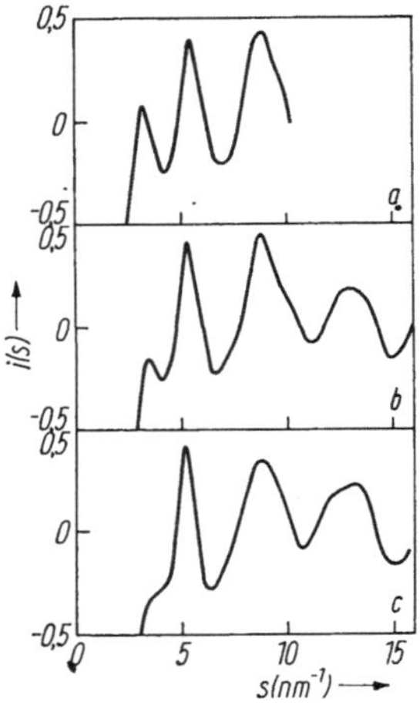

Fig. 7

Fig. 6. Vergleich der Interferenzfunktion des getemperten Ge + 40 At% Cu-Films (ausgezogene Kurve) bei 350 K mit der eines Mikrokristallmodells (gestrichelte Kurve), bei dem eine Mischung von Mikrokristallen des Diamantgitters mit solchen der hexagonalen ζ-Phase angenommen wurde

Fig. 7. Interferenzfunktionen der Filme unmittelbar nach der abschreckenden Kondensation bei $\approx$ 10 K. a) Reines Ge, b) Ge + 23 At% Cu, c) Ge + 40 At% Cu

primitiv kubischer Struktur herangezogen haben, wird kleiner. Ein 4. Maximum wird deutlich sichtbar.

Man wird zunächst versuchen, diese Veränderungen dadurch zu erklären, daß der Anteil der Mikrobereiche mit Diamantstruktur abnimmt, und dafür die dichter gepackten, primitiv kubischen Kristallite zunehmen, indem diese Nahordnung durch die Cu-Atome stabilisiert wird. Die in Fig. 4 dargestellte Interferenzfigur von statistisch verteilten Mikrokristallen mit primitiv kubischer Struktur zeigt unmittelbar, daß eine solche Deutung ausscheiden muß. Die Überlagerung dieser Funktion mit den in Fig. 3 dargestellten kann nicht zu den für die Legierungsfilme beobachteten Interferenzfunktionen führen.

Der Abbau des 1. Maximums und eine Abnahme der Intensität des 3. Maximums läßt sich aber erreichen, wenn man annimmt, daß mit zunehmendem Cu-Gehalt neben

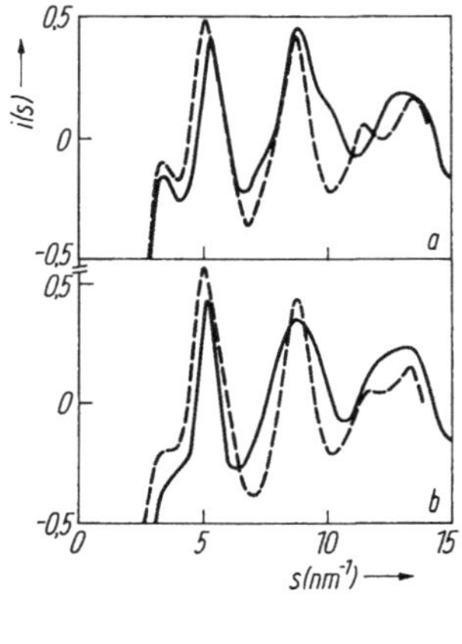

Fig. 8. Vergleich der an frisch kondensierten Ge–Cu-Filmen beobachteten Intererenzfunktionen (ausgezogene Kurven) mit den Interferenzfunktionen, die für bestimmte Mikrokristallmodelle (siehe Tabelle 1) berechnet wurden (gestrichelte Kurven). a) Ge + 23 At% Cu bei 4,2 K, b) Ge + 40 At% Cu bei 4,2 K

den Mikrokristallen in Diamantstruktur und denen in primitiv kubischer Struktur auch schon solche mit hexagonal dichtester Packung auftreten. Diese Annahme erscheint nicht abwegig, wenn man bedenkt, daß nach dem Tempern offenbar alles Cu in solchen Mikrokristallen der ζ-Phase eingebaut ist.

In Fig. 8 sind für die Legierungsfilme die mit dieser Modellvorstellung berechneten Interferenzfunktionen zusammen mit den experimentell bestimmten dargestellt. Die für die Modellrechnungen zugrunde gelegten Parameter sind in Tabelle 1 zusammengefaßt. Die Zahlen in den Spalten geben den prozentualen Anteil der Atome, die in den betreffenden Kristalliten vorliegen. Eine Änderung in einer der drei Komponenten um 10% von der hier gewählten Zusammensetzung gibt vor allem für die Intensität des 1. Maximums eine deutlich schlechtere Anpassung.

Tabelle 1

Struktur der Mikrokristallite	reines Ge	Ge + 23 At% Cu	Ge + 40 At% Cu
Diamantgitter 26 Atome/Kristall $r_0 = 0{,}245$ nm	40	40	40
primitiv kubisch 19 Atome/Kristall $r_0 = 0{,}26$ nm	60	40	30
hex. dichteste Kugelpackung 21 Atome/Kristall $r_0 = 0{,}264$ nm	–	20	30

Die Übereinstimmung ist auch qualitativ nicht sehr befriedigend. Das Modell gibt aber den Trend der Veränderungen richtig wieder. Überraschend ist das Ergebnis, daß offenbar auf der Ge-reichen Seite des Ge–Cu-Systems immer etwa 40 At% des Ge (unabhängig von der Cu-Konzentration) in Mikrokristallen mit Diamantstruktur aufwachsen. Die gleichzeitige abschreckende Kondensation von Cu-Atomen scheint lediglich dazu zu führen, daß die ohnehin nicht in der Nahordnung der Diamantstruktur aufwachsenden Kristallite mehr und mehr in einer dichtesten Packung eingefroren werden. Der wachsende metallische Charakter der Legierungsfilme wird damit verständlich. Man muß aber auf Grund dieser Analyse feststellen, daß die amorphen Filme auf der Ge-reichen Seite dieses Systems offenbar keine homogenen Phasen sind, sondern deutlich Entmischungstendenzen zeigen.

Ein großer Nachteil der Mikrokristallmodelle ist es zweifellos, daß sie die Übergangsbereiche zwischen den Mikrokristallen, die einen wesentlichen Anteil der abschreckend kondensierten Filme ausmachen, nicht richtig erfassen.

Der Deutschen Forschungsgemeinschaft danken wir für die finanzielle Unterstützung und Herrn Prof. G. v. Minnigerode für die Überlassung der Hochspannungsquelle.

Literatur

[1] J. Graczyk und P. Chaudari, phys. stat. sol. (b) **58**, 501 (1973); J. non-crystall. Solids **17**, 299 (1975).

[2] R. Grigorovici, R. Manaila, M. Popescu und C. Stanescu, AIP Conf. Proc. **20**, 200 (1974).

[3] W. E. Spear und P. G. LeComber, Solid State Commun. **17**, 1193 (1975); Phil. Mag. **33**, 935 (1976).
[4] B. Stritzker und H. Wühl, Z. Phys. **243**, 361 (1971).
[5] E. Haug, N. Hedgecock und W. Buckel, Z. Phys. **B 22**, 237 (1975).
[6] W. Buckel, Z. Phys. **138**, 136 (1954).
[7] W. Buckel, Z. Phys. **154**, 474 (1959).
[8] W. Buckel und W. Gey, Z. Phys. **176**, 336 (1963).
L. Bosio, A. Defrain und I. Epelboin, J. Physique **27**, 61 (1966).
[9] J. Bolz und F. Pobell, Z. Phys. **B 20**, 95 (1975).
[10] J. Wittig, Z. Phys. **195**, 215 (1966).
[11] F. Fischer, Z. Phys. **139**, 328 (1954).
[12] S. Steeb, Springer Tracts mod. Phys., **47**, 1 (1968).
[13] Yu. F. Komnik, B. I. Belevtsev und L. A. Yatsuk, Thin Solid Films **21**, 189 (1974).
[14] M. Hansen, Constitution of Binary Alloys, McGraw-Hill Publ. Co. 1958.
[15] E. A. Davis und F. C. Weinstein, J. non-crystall. Solids **13**, 401 (1973).
[16] S. P. Isherwood, R. Manaila und B. R. Orton, J. non-crystall. Solids **8**, 691 (1972).

(Received June 2, 1978)

phys. stat. sol. (a) 49, 83 (1978)

Subject classification: 13.3; 16; 17; 22.2.3

Institut für Angewandte Physik der Technischen Hochschule Darmstadt

Work Function of Irradiated Real and Cleaned Indium Antimonide Surfaces

By

S. LEHR[1]) and H. PAGNIA

In honour of Prof. Dr. F. STÖCKMANN's 60th birthday

Real surfaces of InSb are covered with a thin oxide and normally with adsorbed water molecules on the outer surface forming a surface dipole layer. Irradiation with light or a low electron dose leads to an increase of the work function of maximum 1 eV by means of water desorption. Simultaneously, water dissociation and desorption of only the hydrogen component cause after high electron doses a decrease of about 1 eV compared with the initial state. A modified model of the real surface is presented.

Reale InSb-Oberflächen sind mit einer dünnen Oxidschicht bedeckt und haben unter normalen Bedingungen eine adsorbierte Schicht aus Wassermolekülen, die ein Oberflächen-Dipolmoment erzeugen. Bestrahlungen mit Licht oder einer kleinen Elektronendosis führen zu einem Anstieg der Austrittsarbeit um maximal 1 eV infolge Wasserdesorption. Gleichzeitig findet eine Wasserdissoziation statt und eine Desorption der Wasserstoff-Komponente. Dies führt bei hohen Elektronendosen zu einem Abfall von etwa 1 eV unter den Ausgangszustand. Ein modifiziertes Modell der realen Oberfläche wird vorgeschlagen.

1. Introduction

The influence of electron irradiation on semiconductor surfaces has been of interest in the last twenty years as well for example in connection with irradiation damage studies, as in irradiation-induced desorption phenomena. In the early times real surfaces were the object of investigations in this field. Now, with the development of ultrahigh vacuum (UHV) techniques, clean surfaces are favoured. Without any doubt the latter are of more fundamental importance. Real surfaces, on the other hand, are essential constituents of a lot of technological and chemical applications. But our knowledge of the whole complex "surface" is at least incomplete, and more useful data should be known.

Our interest is of more fundamental nature and joins to previous investigations mainly on the semiconductor–oxide interface of real surfaces [1]. Here we shall report on an expanded surface model including the outer insulator vacuum interface, deduced among others from results of contact potential difference (CPD) and photoelectric work function measurements. Electron irradiation was used to change definitely the surface conditions. A few results have already been published in a short note [2]. In contrast to similar studies on real germanium surfaces with somewhat lower electron energies [3], we found CPD changes two orders higher than those of the surface potential V_s and with no determined correlations of the signs.

2. Experimental

The specimens used were p-doped InSb single crystals (from M.C.P., Ltd.) grown in [211] direction with $N_A \approx 10^{13}$ and [111] with $N_A \approx 6 \times 10^{12}$ and $\approx 10^{16}$ cm^{-3},

[1]) Present address: Brown, Boverie & Cie., Mannheim ZEK/P, BRD.

respectively, mechanically polished and etched in CP4 A. Contacts were prepared by thermocompression of gold and silver wires, and in a few cases with silver paint. Areas of the samples were a few cm^2.

The measurements have been performed in either a high-vacuum (HV) or an 8″ ultra-high vacuum (UHV) system (from V.G.). In both systems the specimens were mounted on a turnable cold finger. In Fig. 1a sectional view of the UHV chamber is shown. The components seen are: an electron gun (EG) for the irradiations with 25 to 100 keV electrons (maximum current 10^{-6} A), an argon ion gun (AG1), a quartz window for optical irradiation, a view window, and the electromechanically driven vibrating stainless steel electrode (in the HV system gold was used) for CPD measurements which could be cleaned by e-gun heating (E2). Also to be seen is the localization of the specimen and the tungsten (W) reference electrode which also could be cleaned by e-gun heating (E1) and, in a lower plane, the quadrupole mass-spectrometer Q 7 (from V.G.).

The measuring signal form CPD was connected triaxially to an operation amplifier (Analog Dev. 311) and phase sensitive amplified (381A Dynatrac) with the 2ω reference signal from the electromagnetic coil of the vibrating electrode assembly driven with an ac voltage of about 12.5 Hz in the HV system, and 23 Hz in the UHV system, respectively. The stationarity of the signal and the automatic compensation of the CPD by the balance voltage was controlled by a double beam oscilloscope. CPD's up to 10 V could be measured to an accuracy of 1 mV. The errors being principally due to the feedback and the chosen overall amplification.

The optical equipments were used as well for the irradiation experiments as for the measurements of the photoelectric work function W_{ph} or the surface photovoltage (SPV), respectively. They consist of grating monochromators with different gratings (B. & L.) ranging from 200 to 1700 nm with (150 and 450 W) xenon light sources and Suprasil windows and lenses. The photoelectric yield was measured with a phase sensitive detector. The light intensity of every measure point was controlled via semiconductor thermocouple to exclude the faults from xenon lamp instabilities.

The SPV was measured either with white light or for every point in the spectre via intensity measurements by aims of gray filters and graphically evaluation the semi-logarithmic plots.

The conductivity of the specimens and their changes (the surface conductance (SC) $\Delta\Sigma$) as well as the field effect of SC was measured with convenient dc equipments.

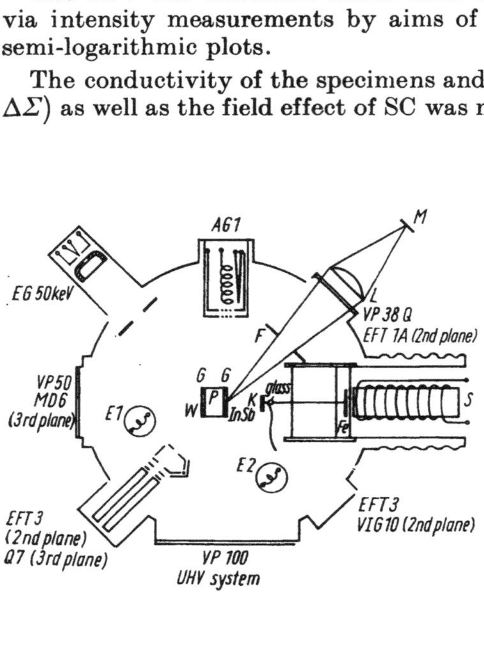

Fig. 1. Sectional view of the UHV system with the different components

3. Experimental Results

3.1 Contact potential

The electron energies between 25 and 100 keV are too low to create point defects in the bulk of a semiconductor [4, 5]. But they are able to change surface parameters as a consequence of electronic excitation, charge trapping or retrapping in surface states, or impact processes with adsorbed gas atoms or molecules, for example.

The changes of CPD, a very sensitive quantity for those parameters, are shown in Fig. 2. Different conditions of the environment are chosen in this case. With increasing electron dose in all cases observed, firstly CPD and the work function W of InSb increases (state II), then decreases and saturates (state III) at a lower value than original (state I). The maximum changes found are ± 1 eV. At temperatures below 100 K the changes are retentive for at least some hours, and can be annealed by heating up to RT [2]. The annealing curves show independently of the starting point after increased or decreased W's the characteristic peaked structure near 150 K, accompanied by a measurable gas outbreak.

Irradiations at RT always lead to increases of positive CPD's. The conversion temperature for the direction of the work function changes ΔW was about 130 K. With the electron energy range chosen no dependence on energy was found. At constant temperature, CPD changes are determined solely by the electron dose.

The influence of light (between 1.7 and 6.2 eV) on W was found independent of temperature in the positive direction, in the limit 0.6 eV at about 80 K. Combined irradiations alternately with electrons or light, show with light always changes in the direction to state II.

3.2 Surface conductance

The conductances of thin enough p-InSb specimens are always increased under irradiation influences, light, X-rays, and electrons as well. On this surface photoconductivity (SPC) we already reported previously [1, 6], also on the spectral distribution of optically induced CPD changes [2].

In Fig. 3 CPD and SC are shown, both measured in irradiation breaks. The positive sign of $\Delta\Sigma$ is obviously independent of the simultaneously observable direction of ΔW. An analysis of field effect curves showed the reduced surface potential v_s after cooling down the specimens in the dark before any irradiation to lie between -5 and -10. Bands were thus always found to be bent up originally, caused by a negative oxide charge. This is in agreement with the observations of other authors on real [7] and even (thin) oxidized InSb [8]. The density of the surface states N_{ss} for the $10^{13}/cm^3$ crystal was about $4 \times 10^{12}/cm^2$ eV in the broad minimum ascending to about 3 to $4 \times 10^{13}/cm^2$ eV near the band edges, deduced from the measured $\Delta\Sigma$ with the help of the Brown concept [9] and the Schrieffer approximation [10] for surface mobility influence. After irradiation, the minima of the field effect curves ($\Delta\Sigma - V$) were shifted

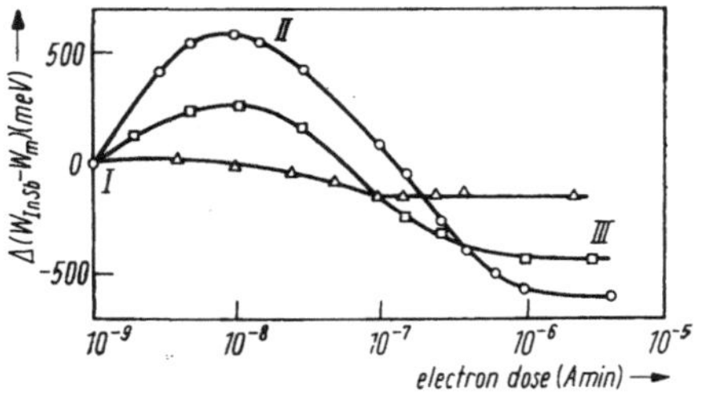

Fig. 2. CPD changes of InSb (IS 261 A, 50 keV electrons) vs. the electron irradiation dose; parameter is the different vapour pressure in the UHV system. △ $p < 5 \times 10^{-11}$ Torr, $T = 90$ K; □ $p = 1 \times 10^{-9}$ Torr, $T = 85$ K; ○ 10^{-6} Torr $> p > 10^{-7}$ Torr, $T = 92$ K

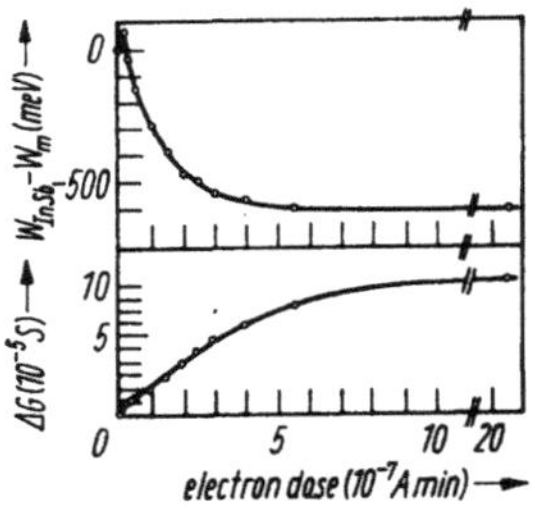

Fig. 3. An example of simultaneous changes of CPD and conductance ΔG of InSb (IS 5) under electron irradiation influence in the HV system; vibrating gold electrode. $T = 80$ K

in the direction of positive gate voltages, referring to an increased negative oxide charge. Field effect curves were found weakly deformed after irradiation. The analyses show increased N_{ss} of some tenths of percentage, largest near the conduction band edge and weakly structured near $v_s = 2$ and 14 in the case, for example, of a specimen with a reduced bulk potential of $u_b = -8.3$.

3.3 Photoemission

Photoelectric yield (y) measurements have been performed to find independently the absolute value of the work function of the InSb and the W foil used for calibration, and also to compare our values of the InSb with those known from literature. The best fit of the measured yield with the theoretical calculations of Kane [11] $y \sim (h\nu - W_{ph})^{n/2}$ was found with $n = 6$ instead of the expected 5 for the low doped semiconductor. This agrees with the findings of Gobeli and Allen [12]. One example of a specimen baked in the UHV system is shown in Fig. 4. In this case the work function is increased compared with the etched untreated specimens to 4.4 eV, but lies below those of cleaved crystals with values between 4.5 and 4.6 eV [13], respectively, 4.77 eV [12]. Initially, we found typically $W_{ph} = 4$ eV similar to our early observation with SPC [1] and Haneman's value of 4.15 eV [13].

The work function of the W foil was found to be 4.04 eV from fitting measured values to a yield $y \sim (h\nu - W_m)^2$ following the Fowler/Du Bridge [14, 15] estimation. With the help of this foil the different Kelvin electrodes were adjusted by CPD measurements. The systematical errors in the estimation of the work functions of InSb are essentially given by this method of fitting and are thus higher than the work function differences from CPD measurements.

A comparison of the thermionic and the photoelectric work function showed in the UHV system the thermally (250 °C) treated specimens to have $v_s = 0$, i.e. bands are then flat up to the semiconductor surface.

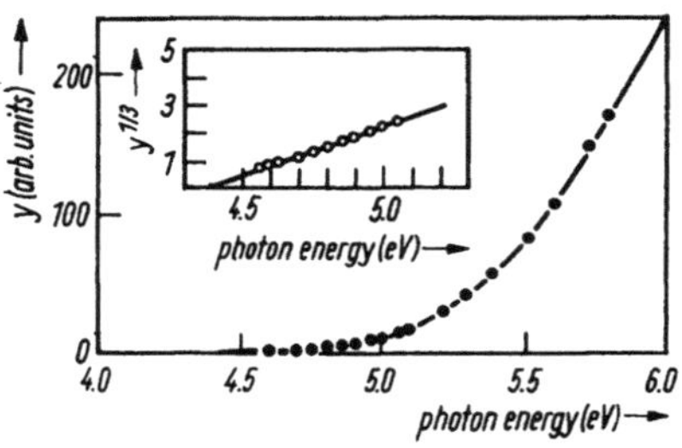

Fig. 4. Photoemission yield (y) from a real (111) InSb (IS 5 3.1) surface, measured in the degased UHV system (5×10^{-11} Torr); values are corrected concerning optical reflection

3.4 *Surface photovoltage*

At least at high light intensities, leading to a high excitation level, and a Dember potential negligible compared to the surface photovoltage (SPV), the SPV δV_s tends to $-V_s$. Comprehensive presentations of this topic have been given in the literature [16 to 18]. Under those conditions, the values of δV_s were found to be about -0.1 V. This is in agreement with the field effect analysis, bands are initially always bent up maximal to 0.1 eV. Thermally degased specimens in UHV environment, on the other hand, showed at RT a zero SPV, referring to flat band conditions (see also Section 3.3).

The spectral dependence of SPV at lower temperatures is shown in Fig. 5. A steep increase at 0.84 eV is followed by at least two structures at higher photon energies. The latter are probably correlated to the interband transitions near 2 and 2.5 eV, also to be found by SPC measurements [1], and which are established as well experimentally as theoretically [19, 20].

3.5 *Adsorption of water molecules*

The mass spectrometric analysis in the two vacuum systems showed the main components of the residual gas to be water vapour (H_2O^+ and OH^+) and hydrocarbons in the HV (10^{-6} Torr) apparature, whilst masses larger than 46 are absent even in the unbaked UHV system (10^{-9} Torr). Other residual gas components like H^+, C^+, CH_3^+, O^+, OH^+, H_2O^+ are then low and of the same order as the N_2^+ contents, H_2^+ somewhat higher. They all vanish in the measuring limit after degasing the system thermally at 250 °C (for about 10 h). In this environment one frequently measured specimen was found to be non-aged after three months with the CPD reproducible within the measuring accuracy of 1 mV (at 100 K).

The influence of the different environments on the behaviour under electron irradiation was shown in Fig. 2 for the UHV system, in [2] for the HV system. No difference is to be observed comparing the latter with Fig. 2. The environments differ in the contents of hydrocarbons, whereas water vapour is present in both. Obviously, a reduction of the water vapour concentration in Fig. 2 drastically reduces the CPD changes under irradiation. Thus, the influence of water was studied. Fig. 6 shows the change of CPD after inlet of water vapour in the degased (10^{-11} Torr) UHV system. Stationary values of CPD follow reversibly the partial pressure of the water vapour. Even in the low-pressure region of unchanged CPD (Fig. 6) a deposition of water molecules must be assumed, for electron irradiation then leads again to CPD changes of for example -0.5 eV (at 10^{-10} Torr) and the starting point can be reached by illumination like in the cases discussed in Section 3.3.

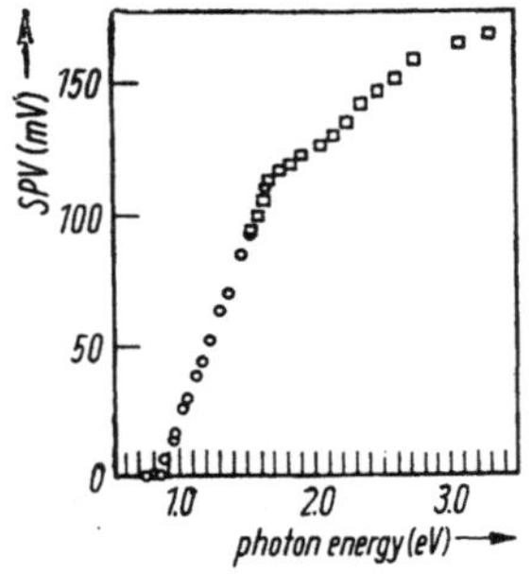

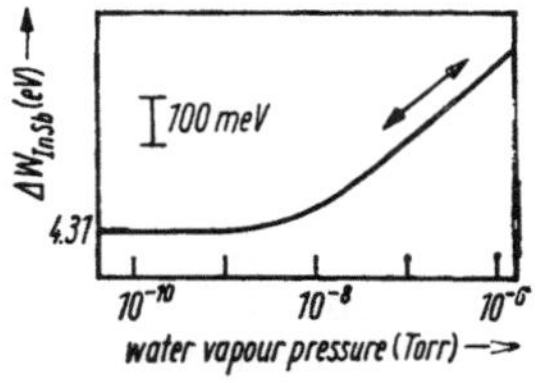

Fig. 6. Reversible changes of the thermionic work function of InSb (IS 261 A) after inlet of pure water vapour in the degased UHVsystem

Fig. 5. Spectral dependence of SPV of InSb (IS 261 A) normalized to a constant photon flux (UHV system 250 °C degased); measure points noted with circles and squares are from different gratings. $p < 5 \times 10^{-11}$ Torr. $T = 100$ K

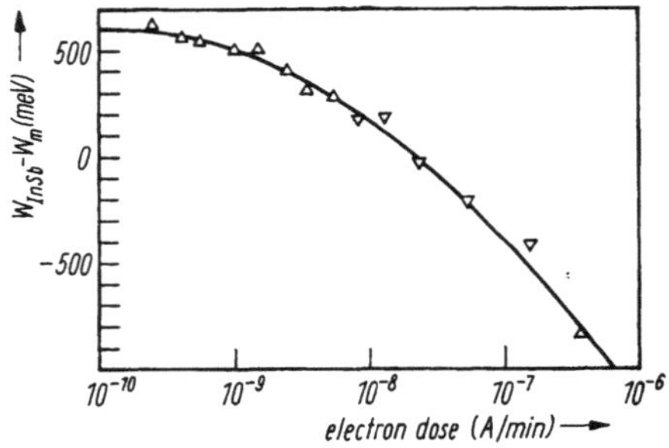

Fig. 7. CPD changes of InSb (IS 261 A) under electron irradiation influence in a moist environment; partial pressure of the water vapour 10^{-8} Torr. $T = 98$ K

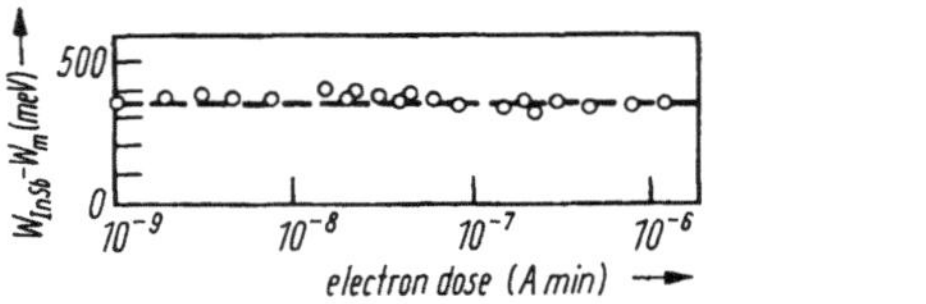

Fig. 8. Electron irradiation influence of InSb (IS 261 A) on a specimen in the UHV (10^{-10} Torr(. The specimen was firstly cleaned by Ar^+ bombardment and thermal treatment, then exposed to a water vapour pressure of 5×10^{-8} Torr. $T = 100$ K

The maximum electron irradiation induced changes after water deposition were found to be -1 eV (Fig. 7). Annealing curves with a constant heating rate then show exactly the same shapes with the characteristic peak like in those measured in HV environment.

Specimens have then be cleaned by argon ion bombardment of some hours duration (196 eV, 5×10^{-6} A/cm^2) and subsequent heating slowly to maximum 350 °C. After that procedure the work function has increased and the behaviour is drastically altered. The influence of electron irradiation is shown in Fig. 8. The cleaned (111) surfaces behave likely before and after water vapour inlet: CPD fluctuates maximum about ± 30 meV, measured in irradiation breaks. The irradiation effects observed with real surfaces are no longer present.

4. Discussion

The result of the investigations show that due to electron irradiation with increasing dose for InSb crystals with real surfaces in a moist environment the surface potential V_s only increases in the order of 10 mV. The work function, on the other hand, firstly increases, but then decreases. The changes are in the order of ± 1 eV (states II and III). Initially, these p-type specimens always had accumulation space charge layers (state I). The influence of light causes changes of the same order independent of the starting conditions in the same direction of increasing band bending and work function.

An estimation for the conditions leading to state II shows the following: A $\Delta W \approx +1$ eV ist correlated to a change of the oxide charge density of about 10^{-7} As/cm^2 (from field effect minima shift). With the realistic assumption of an oxide thickness of about 10^{-8} m and an $\varepsilon \approx 11$ [21] an overall potential drop of only $\Delta U = \Delta Q/C_{ox} \approx 0.1$ V follows. Band bending in the semiconductor was found even one order lower, about 12 meV. A plausible explanation is, that work function changes occurred due to a surface dipole layer caused by adsorption, desorption, or orientation of neutral molecules. The two signs of ΔW claim for at least two processes of this kind, different in magnitude and which saturate both independently or lead to stationary conditions of state III.

Dipoles may either be permanent or induced in the surface field. The work function is always lowered depending on their density n_A and dipole momentum μ. The potential flaw $\Delta\Phi = n_A\mu/\varepsilon_0$ is for non-interacting dipoles (at low densities) simply proportional to n_A. The water molecule has by far the greatest μ (6.1×10^{-30} As m [22]) of the residual gas components. We assume the latter and its neutral radicals to be responsible for the observed phenomena. Electron impact desorption (EID) with electrons of a few or a few hundred eV have been observed by some authors, for example from H_2O and CO on CdS [23] or O_2 on ZnO [24]. Here also a desorption under irradiation influence was found. The whole topic was discussed in review articles by Menzel [25] and Madey and Yates [26]. Impacts on electro-negative adsorbates result in a desorption of neutral or positively charged molecules. Changes of the adsorption state under irradiation influence are also possible.

In the non-irradiated state water dipoles may be adsorbed even at RT, aligned in the surface field with the positive charge on the vacuum side. Irradiations with electrons and light lead to a state II′ (at RT) via EID or photo-desorption of neutral water dipoles. The lower saturation value under illumination can be caused by a lower cross-section σ. Similar differences in the σ's have been found also with other systems [27]. Re-establishment of the adsorption equilibrium state I is to be observed again without irradiation. At low temperatures the two kinds of irradiation have different effects. In the electron irradiation case, a second mechanism must be superimposed causing the work function to decrease at high doses. The electron energy is higher than necessary for a dissociation of water molecules in H and OH radicals. Bonding energies of O–H or OH–H are about 4.3 eV [22] and thus low compared with the transmitted electron impact energy. The increased H^+ contents found in the residual gas after electron irradiation lead to the assumption, that the latter are desorbed, whilst the OH dipoles stay adsorbed. A similar behaviour was observed on other semiconductor surfaces, for example a dissociation in the components H and OH of water adsorbed on germanium followed by a H_2 desorption [28]. Generally it is to state, that water has frequently been found to be adsorbed molecularly and from its radicals hydrogen was the one with the lower adsorption energy [29].

An estimation of the number of the dipoles needed to have the effect on the work function of ± 1 eV shows realistic quantities. A monolayer of water molecules consists of $7.4 \times 10^{14}/\text{cm}^2$ (with an area of 13.4 Å^2 for one single molecule). With the dipole momentum of 6.2×10^{-28} As cm state II requires from these about $1.5 \times 10^{14}/\text{cm}^2$ desorbed water molecules. A portion of about $1.7 \times 10^{14}/\text{cm}^2$ OH dipoles with $\mu \approx 5 \times 10^{-28}$ As cm must have been created for state III. The degased real surface was found to be correlated to flat band conditions. The absence, or at least drastically reduction, of oxide charge and surface field has the consequence of no alignment of in a low density adsorbed of water molecules. The work functions stay unaltered, 4.4 eV photoelectric and 4.31 eV CPD measured. Charge excitation and trapping under electron irradiation influence lead to a surface field again and alignment starts. An increased adsorption density, on the other hand, has the result of dipole orientations by dipole–dipole interaction and leads to reversibly increased work functions.

Electron trapping in oxide states was the main feature leading to the model of the semiconductor–insulator interface of real InSb surfaces [1]. The behaviour found in the course of work function measurements are shown to be correlated to adsorption, desorption, and creation of dipole molecules and radicals on the outer insulator surface. The model of the real surface must thus be modified with an additional potential flaw or an "electron affinity" strongly dependent on the surface dipole state (Fig. 9). In this model the energy difference between the bonding state of the adsorbate and the anti-bonding excited state was assumed to be given by the threshold in the spectral dependence of ΔW [2]. We are not sure that the states H_1 0.84 eV above the semiconductor valence band edge are identical with those found by SPC measurements [1] at 0.6 eV. There is better agreement with the energetic localization of an empty states band of the clean surface proposed by Huijser et al. [30].

A cleaned surface should show under irradiation influence no trapping processes. This was found to be true. With a cleaned surface in a moist environment, also no

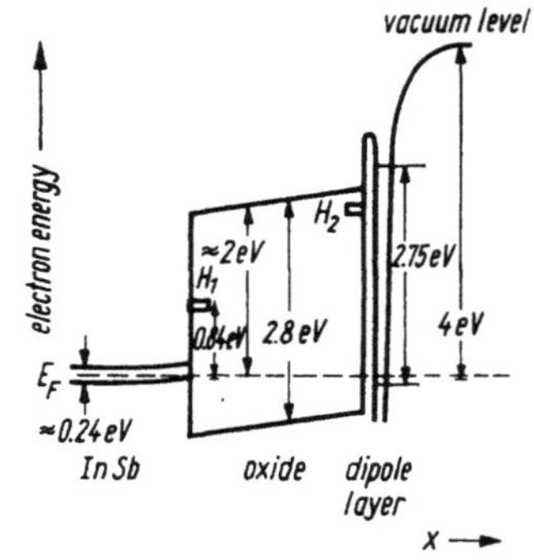

Fig. 9. Band model of the real (CP4-A-etched) InSb surface, describing a system semiconductor–oxide–adsorption layer–vacuum

changes of the work function were observed. Possibly, that is due to either a low surface field or to an adsorption coefficient near zero.

5. Conclusions

The electronic behaviour of InSb with a real surface may well be explained by the assumption of interface states with a U-shaped distribution and an oxide cover with a high density of electron traps. Specimens cooled down to 77 K always show a weak negative oxide charge, with the consequence of the semiconductor bands bent up to the surface about 50 to 100 meV. Work functions were found to be about 4 eV. Any irradiation leading to electron excitation in the semiconductor above the InSb–oxide barrier causes an increase in the negative oxide charge, and bands are bent up further about 10 meV. Simultaneously, but independently, photon or electron-induced processes on an adsorption layer of water molecules take place with the result of an about 1 eV increased work function via desorption of water dipoles or, respectively, an about −1 eV decreased work function via electron impact dissociation of water molecules in the course of which the OH dipoles stay adsorbed and hydrogen desorbs.

In degased specimens the oxide charge is at least very low. Bands are then flat in the semiconductor and no surface field exists. The work function is increased to about 4.4 eV. Under those conditions water molecules can be adsorbed statistically non-directed and with no influence on the work function. Alignment occurs only after charge trapping in the oxide by any event. As was to be expected, cleaned surfaces show none of the discussed processes, obviously correlated to the possibility of charge trapping in an insulator cover on the semiconductor surface.

References

[1] H. PAGNIA, phys. stat. sol. **34**, 121 (1969).
[2] S. LEHR, W. MANTKE, and H. PAGNIA, phys. stat. sol. (a) **13**, K159 (1972).
[3] D. N. BOSE, phys. stat. sol. **27**, 325 (1968).
[4] F. H. EISEN and P. W. BICKEL, Phys. Rev. **115**, 345 (1959).
[5] R. BÄUERLEIN, Z. Phys. **176**, 498 (1963).
[6] E. W. KREUTZ, H. PAGNIA, and W. WAIDELICH, phys. stat. sol. **31**, K147 (1969).
[7] H. R. HUFF, S. KAWAJI, and H. C. GATOS, Surface Sci. **5**, 399 (1966).
[8] A. ETCHELLS and C. W. FISCHER, J. appl. Phys. **47**, 4605 (1976).
[9] W. L. BROWN, Phys. Rev. **100**, 590 (1955).
[10] J. R. SCHRIEFFER, Phys. Rev. **97**, 641 (1955).
[11] E. O. KANE, Phys. Rev. **127**, 131 (1962).
[12] G. W. GOBELI and F. G. ALLEN, Phys. Rev. **137**, A245 (1965).
[13] D. HANEMAN, J. Phys. Chem. Solids **14**, 162 (1960).
[14] R. H. FOWLER, Phys. Rev. **38**, 45 (1931).

[15] L. A. Du Bridge, Phys. Rev. 39, 108 (1932).
[16] C. G. B. Garrett and W. H. Brattain, Phys. Rev. 99, 376 (1955).
[17] D. L. Lile, Surface Sci. 34, 337 (1973).
[18] A. Many, Y. Goldstein, and N. B. Grover, Semiconductor Surfaces, North-Holland Publ. Co., Amsterdam 1965.
[19] K. L. Shaklee, M. Cardona, and F. H. Pollak, Phys. Rev. Letters 16, 48 (1966).
[20] J. R. Chelikowsky and M. L. Cohen, Phys. Rev. B 14, 556 (1976).
[21] E. T. Yon, J. appl. Phys. 41, 2185 (1970).
[22] W. J. Moore and D. O. Hummel, Physikalische Chemie, de Gruyter, Berlin 1976.
[23] S. Baidyaroy, W. R. Bottoms, and P. Mark, Surface Sci. 28, 517 (1971).
[24] T. Kwan and Y. Fujita, Bull. Chem. Soc. Japan 31, 379 (1958).
[25] D. Menzel, Angew. Chem. 82, 263 (1970).
[26] T. E. Madey and J. T. Yates, Surface Sci. 63, 203 (1977).
[27] P. Kronauer and D. Menzel, Adsorption–Desorption Phenomena, Academic Press, New York 1972.
[28] S. Sinharoy and M. Henzler, Surface Sci. 51, 75 (1975).
[29] G. Ertl and T. Giovanelli, Ber. Bunsenges. phys. Chem. 72, 74 (1968).
[30] A. Huijser, J. van Laar, and T. L. van Rooy, Surface Sci. 62, 472 (1977).

(Received June 2, 1978)

phys. stat. sol. (a) 49, 93 (1978)

Subject classification: 10.2; 22.2.1

Institut für Metallphysik der Universität Göttingen and Sonderforschungsbereich 126 der Deutschen Forschungsgemeinschaft, Göttingen/Clausthal

Creep and Dislocation Velocities in Gallium Arsenide

By

H. STEINHARDT and P. HAASEN

In honour of Prof. Dr. F. STÖCKMANN's 60th birthday

The velocities v_i of α, β, and screw dislocations are measured by double etching in GaAs crystals with different doping and at different temperatures. Also the maximum creep rates $\dot{\varepsilon}$ in compression are determined. A method of averaging the v_i is proposed to obtain $\dot{\varepsilon}$ from the theory of multiplication-limited creep proposed by Haasen for Ge and Si. The dependence of v_i on doping is attributed to an effect of the dislocation charge on kink formation energy.

In unterschiedlich dotierten GaAs-Kristallen werden die Geschwindigkeiten v_i der α-, β- und Schraubenversetzungen bei verschiedenen Temperaturen nach der Doppelätzmethode gemessen. Ebenso werden die maximalen Kriechgeschwindigkeiten $\dot{\varepsilon}$ im Kompressionsversuch bestimmt. Um nach der von Haasen für Ge und Si vorgeschlagenen Theorie des durch Versetzungsmultiplikation bestimmten Kriechens einen Zusammenhang zwischen $\dot{\varepsilon}$ und v_i zu gewinnen, wird eine Methode zur Mittelung über die Geschwindigkeiten der Versetzungstypen vorgestellt. Die Abhängigkeit der v_i von der Dotierung wird dem Einfluß einer Versetzungsladung auf die Kinkenbildungsenergie zugeschrieben.

1. Introduction

For gallium arsenide and other III–V compound semiconductors with sphalerite structure, α and β dislocations (positive and negative 60° dislocations) have to be distinguished [1]. The α dislocation (in the shuffle set [2]) has a row of Ga atoms at the end of the extra half plane and the dislocation core resembles a linear arrangement of vacancies in the As sublattice of GaAs while the core of the β dislocation may be compared with a row of Ga vacancies. (The inverse picture would apply to α* and β* dislocations of the glide set of slip planes proposed by Hirth and Lothe [2].) Different velocities for α and β dislocations have been observed in InSb [3 to 5], GaAs [6 to 8], and other III–V compounds [9]. Doping of the crystals leads to quite complicated velocity changes [10 to 12] which may be related to the electrical charge of the dislocations [13]. Plastic flow in gallium arsenide takes place at elevated temperatures [14 to 18] similarly to that in germanium and silicon [19]. The initial macroscopic deformation of these materials depends strongly on temperature and also on the applied shear stress and is correlated directly to the velocities of the dislocations that are moving, according to a theory by Haasen [20, 21]. This theory does not take into account, however, differences in velocity between edge and screw dislocations or between positive and negative edge dislocations. For GaAs, a simultanous investigation of the individual dislocation motion and the macroscopic plastic flow has not been done so far. Here we present experimental results on the creep rates as well as on dislocation velocities in undoped GaAs and in GaAs crystals doped with Zn and Te in the order of 10^{18} cm^{-3}. The results show that the strain rate is determined by a suitable average of the various dislocation types of different mobilities. A theoretical explanation for this is proposed extending the theory for Ge and Si.

2. Experimental

Rectangular samples for uniaxial compression tests were cut from not specifically doped GaAs ($n = 10^{17}\ cm^{-3}$), from a crystal doped with Zn ($p = 2.9 \times 10^{18}\ cm^{-3}$), and from a crystal doped with Te ($n = 2.3 \times 10^{18}\ cm^{-3}$). The creep tests were performed under inert gas atmosphere with samples in ⟨123⟩ orientation suitable for single glide. Dislocation etch pits were produced on (111) surfaces by a solution according to Schell [22] and on $(\bar{1}\bar{1}\bar{1})$ surfaces by an etchant developed by Richards and Crocker [23].

3. Results

In the creep tests, the plastic strain was measured as a function of time. Temperature and stress were kept constant. The maximum strain rate $\dot{\varepsilon}_m$ of the S-shaped creep curve was taken as the quantity characterizing the plasticity of the crystals. Typical data are shown in Fig. 1 as a function of the temperature for constant shear stress $\tau = 10^7\ N\ m^{-2}$. The lines represent Arrhenius equations $\dot{\varepsilon}_m = C \exp(-Q/kT)$. Measured values are $C^{(u)} = 10^6\ s^{-1}$ and $Q^{(u)} = 1.5$ eV for the undoped GaAs, $C^{(p)} = 3 \times 10^5\ s^{-1}$ and $Q^{(p)} = 1.3$ eV for the crystal doped with Zn, $C^{(n)} = 5 \times 10^8\ s^{-1}$ and $Q^{(n)} = 2.2$ eV for the crystal doped with Te. A variation of the applied shear stress at constant temperature showed that the creep rate depends on stress similarly to Ge and Si [19]. Fig. 1 shows that GaAs is softened by doping with Zn and hardened by doping with Te. This is just opposite to the behaviour of Ge [24, 25] which is hardened by p-doping and softened by n-doping. From Fig. 1 it is seen that the change of the maximum creep rate with doping is mainly due to a variation of the activation energy Q. A change of the stress dependence of $\dot{\varepsilon}_m$ with doping has not been observed.

Dislocation velocities were measured by the double etch technique [4]. Fresh dislocations were produced at a scratch on the surface of the sample and were driven away from it by an applied shear stress. At constant stress and temperature, the distances of motion are proportional to the time of stress application. So the ratio of distance and time defines the dislocation velocity. Velocities of α and β dislocations could be measured simultaneously on samples in ⟨123⟩ orientation as used for the creep tests. Velocities of screw dislocations were measured on samples on ⟨011⟩ orientation. The dislocation types were identified by a method developed for InSb

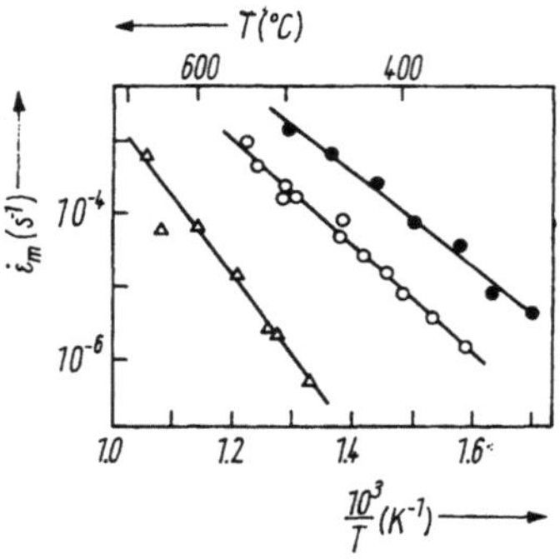

Fig. 1

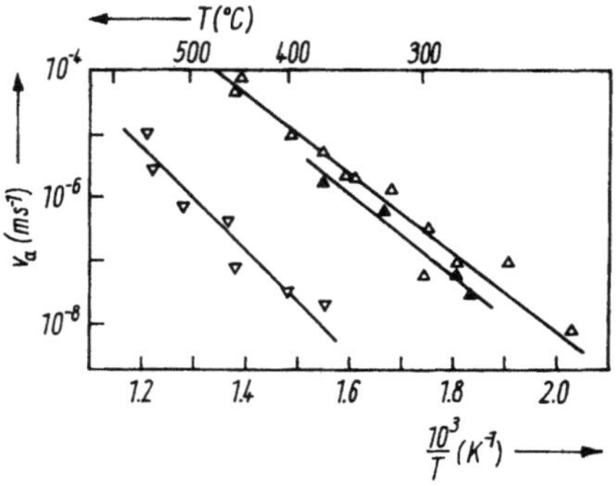

Fig. 2

Fig. 1. Maximum strain rate vs. inverse temperature for differently doped GaAs crystals ($\tau = 10^7\ Nm^{-2}$). ○ undoped; ● doped with Zn, $3 \times 10^{18}\ cm^{-3}$; △ doped with Te, $2 \times 10^{18}\ cm^{-3}$

Fig. 2. Velocity of α dislocations vs. inverse temperature for differently doped GaAs crystals $\tau = 10^7\ Nm^{-2}$). △ undoped GaAs; ▲ doped with Zn, $3 \times 10^{18}\ cm^{-3}$; ▽ doped with Te, $2 \times 10^{18}\ cm^{-3}$

[4]. For the velocity measurements and for the creep tests, the investigated ranges of temperatures overlap and the applied shear stresses coincide. The velocities of the α dislocations are shown in Fig. 2 as a function of temperature at constant stress τ. The lines represent equations $v = B \exp(-Q/kT)$ with $Q^{(u)} = 1.2$ eV for the undoped GaAs, $Q^{(p)} = 1.3$ eV and $Q^{(n)} = 1.8$ eV for the crystals doped with Zn and Te, respectively. Fig. 2 shows that doping with Zn slightly reduces the velocity of the α dislocation and doping with Te strongly reduces it. The activation energy Q varies in a corresponding way.

On notes that doping with Zn increases the creep rate (Fig. 1) whereas it decreases the velocity of α dislocations (Fig. 2). Therefore the maximum creep rate $\dot{\varepsilon}_m$ cannot be directly related to the velocity v_α of the α dislocations alone. Similar comparisons with the velocities v_β of β dislocations and v_s of screw dislocations indicate that the creep rate is not related to any single one of these either. So it must be concluded that all three dislocation types or at least two of them contribute to the initial plastic flow in gallium arsenide.

A comparison of differently doped GaAs yields the smallest differences between the velocities of the three types of dislocations for the case of crystals doped with Zn. Results are shown in Fig. 3. α and screw dislocations are equally fast within the limits of error of the experimental data and β dislocations were observed to be only 2 to 3 times slower than the other types. The velocities v vary with temperature in the same way as the maximum strain rate $\dot{\varepsilon}_m$. So a coincidence of these data is achieved by scaling as $\dot{\varepsilon}_m = v\ 16\ \text{m}^{-1}$ for Fig. 3. This may be compared in more detail to the behaviour of Ge crystals deformed in creep tests at 580 °C [26]. At this temperature, edge and screw dislocations move equally fast in Ge at an applied shear stress $\tau = 10^7\ \text{N m}^{-2}$ [27]. The strain rate $\dot{\varepsilon}$ is related to the dislocation velocity by the Orowan equation $\dot{\varepsilon} = bNv$, where b is the Burgers vector and N is the density of mobile dislocations. At the inflection point of the creep curve, the dislocation density is remarkable independent of temperature for Ge [19]. A relation $\dot{\varepsilon}_m(T) = bNfv(T)$ is obtained. The factor $f \approx 0.3$ takes into account the internal stresses in the deformed crystal [20]. At the applied stress $\tau = 10^7\ \text{N m}^{-2}$, a dislocation density $N \approx 6 \times 10^{11}\ \text{m}^{-2}$ and a scaling factor $bNf = 70\ \text{m}^{-1}$ are obtained for Ge [26, 27]. The scaling factor for p-doped GaAs in Fig. 3 is lower than the value for Ge but of the same order of magnitude. Therefore the macroscopic creep rate in p-GaAs may be rather directly related to the dislocation velocities in the way known from Ge.

For GaAs doped with Te, the α dislocation is about ten times faster than the β dislocation and more than hundred times faster than the screw dislocation (at 550 °C). Results are shown in Fig. 4. For a comparison with the creep rate of n-doped GaAs

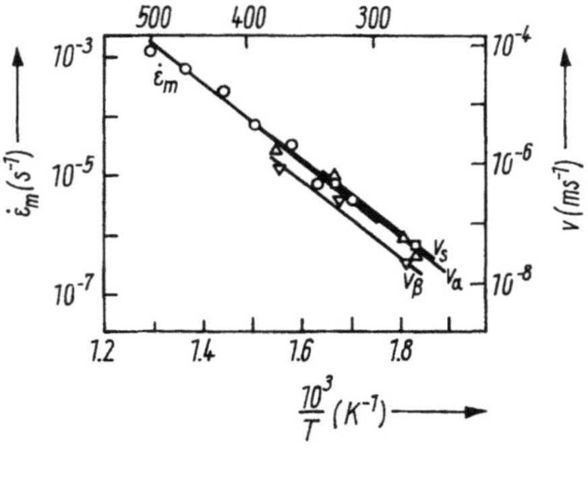

Fig. 3. Maximum strain rate and dislocation velocities vs. inverse temperature for p-GaAs doped with Zn, 3×10^{18} cm^{-3}. (○ $\dot{\varepsilon}_m$, △ v_α, ▽ v_β, □ v_s)

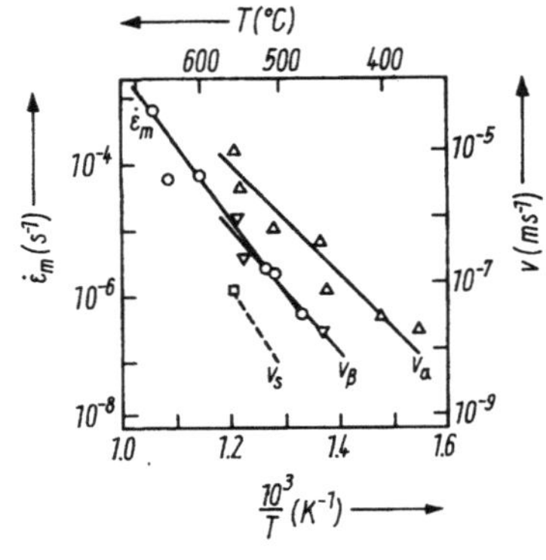

Fig. 4. Maximum strain rate and dislocation velocities vs. inverse temperature for n-GaAs doped with Te, 2×10^{18} cm^{-3}. (○ $\dot{\varepsilon}_m$, △ v_α, ▽ v_β, □ v_s)

the scaling law $\dot{\varepsilon}_m = v\ 16\ m^{-1}$ from Fig. 3 is tried again. It is seen that the α dislocation velocity is higher than expected from the creep rate and the screw dislocation velocity is lower, on this scale. Unfortunately, for n-GaAs it has not been possible to measure the velocity of screw dislocations at temperatures below 550 °C. Therefore the broken line in Fig. 4 indicates only the probable dependence of the screw dislocation velocity on temperature. From Fig. 4 it is concluded that the maximum creep rate corresponds to an averaged dislocation velocity which is neither near to the highest observed velocity of a dislocation type nor to the smallest. The experimental results show that the creep rate of GaAs depends on the velocities of two or three dislocation types in a not quite simple way.

For a further comparison of creep rate and dislocation velocities, experimental data at a fixed temperature shall be analyzed in more detail. In Table 1 values are shown

Table 1
Comparison of creep rates and dislocation velocities at $T = 400$ °C and $\tau = 10^7$ N m^{-2}

	u-GaAs (undoped)	p-GaAs (3×10^{18} Zn cm^{-3})	n-GaAs (2×10^{18} Te cm^{-3})
$\dot{\varepsilon}_m$ (10^{-5} s^{-1})	0.9	10	0.001
v_α (10^{-6} m s^{-1})	12	6	0.03
v_β (10^{-6} m s^{-1})	0.01	2.5	0.002
v_s (10^{-6} m s^{-1})	0.06	7	0.00002

for $T = 400$ °C. At this temperature, the creep rates have been directly measured for undoped and Zn-doped GaAs. A value for Te-doped GaAs is obtained by extrapolation from Fig. 1. The dislocation velocities at 400 °C are either directly measured, or values are obtained by slight extrapolations from Fig. 3 and 4. (Only the screw dislocation velocity in n-GaAs had to be estimated rather crudely from Fig. 4.) Table 1 shows that p-doping increases the velocities of β and screw dislocation stronger (by factors 250 and 100) than the creep rate (factor 11) whereas it decreases slightly the α dislocation velocity (factor 2). On the other hand, a comparison of n-GaAs to the undoped crystal shows that n-doping reduces the velocities of α and β dislocations less than the creep rate while it reduces the screw dislocation velocity more strongly (by factors 400, 20, 900 and 3000, respectively). This leads to the conclusion that α dislocations as well as screw dislocations participate in the average dislocation velocity corresponding to the creep rate. The β dislocations, however, may not in every case contribute to the creep rate of GaAs crystals. Quantitatively, the creep rate changes with doping rather like the geometrical average of the velocities of screw and α dislocations. A theoretical explanation for this is proposed in the following.

4. Theoretical Considerations

For Ge and Si, the initial stage of plastic creep corresponds with the motion and multiplication of dislocations according to a theory by Haasen [19 to 21]. The basic equations of this theory are

$$\dot{\varepsilon} = bNv\,, \tag{1}$$

$$\dot{N} = \frac{\mathrm{d}N}{\mathrm{d}t} = KNv\tau_{\mathrm{eff}}\,, \tag{2}$$

$$\tau_{\mathrm{eff}} = \tau - A\sqrt{N}\,, \tag{3}$$

$$v = B\tau_{\mathrm{eff}}^{m} \exp\left(-\frac{U}{kT}\right), \tag{4}$$

where $\dot{\varepsilon}$ is the strain rate, b the Burgers vector, N the density of mobile dislocations, v the average dislocation velocity, K a multiplication constant, τ_{eff} an effective stress acting upon the dislocations, τ the externally applied stress, A and B are constants, and the exponent m and the activation energy U are parameters characterizing the stress and temperature dependence of the dislocation velocity v. From (1) to (4) one obtains a maximum creep rate

$$\dot{\varepsilon}_{\mathrm{m}} = C\tau^{m+2} \exp\left(-\frac{U}{kT}\right) \tag{5}$$

which relates $\dot{\varepsilon}_{\mathrm{m}}$ to the parameters m and U of the dislocation velocity v. (C contains a combination of the previously introduced constants.) This theory does not yet distinguish between various dislocation types. Furthermore, the empirical equation (4) may be too simple for the description of the dislocation velocities v which have been measured [27, 28]. A generalization of (4) should allow for a more complicated dependence of v on temperature and stress. The following set of equations extends the theory by taking into account the existence of the three differently mobile types of dislocations in gallium arsenide.

The equations for the strain rate extends the Orowan relation (1) to

$$\dot{\varepsilon} = b(N_\alpha v_\alpha + N_\beta v_\beta + N_{\mathrm{s}} v_{\mathrm{s}})\,, \tag{6}$$

where N_i is the density and v_i the velocity of i-type dislocations. It is clear that the mobile dislocation density is given as $N = N_\alpha + N_\beta + N_{\mathrm{s}}$. So we extend (3) into

$$\tau_{\mathrm{eff}} = \tau - A\sqrt{N_\alpha + N_\beta + N_{\mathrm{s}}}\,. \tag{7}$$

For the multiplication of dislocations, we propose that screw dislocations are generated by moving 60° (nearly edge) dislocations and vice versa. Then the following equations replace (2):

$$\dot{N}_{\mathrm{s}} = K_\alpha N_\alpha v_\alpha \tau_{\mathrm{eff}} + K_\beta N_\beta v_\beta \tau_{\mathrm{eff}}\,, \tag{8a}$$

$$\dot{N}_\alpha = \dot{N}_\beta = \tfrac{1}{2} K_{\mathrm{s}} N_{\mathrm{s}} v_{\mathrm{s}} \tau_{\mathrm{eff}}\,. \tag{8b}$$

This allows for various multiplication mechanisms but excludes multiplication as straight dislocations. Finally, (4) is replaced by three equations for the dislocation velocities:

$$v_\alpha = B_\alpha(T)\, F(\tau_{\text{eff}})\,, \tag{9a}$$

$$v_\beta = B_\beta(T)\, F(\tau_{\text{eff}})\,, \tag{9b}$$

$$v_s = B_s(T)\, F(\tau_{\text{eff}})\,. \tag{9c}$$

Here the different velocities of the dislocations are expressed by different temperature functions $B_i(T)$. It is assumed that the stress function $F(\tau_{\text{eff}})$ is the same for the three dislocation types and does not change with doping. This behaviour is in fact observed for Ge [25].

In principle, the creep curve of GaAs can be derived from (6) to (9). The constants K_i and A are not precisely known for GaAs, however. So we shall only derive an expression for the average dislocation velocity v which corresponds to the creep rate ((cf. (1) and (5)). It appears reasonable to set $K_\alpha = K_\beta$ in (8a). We also neglect the initial dislocation densities, because the dislocations multiply by several orders of magnitude during a creep test. Then equations (8) yield

$$N_\alpha = N_\beta\,, \tag{10a}$$

$$N_s \sqrt{K_s v_s} = N_\alpha \sqrt{2K_\alpha(v_\alpha + v_\beta)} \tag{10b}$$

which means that the densities of the dislocation types grow proportional to each other during the initial stage of plastic deformation. Equations (10) combine with (6) to

$$\dot\varepsilon = bN \sqrt{\frac{1}{2}(v_\alpha + v_\beta)\, v_s}\, \frac{\sqrt{2K_\alpha v_s} + \sqrt{K_s(v_\alpha + v_\beta)}}{\sqrt{2K_s v_s} + \sqrt{K_\alpha(v_\alpha + v_\beta)}}\,. \tag{11}$$

This expression means that the average dislocation velocity is given by

$$v = q \sqrt{\tfrac{1}{2}(v_\alpha + v_\beta)\, v_s}\,,$$

where the dimensionless factor q combines the multiplication constants K_α, K_s and the velocities of "edge" dislocations $v_e = \frac{1}{2}(v_\alpha + v_\beta)$ and screw dislocations v_s. If $K_\alpha = K_s$, then $q = 1$ and (11) simplifies to

$$\dot\varepsilon = bN \sqrt{\tfrac{1}{2}(v_\alpha + v_\beta)\, v_s} \tag{12}$$

which may be assumed to apply for GaAs. (A more detailed analysis shows that (11) agrees with the experimental data within the limit of error for the range $0.3\, K_\alpha < K_s < 3K_\alpha$.) For $v_\alpha \gg v_\beta$, (12) reduces further to

$$\dot\varepsilon = bN \sqrt{\tfrac{1}{2} v_\alpha v_s}\,. \tag{13}$$

This correlation seems to be applicable to undoped GaAs as well as to GaAs doped with Te. In these cases, the creep rate averages geometrically over the faster edge dislocation velocity and the screw dislocation velocity, whereas the slower edge dislocation apparently does not contribute to the creep rate.

5. Discussion

A correlation between the macroscopic creep rate and the various dislocation velocities in GaAs has not been achieved previous to this work. Rather it had been assumed earlier that the velocity of one dislocation type determines the plastic strain rate in bending experiments of InSb crystals [29]. For GaAs, Osvenskii et al. [16]

investigated bending creep of undoped crystals and found three times higher creep rates for an orientation where α dislocations in the main slip system are the majority type than when they are the minority type. This anisotropy of bending became even smaller with p-doping or (at higher temperature) with n-doping [16]. In the latter case and in undoped GaAs the velocities of α and β dislocations differ, however, by factors 15 and 1200, respectively, according to Table 5. Therefore, the bending rate cannot be directly related to the velocity of the majority dislocation type. Such a discrepancy has also been found for InSb and has recently been explained convincingly by Shimizu and Sumino [30, 31] as follows: On bending InSb in such a way that the slower β dislocations should be majority dislocations on the main slip system, the crystal in fact deforms at first on the conjugate slip system (having a lower Schmid factor) where the faster α dislocations are again majority dislocations. Only later the primary slip system takes over. Shimizu and Sumino conclude that a forest type interaction between the two slip systems is responsible for the slower creep in this orientation. They believe that the motion of screw dislocations controls the deformation rate in both orientations of bending. The same mechanism should apply to undoped GaAs and to GaAs doped with Te. Instead of a rate control by moving screw dislocations only, however, we propose that the geometrical average of screw and α dislocation velocities determines the bending rate according to the considerations of this work. This means that in the case $v_\alpha \gg v_\beta$ the plastic bending for the introduction of β dislocations is achieved by the motion of α and screw dislocations without much contribution of the moving β dislocations themselves.

Previous measurements of dislocation velocities in GaAs [6 to 8, 11, 12] have been analysed in terms of various theoretical models [32 to 34] which had been proposed earlier for the description of the velocity of undissociated dislocations in germanium and silicon. Investigations by high-resolution transmission electron-microscopy, however, revealed that the dislocations in Si [35] and Ge [36] are dissociated. Furthermore, constrictions of the extended dislocations were observed at various points along the dislocations. Recently a similar observation has been reported for CdTe as well [37]. CdTe is a II–VI compound semiconductor crystallizing in the same structure as GaAs. Therefore, one must assume that dislocations are dissociated in GaAs, too. This has to be taken into account by a theory of dislocation velocities in GaAs and other III–V compounds. A suitable theory is not available, at present, which describes the velocity depending on temperature and applied shear stress as well as the velocity change with doping and the velocity difference between various dislocation types in GaAs. Experimentally, velocity differences between the dislocation types have been observed in InSb, InAs, and GaSb [9] similarly as in GaAs. In undoped crystals of these III–V compounds, α dislocations always move faster than β dislocations.

Various mechanisms have been considered for the influence of doping on the dislocation velocities in GaAs. At first, local obstacles to the motion of kinks along the dislocations have been proposed to explain the effects of doping [11]. Especially an electrostatic interaction between dislocations and charged impurities has been discussed [8]. These types of interaction between point defects and dislocations, however, generally reduce the velocity of a dislocation and cannot explain the observed increase of velocities in GaAs with Zn doping. At second, a variation of the Peierls barrier with doping has been proposed to apply for GaAs [7]. At present, a calculation of the Peierls barrier does not exist for the three dislocation types as a function of doping. For consistency with the experimental observations, the Peierls barriers should vary differently with doping for the three dislocation types. There seems to be no simple argument for such a complicated behaviour.

We think that the influence of doping is mainly due to an effect of the electrical charge on the dislocations upon the formation of double kinks in the dislocations [13]. We conclude from Fig. 1 to 4 that the variation of the dislocation velocities with doping is determined by a corresponding variation of the activation energies, at constant shear stress τ. These energies may be identified with the energies of the formation of kink pairs in the Peierls potential. The kink formation energy decreases with increasing electrical charge of the dislocation [13]. This is because of the higher electrostatic energy of a straight than of a kinked dislocation. Quantitatively one obtains [39] for the decrease of the kink formation energy due to a ratio f of charged dislocation sites:

$$\Delta E = \frac{e^2 f^2}{2\pi\varepsilon b} \, e^{-b/\lambda} \,. \tag{14}$$

Here e is the elementary charge, ε the dielectric constant, d the double kink width, λ the Debye screening length. f is obtained from the position of the Fermi energy E_F relative to the energy level E_0 of the neutral dislocation [38] according to

$$E_F = E_0 + \frac{e^2 f}{2\pi\varepsilon b}\left(\ln\frac{\lambda}{r_0} + \frac{1}{2}\right). \tag{15}$$

Here an entropy contribution is neglected, see [13]. The Fermi level E_F in turn is obtained from the doping concentration N_c (for p-material),

$$E_F = kT \ln \frac{C_v T^{3/2}}{N_c} \,. \tag{16}$$

$C_v T^{3/2}$ is the effective density of states at the valence band edge. Unfortunately the energetic positions E_0 for α, β, screw dislocations are not known from electrical measurements on GaAs so far. Therefore, we cannot yet compare changes in activation energy with doping between theory and experiment. We suggest, however, that the three dislocation types carry a strong positive charge is p-doped GaAs, while they are only weakly positive (α dislocation) or weakly negative (β dislocation) or even neutral (screw dislocation) in n-doped GaAs. Then it is likely that the three dislocation types in undoped GaAs differ in mobility mainly because their different electrical charges give more or less help to the formation of kink pairs.

References

[1] P. Haasen, Acta metall. 5, 598 (1957).
[2] J. P. Hirth and J. Lothe, Theory of Dislocations, Mc Graw-Hill, New York 1968 (p. 353).
[3] M. Mihara and T. Ninomiya, J. Phys. Soc. Japan 25, 1198 (1968).
[4] H. Steinhardt and S. Schäfer, Acta metall. 19, 65 (1971).
[5] M. Mihara and T. Ninomiya, phys. stat. sol. (a) 32, 43 (1975).
[6] S. K. Choi and M. Mihara, J. Phys. Soc. Japan 32, 1154 (1972).
[7] V. B. Osvenskii and L. P. Kholodnyi, Soviet Phys. — Solid State 14, 2822 (1973).
[8] S. A. Erofeeva and Yu. A. Osipyan, Soviet Phys. — Solid State 15, 538 (1973).
[9] S. K. Choi, M. Mihara, and T. Ninomiya, Japan J. appl. Phys. 17, 329 (1978)
[10] H. Steinhardt, Thesis, Göttingen 1972.
[11] V. B. Osvenskii, L. P. Kholodnyi, and M. G. Milvidskii, Soviet Phys. — Solid State 15, 661 (1973).
[12] S. K. Choi, M. Mihara, and T. Ninomiya, Japan, J. appl. Phys. 16, 737 (1977).
[13] P. Haasen, phys. stat. sol. (a) 28, 145 (1975).
[14] N. P. Sazhin, M. G. Milvidskii, V. B. Osvenskii, and O. G. Stolyarov, Soviet Phys. — Solid State 8, 1223 (1966).
[15] V. B. Osvenskii, M. G. Milvidskii, and O. G. Stolyarov, Soviet Phys. — Cryst. 13, 718 (1969).

[16] V. B. OSVENSKII, L. P. KHOLODNYI, and M. G. MILVIDSKII, Soviet Phys. – Doklady **14**, 144 (1969).
[17] V. B. OSVENSKII, O. G. STOLYAROV, and M. G. MILVIDSKII, Soviet Phys. – Solid State **10**, 2540 (1969).
[18] D. LAISTER and G. M. JENKINS, J. Mater. Sci. **8**, 1218 (1973).
[19] H. ALEXANDER and P. HAASEN, Solid State Physics **22**, 27 (1968).
[20] P. HAASEN, Z. Phys. **167**, 461 (1962).
[21] P. HAASEN, Festkörperprobleme III, Ed. F. SAUTER, Vieweg Verlag Braunschweig, 1964 (p. 167).
[22] H. A. SCHELL, Z. Metallk. **48**, 158 (1957).
[23] J. L. RICHARDS and A. J. CROCKER, J. appl. Phys. **31**, 611 (1960).
[24] M. M. SHEA. L. E. HENDRICKSON, and L. A. HELDT, J. appl. Phys. **37**, 4572 (1966).
[25] J. R. PATEE and A. R. CHAUDHURI, Phys. Rev. **143**, 601 (1966).
[26] K. BERNER and H. ALEXANDER, Acta metall. **15**, 933 (1967).
[27] H. SCHAUMBURG, Phil. Mag. **25**, 1429 (1972).
[28] A. GEORGE, C. ESCARAVAGE, G. CHAMPIER, and W. SCHRÖTER, phys. stat. sol. (b) **53**, 483 (1972); A. GEORGE, Thesis, Institut National Polytechnique de Lorraine, Nancy 1977.
[29] E. PEISSKER, P. HAASEN, and H. ALEXANDER, Phil. Mag. **7**, 1279 (1962).
[30] H. SHIMIZU and K. SUMINO, Phil. Mag. **32**, 123 (1975).
[31] K. SUMINO and H. SHIMIZU, Phil. Mag. **32**, 143 (1975).
[32] V. CELLI, M. KABLER, T. NINOMIYA, and R. THOMSON, Phys. Rev. **131**, 58 (1963).
[33] T. NINOMIYA, R. THOMSON, and F. GARCIA-MOLINER, J. appl. Phys. **35**, 3067 (1964).
[34] V. V. RYBIN and A. N. ORLOV, Soviet Phys. – Solid State **11**, 2635 (1970).
[35] I. L. F. RAY and D. J. H. COCKAYNE, Proc. Roy. Soc. **A 325**, 543 (1971).
[36] G. PACKEISER and P. HAASEN, Phil. Mag. **35**, 821 (1977).
[37] E. L. HALL and J. B. VANDER SANDE, Phil. Mag. **37**, 137 (1978).
[38] W. SCHRÖTER, phys. stat. sol. **21**, 211 (1967).
[39] P. HAASEN and W. SCHRÖTER, Verh. DPG (VI) **12**, 264 (1977).

(Received June 2, 1978)

phys. stat. sol. (a) **49**, 103 (1978)

Subject classification: 2; 14.1; 14.3.1; 21.6; 22.1

Physikalisches Institut der Universität Karlsruhe[1])

Electrical Resistivity of Amorphous Sb Films and Sb–Au Films

By

D. MÖCKEL, W. RÖSSLER, and F. BAUMANN

In honour of Prof. Dr. F. STÖCKMANN's 60th birthday

The semiconductor–metal transition is investigated in amorphous, quenched Sb–Au films. The resistivity of pure quenched Sb films depends on the film thickness as well as on vapour composition. This can be seen by studying quenched sputtered films. Their resistivity is smaller by many orders of magnitude compared with films quenched from Sb_4 molecules and will be determined only by variable range hopping. Films doped with more than 5 at% Au do not show any indication for hopping conductivity.

Der Halbleiter–Metall-Übergang wird an amorphen, abschreckend kondensierten Sb–Au-Schichten untersucht. Der spezifische Widerstand reiner Sb-Schichten hängt sowohl von der Schichtdicke als auch von der Dampfzusammensetzung ab. Dies wird durch Messungen an abschreckend kondensierten Sputterschichten gezeigt. Ihr spezifischer Widerstand ist um viele Größenordnungen kleiner als der aus Sb_4-Molekülen abschreckend kondensierten Schichten, wird jedoch ausschließlich durch Hoppingprozesse bestimmt. Die Hoppingleitfähigkeit verschwindet vollständig bei Filmen mit einem Zusatz von etwa 5 At% Gold.

1. Introduction

Among non-crystalline materials one can find insulators, semiconductors, and metals. This demonstrates that it is not the long-range ordering which is immediately responsible for the different properties. These are much more influenced by the nearest neighbourhood. Silicon, germanium, and antimony are for instance metals in the liquid state. By quenching from the vapour onto a cold substrate, thin films of these elements without any long-range order can be produced [1 to 3]. These amorphous, metastable films are semiconducting and subject to many investigations in the field of basic as well as applied research.

Besides this it is possible by quenching condensation simultaneously with noble metals to force the three elements into a metallic phase which is superconducting [4 to 6]. Systematic studies of the structure [7 to 9], the Hall coefficient [10, 11], and the change of the transition temperature to the superconducting state using uniaxial stress and strain [12] make clear that these films with more than 20 at% noble-metal content exist in a phase very similar to that of the liquid alloys. Therefore, the films are often called the liquid-like amorphous ones. Because the pure metastable Si, Ge, and Sb films have a nearest-neighbour order similar to the crystalline elements [13, 14] it is sometimes referred to them as the lattice-like amorphous ones.

Apparently it is not possible to produce pure films of the elements mentioned above without any covalent bonds. This leads to the question whether the covalent bonds which already do exist in the vapour phase are the ones responsible for the nearest-neighbour ordering in these films. As well known the vapour of the three elements consists mostly of molecules and not of atoms [15 to 17] as long as sophisticated

[1]) Engesserstr. 7, D-7500 Karlsruhe 1, BRD.

evaporation conditions are not used. Another question is concerned with the role of the noble metals in building the metal phase in the liquid-like amorphous films.

In this paper we report on investigations on pure antimony films and antimony films with small gold concentrations concerning the two questions just mentioned. The pure films are produced by sputtering antimony atoms from a target using an antimony ion beam as well as by condensation of Sb^+ or Sb_2^+ ions. The experiments performed in a helium cryostat should show the influence of the molecular bonds of the vapour upon the properties of the films. In another series of experiments antimony films with small gold content are investigated using sputtering evaporation technique.

During the experiments resistance measurements as a function of temperature after different annealing stages are carried out. By doing this it is possible to follow the transition from the lattice-like amorphous phase with the very high resistivity to the metallic phase of the liquid-like amorphous films in a sequence of different films quenched at helium temperatures.

2. Experimental

In this part some of the experimental details which are not standard will be sketched. Between the evaporation source and the substrate within the helium cryostat there is an ion source, followed by a magnet to separate the Sb^+ and the Sb_2^+ ions and some electrostatic lenses to take care of the profile of the ion beam. In order to condense ions an immersion lens just in front of the substrate has to slow down the ions and to focus them at the same time to the substrate. There is no need for the immersion lens when the system is used to produce the films by sputtering. In this case the antimony ion beam hits a target from which the atoms are sputtered and condensed at a substrate nearby. It should be emphazised that contrary to normal sputtering techniques, no additional gas is used. The ion source produces an ion current of 25 mA using a voltage of 10 kV. Just in front of the substrate ion current densities up to 80 μA/cm² have been measured corresponding to evaporation rates of 50 Å/min. The average energy of the ions has to be between 20 and 60 eV; otherwise no film can form. The upper limit is caused by self-sputtering while for energies below 20 eV the Coulomb interaction in the beam does not allow any substantial condensation rate of ions.

In order to produce antimony films by sputtering with antimony ions, antimony targets have been used after polishing. In the case of the Sb–Au films the flat surface of the target has been prepared with small evaporated gold strips of equal width and thickness separated by the same distance. From optical measurements using multibeam interferometry before and after sputtering has been carried out the amount of the two elements evaporated can be calculated. The gold concentrations of the films determined in this way agree quite well with the values measured by electron probe microanalysis. The evaporation rate for the sputtered films is about 200 Å/min.

3. Results

In Fig. 1 the electrical resistivity is shown as function of temperature in logarithmic scales. The film indicated as "Sb_4" is produced by evaporating antimony from a heated tungsten band and quenching the vapour onto a quartz substrate held at 4 K. The film is 870 Å thick and is grown up in the lattice-like amorphous phase. The change in resistivity is reversible as long as the film is not annealed above 280 K. Around 280 K crystallization into the stable, rhombohedral phase of antimony takes place. This behaviour is characteristic of films thicker than 900 Å. Thinner films possess a smaller resistivity after quench condensation at 4 K. After annealing up to 90 K

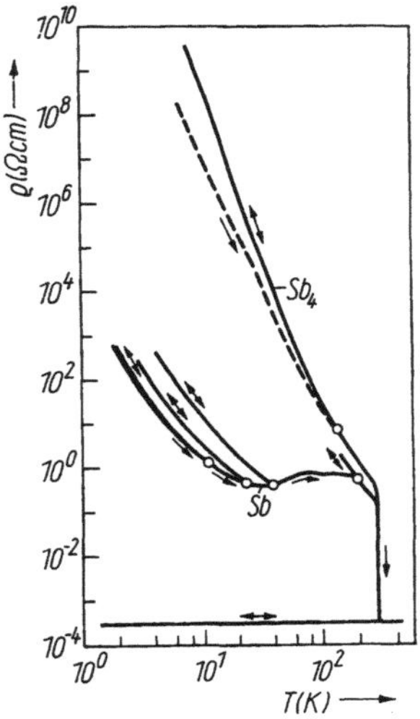

Fig. 1

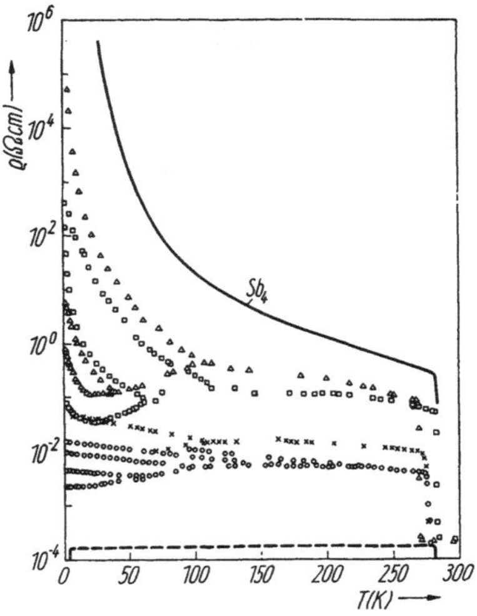

Fig. 2

Fig. 1. Temperature dependence of the resistivity for pure evaporated antimony films ("Sb_4", solid symbols) and sputtered Sb films ("Sb"). Single and double arrows indicate irreversible end reversible changes, respectively

Fig. 2. Temperature dependence of the resistivity for sputtered Sb films with different gold content △ 2, □ 4, × 6, ○ 8, - - - - 44 at% Au

the thinner films show the same resistivity behaviour as indicated by the Sb_4 curve. The dotted curve represents the results for a thin film with a thickness of 460 Å.[2])

In the same figure are also shown some results measured on a film produced by the sputtering technique mentioned above. These results are marked by "Sb". The film is 640 Å thick and its resistivity just after quenching at 4 K is much smaller than the films built up by Sb_4 molecules. After annealing and cooling down again the resistivity of the film has increased irreversibly. Warming up the film to 200 K almost finishes the annealing process and the sputtered film behaves nearly in the same manner as the films produced in the conventional way. Crystallization takes place at 280 K for both films. By sputtering antimony using antimony ions with energy and density chosen in these measurements one can be sure that the material will be carried off by single atoms. Apparently these measurements demonstrate the strong influence of the vapour composition on the resistivity of the films. The resistivity is also strongly dependent on film thickness. Thinner films produced by sputtering show clearly a smaller resistivity than thicker ones. This effect is much more pronounced in the case of the "Sb" films than in the case of the "Sb_4" films and can be seen in the whole range of thickness between 200 and 1200 Å of sputtered films are investigated.

In Fig. 2 some results are shown concerning Sb films containing different amounts of gold. The films are produced by sputtering and the condensation took place always at a substrate temperature of 4 K. With increasing gold content there is a strong decrease in resistivity. While the resistivity of the film with 4 at% gold changes by

[2]) The two Sb_4 films are measured by Mr. S. Grau and Mr. E. Rammelsberg. The authors are grateful to both for allowing to publish the data in this form.

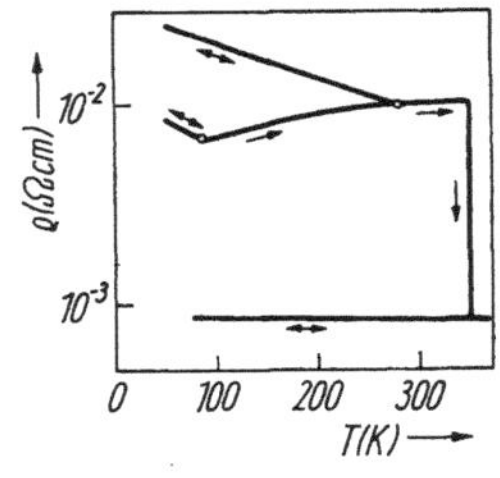

Fig. 3. Temperature dependence of the resistivity for an Sb film produced by quenching Sb^+ ions

some orders of magnitude cooling down from 100 to 1.5 K the film with 8 at% gold shows a change in resistivity by only a factor of two, during the same procedure. It is also interesting to note that films with a gold content above 5 at% do show an increase in resistivity by cooling proportional to $e^{-\alpha T}$ which is quite different from the behaviour of the pure antimony films and the films with a small amount of gold. The behaviour of a sputtered antimony film with 44 at% gold is also shown in Fig. 2. This film, which is metallic and becomes superconducting below 3.3 K, represents the liquid-like amorphous Sb–Au films.

In Fig. 3 the curve represents results of a film produced by quenching Sb^+ ions onto a substrate at 80 K. Films produced in the same way by quenching Sb_2^+ ions show qualitatively a behaviour very similiar to the curve shown in Fig. 3, but their resistivity values are always a little bit larger. The crystallization of the films produced by quenching ions takes place well above room temperature as shown by the irreversible drop in resistivity. This shift of the crystallization temperature to higher values may be caused by the condensation of residual gases during the relatively long evaporation time of these films.

Comparing the curve of Fig. 3 with the behaviour of the film containing 8 at% gold in Fig. 2 one sees a great similarity. From this it cannot be ruled out that part of the ion beam has met some parts of the immersion lens or the slits resulting in sputtering some noble metals. Although the slits and the lenses just mentioned were coated with a thin layer of antimony using a separate evaporation equipment every time before the films under investigation were condensed an uncontrolled doping of the films by noble metal cannot be ruled out completely. This difficulty does not exist in the case of films produced by the sputtering technique.

4. Discussion

The resistivity of pure quenched antimony films depends on film thickness as well as on vapour composition. The films produced by sputtering and the thin Sb_4 films have a smaller resistivity but approach continuously the behaviour of the thicker Sb_4 films during the course of annealing. The latter do not show any irreversible change as long as crystallization does not appear and their values of resistivity are the highest ones. This is shown by the upper branch of the Sb_4 curve shown in the figures so far.

These results can be discussed very naturally in terms of the bonding properties. The covalent bonds between the antimony atoms in the films condensed from Sb_4 molecules are established to a maximum degree. This is not the case in sputtered films built up atom by atom. In these films it happens during annealing that the covalent bonds are more and more established connected with an irreversible increase in resistivity.

There are some other arguments that vapour composition has an influence on the resistivity behaviour of antimony films. For example, the resistance of a film condensed by Sb^+ ions always differs much more with respect to an Sb_4 films of the same thickness than with a film produced by Sb_2^+ ions. Smaller resistivity was also measured

Fig. 4. Temperature dependence of the resistivity for pure evaporated antimony films ("Sb_4") and sputtered Sb films. The parameters indicate the annealing temperatures (T_a) seen by the sputtered films. Sb_4: ○ 240, □ 460, △ 870 Å; Sb: ▲ 240, ■ 640, ● 1200 Å

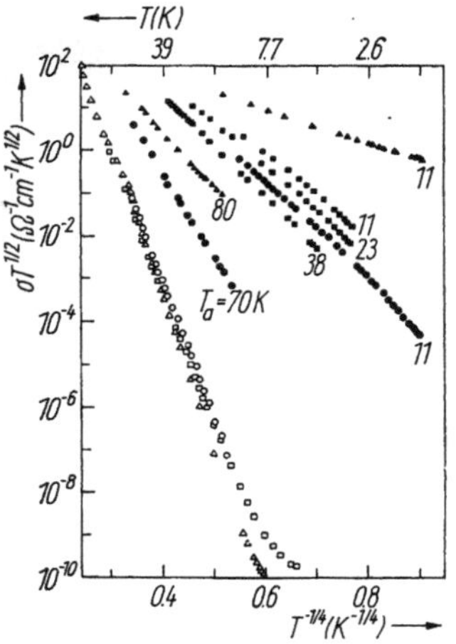

in pure antimony films produced by evaporating antimony from the compound $PtSb_2$ at the melting point [6]. Under these conditions the vapour composition is changed in favour of Sb_2 molecules [17, 18]. From this point of view it also makes sense that quenched Ge and Si films anneal with an irreversible increase in resistivity. In contrast to antimony the vapour of germanium and silicon is a mixture of atoms and various molecules [15, 16]. Therefore, one cannot expect to find the amorphous Ge and Si films just after quenching in a state with a maximum degree of covalent bonds.

At first it seems very surprising that the thinner films have a smaller resistivity. Assuming that the films are almost homogeneous one would be led to the conclusion that the thinner the films the fewer covalent bonds are established.

To be more precise one has to analyse the temperature dependence of the resistivity. Following Mott [19] the hopping conductivity σ as a function of temperature T for variable range hopping is given by

$$\sigma(T) = \sigma_0 T^{-1/2}\, e^{-(T_0/T)^{1/4}}$$

with the constants σ_0 and T_0. The latter is

$$T_0 = \frac{18\alpha^3}{k_B N(E_F)},$$

where α is the coefficient of exponential decay of localized states, $N(E_F)$ the density of localized states at the Fermi level, and k_B the Boltzmann constant. The constant σ_0 depends strongly upon the theoretical model under consideration [19, 20].

In Fig. 4 the representation is chosen in such a way that results which obey the first equation above lie on a straight line. The values marked by "Sb_4" belong to a 870 Å thick antimony film acting as a representative for the thicker films quenched from Sb_4 molecules. Obviously these data can be described very well with the equation mentioned above. Beside these there are also plotted results from pure antimony films quenched by sputtering. As a parameter there is noted the annealing temperature below that the measurement has been carried out. The strong influence of both annealing and film thickness can be seen very easily. The drawing also demonstrates that the reversible changes in resistivity of these films are described very well by the above equation. From this one can conclude that the conductivity of pure antimony films is determined completely by variable-range hopping processes. Despite the large changes this is true for all pure non-crystalline antimony films, independent of vapour composition, annealing stage, and film thickness. Surprisingly attempts made to fit the results of the thinner films with expressions for two-dimensional hopping [21] did not succeed.

It is impossible to determine the values for α and $N(E_F)$ using resistance measurements only. Assuming for α a value of 0.1 Å^{-1}, corresponding to a radius of localized states of 10 Å, the density of localized states at the Fermi level is calculated. These values as well as σ_0 and T_0 are given in Table 1.

Table 1
Properties of pure antimony films

film thickness (Å)	deposition method	annealing temperature (K)	σ_0 (Ω^{-1} cm^{-1} K$^{1/2}$)	T_0 (K)	$N(E_F)$ (eV^{-1} cm^{-3})
240	evaporation	80	3.81×10^9	2.7×10^7	7.7×10^{18}
460	evaporation	80	7.09×10^8	2.4×10^7	8.7×10^{18}
870	evaporation	80	1.76×10^{10}	3.7×10^7	5.6×10^{18}
240	sputtering	11	2.04×10^3	6.6×10^3	3.16×10^{22}
		80	4.75×10^5	8.49×10^5	2.46×10^{20}
640		11	1.82×10^5	1.73×10^5	1.21×10^{21}
		38	9.27×10^5	5.28×10^5	3.95×10^{20}
1200		11	4.21×10^5	3.54×10^5	5.90×10^{20}
		70	1.74×10^7	4.05×10^6	5.16×10^{19}

In order to calculate the resistivity a homogeneous structure of the films has been tacitly assumed. Although inhomogeneities may not be excluded completely they cannot play a dominant role. In particular a sandwich structure with a thin layer of high-conductivity material with a constant thickness cannot be made responsible for the strong thickness dependence of the resistivity observed in the sputtered films. It would be unlikely to find such a simple temperature dependence for sandwich films as shown in Fig. 4. In order to explain the drastic change in resistivity with film thickness one has to assume that during condensation the resistivity increases continuously. With increasing film thickness there is an increase in the density of covalent bonds equivalent to a decrease of the density of localized states at the Fermi level. This should happen in the case of thin Sb_4 films as well as in a much more pronounced manner in the sputtered films. This is not in agreement with measurements carried out on antimony films produced by sputtering in an argon atmosphere and condensed at a substrate temperature of 90 K [22]. The author himself mentioned that his results may be influenced by argon trapped in the films. This may also explain the higher crystallization temperature of these films.

It has to be emphasized explicitily that there was no success in producing metallic films by quenching pure antimony. The smallest resistivity was 0.4 Ω cm measured on a sputtered film with a thickness of 240 Å quenched at 4 K. Thinner films, which probably exist in form of small islands, have a very large resistivity. This was found at a sputtered film with a thickness of 160 Å and at a Sb_4 film with a thickness of 130 Å.

According to Ostwald's step rule [23] quenching from the vapour phase should result in the formation of a liquid-like state at the first step. Molten antimony behaves as a liquid metal and, therefore, one would expect the quenched antimony films to be metallic as seen in the analogous case of bismuth [24]. However, this metallic phase is obviously extremely unstable and for this reason pure antimony films change their properties already during condensation. On the other hand, it is well known that a metastable phase in a thin film is more stable than in thicker ones. In the picture proposed here it is also possible to explain why there is a continuous change in the density of localized states instead of a sudden transformation from a metallic to a semiconducting phase. To our opinion this is a change from one amorphous phase to another one connected with rearrangements only between nearest neighbours. These may be the reason for the thickness dependence of the resistivity measured in pure quenched antimony films.

In comparison with the pure antimony films discussed so far the quenched antimony films with small gold content behave quite differently. First of all the resistivity of the Sb–Au sputtered film is lowered in comparison with pure sputtered film of same thickness. As mentioned above the highest conductivity of pure Sb film was 2.5 $(\Omega\ \text{cm})^{-1}$ measured on a sputtered film 240 Å thick just after condensing at 4 K. The corresponding values are 2.4, 175, and 430 $(\Omega\ \text{cm})^{-1}$ for a 890 Å thick film with 2 at% Au, a 1020 Å thick film with 6 at% Au, and a 900 Å thick film with 8 at% Au, respectively. Also with increasing gold content the temperature dependence of the resistivity is changed. Cooling down the films immediately after condensation the resistivity increases considerably except for the films with a gold content above 5 at%. Also the behaviour of the films after annealing is quite interesting. Films with less than 5 at% Au show hopping conductivity at high temperatures, but at low temperatures their resistivity does not increase as much as one would expect according to the equation mentioned above. Similar results are reported for quenched Si films with small Au concentrations [25]. The resistivity of Sb–Au films with 5 to 10 at% gold increases exponentially with decreasing temperature. For this well-defined temperature dependence no explanation can be given so far. If one were to try to describe these results in the frame of a doped semiconductor neglecting the temperature dependence of the mobility, the activation energy would turn out to increase from 0.1 meV at 10 K to 8 meV at 100 K for the film with 6 at% Au. Quenched Sb–Au films with more than 15 to 20 at% Au but less than 80 at% Au grow up in the liquid-like amorphous phase, which is metallic as shown by its superconductivity [6]. In this phase the resistivity is linearly dependent on temperature with a negative coefficient.

In conclusion it is possible to study the semiconductor–metal transition in quenched Sb–Au films. The transition occurs in sputtered films with a doping of about 5 at% Au but is not accompanied with any change in long-range order. At the transformation the conductivity of the films is about 100 $(\Omega\ \text{cm})^{-1}$ which is in good agreement with the Joffe-Regel condition [26]. In order to freeze in the metallic phase about 15 to 20 at% Au is needed.

Acknowledgements

Stimulating discussions with Prof. W. Buckel, P. Häußler, H. Leitz, and Dr. W. Müller are acknowledged. The authors would also like to thank the members of the Laboratorium für Elektronenmikroskopie der Universität Karlsruhe. This work was supported by the SFB 66 of the Deutsche Forschungsgemeinschaft.

References

[1] S. K. Bahl and S. M. Bhagat, J. non-crystall. Solids **17**, 409 (1975).
[2] Z. A. Walley and A. K. Jonscher, Thin Solid Films **1**, 367 (1967).
A. H. Clark, Phys. Rev. **154**, 750 (1967).
[3] R. Suhrmann and W. Berndt, Z. phys. **37**, 146 (1936).
[4] D. Möckel and F. Baumann, Verh. D. P. G. **7**, 742 (1976).
[5] B. Stritzker and H. Wühl, Z. Phys. **243**, 361 (1971).
[6] W. H.-G. Müller, F. Baumann, and W. Buckel, Thin Solid Films **28**, 83 (1975).
F. Baumann, Thin Solid Films **34**, 87 (1976).
[7] E. Haug, N. Hedgecock, and W. Buckel, Z. Phys. **B 22**, 237 (1975).
[8] H.-J. Nowak, H. Leitz, and W. Buckel, phys. stat. sol. (a) **49**, No. 1 (1978).
[9] H. Leitz, private communication.

[10] G. Bergmann and R. Koepke, private communication.
[11] P. Häussler and W. H.-G. Müller, Verh. D. P. G. **1**, 427 (1977).
[12] R. Blocksdorf, H. Seifert, W. H.-G. Müller, and F. Baumann, Verh. D. P. G. **7**, 743 (1976).
[13] G. Breitling, J. Vacuum Sci. Technol. **6**, 628 (1969).
[14] H. Richter, H. Berckhemer, and G. Breitling, Z. Naturf. **9a**, 236 (1954).
[15] Gmelin, Handbuch der anorganischen Chemie, Si, Vol. 15B, 8th edition, Verlag Chemie GmbH, Weinheim/Bergstr. 1959 (p. 33).
[16] G. Köhl, Z. Naturf. **9a**, 913 (1954).
[17] F. Baumann, J. Kessler, and W. Rössler, J. appl. Phys. **38**, 3398 (1967).
[18] J. Sommer, J. appl. Phys. **37**, 2789 (1966).
G. Rosenblatt, J. appl. Phys. **38**, 888 (1967).
R. M. Lowe, J. appl. Phys. **39**, 2476 (1968).
[19] N. F. Mott, Phil. Mag. **19**, 835 (1969).
[20] W. Brenig, G. H. Döhler, and P. Wölfle, Z. Phys. **258**, 381 (1973).
[21] E. M. Hamilton, Phil. Mag. **26**, 1043 (1967).
W. Brenig, G. H. Döhler, and H. Heyszmann, Phys. Letters **39** A, 175 (1972).
M. L. Knotek, M. Pollak, T. M. Donovan, and H. Kurtzmann, Phys. Rev. Letters **18**, 853 (1973).
[22] J. J. Hauser, Phys. Rev. **B 9**, 2623 (1974).
[23] W. Ostwald, Z. phys. Chem. **22**, 289 (1897).
[24] W. Buckel and R. Hilsch, Z. Phys. **138**, 109 (1954).
[25] N. Kishimoto, K. Morigaki, A. Shimizu, and A. Hiraki, Solid State Commun. **20**, 31 (1976).
[26] A. F. Joffe and A. R. Regel, Progr. Semicond. **4**, 237 (1960).
N. F. Mott, J. non-crystall. Solids **8/10**, 1 (1972).

(Received June 9, 1978)

phys. stat. sol. (a) **49**, 111 (1978)

Subject classification: 14.3.1 and 14.3.3; 13.4; 22.2.1

Institut für Elektrophysik der Technischen Universität Braunschweig (a) and Institut für Halbleitertechnik der Rheinisch-Westfälischen Technischen Hochschule Aachen (b)

Low-Frequency Current Noise of GaAs Epitaxial Layers between 4.4 and 300 K

By

W. HOYER (a), G. LAUTZ (a), and M. HEYEN (b)

In honour of Prof. Dr. F. STÖCKMANN's 60th birthday

From several GaAs epitaxial layers ($n \approx 10^{17}$ cm^{-3}) on semiinsulating substrates MESFET device structures are fabricated and current noise measurements are performed between 1 kHz and 1 MHz in the range 4.4 to 300 K, to examine the influence of different epitaxial technologies on the low frequency noise. Below 20 K the noise is generally white up to 1 MHz with $1/f$-noise superimposed. The white noise can be attributed to warm electron effects. Above 20 K the temperature dependence of the noise, which is mainly determined by generation–recombination processes in the conducting channel, yields several activation energies of impurities interacting with the conduction band. As far as published energy values are available good agreement could be achieved. A chart containing the relative noise intensities connected with the different impurity levels is given.

Auf einer Reihe von GaAs-Epitaxieschichten ($n \approx 10^{17}$ cm^{-3}) mit semiisolierendem Substrat werden MESFET-Probenstrukturen erzeugt und ihr Stromrauschen zwischen 1 kHz und 1 MHz bei Temperaturen von 4,4 bis 300 K gemessen. Unterhalb 20 K wird neben $1/f$-Rauschen ein im allgemeinen bis 1 MHz weißes Rauschen gefunden, das auf Effekte warmer Elektronen zurückgeführt werden kann. Aus der Temperaturabhängigkeit des Rauschens oberhalb 20 K, das als Generations–Rekombinations-Rauschen des leitenden Kanals identifiziert wird, können zahlreiche Aktivierungsenergien von Störstellen bestimmt werden, die mit dem Leitungsband wechselwirken. Die gefundenen Energiewerte befinden sich in guter Übereinstimmung mit den Ergebnissen anderer Methoden und anderer Autoren. In einer Aufstellung wird die relative Rauschintensität der verschiedenen Störniveaus angegeben.

1. Introduction

Recently GaAs MESFETs have gained increasing importance as low noise microwave amplifiers. Since the epitaxial n-GaAs layers for the devices are doped to 10^{17} cm^{-3} or even higher, they show no freeze-out of electrons at low temperatures down to 4.2 K. So GaAs MESFETs appear also to be useful in cooled amplifier applications at low frequencies.

However, some GaAs MESFETs exhibit a poor dc stability and a rather strong low-frequency noise. Barrera [1] attributed this disadvantage to the influence of deep impurity states located at the interface between the epitaxial layer and the semi-insulating substrate, giving rise to a space charge region. To overcome this problem, some manufactures habe grown undoped buffer layers between the substrate and the strongly doped "active" layer.

In this study various n-GaAs epitaxial layers with and without buffer layers were investigated in the temperature range from 4.4 to 300 K by measuring the current noise between 1 kHz and 1 MHz, the Hall effect, and the conductivity. By noise analysis the activation energies E_A of the involved impurities were determined. To get further information about the location of the noise sources, the current distribution in the GaAs layers was influenced by the field effect.

Additional C–U measurements on the Schottky junction gates were performed at 300 and 78 K to control the doping profiles and to look for deep impurity concentrations in excess of about 1% of the doping concentration.

2. Sample Preparation

The epitaxial wafers are listed in Table 1, the shape of the samples is shown in Fig. 1. Six or sometimes more sample structures were produced simultaneously.

Table 1
Epitaxial GaAs wafers

wafer No.	electron concentration n (cm^{-3}) at 300 K	active layer thickness (μm)	manufacturer	Hall mobility at 300 K (10^3 cm^2/Vs)	type of epitaxy
E 397	9.8×10^{14}	9 to 11	Fernmeldetechnisches Zentralamt Darmstadt, BRD	7.58	LPE
D 332	1.4×10^{17}	0.2	BDH Chemicals, Poole Dorset, England	3.6	LPE
He 384	4×10^{17}	≈ 0.2	Technical University	3.2	VPE
411	2×10^{17}	≈ 0.2/2.5*)	Aachen, BRD,	3.0	VPE
413	2×10^{17}	≈ 0.2	Institut f. Halbleiter-	3.0	VPE
417	1×10^{17}	≈ 0.4/2.5*)	technik	3.5	VPE
425	7×10^{16}	≈ 0.5/2.5*)		3.4	VPE
426	7×10^{16}	≈ 0.5		3.7	VPE

HFET 1000, complete microwave MESFET chip (Hewlett-Packard).
*) Buffer layer thickness in μm.

Chips were cut from epitaxial wafers with a wire saw. After cleaning in organic solvents and distilled water, mesa structures were formed by chemical etching in a $NaOH:H_2O_2:H_2O$ solution using Shipley AZ 1350 H photoresist.

Prior to evaporation of metal films for the electric contacts, the semiconductor surface was cleaned in vacuum by argon ion dc sputter etching [2] with an AZ 1350 H photoresist mask. This procedure guaranted good adhesion of the metal films, nondirectional contacts with low-resistivity and nearly ideal Schottky diodes.

The ohmic contacts were obtained by photoresist lift-off techniques after vacuum evaporation of 0.005 μm of Ni and 0.09 μm of Au/Ge eutectic and a subsequent anneal for 30 s at 450 °C under dry H_2 atmosphere. To ensure homogeneous processing over the contact areas the heat-up time was kept short (about 14 s) by pushing the samples into a preheated zone of the quartz oven.

Schottky junction gates were formed by evaporation of 0.005 μm of Cr, 0.1 μm of Ag, 0.03 μm of Au in a similar way, the annealing time and temperature beeing 30 s and 400 °C. The n values in the current voltage characteristics $I = I_0\,(\exp qU/(nkT) - 1)$ of these Schottky diodes were as low as 1.04 at 295 K.

Finally the samples were gold wire bonded.

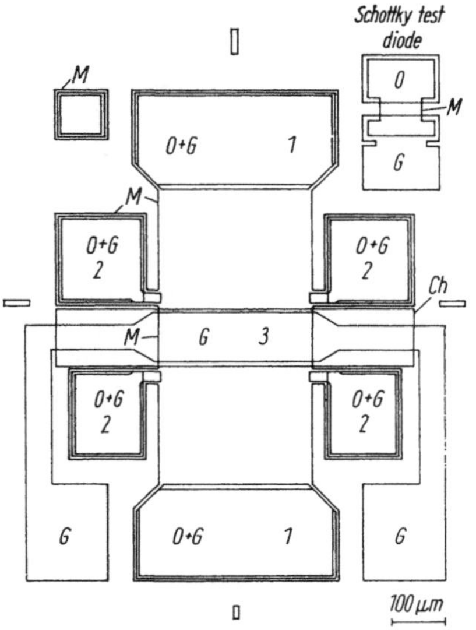

Fig. 1. Sample structure. O ohmic contact, G gate metal layer, M edge of mesa, Ch edge of etche channel zone; bond pads: 1 current contact, 2 potential contact, 3 gate

Three kinds of devices were prepared of each wafer:

1. Devices with ohmic contacts only for Hall effect and conductivity measurements,
2. devices with ohmic contacts and a chemically etched zone between the potential contacts to obtain sample resistances of about 1 kΩ convenient for current noise measurements,
3. devices like those under 2. but with an additional Schottky gate.

3. Experimental

An electronically regulated helium cryostat was used to supply temperatures from 4.4 to 300 K. Hall effect and resistivity measurements were performed with the usual dc methods. Additional control measurements with ac [3, 4] using silver paint contacts on stripes on the epitaxial wafers were employed to make sure that the sample preparation processes do not change the wafer properties.

For noise measurements the apparatus of Fig. 2 was used. To obtain a high sensitivity, the current noise is analysed with a correlation method and compared with the output of a calibrated noise generator. The current noise is characterized by the spectral mean square noise voltage S_U (V^2/Hz) that the noise generator must feed into the 1 Ω resistor, to obtain the same meter deflection with the dc current I switched off. In this way the thermal noise of the contacts will be eliminated. The sensitivity is better than 6×10^{-19} V^2/Hz at 1 kHz and better than 2.5×10^{-19} V^2/Hz from 10 kHz to 1 MHz.

Searching for the connection between noise intensities and the impurity content of specimens it is convenient not to employ the usual presentation of spectra as a function of frequency but to perform the measurements by observing the temperature variation at fixed frequencies, currents I, and gate reverse voltages U_G. In case of dominating generation-recombination (g-r) processes the spectral noise S_U should be

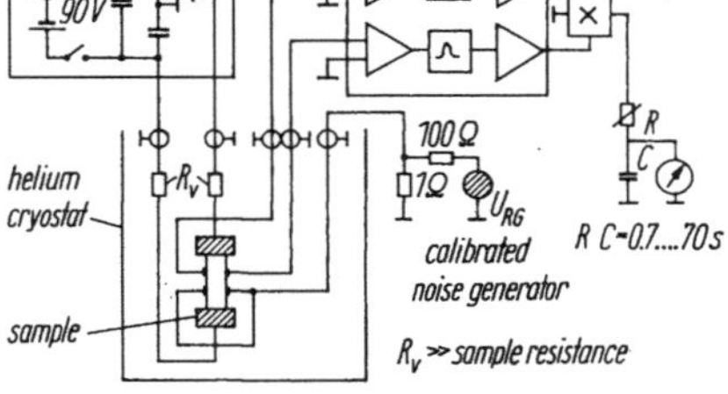

Fig. 2. Noise measurement set-up

proportional to $\tau/(1 + \omega^2\tau^2)$ with the noise time constant τ. τ is related to the activation energies of impurities mainly responsible for the fluctuation process under consideration and depends on the temperature. $S_U(T)$ shows maxima at those temperature values (T_m) where $\omega\tau = 1$; for different frequencies $\tau(T_m)$ can be obtained.

4. Results

The typical behaviour of the samples listed in Table 1 in respect of the resistivity, the Hall coefficient, and the Hall mobility as a function of temperature (Fig. 3) can be understood by the assumption of impurity band conduction at low temperatures. Unlike the dc measurements, which give mean values the ac results are related to the local properties of the samples between the capacitive electrodes; the differences between the two experimental methods give information on the variation of layer thickness and concentration inhomogeneity.

The non-ohmic behaviour of the samples below 20 K (Fig. 4) is connected with the existence of an impurity band, which partially overlaps the conduction band; this is to be expected in n-GaAs for carrier concentrations greater than 10^{16} cm^{-3} [5]. At low electric fields the charge transport takes place mainly in the impurity band with a low electron mobility. At higher fields the "electron temperature" is rising above the lattice temperature and more carriers move into the conduction band achieving a higher mobility. This causes a decrease in differential sample resistence with increasing electric field. At even higher fields the FET samples begin to saturate.

Representative results of noise in the unsaturated part of the dc characteristics (Fig. 5) show the expected maxima in the spectral mean square voltage S_U. With $\omega\tau = 1$ corresponding values of τ and T_m are obtainable. A plot of $\log \tau$ versus $1/T_m$ yields a straight line with a slope proportional to the activation energy E_A of the impurity involved in the g-r process [6 to 8]. For all samples several activation energies were found this way (Fig. 6).

Further details on the noise mechanism are supplied by the noise intensity–current relation (Fig. 7). Above 20 K a strong proportionality to I^2 was found; this means S_U should be proportional to the dc power consumed in the samples. Below 20 K, where no maxima in $S_U(T)$ were observed, the intensity was no longer proportional to IU_{DS}. In addition noise spectra with T and I as parameters tend to be "white" up to 1 MHz with some $1/f$ noise superimposed at low frequencies; they exhibit a non-monotonic current dependence.

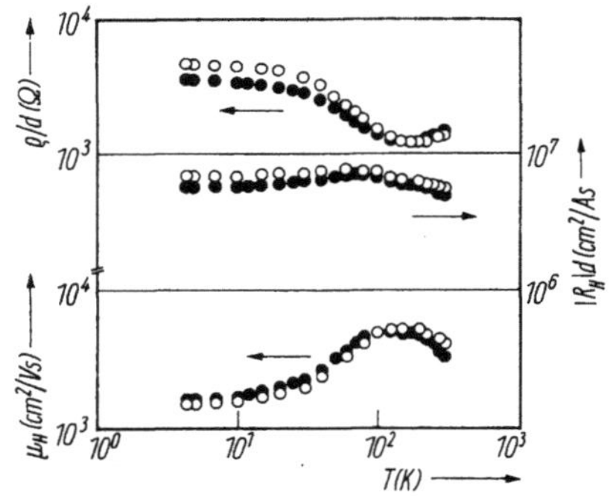

Fig. 3. Temperature dependences of resistivity, Hall coefficient, and Hall mobility, sample D 332; ● dc, ○ ac, d is the thickness of epitaxial layer

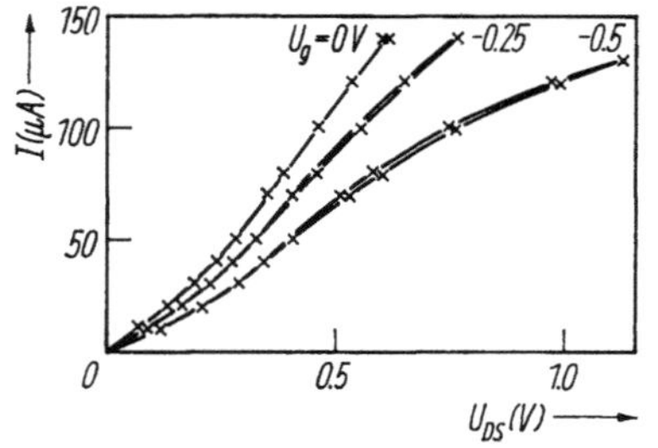

Fig. 4. Current voltage characteristics for sample D 332/2 at 4.4 K. U_{DS} drain source voltage, U_G gate reverse voltage

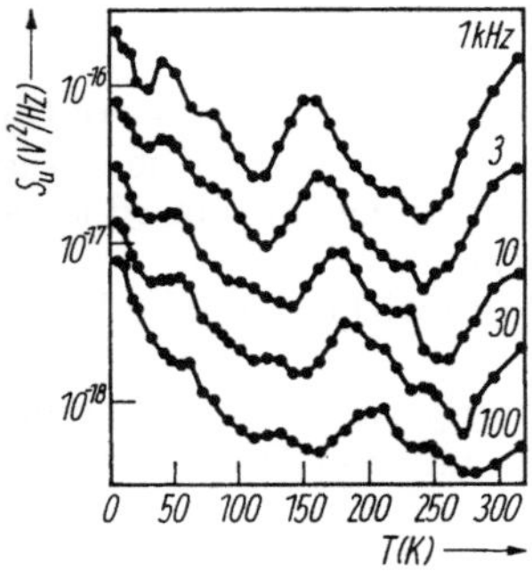

Fig. 5

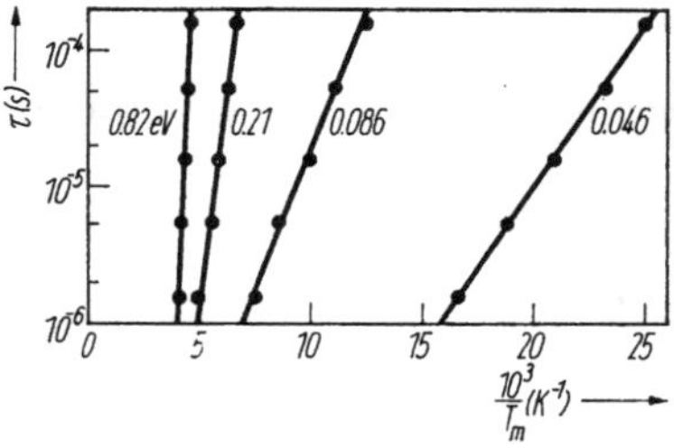

Fig. 6

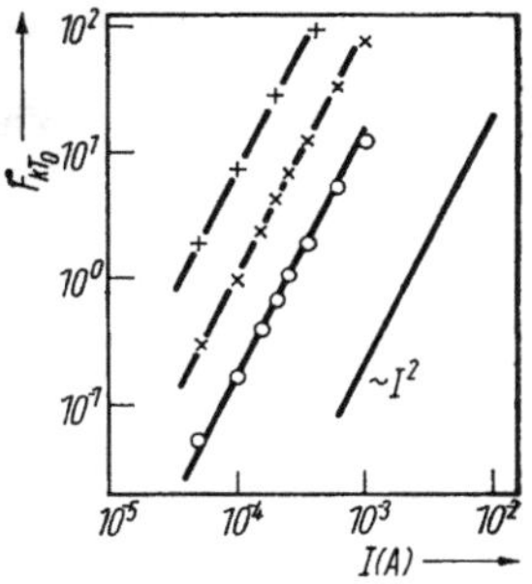

Fig. 7

Fig. 5. Temperature dependence of spectral mean square noise voltage S_U for sample 413/1 at fixed frequencies; $I = 2$ mA

Fig. 6. Semilogarithmic plot of noise time constants as a function of the reciprocal matching temperatures from Fig. 5. Values of activation energies are shown

Fig. 7. Current dependence of noise figure $F_{kT_0} = S_U\,\overline{U_{\text{th}}^2}$; sample E 397, $T = 300$ K. Parameter: noise frequency, + 1 kHz, × 10 kHz, ○ 100 kHz

Noise investigations on samples from the same material with and without Schottky gate show with very few exceptions identical energy levels. Therefore the gates can be considered as free from contact noise.

5. Discussion

Above 20 K and in the ohmic region of the samples g-r noise of the conducting channel indeed can be supposed to be the dominating effect [9]. The noise analysis yields several activation energies (Fig. 8). Some materials within the limits of accuracy exhibit the same energy levels, but large differences in the associated normalized noise intensities.

The noise power of the liquid phase epitaxy (LPE) materials D 332 and E 397 and of the HFET 1000 is more than two orders of magnitude higher than that of the vapour phase epitaxy (VPE) materials fabricated in Aachen.

Since it was not possible to find impurities in the LPE materials by the *C–U* methods, it can be concluded that the deep impurity concentrations are more than two orders of magnitude below the doping concentration in the LPE materials and more than three orders of magnitude below the doping concentration in the VPE materials investigated.

Comparing noise with and without a reverse gate voltage U_G gives information about the spatial distribution of the impurities. With U_G increasing the current flows

Fig. 8. Relative noise power of impurity levels in n-GaAs epitaxial layers. P_{rel} spectral noise power at 10 kHz/dc power in the channel. ● samples without gate, ○ $U_G = 0$ samples with gate, × $U_G \neq 0$ samples with gate. (On the right-hand side there are shown the published E_A values)

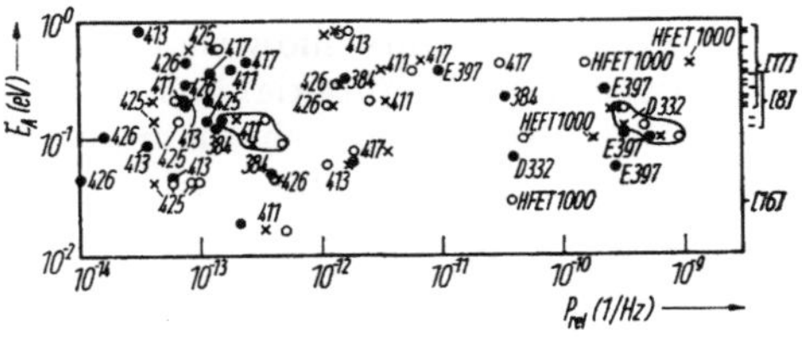

more and more in the transition region between the active layer and the substrate or the buffer layer.

Under this point of view, only the HFET 1000 exhibits a significant influence of the transition region for the 0.45 and 0.1 eV levels, while for the 0.45 eV level of material 417 surface effects seem to be responsible. All other impurities can be considered to be nearly homogeneously distributed throughout the active layer.

No characteristic differences in the noise of samples with and without buffer layers could be found in the investigated frequency range.

While all VPE materials showed good dc stability and a fairly good reproducibility of the noise after several low-temperature cycles, the LPE material D 332 and the HFET 1000 were different.

D 332 and HFET 1000 both exhibited irreproducible noise after two full temperature cycles. At temperatures below 100 and 270 K respectively it was possible to change the noise intensity permanently by applying a short pinch-off pulse to the gate. Further the HFET 1000 was unstable in the dc output characteristic below 270 K. It is interesting in this respect that both samples are similar with regard to their activation energies and noise intensities.

Below 20 K, however, because of the different current dependence the onise cannot be explained by the g-r mechanism in the usual sense. The noise temperatures formally introduced with the Nyquist relation for the frequency independent part of the spectra are so high, that it would be unrealistic to take these as "electron temperatures" (e.g. 44 K for 4.4 K lattice temperature and 40 μW dc power). But as the noise temperatures exhibit maxima at current values at which the decrease of the differential resistance shows a maximum transitions of warm electrons between the impurity and conduction bands may offer a possible interpretation of the experimental data.

6. Comparison with Published Values of Activation Energies

Above 0.12 eV all activation energies found by noise analysis coincide with values already published [8, 16, 17]. Little is known however about the nature of the impurities with exception of the 0.82 eV level. This level is attributed to Ga vacancies in VPE materials by Hasegawa and Majerfeld [10, 11], while measurements by Copeland [12] on LPE materials and by Adlerstein [13] on VPE materials indicate that it is caused by oxygen. Chromium is also discussed [14]. Finally, Lum et al. [15] found an activation energy of 0.837 eV for Cr by photoluminescence (PL) techniques.

Below 0.12 eV only few E_A values are published. For instance Lum et al. [15, 16] performed PL measurements at 20 K on different semiinsulating and doped GaAs layers before and after heat treatment. PL peaks found at 1.491 eV are attributed to carbon in Ga vacancies acting as donors ($E_A \approx 0.027$ eV), while it is believed that PL peaks at 1.412 eV belong to carbon acceptor arsenic vacancy complexes ($E_A \approx 0.10$ eV).

However the 0.1 eV level, which strongly effects the noise of the materials D 332, E 397, and the HFET 1000 in our experiments, should be connected with donor states.

7. Conclusions

The presented results show that the noise method is very useful for sensitive impurity analysis in GaAs epitaxial layers above 20 K. Significant improvement of the GaAs MESFET low-frequency behaviour appears to be possible by eliminating those steps in production which contribute predominantly to the impurity content. Systematic and correlated work in physics and technology with the aid of noise measurements can make a substantial contribution to this goal.

Acknowledgement

We are grateful to the Deutsche Forschungsgemeinschaft for financial support.

References

[1] I. Barrera, Proc. Active Semicond. Dev. for Microwaves and Integr. Optics, Cornell University, Ithaca New York 1975 (p. 135 to 144).
[2] K. Heime, U. König, E. Kohn, and A. Wortmann, Solid State Electronics **17**, 835 (1974).
[3] W. Hoyer, Arch. Elektrotech. **57**, 93 (1975).
[4] W. Hoyer, Arch. Elektrotech. **57**, 215 (1975).
[5] O. V. Emelyanenko, T. S. Lagunova, and D. N. Nasledov, Soviet Phys. – Solid State **3**, 144 (1961).
[6] C. F. Hiatt, A. van der Ziel, and K. M. van Vliet, IEEE Trans. Electron Devices **22**, 614 (1975).
[7] R. A. Spaulding, Proc. IEEE **56**, 886 (1968).
[8] D. Sodini, A. Touboul, G. Lecoy, and M. Savelli, Electronics Letters **12**, 42 (1976).
[9] J. Graffeuil and J. Caiminade, Electronics Letters **10**, 266 (1974).
[10] F. Hasegawa and A. Majerfeld, Electronics Letters **11**, 286 (1975).
[11] F. Hasegawa and A. Majerfeld, Electronics Letters **12**, 52 (1976).
[12] J. A. Copeland, IEEE Trans. Electron Devices **18**, 50 (1971).
[13] M. G. Adlerstein, Electronics Letters **12**, 297 (1976).
[14] G. A. Aleksandrova, G. G. Koleshnikova, A. P. Korovin, N. V. Korotkova, G. F. Lymar, and D. P. Sarnatskii, Soviet Phys. – Semicond. **10**, 1162 (1976).
[15] W. Y. Lum, H. H. Wieder, W. H. Koschel, S. G. Bishop, and B. D. Mc Combe, Appl. Phys. Letters **30**, 1 (1977).
[16] W. Y. Lum and H. H. Wieder, Appl. Phys. Letters **31**, 213 (1977).
[17] G. M. Martin, A. Mitonneau, and A. Mircea, Electronics Letters **13**, 191 (1977).

(Received June 23, 1978)

phys. stat. sol. (a) 49, 119 (1978)

Subject classification: 1.6

Institut für Kristallographie und Mineralogie der Universität München[1])

Systematic Absences in Low Energy Electron Diffraction

By

H. JAGODZIŃSKI

In honour of Prof. Dr. F. STÖCKMANN's 60th birthday

The "cluster-embedded chain"-method is applied to calculate extinction rules valid for sharp and diffuse LEED-reflections in ordered and disordered crystal surfaces. Besides the wellknown absences due to glide planes with a translational component parallel to the surface, there are others due to translations vertical to it. Since the crystal surface has no translation symmetry in this direction, these extinctions are no longer strictly fulfilled. The behaviour of sharp and diffuse reflections of a stepped surface is discussed with respect to the differences between a kinematic and a dynamic calculation.

Die „cluster-embedded chain"-Methode wird für die Berechnung von scharfen und diffusen LEED-Reflexen geordneter und fehlgeordneter Kristalloberflächen angewendet. Neben den bekannten Auslöschungsregeln für Gleitspiegelebenen mit paralleler Gleitkomponente zur Oberfläche gibt es auch Auslöschungen, die auf Translationen senkrecht dazu zurückzuführen sind. Da die Kristalloberfläche keine Translationssymmetrie in dieser Richtung besitzt, werden diese Auslöschungen nicht streng erfüllt. Das Verhalten von scharfen und diffusen Reflexen einer gestuften Oberfläche wird im Zusammenhang mit den Unterschieden zwischen einer dynamischen und kinematischen Berechnung diskutiert.

1. Introduction

There are four different types of extinction rules in the kinematic theory of diffraction [1]. The first one influences all reflections (integral extinction law), and it is caused by a non-primitive setting of the unit cell often prescribed by the symmetry of the crystal. This group of absences is of general validity for all diffraction methods because of the fact that a physical observation should be independent of the choice of the unit cell. The second group of extinctions refers to reflections of a zone (a plane passing through the origin of the reciprocal lattice) or a series ($\nu h, \nu k, \nu l$; ν integer), respectively. They are characteristic for a glide plane or a screw axis, and contain valuable information on the space group of the crystal. The third type of absences is often realized if special positions of the space group are occupied only; they are caused by the assumption of spherical atoms and often violated for this reason. But there is another process breaking this extinction rule, which has been studied very carefully in the case of the forbidden (222) reflection of diamond [2]: the dynamical multiple scattering (Umweganregung) which may also take place in an incoherent way (kinematic multiple scattering). These two types of multiple scattering occur if at least two Bragg reflections are scattered at the same time, e.g. (311) and (222) in the case of diamond. It may easily be shown then that the reflection with indices corresponding to the difference — $(1\bar{1}\bar{1})$ in our case — is scattered into the direction of (222). The fourth type of system aticabsences is caused by glide operations just being realized for the contents of the unit cell, but not obeying the symmetry of the space group. This type of extinction rules is often correlated with order–disorder phenomena.

[1]) Theresienstr. 41, D-8000 München 2, BRD.

Holland and Woodruff [3] have shown that systematic absence are also observed when a glide plane vertical to the surface with a translation parallel to it is realized even in the case of low energy electron diffraction in crystal surfaces where diffraction is governed by coherent multiple scattering. The main difference between the kinematic and the dynamic calculation is, that the wave vectors of the incident and the diffracted beam, respectively, must not have a component vertical to the glide plane under investigation. Other absences could be indicated [3] if a spherically symmetric scattering (s-wave scatterer) is assumed. Although the statement that a glide plane — having a glide component vertical to the surface — does not cause extinction rules is correct for diffraction in two dimensions, it has to be kept in mind that all physical processes are three-dimensional ones, and extinction rules corresponding with the third dimension cannot be excluded. It will be shown that a new type of absences may occur under certain geometrical conditions. Their existence generally leads to a simple geometrical interpretation, thus contributing essentially to the solution of the structure.

In the kinematic treatment of diffraction the intensity of the sharp reflections is completely determined by the so-called averaged structure factor, and the diffuse scattering by their variance. According to a recent approach to the formal solution of the multiple scattering problem [4 to 6] a similar expression may be derived in the case of LEED, but it could also be shown that typical differences of the behaviour of sharp and diffuse reflections may exist. It will be shown in this paper how the experimental observation of extinction of sharp or diffuse scattering may be interpreted from a structural point of view.

2. The Cluster-Embedded Chain in Multiple Diffraction

As has been shown recently, a formal solution of the problem of multiple diffraction may be given if the cross-section of absorption is large, such, that any multiple scattering process ends at a distance relatively short. Under these circumstances an area of multiple scattering (AMS) may be defined in the following way: The whole crystal and the surface are subdivided into chains extending from the surface into the interior of the crystal. A certain complex of surrounding chains is defined as AMS, as shown in Fig. 1. Within the said area all absorption and multiple scattering processes are calculated accurately, multiple scattering processes having left the AMS at least once are neglected; they are taken into account together with the influence of the chains outside the AMS by averaging. What we need for a solution of the diffraction problem is the contribution of the central chain to the various diffracted beams; it has been shown recently [5] that it may be calculated at least in principle. In this approximation all chains having the same structural surrounding within the AMS contribute identically to the diffracted beam, a phase factor caused by the position of the central chain excepted.

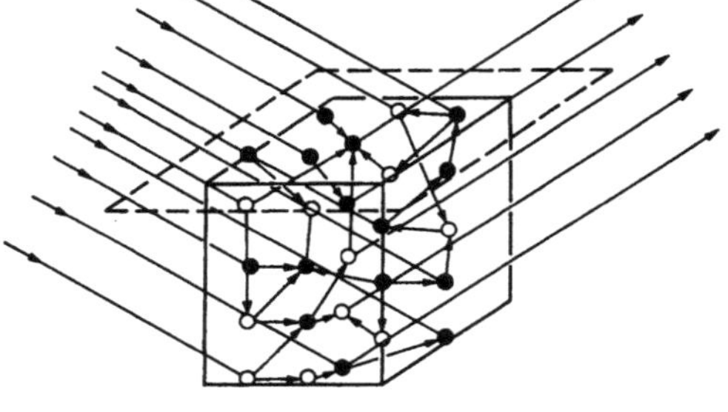

Fig. 1. Definition of the area of multiple scattering (AMS). All scattering processes ending at the central chain before being scattered into the direction $\boldsymbol{k}'$ are taken into account, if their ways lie within the area indicated. The remaining scattering processes are only considered by averaging. Each chain appears just once as central chain

The incident and the diffracted electron beams are described as plane waves (outside the crystal) and determined by their wave vectors $\boldsymbol{k}'$, $\boldsymbol{k}$.

Since we are mainly concerned with phase shifts of plane waves $\boldsymbol{k}'$ and $\boldsymbol{k}$ which are described with the aid of the reciprocal lattice used in the kinematic theory of diffraction we introduce

$$\boldsymbol{k} = 2\pi(k_a\boldsymbol{a}^* + k_b\boldsymbol{b}^* + k_c\boldsymbol{c}^*); \qquad \boldsymbol{k}' = 2\pi(k_a'\boldsymbol{a}^* + k_b'\boldsymbol{b}^* + k_c'\boldsymbol{c}^*)\,, \tag{1}$$

where $\boldsymbol{a}^*$, $\boldsymbol{b}^*$ are (not necessarily) parallel and $\boldsymbol{c}^*$ vertical to the surface. Let us assume that the number of clusters realized in the crystal and its surface be limited to n; the contribution of the central chain of the ν-th cluster at a distance $\boldsymbol{r}_{\nu\mu}$ to the diffracted beam $\boldsymbol{k}$ is then given by

$$F_{\nu\mu}(\boldsymbol{k}, \boldsymbol{k}') = \mathrm{e}^{i(\boldsymbol{k}-\boldsymbol{k}')\boldsymbol{r}_{\nu\mu}}\, F_{\nu 0}(\boldsymbol{k}, \boldsymbol{k}')\,, \tag{2}$$

where $F_{\nu\mu}$ is the contribution of the same cluster placed at the origin. For an ordered crystal with chains in the unit cell the resulting amplitude of the diffracted electron beam is

$$\begin{aligned} F(\boldsymbol{k}, \boldsymbol{k}') &= \sum_{\nu}\sum_{\mu} \mathrm{e}^{i(\boldsymbol{k}-\boldsymbol{k}')\boldsymbol{r}_{\nu\mu}}\, F_{\nu 0}(\boldsymbol{k}, \boldsymbol{k}') \\ &= \frac{\sin \pi N_1 h}{\sin \pi h}\, \frac{\sin \pi N_2 k}{\sin \pi k} \sum_{\nu} F_{\nu 0}(\boldsymbol{k}, \boldsymbol{k}')\, \mathrm{e}^{i(\boldsymbol{k}-\boldsymbol{k}')\boldsymbol{r}_{\nu 0}}\,, \end{aligned} \tag{3}$$

where $\boldsymbol{h} = h\boldsymbol{a}^* + k\boldsymbol{b}^* + l\boldsymbol{c}^* = \dfrac{1}{2\pi}(\boldsymbol{k} - \boldsymbol{k}')$, μ, ν stand for the summation over the lattice and the unit cell, respectively. Although equation (3) has a very similar structure compared with a kinematic calculation, it is fully dynamic because of the different definition of $F_{\nu_0}(\boldsymbol{k}, \boldsymbol{k}')$, which can no longer be described with the aid of $\boldsymbol{k} - \boldsymbol{k}'$, it is sensitively dependent on $\boldsymbol{k}$ and $\boldsymbol{k}'$.

As a consequence of the similar structure of the scattering amplitudes of all diffracted beams $\boldsymbol{k}$ associated with the incident beam $\boldsymbol{k}'$, we again can derive formulae for a random arrangement of cluster-embedded chains. Since chains being equal from a topological point of view may be embedded in different clusters, the atomic coordinates within the chain are determined by the cluster. Consequently the cluster, as defined for multiple diffraction, has an important structural meaning as long as the chain at the boundary of the AMS has an influence on the positions of the central chain. LEED calculations of two arrangements may be evaluated easily:

a) the ordered arrangement of chains according to equation (3),

b) the random distribution of chains embedded in different clusters, in the latter case we get "sharp" reflections with the intensity I_s given by

$$I_\mathrm{s} \sim \langle F(\boldsymbol{k}, \boldsymbol{k}')\rangle^2\,, \tag{4}$$

and diffuse intensity I_d

$$I_\mathrm{d} \sim \langle |F(\boldsymbol{k}, \boldsymbol{k}')|^2\rangle - |\langle F(\boldsymbol{k}, \boldsymbol{k}')\rangle|^2\,. \tag{5}$$

The diffraction theory of the usual statistical distribution of chains may often be approximated with a kind of interpolation between the formulae given in (3) and (4), (5).

3. Calculation of Systematic Absences

Fig. 2 shows a top view on a surface with a glide plane vertical to $\boldsymbol{b}$. According to the results discussed in the preceding chapter we have to look after chains generated by translation or other symmetries correlated with translation. The unit cell in Fig. 2 has

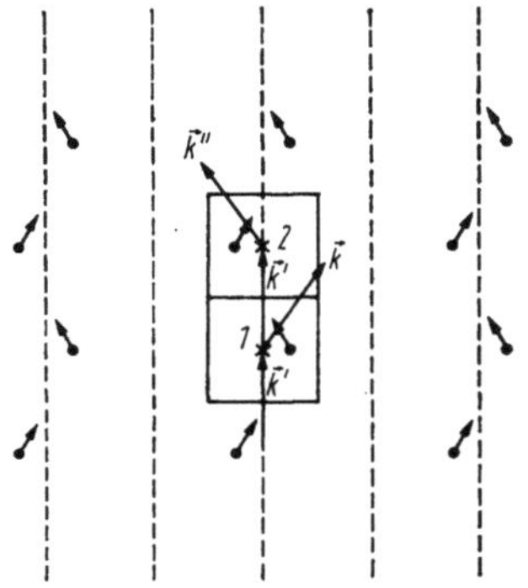

Fig. 2. Extinction by a glide plane in mutliple scattering. According to the explanation given for Fig. 1 there are two contributions originating from the chains of the subcells designated 1, 2. The two contributions are symmetrically equivalent with respect to $\boldsymbol{k}$ and $\boldsymbol{k}''$, if $\boldsymbol{k}'$ lies in the glide plane

two subcells containing the object (circle and arrow) and its picture caused by a mirror plane, and displaced by $\frac{1}{2}\,\boldsymbol{a}$. With respect to the origin of the two subcells the two contributions shall be designated as

$$F_a(\boldsymbol{k}, \boldsymbol{k}') \quad \text{and} \quad \overline{F}_a(\boldsymbol{k}, \boldsymbol{k}')$$

and the structure amplitude $F(\boldsymbol{k}, \boldsymbol{k}')$ is according to equation (3)

$$F(\boldsymbol{k}, \boldsymbol{k}') = F_a(\boldsymbol{k}, \boldsymbol{k}') + e^{\pi i h}\, \overline{F}_a(\boldsymbol{k}, \boldsymbol{k}') \, .$$

Obviously, there is an extinction rule, if $h =$ odd, and $F_a(\boldsymbol{k}, \boldsymbol{k}') = \overline{F}_a(\boldsymbol{k}, \boldsymbol{k}')$. Checking all possible multiple scattering events with the aid of Fig. 2, it is easily to be seen that equal amplitudes may only be observed if $\boldsymbol{k}'$ lies in the glide plane

$$\boldsymbol{k}' = \boldsymbol{k}_a \boldsymbol{a}^* + \boldsymbol{k}_c \boldsymbol{c}^* \, .$$

This results from the fact that all scattering events ending in the two subcells have analogous ones generated by the mirror plane and translation. Consequently we have

$$F_a(\boldsymbol{k}, \boldsymbol{k}') = \overline{F}_a(\boldsymbol{k}'', \boldsymbol{k}') \, , \qquad F_a(\boldsymbol{k}'', \boldsymbol{k}) = \overline{F}_a(\boldsymbol{k}, \boldsymbol{k}') \, ,$$

where

$$\boldsymbol{k}'' = k_a \boldsymbol{a}^* - k_b \boldsymbol{b}^* + k_c \boldsymbol{c}^* \, ,$$

and

$$\boldsymbol{k} = \boldsymbol{k}_a \boldsymbol{a}^* + k_b \boldsymbol{b}^* + \boldsymbol{k}_c \boldsymbol{c}^* \, .$$

Now it can be shown easily that $\boldsymbol{k} = \boldsymbol{k}''$ is a necessary and sufficient condition to obey the equation

$$F_a(\boldsymbol{k}, \boldsymbol{k}') = \overline{F}_a(\boldsymbol{k}, \boldsymbol{k}') \left(= F_a(\boldsymbol{k}'', \boldsymbol{k}')\right) .$$

$\boldsymbol{k}'' = \boldsymbol{k}$ means $k_b = 0$. The reflected beam obeys the extinction rule $k =$ odd if $\boldsymbol{k} = \boldsymbol{k}''$, consequently $\boldsymbol{k}$, $\boldsymbol{k}''$ lie in the glide plane. This derivation is based on the same principle as the one given in [2], other extinction rules may easily be found by demanding $F(\boldsymbol{k}, \boldsymbol{k}') = 0$, and there are many of type 4, but they shall not be discussed here because their systematic discussion is tedious. Therefore, it seems to be more efficient to give some important examples. It may be shown in the same way that no extinction rules are observable in the case of a screw axis vertical and parallel to the surface, because $\boldsymbol{k}'$, and $\boldsymbol{k}$ as well, should lie parallel to the screw axis, which cannot be realized experimentally. But it will be shown below that pseudo-extinction rules may occur in the case of domain ordering on the surface.

4. Absences of Sharp and Diffuse Reflections

According to (4) and (5), describing diffraction in disordered surfaces, the condition $\langle F \rangle = 0$ may easily be realized if clusters differing by translation do occur with the same a priori probability. No condition has to be fulfilled with respect to their sequence, since the averaging given in (4) and (5) has to be done independently. Many typical cases obeying $\langle F \rangle = 0$ may be suggested, for a displacement vector being a fraction of a translation vector of the averaged lattice (basic vector) of the plane unit cell parallel to the surface, e.g.

$$\boldsymbol{r}_{\nu\mu} = \frac{\nu}{n}\,\boldsymbol{a} + \frac{\mu}{m}\,\boldsymbol{b} \qquad (\nu = 0, \dots, n-1;\ \mu = 0, \dots, m-1)\,, \tag{6}$$

if all displaced AMS occur with the same a priori probability $p_{\nu\mu} = \dfrac{1}{n+m}$.

Then

$$\left.\begin{aligned} \langle F(\boldsymbol{k}, \boldsymbol{k}')\rangle &= F(\boldsymbol{k},\boldsymbol{k}') \sum_{\nu,\mu} p_{\nu\mu}\, \mathrm{e}^{i(\boldsymbol{k}-\boldsymbol{k}')\left(\frac{\nu}{n}\boldsymbol{a}+\frac{\mu}{m}\boldsymbol{b}\right)} = F(\boldsymbol{k},\boldsymbol{k}') \sum_{\nu\mu} p_{\nu\mu}\, \mathrm{e}^{2\pi i\left(h\frac{\nu}{n}+k\frac{\mu}{m}\right)} \\ &= F[h \equiv 0 \pmod{n}\,, \quad k \equiv 0 \pmod{m}]\,, \\ &= 0 \text{ otherwise.} \end{aligned}\right\} \tag{7}$$

In the same way the complementary extinction rules for the diffuse scattering may be given:

$$\begin{aligned} \langle |F|^2\rangle - |\langle F\rangle|^2 &= 0 \quad \text{if} \quad h \equiv 0 \pmod{n},\ k = 0 \pmod{m}\,, \\ &= |F|^2 \text{ otherwise}\,. \end{aligned}$$

These relations are in complete agreement with the analogous kinematic calculation.

Similar equations may be set up for glide planes correlated with normal translations, the result is identical with the one given in the preceding chapter; only if $\boldsymbol{k}$ and $\boldsymbol{k}'$ fall into the glide plane systematic extinction rules may result.

There is one difficulty which should be discussed briefly: In a statistical arrangement of chains many clusters have to be taken into account and it seems to be very difficult to derive a simple expression for the extinction rules because of the great many probabilities involved. Often this difficulty can be overcome by remembering that each cluster will occur in the superstructure with the same probability in subcells generated by the translations of the bulk crystals parallel to the surface.

Another difficulty arises from the fact that a single defect chain embedded in a cluster causes as many clusters as given by the number of chains within that cluster [1]. Their distribution is strictly correlated with the cluster containing the defect chain as central one. Once a defect chain has occurred the sequence of the remaining clusters containing the defect chain is strictly fixed for a certain set of neighbouring clusters. (The defect occurs once in each position of chains within the AMS with normal chains as central ones). A more complicated situation arises if more than one defect chain has to be considered within the same cluster, which shall not be discussed here. All these considerations are not simple and it may be guessed that solution of the multiple diffraction problem for disordered systems must be very complicated, although the huge number of areas of multiple scattering involved favours the averaging necessary for intensity calculations [4, 5].

5. Non-Trivial Extinctions

A completely new type of extinctions, which has not been discussed explicitly yet, is not easily seen in the LEED picture. These extinctions are generally realized if there is a group of AMS, basically with the same structure, but differing by a translation vector involving a component vertical to the surface. A periodically stepped surface may also be described as a vicinal face slightly inclined with respect to the normal of the low-indexed plane, (001) in Fig. 3. With $\boldsymbol{a}, \boldsymbol{b}, \boldsymbol{c}$ as basic vectors of the bulk crystal we

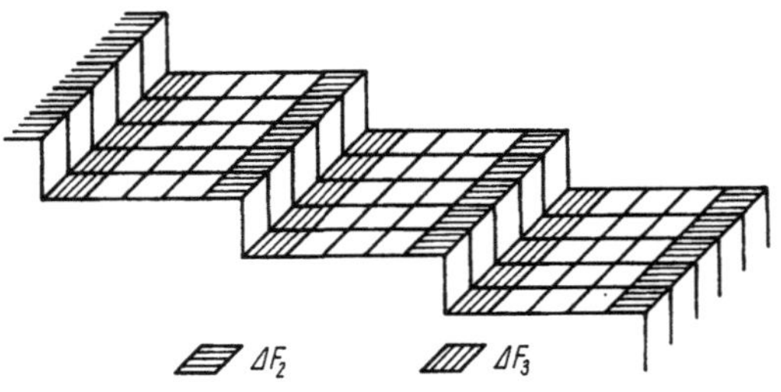

Fig. 3. Periodically stepped surface and its LEED pattern. The cells are equivalent from a structural point of view, but the chains at the boundary contribute differently to the diffracted beam. The deviations are designated ΔF_1, ΔF_2

introduce the set of new vectors

$$\boldsymbol{a}' = N\boldsymbol{a} + z\boldsymbol{c}\,, \qquad \boldsymbol{b}' = \boldsymbol{b}\,, \qquad \boldsymbol{c}' = \boldsymbol{c}\,.$$

The transformation of the new reciprocal vector is then given by

$$\boldsymbol{a}'^* = \frac{1}{N}\,\boldsymbol{a}^*\,, \qquad \boldsymbol{b}'^* = \boldsymbol{b}^*\,, \qquad \boldsymbol{c}'^* = -\frac{z}{N}\,\boldsymbol{a}^* + \boldsymbol{c}^*\,.$$

Since the indices of the reciprocal vector

$$h\boldsymbol{a}^* + k\boldsymbol{b}^* + l\boldsymbol{c}^*$$

are transformed according to

$$h = \frac{1}{N}\,(h' - zl)\,, \qquad k = k'\,, \qquad l = l'\,.$$

The sharp spots (beams) of the ideal (001) surface are given by h integer, consequently

$$(h' + zl) = \nu N \qquad (\nu = \text{integer})\,.$$

Since h' integer (μ) has also to be obeyed because of the periodicity of the steps, we get

$$-zl = \nu N - h' = \text{integer} \tag{8}$$

as condition for appearance or extinction of the beams of the ideal surface. On the other hand, no condition for l and l' can be given for LEED. Consequently the equation is solvable and there are no extinctions other than those using a non-primitive setting, which are trivial absences. But there is a very important physical meaning in this result if the structure amplitudes of the different blocks repeated periodically are taken into account. As shown in Fig. 3, there is an "almost" periodicity within the block, and a deviation at its border. Assuming that the AMS may be restricted to next-nearest neighbours, only the central chains hatched in Fig. 3 have a different contribution to the scattered beam other than those produced by (pseudo-) translation. Since these differences may be small they are given by

$$\overline{F}(\boldsymbol{k}, \boldsymbol{k}') + \Delta F_i \qquad (i = 1, 2)\,,$$

where $\overline{F}$ represents the averaging multiple scattering for the cells within one block. The structure amplitude of the structure becomes

$$F_b(\boldsymbol{k}, \boldsymbol{k}') = \overline{F}(\boldsymbol{k}, \boldsymbol{k}')\; e^{i\pi N}\,\frac{\sin \pi N h}{\sin \pi h} + (e^{-\pi i h}\,\Delta F_2 + e^{\pi i h}\,\Delta F_3)\,,$$

N number of cells in one block. The resulting amplitude F_s for the periodical stepped surface is then

$$F_s(\boldsymbol{k}, \boldsymbol{k}') = \frac{\sin \pi M(Nh + zl)}{\sin \pi(Nh + zl)} \left\{ e^{i\pi N} \frac{\sin \pi Nh}{\sin \pi h} \overline{F}(\boldsymbol{k}, \boldsymbol{k}') + \right.$$

$$\left. + e^{-i\pi h} \Delta F_2(\boldsymbol{k}, \boldsymbol{k}') + e^{i\pi h} \Delta F_3(\boldsymbol{k}, \boldsymbol{k}') \right\}, \qquad (9)$$

where M is the number of steps.

Equation (9) describes strong intensities zl integers, but very weak ones if this condition is violated (they do not appear in the kinematic theory). Furthermore, it may easily be seen that in the case of large blocks ($N \gg 1$) a typical diffraction pattern of satellites will be produced. Additional extinction rules which are approximately valid are given by the zeros of the term

$$\frac{\sin \pi Nh}{\sin \pi h} \quad \text{for} \quad Nh + zl = \nu = (\text{integer}) ,$$

$$h = \frac{\mu}{N} \; [\mu \text{ integer} \neq 0 \; (\text{mod } N)] .$$

These pseudo-extinction rules are very common in LEED diffraction pictures, a detailed discussion will be given in a later paper where also some examples will be outlined.

Finally, let us discuss the multiple diffraction problem of a surface stepped statistically, as shown in Fig. 4. According to equations (4), (5) we have to form the average of a statistical arrangement of blocks at different heights. According to Fig. 4 we have four different types of blocks:

a) "top" blocks with two lower neighbouring blocks,

b) "base" blocks with two higher neighbouring blocks,

c) "right" blocks with a high neighbour on the right-hand side,

d) "left" blocks with a high neighbour on the left-hand side. The four structure amplitudes of the blocks are

$$F_a(\boldsymbol{k}, \boldsymbol{k}') = \frac{\sin \pi N_a h}{\sin \pi h} F(\boldsymbol{k}, \boldsymbol{k}') \left[e^{-i\pi(N_a-1)h} \Delta F_1 + e^{i\pi(N_a-1)h} \Delta F_2\right] ,$$

$$F_b(\boldsymbol{k}, \boldsymbol{k}') = \frac{\sin \pi N_b h}{\sin \pi h} F(\boldsymbol{k}, \boldsymbol{k}') \left[e^{-i\pi(N_b-1)h} \Delta F_3 + e^{i\pi(N_b-1)h} \Delta F_4\right] ,$$

$$F_c(\boldsymbol{k}, \boldsymbol{k}') = \frac{\sin \pi N_c h}{\sin \pi h} F(\boldsymbol{k}, \boldsymbol{k}') \left[e^{-i\pi(N_c-1)h} \Delta F_1 + e^{i\pi(N_c-1)h} \Delta F_4\right] ,$$

$$F_d(\boldsymbol{k}, \boldsymbol{k}') = \frac{\sin \pi N_d h}{\sin \pi h} F(\boldsymbol{k}, \boldsymbol{k}') \left[e^{-i\pi(N_d-1)h} \Delta F_3 + e^{i\pi(N_d-1)h} \Delta F_2\right] .$$

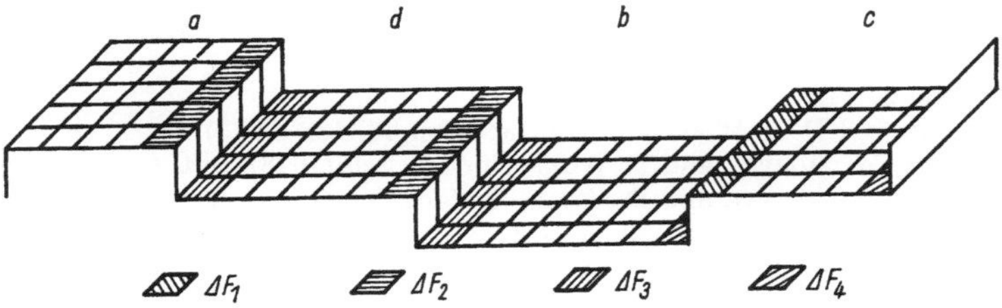

Fig. 4. Stepped surface with statistically ordered steps; note the four different scattering amplitudes ΔF_1, ΔF_2, ΔF_3, ΔF_4

For the sake of simplicity, and without lack of qualitative validity of the result, we may assume that all blocks are distributed at random with the same size ($N_a = N_b = N_c = N_d$) and at three different heights $-\boldsymbol{c}$, 0, $+\boldsymbol{c}$.

Inserting these conditions in equations (4) and (5) we get

$$\left.\begin{aligned} |\langle F(\boldsymbol{k}, \boldsymbol{k}')\rangle|^2 &= \left[\frac{1}{3}\,\frac{\sin \pi 3l}{\sin \pi l}\right]^2 \left|\bar{F}(\boldsymbol{k}, \boldsymbol{k}')\,\frac{\sin \pi N_a h}{\sin \pi h} + \Delta C(\boldsymbol{k}, \boldsymbol{k}')\right|^2, \\ \langle |F(\boldsymbol{k}, \boldsymbol{k}')|^2\rangle - |\langle F(\boldsymbol{k}, \boldsymbol{k}')\rangle|^2 &= \left[1 - \left(\frac{1}{3}\,\frac{\sin \pi 3l}{\sin \pi l}\right)^2\right] \left|\bar{F}(\boldsymbol{k}, \boldsymbol{k}')\,\frac{\sin \pi N_a h}{\sin \pi h} + \Delta C(\boldsymbol{k}, \boldsymbol{k}')\right|^2 \end{aligned}\right\} \tag{10}$$

with

$$\Delta C(\boldsymbol{k}, \boldsymbol{k}') = 2\mathrm{e}^{-i\pi(N_a-1)h}\,[\Delta F_1(\boldsymbol{k}, \boldsymbol{k}') + \Delta F_3(\boldsymbol{k}, \boldsymbol{k}')] + {} + 2\mathrm{e}^{\,i\pi(N_a-1)h}\,[\Delta F_2(\boldsymbol{k}, \boldsymbol{k}') + \Delta F_4(\boldsymbol{k}, \boldsymbol{k}')]\,.$$

Equation (10) shows that there are sharp reflections having maximum intensity, if l is integer, they have zero intensity, if $l = \frac{1}{3} + \nu$ and $l = (\frac{2}{3} + \nu)$. Again we have found a completely new type of extinction rules, depending on the index l. Complementary extinctions exist for the diffuse scattering.

According to my knowledge, experimental evidence of this type of extinctions is not yet available. But this fact is not surprising, because authors deal with their experimental measurements as a function of energy, where systematic absences dependent on l cannot be detected easily, but there are many indications for the existence of such effects. The diffuse scattering should have its maximum at diffraction conditions, where the sharp reflections vanish, consequently, the extinction rules are not strictly obeyed.

There are many possibilities to find other structural models with typical extinction rules for LEED. In domain structures caused by translation and rotation, or by a glide plane, which needs not necessarily be a glide plane of the bulk crystals, they should also occur occasionally, but as has been shown in this paper they cannot be fulfilled exactly. Their study is essential because of the strong influence on the diffraction pattern. It is this influence which may hamper structure determination of the surface considerably.

Acknowledgements

The author thanks the Deutsche Forschungsgemeinschaft for financial assistance through the SFB 128. Helpful discussions with Dr. W. Moritz are gratefully acknowledged.

References

[1] International Tables for X-Ray Cristallography, Vol. I, The Kynoch Press, Birmingham (England) 1952.
[2] M. RENNINGER, Z. Phys. **106**, 141 (1937).
[3] B. W. HOLLAND and D. P. WOODRUFF, Surface Sci. **36**, 488 (1973).
[4] H. JAGODZIŃSKI, W. MORITZ, and D. WOLF, Surface Sci. (1978), in the press.
[5] W. MORITZ, H. JAGODZIŃSKI, and D. WOLF, Surface Sci. (1978), in the press.
[6] H. JAGODZIŃSKI, in: Festkörper, Vol. XVIII, Westdeutscher Verlag, Vieweg, Wiesbaden 1978.
[7] H. JAGODZIŃSKI, Proc.VII. Internat Vac. Congress, and III. Internat Conf. Solid Surfaces 1977 (p. 2391).

(Received June 23, 1978)

phys. stat. sol. (a) **49**, 127 (1978)

Subject classification: 14.3.4; 14.3.2; 22

Institut für Angewandte Physik der Universität Karlsruhe

Electrical Conductivity of Passivated Si_2Te_3 Single Crystals[1])

By

H. P. BAUER and U. BIRKHOLZ

In honour of Prof. Dr. F. STÖCKMANN's 60th birthday

In order to obtain reliable data of the dark conductivity for Si_2Te_3 single crystals it is necessary to use samples with ohmic contacts and low conductivity surface layer. It is shown that these conditions can be met by evaporation of gold contacts after vacuum cleaving and a subsequent heat treatment, which produces a SiO_2 surface layer. The results of dc and ac conductivity measurements indicate a thermal activation of the carrier mobility parallel to the *c*-axis.

Um zuverlässige Werte für die Dunkelleitfähigkeit von Si_2Te_3-Einkristallen zu erhalten, müssen Proben mit Ohmschen Kontakten und schlecht leitender Oberflächenschicht verwendet werden. Diese Bedingungen können durch Aufdampfen von Goldkontakten nach Vakuum-Spaltung und anschließender Wärmebehandlung erfüllt werden. Die Ergebnisse von Gleich- und Wechselstrommessungen der Leitfähigkeit zeigen, daß die Beweglichkeit der Ladungsträger parallel zur *c*-Achse thermisch aktiviert ist.

1. Introduction

In the presence of water vapour a chemical reaction takes place at the surface of Si_2Te_3 crystals according to the formula

$$Si_2Te_3 + 4\,H_2O \rightarrow 2\,SiO_2 + Te + 2\,H_2Te + 2\,H_2\,.$$

As a result of this reaction SiO_2 and free tellurium is deposited on the sample surface [1]. For electrical or optical measurements on the high resistivity Si_2Te_3 it is necessary either to avoid the reaction or to restore the original low conductivity of the surface. As reported by Ziegler and Birkholz [2] sample preparation in a glove box containing dry argon gives satisfactory results in the case of photoconductivity measurements. According to Ziegler et al. [3] the conductivity of a slightly contaminated surface can be strongly reduced by a vacuum heat treatment. Both methods, however, turned out to be insufficient for sample preparation in the case of dark conductivity measurements. As the dark conductivity of Si_2Te_3 is of the order $10^{-17}\ \Omega^{-1}\ cm^{-1}$ at room temperature, the measurements are extremely sensitive to surface and contact properties.

Using Auger electron spectroscopy (AES) Erlandsson et al. [4, 5] investigated the surface layer structure of Si_2Te_3 samples after exposition to humid air and also after heat treatment in vacuum. The results for an exposed sample are reproduced in Fig. 1, a scheme of the proposed layer structure is added. According to the AES results the layer structure changes drastically with heat treatment. After a tempering process in vacuum at 600 K the tellurium is evaporated, and the remaining layer consists of SiO_2 only. It may be assumed that contacts applied to samples after such a heat treatment will show non-ohmic behaviour because of the presence of the SiO_2 layer. On the other hand, it seems reasonable to assume that this layer will protect the sample against further reaction with water, i.e. the heat treatment should produce a passivated sur-

[1]) This work was supported by Deutsche Forschungsgemeinschaft, Bonn-Bad Godesberg.

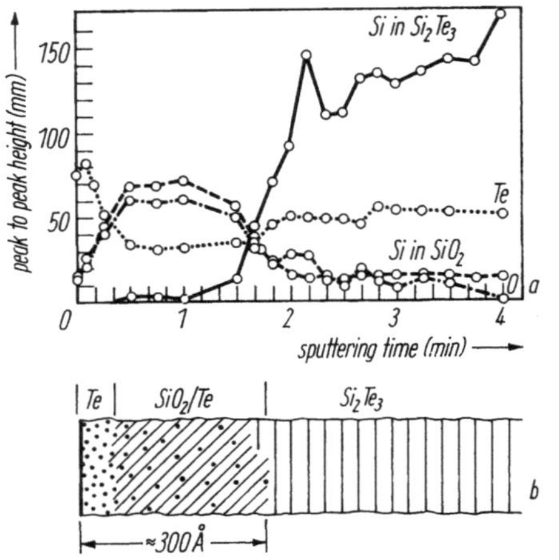

Fig. 1. Surface layer structure of a Si_2Te_3 crystal exposed to air. a) AES depth profile after Erlandsson et al. [4, 5]; b) scheme of the surface layer structure

face. In this paper methods for producing an ohmic metal contact on Si_2Te_3 are investigated, and the optimum conditions for the passivation are studied. Results of dc and a cconductivity measurements, which were performed on samples with passivated surfaces and non-blocking contacts, are reported.

2. Experimental

2.1 Sample preparation

Single crystals were grown by transport in the vapour phase. Platelets with area 7 to 40 mm² and thickness 30 to 100 μm were prepared. As mentioned before no satisfactory sample preparation was possible by using a dry argon atmosphere. A very low water vapour concentration of the order 50 ppm was sufficient to affect the surface conductivity seriously. Therefore it was necessary to perform the preparation in high vacuum. The samples were cleaved perpendicular to the *c*-axis. Gold contacts were applied to the cleavage planes by evaporation. The time delay between cleavage and evaporation was a few second only. The samples were then taken out of the evaporation system and transferred to the conductivity apparatus. During this transfer the samples were exposed to the laboratory atmosphere for some minutes. After vacuum tempering at 520 K for some hours inside the conductivity apparatus the samples were ready for the conductivity measurements.

To study the influence of exposition to water vapour on the properties of the contact, two samples, PZ1 and PZ2, were prepared from the same crystal using the same cleavage plane. Sample PZ1 was prepared as described before, whereas sample PZ2 was exposed to laboratory atmosphere for some seconds before the contacts were applied.

2.2 Conductivity measurements

Because of the very low dark conductivity of Si_2Te_3 the circuits had to be designed with good electrical isolation and low stray capacitance. The samples were mounted on a quartz plate. The electrical contact was obtained by pressing gold wires on the evaporated gold contacts. For temperature control a P.I.D. regulator was used to keep the temperature constant with high long-time stability or to obtain a linear change of temperature with time.

The dc conductivity was measured with help of Keithley electrometers 602 and 616 used in a fast-mode as a feedback-type picoammeter. In order to obtain equilibrium states the heating rate was 0.04 K/min below 470 K and 0.4 K/min above that temperature.

The ac conductivity was measured at constant temperature. In the frequency range 10^{-5} to 0.5 Hz the real and imaginary part of the conductivity were obtained by measuring current and voltage with *X–Y* recorder. In the range 0.1 to 100 Hz a storage oscilloscope was used. A two-channel lock-in amplifier Brookdeal 9502 with current preamplifier was applied in the range 1 Hz to 100 kHz. For all ac measurements it was necessary to control the phase shift of the instruments very carefully.

3. Results and Discussion

3.1 Passivation process

The change of the dc conductivity of some samples, which were exposed to humid air after the preparation, during a heat treatment is shown in Fig. 2. The curves a and b exhibit maxima near 400 K. It is assumed that the uppermost Te layer (Fig. 1) is evaporated in this temperature range. The curves b and c each show a second maximum near 520 K. Probably these maxima correspond to the removal of Te from the second surface layer, which consists of Te and SiO_2. After the samples had been heated above 520K, the conductivity curves turned out to be reversible. It was also observed that samples treated in this way did no longer change their conductance when exposed to humid air. It can be concluded that a heat treatment of exposed samples results in a passivation of the surface. A scheme of the surface layer in the three stages is shown in Fig. 3.

3.2 dc conductivity

To test the contact properties, the current voltage characteristics of the samples PZ1 and PZ2 were investigated for 296 and 575 K (Fig. 4 a and b). As described before the gold contact of sample PZ1 was applied to a clean cleavage plane. This sample shows an ohmic behaviour ($I \sim U$) at low fields and space charge limited current ($I \sim U^2$) at higher fields. For sample PZ2, the gold contact of which was applied to an exposed surface, a lower conductance and non-ohmic behaviour ($I \sim U^{0.7}$) was observed at 296 K. At 575 K sample PZ2 exhibits an ohmic characteristic too, but the conductance still is reduced.

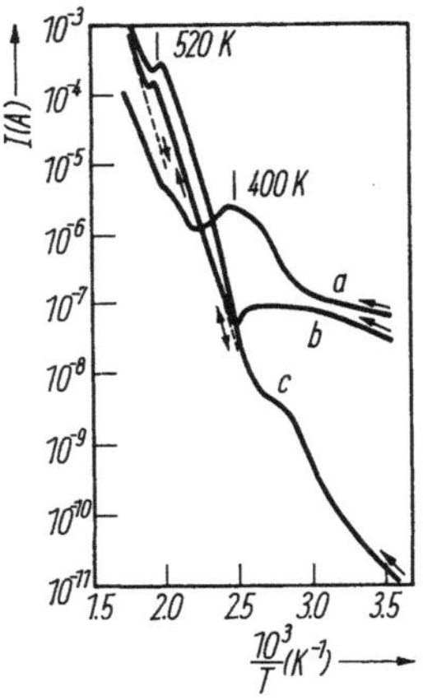

Fig. 2

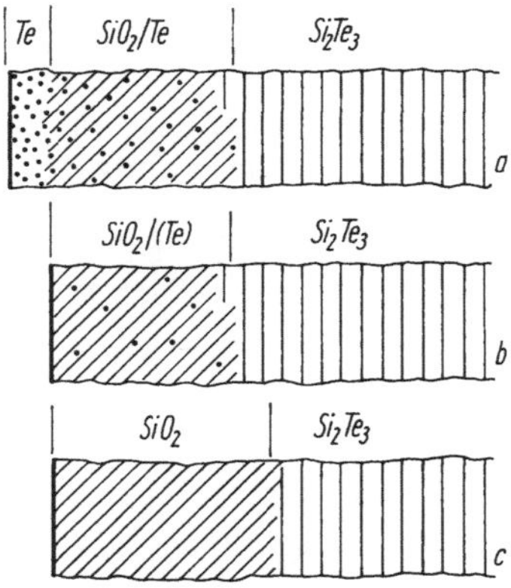

Fig. 3

Fig. 2. Conductivity changes during the heat treatment of Si_2Te_3 crystals, which had been slightly exposed to humid air. Heating rates: (a) 3, (b) 0.7, (c) 0.7 K/min. Expositions times: (a) 1 min, (b) 5 min, (c) 1 h

Fig. 3. Scheme of the surface layer structure of Si_2Te_3 crystals a) after exposition to humid air for some weeks, b) after 2 h heat treatment of the exposed sample at 500 K, c) after 4 h heat treatment of the exposed sample at 650 K

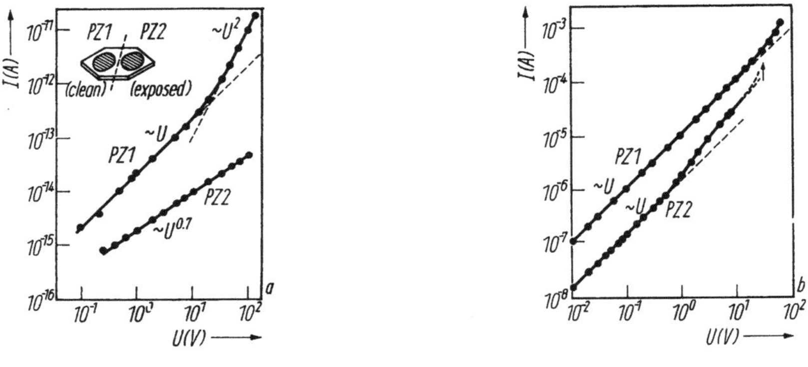

Fig. 4. Current–voltage characteristic of Si_2Te_3 at a) 296 K and b) 575 K (parallel c-axis). PZ1: ohmic behaviour, PZ2: non-ohmic behaviour

The temperature dependence of the dc conductivity parallel to the c-axis of samples PZ1 and PZ2 is shown in Fig. 5. The data for sample PZ1 will be discussed only. The activation energy of the conductivity is 1.04 eV below 620 K and 1.43 eV above 620 K. As shown in [2] the hole mobility parallel to the c-axis is very low ($\mu_{||} \approx \approx 2 \times 10^{-3}$ cm²/Vs at 283 K). Therefore, it may be concluded that the observed activation energy of the conductivity has its origin both in carrier activation and an activated mobility $\mu_{||}$ due to a hopping process. Assuming that intrinsic conduction is dominant above 620 K and using a value for the band gap $E_g = 2.2$ eV (1), an activation energy of the mobility $E_\mu \approx 0.3$ eV can be estimated. Therefore, below 620 K the carriers (holes) are assumed to be activated from deep impurity levels lying 0.7 eV above the valence band. This result is in good agreement with data obtained from the analysis of thermally stimulated currents. The corresponding experiments were performed by cooling the crystal with a voltage bias applied and subsequent heating without bias.

At a certain electrical field strength parallel to the c-axis switching effects have been observed. An example is shown in Fig. 6. The switching takes place from a high resistivity to a low resistivity state. If the voltage is reversed the opposite effect occurs. It has not yet been established whether the transition is controlled by traps or by ionic conduction due to the defect structure of the lattice.

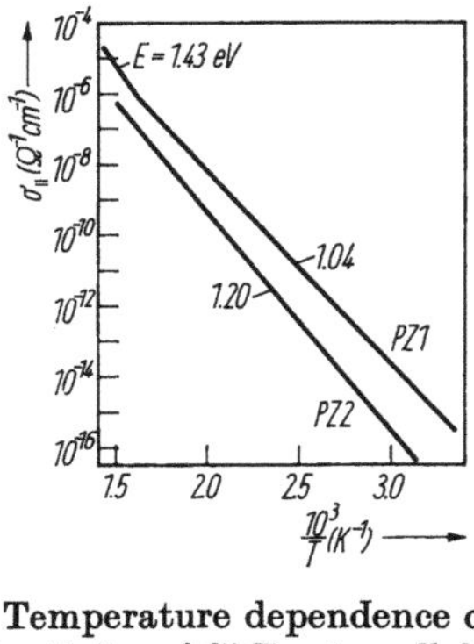

Fig. 5. Temperature dependence of the dc conductivity of Si_2Te_3 (parallel c-axis)

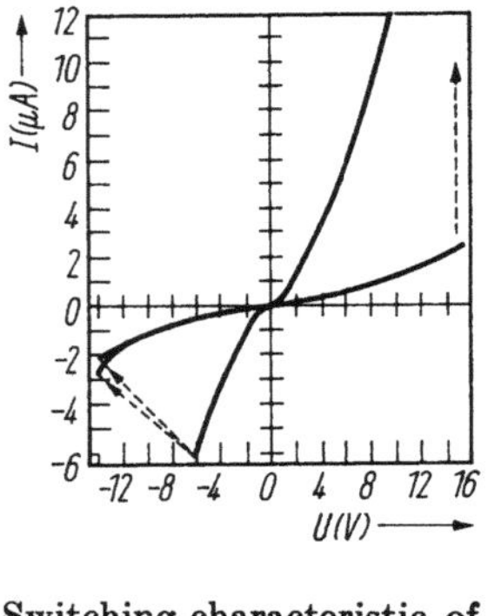

Fig. 6. Switching characteristic of a Si_2Te_3 device (430 K)

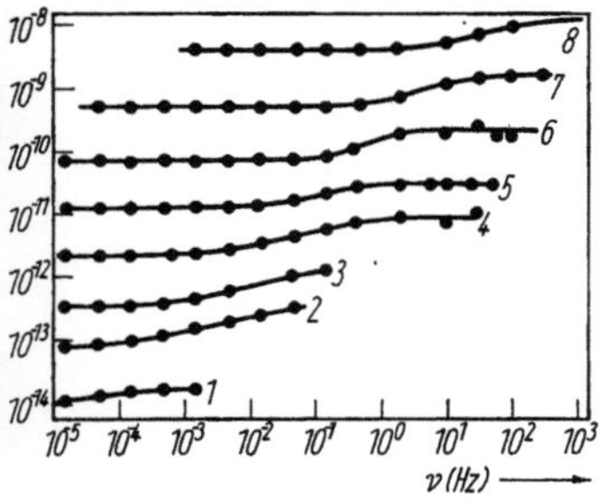

Fig. 7

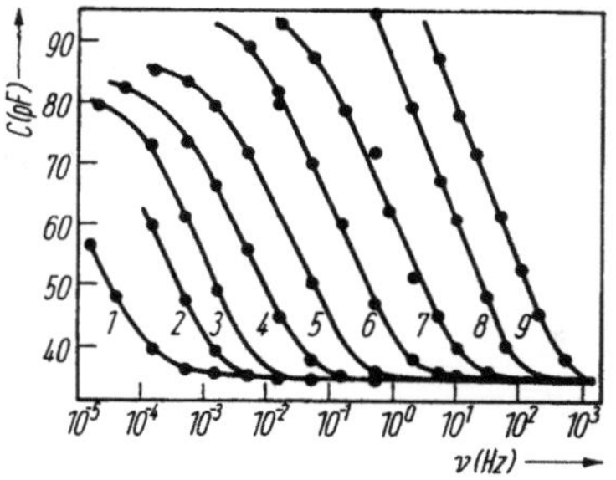

Fig. 8

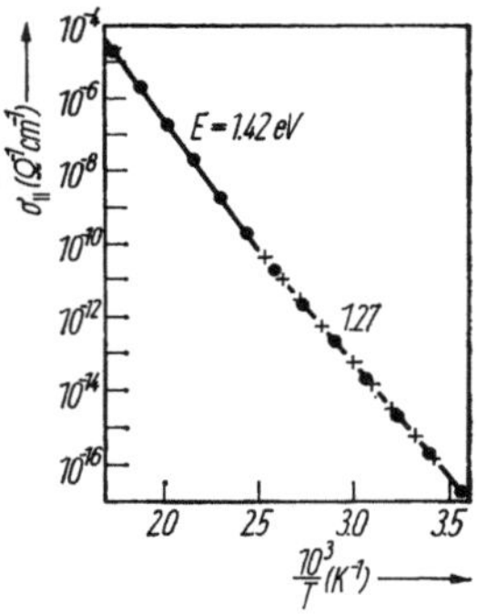

Fig. 9

Fig. 7. Frequency dependence of the real part of the electrical conductance G of Si_2Te_3 at different temperatures (parallel c-axis, Li doping). (1) $T = 294$, (2) 307, (3) 317, (4) 331, (5) 346, (6) 361, (7) 378, (8) 397 K

Fig. 8. Capacity of Si_2Te_3 versus frequency at different temperatures (parallel c-axis, Li doping). (1) $T = 294$, (2) 307, (3) 317, (4) 331, (5) 346, (6) 361, (7) 378, (8) 397, (9) 415 K

Fig. 9. Temperature dependence of the dc conductivity (———) of Si_2Te_3 (parallel c-axis, Li doping + indicates extrapolated dc conductivity from Fig. 7 $\left(\lim\limits_{\omega \to 0} \sigma(\omega)\right)$

3.3 ac conductivity

Real and imaginary part of the electrical conductance of Si_2Te_3 doped with Li is plotted versus frequency in Fig. 7 and 8. The real part of the conductivity increases with frequency approximately by a factor 2. The characteristic frequency for this process increases exponentially with temperature. An activation energy of 1.27 eV is obtained. Further investigations are necessary to decide whether traps or ionic conduction contribute significantly to the observed effects.

The temperature dependence of the dc conductivity for Li-doped Si_2Te_3 is shown in Fig. 9. The activation energy in the extrinsic range is enhanced by Li-doping. As indicated the extrapolated dc conductivity data from Fig. 7 are in good agreement with these results.

References

[1] J. W. Rau and C. R. Kannewurf, J. Phys. Chem. Solids **27**, 1097 (1966).
[2] K. Ziegler and U. Birkholz, phys. stat. sol. (a) **39**, 467 (1977).
[3] K. Ziegler, H.-D. Junker, and U. Birkholz, phys. stat. sol. (a) **37**, K 97 (1976).
[4] R. Erlandsson, U. Birkholz, and S.-E. Karlsson, phys. stat. sol. (a) **41**, K 163 (1977).
[5] R. Erlandsson, U. Birkholz, and S.-E. Karlsson, phys. stat. sol. (a) **47**, 85 (1978).

(Receveid June 23, 1978)

phys. stat. sol. (a) 49, 133 (1978)

Subject classification: 1.1 and 14.4; 15; 22.9

Laboratories RCA Ltd., Zürich[1]) *(a) and J. Franck Institute, The University of Chicago*[2]) *(b)*

Pyroelectric Effect in PTS

By

H. Kiess (a) and R. Clarke (b)

In honour of Prof. Dr. F. Stöckmann's 60th birthday

Pyroelectricity is observed and measured in poly-2,4-hexadiyne-1,6-diol bis(p-toluene sulphonate), (abbreviated PTS), between 76 and 300 K. PTS is claimed to be monoclinic and to belong to the space group $P2_1/c$, i.e. it should have a center of symmetry, so that the observation of pyroelectricity is in conflict with the structure determination. Since the measurements indicate that no other mechanisms such as charge injection, orientation or impurity caused dipoles are involved it is concluded that PTS must have a non-centrosymmetric crystal structure.

An Poly-2,4-Hexadiin-1,6-diol-bis(p-Toluolsulphonat)-Kristallen (abgekürzt PTS) wird Pyroelektrizität beobachtet und der Effekt im Temperaturbereich zwischen 76 und 300 K gemessen. PTS ist monoklin und soll die Raumgruppe $P2_1/c$ haben, d. h. ein Symmetriezentrum besitzen. Diese Aussage steht im Widerspruch zum beobachteten pyroelektrischen Effekt. Da die Messungen andere Mechanismen ausschließen (z. B. Ladungsinjektion, Orientierung von Dipolmomenten von Verunreinigungen des PTS, etc.), wird gefolgert, daß PTS eine Kristallstruktur ohne Symmetriezentrum besitzen muß.

1. Introduction

There has been considerable practical as well as theoretical interest in recent years in the pyroelectric effect of polymers [1, 2]. Since the vectorial pyroelectric effect is a crystal property and since it is restricted to non-centrosymmetric materials [3], actually no pyroelectric effect is observed in the amorphous or at best (partially) microcrystalline polymer films unless films of optically active polymers are uniaxially oriented by mechanical deformation or unless the material has been polarized electrically at elevated temperatures. The dc bias applied at high temperatures induces a polarization of the material either by orientation of dipoles or by injection of charge carriers, a state which can be frozen-in at low temperatures. Thus, in order to observe pyroelectricity in polymer films some manipulation of the material is required and also annealing at elevated temperatures destroys usually the polarization.

There are polymers, however, which polymerize in the solid state to give macroscopic single crystals whose properties differ dramatically from those of conventional amorphous polymers reflecting the high structural perfection and extended chain morphology. The observation of pyroelectricity is then, of course, primarily a question of the symmetry of the crystal structure and only accidental preparation conditions might induce a polarization in a way as was briefly discussed for the polymer films.

2. Experimental

The single crystals of poly-2,4-hexadiyne-1,6-diol bis(p-toluene sulphonate), referred to hereafter as PTS, were obtained by growing monomer crystals from acetone solutions. The monomer crystals were then thermally polymerized at 60 °C for 40 h.

---

[1]) Badenerstr. 569, CH-8048 Zürich, Switzerland.

[2]) Chicago, Illinois, 60637, USA.

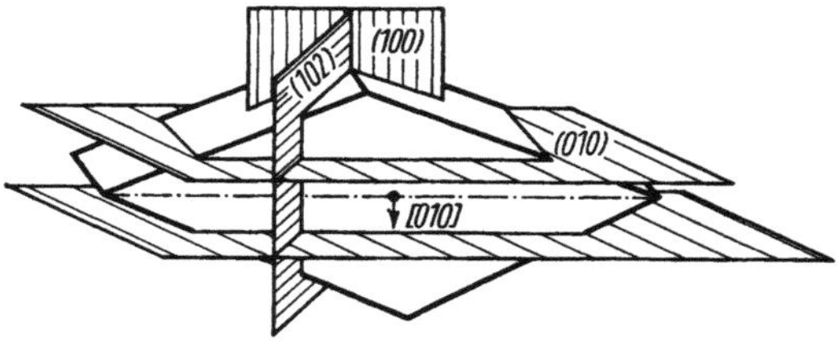

Fig. 1. The configuration used in this series of measurements. In terms of the unit cell reported by Kobelt and Paulus [4] the crystal was cut parallel to the (010) planes onto which the electrodes were placed. The arrow indicates the chain direction. The (102) and (100) planes are cleavage planes

Au electrodes or occasionally silver paste electrodes were placed on the crystal in the orientation illustrated in Fig. 1. In order to obtain this configuration the crystal had to be cut and the surfaces polished with analytical grade acetone prior to polymerization. The polymer chains are thus extended perpendicular to the contact area and all measurements which are reported upon are made in the crystallographic *b*-direction. Typically the contact area was between 10^{-2} and 10^{-1} cm^2, the thickness of the crystals was varied systematically between 2×10^{-3} cm and 10^{-1} cm.

For measuring the pyroelectric effect a quasi-static method was used: The crystal was heated or cooled at a constant rate and the pyroelectric current was recorded continuously. The electric field in the crystal remained low (< 1 V/cm). The currents were measured with Keithley Electrometers Model 602 and 614, the rise times (10% to 90%) of which were $\leqq 1$ s in the current ranges used.

The crystals were placed on a copper holder which could be mounted onto a cooling (or heating) finger. The sample holder itself was electrically insulated from the cooling finger but in good thermal contact with it. Special care was taken to eliminate any spurious electrical signals which were caused by bad insulating materials and at pressures of the vacuum system $> 10^{-6}$ Torr. As a rule measurements were only made when the pressure in the system was below 5×10^{-7} Torr.

The temperature and its gradient with time $\mathrm{d}T/\mathrm{d}t$ were monitored with a thermocouple. Typically, the temperature was changed at a rate of 1 K/s. By increasing the thermal mass of the cooling finger, the cooling rate could be reduced to 0.2 K/s. Obviously, the heating rate could be easily changed and the ratio could be varied between 0.1 and 1.5 K/s.

3. Results

The pyroelectric signal measured in *b*-direction (i.e. in chain direction) as a function of temperature is shown in Fig. 2. The curves are not corrected for constant rate of change of the temperature. Between 170 and 210 K the pyroeffect changes sign and in addition the signal oscillates with temperature. The positions of the two relative extrema in this temperature range are well defined with respect to temperature. Varying the heating rate and extrapolating to zero heating rate the extrema could be positioned at 193 and 187.5 K, respectively (Fig. 3). The mean values of the extrema positions obtained by averaging over measurements on five crystals gave temperatures of 194.0 and 187.5 K. The strength of the oscillations varied from crystal to crystal and typically the extrema were more pronounced in the heating than in the cooling cycle even if corrected for equal $\mathrm{d}T/\mathrm{d}t$.

The component π_b of the pyroelectric constant in *b*-direction was found to be about 2×10^{-11} C/cm^2 K above 210 K and below 170 K. An accurate determination of π_b proved to be difficult since the crystals crack easily along the (100) cleavage plane which gives then rise to noisy signals and to an inaccurate determination of the relevant contact area. In the temperature regions below 170 K and above 210 K the pyroelectric constant is practically independent of temperature.

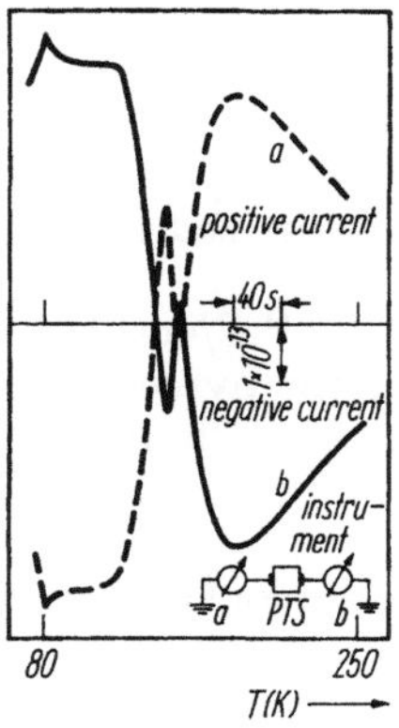

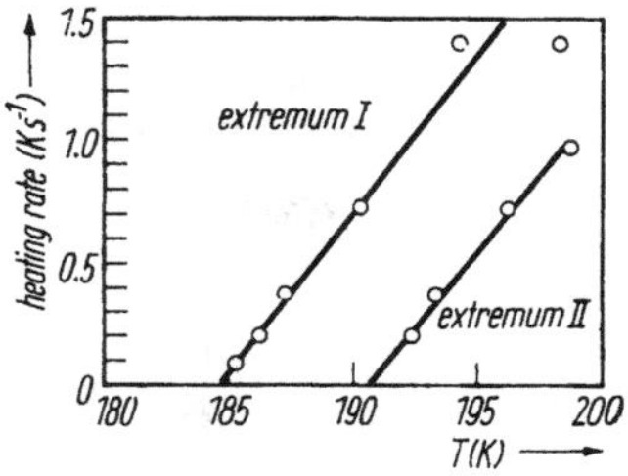

Fig. 3. Variation of the positions of the extrema with heating rate

Fig. 2. Recording traces of the pyroelectric polarization current upon cooling. Curve a and b were measured simultaneaously and represent the currents measured on opposite faces (see insert). The increase of the current at 80 K reflects the higher cooling rate close to liquid nitrogen temperature

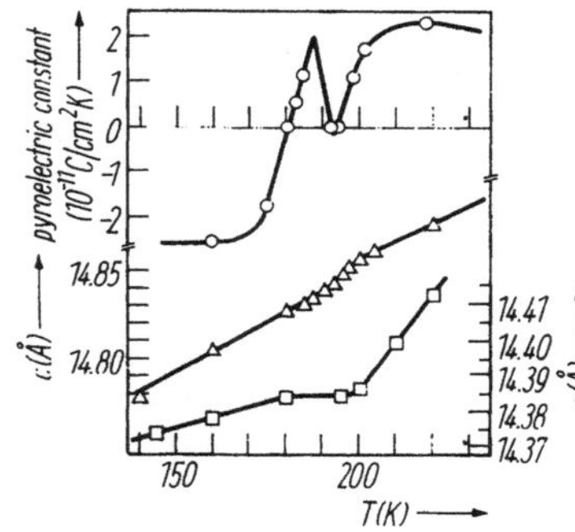

Fig. 4. The upper part of the figure represents the pyroelectric constant π_b as a function of temperature, the lower part gives the variation of the lattice spacings with temperature. There is a clear correlation of the anomalous behaviour of π_b with the phase transition

The temperature in which the pyroelectric signal behaves anomalously coincides exactly with the anomalous change with temperature of the lattice spacings a and c as measured by a method specially designed for this type of study [5] (Fig. 4). This correlation indicates clearly that the change in sign of the pyroelectric signal and its oscillations with temperature are connected with the phase transition which PTS is known to undergo in this temperature range [6].

PTS crystals of different provenience were measured and the pyroelectric effect was found to vary only within the limits of the accuracy of the present measurements. Also, neither variation in polymerization temperatures and times nor polymerizing the crystals by γ-irradiation did have an influence on the pyroelectric effect in any significant way. Similarly, poling of the crystals or annealing at 70 °C for several hours did not affect the signal.

4. Discussion

The annealing and poling experiments prove that the pyroelectricity observed in PTS cannot be due to any of the mechanisms which are normally responsible for the pyroelectric effect in polymer films. Similarly, tensorial pyroelectricity, theoretically predicted by Voigt [7, 8], can also be excluded since the signal does not depend on the crystal shape and since it is perfectly symmetrical with respect to the top and bottom electrodes. Therefore we conclude that vectorial pyroelectricity is observed which implies either that the space group of the crystal must lack a center of symmetry or that impurities with dipole moments must be incorporated in the crystal in a preferred orientation. Since no significant differences between individual crystals, between crystals of different provenience and crystals which were polymerized in different ways could be found we exclude the latter possibility.

The pyroelectric constant π_b, being approximately 2×10^{-11} C/cm² K, is about one order in magnitude smaller than that of non-ferroelectric pyroelectric materials (Table 1). The value of 2×10^{-11} C/cm² K represents a lower limit since the polar

Table 1

Pyroelectric constants of various materials at room temperature for comparison

polymers PVF_2	PTS	II–VI compounds ZnO	CdS	ferroelectricum $LiTaO_3$	
4×10^{-10} to 1.8×10^{-9}	$> 2 \times 10^{-11}$	9×10^{-10}	4×10^{-10}	1.7×10^{-8}	C/cm² K

axis of the material is unknown and since only the component in *b*-direction was determined. Preliminary measurements perpendicular to the (100) cleavage plane indicate that the component of the pyroelectric constant in this direction is of similar magnitude. Therefore the polar axis is not in *b*-direction, i.e. parallel to the chains. Further measurements are under way which will allow to determine the polar axis.

The change of sign of the pyroelectric constant and the oscillation with temperature in the range of the phase transition are not understood at the present time. The available information about different properties of PTS, in particular also about the phase transition, is not sufficient to make a microscopic model. Further measurements on the temperature dependence of the pyroelectric constant at low temperatures and measurements on monomers and partially polymerized PTS are planned and will be reported elsewhere.

Acknowledgements

The authors thank J. Kane and V. Enkelmann for numerous discussions and for the preparation of the crystals. They are also indebted to W. J. Merz and G. Harbeke for their continuous interest in this work and to R. Keller for performing many of the experiments.

References

[1] Y. Wada and R. Hayakawa, Japan. J. appl. Phys. **15**, 2841 (1976).
[2] M. G. Broadhurst, G. T. Davis, and R. E. Collins, Bull. Amer. Phys. Soc. **22**, 282 (1977).
[3] See e. g. W. G. Cady, in: Piezoelectricity, Dover Publ., Inc., New York 1964.
[4] D. Kobelt and E. F. Paulus, Acta cryst. **B 30**, 232 (1974).
[5] R. Clarke and R. E. Morley, J. appl. Cryst. **9**, 481 (1976).
[6] V. Enkelmann and G. Wegner, Makromol. Chem. **178**, 635 (1977).
[7] W. Voigt, Göttingen Nachr., **394** (1905).
[8] See also H. Falkenhagen, Hdb. Phys., Vol. XIII, Springer-Verlag,1928 (p. 291).

(Received June 23, 1978)

phys. stat. sol. (a) 49, 137 (1978)

Subject classification: 9 and 14.3; 10.2; 22.1.2

Semiconductor Division, AEG-Telefunken, Heilbronn [1])

Degradation of Carrier Lifetime in Silicon Crystals at Room Temperature

By

K. GRAFF and H. PIEPER

In honour of Prof. Dr. F. STÖCKMANN's 60th birthday

After quenching high-purity n-type silicon samples from about 1000 °C to room temperature the minority carrier lifetime is measured as a function of the time elapsed after this heat treatment. A more or less steep decrease of the carrier lifetime is observed within the first 100 h. This decrease depends on the purity of the samples before annealing and the cleanness of the furnace tube. Samples intentionally doped with heavy metal impurities show a similar decrease in carrier lifetime if the impurity is gold. A decrease in carrier lifetime is also observed after cutting samples from Czochralski-grown silicon crystals without application of any heat treatment. These results can be explained by the diffusion of vacancies which were frozen in during quenching the samples or generated during the sawing process. The vacancies subsequently react with interstitial gold, a residual impurity in the crystal, forming recombination centers and reducing the carrier lifetime.

Nach dem Abschrecken von hochreinen n-Siliziumproben von ca. 1000 °C auf Raumtemperatur wird die Minoritäts-Trägerlebensdauer als Funktion der Zeit nach der Wärmebehandlung gemessen. Innerhalb der ersten 100 Stunden wird ein mehr oder weniger steiler Abfall der Trägerlebensdauer beobachtet. Dieser Abfall hängt von der Reinheit der Proben vor dem Temperaturprozeß und von der Sauberkeit des Ofenrohres ab. Proben, die absichtlich mit Schwermetallverunreinigungen dotiert wurden, zeigen einen ähnlichen Abfall der Trägerlebensdauer, wenn zur Verunreinigung Gold verwendet wurde. Ein Abfall der Trägerlebensdauer kann auch nach dem Sägen von Proben aus Czochralski-gezogenen Siliziumkristallen beobachtet werden, ohne daß sie einem Temperaturprozeß unterzogen werden. Diese Ergebnisse können durch die Diffusion von Leerstellen erklärt werden, die während des Abschreckens der Proben eingefroren, oder durch den Sägeprozeß erzeugt werden. Diese Leerstellen reagieren anschließend mit Zwischengitter-Gold als Restverunreinigung im Kristall, wobei sie Rekombinationszentren bilden, die die Trägerlebensdauer verringern.

1. Introduction

One of the problems in semiconductor device production not fully resolved so far is the preservation of the high purity of floating-zone starting material during the various high-temperature processes which are needed for oxidation of and diffusion in silicon wafers. In general this problem is avoided by adding a getter process as a final step such as phosphorus diffusion [1] or ion implantation followed by an annealing treatment [2]. In this way the impurities unintentionally diffused into the semiconductor material will be restricted to regions where they do not deteriorate the electrical properties of the device.

On the other hand, there are many efforts to keep semiconductor material as clean as possible usually by applying highly sophisticated cleaning procedures immediately before heat treatment of the wafers [3] and by application of specially purified tubes [4]. Some work was done to avoid impurities getting into the tube by evaporation from the heat-windings and diffusion through the tube material. For this purpose silicon tubes partly doped and directly heated were developed [5].

---

[1]) Theresienstr. 2, D-7100 Heilbronn, FRG.

Most of the problems arise from the presence of heavy metal impurities during handling and treatment of the wafers. They tend to be adsorbed on the silicon surfaces. They are highly mobile in the silicon host lattice at elevated temperatures because of a fast diffusion process via interstitial sites. They strongly decrease recombination carrier lifetime since most of them form deep level impurities in silicon. As pointed out in the literature it is rather difficult to clean silicon surfaces if heavy metals are once adsorbed especially in the case of noble metals such as gold or copper [3]. The adsorption takes place during cutting the crystal or handling the wafers or even during cleaning processes, especially HF-dip, if the purity of the chemicals is not extremely high. Large efforts are necessary to purify the water used in rinsing the wafers at the end of the process [6].

Even small concentrations of heavy metals strongly decrease the carrier lifetime. As a result of a calculation using recombination parameters for gold from the literature [7] a concentration of $3 \times 10^{10}\ cm^{-3}$ yields a low level minority carrier lifetime of 300 μs in n-type silicon. This value, considerably below carrier lifetime of as-grown floating-zone silicon (several ms), however, is caused by an impurity concentration which can be detected only by activation analysis. Even with the DDLTS-technique [8] enhanced to high sensitivity the detection limit is of the same order of magnitude. So carrier lifetime measurements are extremely sensitive to detection of deep level impurities and for investigation of their reactions.

In this paper reactions of deep level impurities are discussed which were observed by means of carrier lifetime measurements on n-type silicon samples of high-purity crystals. These reactions take place at room temperature. The concentration of the impurities responsible for the change in the electrical parameter is of the order of magnitude of $10^{10}\ cm^{-3}$. Further investigations by means of other techniques to get more information about the chemical properties of the impurities are not possible because this concentration is beyond the detection limit of usual techniques.

2. Experimental Methods

Samples $5 \times 5 \times 25\ mm^3$ in size were cut from the seed-end of dislocation-free floating-zone n-type silicon crystals with specific resistivities of 60 to 70 Ωcm. After a first cleaning the samples were etched twice in a usual mixture of HNO_3, HF, and acetic acid. Subsequently they were rinsed in de-ionized water, boiled in aqua regia and rinsed once more. The samples were dried by means of compressed air.

Heat treatments were carried out using a furnace with a silica tube and a silica sample holder. To clean the tube the temperature was elevated about 100 °C above the usual working point. The samples were isolated from the silica sample holder by mounting them on top of two other samples. Usually heat treatment was carried out at a temperature of 1000 °C for 1 h in dry oxygen which resulted in an oxide layer of about 900 Å in thickness. To finish the heat treatment the sample holder was pulled out of the furnace and the sample was allowed to cool down to about 200 °C rather quickly. Slow cooling of the samples was accomplished in the furnace by turning off electrical heating.

For some experiments samples were doped intentionally with heavy metals. They were etched and subsequently treated with HF after adding a small amount of a solution containing gold, copper, or iron. In a proceeding cleaning process the samples were boiled in aqua regia etched once or twice, rinsed, and dried as described above.

After heat treatment the oxide layer was removed and the two smallest opposing sides of the samples were lapped in order to prepare electrical contacts by rubbing gallium on the surfaces.

For the measurement of the minority carrier lifetime the photoconductive decay (PCD) method was applied using special IR-emitting diodes for the generation of excess carriers. The change in conductivity of the sample caused by illumination with light pulses was amplified, stored in a Waveform Eductor (PAR) and recorded as published previously [9]. Higher values of carrier lifetime were corrected for specimen size following ASTM-Standards [10].

3. Experimental Results

3.1 Annealing under clean conditions

The carrier lifetime of samples cleaned and heat treated as described above is plotted versus the time after heat treatment on a double-logarithmic scale as shown in Fig. 1. Quenched samples usually show variations in carrier lifetime during about 100 h following the heat treatment. Generally there is a decay in carrier lifetime. Sometimes, however, the values increase slightly during the starting period and reach a maximum several hours after quenching. Whereas the accuracy of the measurement is within $\pm 10\%$ the variations in carrier lifetime may amount to a factor of 10 thus being far larger than the measurement error. The variations in the carrier lifetime represent variations of the properties in the bulk of the silicon crystal since measurements after lapping all surfaces of the sample yield the same results.

After heat treatments in atmospheres of oxygen, argon, as well as oxygen with an additional small amount of HCl the same results were obtained. Slow cooling of the samples after heat treatment, however, does not cause any decrease in carrier lifetime. The measured values start at a lower level equal or near to the final results of the quenched samples as demonstrated by the dashed lines in Fig. 1. It can be deduced from the experimental findings that the decay of the carrier lifetime is caused by the quenching of the sample after heat treatment. But there must be another property of the individual sample which fix the amount of the decay since samples annealed simultaneously may differ.

As can be seen in Fig. 1 the values of the carrier lifetime soon after quenching the samples nearly equal to those of the as-grown crystals. This is still more pronounced in the case of adding HCl to the oxygen atmosphere which is assumed to shift the equilibrium of vacancies and self-interstitials in silicon. The final carrier lifetime, however, seems to be the same with or without addition of HCl.

Differences are observed in the slopes of the curves of carrier lifetime after quenching the samples. Sometimes an abrupt decay changes to a constant value whereas the approach of the final value in other cases is performed gradually. Decay curves may cross one another. An increase in the carrier lifetime during the starting period is only observed if the following decay is not strongly marked. Because it was suspected that residual impurities introduced during handling and cleaning may be responsible for the individual properties of the samples, some of them were intentionally doped with heavy metals.

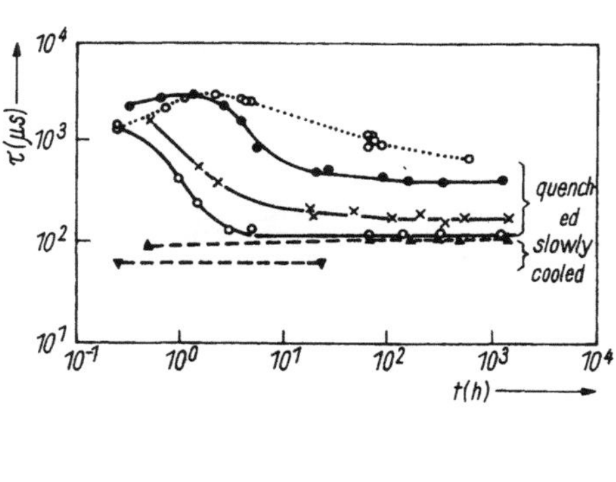

Fig. 1. Carrier lifetime (τ) of various floating-zone n-type silicon samples versus time after heat treatment (t) at 1000 °C

3.2 Annealing of intentionally doped samples

After annealing and quenching samples intentionally doped with *gold*, results resembling those discussed above were obtained. They are shown in Fig. 2. The smaller values of carrier lifetime demonstrate that a certain amount of gold has penetrated the silicon by diffusion. Maximum variation of the carrier lifetime with elapsed time rises by a factor of 3 which again is well above measurement error. The beginning of carrier lifetime decay in some cases seems to come later compared to the results on undoped samples.

As a consequence of variations in the contamination of the samples before heat treatment there are differences in the results. After deposition of gold all samples were boiled in aqua regia to remove most of the contamination. Samples annealed without any further cleaning exhibit carrier lifetimes in the region 2 to 3 μs which corresponds to a gold concentration of 4×10^{12} cm^{-3} in the bulk of the silicon. Samples subsequently etched in KOH or CP4-like solution as mentioned above yield carrier lifetimes of 16 to 18 μs soon after quenching. Samples etched twice after treatment in aqua regia have 35 μs corresponding to a gold content of about 3×10^{11} cm^{-3} after oxidation. Lapping all surfaces twice followed by two etching processes finally results in rather high values of the carrier lifetime after heat treatment of about 800 μs corresponding to a gold concentration of 10^{10} cm^{-3}.

Samples doped with *copper* and subsequently boiled in aqua regia or treated in HNO_3 again yield small values of carrier lifetime after annealing but no decay with time was observed as demonstrated by the dashed lines in Fig. 2. On the other hand, traps were formed in the samples treated in HNO_3 at a certain time after quenching the sample. The appearance of traps is indicated by an non-exponential decay of photo-conductivity. The concentration of these traps increased with time. In samples boiled in aqua regia no traps were observed.

Results obtained on samples doped with *iron* in HF-solution did not differ significantly from those reported on clean samples. Obviously iron cannot be deposited on silicon in this way. From earlier experiments it is known, however, that silicon samples contain considerable concentrations of iron after heat treatment in hydrogen atmosphere. Concentrations of neutral, interstitial iron up to 10^{14} cm^{-3} were determined by ESR spectroscopy after applying an epitaxial process [13]. Therefore, this process was applied to our samples in order to dope them with iron. The results as indicated as dotted lines in Fig. 2 show constant carrier lifetimes which are of the same order of magnitude as those in the other doped specimens. These results are independent of the time after annealing and quenching the samples as in the case of copper. From the investigations carried out so far it can be deduced that only gold may cause a carrier lifetime decay with elapsed time similar to that originally observed on undoped samples.

3.3 Multiple heat treatments

For a further clarification of the impurity reactions taking place at room temperature some samples were exposed to repeated heat treatments at various temperatures. Samples quenched from 1000 °C were kept at room temperature until the carrier

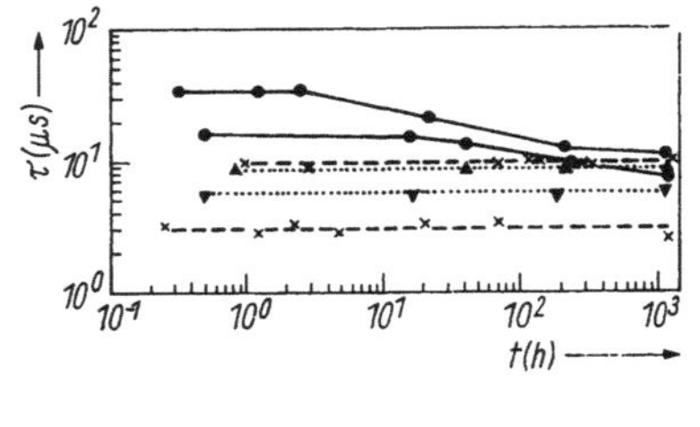

Fig. 2. Carrier lifetime of floating-zone n-type silicon samples doped with various heavy metal impurities versus time after heat treatment and quenching of the samples. Intentionally doped with: ● Au, × Cu, ▲ Fe

lifetime showed its final constant value. Subsequently they were annealed at 450 °C and quenched once more. As a result the carrier lifetime again starts at high values and decreases with time. This procedure was repeated at 600, 800, and 1000 °C. The tendency of the results again was the same whereas the final carrier lifetime in general decreased more or less because of increasing total impurity content of the samples after multiple annealing processes.

Samples which were slowly cooled during the first step did not change considerably in carrier lifetime until the temperature of the subsequent heat treatment was raised to about 800 °C. After quenching the carrier lifetime starts at a higher level and decreases with time. Finally quenching these samples from 1000 °C yields results identical to those discussed above.

3.4 Test for diffusion-limited reaction

An experiment to decisively determine whether the decay of the lifetime after heat treatment is diffusion-limited was carried out in which some samples were quenched from 1000 °C, the carrier lifetime was determined as usual, and subsequently the samples were kept in liquid nitrogen. After an elapsed time sufficiently long for the carrier lifetime to approach a constant value, the samples were warmed up to room temperature and the carrier lifetime was remeasured in dependence on the time. The results are shown in Fig. 3 where the carrier lifetime versus the time after oxidation is plotted linearly. During the cold period the value remains unchanged at a high level. After warming up, however, the carrier lifetime steeply decreases in the same way as reported for samples at room temperature immediately after quenching. For comparison both results are indicated in Fig. 3. It can be seen that the reaction is highly dependent upon temperature and can be interrupted by cooling the sample.

3.5 Decay of carrier lifetime after cutting samples

In crystalline silicon rods local variations in the carrier lifetime can be observed. From measurements on a rod only a mean value is obtained which depends on the radial profile of the carrier lifetime and the penetration depth of the pulsed light generating excess carriers. To determine more exact values samples must be cut from the crystal rod. During examinations of the radial profiles of the carrier lifetime on various crystals, samples were found which demonstrated smaller carrier lifetimes when measured a second time. These results are demonstrated in Fig. 4. Here the radial profile of carrier lifetime of a 1 Ωcm Czochralski-grown p-type silicon crystal which was cut in samples 5 × 5 mm² in cross-section is recorded. The profile is a composition of results obtained on eleven samples forming one diameter of the crystal. The first measurements were carried out three days and the second (dotted line) 13 days after sawing the samples. In general the tendency of the two profiles are similar whereas the carrier lifetimes of the individual samples diminished more or less in the meantime.

From measurements on numerous crystals it was deduced that a decay of carrier lifetime only takes place in Czochralski-grown crystals. Although the higher values in

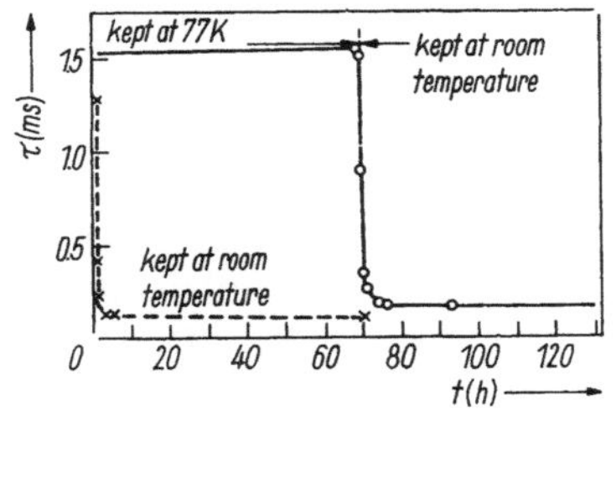

Fig. 3. Carrier lifetime of floating-zone n-type silicon samples kept for a time at 77 K versus time after heat treatment and quenching of the samples

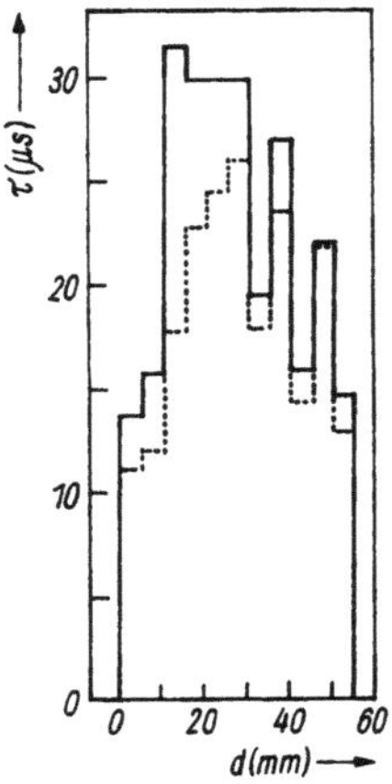

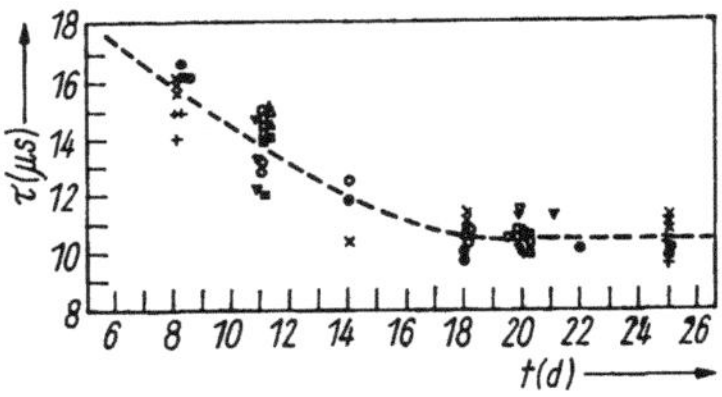

Fig. 4. Radial profile of carrier lifetime of Czochralski-grown p-type silicon measured 3 (———) and 13 (······) days after sawing the samples (*d* is the distance from rim of crystal)

Fig. 5. Carrier lifetime of eight Czochralski-grown p-type silicon crystals versus time after sawing the samples

carrier lifetime of floating-zone silicon enables higher sensitivity for detection of any impurity reaction it could not be observed in this material. The decay starts after the cutting process and saturates after about 20 days as demonstrated in Fig. 5 in which measurements on various 1 Ωcm Czochralski-grown p-type silicon crystals are reproduced. A particular symbol represents a partical crystal. Usually several samples from one crystal were measured. All samples were cut in succession. As can be seen the variation in carrier lifetime again is well above experimental error. There are some differences in decay from crystal to crystal (Fig. 5) and even from sample to sample (Fig. 4).

4. Discussion

From the experimental findings it seems likely that the carrier lifetime decay caused by quenching the samples from high temperatures as well as sawing the crystals are based on the same impurity reaction. In both cases a certain process initiates the decay and individual properties of the samples are involved which are probably the result of a special impurity content namely the gold concentration of the crystal. The reaction time in both cases is of the same order of magnitude. In quenched samples the reaction is proved to be highly dependent on the temperature, so it might be assumed that it is diffusion-limited.

To explain the impurity reaction causing the decay of the carrier lifetime with elapsed time, a defect must be found which diffuses quickly at room temperature since diffusion coefficients of usual impurities are extremely low. At higher temperatures vacancies are created in equilibrium with self-interstitials. The existence of these vacancies may be maintained by quenching the samples. It is known that they are highly mobile even at low temperatures. It may be imagined that vacancies are also created during sawing a silicon crystal.

Since the size of the samples enables the vacancies to diffuse through the crystal for rather long periods of time they can collide with interstitial gold atoms which are known to be electrically inactive. Most of the gold may originate from residual impurities on the surfaces before heat treatment. A reaction between vacancies and interstitial gold takes place which creates a recombination centre and thus lowers the carrier lifetime. The total decay of the carrier lifetime depends on the quantity of vacancies as well as on the concentration of interstitial gold. Assuming the validity of this model the following experimental details may be explained:

If the samples were cooled slowly all reactions take place during the cooling period. The carrier lifetime starts at a lower level and remains constant.

In quenched samples reactions take place as long as vacancies and interstitial gold atoms are present. In quenched samples with very low gold concentrations the time needed for collisions and reactions will be longer. Vacancies partly diffuse to the surfaces of the sample where they recombine. If the vacancy itself operates as a recombination center the carrier lifetime may increase during outdiffusion of vacancies.

At low temperatures the diffusion coefficient of the vacancies is considerably reduced. The vacancies remain nearly immobile until the sample is warmed up. So the reaction is interrupted during the cold period.

After sawing Czochralski-grown silicon crystals the reactions take place as described above. In the case of cutting floating-zone crystals reactions may be prevented because of the lack of interstitial gold atoms in this material. The vacancy–gold complex may be disturbed by annealing the sample once more and the gold is resolved to interstitial sites. So the process can be reproduced by repeated heat treatments. There might be some differences in binding energy between the assumed complexes and substitutional gold since samples cooled slowly during a first annealing do not exhibit a change in carrier lifetime until the temperature of a second annealing process is raised to about 800 °C whereas quenched samples already react at a temperature of 450 °C.

Some quantitative estimations were performed to determine whether the model is realistic. Assuming the carrier lifetime decay is only caused by changing gold atoms from interstitial sites to the vacancy complex (with carrier cross sections of the substitutional gold) the needed gold concentration can be calculated. In our experiments the maximum concentration was found for intentionally doped gold without a subsequent etch process. The change in carrier lifetime from 4 to about 2.8 μs results in a gold concentration of 10^{12} cm^{-3}. In this case the relative experimental error may be higher than usual because 2.8 μs begins to approach the lower limit of the electronic equipment [14]. Since each reaction needs at least one vacancy the minimum vacancy concentration is 10^{12} cm^{-3}. The concentration of vacancies in thermal equilibrium is not well-known. Published data for 1000 °C, however, are about 10^{12} to 10^{13} cm^{-3} which is the needed order of magnitude [11].

To estimate diffusion lenghts of vacancies their electrical charge must be specified because the diffusion coefficients published in the literature depend to a large extent on it. Since the activation energy of vacancies in the silicon band gap also depends on the charge, the occupation probability of the different levels is determined by the Fermi level of the silicon crystal investigated. In the present case most of the vacancies should have a single negative charge at room temperature whereas neutral vacancies should predominate during oxidation process according to the data of Fair [12]. For singly negatively charged vacancies a diffusion coefficient of $1.38 \times \times 10^{-4} \exp(-0.33\ \text{eV}/kT)$ has been published [11] resulting in 4×10^{-10} cm^{2}s^{-1} at room temperature. The time needed for the diffusion length to equal one half of a sample diameter is thus about 10^4 h which is nearly one order of magnitude higher than the time during which changes in the carrier lifetime were observed.

For the carrier lifetime decay usually observed in "clean" samples a gold concentration of about 5×10^{10} cm^{-3} is needed corresponding to a mean distance of 3 μm between individual gold atoms. Most of the decay takes place in the first hours after quenching the sample. For a period of two hours, for example a diffusion length of 30 μm is calculated for the vacancies at room temperature. This diffusion lenght is one order of magnitude higher than the mean distance of gold atoms and may be sufficient to enable reactions. The corresponding diffusion lenght at 77 K amounts to 3×10^{-11} cm which is small compared to the mean distance of gold atoms. So reactions are prevented.

Doubly negatively charged vacancies diffuse considerably faster [11]. They might be present in minor concentrations since their activation energy is above the Fermi level at room temperature [12]. The time needed for these vacancies to diffuse from the center of the sample to the surface amounts to about 30 h which also could be sufficient to enable reactions with gold. The diffusion length calculated for 77 K and 2 h is about 800 Å which again is small compared to the mean distance between gold atoms.

The diffusion coefficient of divacancies [11] finally is too small to enable reactions at room temperature. Assuming that the formation of divacancies after quenching the sample would cause the decay of the carrier lifetime is in contrast to the experimental finding that only gold doping results in a decay of lifetime whereas doping with copper or iron does not show the same effect.

A direct proof of the model by detecting vacancies after quenching the samples seems quite difficult because of the low concentrations of the vacancies compared to the detection limit of for example ESR-spectroscopy which is in the region of 10^{13} to 10^{14} cm^{-3} [13]. The low concentration of vacancies also prevents the formation of divacancies to affect significantly the carrier lifetime. The mean distance between vacancies after quenching the samples is about 1 μm compared to 3 μm for the gold atoms which is of the same order of magnitude. The carrier cross sections of gold, however, seem to be the highest known so far.

5. Conclusions

The observed decay of carrier lifetime with time elapsed after quenching silicon samples from high temperatures and after cutting samples from Czochralski-grown crystals gives us some interesting information about impurity reactions in semiconductor material which take place even at room temperature. The experimental results of the carrier lifetime measurements in dependence of time is quite useful for furnace control or characterization of materials and processes. If the proposed model is valid it may be possible to seperate the effect of gold from that of other impurities by means of highly sensitive carrier lifetime measurements. Furthermore the importance of vacancies and self-interstitials for impurity reactions in silicon is emphasized again.

Acknowledgement

This work was supported by the Bundesministerium für Forschung und Technologie of the FRG.

References

[1] A. GOETZBERGER and W. SHOCKLEY, J. appl. Phys. **31**, 1821 (1960).
[2] T. M. BUCK, K. A. PICKAR, J. M. POATE, and C. M. HSIEH, appl. Phys. Letters **21**, 485 (1972).
[3] W. KERN and D. A. PUOTINEN, RCA Rev. **31**, 187 (1970).
[4] R. J. KRIEGLER, Y. C. CHENG, and D. R. COLTON, J. Electrochem. Soc. **119**, 388 (1972).
[5] W. DIETZE, Spring Meeting Electrochem. Soc. Philadelphia, May 1977.
[6] A. C. BONARA, Semiconductor Silicon, Ed. H. R. HUFF and E. SIRTL, The Electrochemical Society, Princeton N. J. 1977 (p. 154).
[7] J. M. FAIRFIELD and B. V. GOKHALE, Solid State Electronics 8, 685 (1965).
[8] M. SCHULZ and H. LEFEVRE, see [6] (p. 142).
[9] K. GRAFF, H. PIEPER, and G. GOLDBACH, Semiconductor Silicon, Ed. H. R. HUFF and H. R. BURGESS, The Electrochemical Society, Princeton N. J. 1973 (p. 170).
[10] Annual Book of ASTM Standards, Philadelphia 1976 (F 28 to 75).
[11] A. J. R. DEKOCK, Philips Res. Rep., Suppl. **1**, 9 (1973).
[12] R. B. FAIR, see [6] (p. 968).
[13] H. NEUBRAND, private communication.
[14] K. GRAFF and H. PIEPER, phys. stat. sol. (a) **30**, 593 (1975).

(Received April 10, 1978)

phys. stat. sol. (a) 49, 145 (1978)

Subject classification: 3; 8; 22.8.1

Kristall- und Materiallabor der Fakultät für Physik der Universität Karlsruhe

Solubility Curves in High-Temperature Melts for the Growth of Single Crystals of Rare Earth Vanadates and Phosphates

By

W. EIGERMANN, G. MÜLLER-VOGT, and W. WENDL

In honour of Prof. Dr. F. STÖCKMANN's 60th birthday

With thermoanalytical methods the solubility of rare earth salts in fluxes used for the growing of single crystals are investigated and the systems YVO_4–$NaVO_3$ and $REPO_4$–$Pb_2P_2O_7$ (RE = Tb to Lu, Y) are measured. The various $REPO_4$ show different solubilities. Borax and lead oxide as additions to the fluxes increase the solubilities of the vanadates and phosphates respectively. The results allow the exact determination of starting temperatures and compositions for the flux growth of better crystals of these compounds.

Mit Hilfe thermoanalytischer Methoden werden die Löslichkeiten von RE-Salzen in Flußmitteln untersucht, die zur Kristallzüchtung dieser Verbindungen dienen. Gemessen werden die Systeme YVO_4–$NaVO_3$ und $REPO_4$–$Pb_2P_2O_7$ (RE = Tb bis Lu, Y). Die verschiedenen $REPO_4$ zeigen unterschiedliche Löslichkeiten. Borax und PbO-Zusätze zu den Flußmitteln für Vanadate und Phosphate erhöhen deren Löslichkeiten in den betreffenden Flußmitteln. Aus den Meßergebnissen lassen sich die Versuchsparameter, wie Anfangstemperaturen und Zusammensetzung, zur Züchtung besserer Kristalle dieser Verbindungen entnehmen.

1. Introduction

The vanadates, phosphates, arsenates, and chromates of the rare earths (RE) are an isostructural series of compounds with highly interesting magnetic and crystallographic properties at low temperatures [1 to 3]. Crystal growth of the vanadates is possible from the melt [4, 5] with some difficulties because of the reduction of the V^{5+}-ion at high temperatures. Another possibility is the growing from high-temperature solutions [6 to 9]. For the system Y_2O_3–V_2O_5 there exists a phase diagram [10] and for YVO_4–$NaVO_3$ there is a report on the solubility curve [9]. Because of their thermal instability, the phosphates and arsenates are only reported to grow from high-temperature solutions. For the phosphates the flux is $Pb_2P_2O_7$ either pure [8, 11] or with additions of PbO [12] and for the arsenates $Pb_2As_2O_7$ [8, 13]. No solubility curves are measured for the last three systems. The flux $Pb_2As_2O_7$ decomposes at higher temperatures so that the investigation with thermoanalytical methods in the interesting temperature region is difficult. There is no information about a successful growing of the $RECrO_4$. The use and optimization of such high-temperature solutions for crystal growth depends strongly on the knowledge of the solubility curves. They must be known exactly for studying the crystal growing mechanism. Therefore, we determined the solubility curves for the $REVO_4$ and $REPO_4$ by means of thermoanalytical techniques.

2. Experimental Methods

The solubility curves were determined with an apparatus for differential thermal analysis, which was built up with commercial components and self-made parts to an automatically working equipment allowing the investigation throughout any temperature interval from room temperature to 1700 K with 12 heating rates from 1.5 to 25 K/min.

The measured systems were examined in the region between 800 and 1650 K with concentrations of the crystallizing component from 1 to 20 mol%. The temperature accuracy was estimated from measurements of the melting points of various fluxes and the temperature spread of the starting temperatures of crystallization confirmed by repeated runs.

The platinum crucibles were filled with about 200 mg of the mixture of rare earth salt and flux; the same quantity of Al_2O_3 powder served as reference material. A more exact determination of the saturation temperature was made in the system YPO_4–$Pb_2P_2O_7$ by the method of immersing crystals [14]. At various temperatures, small YPO_4 crystals were exposed to a solution with constant concentration of YPO_4. A decrease or increase of weight of the crystals during the procedure indicated that the applied temperature lay above or below the starting temperature of crystallization (the saturation temperature).

3. Experimental Results

3.1 The system YVO_4–($NaVO_3$ + $xNa_2B_4O_7$)

The solubilities of YVO_4 in $NaVO_3$ either pure or with additions of borax are plotted in Fig. 1 and 2. In Fig. 1 the reported values of Phillips and Pressley [9] for pure $NaVO_3$ and an addition of 6 mol% borax are shown, too. These values for the saturation temperature are in both cases lower than ours for equal concentrations. Fig. 2 shows that the solubility of YVO_4 increases with increasing amount of borax.

3.2 The system $REPO_4$–($Pb_2P_2O_7$ + $xPbO$)

The $REPO_4$ from Tb to Lu and Y crystallize in the tetragonal zircon structure, whereas the others appear in the monoclinic huttonite structure [15]. Because of their physical properties we grow only crystals with the zircon structure. We measured therefore the solubility of all these $REPO_4$ in the flux $Pb_2P_2O_7$. Fig. 3 shows the results. It can be seen that there are significant differences in the solubility of the different $REPO_4$. At the same temperature the solubility of $TbPO_4$ is 1.5 times the one of $LuPO_4$. The solubility decreases approximately with increasing atomic number of the RE, but the differences in the solubilities between neighboured RE are rather small. The solubility of YPO_4 is near that of $ErPO_4$.

Wanklyn [12] reports about the influence of additions to various fluxes on the results of crystal growth experiments. Admixture of PbO to the flux $Pb_2P_2O_7$ should improve the crystal quality and dimensions. Our solubility determinations showed quantitatively that the solubility increases with an increasing content of PbO in the flux (Fig. 4).

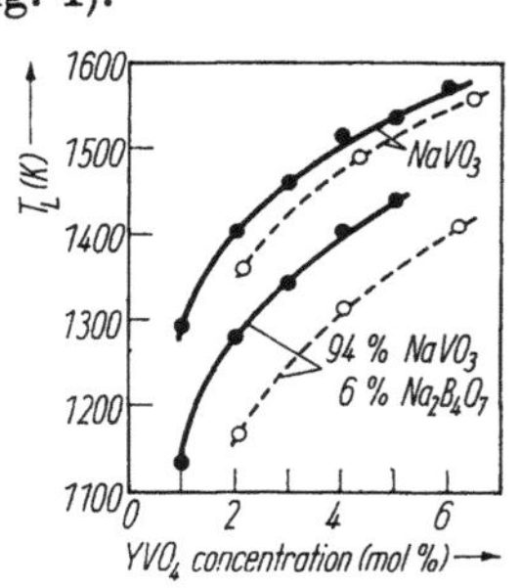

Fig. 1. Solubility curves of YVO_4 in $NaVO_3$ pure and $NaVO_3$ with addition of borax (○ and dashed line from [9])

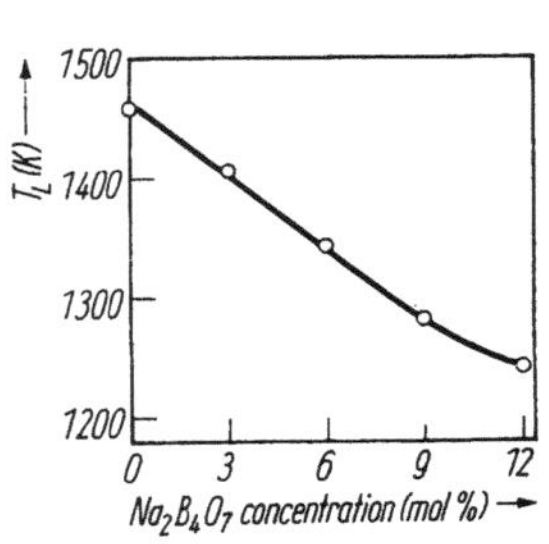

Fig. 2. Saturation temperature of 3 mol% YVO_4 as function of borax concentration in $NaVO_3$

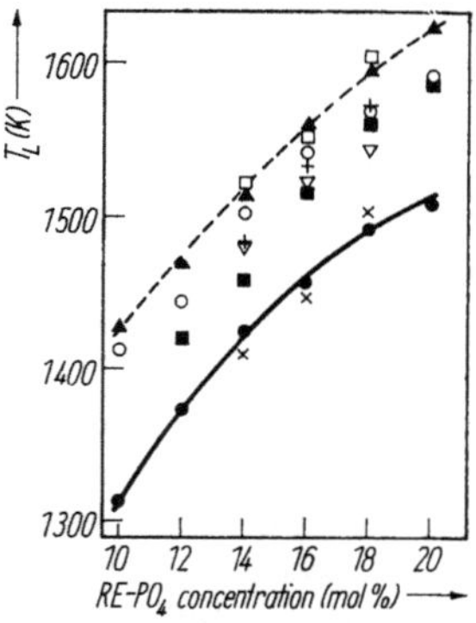

Fig. 3. Solubility curves of $REPO_4$ in $Pb_2P_2O_7$. □ $LuPO_4$, ▲ $YbPO_4$, ▽ $TmPO_4$, + $ErPO_4$, ■ $HoPO_4$, × $DyPO_4$, ● $TbPO_4$, ○ YPO_4

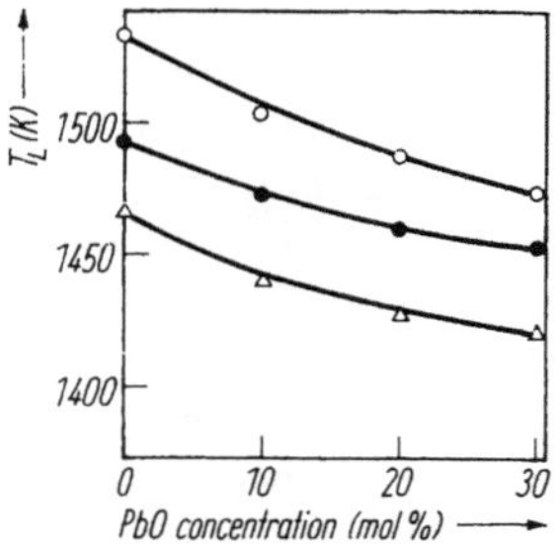

Fig. 4. Saturation tempertures of various YPO_4 concentrations as a function of additional PbO in $Pb_2P_2O_7$. ○ 15, ● 14, and △ 12 mol% YPO_4

4. Discussion

Melting point and fusion enthalpy of the pure components can be estimated from the solubility curves under certain assumptions.

For the thermodynamic ideal mixture of two components A and B forming an eutectic system, the following relation is valid:

$$\ln x_B = -\frac{\Delta H_B}{R}\left(\frac{1}{T_L} - \frac{1}{T_B}\right),$$

where T_L is the equilibrium temperature, at which solid B is in equilibrium with a melt of A and B,

x_B the corresponding mole fraction of B in the melt,

ΔH_B the molar fusion enthalpy of B, and

T_B the melting point of B.

Plotting $\ln x_B$ against $1/T_L$, we get ΔH_B from the slope of the straight line, and T_B from the section of the $1/T_L$ axes. In this way, we found for YPO_4 a melting point of 2500 K and a molar fusion enthalpy of 60 kJ/mol. The values for YVO_4 are 2300 K and 115 kJ/mol, respectively. Our value for the melting point of YVO_4 shows a fairly good agreement with that reported in literature as 2083 K [5].

a

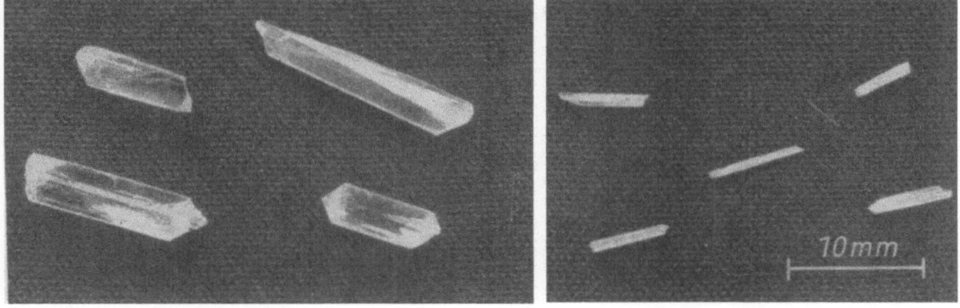

b

Fig. 5. Crystals of YPO_4. a) Results of a run with 12 mol% and b) with 7 mol% YPO_4 in $Pb_2P_2O_7$ with starting temperature 1500 K

We used the knowledge of the measured solubility curves to grow crystals, here for example YPO_4. The recommended starting composition and temperature in the literature was 7 mol% YPO_4 and 1500 K, respectively. Runs with 12 mol%, the above-mentioned starting temperature, and the same crystallization region yielded crystals with greater dimensions and better quality. To the left, Fig. 5 shows crystals grown with 12 mol% YPO_4 and to the right the same with 7 mol%.

5. Conclusions

The knowledge of the solubility curves, measured for a series of various RE compounds improved the results of our crystal growth from high-temperature solutions. The determination of the saturation temperatures in high-temperature solutions is a first step to find out quantitative experimental data for our systems. These data are necessary in a crystal growing field partly with a great amount of successful empiric work.

Acknowledgement

We are indebted to the Deutsche Forschungsgemeinschaft for financial support.

References

[1] G. A. GEHRING and K. A. GEHRING, Rep. Progr. Phys. **38**, 1 (1975).
[2] M. SCHWAB and H. G. KAHLE, phys. stat. sol. (b) **84**, 167 (1977).
[3] H. KALBFLEISCH, H. KUMBERG, G. MÜLLER-VOGT, and M. STENGEL, phys. stat. sol. (b) **52**, 499 (1972).
[4] J. J. RUBIN and L. G. VAN UITERT, J. appl. Phys. **37**, 2920 (1966).
[5] R. C. ROPP, Mater. Res. Bull. **10**, 271 (1975).
[6] L. H. BRIXNER and E. ABRAMSON, J. Electrochem. Soc. **112**, 70 (1965).
[7] R. S. FEIGELSON, J. Amer. Ceram. Soc. **51**, 538 (1968).
[8] W. HINTZMANN and G. MÜLLER-VOGT, J. Crystal Growth **5**, 274 (1969).
[9] W. PHILLIPS and R. J. PRESSLEY, J. Amer. Ceram. Soc. **46**, 366 (1967).
[10] E. M. LEVIN, J. Amer. Ceram. Soc. **50**, 381 (1967).
[11] R. S. FEIGELSON, J. Amer. Ceram. Soc. **47**, 257 (1964).
[12] B. M. WANKLYN, J. Crystal Growth **37**, 334 (1977).
[13] R. S. FEIGELSON, J. Amer. Ceram. Soc. **50**, 437 (1967).
[14] V. A. TIMOFEEVA and J. KVAPIL, Soviet Phys. — Cryst. **11**, 263 (1966).
[15] H. SCHWARZ, Z. anorg. allg. Chem. **323**, 44 (1963).

(Received June 9, 1978)

phys. stat. sol. (a) **49**, 149 (1978)

Subject classification: 10.1 and 14.1; 1.3; 21.1

Gesamthochschule Kassel

Zum Einfluß von Nickel-Clustern auf den spezifischen elektrischen Widerstand von paramagnetischen Kupfer–Nickel-Legierungen

Von

H. GÄRTNER und W. AUER[1])

Herrn Prof. Dr. F. STÖCKMANN zum 60. Geburtstag gewidmet

Es wird der spezifische elektrische Widerstand von drei $Cu_{1-x}Ni_x$-Legierungen mit $x = 0{,}36$; 0,40; 0,42 im Temperaturbereich von 1,5 bis 250 K gemessen. Von jeder Legierung werden zwei Proben nach unterschiedlicher Wärmebehandlung untersucht. Langsames Abkühlen führt zur verstärkten Bildung von Nickel-Clustern, die magnetische Nickelatome enthalten. Das Tieftemperaturverhalten der Proben kann mit einem phänomenologischen Modell von Iha und Jericho erklärt werden.

The resistivity of three $Cu_{1-x}Ni_x$-alloys with $x = 0.36$, 0.40, 0.42 is measured from 1.5 to 250 K. From each alloy two samples are investigated after different heat treatment. Slow cooling causes an enhanced formation of nickel clusters containing magnetic nickel atoms. The low temperature behaviour of the samples can be explained with a phenomenological modell by Iha and Jericho.

1. Einleitung

Messungen der magnetischen Suszeptibilität [1] und der diffusen Neutronenstreuung [2] an Kupfer–Nickel-Legierungen haben ergeben, daß es im paramagnetischen Konzentrationsbereich, also bei Nickelgehalten unter 50%, zur Bildung von Nickel-Agglomeraten kommt. Diese Cluster haben ihre Ursache in der Tendenz der Nickelatome, sich mit gleichartigen Nachbarn zu umgeben. Nach Robbins et al. [3] besitzt ein Nickelatom in Kupfer–Nickel-Legierungen ein magnetisches Moment, wenn es zwischen acht und zwölf Nickelatome als nächste Nachbarn hat.

Neben den genannten Messungen deuten auch Anomalien in der Temperaturabhängigkeit des spezifischen elektrischen Widerstandes auf das Vorhandensein von Nickel-Clustern hin. Houghton et al. [4, 5] fanden im Bereich 30 bis 50% Nickel ausgeprägte Minima in den Widerstandskurven, die sie auf die Streuung von Leitungselektronen an Nickel-Clustern zurückführten. In einer Untersuchung des Kondo-Effektes an Eisen in Kupfer–Nickel-Legierungen [6] zeigten die eisenfreien Proben mit 12 und 23% Nickel schwache Minima, die selbst bei diesen geringen Konzentrationen auf Abweichungen von der regellosen Verteilung der Nickelatome hinwiesen.

In der vorliegenden Arbeit berichten wir über den Temperaturverlauf des spezifischen elektrischen Widerstandes von drei $Cu_{1-x}Ni_x$-Legierungen mit $x = 0{,}36$, 0,40 und 0,42 zwischen 1,5 und 250 K. Ziel der Untersuchung war es, den Einfluß der Wärmebehandlung der Proben auf die Clusterbildung und insbesondere das Verhalten der Proben bei tiefen Temperaturen zu studieren.

2. Präparation und Messung

Die Proben wurden in einem Argon-Bogen-Ofen nach Geach und Jones [7] aus Kupfer (5N) und Nickel (4N8) der Firma Metals Research erschmolzen. Um eine mög-

[1]) Jetzt: Standard Elektrik Lorenz Forschungszentrum Stuttgart, BRD.

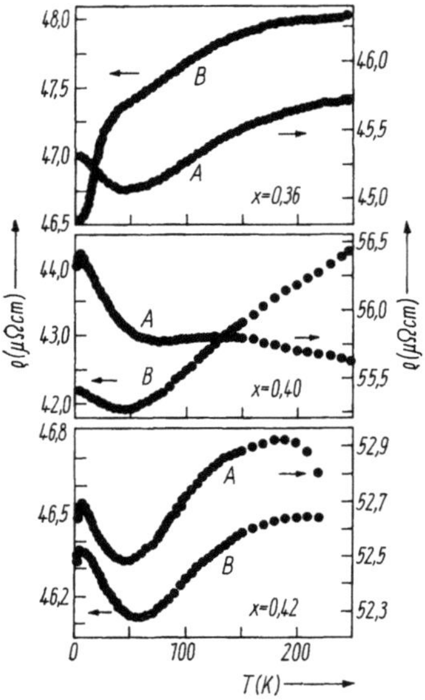

Fig. 1. Spezifischer elektrischer Widerstand der $Cu_{1-x}Ni_x$-Legierungen ($x = 0{,}36$, 0,40, 0,42) in Abhängigkeit von der Temperatur. Nach dreitägiger Temperung bei 1000 °C wurden die Proben A in Eiswasser abgeschreckt, die Proben B langsam im Ofen abgekühlt

lichst gute Durchmischung der Partner zu erreichen, wurden die Barren mehrere Male zerschnitten und erneut zusammengeschmolzen. Aus jeder Legierung wurden zwei Proben mit quadratischem Querschnitt von 2×2 mm² und einer Länge von 20 mm herausgeschnitten. Anschließend wurden die Proben in Quarzampullen unter Vakuum bei 1000 °C drei Tage getempert. Von jeder Legierung wurde eine Probe in Eiswasser abgeschreckt (A), die jeweils andere langsam im Ofen abgekühlt (B).

Die Messungen des elektrischen Widerstandes wurden am Institut für Festkörperforschung der Kernforschungsanlage Jülich durchgeführt. Die Widerstände der Proben lagen bei einigen Milliohm. Sie konnten nach der Vier-Punkte-Methode mit Hilfe einer Konstantstromquelle (North Hills Tc 602 CR) und eines hochauflösenden Digitalvoltmeters (Nanomat Typ 920) mit Nanovoltverstärker (Amplispot Typ 9201) mit großer Genauigkeit gemessen werden. Die Ergebnisse der Widerstandsmessungen an den drei $Cu_{1-x}Ni_x$-Legierungen sind in Fig. 1 für die verschiedenen Konzentrationen und Wärmebehandlungen zusammengestellt. Ein Vergleich der Kurven zeigt den drastischen Einfluß der unterschiedlichen Wärmebehandlungen insbesondere bei den Proben $x = 0{,}36$ und 0,40. Alle Proben, außer der Probe $x = 0{,}36$ B, haben ein Tieftemperaturminimum und weisen, bis auf die Probe $x = 0{,}40$ B, bei höheren Temperaturen ein Maximum auf oder zeigen die Tendenz in ein Maximum einzulaufen. Außerdem besitzen die Proben $x = 0{,}42$ A und B und $x = 0{,}40$ B ein weiteres Maximum bei etwa 10 K, das im folgenden eingehender diskutiert werden soll.

3. Auswertung und Diskussion

Die auch von anderen Autoren [4, 5] beobachteten Maxima bei hohen Temperaturen können zumindest qualitativ im Rahmen des von Levin und Mills [8] entwickelten Modells erklärt werden. Mit diesem Modell wird für Nickel-Cluster, die jeweils 50 Atome enthalten sollen, unter stark vereinfachenden Annahmen der elektrische Widerstand berechnet.

Im Temperaturbereich zwischen 20 und 40 K wird der Verlauf des Widerstandes durch eine lineare Abhängigkeit von $\ln T$ gekennzeichnet (Fig. 2). Es herrschen also Streuprozesse vor, wie sie von der Kondo-Theorie beschrieben werden.

Die Widerstandskurven der Proben $x = 0{,}42$ A und B und $x = 0{,}40$ A haben bei $T = 10$ K ein Maximum, an das sich zu tieferen Temperaturen hin ein steiler Widerstandsabfall anschließt. Für Legierungen des Kupfer–Nickel-Systems im Konzentrationsbereich dicht unterhalb des Überganges zum Ferromagnetismus gibt es noch keine theoretische Behandlung des Tieftemperaturverhaltens. Zur Interpretation der hier vorgelegten Ergebnisse wird deshalb auf ein phänomenologisches Modell zurück-

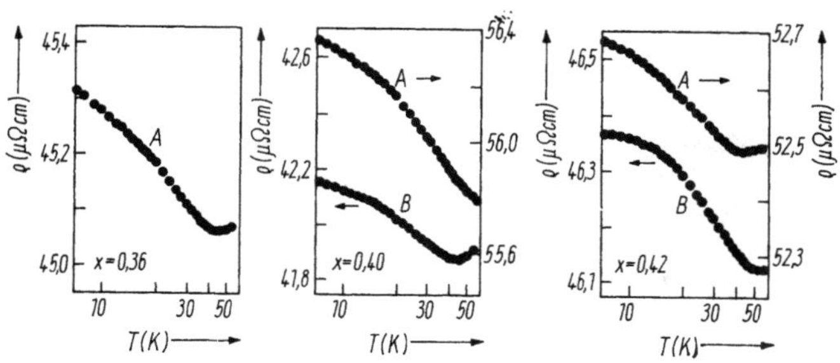

Fig. 2. Spezifischer elektrischer Widerstand der $Cu_{1-x}Ni_x$-Legierungen ($x = 0{,}36,\ 0{,}40,\ 0{,}42$) im Bereich der Kondo-Streuung

gegriffen, das Iha und Jericho [9] im Anschluß an eine Arbeit von Klein und Brout [10] zur Deutung von Widerstandsmessungen an Ag(Mn)-Legierungen verwendet haben, die ähnliche Widerstandscharakteristika wie die hier untersuchten Kupfer-Nickel-Legierungen hatten. Ein Hinweis auf die Anwendung des Modells auf Kupfer-Nickel-Legierungen findet sich schon in der gennanten Arbeit von Klein und Brout [10].

Im Einklang mit den Ergebnissen der diffusen Neutronenstreuung [2] nehmen wir an, daß in den Clustern unserer Proben nur ein kleiner Teil der Nickelatome magnetisch wirksam ist. Obwohl also nominell ein hoher Nickelgehalt vorliegt, behandeln wir unsere Proben im Hinblick auf die Elektronenstreuung an magnetischen Momenten als verdünnte Systeme. Wie Iha und Jericho im Falle des Ag(Mn) gehen wir von einer Wahrscheinlichkeitsverteilung der Energie der magnetischen Nickelatome aus, die, wegen der Annahme eines Ising-Modells in dem Ansatz von Klein und Brout [10], durch eine Lorentz-Kurve

$$W(g\mu_B H) = \frac{\Delta}{\pi\left(\Delta^2 + (g\mu_B H)^2\right)} \tag{1}$$

angegeben werden kann, wobei g der Landésche Faktor, μ_B das Bohrsche Magneton, H das effektive Feld am Ort des magnetischen Nickelatoms und Δ die Halbwertsbreite der Verteilung ist. Nur Leitungselektronen mit einer thermischen Energie $k_B T$, die größer als die Energie des magnetischen Nickelatoms im effektiven Feld H ist, können einen Nickelspin umklappen und so zum Kondo-Effekt beitragen. In einem Bereich der Breite $2k_B T$ um den Nullpunkt der Energieverteilung (1) wird ein Verhalten des Widerstandes nach Kondo angenommen, während der Widerstand außerhalb dieses Bereiches konstant sei. Unter Berücksichtigung, daß die Breite der Verteilung (1) der Konzentration c der magnetischen Momente proportional ist und die nichtmagnetischen Atome einen Beitrag ϱ_1 zum Widerstand liefern, folgt für die

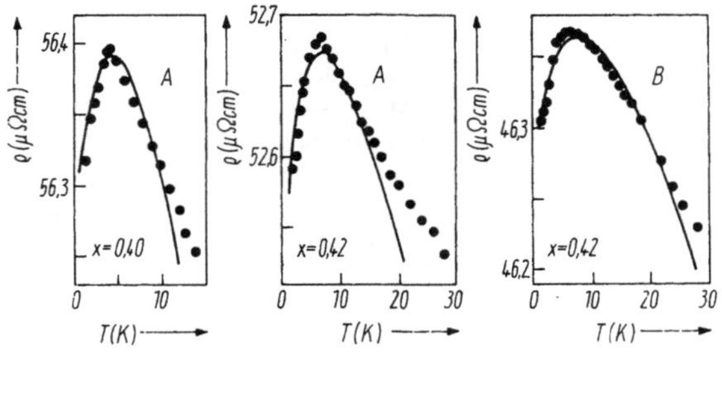

Fig. 3. Spezifischer elektrischer Widerstand der $Cu_{1-x}Ni_x$-Legierungen ($x = 0{,}40,\ 0{,}42$) in der Umgebung des Tieftemperaturmaximums. Die ausgezogenen Kurven entsprechen der mit den Parametern der Tabelle 1 angepaßten Gleichung (2)

Temperaturabhängigkeit des Gesamtwiderstandes (Gleichung (6) in [9]):

$$\varrho(T) = \varrho_1 \left[1 - \frac{2}{\pi} a_1 (\ln T - a_3) \operatorname{arctg}\left(\frac{a_2 T}{c}\right)\right]. \tag{2}$$

Mit den aus Fig. 1 durch lineare Extrapolation nach $T = 0$ K gewonnenen Werten für ϱ_1 können aus den Steigungen der Kurven in Fig. 2 im Bereich der Kondo-Streuung die Werte a_1 ermittelt werden. Neben diesen Größen sind in Tabelle 1 die Werte für die Parameter a_2/c und a_3 aufgeführt, mit denen es gelang, die durch (2) gegebene Kurve unterhalb der Temperatur T_m recht gut an die experimentellen Werte anzu-

Tabelle 1

Parameter zur Anpassung von Gleichung (2) an die Meßwerte

x	ϱ_1 (μΩ cm)	a_1	a_2/c	a_3	T_m (K)
0,40 A	56,17	0,0235	0,055	2,6	9
0,42 A	52,486	0,0096	0,075	3,15	15
0,42 B	46,25	0,0147	0,031	3,15	22

ϱ_1 und a_1 wurden aus den Widerstandskurven (Fig. 1 und 2) entnommen. a_2/c und a_3 sind die Parameter für optimale Anpassung unterhalb der Temperatur T_m.

passen, wie Fig. 3 zeigt. Wählt man für den Parameter a_1, der eigentlich konzentrationsunabhängig sein sollte, den arithmetischen Mittelwert 0,0159 der drei Proben in Tabelle 1, so lassen sich mit $a_3 = 3{,}15$ für die Proben $x = 0{,}42$ A und B die Parameter $a_2/c_{x=0,42\,\mathrm{B}} = 0{,}03$ und $a_2/c_{x=0,42\,\mathrm{A}} = 0{,}046$ bei optimaler Anpassung der Kurve (2) bestimmen. Das Verhältnis dieser beiden Konstanten liefert einen Hinweis auf die Konzentration der magnetischen Momente in den Proben $x = 0{,}42$ A und B, die unterschiedliche Wärmebehandlungen erfahren haben. Es ergibt sich $c_{x=0,42\,\mathrm{B}} = 1{,}53 c_{x=0,42\,\mathrm{A}}$. Das langsame Abkühlen der Probe B führt offenbar zur verstärkten Bildung von Clustern, die magnetisch wirksame Nickelatome enthalten, während in der abgeschreckten Probe A die durch den Temperprozeß erzielte gleichmäßigere Verteilung der Nickelatome vorliegt. Diese Vorstellung wird bestätigt durch den Verlauf der Widerstandskurven der Proben $x = 0{,}42$ A und B in Fig. 1. Nimmt man als Maß für die Konzentration der magnetischen Momente die Tiefe der beobachteten Minima $[\varrho(0\,\mathrm{K}) - \varrho(T_{min})]/\varrho(0\,\mathrm{K})$, so zeigt die Probe B mit 6,51% einen deutlich höheren Wert als die Probe A mit 4,92%. Das Verhältnis dieser beiden Konzentrationen ist 1,32, was zumindest größenordnungsmäßig mit dem oben ermittelten Wert 1,53 übereinstimmt.

Literatur

[1] J. S. Kouvel und J. B. Comly, Phys. Rev. Letters **24**, 598 (1970).
[2] T. J. Hicks, B. Rainford, J. S. Kouvel und G. G. Low, Phys. Rev. Letters **22**, 531 (1969).
[3] C. G. Robbins, H. Claus und P. A. Beck, Phys. Rev. Letters **22**, 1307 (1969).
[4] R. W. Houghton, M. P. Sarachik und J. S. Kouvel, Solid State Commun. **8**, 943 (1970).
[5] R. W. Houghton, M. P. Sarachik und J. S. Kouvel, Phys. Rev. Letters **25**, 238 (1970).
[6] H. Gärtner, D. R. Zrudsky und S. Legvold, Solid State Commun. **8**, 913 (1970).
[7] G. A. Geach und F. O. Jones, J. less-common Metals **1**, 56 (1959).
[8] K. Levin und D. L. Mills, Phys. Rev. B **9**, 2354 (1974).
[9] D. Iha und M. H Jericho, Phys. Rev. B **3**, 147 (1971).
[10] M. W. Klein und R. Brout, Phys. Rev. **132**, 2412 (1961).

(Received June 9, 1978)

phys. stat. sol. (a) **49**, 153 (1978)

Subject classification: 13.4 and 14.3.4; 14.3.3; 22.1.2

Institut für Angewandte Physik der Universität Münster

Determination of the Temperature Dependence of the Capture Cross-Sections of the Gold Acceptor Level and of the Temperature of Current Filaments in Silicon pin Diodes

By

I. Dudeck and R. Kassing

In honour of Prof. Dr. F. Stöckmann's 60th birthday

The temperature dependence of the capture cross section of negatively charged gold centers in n-type silicon for holes σ_p^- is determined in the temperature range of 55 K $\leqq T \leqq$ 340 K from the temperature dependent breakdown voltage U_{BD} of gold compensated pin diodes showing a negative differential resistance region. The temperature dependence of the capture cross section of neutral gold centers in the range of 193 K $\leqq T \leqq$ 320 K, which was obtained in an earlier paper, is confirmed by means of small signal pulse measurements in the high injection regime of the same double injection diodes. The over-temperature caused by Joule heating in current filaments occurring in the high injection region of the stationary S-shaped I–U characteristic is measured for the first time using a large signal pulse method.

Es wird die Temperaturabhängigkeit des Einfangquerschnitts von negativ geladenen Goldzentren in n-leitendem Si für Löcher, σ_p^-, im Temperaturbereich 55 K $\leqq T \leqq$ 340 K aus der temperaturabhängigen Durchbruchsspannung U_{BD} von goldkompensierten pin-Dioden, die einen Bereich negativen differentiellen Widerstandes aufweisen, bestimmt. Die Temperaturabhängigkeit des Einfangquerschnitts von neutralen Goldzentren im Bereich von 193 K $\leqq T \leqq$ 320 K, die in einer früheren Arbeit erhalten wurde, wird mittels einer Kleinsignal-Impulsmessung im Hochinjektionsregime von Doppelinjektionsdioden bestätigt. Die durch Joulsche Wärme verursachte Übertemperatur in Stromfäden, die im Hochinjektionsbereich der stationären S-förmigen I–U-Charakteristiken auftreten, wird zum erstenmal mit einer Großsignal-Impulsmethode gemessen.

1. Introduction

Despite the fact that gold is one of the most investigated doping materials, up to now the values of the capture cross-sections, especially that of the gold acceptor level for electrons σ_n, are very controversial (see, e.g., Table 1 in [1]). Therefore, the capture cross-sections of the gold acceptor level have been and still are the subject of many investigations [2 to 14]. In February 1978 Brotherton and Bicknell [14] reported on measurements yielding $\sigma_n = (9 \pm 2) \times 10^{-17}$ cm², temperature independent in the range of 77 to 280 K. This value agrees with those given by Fairfield and Gokhale [7] and Barbolla et al. [13]. In 1975 we reported, however, on a pulse and small signal capacitance method [11] for a direct determination of the electron capture time constant τ_{c_n} in Au-compensated n-type silicon yielding a temperature independent value of $\sigma_n = 2.6 \times 10^{-15}$ cm² in the range of 172 K $\leqq T \leqq$ 316 K. This value was 1977 confirmed for room temperature by a numerical evaluation of the dc I–U characteristic of silicon pin diodes [12]. Our value is about ten times larger than that given in [7, 13].

Further, we obtained, in agreement with results of Engström and Grimmeiss [15], the gold acceptor level $(E_c - E_{Au})$ (T) to be pinned to the conduction band contrary to the result of [14] where the acceptor level is found to be pinned to the valence band and, therefore, the unusual room temperature value of $(E_c - E_{Au}) = 0.492$ eV is given.

The low-injection region of the S-shaped I–U characteristic of these pin diodes, especially the breakdown voltage U_{BD}, i.e. the onset of the negative differential resistance (n.d.r.) region, also yields information about the capture cross-section of negatively charged gold centers for holes σ_p^- [12]. Therefore, from the temperature dependence of the dc I–U characteristic the temperature dependence of σ_p^- can be determined. This is the subject of Section 2. For temperatures above 240 K a numerical evaluation of the breakdown voltage U_{BD} is necessary to achieve correct results, whereas for lower temperatures ($T \leqq 240$ K) the values of σ_p^- calculated with a formula of Ashley and Milnes [16],

$$U_{BD} = \frac{1}{2} L^2 n_{R_0} \sqrt{\frac{q p_{R_0} v_{th}}{\pi \varepsilon \mu_p N_R}} \sqrt{\sigma_p^-}\,, \tag{1}$$

are in excellent agreement with the numerical results. (L is the sample length, n_{R_0}, p_{R_0} the density of gold centers occupied and unoccupied by electrons, respectively; v_{th} the thermal velocity of holes, μ_p the hole mobility, and N_R the total recombination center density.) This yields a very effective and easy method to determine $\sigma_p^-(T)$.

Section 3 is concerned with the determination of $\sigma_n(T)$ from the high-injection lifetime by means of small-signal pulse measurements within the high-injection square-law regime ("semiconductor regime"). Baron et al. [17] were the first to deal with this problem. However, until now this method has never been applied to compensated pin diodes and used for the determination of c_n, presumably because proper samples showing the high-injection square-law regime in the dc I–U characteristics were lacking. In [12] we reported for the first time on ideal non-filamentary pin diodes showing the high-injection square-law regime. Therefore, with these diodes it is possible, for the first time, to determine σ_n and its temperature dependence from the high-injection lifetime by means of small-signal pulse measurements.

A problem arises from the Joule heating of the samples when biased in the high-injection regime which can falsify the determination of the actual sample temperature. Therefore, it is necessary to use a pulse technique for biasing the sample in the high-injection regime. This is possible because the time constant for reaching the steady high-injection state is in the order of the high-injection lifetime, i.e. in the range of μs, while the thermal time constant has been measured to be in the range of 1 s. This is described in Section 4 where also the over-temperature within the sample as a function of the applied bias is determined. These measurements allow conclusions on the temperature within current filaments occurring in large-contact-area pin diodes.

2. Temperature Dependence of σ_p^-

The samples used for the measurements were performed of high-resistivity n-type silicon with a shallow-donor concentration of $N_D = 2.6 \times 10^{13}$ cm^{-3} compensated with gold of a total density $N_R = 4 \times 10^{13}$ cm^{-3}. In [12] we have shown that the low-injection regime of the I–U characteristic of pin diodes with these sample data is mainly affected by the capture cross-section of negatively charged gold centers for holes σ_p^- and that the high-injection regime is mainly influenced by the capture cross-section of neutral gold centers for electrons σ_n. Therefore, it is principally possible to determine σ_p^- from the breakdown voltage U_{BD} and σ_n from the minimum voltage U_{min}. In [12] we have also shown that pin diodes with contact diameters smaller than about 500 μm and a sample length of about 200 μm show dc I–U characteristics that can be compared in all respects to theoretical I–U characteristics and are, therefore, ideally suited for a quantitative evaluation by fitting the experimental curves. This was shown in Fig. 7 of [12] for room temperature, the fitting parameter were $c_n = 3.4 \times 10^{-8}\ \text{cm}^3\ \text{s}^{-1}$, $c_p^- = 1 \times 10^{-7}\ \text{cm}^3\ \text{s}^{-1}$. Fig. 1 shows a set of theoretical I–U characteristics for different temperatures obtained by numerically solving the equations that define the problem of double injection in semiconductors with one deep level (with the data of our samples the influence of the gold donor level is negligible):

$$j = q(\mu_n n + \mu_p p)\, E\,, \tag{2}$$

$$j' = (j_n + j_p)' = 0\,, \tag{3}$$

$$E' = \frac{q}{\varepsilon}(p - p_0 - n + n_0 - n_R + n_{R_0})\,, \tag{4}$$

$$n_R = N_R \frac{c_n n + c_p p_1}{c_n(n + n_1) + c_p(p + p_1)}, \tag{5}$$

$$\left.\begin{aligned} \frac{1}{q} j_n' &= R = c_n[n(N_R - n_R) - n_1 n_R]\,, \\ &= -\frac{1}{q} j_p' = c_p[p n_R - p_1(N_R - n_R)]\,, \end{aligned}\right\} \tag{6}$$

where n, p denote the free electron, hole concentration, n_0, p_0 the corresponding values in thermal equilibrium, μ_n, μ_p the electron, hole mobility, n_R the density of negatively charged gold centers; R the recombination rate density, and $n_1 = N_c \exp\{-(E_c - E_R)/kT\}$, $p_1 = N_v \exp\{-(E_R - E_v)/kT\}$. The beginning and the end of the i-region length is defined by the boundary condition that the field currents are larger than or equal to the diffusion currents $kT\mu_n n'$, $kT\mu_p p'$ that would arise from the gradients of free-i carrier concentrations n', p' rather than by $E(0) = E(L) = 0$, as used by nearly all other authors. Since in our samples the length of the diffusion dominated regions is negligibly small compared with the sample length L, these regions are neglected in the numerical calculations.

The curves in Fig. 1 show that the I–U characteristics are strongly temperature dependent. Therefore, a new method arises for the determination of the temperature dependence of σ_n, σ_p^- by fitting the experimentally obtained minimum and breakdown voltages U_{min} and U_{BD}, respectively. It is worthwile to note, that the capture cross-sections σ_n, σ_p as dynamical parameters can in principle both be determined from the dc I–U characteristic.

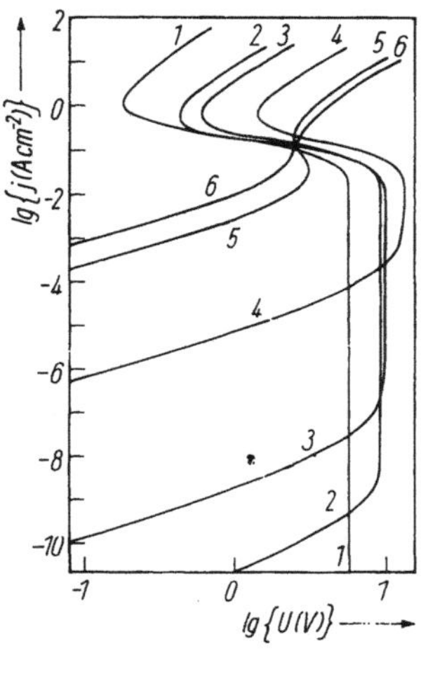

Fig. 1. Calculated I–U characteristics of Au-compensated pin diodes for different temperatures. (1) $T = 100$ K, $\sigma_p^- = 1.3 \times 10^{-13}\ \text{cm}^2$; (2) $T = 160$ K, $\sigma_p^- = 6.5 \times 10^{-14}\ \text{cm}^2$; (3) $T = 180$ K, $\sigma_p^- = 4.6 \times 10^{-14}\ \text{cm}^2$; (4) $T = 240$ K, $\sigma_p^- = 2.9 \times 10^{-14}\ \text{cm}^2$; (5) $T = 314$ K, $\sigma_p^- = 6 \times 10^{-15}$; (6) $T = 340$ K, $\sigma_p^- = 4.3 \times 10^{-15}\ \text{cm}^2$. Other parameters are $L = 230\ \mu\text{m}$, $N_D = 2.9 \times 10^{13}\ \text{cm}^{-3}$, $N_R = 3.5 \times 10^{13}\ \text{cm}^{-3}$, $\sigma_n = 2.6 \times 10^{-15}\ \text{cm}^2$

The numerical calculations yield the voltage drops U_i across the i-region. The measured voltage drop, however, is given by

$$U = U_i + U_j ,$$

where $U_j = U_{n^+i} + U_{ip^+} = (kT/q) \ln (n_K p_A / n_i^2)$ is the sum of the voltage drops across the junctions, n_K, p_A are the non-equilibrium free carrier densities at the n^+ and p^+-junctions, respectively, as calculated by the computer for the boundaries of the i-region.

Fig. 1 shows that U_{min} strongly decreases with temperature to values of some hundred mV, while U_j increases with decreasing temperature approaching the value of $(1/q)\,(E_c - E_v)$, see Table 1. This means that the determination of $\sigma_n(T)$ from the minimum voltage $U_{min}(T)$ is not very sensible because the correction U_j is large compared with the actual quantity to be measured, U_i, for temperatures below $T \approx$ ≈ 260 K. Therefore, $\sigma_n(T)$ could not be determined this way in the low-temperature range. However, as far as the breakdown voltage is concerned, these difficulties do not arise, because $U_{BD}(T)$ is always by factors larger than the "correction" U_j as shown in Table 1. The optimum range where the correction is negligible considering

Table 1

T (K)	340	314	294	240	180	100	55
U_{BD} (V)	2.5	3.1	5.8	13.5	10	5.8	46.
U_j (V)	0.2	0.2	0.3	0.3	0.5	0.8	1.1

an error of about 30% for the determination of σ_p^- arising from uncertainties in N_R, N_D and the spread of identically fabricated devices, lies between 250 and 300 K.

Fig. 2 shows the measured breakdown voltages $U_{BD}(T)$ (mean values of four typical samples) together with the corrected values $U_i = U_{BD} - U_j$ which were numerically fitted by variation of $c_p^- = \sigma_p^- v_{th}$ in (3) and (4). For the mobilities $\mu_n \sim T^{-2.5}$, $\mu_p \sim T^{-2.7}$ was used. This fit yields $\sigma_p^-(T)$ in the range of $55\text{ K} \leq T \leq 340$ K as shown in Fig. 3. In the range $160\text{ K} \leq T \leq 340$ K we found $\sigma_p^- \sim T^{-4}$, in agreement with measurements of Bemski [6] in the range of $200\text{ K} \leq T \leq 500$ K. Below 160 K no power law can be given. It is important to note that for the calculation of $\sigma_p^-(T)$ the hole mobility $\mu_p(T)$ was assumed to be $\mu_p \sim T^{-2.7}$ in the whole temperature range. However, to our knowledge, this has not been established, yet.

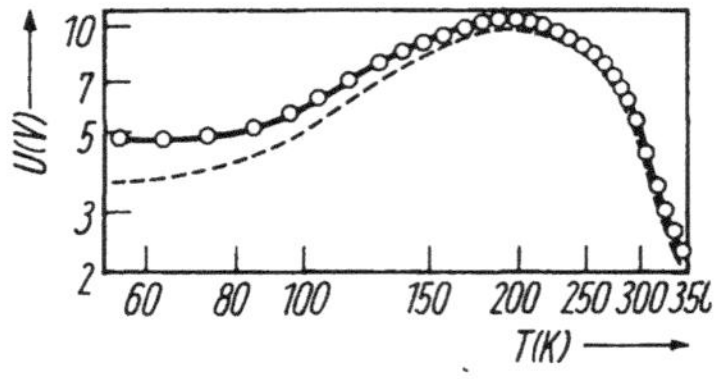

Fig. 2. Temperature dependence of breakdown voltage U_{BD} ($L = 200$ μm). Solid line: values corrected for junction voltage drops $U_i = U_{BD} - U_j$

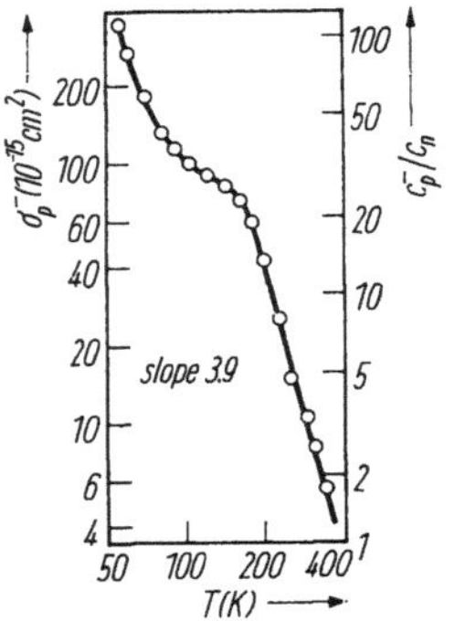

Fig. 3. Temperature dependence of capture cross-section for holes σ_p^- of the negatively charged gold centers

3. Determination of $\sigma_n(T)$

The square-law high-injection regime in the I–U characteristic of pin diodes as numerically calculated in Fig. 1 with the sample data used here, can be excellently approximated by an analytical expression derived for high-injection conditions:

$$n \approx p \gg n_0, p_0, \frac{c_p^-}{c_n} p_1, \frac{c_n}{c_p^-} n_1 . \tag{7}$$

The assumption of qasineutrality yields

$$n - p \approx -p_0 + n_0 + n_{R_0} - N_R \frac{1}{1 + c_p^-/c_n} = Q_F \tag{8}$$

and the "quasicontinuity" of the electrons and hole current (i. e. the recombination current $q \int_0^L R(x)\, dx$ small compared with both electron and hole current)

$$0 \approx (nE) . \tag{9}$$

The recombination rate density is given by

$$R = c_n n N_R \frac{c_p^-/c_n}{1 + c_p^-/c_n} = j_n' = -j_p' . \tag{10}$$

The derivative of the identity

$$\frac{j_n}{j_p} = \frac{\mu_n}{\mu_p} \frac{n}{p} \tag{11}$$

yields together with (7) and (8)

$$n' \approx p' = \frac{j_n'}{\mu_n E} \frac{p + n\mu_n/\mu_p}{p - n} . \tag{12}$$

With (9) and (10) we get a proportionality of E' and n:

$$n = E' \left(\frac{N_R c_n}{\mu_n} \frac{1}{1 + c_n/c_p^-} \frac{\mu_n/\mu_p + 1}{Q_F} \right)^{-1} \tag{13}$$

yielding with (2) the square-law regime

$$j = \frac{9}{8} q \mu_n \mu_p Q_F \tau \frac{U^2}{L^3} ; \tag{14}$$

$$\tau = \left(\frac{N_R c_n}{c_n/c_p^- + 1} \right)^{-1} . \tag{15}$$

For $c_p^- \gg c_n$ and negligible concentration of recombination centers equation (14) is the well-known formula for the semiconductor regime as first derived by Lampert and Rose in 1961 [18]:

$$j = \frac{9}{8} q \mu_n \mu_p (n_0 - p_0)\, \tau_H \frac{U^2}{L^3} ; \tag{16}$$

$$\tau_H = \frac{1}{c_n N_R} . \tag{17}$$

In the case of our sample data n_0, p_0 are negligible and with $c_p^- \gg c_n$ (which holds at lower temperatures) we get

$$j = \frac{9}{8} q \mu_n \mu_p N_D \tau \frac{U^2}{L^3} ; \tag{18}$$

$$\tau = (c_n N_R)^{-1} . \tag{19}$$

This shows that a measurement of the lifetime τ appearing in (19) allows the determination of c_n. If c_p^- is known, this is also possible for higher temperatures with the lifetime τ given in (15). Baron et al. [17] have analytically treated this problem of determining τ by means of small-signal pulse measurements. Therefore, we will restrict ourselves to shortly repeating their physical arguments:

If a small voltage step is applied to a sample biased in the semiconductor regime, free charge will move around so as to reduce the displacement current to zero within about the dielectric relaxation time. Since $n, p \gg (\varepsilon/q) E'$ holds, this small change in the electric field can practically be achieved without altering n and p. Therefore, the current response for time $t = 0^+$ is "ohmic"

$$\Delta j = j \frac{\Delta U}{U}. \tag{20}$$

Now the Maxwell relations are satisfied and the electric field does not change anymore, but recombination statistics are not, which may be seen from (13): with higher electric field gradients, n and p have to increase in order to establish a stationary state. Therefore, after the first fast capacitive rise the small signal current increases with a time constant τ, due to the generation of free carriers Δn, Δp by the recombination centers if a positive voltage step is applied, while it decreases with τ, due to recombination of excessive free carriers if a negative voltage step is applied. Since the electric field remains constant for $t > 0^+$, the change in the free-carrier density is directly reflected by the current change. Fig. 4 shows the agreement between theoretical and experimental results for our Au-compensated diodes biased in the high-injection square-law regime. Therefore, these samples are suited for the determination of $c_n(T)$ by measuring the mentioned time constant τ, equation (15). The capture cross-section σ_n can then be calculated:

$$\sigma_n = \frac{1}{v_{th}(\tau N_R - 1/c_p^-)}.$$

As long as $c_p^- \gg c_n$ is valid, this method is independent of the knowledge of c_p^-.

In Fig. 5 the experimental results are shown in comparison with the values of σ_n obtained from large-signal pulse and small-signal admittance measurements at n–i–n SCLC diodes performed of the same material [11]. In [11] we concluded that σ_n is temperature independent within the range of $173\ \text{K} \leqq T \leqq 316\ \text{K}$. Together with

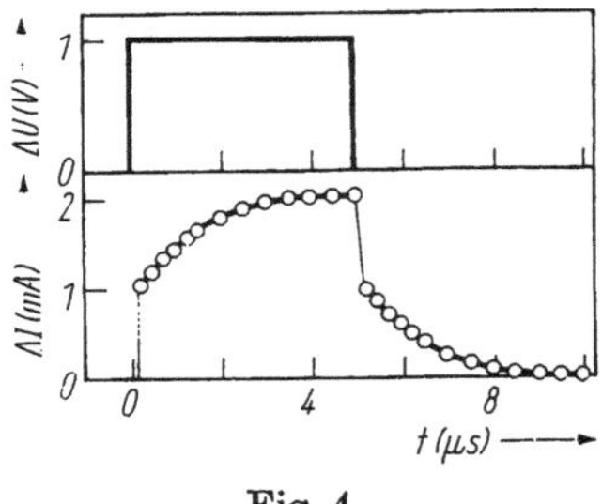

Fig. 4

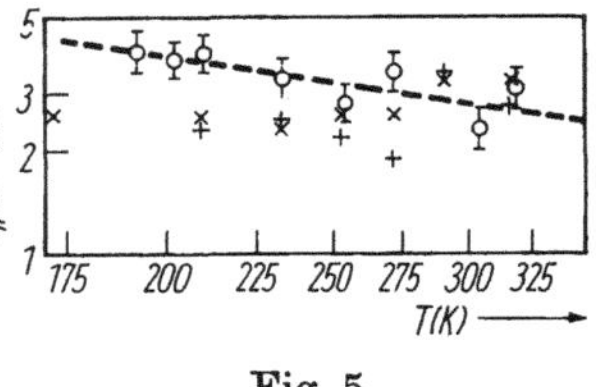

Fig. 5

Fig. 4. Voltage step (upper curve) and current response (lower curve) for a sample biased in the high-injection square-law regime ($U_{bias} = 22$ V, $I_{bias} = 20$ mA, $L = 350$ μm, $\tau = 1.3$ μs)

Fig. 5. Temperature dependence of capture cross-section for electrons σ_n determined by means of small-signal pulse measurements in pin diodes (●) in comparison with earlier results obtained in n–i–n diodes (large-signal pulse (×) and capacitance measurements (+))

the new results obtained with the above described method there seems to be a tendency of σ_n slightly increasing with decreasing temperature. Within the limit of error of all methods of bout 30%, all the results agree and a value of

$$\sigma_n = 2.8 \times 10^{-15} + \left(1 - \frac{T}{300}\right) 3.2 \times 10^{-15}\ \text{cm}^2$$

is obtained. The temperature limitations of this method are given by the fact that at higher temperatures the relation (7) is no more valid and that at lower temperatures the shape of the current pulse deviates from the shape shown in Fig. 4 and therefore does not allow a correct determination of the time constant τ anymore.

4. Determination of Over-Temperature in Current Filaments

In the high-injection regime in which the small-signal pulse measurements were performed, the power density approaches the KW/cm² range, see Fig. 1. Therefore, at higher voltages considerable Joule heating of the samples occurs leading to a thermal hysteresis in the dc I–U characteristics or even thermal destruction of the sample when dc biased. Therefore, to ensure the agreement of the measured bath temperature with that of the sample, the diode was biased using a voltage pulse of suitable duration. The small-signal pulse was applied 50 μs after applying the bias, see Fig. 6. Since the large-signal time constant for reaching the steady-state value within the semiconductor regime is in the μs range, this ensures steady-state conditions when the small-signal voltage step is applied, but prevents heating of the sample, because the thermal time constant lies in the 1s range. This and the temperature of the sample under thermally stationary conditions was determined using the sample itself as a thermometer: In the ohmic regime the conduction of our compensated samples is given by

$$\sigma = q\mu_n n_0 \approx q\mu_n N_c \exp\left\{-\frac{(E_c - E_R)}{kT}\right\}, \tag{21}$$

because the Fermi level lies nearly on the recombination center level $(E_c - E_R)$. For applied voltages of about 1 V the sample is biased in the ohmic regime for all temperatures, see Fig. 1. The basic idea of measuring the temperature within the sample biased in the semiconductor regime, therefore, is to short-circuit the sample for several free-carrier time constants (thus all excess free carriers have recombined, but the thermal situation has not changed) and then apply a voltage within the ohmic regime and measure the corresponding current as a function of time. According to (21) from $j(t)$ the over-temperature in the sample can be calculated, since for $j(t \to \infty)$ the temperature, the bath temperature, is known. Furthermore, if the theoretical expectations are experimentally verified, the thermal time constant can be determined and

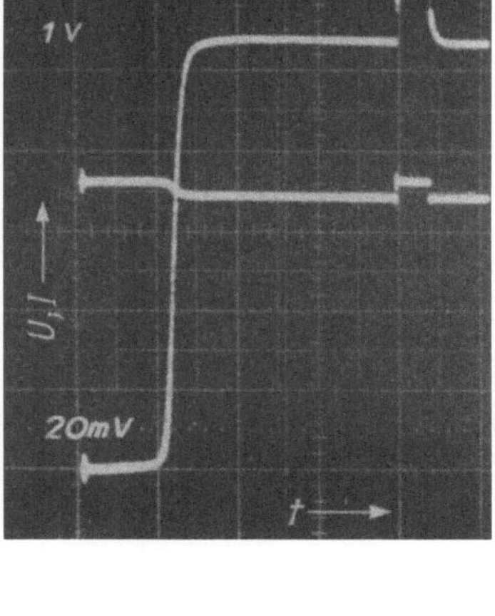

Fig. 6. Bias and voltage step (lower trace) and corresponding current response (upper trace) in the high-injection square-law regime (time scale 10 μs/small div; voltage scale 1 V/div; current scale 1.1 mA/div)

an extrapolation to $t = 0$ yields the temperature in a sample continuously biased in the semiconductor regime.

If the cooling of the sample obeys a simple exponential law

$$T(t) = T_0 + \Delta T \exp\left(-\frac{t}{\tau_{th}}\right) \tag{22}$$

with T_0 the bath temperature and τ_{th} the thermal time constant, then together with (21) and $(E_c - E_R) = 0.54$ eV one obtains for the ohmic conductivity as a function of time:

$$\sigma(t) = q\mu_n N_c \exp\left[-\frac{0.54}{kT_0}\,\frac{1}{1 + (\Delta T/T_0)\exp(-t/\tau_{th})}\right]. \tag{23}$$

For $\Delta T \ll T_0$ equation (23) yields

$$\ln\left[\ln\frac{\sigma(t)}{\sigma_\infty}\right] = \ln\left(\frac{0.54}{kT_0}\,\frac{\Delta T}{T_0}\right) - \frac{t}{\tau_{th}} \tag{24}$$

which allows the determination of τ_{th}. Fig. 7 shows the experimental results yielding $\tau = 2.3$ s for the corresponding sample. This sample only had thermal contact with the sample holder by the two contact wires. (The value of τ_{th} depends of course on the quality of the thermal contact of the chip with the bath.) In the semiconductor regime the power dissipation is given by

$$P = AjU \approx Aq\,\frac{\mu_n\mu_p}{c_n}\,\frac{N_D}{N_R}\left(\frac{U}{L}\right)^3 \tag{25}$$

with A the contact area, therefore Fig. 8 shows the overtemperature determined that way as a function of the normalized bias and the power, respectively.

Since in large-contact-area pin diodes with an n.d.r. region the occurring current filaments ($A \approx 10^{-3}$ cm^2) correspond to our small-contact-area samples, these measured over-temperature values correspond to those within the current filaments as a function of the normalized voltage if the filaments are assumed to be biased in the semiconductor regime. Therefore, this may be used as a new method for the determination of the temperature in current filaments. It is, however, an integral method which does not allow the determination of temperature distributions.

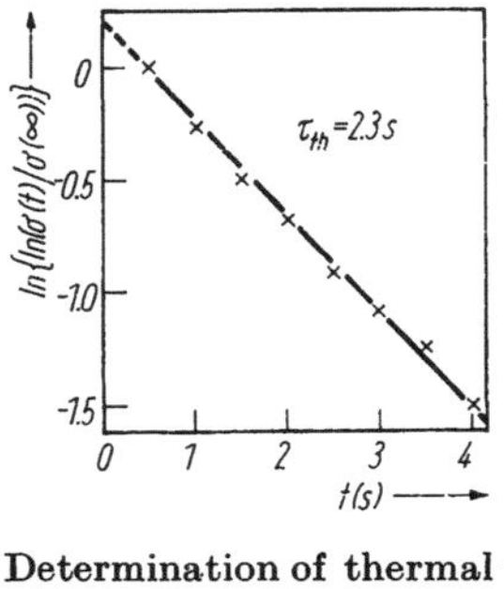

Fig. 7. Determination of thermal time constant τ_{th} of a cooling pin diode, $\tau_{th} = 2.3$ s

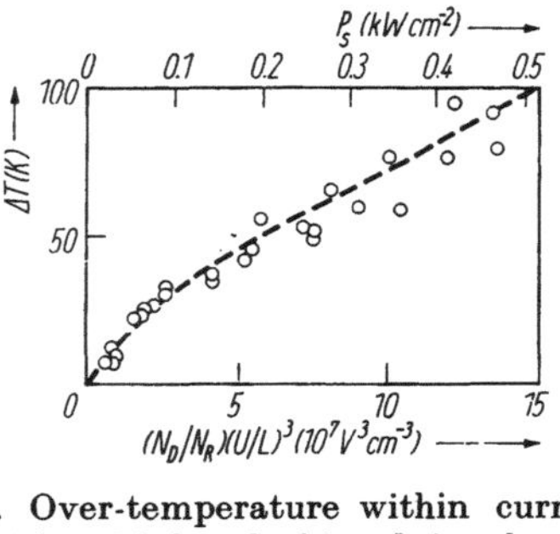

Fig. 8. Over-temperature within current filament ($A = 10^{-3}$ cm^2) biased in the semiconductor regime as a function of normalized voltage or power density

5. Conclusion

The gold acceptor cross-sections for electrons and holes σ_n and σ_p^-, respectively, were measured in Au-compensated pin diodes using a small-signal pulse method and evaluating the dc I–U characteristics of pin diodes.

In the range 193 K $\leqq T \leqq$ 316 K it is

$$\sigma_n = \left[2.8 + 3.2\left(1 - \frac{T}{300}\right)\right] 10^{-15}\ \text{cm}^2$$

which agrees with earlier results obtained with other methods.

For σ_p^- we found in the range of 160 K $\leqq T \leqq$ 340 K

$$\sigma_p^- = 8 \times 10^{-15} \left(\frac{300}{T}\right)^{3.9} \text{cm}^2$$

in agreement with measurements of Bemski (200 K $\leqq T \leqq$ 500 K) and Fairfield and Gokhale at room temperature.

The Joule heating within current filaments was measured with a pulse method using the temperature dependent ohmic regime of the sample for the temperature determination.

References

[1] R. Kassing and H. Lenz, phys. stat. sol. (a) **26**, 155 (1974).

[2] C. T. Sah, L. Forbes, L. I. Rosier, A. E. Tasch, Jr., and A. B. Tole, Appl. Phys. Letters **15**, 145 (1969).

[3] H. Klose, Probleme der Festkörperelektronik **3**, 91 (1971).

[4] Y. Zohta, J. appl. Phys. **43**, 1713 (1972).

[5] R. R. Senechal and J. Basinki, J. appl. Phys. **39**, 3723 (1968).

[6] G. Bemski, Phys. Rev. **111**, 1515 (1958).

[7] J. M. Fairfield and B. V. Gokhale, Solid State Elektronics **8**, 685 (1965).

[8] W. D. Davis, Phys. Rev. **114**, 1006 (1959).

[9] L. C. Parrillo and W. C. Johnson, Appl. Phys. Letters **20**, 104 (1972).

[10] R. Kassing and H. Lenz, phys. stat sol. (a) **25**, 131 (1974).

[11] R. Kassing, E. Kähler, and I. Dudeck, phys. stat. sol. (a) **30**, 141 (1975).

[12] I. Dudeck and R. Kassing, J. appl. Phys. **48**, 4786 (1977).

[13] J. Barbolla, M. Pugnet, J. C. Brabant, and M. Brousseau, phys. stat. sol. (a) **36**, 495 (1976).

[14] S. D. Brotherton and I. Bicknell, J. appl. Phys. **49**, 667 (1978).

[15] O. Engström and H. G. Grimmeiss, Appl. Phys. Letters **25**, 413 (1974).

[16] K. L. Ashley and A. G. Milnes, J. appl. Phys. **35**, 369 (1964).

[17] R. Baron, O. J. Marsh, and J. W. Mayer, J. appl. Phys. **37**, 2614 (1966).

[18] M. A. Lampert and A. Rose, Phys. Rev. **121**, 26 (1961).

(Received June 15, 1978)

phys. stat. sol. (a) 49, 163 (1978)

Subject classification: 16 and 20.3; 22.2.1

Forschungslaboratorien der Siemens AG, München

Effect of Reabsorbed Recombination Radiation on Photoluminescence and Photoconductivity in a Semi-Infinite Direct-Gap Semiconductor

By

K. METTLER

In honour of Prof. Dr. F. STÖCKMANN's 60th birthday

A one-dimensional steady-state continuity equation for optically generated excess minority carriers is derived in which the effect of reabsorbed recombination radiation on the excess carrier profile in a semiinfinite direct-gap semiconductor is taken into account. The influence of reexcited carriers is illustrated by means of calculated photoconductivity and photoluminescence data using optical absorption data of a highly doped GaAs sample as an example. The results show that the spatial carrier distribution, the photoluminescent intensity, and the photocurrent are appreciably affected by reabsorbed luminescence only for internal quantum efficiency values of at least a few tens of percent.

Für den Fall der optischen Erzeugung von Überschußminoritätsträgern im halbunendlich ausgedehnten direkten Halbleiter wird eine eindimensionale stationäre Kontinuitätsgleichung hergeleitet, die den Einfluß reabsorbierter Rekombinationsstrahlung auf das Profil der Überschußträger berücksichtigt. Der Einfluß der durch Wiederanregung erzeugten Träger wird am Beispiel einer hochdotierten GaAs-Probe an Hand berechneter Photoleitungs- und Photolumineszenzdaten veranschaulicht. Die Ergebnisse zeigen, daß räumliche Trägerverteilung, Photolumineszenzintensität und Photostrom nur bei internen Quantenwirkungsgraden von mindestens einigen zehn Prozent durch reabsorbierte Lumineszenz nennenswert beeinflußt werden.

1. Introduction

Quantitative studies of photoluminescence (PL) and photoconductivity (PC) often require the solution of a continuity equation describing the dependence of the non-equilibrium carrier density on the rate of excitation [1]. It is usually assumed that the resulting excess carrier density does not react upon the carrier excitation rate itself. If in a semiconductor excess carriers recombine radiatively, however, such an influence may result from the creation of new electron–hole pairs due to self-absorption of recombination radiation [2]. The magnitude of this effect mainly depends on the internal quantum efficiency of the radiative recombination process. Thus, a most suitable material for studying the effect of reexcited carriers on other semiconductor properties is liquid-phase epitaxial (LPE) gallium arsenide since it exhibits internal radiative efficiencies of nearly up to unity [3 to 5]. An influence of reexcitation on the minority carrier diffusion length [6, 7] and lifetime [8] and on the external quantum efficiency [6] has been predicted from theoretical studies and from experimental investigations on thin LPE-grown GaAs layers bounded by (Ga, Al)As layers.

In this paper the effect of reabsorbed recombination radiation on the external photoluminescent intensity and on the photocurrent is treated for the case of a semi-infinite sample. The results are illustrated using a highly doped GaAs sample, as an example. This work was also initiated by the need for explaining experimental results of PL measurements which gave obviously wrong internal quantum efficiencies of up to 135% due to the contribution of reexcited carriers [9].

2. Continuity Equation

Large area excitation of a semi-infinite sample by external light is assumed in the present calculation. In this case it is often found convenient to use a one-dimensional continuity equation to calculate the density of excess minority carriers. Thus, following [6], the continuity equation which includes the effect of reexcitation is written

$$D\frac{\mathrm{d}^2n}{\mathrm{d}x^2}-\frac{n}{\tau}+G_\mathrm{e}+G_\mathrm{r}=0\,. \tag{1}$$

In (1) G_e and G_r are the generation rates due to the external optical excitation and to the reabsorbed recombination radiation, respectively, n denotes the density, D the diffusion constant, and τ the lifetime of excess carriers. When the generation rate for carriers is governed by Lambert's law of photon absorption, G_e is simply given by

$$G_\mathrm{e}(x)=F_\mathrm{e}(1-R_\mathrm{e})\,\alpha_\mathrm{e}\exp\left(-\alpha_\mathrm{e}x\right), \tag{2}$$

where F_e is the number of incident photons per unit time and unit area, R_e the reflectivity, and α_e the optical absorption coefficient at the energy of the excitation. However, with the assumption of a one-dimensional carrier profile difficulties arise from the necessity to find also a one-dimensional expression for G_r, although the emission and reabsorption of internal luminescence is a priori a three-dimensional effect. It is shown in a physically illustrative approach different from the derivation in [6] that a function $G_\mathrm{r}(x)$ can indeed be found which only depends on x and which is equivalent to the three-dimensional reexcitation rate G_r.

The half-space representing the semiconductor bulk (cf. Fig. 1) is partitioned into an infinite number of layers being situated at various distances x from the surface at $x=0$ and having the same thickness $\mathrm{d}x$. The generation rate $G_\mathrm{r}(x)$ at the depth x is then obtained from the total number of photons which are reabsorbed within that specific layer. It is assumed that (i) the generated excess carriers — also those that are carried over a larger distance by the photon transfer — thermalize before recombining and thus the spectral distribution of the internal luminescence remains unaffected by reexcitation, and that (ii) the emission and absorption of the recombination radiation is spatically isotropic. The rate of carriers generated by absorption of radiation emitted only from the fixed point source S into the elemental volume

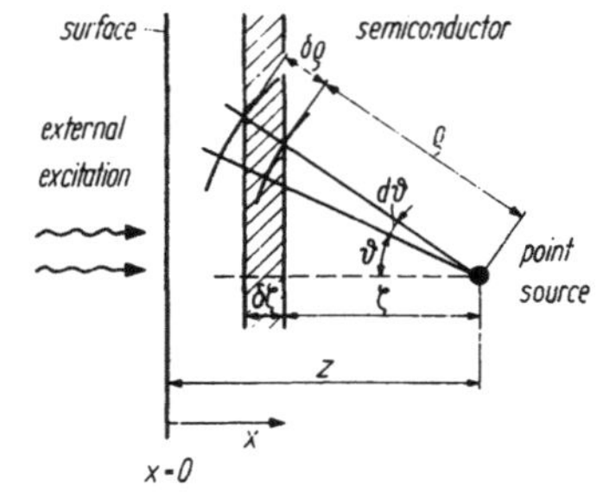

Fig. 1. Configuration for the absorption of luminescence emitted from a point source

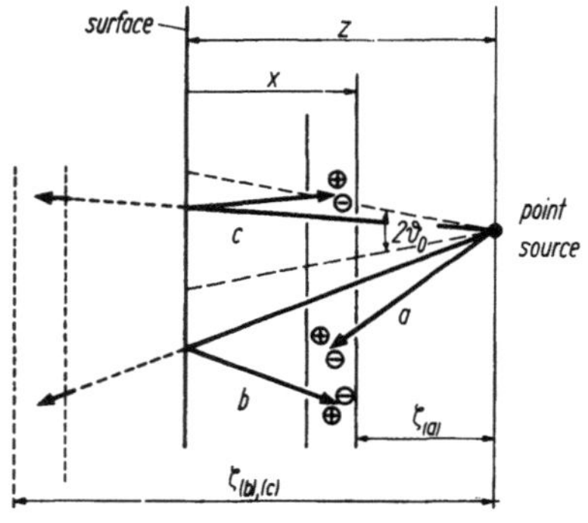

Fig. 2. Possible modes of propagation of light which is absorbed in the hatched cross-section of a layer

$\mathrm{d}\varrho \sin \vartheta \, \mathrm{d}\vartheta \, \mathrm{d}\varphi$ (Fig. 1) is

$$r_s = \frac{1}{4\pi} \int_0^\infty \mathrm{d}(h\nu) \, \eta_q \frac{n(z)}{\tau} \gamma(h\nu) \, \alpha(h\nu) \exp\left[-\alpha(h\nu) \, \varrho\right] \mathrm{d}\varrho \sin \vartheta \, \mathrm{d}\vartheta \, \mathrm{d}\varphi \,. \tag{3}$$

In (3) η_q is the internal quantum efficiency of the recombination radiation, $\gamma(h\nu)$ and $\alpha(h\nu)$ are the normalized internal emission spectrum and the absorption coefficient, respectively, at the photon energy $h\nu$, and are mutually related through the van Roesbroeck-Shockley relation [10]; ϱ and ϑ are geometrical quantities as defined in Fig. 1 and φ is the azimuth angle. The carrier generation rate for the whole layer due to luminescence from S is given by

$$\frac{1}{4\pi} \eta_q \frac{n(z)}{\tau} \int_0^\infty \mathrm{d}(h\nu) \, \gamma(h\nu) \, \alpha(h\nu) \int_0^{2\pi} \mathrm{d}\varphi \int_0^{\pi/2} \sin \vartheta \, \mathrm{d}\vartheta \int_\varrho^{\varrho+\delta\varrho} \exp\left[-\alpha(h\nu) \, \varrho\right] \mathrm{d}\varrho \tag{4}$$

or, after integrating for ϱ from $\varrho = \zeta/\cos \vartheta$ to $\varrho = (\zeta + \delta\zeta)/\cos \vartheta$ and then integrating for ϑ, by

$$\frac{1}{2} \eta_q \frac{n(z)}{\tau} \int_0^\infty \mathrm{d}(h\nu) \, \gamma(h\nu) \, \alpha(h\nu) \, E_1[\alpha(h\nu) \, \zeta] \, \delta\zeta \,, \tag{5}$$

where

$$E_1(s) = \int_s^\infty \frac{\exp(-t)}{t} \, \mathrm{d}t \,. \tag{6}$$

In (5) ζ denotes the shortest possible distance a photon may travel until it is absorbed in the layer under consideration. The factor $\frac{1}{2}$ results from half of the luminescence being emitted into the half-space behind the point source ($\vartheta > \pi/2$). The total generation rate for a layer at the depth x is obtained by integrating (5) for z in the limits $0 \leqq z \leqq \infty$, i.e. for the source varying over the total half-space. According to Fig. 2 three different modes of light propagation thereby must be taken into account corresponding to a) direct rays, b) totally reflected light with an angle of incidence $\vartheta \geqq \vartheta_0$ from the point source to the sample surface (refractive index n^*), and c) partially reflected light with an incidence angle $\vartheta < \vartheta_0$ where

$$\vartheta_0 = \arcsin\left(\frac{1}{n^*}\right). \tag{7}$$

It must also be considered that in (5) ζ has to be set equal to $(z + x)$ or to $|z - x|$ depending on whether or not the emitted light is totally reflected at the surface before becoming absorbed in a specific layer. Integration for z then leads to

$$G_r(x) = \frac{\eta_q}{2} \int_0^\infty \int_0^\infty \gamma(h\nu) \, \alpha(h\nu) \frac{n(z)}{\tau} \Bigg\{(1 + R) \, E_1[\alpha(h\nu) \, |z - x|] + \\ + (1 - R) \, E_1\left[\frac{\alpha(h\nu) \, (z + x)}{\cos \vartheta_0}\right]\Bigg\} \mathrm{d}z \, \mathrm{d}(h\nu) \,, \tag{8}$$

where R is the reflectivity of luminescence at the surface. Equation (1) together with (2) and (8) represents the one-dimensional steady-state continuity equation for excess minority carriers with a generation rate including both excitation by external light and reexcitation due to internally reabsorbed luminescence.

3. Photoluminescence and Photoconductivity

Numerical data of a highly doped p-type GaAs sample were used to demonstrate by means of an example the effect of reabsorbed recombination radiation on the minority carrier profile, on the external photoluminescence, and on the photoconductivity of a semiconductor sample. A diffusion length of $L = 8.5 \times 10^{-5}$ cm and data for the absorption coefficient $\alpha(h\nu)$ were taken from [11] for $p = 1 \times 10^{19}$ cm^{-3}; further a value of $n^* = 3.59$ [12] was used for the refractive index at the wavelength of the luminescence peak and a value of $\alpha_e = 4.4 \times 10^4$ cm^{-1} for the absorption coefficient at a wavelength corresponding to excitation by a HeNe laser [13].

Numerical results were obtained from the integral equation (1) by an iteration procedure described in [6]. Accordingly, both $G(x)$ and $n(x)$ are developed in a series:

$$G(x) = G_e(x) + G_{r,0}(x) + \dots + G_{r,k-1}(x) + \dots, \tag{9}$$

$$n(x) = n_0(x) + n_1(x) + \dots + n_k(x) + \dots, \tag{10}$$

where $n_k(x)$ is proportional to η_q^k and represents the density of excess minority carriers generated at a rate $G_{r,k-1}$ in the $(k-1)$-th reexcitation process ($k = 1, 2, \dots$). The carrier density $n_0(x)$ without reexcitation is obtained from (1) for $G_r = 0$. With $n_0(x)$ and $n_k(x)$ being known, the single terms $G_{r,k-1}(x)$ and $n_k(x)$ are then alternately calculated from (8) to obtain $G_{r,k-1}(x)$, and using

$$n_k(x) = \frac{L}{2D}\left\{\exp\left(-\frac{x}{L}\right)\left[\int_0^x \exp\left(\frac{\xi}{L}\right) G_{r,k-1}(\xi)\,d\xi - \right.\right.$$
$$\left.\left. - \frac{s_r - 1}{s_r + 1}\int_0^\infty \exp\left(-\frac{\xi}{L}\right) G_{r,k-1}(\xi)\,d\xi\right] + \exp\left(\frac{x}{L}\right)\int_x^\infty \exp\left(-\frac{\xi}{L}\right) G_{r,k-1}(\xi)\,d\xi\right\} \tag{11}$$

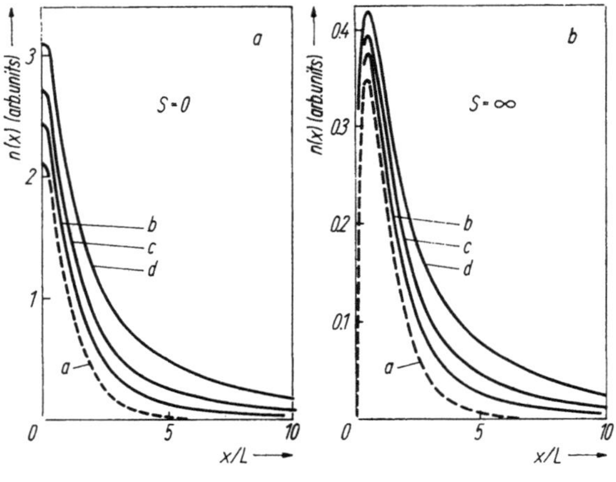

Fig. 3. Calculated spatial profiles of the excess minority carrier density (a) without reexcitation, (b) to (d) with reexcitation and for various values of $\eta_q = 0.4$ (b), 0.6 (c), 0.8 (d). S is the surface recombination velocity

to determine $n_k(x)$. Equation (11) follows by solving (1) for $G_e = 0$ and $k = 1, 2, \dots$ and with the boundary conditions

$$n_k(x = \infty) = 0\,, \tag{12}$$

$$\left.\frac{\mathrm{d}n_k}{\mathrm{d}x}\right|_{x=0} = \frac{s_r}{L}\, n_k(x = 0)\,, \tag{13}$$

where $s_r = S\tau/L$ and S is the surface recombination velocity. Different from [6] where $\gamma(h\nu)$ was approximated by a rectangular distribution, the emission spectrum determined from $\alpha(h\nu)$ was used here on evaluating (8) and (11).

From $n(x)$, as shown in Fig. 3, values for the luminescence intensity I_{PL} and the photoconductivity I_{PC} may be obtained. These were calculated from the expressions

$$I_{PL} = F_e(1 - R_e)\, \alpha_e \eta_q \int_0^\infty \int_0^\infty \frac{n(x)}{\tau}\, \gamma(h\nu) \exp\left[-\alpha(h\nu)\, x\right] \mathrm{d}x\, \mathrm{d}(h\nu) \tag{14}$$

and

$$I_{PC} \sim F_e(1 - R_e)\, \alpha_e \int_0^\infty n(x)\, \mathrm{d}x\,, \tag{15}$$

respectively, and are shown in Fig. 4 and 5 as functions of the quantum efficiency η_q. The surface recombination velocity has been assumed to be zero and infinite in the calculations.

The results in Fig. 3 to 5 show that the effect of reabsorbed recombination radiation becomes significant already for quantum efficiency values of a few tens of a percent. It is seen that the increase in signal with increasing η_q is more pronounced for the photoconductivity (Fig. 5) than for the external photoluminescence (Fig. 4) of the sample. This difference is due to the fact that the carrier profile extends appreciably to larger x, for higher η_q. Therefore, an increasing part of the recombination radiation originates at greater depths in the sample and, according to (14) becomes "filtered out" by reabsorption before reaching the surface, whereas after (15) the photoconductivity remains unaffected by the spatial broadening of the carrier distribution in this model.

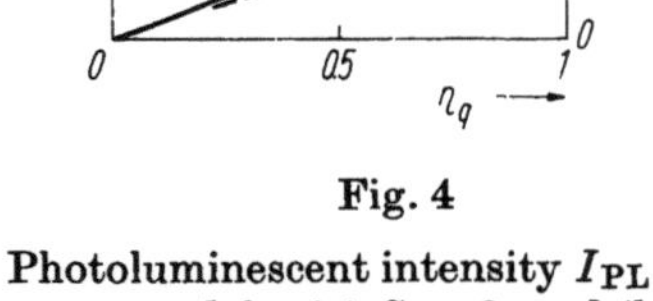

Fig. 4

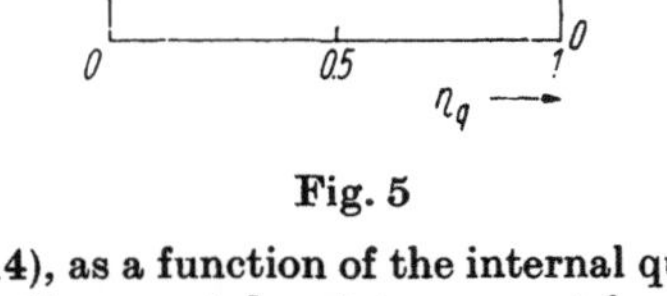

Fig. 5

Fig. 4. Photoluminescent intensity I_{PL} calculated from (14), as a function of the internal quantum efficiency η_q, and for (a) $S = 0$, and (b) $S = \infty$. Reexcitation was taken into account for the full curves. The dashed line gives I_{PL} in the absence of reexcitation. The curves have only been drawn for η_q up to 0.8 because of the slow convergence of (10) in the present numerical evaluation

Fig. 5. Photoconductivity I_{PC} calculated from (15), as a function of the internal quantum efficiency η_q and for (a) $S = 0$ and (b) $S = \infty$. Reexcitation was taken into account for the full curves. The dashed line gives I_{PC} in the absence of reexcitation

It should be noted, however, that the doping concentrations in conventional photoconductors are usually much lower than the value taken in the present example of GaAs where a high value of the internal quantum efficiency is only obtained at high doping densities.

Acknowledgements

The author wishes to acknowledge the many helpful discussions with K.-H. Zschauer. He would also like to thank G. Winstel for his interest in this work.

References

[1] H. B. DEVORE, Phys. Rev. **102**, 83 (1956).
[2] W. P. DUMKE, Phys. Rev. **105**, 139 (1957).
[3] ZH. I. ALFEROV, V. M. ANDREEV, D. Z. GARBUZOV, and V. D. RUMYANTSEV, Soviet Phys. — Semicond. **9**, 305 (1975).
[4] M. ETTENBERG and H. KRESSEL, J. appl. Phys. **43**, 1204 (1972).
[5] E. PINKAS, B. I. MILLER, I. HAYASHI, and P. W. FOY, J. appl. Phys. **43**, 2827 (1972).
[6] T. KURIYAMA, T. KAMIYA, and H. YANAI, Japan. J. appl. Phys. **16**, 465 (1977).
[7] M. ETTENBERG, Appl. Phys. Letters **30**, 207 (1977).
[8] P. ASBECK, J. appl. Phys. **48**, 820 (1977).
[9] K. METTLER, Siemens Forsch.- u. Entwickl.-Ber. **6**, 152 (1977).
[10] W. VAN ROESBROECK and W. SHOCKLEY, Phys. Rev. **94**, 1558 (1954).
[11] H. C. CASEY, Jr., B. I. MILLER, and E. PINKAS, J. appl. Phys. **44**, 1281 (1973).
[12] D. D. SELL, H. C. CASEY, JR., and K. W. WECHT, J. appl. Phys. **45**, 2650 (1974).
[13] D. D. SELL and H. C. CASEY, JR., J. appl. Phys. **45**, 800 (1974).

(Received June 9, 1978)

phys. stat. sol. (a) 49, 169 (1978)

Subject classification: 16; 22.4.1

Institut für Angewandte Physik der Universität Karlsruhe

Photovoltage at the Metal–CdS Schottky Contact

By

R. Butendeich and W. Ruppel

In honour of Prof. Dr. F. Stöckmann's 60th birthday

Photovoltages at a metal–CdS Schottky contact occur at photon energies that are smaller than the CdS band gap. They may be due either to photoemission of electrons from the metal into the CdS or to hole generation in the CdS. The first mechanism is shown to occur in conducting CdS. From its observation values for the attenuation length of the photoexcited electrons in Au are derived as a function of photon energy. The second mechanism is shown to occur in insulating CdS. At intermediate conductivities of the CdS photoemission is found for small photon energies, while minority carrier generation in the CdS is responsible for the generation of a photo-e.m.f. at somewhat higher photon energies.

Photospannungen an einem Metall–CdS-Schottky-Kontakt entstehen bei Belichtung auch mit Photonen, deren Energie kleiner ist als der Bandabstand des CdS. Diese Photospannungen beruhen entweder auf Photoemission von Elektronen aus dem Metall in das CdS oder auf Löchererzeugung im CdS. Bei CdS hoher Leitfähigkeit tritt der erstgenannte Mechanismus auf. Aus seiner Beobachtung werden Werte für die Reichweite der optisch erzeugten Elektronen in Au als Funktion der Photonenenergie gewonnen. Der zweitgenannte Mechanismus tritt in isolierendem CdS auf. Im Zwischengebiet der Leitfähigkeit des CdS wird für kleine Photonenenergien Photoemission beobachtet, während die Erzeugung von Minoritätsträgern im CdS bei größeren Photonenenergien die Photospannung verursacht.

1. Introduction

The occurrence of a photoelectromotive force (photo-e.m.f. or photovoltage) at a metal–CdS Schottky contact is observed also for photons the energy of which is smaller than the CdS band gap [1]. The origin of this photo-e.m.f. has been explained in two different ways:

1. Excitation of the metal adjacent to the semiconductor and emission of electrons from the metal into the semiconductor [2 to 5]. This operation is analogous to the vacuum photocell.

2. Excitation in the semiconductor space charge layer and creation of mobile minority carriers [6 to 12].

Both excitation mechanisms have in common that the band gap energy does not present a low-energy limit for the photon energy at which a photo-e.m.f. is observed (Fig. 1). Electrons may be emitted from the metal when their excitation energy exceeds that of the Schottky potential barrier at the contact which may be considerably less than the band gap. Free minority carriers in the semiconductor may as well be created by photons of smaller energy than that of the band gap.

Conclusions about the concentration and the energetic distribution of traps and recombination centres from measurements of the photoconduction under applied voltage for majority carriers and from measurements of the photo-e.m.f. for minority carriers have been the aim of the work of Stöckmann and his coworkers in the last few years. It is evident that conclusions on the properties of minority carrier traps and recombination centres may only be drawn from measurements of the photo-e.m.f.

if it is asserted that the observed photovoltage is due to minority carrier generation in the semiconductor. On the other hand, detection of the photovoltage due to electron excitation in the contact metal may lead to the determination of properties of the excited metal electron gas.

It will be shown in the present paper that the parameter which delineates the occurrence of either of these mechanisms is the position of the Fermi level in the dark and hence the dark conductivity of the CdS. As we will see, for a low-lying Fermi level, namely insulating CdS, excitation of minority carriers prevails as the mechanism responsible for the photo-e.m.f., while for a shallow Fermi level, namely rather conducting CdS, it is electron emission from the metal into the CdS.

2. Origin of the Photovoltage

2.1 Photoemission into CdS

Electrons excited in the metal contact layer and crossing the metal–CdS boundary generate an e.m.f. if they penetrate into the CdS at least by the distance x_m (Fig. 1). x_m denotes the maximum of the potential. The distribution of the electrical potential in the CdS results from the superposition of a contribution from the difference in the electron chemical potential at the metal–semiconductor boundary and a contribution from the image force. The distance x_m from the metal–CdS boundary is [13]

$$x_m = 1.9 \times 10^{-4} \left(\frac{\varepsilon F_0}{\text{V/cm}}\right)^{1/2} \text{cm}\,, \tag{1}$$

where F_0 denotes the field strength at the boundary without consideration of the image force, and ε the dielectric constant.

Assuming an electron mean free path in the CdS in the order of 1 to 5 nm and considering that the hot electrons emitted from the metal may pass the barrier maximum

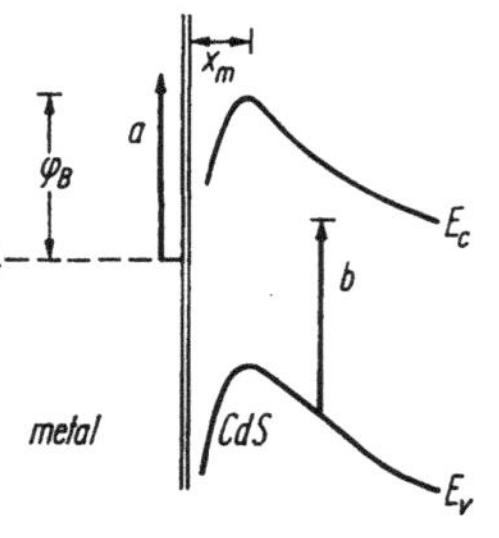

Fig. 1

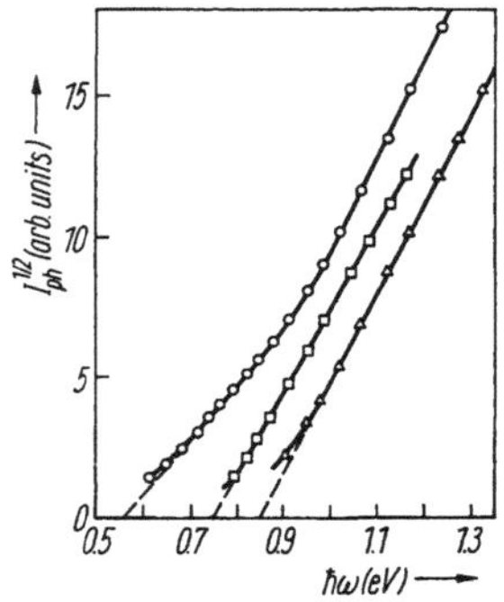

Fig. 2

Fig. 1. Metal–CdS Schottky contact with the contribution of the image force. E_F denotes the Fermi level, E_c and E_v the band edges of conduction and valence band. x_m is the distance of the potential maximum from the metal–CdS interface. φ_B is the barrier height. The transition a shows an excitation of the metal electron gas beyond φ_B. The transition b shows an excitation with equal photon energy as transition a leading to a mobile hole in the CdS

Fig. 2. Square root of the short-circuit photocurrent as a function of photon energy (Fowler plot) for various metal–CdS contacts. The extrapolation to the abscissa yields the barrier height φ_B. Dark conductivity of the CdS $\sigma = 10^{-1}$ $(\Omega\ \text{cm})^{-1}$. For the Au and Pt contact $T = 295$ K. For the Ag contact $T = 110$ K. ○ Ag–CdS, △ Pt–CdS, □ Au–CdS at x_m only after several collisions with LO-phonons, one is to expect the generation of a photo-e.m.f. by photoemission into the CdS for values of x_m in the range

$$x_m \lessapprox 10 \text{ to } 50 \text{ nm}.$$

The boundary field strength F_0 of a Schottky electron depletion layer

$$F_0 = \left(\frac{2\sigma U_D}{\mu_n \varepsilon \varepsilon_0}\right)^{1/2} \tag{2}$$

yields in combination with (1) for values of the diffusion potential $U_D = 0.6$ V and the electron mobility $\mu_n = 10$ cm²/Vs a conductivity range for CdS of

$$\sigma \gtrapprox 10^{-5} \, (\Omega \text{ cm})^{-1} \tag{3}$$

in which an e.m.f. due to photoemission of electrons from the metal into CdS should be observable.

Experimentally, Fig. 2 shows a Fowler plot for an illuminated CdS contact with Au, Ag, and Pt. The samples were prepared by evaporating the metal contact on to the CdS crystal while the crystal was cleaved in high vacuum under the metal vapour stream. The fact that the square root of the short-circuit photocurrent depends linearly on the photon energy of the incident photons above the current threshold, points to generation of the photocurrent by photoemission. The values for the potential barrier heights φ_B of the Schottky barrier, as derived from the threshold photon energies in Fig. 2, are in good agreement with observations reported in the literature. A review of φ_B of various material combinations is presented in [14]. Lepley and Ravelet find for Au on polycrystalline CdS films of 1 to 100 Ω cm that φ_B lies in the range 0.6 to 0.9 eV according to the treatment of the CdS surface before the Au deposition [15]. φ_B is related to the photo-e.m.f. and to the thickness of an insulating interface layer which stems either from chemisorbed impurities or oxide formation. The existence of

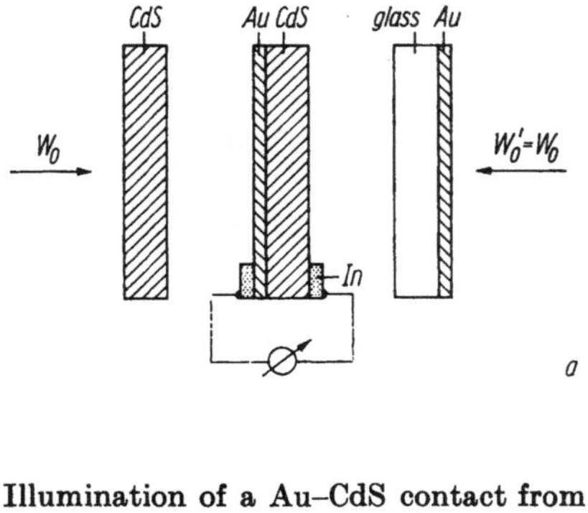

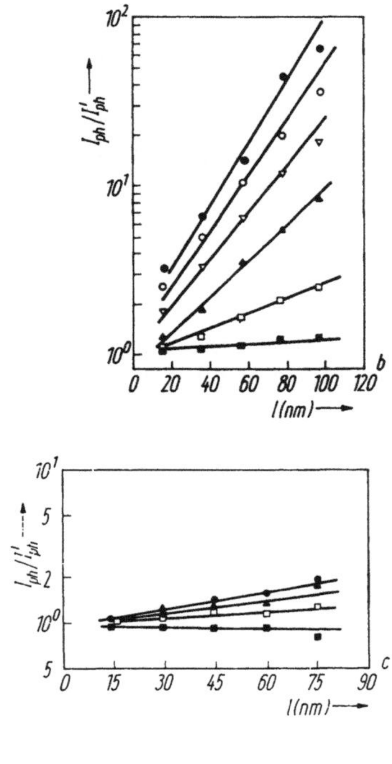

Fig. 3. a) Illumination of a Au–CdS contact from two sides with equal light intensities $W_0 = W'_0$. An additional CdS crystal and Au layer provide equal light intensities incident on the Au–CdS interface. b) Ratio I_{ph}/I'_{ph} of the short-circuit photocurrents measured with the arrangement shown in a) as a function of the Au-layer thickness l at different photon energies $\hbar\omega$ of monochromatic light of the intensity $W_0 = W'_0$ in a). Dark conductivity of the CdS crystal $\sigma = 10^{-1}$ $(\Omega \text{ cm})^{-1}$. ● $\hbar\omega = 1.13$, ○ 1.27, ▽ 1.44, ▲ 1.69, □ 1.92, ■ 2.27 eV. c) As b), but for a CdS crystal of the dark conductivity $\sigma = 10^{-12}$ $(\Omega \text{ cm})^{-1}$ (values of $\hbar\omega$ as under b))

chemically reacted interface layers has recently been demonstrated by the energy loss spectra of low-energy electrons [16]. Other observations on Au contacts on CdS single crystals or polycrystalline films yield also values in this range, namely $\varphi_B = (0.67 \pm 0.03)$ eV [17] and $\varphi_B = 0.7$ to 0.8 eV [18].

Values for Cu have not been included in Fig. 2. In agreement with Lepley et al. [19] who observe a change of the short-circuit photocurrent upon heat treatment for a Cu contact in contrast to a Au contact, we find a time variation of φ_B for Cu even without heat treatment in the course of several weeks. Photoemission, originally observed at the Cu–CdS contact, weakens and at the same time the characteristic features of a photo-e.m.f. as due to minority carrier generation in the CdS become more and more pronounced. Also, the resistivity of the CdS close to the Cu layer increases. These observations are accounted for by diffusion of Cu from the contact into the CdS. It appears that certain contradictory observations on the photovoltaic behaviour of Cu layers deposited on CdS may be readily reconciled if a different degree of diffusion of Cu into the CdS at different times after the Cu deposition is considered.

A direct proof for the contribution of photoemission to the generation of the e.m.f. at a metal–CdS contact for conducting CdS, and for an origin of the photo-e.m.f. different from photoemission for insulating CdS is constituted by the following experiment: The Au–CdS interface is illuminated from both sides in a completely symmetrical way (Fig. 3a). When illuminated from the Au side (left in Fig. 3a) through an additional CdS crystal, a short-circuit photocurrent I_{ph} results. If illuminated from the CdS side through an additional Au layer (right in Fig. 3a), a short-circuit photocurrent I'_{ph} is obtained. The number of photons arriving at the Au–CdS interface is the same for both illuminations. The observed ratio I_{ph}/I'_{ph} is shown in Fig. 3b and c as a function of the thickness l of the Au contact layer. It is seen that for CdS of high conductivity ($\sigma = 10^{-1}\ (\Omega\ \text{cm})^{-1}$) in Fig. 3b the ratio I_{ph}/I'_{ph} does not equal unity. Particularly for photons of high energy it strongly increases with increasing thickness l of the Au contact layer. An indicator for the optimum thickness of evaporated metal layers in Schottky barrier solar cells is described in [20]. For CdS of low conductivity ($\sigma = 10^{-12}\ (\Omega\ \text{cm})^{-1}$) the ratio I_{ph}/I'_{ph}, shown in Fig. 3c, is almost equal to unity independently of the incident photon energy and of the Au layer thickness.

The dependence of I_{ph}/I'_{ph} on l and on $\hbar\omega$ in Fig. 3b may readily be understood if the origin of both I_{ph} and I'_{ph} is taken into account as being due to photoemission. Let L denote the attenuation length or range of the optically excited electrons in the Au metal. Provided that for the penetration depth α^{-1} of the light into the layer holds that $\alpha^{-1} < l$ and $\alpha^{-1} < L$, the ratio I_{ph}/I'_{ph} is given by

$$\frac{I_{ph}(l, \hbar\omega)}{I'_{ph}(l, \hbar\omega)} = \exp\left[l\left(\alpha - \frac{1}{L}\right)\right]. \tag{4}$$

Evaluating the curves in Fig. 3b and considering $\alpha = \alpha(\hbar\omega)$ for Au as shown in Fig. 4, one finds for the attenuation length L of the photoexcited electrons in the Au

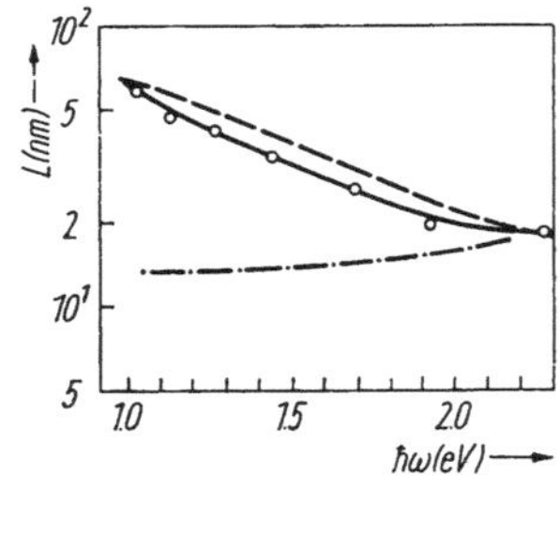

Fig. 4. Attenuation length of optically excited hot electrons in Au at $T = 295$ K as a function of excitation energy. The open circles are deduced from the measurements shown in Fig. 3b. For comparison, results of Sze et al. [21] (– – – –) are given. The dash-dotted line shows the penetration depth of light in Au

layer as a function of the excitation energy $\hbar\omega$ the values shown in Fig. 4. They are in close agreement with values obtained by Sze et al. [21] from measurements on the collection of photoexcited electrons across Schottky barriers in Au–GaP and Au–Si structures.

2.2 *Minority carrier generation in the CdS*

With decreasing σ of the CdS the distance x_m of the potential barrier maximum from the metal–CdS interface increases (Fig. 1). For large values of x_m photoemission of electrons from the metal cannot lead any more to the generation of a photo-e.m.f. This expectation is borne out by the results shown in Fig. 3c. The fact that for insulating CdS the photovoltage has the same value irrespective of the side from which the CdS–metal interface is illuminated, demonstrates that the excitation leading to the generation of the photo-e.m.f. may not be located in the metal but in the CdS adjacent to the interface. If the excitation takes places in the CdS, it must be the generation of mobile holes as the minority carriers in the CdS that leads to the observed photovoltage.

The generation of minority carriers in insulating CdS as the source of the photo-e.m.f. in a CdS–metal contact is demonstrated by its correlation with the quenching of the photocurrent under externally applied voltage [11]. Fig. 5 shows the spectral dependence of the photovoltage under two-photon-beam excitation (Fig. 5a) and the quenching of the photocurrent under applied voltage (Fig. 5b). In both cases one photon beam of 2.0 eV photons is incident on the Au–CdS contact of CdS with a dark conductivity $\sigma = 10^{-12}$ $(\Omega\ \text{cm})^{-1}$. The abscissa in Fig. 5 shows the photon energy of the second photon beam. The maxima of the photovoltage are seen to coincide in photon energy with the minima of the photocurrent under applied voltage. They occur at photon energies following from a model by Hemila and Bube [22]. The quenching of the photocurrent under applied voltage by additional illumination with photons of lower energy is well known to be due to the excitation of minority carriers. This exci-

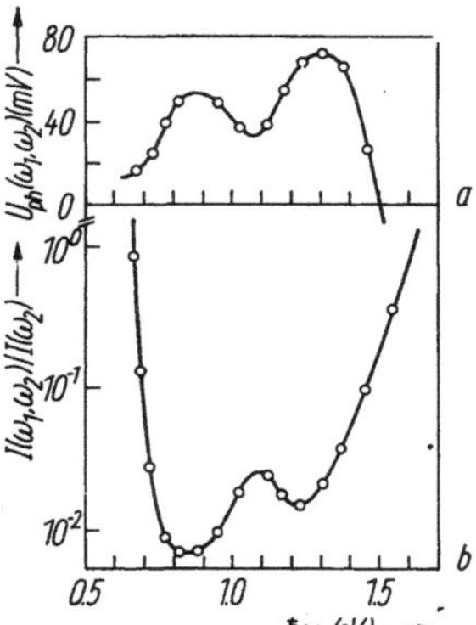

Fig. 5

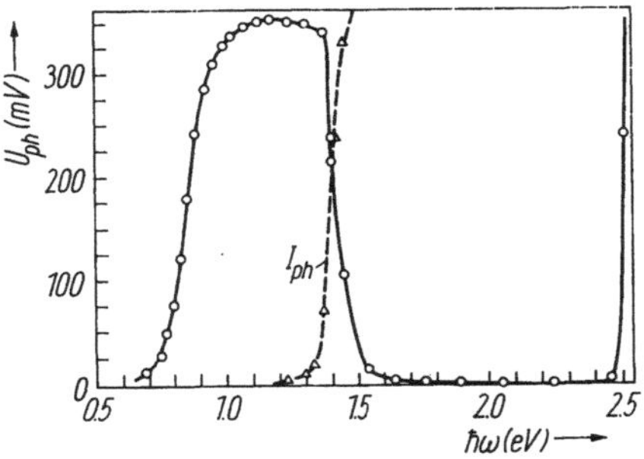

Fig. 6

Fig. 5. a) Photovoltage U_{ph} at Au–CdS contact under two-beam illumination as a function of the photon energy of the second beam. The photon energy of the first beam is 2.0 eV. Dark conductivity of the CdS crystal $\sigma = 10^{-12}$ $(\Omega\ \text{cm})^{-1}$. b) Photocurrent under externally applied voltage for the same illumination and the same sample as a). The quenching extrema occur at the same photon energies as the maxima of the photovoltage

Fig. 6. Photovoltage U_{ph} and short-circuit photocurrent I_{ph} (dashed line) of a Au–CdS contact at $T = 110$ K as a function of photon energy. Dark conductivity of the CdS crystal $\sigma = 10^{-9}$ $(\Omega\ \text{cm})^{-1}$

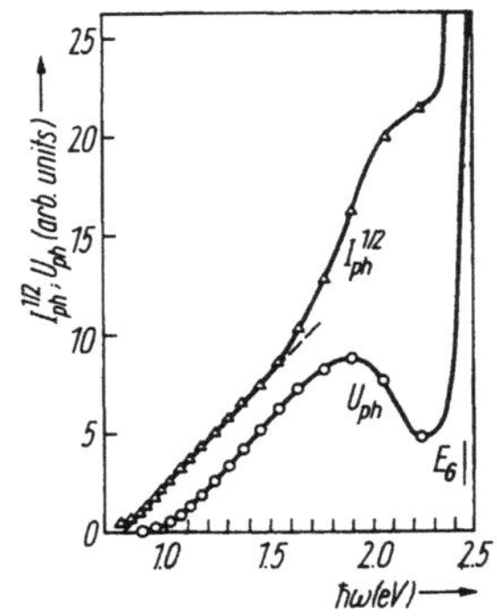

Fig. 7. Photovoltage U_{ph} and square root of the short-circuit photocurrent $I_{ph}^{1/2}$ (Fowler plot) of a Au–CdS contact at $T = 295$ K. Dark conductivity of the CdS crystal $\sigma = 10^{-5}$ $(\Omega \text{ cm})^{-1}$

tation causes an enhanced recombination of majority carriers and hence a reduction of the photocurrent.

On the other hand, the holes as minority carriers can be quenched as well by the generation of electrons as the majority carriers. An example for this effect is shown in Fig. 6. At a Au–CdS contact at 110 K with CdS of a dark conductivity of 10^{-9} $(\Omega \text{ cm})^{-1}$ a maximum of the photovoltage is observed at about half the band-gap value due to minority carrier generation in the CdS by two-step excitation of electrons from the valence band into the conduction band. For the photovoltage at still lower photon energies a small contribution of photoemission from the Au contact may not be excluded. Around 1.4 eV, however, a strong generation of electrons is indicated by a steep rise of the short-circuit photocurrent due to a reduction of the series resistance of the CdS. As Fig. 6 shows, the rise of the short-circuit photocurrent is accompanied by a pronounced decrease of the photovoltage over the spectral region up to the CdS band gap. This decrease of the photo-e.m.f. is due to the quenching of the holes by photoexcited electrons.

In a Au–CdS contact with CdS of an intermediate value of the conductivity, both mechanisms for the generation of a photo-e.m.f., photo emission from the Au layer and hole generation in the CdS, may be observed. This is shown in Fig. 7 for CdS of a dark conductivity $\sigma = 10^{-5}$ $(\Omega \text{ cm})^{-1}$. At small photon energies photoemission is observed in agreement with Fig. 2. With increasing photon energies the short-circuit photocurrent increases somewhat more strongly than expected from a Fowler plot. This effect could be due to a reduction of the series resistance in the CdS and is not necessarily incompatible with predominant electron emission from the Au layer as the source of U_{ph} and I_{ph} in Fig. 6. With further increase of the incident photon energy the photovoltage is, however, observed to decrease. This behaviour cannot be accounted for by photoemission but points to hole generation in the CdS as the predominant source of the photo-e.m.f.

3. Conclusions

In conclusion, it may be stated that the mechanism of the generation of a photo-e.m.f. at photon energies smaller than the band gap in a metal–CdS Schottky contact depends on the conductivity of the CdS. For CdS with a dark conductivity $\sigma >$ $> 10^{-5}$ $(\Omega \text{ cm})^{-1}$, the photo-e.m.f. is almost entirely due to the emission of photoexcited electrons from the metal into the CdS in close analogy to the emission of electrons from the metal into vacuum in a vacuum photocell. Into CdS with a dark conductivity $\sigma < 10^{-9}$ $(\Omega \text{ cm})^{-1}$ no photoemission of electrons occurs beyond the potential maximum in the CdS. The observed photo-e.m.f. is due to the optical generation of mobile holes as the minority carriers in CdS. In metal–CdS contacts with conductivity values between the extremes of the conducting and the insulating type both mechanisms may occur. While at the tail of small photon energies the photo-e.m.f. is predominantly due to photo emission, at higher photon energies, but still at values below that of the CdS band gap, hole generation in the CdS is observed to be the prevailing mechanism for the generation of a photo-e.m.f.

References

[1] D. C. Reynolds, G. Leies, L. L. Antes, and R. E. Marburger, Phys. Rev. **96**, 533 (1954).
[2] R. Williams and R. H. Bube, J. appl. Phys. **31**, 968 (1959).
[3] C. A. Mead and W. G. Spitzer, J. appl. Phys. **34**, 3061 (1963).
[4] C. A. Mead, Solid State Electronics **9**, 1023 (1966).
[5] A. M. Goodman, J. appl. Phys. **35**, 573 (1964).
[6] H. G. Grimmeiss and R. Memming, J. appl. Phys. **33**, 2217 (1961).
[7] E. D. Fabricius, J. appl. Phys. **33**, 1597 (1962).
[8] J. Woods and I. A. Champion, J. Electronics **7**, 243 (1959).
[9] N. Duc Cuong and J. Blair, J. appl. Phys. **37**, 1660 (1966).
[10] P. N. Keating, J. Phys. Chem. Solids **24**, 1101 (1963).
[11] W. Palz and W. Ruppel, phys. stat. sol. **15**, 649 (1966).
[12] S. Yu. Pavelets and G. A. Fedorus, Ukr. fiz. Zh. **14**, 620 (1969).
[13] R. Williams, in: Semiconductors and Semimetals, Vol. 6, Academic Press, New York/London 1970.
[14] J. O. McCaldin, T. C. McGill, and C. A. Mead, J. Vacuum Sci. Technol. **13**, 802 (1976).
[15] B. Lepley and S. Ravelet, phys. stat. sol. (a) **33**, 517 (1976).
[16] L. J. Brillson, Phys. Rev. Letters **40**, 260 (1978).
[17] G. Lubberts, J. appl. Phys. **47**, 365 (1976).
[18] V. M. Marakhonov, V. M. Busov, E. L. Seisyan, and M. M. Shulinskaya, Zh. tekh. Fiz., Pisma **2**, 356 (1976).
[19] B. Lepley, S. Ravelet, and P. Renard, phys. stat. sol. (a) **22**, K191 (1974).
[20] Juh Tzeng Lue, J. Phys. E **11**, 84 (1978).
[21] S. M. Sze, I. L. Moll, and T. Sugano, Solid State Electronics **7**, 509 (1964).
[22] S. O. Hemila and R. H. Bube, J. appl. Phys. **38**, 5258 (1967).

(Received June 23, 1978)

phys. stat. sol. (a) 49, 177 (1978)

Subject classification: 13.3; 16; 20.1; 22.4.1

Institut für Angewandte Physik der Universität Karlsruhe

Investigation of Surface Properties of CdS by Means of Photoreflectance in the Exciton Region

By

L. RISCH[1])

In honour of Prof. Dr. F. STÖCKMANN's 60th birthday

CdS crystals exhibit two kinds of photoreflectance spectra which are classified as type A and type B. Surface-sensitive experiments show these spectra to be induced or influenced by surface electric fields. Type A spectra are favoured if an enhancement layer is present and type B spectra if there is a depletion layer. These spectra can be transformed from one type into the other by additional electric fields or temperature treatment of the crystals. Although a similar behaviour of CdS is known from photoconductivity experiments, a direct relation between photoconductivity and photoreflectance is questionable.

An CdS-Kristallen treten zwei Typen von Photoreflexionsspektren auf, die mit Typ A und Typ B bezeichnet werden. Mit Hilfe von oberflächenempfindlichen Experimenten wird bewiesen, daß die Photoreflexionstypen empfindlich auf elektrische Randschichten reagieren. Eine Anreicherungsrandschicht liefert bevorzugt Typ A-Spektren während eine Verarmungsrandschicht Typ B-Spektren zeigt. Diese Spektren lassen sich durch zusätzliche elektrische Felder oder durch Temperaturbehandlung der Kristalle ineinander umwandeln. Ein sehr ähnliches Verhalten ist bei CdS aus Photoleitungsexperimenten bekannt. Eine direkte Korrelation der Photoleitungstypen und Photoreflexionstypen erscheint jedoch fraglich.

1. Introduction

Photoreflectance measurements in the exciton region of CdS platelets with high resistivity exhibit two different kinds of spectra classified as type A and type B [1]. The intercomparison of photoreflectance and electroreflectance spectra showed type A modulation spectra to result from an increase in electric field strength through illumination. This increase in electric field strength increases the thickness of the exciton-free layer and the half-width of the exciton line. Type B modulation spectra result from a decrease in strength of the built-in electric field through illumination. This decrease in electric field strength decreases the thickness of the exciton-free layer and the half-width of the exciton line.

The appearance of type A or type B photoreflectance is due to the CdS crystals having different electronic properties although their growing conditions were kept as constant as possible. Even crystals from the same batch may differ in surface properties. It has as yet not been possible to discover any relation between their electronic properties and the growing parameters such as the gradient of temperature across the furnace.

CdS crystals with different electronic properties have long been known from measurements of spectral photoconductivity. Gross and Novikov [2] have classified these as type I and type II crystals. It is further known that the properties of such crystals can be changed for example by temperature treatment and that such change is partially reversible.

[1]) Present address: Siemens AG, FL HLT 21, Otto-Hahn-Ring 6, D-8000 München 83, BRD.

An investigation of whether photoreflectance and photoconductivity are related is of great interest as a means of arriving at a better understanding of the electronic properties of the CdS surface.

Among various interpretations of this phenomenon, the explanation proposed in [3, 4], which is based upon different electric space charge layers at the crystal surface, seems most convincing.

Since the modulation mechanism of photoreflectance depends on the change in the electric surface layers [1] it should be possible to induce or change an electric field at the surface and consequently to observe its effect on the photoreflectance spectra.

2. Experimental Details

The apparatus and CdS crystals used have been described in [1]. The crystals which had a temperature treatment (see Section 3.3) were heated at a rate of about 2 K/min in a vacuum chamber. After the final temperature was reached, it was kept constant for about 30 min before switching off the heating. The crystals took several hours to cool down to room temperature. The photoreflectance spectra were measured before and after each heat treatment at about 105 K.

3. Experimental Results

3.1 Photoreflectance with external electric field

Colbow et al. [5] demonstrated in a photoconductivity experiment that type I and II properties are related to different electric space charge layers on the crystal surface. Using a field-effect set-up, they observed that the photocurrent changes to type I under the influence of an enhancement field and to type II under the influence of a depletion field.

If the kind of space charge layer also determines the photoreflectance, then the line profiles should also change when an external electric field is applied. The crystal surface was prepared with a semi-transparent Au electrode. A dc voltage changed the electric field on the semiconductor surface.

With the voltage removed the crystal showed photoreflectance spectra of type A (Fig. 1) both under the Au electrode and on the free surface. When a positive voltage (enhancement layer) is applied, the type A properties remain. When a negative voltage (depletion layer) is applied, the line profile changes with increasing voltage into type B until at about −6 V the larger type B signal is clearly present (Fig. 1). (The energy shift is caused by the current which heated up the crystal.)

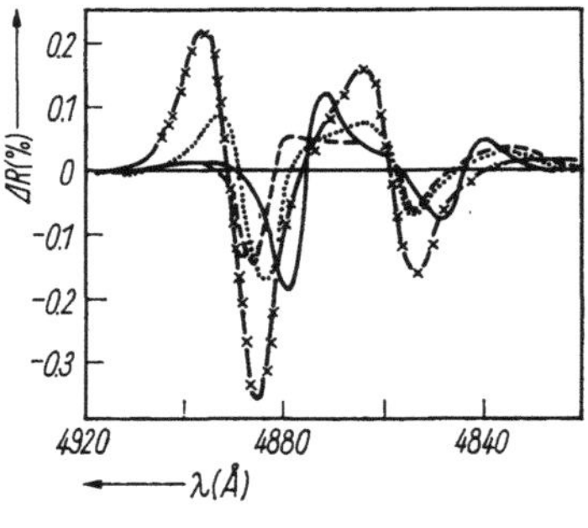

Fig. 1. Photoreflectance spectra with an external electric field. ——— $U = 0$, - - - - −5, ·········· −5.5, ×–×–× −6 V

3.2 Photoreflectance on enhancement layer and on depletion layer

The preceding experiments established the sensitivity of the photoreflectance on space charge layers. The creation of a depletion layer changed the original type A spectrum into type B. The following experiment investigates whether an original type B spectrum, which seems to be related to a depletion layer will change into a type A spectrum under the influence of the field of an enhancement layer.

The enhancement layer was produced by evaporating a thin In layer onto one half of the crystal surface while the other half remained free. Two different photoreflectance spectra (Fig. 2) type A under the In layer, type B on the free surface were measured on the crystal plane which had been thus prepared.

On the unprepared CdS surface a depletion layer is usually formed by negatively charged surface states, probably oxygen. The built-in electric field of this depletion layer decays under the influence of the modulating light beam (inferred from line profile of type B modulation, see [1]).

The conditions under the In layer are more complex. Since the electron density in the contact zone is increased by the metal with a low work function, the original depletion layer changes in the dark into an enhancement layer. Illumination of the In–CdS junction increases the density of free charge carriers in the semiconductor bulk. The number of charge carriers in the contact zone is however determined by the metal and changes only slightly with illumination. When the surface is illuminated with high intensity it is therefore possible that a depletion layer may form on the surface [6].

Fig. 2 shows a type A photoreflectance spectrum to be present after the electronic structure of the surface has been changed by the metal with a low work function (corresponding to an enhancement layer) and that the change can be detected by photoreflectance spectroscopy.

As the excitons are only sensitive to the electric field strength, the sign of the field cannot be the main reason for the different modulation behaviour. The difference is due to a depletion layer forming under the influence of the illumination. It may be inferred from the type A photoreflectance spectrum that the electric field strength of this depletion layer is greater than that of the original enhancement layer.

3.3 Changes of type of photoreflectance under temperature treatment

The behaviour of the type A and B line profiles in response to changes in the space charge layer is very similar to that of type I and type II photoconductivity. If such a relation exists the change in the type of photoconductivity under temperature treatment reported by many authors should be accompanied by a change in the type of photoreflectance.

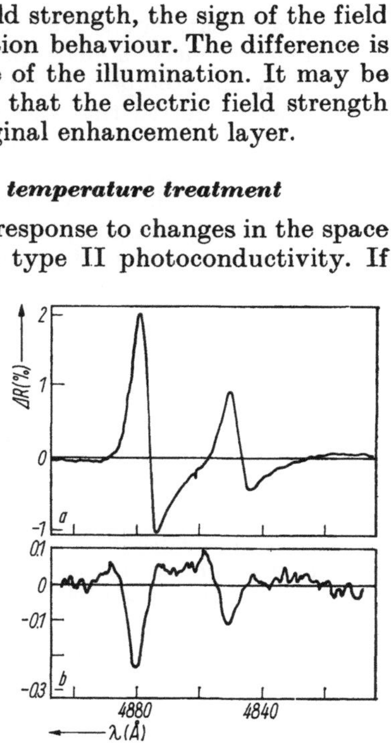

Fig. 2. a) Type B photoreflectance spectrum of the free surface and b) type A photoreflectance spectrum with an In layer

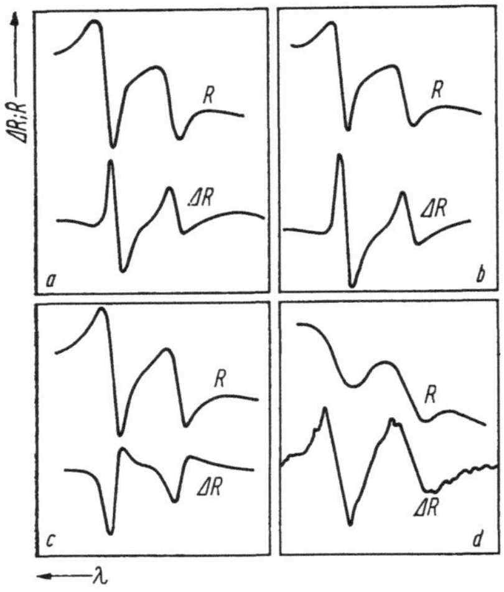

Fig. 3. Reflectance and photoreflectance spectra after different temperature treatments. a) Untreated crystal surface, type B photoreflectance; b) heating to 120 °C, type B photoreflectance; c) heating to 340 °C, type A photoreflectance; d) heating to 490 °C, type A → B

Fig. 3 shows the reflectance and photoreflectance spectra of a crystal after different temperature treatments. The changes in the reflectance spectra are not very pronounced and therefore do not allow any inference to be drawn concerning changes in the properties of the space charge layer except by means of sophisticated calculations. However, the photoreflectance spectra clearly show a change in the electronic structure at the surface.

Since the untreated crystal surface (Fig. 3a) exhibits a type B photoreflectance spectrum, an electric field which decays with illumination must be present on the crystal surface. After the first heating to 120 °C (Fig. 3b) the crystal did not change in type, but there was some change in height of the peaks. Further heat treatment at 340 °C (Fig. 3c) changed the photoreflectance spectrum into type A. The modulating light beam now increases the electric field strength in the space charge layer. After heat treatment to 490 °C (Fig. 3d) the reflectance signal deteriorated. Using a mass spectrometer a sharp peak with the mass 32 was detected in the residual gas analysis at this temperature which was not present at room temperature. It may therefore be inferred that the reflectance signal was reduced by the disturbance of the crystal lattice by the evaporation of S and perhaps Cd (it was not possible to detect such a large mass with the mass spectrometer used). The photoreflectance signal also diminished, but clearly exhibits a reversion from type A to type B.

Similar changes in the photoreflectance signals were observed for many CdS crystals, so confirming that the types of photoreflectance change in the same way as the types of photoconductivity.

3.4 Relation between types of photoconductivity and photoreflectance

Measurements of the photoconductivity showed the crystals used to be of both type I and II. Often they showed a mixed behaviour, making classification difficult.

Classification as type I or II was done on the basis of three accepted assumptions [2, 3, 4]. (i) Type I crystals have photocurrent maxima at the exciton energies, while type II crystals have minima. (ii) Type I crystals become photoconductive at a photon energy which corresponds approximately to the band gap; at higher photon energies the photocurrent remains constant. Type II crystals on the other hand already become photoconductive at a photon energy which is appreciably lower than the band gap energy. The photocurrent has a maximum near the band gap and decreases at high photon energies. (iii) Adsorption of electronegative gases greatly increases the resistivity of type I crystals, whereas type II crystals are almost insensitive to such adsorption.

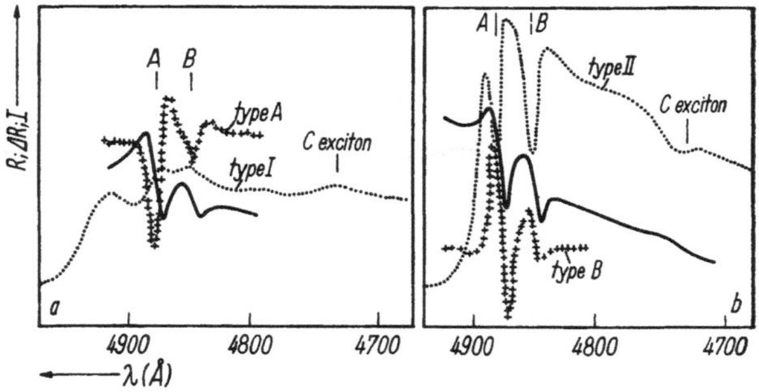

Fig. 4. Comparison of the types of photoconductivity and photoreflectance at two different crystals. ······ photocurrent I, ——— reflectance R, ++++ photoreflectance ΔR

Simultaneous measurements of photoconductivity and photoreflectance indicated a relationship between the crystal types [7]. Type I photoconductivity was favoured with type A photoreflectance (Fig. 4a) and type II photoconductivity with type B photoreflectance (Fig. 4b), although there was no 1:1 relationship.

Later experiments by Aures [10] with about twenty crystals from different batches showed no favorization of type I/type A and type II/type B properties among the crystals. Type II photoconductivity crystals exhibited type A photoreflectance as well.

4. Discussion

The assumed relationship between the types of photoconductivity and photoreflectance agrees very well with a model of the CdS surface given by Bragagnolo et al. [8, 9] for photoconductivity. The explanation of type I and II properties is based on different recombination velocities of the charge carriers in the semiconductor bulk and in the surface region. Type II crystals have a high surface and a low bulk recombination velocity, whereas type I crystals have a higher recombination velocity in the bulk than at the surface. Bragagnolo et al. [8, 9] propose a model assuming an overstoichiometric Cd layer at the surface which produces type I properties. If there is no Cd-rich layer at the surface, the crystal has type II properties. This Cd layer can originate from crystal growth and is easily changed by heat or surface treatments.

This Cd layer at the surface of type I crystals causes a great number of donor states in the first few 100 Å of the crystal, providing a layer of high conductivity beneath the surface. The normal bulk material of high resistivity is below the Cd-doped enhancement layer. In equilibrium an n^+–n junctions is formed with a lower recombination velocity of the electrons in the n^+-region. Directly under the surface this high conductivity layer is more or less depleted by adsorption of oxygen.

In type II crystals there is only a depletion layer at the surface extending a few 1000 Å into the crystal. The recombination velocity is high in this space charge layer.

The Cd-rich layer proposed in [8, 9] can be adapted to the model of photoreflectance proposed in [1], which assumed type A modulation for small and type B modulation for large band bending at the surface especailly under the condition $n_s\sigma_n > p_s\sigma_p$ and $n_s\sigma_n < p_s\sigma_p$, respectively ($n_s$, p_s density of electrons, holes at the surface, σ_n, σ_p capture cross-sections for electrons, holes).

The Cd-rich surface layer of type I crystals explains the preferred appearance of type A photoreflectance in the following way: As a result of the n^+-region the vacant surface state af oxygen, ≈ 0.8 eV below the conduction band, lies closer to the Fermi level (see Fig. 5). An additional occupation with electrons through illumination, which increases the electric field becomes more likely than when this enhancement layer is missing. This would also explain the favorization of type A photoreflectance when an enhancement layer is present at the surface (see Sections 3.1 and 3.2).

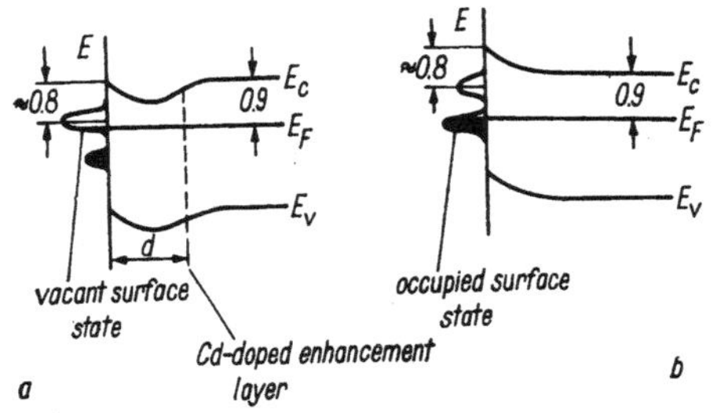

Fig. 5. Model of the CdS surface with and without a Cd enhancement layer. a) Type A, b) type B photoreflectance

A possible alternative explanation for type A photoreflectance is the mechanism discussed in Section 3.2. The Cd layer pins the Fermi level at the surface, while the density of electrons increases below the surface through illumination and so the electric field.

This mechanism also seems to be possible without the Cd layer as the result of a high recombination velocity in the region near the surface, which could explain the appearance of the type II photoconductivity and the type A photoreflectance reported in [10].

Type I photoconductivity crystals which have a strongly depleted Cd layer with high electric field strength are expected to have type B photoreflectance. Thus they can exhibit both type A and B photoreflectance.

Since the behaviour of photoconductivity and that of photoreflectance are very similar, a common mechanism seems to exist. Both effects reinforce each other because they are sensitive to the surface properties at different depths.

It has been found for example that the change in the properties does not take place at the same temperature. For a more sophisticated model of the electronic structure of the CdS surface an additional experimental investigation of the Cd layer using a surface analytic method such as that of Auger or Sims is necessary.

Acknowledgement

I wish to thank Prof. Dr. W. Stößel for helpful suggestions.

References

[1] L. Risch, phys. stat. sol. (b) **88**, 111 (1978).
[2] E. F. Gross and B. V. Novikov, J. Phys. Chem. Solids **22**, 87 (1961).
[3] E.-H. Weber, phys. stat. sol. **28**, 649 (1968); **36**, 175 (1969); phys. stat. sol. (a) **1**, 665 (1970).
[4] R. Schubert and K. W. Böer, J. Phys. Chem. Solids **32**, 77 (1971).
[5] K. Colbow, A. Imaeff, and K. Yuen, Canad. J. Phys. **48**, 57 (1970).
[6] W. Ruppel and H. G. Schlaile, phys. stat. sol. **15**, 675 (1966).
[7] L. Risch, Thesis, TU Karlsruhe, 1976.
[8] J. A. Bragagnolo and K. W. Böer, phys. stat. sol. (a) **21**, 291 (1974).
[9] J. A. Bragagnolo, C. Wright, and K. W. Böer, phys. stat. sol. (a) **24**, 147 (1974).
[10] R. Aures, Diploma Work, TU Karlsruhe, 1978.

(Received June 23, 1978)

phys. stat. sol. (a) 49, 183 (1978)

Subject classification: 13.4 and 20.1; 22.2.1; 22.2.3

A. F. Ioffe Physico-Technical Institute, Academy of Sciences of the USSR, Leningrad

Absorption in Indium Antimonide and Arsenide Crystals Doped with Various Admixtures

By

A. S. FILIPCHENKO, L. P. BOLSHAKOV, and A. NAURIZBAEV

Absorption spectra in indium arsenide and indium antimonide crystals are investigated at temperatures 77 and 300 K. Theoretical calculations are made in strict compliance with Kane's band model and the influence of the doping degree on the forbidden band width. The investigations performed show that the spectral dependence of the absorption coefficient is in good agreement with Kane's band model for the whole interval of the concentrations investigated.

Absorptionsspektren in Indiumarsenid- und Indiumantimonidkristallen werden bei Temperaturen von 77 und 300 K untersucht. Theoretische Berechnungen werden unter strenger Beachtung des Bandmodells von Kane und des Einflusses des Dotierungsgrades auf die Breite der verbotenen Zone durchgeführt. Die Untersuchungen zeigen, daß die spektrale Abhängigkeit des Absorptionskoeffizienten sich in guter Übereinstimmung mit dem Bandmodell von Kane für den gesamten Bereich der untersuchten Konzentrationen befindet.

1. Introduction

The optical absorption in indium antimonide and indium arsenide crystals has been studied in a series of papers [1, 2]. The most interesting results were obtained in [3, 4], which investigated the influence of the doping degree of indium arsenide on the structure of the energy bands. However, the majority of authors in the theoretical calculations of the optical absorption spectrum used various approximations sometimes not quite justified. Thus practically in no single work on indium arsenide one can find strict calculations of the absorption spectra. This could be the source of an uncertainty in the interpretation of the experimental results. Moreover, in the last few years a series of investigations has been published showing that many effects in $A^{III}B^{V}$ compounds substantially depend on the kind of the doping admixture [5, 6]. In connection with this we undertook an investigation of the optical absorption spectra near the fundamental edge of indium antimonide and arsenide crystals, doped with various admixtures,

InSb (Te, Se, S) ,

InAs (Te, Sn, S) ,

in the largest possible interval of concentrations. The theoretical calculations were carried out in strict compliance with Kane's band model [7]; the influence of doping on the gap width and the position of the Fermi level were also, as far as possible, taken into account.

2. Theory

It is known that the optical absorption coefficient in a non-degenerate semiconductor [7] can be represented by the expression

$$\alpha_0 = \frac{4\pi^2 e^2 \hbar}{m^2 c n \hbar \omega} \sum_j M_j^2 \varrho_j \,, \tag{1}$$

where the summation is carried out over all possible transitions between electron states, separated by the energies $\hbar\omega$. The square of the optical matrix element M_j^2 depends on the wave vector $\boldsymbol{k}$ through the wave function coefficients

$$M_j^2 = \frac{2m^2P^2}{3\hbar^2}\,[(a_c c_j + c_c a_j)^2 + (a_c b_j - b_c a_j)^2] \tag{2}$$

and

$$\left.\begin{aligned} a_j &= \frac{1}{N}\,kP\left(E_j' + \frac{2}{3}\,\Delta\right), \qquad b_j = \frac{1}{N}\frac{\sqrt{2}}{3}\,\Delta(E_j' - E_g)\,, \\ c_j &= \frac{1}{N}\,(E_j' - E_g)\left(E_j' + \frac{2}{3}\,\Delta\right), \end{aligned}\right\} \tag{3}$$

where N is a normalizing factor determined from the condition

$$a_i^2 + b_i^2 + c_i^2 = 1\,.$$

In the case of a simple parabolic band $b = 1$, $a = c = 0$. The matrix element P is connected with the width of the energy gap E_g, the energy of the spin–orbit splitting Δ, and the electron effective mass at the bottom of the conduction band m_n by the relation

$$P^2 = \frac{3\hbar^2}{2}\left(\frac{1}{m_n} - \frac{1}{m}\right)\frac{E_g(E_g + \Delta)}{3E_g + 2\Delta}\,. \tag{4}$$

The density of states ϱ_j is given by the expression

$$\varrho_j = \frac{k^2}{2\pi^2}\left|\frac{dE_c}{dk} - \frac{dE_j}{dk}\right|^{-1}\,. \tag{5}$$

In this way, the specific form of the fundamental optical absorption spectrum is determined ultimately by the dependence of the electron energy on its wave vector, i.e. the band structure of the semiconductor. According to Kane's theory [1], the energy band structure in $A^{III}B^{V}$ semiconductor compounds is described by a cubic equation:

$$E'(E' - E_g)\,(E' + \Delta) - k^2P^2(E' + \tfrac{2}{3}\,\Delta) = 0\,, \tag{6}$$

where

$$E' = E - \frac{\hbar^2k^2}{2m}\,.$$

In the majority of papers the approximation $\Delta \gg kP$, E_g has been used which enables to obtain from (6) the explicit form of the $E(\boldsymbol{k})$ dependence. If such an approximation is acceptable for indium antimonide, in the case of indium arsenide, however, it is completely unjustified ($E_g \approx \Delta$). That is why in our theoretical calculations of the absorption spectra of InSb and InAs we, without resorting to approximations, have numerically solved equation (6) using computer techniques. Equation (6) describes three energy bands: the conduction band E_c, the valence band of "light" holes E_{V2}, and the E_{V3} band, which is split due to the spin–orbit interaction. The valence band of "heavy" holes E_{V1} is considered to be parabolic with a curvature determined by the effective mass of the density of states m_p. The dependence of the wave function coefficients a, b, c on the square of the wave vector was calculated according to (3). In the case of InSb these relations agree with those obtained previously by Kane [7] within the limits of the different values used for the initial parameters. The values of the characteristic parameters which were used in our calculations are shown in the following table:

Table 1

	E_g (eV)	Δ (eV)	m_n/m	m_p/m
InSb	0.23 (77 K) 0.18 (300 K)	0.9	0.0152 0.0116	0.5732
InAs	0.41 (77 K) 0.36 (300 K)	0.43	0.026 0.023	0.6736

In the presence of degeneracy a shift of the optical absorption band edge is observed (Moss-Burstein effect) and the dependence of the absorption coefficient on the energy is given by the expression [9]

$$\alpha = \frac{\alpha_0}{\exp \frac{1}{k_0 T}\left[E_F - \frac{\hbar\omega - E_g}{1 - E_{Vj}/(E_c - E_g)}\right] + 1}, \tag{7}$$

where E_F is the Fermi level measured from the top of the valence band, $\alpha_0(\boldsymbol{k})$ the absorption coefficient in the non-degenerate semiconductor calculated from (1).

The spectral dependence of the absorption coefficient was corrected for the narrowing of the band gap due to the high doping level of the samples. The value of this narrowing was assumed to depend on the concentration of the admixture according to the relation [8]

$$\Delta E \approx -\frac{1.5e^2}{\varepsilon r_s}; \qquad r_s = \left(\frac{4}{3}\pi n\right)^{-1/3}. \tag{8}$$

3. Experimental Results and Discussion

About 40 single-crystal indium antimonide and indium arsenide samples grown by the Czochralski method were investigated. The electron concentration determined from Hall effect measurements was in the limit 1.6×10^{17} cm^{-3} to 4.8×10^{18} cm^{-3} for the indium antimonide samples and 1.5×10^{16} cm^{-3} to 5.3×10^{18} cm^{-3} for the indium arsenide samples. The indium arsenide crystals with electron concentration of the order of 10^{16} cm^{-3} were not specially doped. The transmission spectra were observed on an IKS-14A spectrograph with variable prisms. The absorption coefficient was calculated from transmission spectra taking into account the reflection coefficient by the usual methods. The surface area of the samples was not less than 5×5 mm^2. The absorption coefficient for each sample was measured for two different sample thicknesses. Fig. 1 and 2 show the spectral dependence of the absorption coefficients at temperatures 77 and 300 K. The solid lines represent the spectral curves calculated theoretically according to the technique described above. In the investigated range of concentrations the agreement between theoretical and experimental results is satisfactory. The shift of the absorption band edge caused by the filling of the energy bands with carriers (Moss-Burstein effect) is clearly observed. The discrepancy observed in the purest samples can apparently be connected with some compensation of impurities and an appreciable influence of electron transitions impurity level–conduction band. In general, the small deviations from the calculated relations which occur in the samples with higher electron concentrations can be included in the errors introduced by the use of the approximate formula (8). At the same time, the absorp-

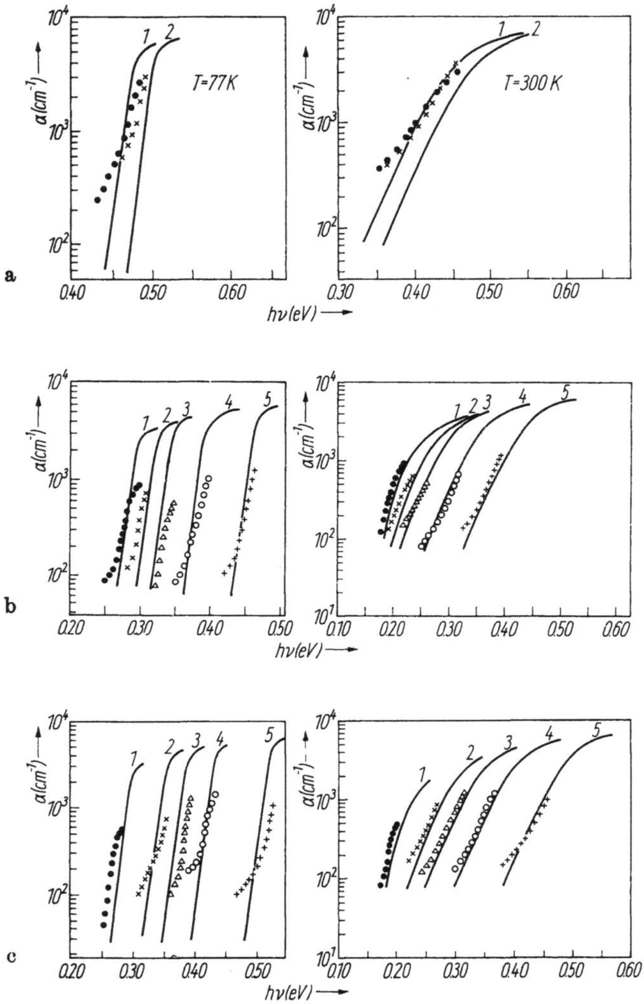

Fig. 1. The spectral dependence of the absorption coefficient for indium antimonide at $T = 77$ K (left-hand side) and $T = 300$ K (right-hand side) doped with a) S; ● $n = 3.4 \times 10^{18}$ and × $4.26 \times 10^{18}{}_3$cm^{-3}. b) Te; ● $n = 2.1 \times 10^{17}$, × 4.28×10^{17}, △ 6.77×10^{17}, ○ 1.45×10^{18}, × 2.97×10^{18} cm^{-3}. c) Se; ● $n = 1.67 \times 10^{17}$, × 7.54×10^{17}, + 4.84×10^{18}, △ 1.27×10^{18}, ○ 2.18×10^{18} cm^{-3}. Solid curves: theory

tion of the photons with energy deficiencies ($E_g - \hbar\omega > 0$) can be connected with the absorption of free carriers and with the existence of fluctnations in the concentration of the charged impurities, which leads to the formation of a potential relief at the bottom of the conduction band. The importance of these effects must naturally grow with increasing doping level of the samples.

The presented data do not permit to make positive conclusions on the influence of the type of the doping admixture on the optical absorption spectra. One can hope that such a dependence can be found from investigations of crystals with maximum concentrations of admixtures.

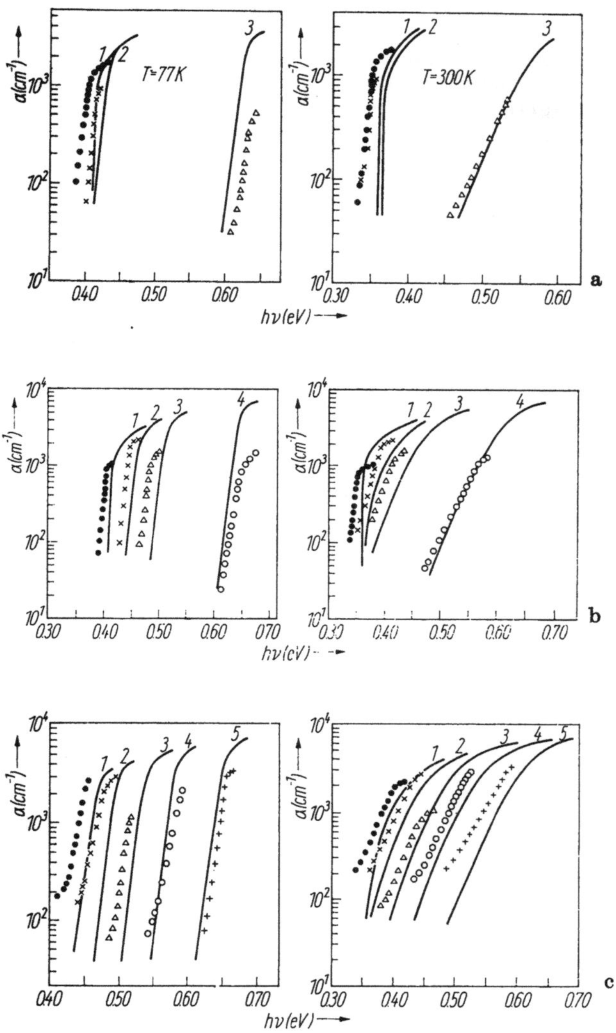

Fig. 2. The spectral dependence of the absorption coeffcient for indium arsenide at $T = 77$ K (left-hand side) and $T = 300$ K (right-hand side) doped with a) S; ● $n = 1.8 \times 10^{16}$, × 1.6×10^{17}, △ 4.78×10^{18} cm^{-3}. b) Te; ● 1.5×10^{16}, × 4.1×10^{17}, △ 4.17×10^{18}, ○ 5.12×10^{18} cm^{-3}. c) Sn; ● $n = 4 \times 10^{17}$, × 8.4×10^{17}, △ 1.65×10^{18}, ○ 2.82×10^{18}, + 5.25×10^{18} cm^{-3}. Solid curves: theory

References

[1] M. Ya. Dashevskii, L. V. Kovalchuk, and L. S. Okun, Fiz. Tekh. Poluprov. 4, 959 (1970).
[2] N. V. Zotova, D .N. Nasledov, and L. D. Neuimina, Fiz. Tekh. Poluprov. 5, 2111 (1971).
[3] N. V. Zotova, D. N. Nasledov, L. D. Neuimina, and A. V. Pentsov, Fiz. Tekh. Poluprov. 6, 558 (1972).

[4] N. V. Zotova, G. Garyagdiev, and D. N. Nasledov, Fiz. Tekh. Poluprov. 8, 921 (1974).
[5] A. S. Filipchenko and D. N. Nasledov, phys. stat. sol. (a) 27, 11 (1975).
[6] M. A. Ambartsumyan and V. G. Karnaukhov, Fiz. Tekh. Poluprov. 2, 734 (1968).
[7] E. O. Kane, J. Phys. Chem. Solids 1, 249 (1957).
[8] A. A. Rogachev, Fiz. tverd. Tela 9, 369 (1967).
[9] E. J. Johnson, in: Optical Properties of $A^{III}B^{V}$ Compounds, Izd. Mir, Moscow 1970 (in Russian).

(Received February 13, 1978)

phys. stat. sol. (a) 49, 189 (1978)

Subject classification: 1.1 and 1.4; 22.8.1

Laboratoire de Cristallographie et Chimie du Solide, associé au C.N.R.S. n⁰ 251, Université de Caen[1])

Etude par microscopie électronique des phénomènes d'intercroissance dans les siliconiobates de baryum

II. Intercroissances multiples et microdomaines[2])

Par

F. STUDER et B. RAVEAU

Des intercroissances multiples, successions périodiques de plusieurs termes d'intercroissances simples de la famille structurale $(Ba_3Nb_6Si_4O_{26})_n \cdot Ba_3Nb_4Ti_4O_{21}$, ont été observées par diffraction et microscopie électronique. Des intercroissances doubles (I^{II}), triples (I^{III}) et quadruples (I^{IV}) ont été identifiées sur les clichés de résolution directe par microphotométrie. Des microdomaines, constitués par des intercroissances simples $||I \mid J||$ telles que $|I - J| = n'$ soit égal à 0,1 ou 2, ont été identifiés. Deux types de parois de domaines ont été observés: parois longitudinales parallèles au plan (00.1), parois transversales contenant l'axe c. Suivant la composition des microdomaines, on distingue les parois d'antiphase, les parois de domaines et les zones de transition.

Multiple intergrowths, periodic series of many simple intergrowths in the structural family $(Ba_3Nb_6Si_4O_{26})_n \cdot Ba_3Nb_4Ti_4O_{21}$, are observed by electron diffraction and microscopy. Double (I^{II}), triple (I^{III}), and quadruple (I^{IV}) intergrowths are identified by microphotometry. Microdomains made up with simple intergrowths $||I \mid J||$ such as $|I - J| = n' = 0$, 1, or 2 determine two types of domain boundaries: longitudinal boundaries parallel to the (00.1) plane and transversal boundaries normal to the (00.1) plane. Depending on the composition of the microdomains one can identify antiphase boundaries, microdomain boundaries, and transition zones.

1. Introduction

Les mécanismes de non-stoechiométrie par intercroissance entre les structures à tunnels de section pentagonale représentées par les siliconiobates $A_{6-x}M_6Si_4O_{26}$ [1 à 3] et les structures en cages de formulation $A_3M_8O_{21}$ [4 à 6] ont été mis en évidence pour la première fois au laboratoire [7]. Les premiers termes d'une série de phases multiples, $(Ba_3Nb_6Si_4O_{26})_n \cdot Ba_3Nb_4Ti_4O_{21}$ ont ainsi été isolés et observés par résolution directe ($n = 1$; 2; 3; 4; 4,5). Dans une étude plus approfondie [8], de nombreux termes ont été obtenus, correspondant à des valeurs entières de n, comprises entre 1 et 10, et fractionnaires variant entre $\frac{3}{2}$ et $\frac{29}{2}$. Il est également apparu que les intercroissances simples correspondant aux premiers termes de la série ($n = 1$; 1,5; 2) étaient susceptibles de se grouper pour donner des intercroissances doubles bien ordonnées et présentant de très importants motifs périodiques.

Nous décrivons ici de nouvelles intercroissances doubles correspondant à des valeurs de n plus élevées, et des mécanismes plus complexes caractérisés par des intercroissances triples et quadruples. Nous abordons d'autre part l'étude des parois et des zones séparant des microdomaines de composition différente.

---

[1]) 14032 Caen Cédex, France.

[2]) Part I see phys. stat. sol. (a) 48, 301 (1978).

2. Techniques expérimentales

Les techniques expérimentales ont été largement décrites dans les publications précédentes [7, 8]. Soulignons cependant que la principale difficulté technique rencontrée est l'identification des termes de l'intercroissance par la mesure d'une interfrange. Le microphotomètre que nous avons transformé pour obtenir un défilement aussi régulier que possible des négatifs, donne une précision sur l'interfrange qui est de 3,5 Å. Sachant que le plus petit écart séparant deux demi-mailles de l'intercroissance est égal au paramètre c d'une maille siliconiobate $Ba_3Nb_6Si_4O_{26}$, soit 7,9 Å, on voit qu'il est possible par cette méthode de séparer toutes les composantes de l'intercroissance quand la résolution initiale du cliché est correcte. Nous avons appliqué systématiquement cette méthode pour le dépouillement des clichés de résolution directe.

3. Résultats et interprétation

3.1 Aspects structuraux et code de description

Nous nous limitons ici au rappel des données structurales et du code de description, strictement indispensables à la compréhension des termes de l'intercroissance; la description détaillée des termes de la série est donnée dans les précédentes publications [7, 8].

Compte tenu de la similitude des deux structures $A_3M_8O_{21}$ et $A_3M_6Si_4O_{26}$, chaque terme de l'intercroissance simple est caractérisé par une maille hexagonale de paramètre a_n voisin de 9 Å et un paramètre c_n relié à ceux des structures $A_3M_8O_{21}$ et $A_3M_6Si_4O_{26}$ de la façon suivante:

$$c_n \approx c_{A_3M_8O_{21}} + 2nc_{A_3M_6Si_4O_{26}} .$$

La structure de chacun de ces termes peut être décrite par l'empilement suivant c de feuillets siliconiobates $A_3M_6Si_4O_{26}$ (Fig. 1a) et de feuillets simples $A_3M_8O_{21}$ (Fig. 1b): des feuillets multiples constitués respectivement de I et J feuillets simples siliconiobates, séparés par des feuillets simples $A_3M_8O_{21}$ alternent donc suivant c. Chaque terme de l'intercroissance simple est alors défini par la valeur de n, telle que $I + J = 2n$ et $|I - J| = 0$ ou 1, respectivement pour n entier ou $n =$ entier $+ \frac{1}{2}$, comme le montre à titre d'exemple la Fig. 2 pour quatre valeurs de n. Pour plus de commodité, nous représenterons le terme n d'une intercroissance simple par la séquence d'empilement suivante:

$$||I \mid J||I \mid J||I \mid J||I \mid J|| \ldots .$$

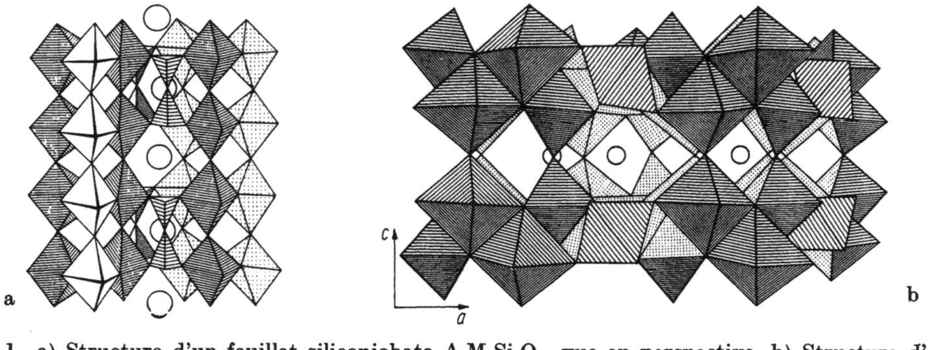

Fig. 1. a) Structure d'un feuillet siliconiobate $A_3M_6Si_4O_{26}$ vue en perspective. b) Structure d'un feuillet $A_3M_8O_{21}$ vue en perspective

Fig. 2. Représentation schématique des quatre premiers termes de la série $(Ba_3Nb_6Si_4O_{26})_n \cdot Ba_3Nb_4Ti_4O_{21}$ par le nombre n et par la description

$$||I \mid J|| = \begin{cases} n = 1 & \text{ou} \quad ||1 \mid 1|| \\ n = 1{,}5 & \text{ou} \quad ||2 \mid 1|| \\ n = 2 & \text{ou} \quad ||2 \mid 2|| \\ n = 2{,}5 & \text{ou} \quad ||3 \mid 2|| \end{cases}$$

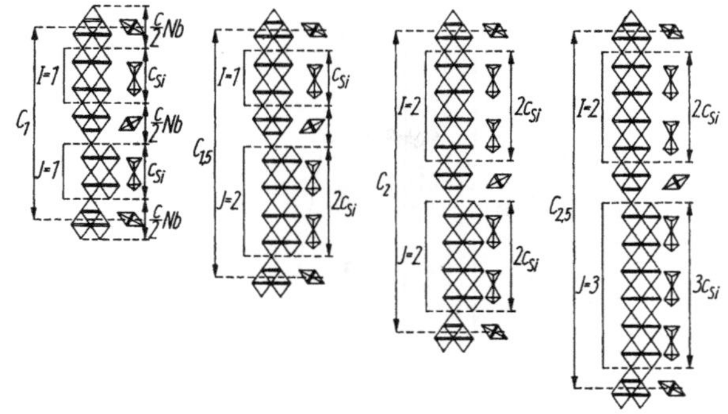

Dans ce code de description, les feuillets simples $A_3M_8O_{21}$ d'épaisseur voisine de 6 Å ($c_{A_3M_8O_{21}}/2$) sont représentés par deux barres verticales lorsqu'ils sont situés à l'origine ($z = 0$), et une barre verticale lorsqu'ils sont à la cote $z = \frac{1}{2}$. Les épaisseurs des feuillets multiples siliconiobates sont alors voisins de $I \times 7{,}9$ Å pour le feuillet I et $J \times 7{,}9$ Å pour le feuillet J. La maille de l'intercroissance est alors représentée par la notation $||I \mid J||$.

Dans les intercroissances multiples des zones peuvent apparaître dans lesquelles $|I - J| = 0$ ou 1. Il semble que dans le cas d'une bonne cristallisation, la différence $n' = |I - J|$ soit limitée aux trois nombres 0, 1 et 2.

3.2 Intercroissances multiples

Plusieurs intercroissances simples, de composition différente, peuvent se succéder de façon périodique, donnant ainsi naissance à des intercroissances multiples. Trois types d'intercroissances ont été observées respectivement: doubles, triples et quadruples.

3.2.1 Intercroissances doubles (I^{II})

Les premières intercroissances doubles ont été décrites précédemment [8]; elles étaient construites à partir des premiers termes de la série $(Ba_3Nb_6Si_4O_{26})_n \cdot Ba_3Nb_4Ti_4O_{21}$ pour $n = 1$, $\frac{3}{2}$ et 2. Nous rappelons ici les quatre principales séquences observées:

$$I_1^{II};\ (||2 \mid 1||2 \mid 1||1 \mid 1||1 \mid 1||)_{n_1}; \qquad c_{I_1^{II}} = 127{,}0 \text{ Å},$$

$$I_2^{II};\ (||2 \mid 1||2 \mid 1||2 \mid 1||1 \mid 1||1 \mid 1||)_{n_2}; \qquad c_{I_2^{II}} = 162{,}7 \text{ Å},$$

$$I_3^{II};\ (||2 \mid 2||2 \mid 1||)_{n_3}; \qquad c_{I_3^{II}} = 79{,}3 \text{ Å},$$

$$I_4^{II};\ (||1 \mid 2||2 \mid 1||)_{n_4}; \qquad c_{I_4^{II}} = 71{,}4 \text{ Å}.$$

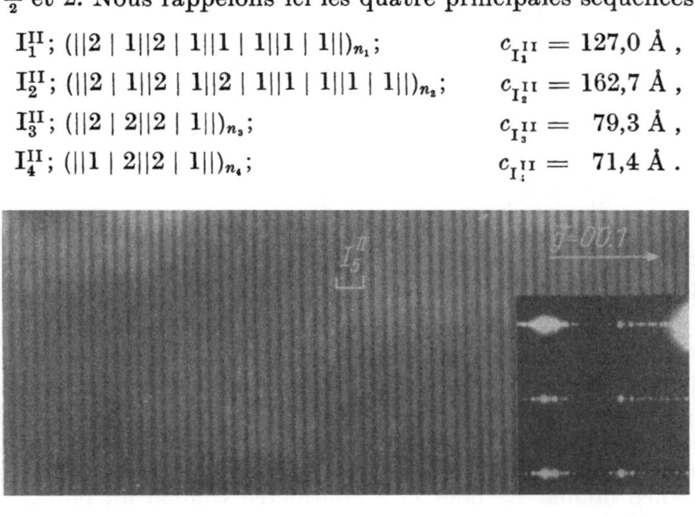

Fig. 3. Résolution directe et diffraction électronique de l'intercroissance double I_5^{II}

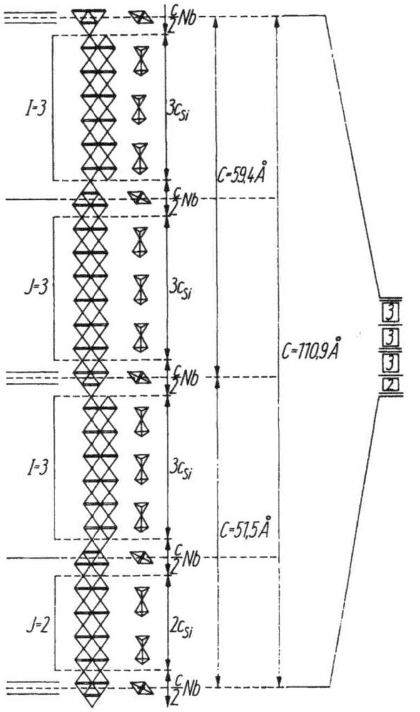

Fig. 4. Représentation schématique de l'intercroissance double I_5^{II} et introduction d'un schématisme plus poussé basé sur les nombres I et J

Les Fig. 3, 4, 5, 6 mettent en évidence trois nouveaux types d'intercroissances doubles construites à partir de termes supérieurs de la série. Le premier type d'intercroissance double, noté I_5^{II}, correspond à la juxtaposition d'une maille $||3 \mid 2||$ et d'une maille $||3 \mid 3||$ conduisant à une maille hexagonale de paramètre $c = 110{,}9$ Å. Des cristaux entiers étaient constitués par cette intercroissance: les clichés de diffraction électronique correspondants (Fig. 3) permettent d'exclure la possibilité d'une intercroissance simple dont les plans (00.1) et (00.2) auraient été résolus.

Fig. 3: I_5^{II}; $(||3 \mid 3||3 \mid 2||)_{n_1}$.

Il faut remarquer que la différence $|I - J|$ est égale à 0 ou 1. La Fig. 4 représente le schéma de la structure de l'intercroissance double I_5^{II} avec les relations entre cette structure et le nombre n dans la formule $(Ba_3Nb_6Si_4O_{26})_n \cdot Ba_3Nb_4Ti_4O_{21}$ d'une part, et la notation $||I \mid J||$ d'autre part. Elle présente aussi une schématisation plus poussée basée sur la notation $||I \mid J||$ qui sera nécessaire pour la description des parois de domaines.

Le deuxième type d'intercroissance double est en fait constitué par la succession dans un même cristal de trois types d'intercroissances doubles bâties sur les deux termes $||4 \mid 4||$ et $||4 \mid 5||$. Les rangées $h0.l$ sur le plan de diffraction de ce cristal ne permettent plus de séparer des réflexions précises. Les intercroissances doubles observées sont les suivantes:

Fig. 5: I_6^{II}; $(||4 \mid 5||4 \mid 5||4 \mid 5||4 \mid 4||)_{n_2}$; $c_{I_6^{II}} = 324{,}5$ Å,

Fig. 5: I_7^{II}; $(||4 \mid 5||4 \mid 5||4 \mid 4||4 \mid 4||)_{n_3}$; $c_{I_7^{II}} = 316{,}6$ Å,

Fig. 5: I_8^{II}; $(||4 \mid 5||4 \mid 4||)_{n_4}$; $c_{I_8^{II}} = 158{,}3$ Å.

Il est à remarquer que la jonction entre I_7^{II} et I_8^{II} se fait par l'intermédiaire d'un défaut dont la séquence est composée de termes I_6^{II} et d'un autre terme:

Fig. 5a: $||4 \mid 5||4 \mid 5||4 \mid 5||4 \mid 5||4 \mid 4||$ ... (A).

Sur la Fig. 5, l'intercroissance double, notée I_6^{II}, peut également être considérée comme un défaut de transition entre le microdomaine contenant l'intercroissance simple $||4 \mid 5||$, notée I_1^{I} sur le cliché, et la zone contenant l'intercroissance double I_7^{II}. On retrouve également l'intercroissance double I_7^{II} de part et d'autre du défaut précédent (A) qui apparaît alors comme bâti sur l'intercroissance double I_6^{II} par l'adjonction d'un terme $||4 \mid 5||$.

L'intercroissance double I_7^{II} peut être considérée comme un polytype de l'intercroissance notée I_8^{II}. Le paramètre c de la maille hexagonale de la séquence I_8^{II} est en

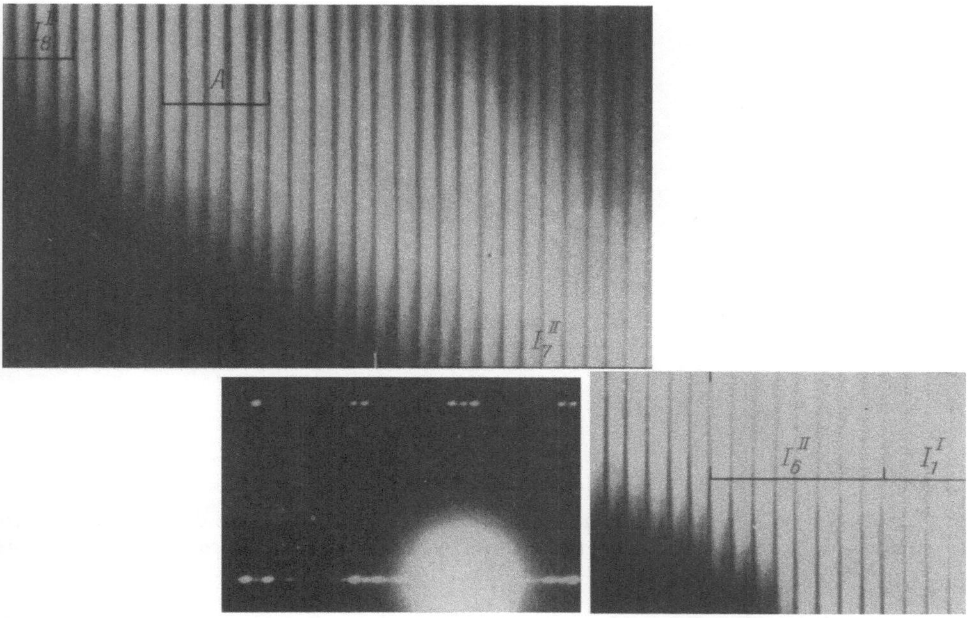

a b

Fig. 5. Résolution directe et diffraction électronique de plusieurs intercroissances doubles I_6^{II}, I_7^{II} et I_8^{II} et de la séquence A: $||4 \mid 5 || 4 \mid 5 || 4 \mid 5 || 4 \mid 5 || 4 | 4 ||$

effet la moitié du paramètre c de la séquence I_7^{II} sans altération globale de la composition. Le troisième type d'intercroissance double, noté I_9^{II}, a été observé dans un microcristal dont le cliché de diffraction permet de distinguer le terme $||9 \mid 9||$ en phase forte. Cette intercroissance double peut être considérée à la limite comme l'apparition d'un défaut localisé mais périodique, le terme $||10 \mid 8||$, dans une séquence régulière du terme $||9 \mid 8||$. La bonne résolution des plans (00.2) permet de déterminer sans ambiguité le nombre de feuillets siliconiobates compris entre deux mailles $A_3M_8O_{21}$ (Fig. 6b). L'intercroissance double observée a pour paramètre $c_{I_9}^{II} = 446{,}8$ Å et correspond à la séquence:

Fig. 6a: I_9^{II}; $(||10 \mid 8||9 \mid 8||9 \mid 8||)_{n_8}$.

Remarquons que le terme de dissymétrie $n' = |I - J|$ prend ici les valeurs 1 et 2. Cette séquence régulière apparaît au milieu d'un cristal présentant une succession désordonnée de termes voisins du terme $||9 \mid 9||$.

3.2.2 Intercroissances triples (I^{III})

Deux types d'intercroissance triples ont été observés. Ils correspondent aux deux premiers empilements possibles, c'est-à-dire la séquence (ABC) et la séquence (ABCB) qui répète un des termes.

La première intercroissance triple, notée I_1^{III}, est bâtie sur les termes $||5 \mid 5||$, $||4 \mid 5||$ et $||4 \mid 4||$ suivant la séquence:

Fig. 7a: I_1^{III}; $(||5 \mid 5||5 \mid 4||4 \mid 4||)_{n_8}$; $c_{I_1^{III}} = 249{,}3$ Å .

Le long de cette séquence, la variation du nombre $n' = |I - J|$ reste égal à 0 ou 1.

La deuxième intercroissance triple I_2^{III} correspondant à la séquence (ABCB), est bâtie sur les termes $||6 \mid 6||$, $||4 \mid 5||$ et $||3 \mid 4||$:

Fig. 7b: I_2^{III}; $(||6 \mid 6||3 \mid 4||4 \mid 5||4 \mid 3||)_{n_7}$; $c_{I_2^{III}} = 324{,}5$ Å .

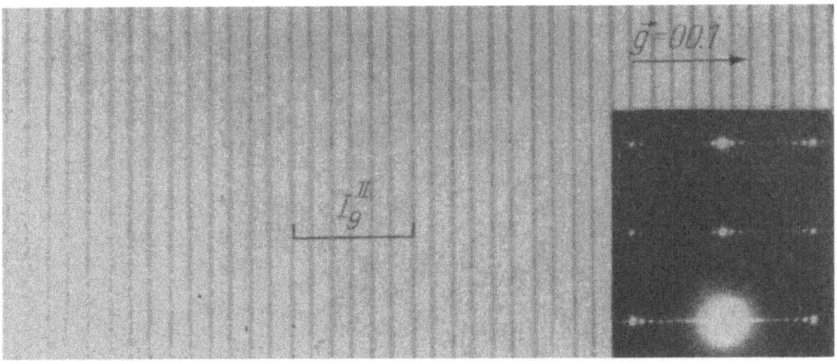

a

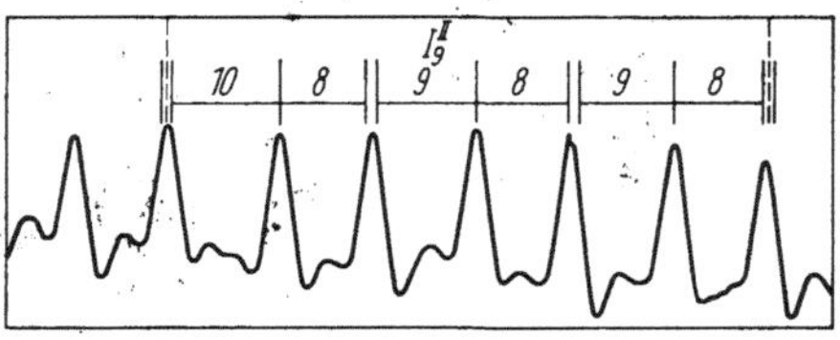

b

Fig. 6. a) Résolution directe et diffraction électronique de l'intercroissance double I_9^{II}. b) Microphotogramme de la résolution précédente mettant en évidence la séquence de l'intercroissance double I_9^{II}

Le terme de dissymétrie $n' = |I - J|$ prend les valeurs 0 ou 1. Pour ces deux intercroissances triples, la résolution directe ne montre pas les plans (00.2) des différents termes (Fig. 7a et b) par suite de l'absorption due à l'épaisseur du cristal; la succession des nombres I et J de feuillets siliconiobates a donc été obtenue en minimisant les différences $|I - J|$ d'une maille à la suivante. Le cliché de diffraction électronique de la Fig. 7b, bien résolu, correspond au mélange des trois termes de base $||6 \mid 6||$, $||3 \mid 4||$ et $||4 \mid 5||$ et de la période de l'intercroissance triple I_2^{III}.

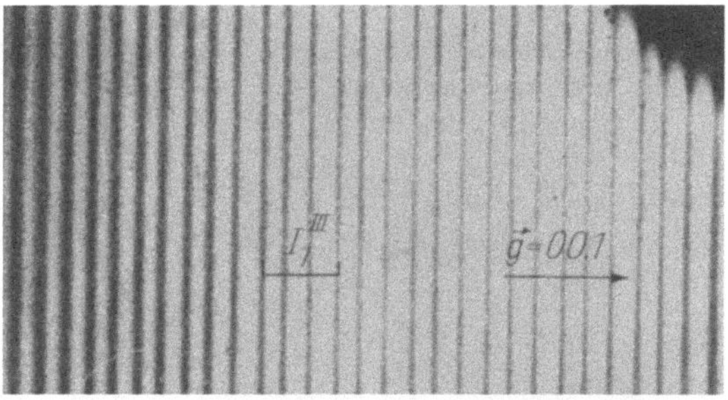

a

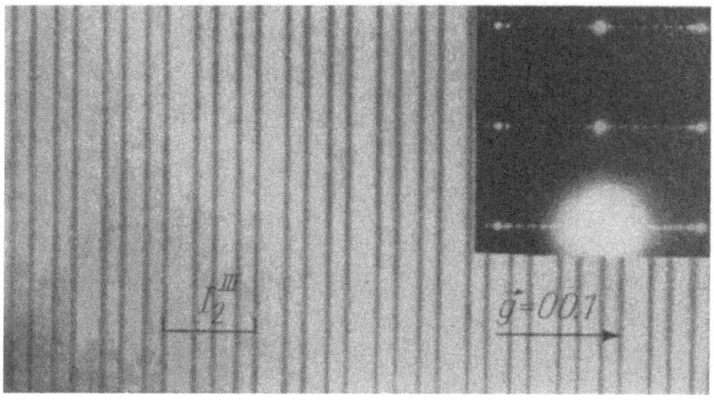

b

Fig. 7. a) Mise en évidence de l'intercroissance triple I_1^{III} dont la séquence est $||5 \mid 5 || 5 \mid 4 || 4 \mid 4||$ b) Résolution directe et diffraction électronique de l'intercroissance triple I_2^{III}

3.2.3 *Intercroissance quadruple* (I^{IV})

Un seul type d'intercroissance quadruple a été observé; il correspond à l'empilement le plus simple de quatre intercroissances simples (ABCD). Ces quatre intercroissances ont été identifiées au microphotomètre (Fig. 8b) ainsi que l'ordre de la séquence:

Fig. 8a: I_1^{IV}; $(\|6 \mid 5\|5 \mid 4\|4 \mid 3\|4 \mid 4\|)_{n_s}$;

le paramètre $c_{I_1^{IV}}$ de cette séquence est égal à 324,5 Å. Le terme de dissymétrie $n' = |I - J|$ est égal à 0 ou 1, tout au long de la séquence. Cette séquence se répète dans la plus grande partie du microcristal observé et les rangées $|h0.l|$ sur le cliché de diffraction électronique ne permettent pas de séparer des réflexions précises compte tenu de la valeur élevée du paramètre.

La stabilisation des intercroissances multiples qui viennent d'être présentées avec une séquence régulière et périodique de plusieurs intercroissances simples, s'étendant sur de grands domaines, n'a pu être obtenue que pour des temps de recuit très longs, allant jusqu'à six semaines. Les compositions de départ étant toujours calculées pour obtenir la synthèse d'une intercroissance simple, les microcristaux présentant des intercroissances multiples correspondent à un écart à la stoechiométrie de départ et sont donc le signe que l'équilibre thermodynamique n'a pas été atteint. Des temps de recuits élevés de plusieurs mois, permettraient sans doute de synthétiser des phases pures ne contenant que l'intercroissance simple prévue par la stoechiométrie.

Le terme de dissymétrie $n' = |I - J|$, permet de caractériser le stade atteint par l'échantillon observé vers l'équilibre thermodynamique. Des valeurs de n' inférieures ou égales à 2, permettent d'obtenir des intercroissances multiples régulières et périodiques. Des valeurs de n' supérieures à 3 caractérisent un désordre important ainsi que le montrent la résolution présentée sur la Fig. 9a et l'exploitation correspondante au microphotomètre (Fig. 9b).

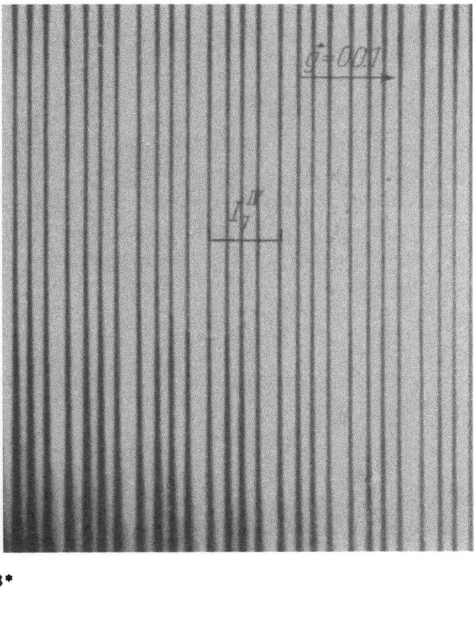

a

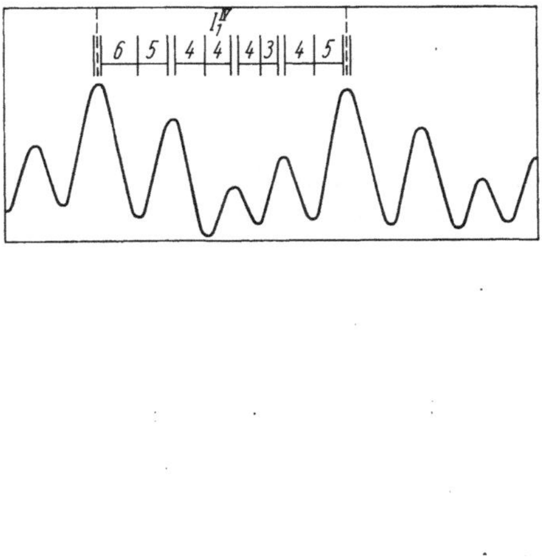

b

Fig. 8. a) Résolution directe et b) microphotogramme correspondant mettant en évidence la séquence de l'intercroissance quadruple I_1^{IV}

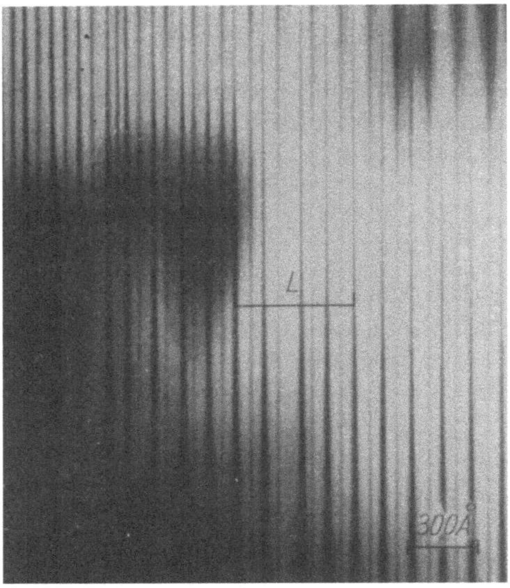

a

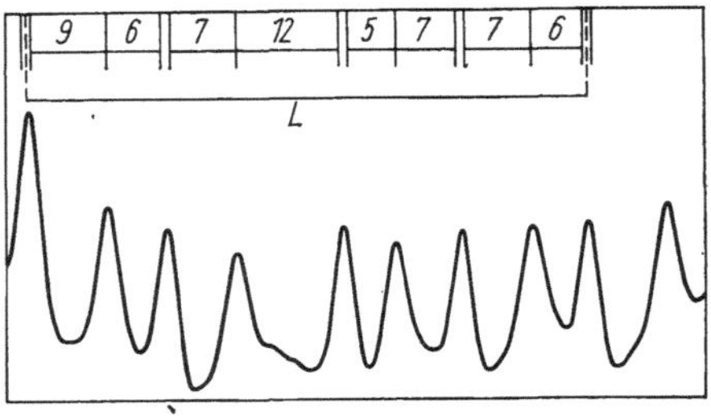

b

Fig. 9. a) Zone désordonnée contenant de nombreux termes différents de l'intercroissance et b) identification au microphotomètre

3.3 Microdomaines et parois de domaines

La mise en ordre des intercroissances simples et multiples s'effectue dans certains cristaux sous la forme de microdomaines ne contenant qu'un seul type d'intercroissance. Les jonctions entre ces domaines ou parois de domaines sont rarement bien localisées dans l'espace; elles s'étendent au contraire sur plusieurs dizaines d'Angstroms et sont le siège d'un désordre important correspondant au passage d'une séquence régulière à une autre. Deux types de parois de domaines ont été observés dans ce travail: les parois longitudinales et les parois transversales.

3.3.1 Parois d'antiphase

Des domaines caractérisés par des compositions et des sous-réseaux identiques et possédant des mailles de même taille ont été observés. La paroi d'antiphase qui sépare ces microdomaines, est définie par un vecteur ***R***, qui est un vecteur de translation de leur sous-réseau.

3.3.1.1 Parois d'antiphase longitudinales

Les parois d'antiphase longitudinales séparent deux domaines de structure différente. Elles sont parallèles aux plans des feuillets $Ba_3M_8O_{21}$ et $Ba_3Nb_6Si_4O_{26}$, soit au plan (00.1); elles sont donc parallèles aux franges de résolution sur les clichés (Fig. 10 et 11). Ces parois sont constituées par une couche $Ba_3M_8O_{21}$ parallèle au plan (00.1) de la structure. Les microdomaines notés M_i ont été repérés sur les clichés des Fig. 10 et 11 et les séquences d'empilement sont les suivantes:

Fig. 10: zone M_1; ... ||6 | 6||6 | 6||6 | 6|| ... .

Le premier domaine correspond au terme $n = 6$ de la série avec des défauts localisés notés B sur la Fig. 11 a identifiés comme le terme ||7 | 6||:

Fig. 10: zone M_2; ... ||5 | 7||5 | 7||5 | 7|| ... .

Le terme ||5 | 7|| de cette zone est identique en composition au terme ||6 | 6||: le paramètre c de la maille est le même $(I + J = 12)$ mais le terme de dissymétrie

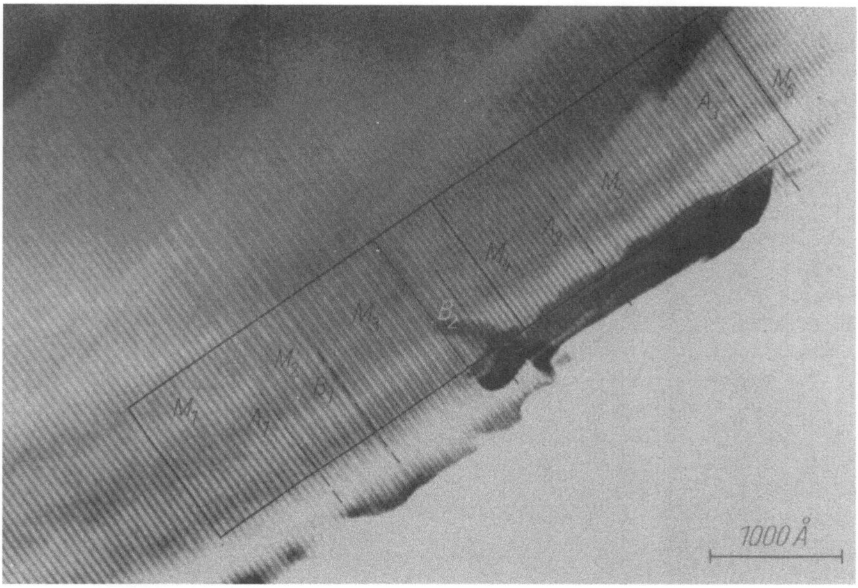

Fig. 10. Mise en évidence de microdomaines M_i séparés par des parois longitudinales: parois d'antiphase A_1, A_2 et A_3, et parois présentant une variation locale de stoechiométrie, B_1 et B_2

$n' = |I - J|$ prend la valeur 2. La résolution des plans (00.2) des structures des domaines M_1 et M_2 ont permis de différencier les mailles $||5 \mid 7||$ des mailles $||6 \mid 6||$. Les structures des microdomaines M_1 et M_2 se déduisent l'une de l'autre par translation suivant c d'un feuillet $Ba_3Nb_8O_{21}$ d'une longueur correspondant à l'épaisseur d'un feuillet simple $Ba_3Nb_6Si_4O_{26}$. Le vecteur qui caractérise la paroi d'antiphase notée A_1 sur les Fig. 10 et 11a est donc donné par $R_{A_1} = c_{Ba_3Nb_6Si_4O_{26}}$.

La deuxième paroi d'antiphase longitudinale notée A_2 (Fig. 10 et 11b) sépare les deux domaines suivants:

$$\text{zone } M_4;\ \ldots\ ||5 \mid 5||5 \mid 5||5 \mid 5||\ \ldots,$$

$$\text{zone } M_5;\ \ldots\ ||6 \mid 4||6 \mid 4||6 \mid 4||\ \ldots.$$

Cette paroi est également caractérisée par un vecteur déplacement identique $R_{A_2} = c_{Ba_3Nb_6Si_4O_{26}}$.

La troisième paroi d'antiphase longitudinale notée A_3 (Fig. 10 et 11b) sépare les domaines M_5 et M_6:

$$\text{zone } M_5;\ \ldots\ ||6 \mid 4||6 \mid 4||6 \mid 4||\ \ldots,$$

$$\text{zone } M_6;\ \ldots\ ||5 \mid 5||5 \mid 5||5 \mid 5||\ \ldots.$$

Le vecteur déplacement R_{A_3} identique à celui de la paroi A_2 est égal à $c_{Ba_3Nb_6Si_4O_{26}}$.

3.3.1.2 Parois d'antiphase transversales

Les parois d'antiphase transversales séparent deux domaines de structure identique. Elles contiennent l'axe c de la maille hexagonale de base et apparaissent normales aux franges de résolution des intercroissances.

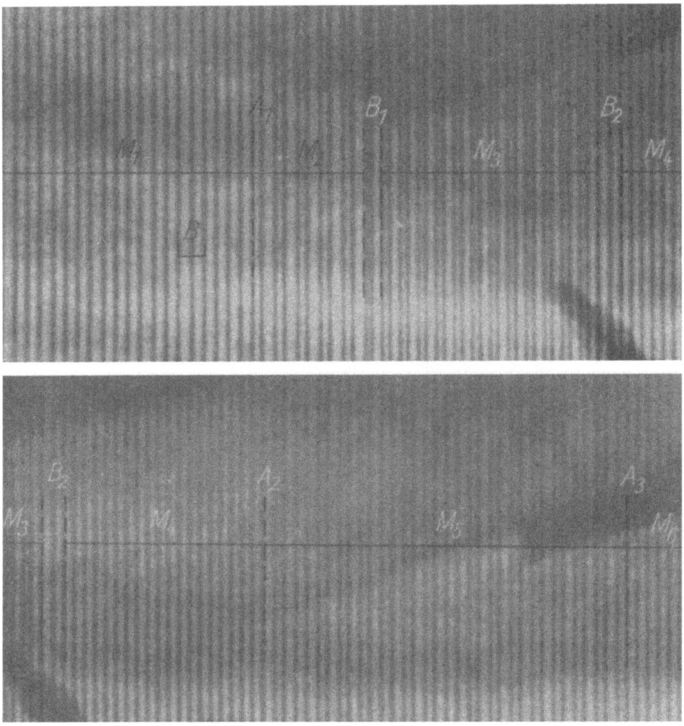

a

b

Fig. 11. Agrandissements de la micrographie de la Fig. 10 montrant les détails des parois

La Fig. 12 montre une résolution directe mettant en évidence une paroi d'antiphase séparant deux domaines contenant le terme $n = 4{,}5$ de la série $(Ba_3Nb_6Si_4O_{26})_n \cdot$ $\cdot\ Ba_3Nb_5Ti_4O_{21}$. Dans les limites de la résolution pratique de notre microscope électronique (≈ 20 Å), la paroi semble parfaitement définie par la rencontre des deux domaines. La paroi étant perpendiculaire au plan du cliché (Fig. 12a), le cliché de diffraction électronique correspondant (Fig. 12b) permet de déterminer le plan de la paroi: c'est un plan (10.0) qui contient l'axe *c*. Sur le cliché de la Fig. 12a, les plans (00.1) et (00.2) de l'intercroissance ||4 | 5|| ont été résolus; ils permettent de déterminer le déplacement qu'a subi la structure au passage de la paroi $\boldsymbol{R} = \frac{1}{2}\langle 00.1\rangle$.

La représentation de la paroi d'antiphase (Fig. 12c) obtenue en utilisant le symbolisme défini à la Fig. 2, montre que le vecteur déplacement $\boldsymbol{R} = \frac{1}{2}\langle 00.1\rangle$ permet de retrouver le contraste observé.

Cependant, la résolution obtenue ne permet pas l'observation de la jonction entre les domaines au niveau de la structure dans le plan hexagonal.

3.3.2 Parois de domaines et zones de transition

Les parois de domaines étudiées ici séparent des microdomaines dont la structure et la composition sont différentes. La résolution des plans (00.1) et parfois (00.2) des intercroissances, permet l'identification des différents termes au microphotomètre. Ces parois longitudinales ou transversales ont une épaisseur variable. A la limite, certains cristaux présentent des domaines bien ordonnés séparés par une zone de transition très large, constituée par un nombre variable d'intercroissances différentes; la notion de paroi disparait alors au profit de la notion de zone de transition.

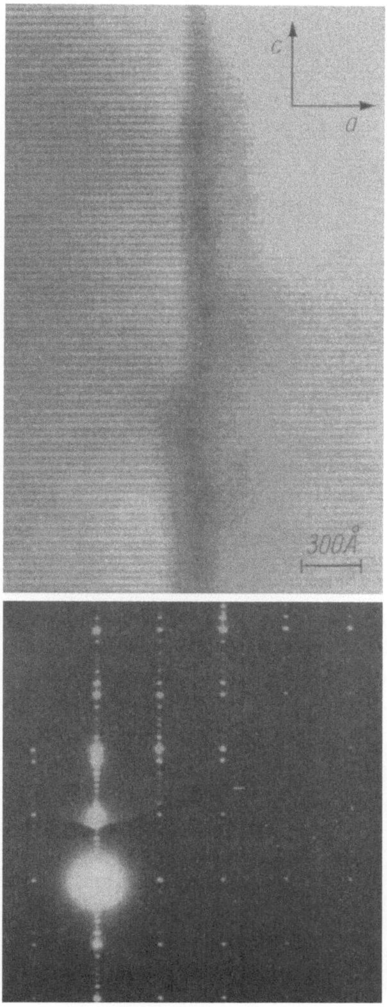

a

b

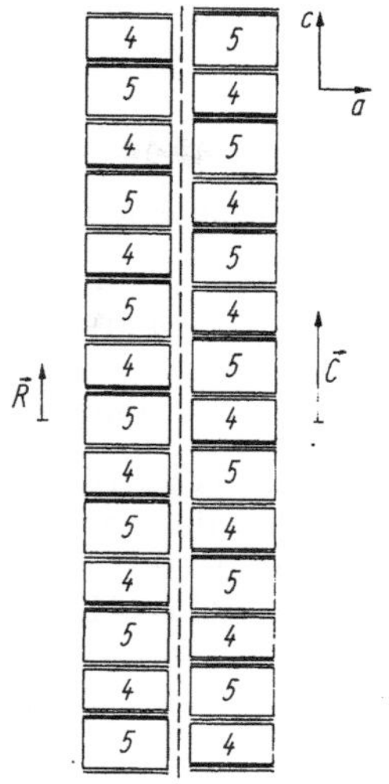

c

Fig. 12. a) Mise en évidence dans un microcristal du terme ||4 | 5|| d'une paroi d'antiphase transversale. b) La diffraction électronique permet de déterminer le plan de la paroi: c'est un plan (10.0). c) Représentation schématique complète par le modèle des blocs définis sur la Fig. 4 de la paroi d'antiphase transversale observée sur la résolution directe de la Fig. 12a. Le vecteur déplacement de la paroi $\boldsymbol{R} = \frac{1}{2}\langle 00.1\rangle$ permet de retrouver le contraste observé

3.3.2.1 Parois longitudinales

3.3.2.1.1 Paroi à stoechiométrie intermédiaire

La paroi est constituée par une intercroissance de composition intermédiaire entre les compositions des deux domaines (Fig. 13 et 14a). Les microdomaines observés en résolution directe ont été identifiés au microphotomètre et numérotés de I à III (Fig. 13):

domaine I; ... ||6 | 6||6 | 6||6 | 6|| ...; $n' = |I - J| = 0$,

domaine II; ... ||6 | 4||6 | 4||6 | 4|| ...; $n' = |I - J| = 2$,

domaine III; ... ||6 | 6||6 | 6||6 | 6|| ... .

La paroi notée C_1 sur les clichés (Fig. 13 et 14b) et sur le schéma de la Fig. 16 entre les domaines II et III, est constituée par les termes ||7 | 3|| ($n = 5$; $n' = 4$) et ||5 | 6||

($n = 5{,}5$; $n' = 1$). La séquence est donc:

$$\underbrace{\ldots\ \|6\,|\,4\|6\,|\,4\|}_{\mathrm{II}}\underbrace{7\,|\,3\|5\,|\,6\|}_{\mathrm{C_1}}\underbrace{6\,|\,6\|\ \ldots}_{\mathrm{III}},$$

où la paroi C_1 présente une composition intermédiaire entre les domaines II et III:

domaine II: $I + J = 10$;
paroi C_1: $I + J = 10$ pour $\|7\,|\,3\|$,
$I + J = 11$ pour $\|5\,|\,6\|$;
domaine III: $I + J = 12$.

La paroi notée C_2 (Fig. 13, 14c et 16) permet l'ajustement des distances entre les plans $Ba_3M_8O_{21}$ décalés par la paroi transversale F. La paroi C_2 est constituée par la répétition du terme $\|6\,|\,5\|$ ($n = 5{,}5$; $n' = 1$) dont la composition est intermédiaire entre les domaines II et III suivant la séquence:

$$\underbrace{\ldots\ \|6\,|\,4\|6\,|\,4\|}_{\mathrm{II}}\underbrace{6\,|\,5\|5\,|\,6\|}_{\mathrm{C_2}}\underbrace{6\,|\,6\|\ \ldots}_{\mathrm{III}}.$$

3.3.2.1.2 Paroi avec effondrement de la stoechiométrie

La composition du ou des termes constituant la paroi est très différente de celles des domaines séparés.

Le premier exemple noté B_1 sur les clichés (Fig. 10 et 11a) montre une paroi longitudinale constituée par le terme $\|2\,|\,5\|$ de l'intercroissance ($n = 3{,}5$ et $n' = 3$) séparant deux domaines de composition $\|5\,|\,7\|$ ($n = 6$, $n' = 2$) et $\|4\,|\,5\|$ ($n = 4{,}5$; $n' = 1$) suivant la séquence:

$$\ldots\ \|5\,|\,7\|5\,|\,7\|\underbrace{2\,|\,5\|}_{\mathrm{B_1}}4\,|\,5\|4\,|\,5\|\ \ldots.$$

Par rapport au terme $\|4\,|\,5\|$, la paroi correspond à un appauvrissement local de deux feuillets siliconiobates, et de cinq feuillets siliconiobates par rapport au terme $\|5\,|\,7\|$.

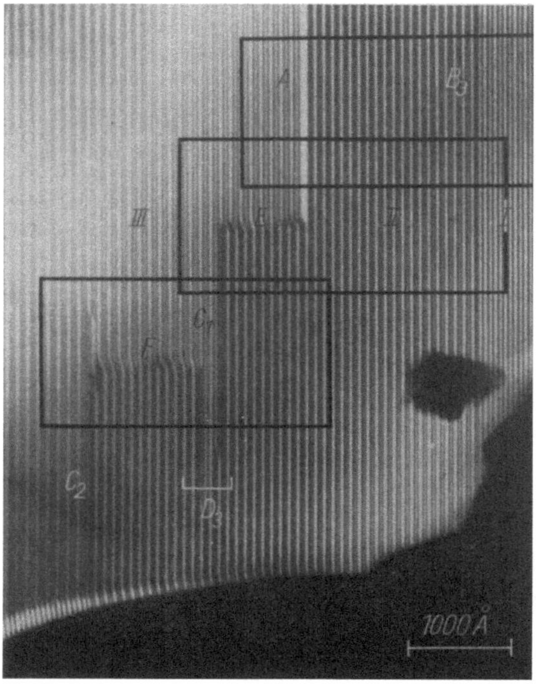

Fig. 13. Mise en évidence de parois longitudinales C_1, C_2, et C_3, de parois transversales E et F et de la zone de transition D_3 séparant trois domaines dans un même cristal

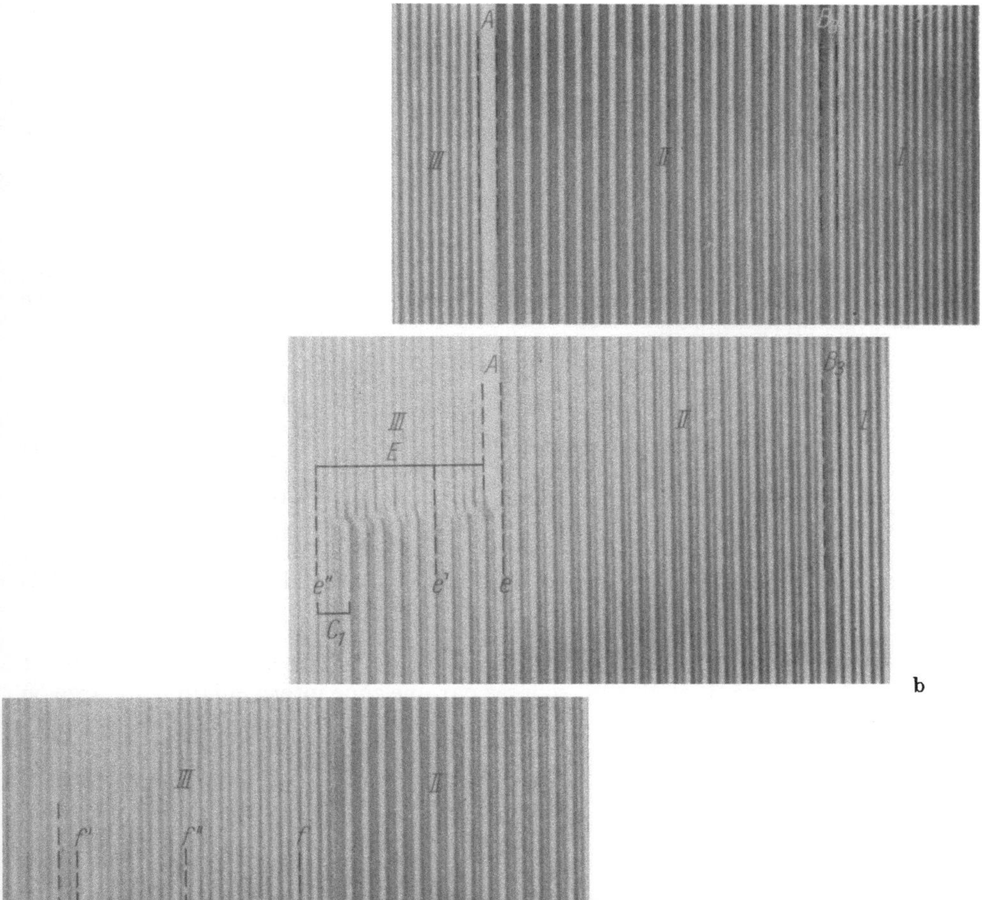

a

b

c

Fig. 14. Agrandissements de la micrographie de la Fig. 13 montrant les perturbations subies par les franges d'interférences au passage des parois transversales E et F

La deuxième paroi notée B_2 (Fig. 10 et 11c) est constituée par un terme $\|6 \mid 5\|$ ($n = 5{,}5$; $n' = 1$) séparant deux domaines de composition $\|4 \mid 5\|$ ($n = 4{,}5$; $n' = 1$) et $\|5 \mid 5\|$ ($n = 5$; $n' = 0$) suivant la séquence:

$$\underline{\ldots\ \|4 \mid 5\|4 \mid 5\|}\underset{B_2}{\underline{\|6 \mid 5\|}}\underline{\|5 \mid 5\|5 \mid 5\|\ \ldots}\,.$$

La paroi correspond ici à un enrichissement local d'un feuillet siliconiobate par rapport au terme $\|5 \mid 5\|$ et de deux feuillets siliconiobates par rapport au terme $\|4 \mid 5\|$.

La troisième paroi notée B_3 (Fig. 13 et 14a) est constituée par un terme $\|5 \mid 3\|$ ($n = 4$; $n' = 2$) séparant deux domaines de composition $\|6 \mid 6\|$ ($n = 6$; $n' = 0$) et

||6 | 4|| ($n = 5$; $n' = 2$) suivant la séquence:

$$\underbrace{\ldots ||6 \mid 6||6 \mid 6||}\underbrace{5 \mid 3||}_{B_3}\underbrace{6 \mid 4||6 \mid 4|| \ldots}.$$

La variation locale de stoechiométrie correspond à un appauvrissement de deux feuillets siliconiobates par rapport au terme ||6 | 4|| et de 4 feuillets siliconiobates par rapport au terme ||6 | 6||.

3.3.2.1.3 Paroi entre intercroissance simple et multiple

La Fig. 15 donne un exemple de paroi séparant deux microdomaines contenant l'un une intercroissance simple longitudinale notée I_1^I, l'autre une intercroissance multiple étudiée précédemment:

zone M_1; (||6 | 5||5 | 4||4 | 3||4 | 4||)$_{m_1}$ (I_1^{IV}) ,
zone M_2; ... ||4 | 5||4 | 5||4 | 5|| ... ($n = 4{,}5$; $n' = 1$) .

Le paroi C est directement constituée par un feuillet $Ba_3M_8O_{21}$. La variation de la composition de part et d'autre de C ne peut être estimée qu'à partir de la composition moyenne de l'intercroissance quadruple qui correspond à $n = 4{,}35$.

3.3.2.1.4 Zones de transition

Plusieurs zones de transition de largeur importante ont été identifiées sur les clichés de résolution:

Fig. 15: zone M_3; (||4 | 5||4 | 5||4 | 4||4 | 4||)$_{m_2}$

Le troisième domaine contient une intercroissance double déjà décrite précédemment (I_7^{II}). Les deux domaines M_2 et M_3 sont séparés par une zone intermédiaire très large où

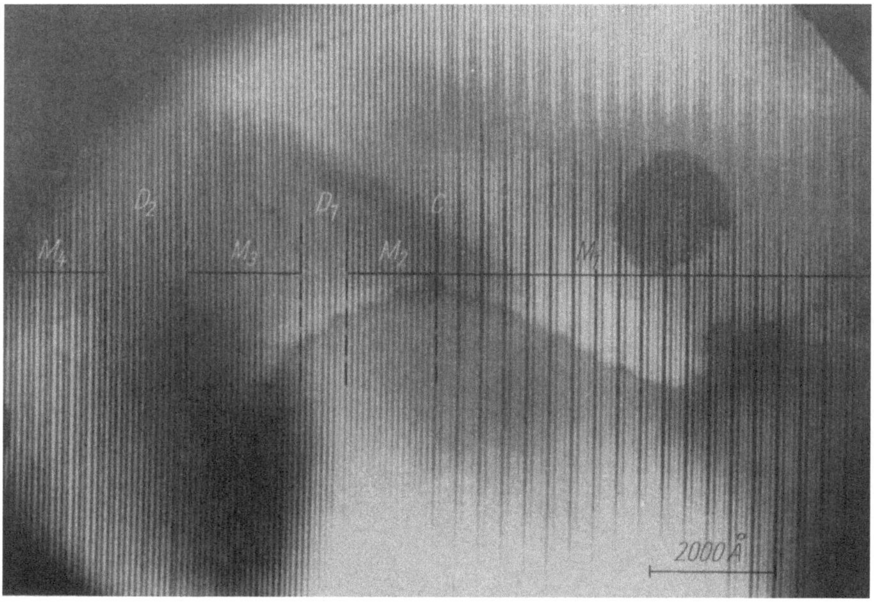

Fig. 15. Mise en évidence dans un microcristal de plusieurs domaines contenant des intercroissances simples I^I, doubles I^{II}, triples I^{III} et quadruples I^{IV}, séparés par de larges zones de transition D_1 et D_2

nous avions mis en évidence deux périodes de l'intercroissance I_6^{II} : $(\|4|5\|4|5\|4|5\|4|4\|)_2$. Cette zone D_1 a pour épaisseur $2c_{I_6^{II}}$, soit 649 Å, et il devient difficile de l'assimiler à une paroi de domaine.

Fig. 15: zone M_4; $(\|5 \mid 5\|5 \mid 4\|4 \mid 4\|)_{m_3}$.

Le quatrième domaine M_4, contenant l'intercroissance triple I_1^{III}, est séparé du domaine M_3 par une très large zone de transition D_2 contenant plusieurs intercroissances doubles suivant la séquence:

$$I_6^{II} \quad A \quad I_6^{II} \quad I_8^{II} \quad I_8^{II},$$

où A est le défaut noté sur la Fig. 5.

Un autre type de zone de transition D_3 est fourni par les clichés des Fig. 13 et 14b. Entre deux parties du domaine II séparées par une maille $\|6 \mid 6\|$ du domaine III, apparaît une zone de transition où la séquence des intercroissances est

$$\underbrace{\ldots \|6 \mid 4\|}_{II}\underbrace{7 \mid 3\|5 \mid 6\|}_{C_1}\underbrace{6 \mid 6\|5 \mid 4\|}_{III}\underbrace{6 \mid 4\| \ldots}_{II}.$$

Cette importante zone de transition large de 364 Å permet le rattrapage des plans $Ba_3M_8O_{21}$ entre les domaines II et III décalés le long des parois transversales E et F.

3.3.2.2 Parois transversales

La première paroi transversale notée E sur les clichés (Fig. 13 et 14b) fait correspondre le terme $\|6 \mid 4\|$ du domaine II et le terme $\|6 \mid 6\|$ du domaine III. Son épaisseur, estimée à partir des perturbations subies par les franges de résolution, est égale à 50 Å. La maille hexagonale des intercroissances est donc le siège d'importants désordres au passage de cette paroi. En prenant pour origine la frange d'interférence notée "e" sur les clichés (Fig. 13 et 14b), nous avons représenté schématiquement (Fig. 16) la disposition des intercroissances le long de la paroi. Il est alors possible de faire apparaître le demi-terme $|6|$ commun aux deux domaines limité à droite par la frange "e'" (Fig. 13, 14b), et de calculer le décalage entre les deux domaines au niveau de la zone de transition D_3.

Dans le domaine II, la séquence d'empilement de la paroi transversale E débute par un plan $Ba_3M_8O_{21}$ noté "e" et représenté par deux barres (Fig. 16). Dans le domaine III, la frange "e" est séparée de l'empilement $\|6 \mid 6\|$ par un important empilement de feuillets siliconiobates noté A (Fig. 14a et b). Ce défaut constitue une paroi longitudinale entre les domaines II et III.

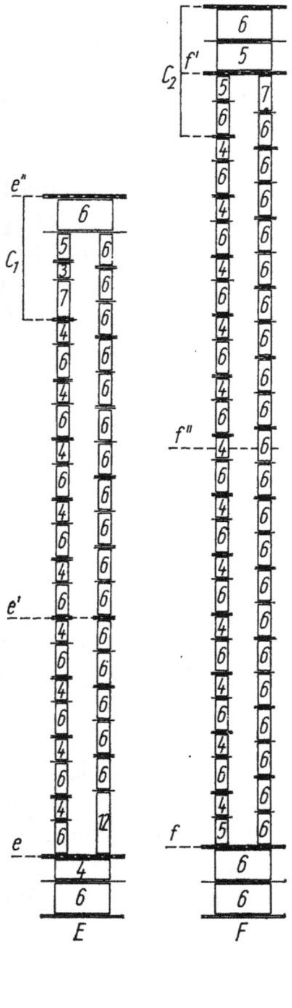

Fig. 16. Représentation schématique complète par le modèle des blocs définis sur la Fig. 4 des parois transversales E et F mises en évidence sur les clichés des Fig. 13 et 14. Les zones de perturbation des franges qui constituent les parois proprement dites sont laissées en blanc entre les domaines II et III. La comparaison avec les clichés des Fig. 14b et c montrent qu'au passage des parois E et F les franges suivent des plans $A_3M_8O_{21}$ identiques des deux cotés de la paroi

L'identification de ce défaut A tient compte de la longueur des séquences dans les domaines II et III le long de la paroi transversale E: Dans le domaine II, la séquence entre les plans "e" et "e'''" est donnée par 9||6 | 4|| + ||7 | 3|| + ||5 | 6||, soit 1008,9 Å et dans le domaine III, la séquence entre "e" et "e'''": 8,5||6 | 6|| + A, permet de déduire l'épaisseur théorique de |A|, soit 101,1 Å. |A| correspond donc à un terme ||4 | 7|| ($c_{th} \approx$ 98,9 Å) ou à un demi-terme |12| ($c_{th} \approx$ 101 Å). Cependant, les franges d'interférences, communes aux domaines II et III, e, e' et e'', correspondent à des plans $Ba_3M_8O_{21}$ identiques repérés par deux barres verticales (Fig. 16). La seule façon de disposer les intercroissances dans le domaine III est la formation d'un terme ||6 | 12|| où le défaut A apparaît comme un demi-terme |12|.

La deuxième paroi transversale notée F est plus complexe. La perturbation des franges d'interférence au passage de la paroi permet d'évaluer son épaisseur à au moins 100 Å. Le schéma complet de cette paroi F est représenté sur la Fig. 16 entre les franges notées f et f' qui sont communes aux domaines II et III. A partir de cette frange f, onze termes ||6 | 6|| dans le domaine III correspondent à treize termes ||6 | 4|| dans le domaine II. Cependant, au niveau de la paroi longitudinale C_2, un rattrapage se fait par un terme ||6 | 5|| dans le domaine II et un terme ||6 | 7|| dans le domaine III (Fig. 16), de telle sorte que: $11c_{||6|4||} + c_{||6|5||} + c_{||5|4||} = 10c_{||6|6||} + c_{||6|7||} \approx 1183$ Å. Le seul point particulier de la paroi transversale F est constitué par le point f'' où un terme |6| et un terme |4| sont alignés.

Nous avons essayé de trouver une relation entre les nombres des mailles de part et d'autre d'une paroi transversale. Considérons les deux intercroissances simples $||I_1 | J_1||$ et $||I_2 | J_2||$ de paramètre $c_1 = (I_1 + J_1)\,c' + c''$ et $c_2 = (I_2 + J_2)\,c' + c''$, où $c' = c_{Ba_3Nb_6Si_4O_{26}}$ et $c'' = c_{Ba_3M_8O_{21}}$. Ces deux intercroissances s'étendent à partir d'un plan origine commun et déterminent une paroi transversale. Le nombre de mailles $||I_1 | J_1||$ désigné par N_1 et le nombre N_2 de mailles $||I_2 | J_2||$ sont reliés par la relation $N_1c_1 = N_2c_2$ qui caractérise l'alignement exact des plans origines. Entre une maille $||I_1 | J_1||$ et une maille $||I_2 | J_2||$, un décalage apparaît égal à $C = [(I_1 - I_2) + (J_1 - J_2)]\,c'$. c_1 et c_2 étant des nombres irrationnels, il est théoriquement impossible de trouver des nombres entiers finis solutions de la relation $N_1c_1 - N_2c_2 = 0$. Cependant, en se bornant à un décalage d'une unité entre N_1 et N_2, nous pouvons résoudre le système suivant par valeurs entières:

$$N_1 - N_2 = 1\,,$$

$$N_1c_1 - N_2c_2 = 0\,.$$

Nous obtenons les solutions

$$N_1 = \frac{c'' + 2n_2c'}{C}, \qquad N_2 = \frac{c'' + 2n_1c'}{C},$$

où $2n_1 = I_1 + J_1$ et $2n_2 = I_2 + J_2$. Ces relations appliquées aux termes ||6 | 6|| et ||6 | 4|| des Fig. 13 et 14b, donnent les résultats suivants: $N_1 = 5{,}76$ et $N_2 = 6{,}76$ avec $C = 15{,}8$; $I_1 = 6$; $J_1 = 6$; $I_2 = 6$; $J_2 = 4$. La solution la plus approchée correspond donc à 7 mailles ||6 | 4|| et 6 mailles ||6 | 6||. Ce résultat a été observé précédemment sur le cliché de la Fig. 14b entre les plans notés e' et e'' de la paroi transversale E.

Dans le cas de la paroi F (Fig. 14b), le système à résoudre: $N_1 - N_2 = 2$; $N_1c_1 - N_2c_2 = 0$, conduit aux solutions approchées $N_1 = 11$ et $N_2 = 13$, qui donnent $11c_{||6|6||} = 1175$ Å et $13c_{||6|4||} = 1183$ Å. L'écart de ces deux valeurs égal à 8 Å correspond à un défaut, d'une maille siliconiobate qui est rattrapée par le terme ||6 | 7|| qui constitue avec le terme ||5 | 6|| la paroi longitudinale C_2 dans le domaine III.

4. Conclusion

Au cours de cette étude, de nouvelles intercroissances multiples entre structures à tunnels et structures en cages ont été observées. Ces dernières sont caractérisées par des mailles géantes, dont le paramètre *c* peut atteindre jusqu'à 450 Å. La grande variété des parois de domaine traduit la souplesse de ces types structuraux, qui peuvent présenter une gamme de compositions très étendue. Une étude systématique des phénomènes d'ordre–désordre dans ces oxydes, en faisant varier les temps et les températures des recuits, devrait permettre d'expliquer l'existence et la nature de défauts plus complexes qui n'ont pas été décrits dans ce travail.

Bibliographie

[1] J. Choisnet, N. Nguyen, D. Groult et B. Raveau, Mater. Res. Bull. **11**, 887 (1976).
[2] D. M. Evans et L. Katz, J. Solid State Chem. 8, 150 (1973).
[3] J. R. Shannon et L. Katz, J. Solid State Chem. **1**, 399 (1970).
[4] D. Groult, J. M. Chailleux, J. Choisnet et B. Raveau, J. Solid State Chem. **19**, 235 (1976).
[5] D. Groult, J. M. Chailleux, B. Raveau et A. Deschanvres, Rev. Chim. Minér. **14**, 1 (1977).
[6] C. Saurel, D. Groult et B. Raveau, Mater. Res. Bull. 12, 629 (1977).
[7] N. Nguyen, F. Studer, D. Groult, J. Choisnet et B. Raveau, J. Solid State Chem. **19**, 369 (1976).
[8] F. Studer et B. Raveau, phys. stat. sol. (a) 48, 301 (1978).

(Received April 24, 1978)

phys. stat. sol. (a) 49, 207 (1978)

Subject classification: 18.1 and 19; 21.1

Faculty of Physics, Cluj-Napoca University[1]) *(a)* and
Department of Chemistry, Pittsburgh University[2]) *(b)*

NMR and Magnetic Susceptibility of Intermetallic Compounds ZrV_2, $Zr_{0.99}Pt_{0.01}V_2$, and $Zr_{0.9}Fe_{0.1}V_2$

By

I. Pop (a), M. Coldea (b), and V. U. S. Rao (b)

The Knight shift and the magnetic susceptibility of ZrV_2, $Zr_{0.99}Pt_{0.01}V_2$, and $Zr_{0.9}Fe_{0.1}V_2$ intermetallic compounds are investigated. The obtained effective magnetic moments show that in these compounds the d electrons of V and Fe atoms are well localized and that in the compound $Zr_{0.99}Pt_{0.01}V_2$ a giant magnetic moment on the Pt atoms is induced by strong conduction electron polarization. The hyperfine field arising from the d core polarization is estimated as $H^d_{HF} = -0.2 \times 10^6$ Oe/spin for all compounds. The phenomenological exchange constant between the localized d electrons of Pt atoms and the conduction electrons is evaluated as $J_{sd} = 0.0242$ eV.

Die Knight-Verschiebung und die magnetische Suszeptibilität der intermetallischen Verbindungen ZrV_2, $Zr_{0,99}Pt_{0,01}V_2$ und $Zr_{0,9}Fe_{0,1}V_2$ werden untersucht. Die beobachteten effektiven magnetischen Momente zeigen, daß in diesen Verbindungen die d-Elektronen der V- und Fe-Atome wohl lokalisiert sind und daß in der Verbindung $Zr_{0,99}Pt_{0,01}V_2$ ein großes Magnetmoment an den Pb-Atomen durch eine starke Polarisation der Leitungselektronen induziert wird. Das Hyperfeinfeld, das von der d-Rumpfpolarisation herrührt, wird zu $H^d_{HF} = -0,2 \times 10^6$ Oe/Spin für alle Verbindungen bestimmt. Die phänomenologische Austauschkonstante zwischen den lokalisierten d-Elektronen der Pt-Atome und der Leitungselektronen wird zu $J_{sd} = 0,0242$ eV berechnet.

1. Introduction

The intermetallic compound ZrV_2 crystallizes in the cubic lattice of Cu_2Mg-type with the lattice parameter $a = 7.445$ Å, belonging to Fd3m space group [1]. The $Zr_{0.99}Pt_{0.01}V_2$ and $Zr_{0.9}Fe_{0.1}V_2$ compounds are isostructural with ZrV_2 and their lattice parameters are nearly the same as that of ZrV_2. These intermetallic compounds are interesting because they are superconducting materials. From the magnetic point of view these compounds were not investigated so far. The elements V and Zr have weak paramagnetic properties. The interaction between these elements and those with other doped transition elements can give rise to interesting magnetic properties.

The aim of this paper is to elucidate the magnetic behaviour of these compounds from the measurements of the temperature variations of Knight shift and magnetic susceptibility.

2. Experimental

The compounds were prepared by melting stoichiometric amounts of the elements with high purity in either a high frequency furnace or an arc melting method in high purity argon atmosphere.

The formation of the compounds was verified by X-ray analysis.

The magnetic susceptibility was measured between 100 and 1100 K at 9450 Oe, with a Faraday type magnetic balance, having 10^{-8} e.m.u./g sensitivity.

---

[1]) 3400 Cluj-Napoca, Roumania.

[2]) Pittsburgh, USA.

The NMR spectra of V^{51} nuclei were taken by a JNM-3 spectrometer at 9.212 MHz, improved by a broad-line attachment JNM-BH-2 at several temperatures in the range from 140 to 475 K in a specially designed dewar. The Knight shifts were determined with reference to the V^{51} resonance in $NaVO_3$ solution.

3. Results and Discussion

The temperature dependence of the reciprocal magnetic susceptibility χ for all investigated compounds is shown in Fig. 1, where one sees that χ follows the Curie-Weiss law

$$\chi = \frac{C}{T - \theta_p}$$

with $\theta_p < 0$. By a partial substitution of Zr atoms with Fe and Pt atoms the magnetic susceptibility increases. Although the concentration of Pt atoms is smaller than that of Fe atoms, the increase in χ is more pronounced in the $Zr_{0.99}Pt_{0.01}V_2$ compound, indicating a very large contribution of Pt atoms to the bulk magnetic susceptibility. From the slope of $1/\chi(T)$ we have determined the following effective magnetic moments per molecule: $5.49\mu_B$ for ZrV_2, $3.58\mu_B$ for $Zr_{0.9}Fe_{0.1}V_2$ and $4.14\mu_B$ for $Zr_{0.99}Pt_{0.01}V_2$.

The effective magnetic moment of ZrV_2 would be attributable only to vanadium atoms, because Zr in the metallic state has a Pauli paramagnetic behaviour [2]. The effective magnetic moment per vanadium atom is estimated as $3.89\mu_B$, which is very close to the value $\mu_v = 3.87\mu_B$ corresponding to the case in which all $3d^3$ electrons of the vanadium atom are localized [3]. In order to determine the effective magnetic moment per Pt and Fe atoms we have subtracted the molar susceptibility of ZrV_2 from that of $Zr_{0.99}Pt_{0.01}V_2$ and $Zr_{0.9}Fe_{0.1}V_2$. The temperature dependences of the reciprocal susceptibility, obtained in this way, is shown in Fig. 2. From the slope of the linear dependences $1/\Delta\chi_{Pt}(T)$ and $1/\Delta\chi_{Fe}(T)$ we have obtained for the effective magnetic moment the very high values $17.8\mu_B$/Pt and $4.18\mu_B$/Fe, respectively. The giant value of magnetic moment per Pt atom in the compound $Zr_{0.99}Pt_{0.01}V_2$ could be explained if we assume the existance of a conduction electron polarization caused by the Pt atoms. This assumption is supported by Knight shift measurements.

The effective magnetic moment per Fe atom in the compound $Zr_{0.9}Fe_{0.1}V_2$ is very close to the experimental value $\mu_{Fe} = 4.2\mu_B$ obtained in the paramagnetic state of metallic iron [4].

Fig. 3 shows the temperature dependence of the Knight shift for three intermetallic compounds. One can see that $dK/dT > 0$ and the Knight shift increases with partial substitution of Zr atoms by magnetic impurities of Fe and Pt. Although the Pt concentration is ten times smaller then the Fe content, the Knight shift for Pt substitution is much larger than in the case of the magnetic susceptibility. This means that Pt atoms play a decisive roll in determining the strong effect of conduction electron polarization.

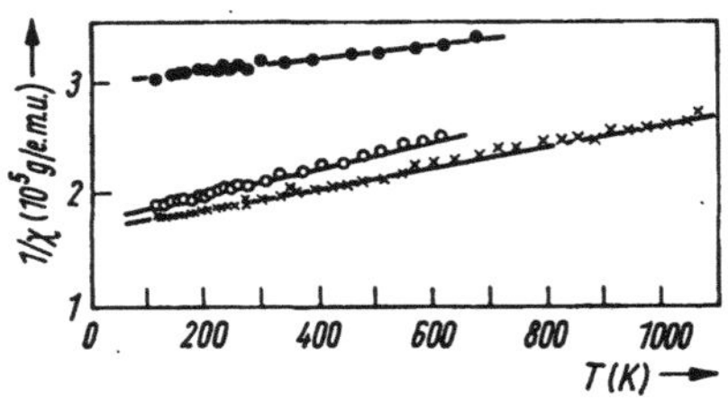

Fig. 1. Reciprocal susceptibility versus temperature for ZrV_2 (●), $Zr_{0.9}Fe_{0.1}V_2$ (○), and $Zr_{0.99}Pt_{0.01}V_2$ (×) compounds

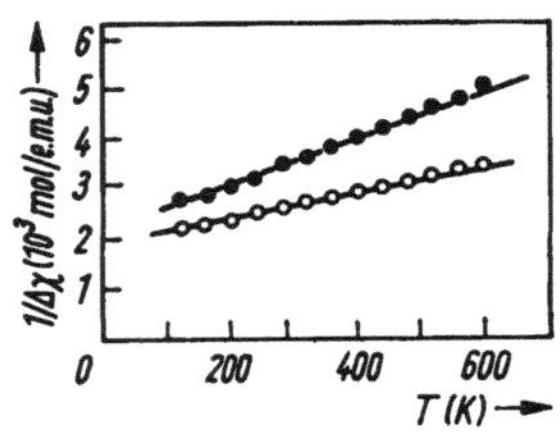

Fig. 2. Reciprocal susceptibility of Pt and Fe atoms versus temperature for $Zr_{0.99}Pt_{0.01}V_2$ (○) and $Zr_{0.9}Fe_{0.1}V_2$ (●) compounds

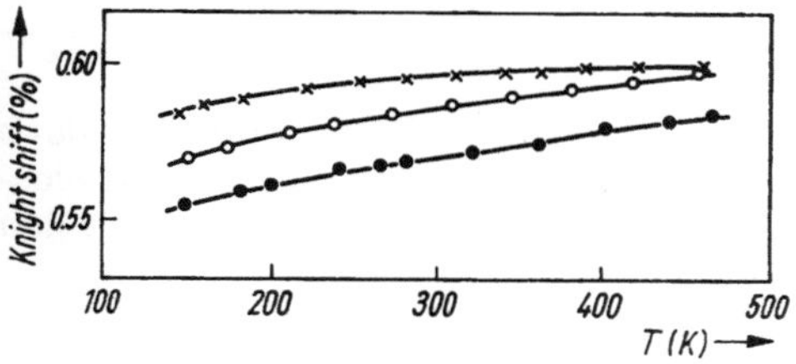

Fig. 3. Temperature dependence of the V^{51} Knight shift in ZrV_2 (●), $Zr_{0.9}Fe_{0.1}V_2$ (○), and $Zr_{0.99}Pt_{0.01}V_2$ (×) compounds

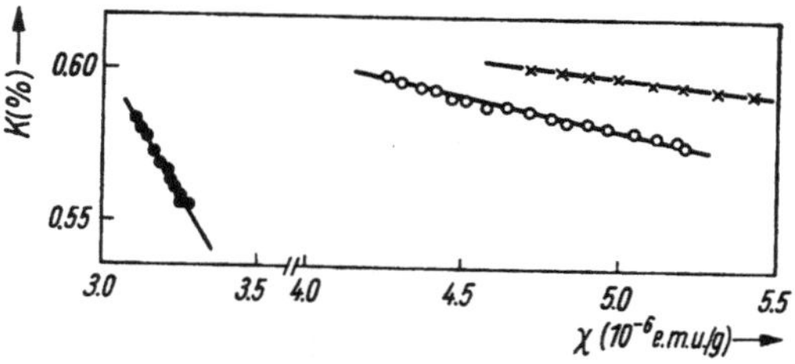

Fig. 4. V^{51} Knight shift versus magnetic susceptibility for ZrV_2 (●), $Zr_{0.9}Fe_{0.1}V_2$ (○), and $Zr_{0.99}Pt_{0.01}V_2$ (×) compounds

The Knight shift changes linearly with magnetic susceptibility for all investigated compounds as shown in Fig. 4.

The bulk Knight shift for ZrV_2 may be expressed as [5]

$$K(T) = K_s + K_d(T) + K_{orb} = a_s\chi_P^s + a_d\chi_P^d(T) + \beta\chi_{orb},$$

where K_s, $K_d(T)$, K_{orb} are the Knight shifts due to the conduction electrons polarization, core polarization, and orbital, respectively. χ_P^s and $\chi_P^d(T)$ are the Pauli spin susceptibilities of s and d electrons and χ_{orb} is the orbital susceptibility, a_s and a_d are the proportionality constants whose values are 0.895×10^{-4} times the hyperfine fields per spin for the s contact and d core polarization and $\beta = (2/A)\langle 1/r^3\rangle$. From the linear dependence of $K(\chi)$ we have obtained $a_d = -8.85$ (e.m.u./mol)$^{-1}$ leading to $H_{HF}^d = -0.2 \times 10^6$ Oe/spin. This value is in good agreement with those obtained for Pt [6], Pd [5], Rh [7]. Because the number of V atoms in all compounds is constant, the obtained value a_d is the same.

In order to know the contribution of the Fe and Pt impurity atoms to the Knight shift we subtracted the ZrV_2 Knight shift from those of $Zr_{0.9}Fe_{0.1}V_2$ and $Zr_{0.99}Pt_{0.01}V_2$, as shown in Fig. 5.

The localized d electrons of Fe and Pt impurities make three contributions to the supplementary Knight shift $\Delta K = K_s'(T) + K_d'(T) + K_{orb}'$, $K_s'(T)$ being positive Knight shift due to the conduction electron polarization, $K_d'(T)$ the negative Knight shift due to core polarization and K_{orb}' is the orbital temperature independent term. As is seen from Fig. 5 (open circles), ΔK for $Zr_{0.9}Fe_{0.1}V_2$ compound is temperature independent. This means that the first two opposite contributions $K_s'(T)$ and $K_d'(T)$ are canceled, remaining only the orbital contribution K_{orb}'.

In the case of $Zr_{0.99}Pt_{0.01}V_2$ compound ΔK is temperature dependent with $\delta(\Delta K)/\delta T < 0$. (Fig. 5, crosses) This effect may be explained in terms of a strong conduction electron polarization by Pt atoms. This effect is supported also by a high value

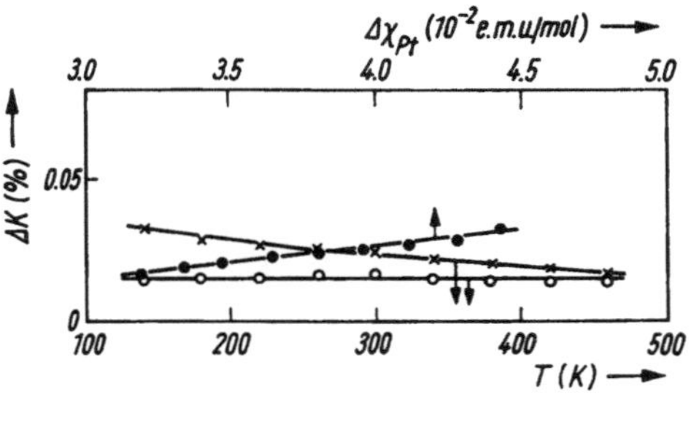

Fig. 5. The contribution of Fe and Pt atoms to the V^{51} Knight shift in $Zr_{0.9}Fe_{0.1}V_2$ (○) and $Zr_{0.99}Pt_{0.01}V_2$ (×) as a function of temperature and in $Zr_{0.99}Pt_{0.01}V_2$ compounds versus the magnetic susceptibility of Pt atoms (●)

of magnetic moment per Pt atom in the compound. Therefore $K_s'(T)$ is much larger than $K_d'(T)$.

ΔK changes linearly with $\Delta\chi_{Pt}$ as shown in Fig. 5 by closed circles. From the slope of the curve we have calculated the exchange integral between the d localized electrons of Pt atoms and conduction electrons as $J_{sd} = 0.0242$ eV, using a similar relation as in the case of rare-earth intermetallic compounds [8, 9]

$$\Delta K(T) = K_0 \left[1 + \frac{J_{sd}(g_d - 1)\, \Delta\chi_{Pt}(T)}{g_s g_e \mu_B^2}\right],$$

where K_0 in the temperature independent Knight shift, g_s and $g_d = 2$ are the Landé factors of conduction electrons and $5d^9$ localized electrons of Pt atoms. The estimated value of J_{sd} is in good agreement with the values for other transition elements.

4. Conclusions

The magnetic susceptibility measurements indicate a strong localization of d electrons of V and Fe and Pt impurity atoms in all investigated compounds. For the $Zr_{0.99}Pt_{0.01}V_2$ compound the magnetic measurements reveal the existence of the giant magnetic moments of $17.8\mu_B$ on Pt atoms caused by the conduction electron polarization.

The comparative study of the three intermetallic compounds leads to the explanation of the magnetic behaviour of these compounds. Assuming that the $3d^3$ electronic states of V atoms in the investigated compounds are the same, we used the compound ZrV_2 as a standard one for the explanation of the complicated picture of magnetic interactions in the isostructural compounds $Zr_{0.9}Fe_{0.1}V_2$ and $Zr_{0.99}Pt_{0.01}V_2$.

By subtracting the Knight shift of the standard ZrV_2 compound from those of $Zr_{0.9}Fe_{0.1}V_2$ and $Zr_{0.99}Pt_{0.01}V_2$ isostructural compounds we have obtained the contribution of Fe and Pt atoms to the bulk Knight shift.

For the compound $Zr_{0.99}Pt_{0.01}V_2$ this method allowed us to determine the exchange integral J_{sd} between the localized d electrons of Pt atoms and the conduction electrons.

References

[1] W. B. Pearson, A Handbook of Lattice Spacings and Structures of Metals and Alloys, Vol. 2, Pergamon Press, 1967 (p. 401).
[2] Ia. H. Dorfman, in: Magnetnîe svoistva i stroenie vestcestva, Gostekhizdat, Moskva 1955 (p. 272).
[3] S. V. Vonsovski, Magnetism, Izd. Nauka, Moskva 1971 (p. 124).
[4] M. Fallot, Ann. Phys. (Paris) **6**, 305 (1936).
[5] J. A. Seitchik, A. C. Gossard, and V. Jaccarino, Phys. Rev. **136**, A1119 (1964).
[6] A. M. Clogston, V. Jaccarino, and Y. Yafet, Phys. Rev. **134**, A 650 (1964).
[7] J. Seitchik, V. Jaccarino, and J. H. Wernick, Phys. Rev. **138**, A148 (1965).
[8] V. Jaccarino, B. T. Matthias, M. Peter, H. Shul, and J. H. Wernick, Phys. Rev. Letters **5**, 251 (1960).
[9] V. Jaccarino, J. appl. Phys. Suppl. **32**, 102 (1961).

(Received February 13, 1978)

phys. stat. sol. (a) 49, 211 (1978)

Subject classification: 14.3.3 and 20.3; 11; 22.2.2

A. F. Ioffe Physico-Technical Institute, Academy of Sciences of the USSR, Leningrad

The Influence of γ-Irradiation on the Degradation of Red-Emitting GaP LED's

By

T. D. Dzhafarov, M. N. Kolotov, and A. A. Litvin

The effect of γ-irradiation on the degradation of p-GaP–n-GaP red light emitting diodes (LED's) is studied. It is shown that after the irradiation to a dose of 1.5×10^{17} γ/cm² the LED's degrade, a decrease of red electroluminescence maximum ($\lambda = 0.675$ μm, $T = 77$ K) by a factor of about 3 to 6 is observed. A following isochronal annealing of the irradiated diodes over the temperature interval from 100 to 300 °C leads to a partial recovery of the starting luminescence properties of the diodes. Similar effect is also observed when forward currents flow through the diode ($j = 10^{-1}$ to 10^{-2} A/cm²) at temperatures 77 and 300 K. The data obtained are explained by defects (Frenkel type) which act as non-radiative centres.

Исследовано влияние γ-облучения на деградацию p-GaP–n-GaP-красных светодиодов. Показано, что при облучении (доза $1{,}5 \times 10^{17}$ γ/cm²) светодиоды деградируют — наблюдается уменьшение максимума интенсивности красной электролюминесценции (при $\lambda = 0{,}675$ μm, $T = 77$ K) приблизительно в 3 до 6 раз. В результате последующего изохронного отжига облучённых диодов от 100 до 300 °C наблюдается частичное восстановление исходных люминесцентных свойств диодов. Аналогичный эффект обнаружен также при пропускании прямых токов через диод ($j = 10^{-1}$до 10^{-2} A/cm²) при температуре 77 и 300 K.

1. Introduction

Only few papers deal with the degradation of GaP LED's due to γ-irradiation [1, 2]. Logan et al. [1] have studied the effect of γ-radiation on diffusion-doped red- and green-emitting GaP LED's while Epstein et al. [2] have investigated the degradation of green GaP and yellow GaAsP light emitting diodes resulting from irradiation with γ-rays.

This paper presents results on the degradation of red-emitting GaP LED's resulting from γ-irradiation, and the effect of a thermal and electrostimulated annealing on the recovery of these diodes.

2. Experimental

Epitaxial p-GaP(Zn, O)–n-GaP(Te) diode structures were grown by liquid phase epitaxy on GaP n-type wafers with electron concentration $n = 3 \times 10^{17}$ cm^{-3}. Thicknesses of the epitaxial p- and n-layers were about 15 and 20 μm, respectively, and their free carrier concentration was about 10^{18} cm^{-3}.

A number of LED's were used for measurements of the electro- and photoluminescence, the spectral dependence of photoresponse (short circuit photocurrent), luminescence brightness, current–voltage and capacitance–voltage characteristics at $T = 80$ K and 300 K. Optical transmittance spectra of the starting p–n-structures were measured over the wavelength range 1 to 25 μm.

The measuremgnts of electro- and photoluminescence and short circuit photocurrent were performed using a monochromator with a 600 grooves/mm grating. A photomultiplier was used for registering the luminescence spectra. Photoluminescence from the p-GaP surface was induced by a helium–cadmium laser (wavelength 0.441 μm) with power output 5 mW. The change of the integrated light intensity (over a wide spectral range) was registered with a photoresistor. The I–U characteristics were measured by a 4-digit voltmeter and the C–U characteristics using a standard capacitance bridge.

After the measurements the starting LED's were irradiated by γ-quanta with 1.3 MeV energy from a ^{60}Co source. During irradiation the diodes were heated up to temperatures not exceeding 50 °C. The irradiation dose was 1.5×10^{17} γ/cm². After that the whole measurement cycle was repeated.

Isochronal annealing of the irradiated diodes was performed in evacuated and argon-filled ampoules over the temperature range 100 to 400 °C for 30 min. After each isochronal annealing the characteristics of the diodes were measured again.

3. Experimental Results

The electroluminescence spectra before (1) and after the irradiation (2) are shown in Fig. 1. It is seen that the maximum of the red electroluminescence band due to recombination via Zn–O complexes lies approximately at 1.84 eV ($\lambda = 0.675$ μm). Besides, in the shorter wavelength part of the spectrum two weak peaks are also observed at 2.19 eV ($\lambda = 0.565$ μm) and 2.14 eV ($\lambda = 0.579$ μm) corresponding to residual impurities (probably to sulphur [3]). With an irradiation dose of 1.5×10^{17} γ/cm² the intensity of the red band decreases approximately by a factor of 3 to 6 (see Fig. 2, curve 2). The intensity of the green band after the irradiation decreases in the same proportion.

In Fig. 2 the relative change of the integrated luminescence brightness versus γ-irradiation dose is shown. The most abrupt fall of the intensity is observed in the dose interval (0.2 to 2) $\times 10^{17}$ γ/cm².

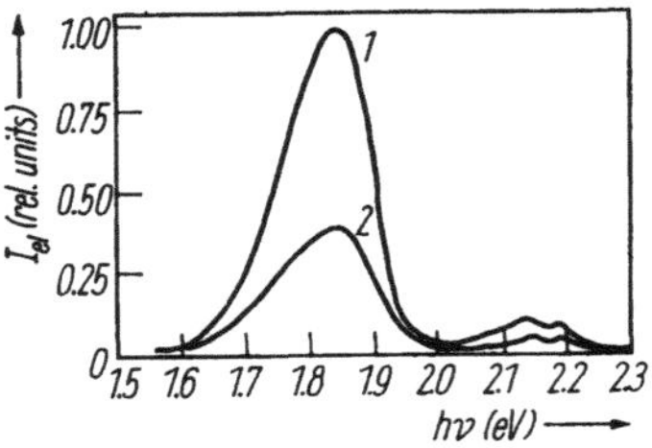

Fig. 1. Red GaP LED electroluminescence spectra ($T = 77$ K). (1) Before, (2) after the irradiation to $\varphi = 1.5 \times 10^{17}$ γ/cm²

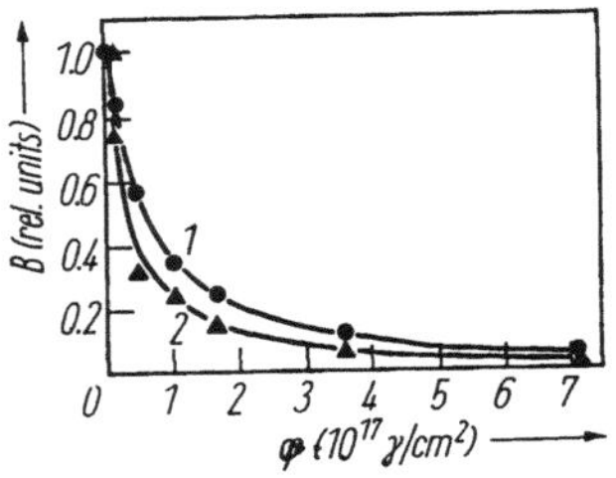

Fig. 2. Red GaP LED's brightness versus γ-irradiation dose. (1), (2) two extreme cases

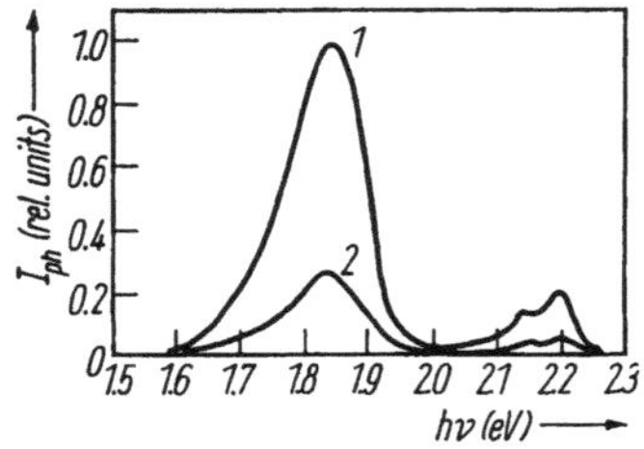

Fig. 3. Red GaP LED photoluminescence spectra of the p-side ($T = 78$ K). (1) Before, (2) after the irradiation to $\varphi = 1.5 \times 10^{17}$ γ/cm²

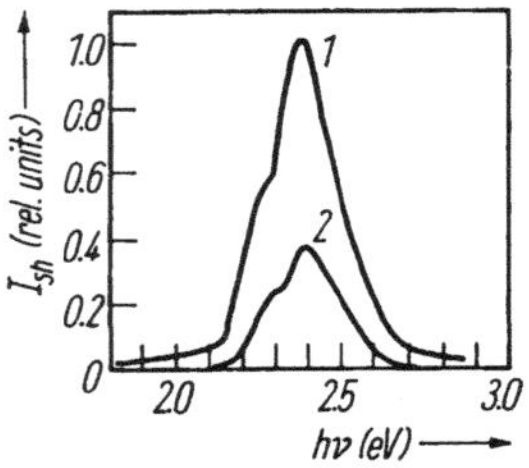

Fig. 4. Spectral dependence of the short circuit photocurrent of red GaP LED's ($T = 300$ K). (1) Before, (2) after the irradiation to $\varphi = 1.5 \times 10^{17}$ γ/cm²

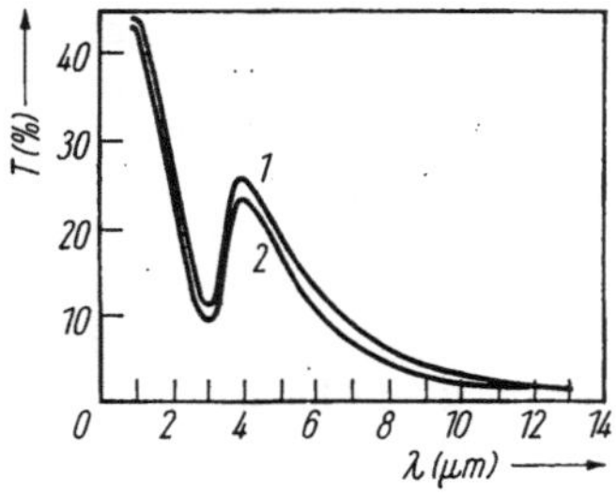

Fig. 5. IR absorption spectra of GaP LED structures ($T = 300$ K). (1) Before, (2) after the irradiation to 10^{18} γ/cm²

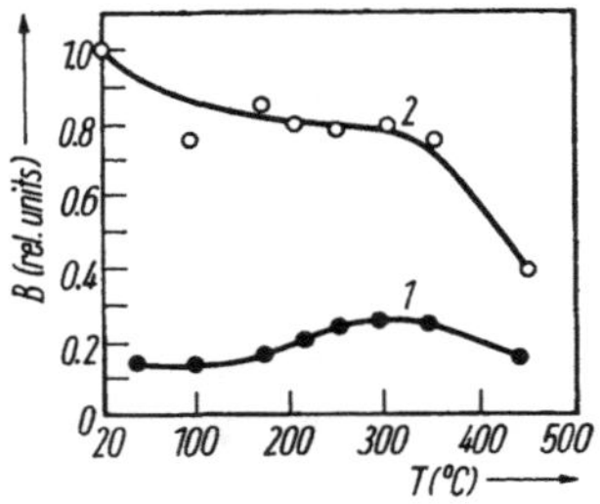

Fig. 6. Isochronal annealing curves of the (1) irradiated and (2) test diodes. $\varphi = 1.5 \times 10^{17}$ γ/cm²

The photoluminescence spectra and their change after the irradiation are similar to those of electroluminescence (Fig. 3).

The spectral dependence of the short circuit photocurrent is shown in Fig. 4. At 2.25 eV there is a band corresponding to indirect interband transitions and the maximum is observed at higher energy and is due to direct transitions. After the irradiation, as it is seen in Fig. 4, the short circuit photocurrent decreases but the pattern of the spectrum does not change.

The optical transmittance of epitaxial p-GaP–n-GaP–n-GaP structures which were used for diodes fabrication after the irradiation as it may be seen from the transmittance spectra (Fig. 5). No new light absorption centres were observed in the irradiated structures over the wavelength range considered.

Capacitance versus reverse bias dependences for the non-irradiated samples are approximated by the equation $C^{-n} = AV$ (with $2 < n < 3$) and change negligibly after γ-irradiation. Forward bias currents increase slightly with irradiation.

Isochronal annealing of the irradiated diodes results in a partial recovery of the luminescent properties. In Fig. 6 the isochronal annealing curve of luminescence brightness (1) is shown, the irradiation dose being 1.5×10^{17} γ/cm². Curve for the test (non-irradiated) LED's is also shown. It is seen that in the beginning of annealing up to 300 °C the radiation brightness increases and at tempeartures higher than 300 °C a brightness decrease is observed. The high temperature effect may be explained by the redistribution of fast diffusing impurities, which create extra non-radiative recombination centres. The test LED's which had not undergone γ-irradiation show a regular decrease of brightness (Fig. 6, curve 2). From measurements of the effect of isochronal and isothermal annealings of the irradiated LED's on the luminescence properties it was found that the activation energy for annealing of the defect centres arising during γ-irradiation and being responsible for the red luminescence degradation is about 0.6 eV.

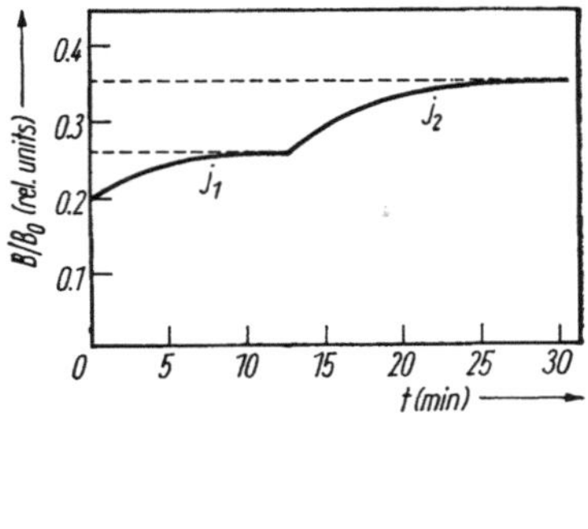

Fig. 7. Change of the irradiated LED's brightness versus time under forward bias. $\varphi = 1.5 \times 10^{17}$ γ/cm², $j_1 = 2.2 \times 10^{-2}$ A/cm² for $t = 0$ to 12 min and $j_2 = 1.3 \times 10^{-1}$ A/cm² for $t > 12$ min

We observed a luminescence brightness recovery of the irradiated diodes at forward current flow. The brightness of a diode as a function of time of forward current flow is shown in Fig. 7. It is seen that the intensity of the diode brightness after γ-irradiation (at the moment $t = 0$ in Fig. 7) was 5 times less than that before the γ-irradiation. When a forward bias is applied ($j_1 = 2.2 \times 10^{-2}$ A/cm^2 during 10 min) as it is seen from Fig. 7, the diode brightness increases from 0.20 to 0.26 and further is saturated.

The increase of the forward current (up to $j_2 = 1.3 \times 10^{-1}$ A/cm^2) results in further brightness growth. The recovery of luminescence brightness of the irradiated diodes is irreversible: after the diode had undergone a forward current flow, a reverse current of the same value does not change its brightness. The reverse current just after the irradiation also does not affect brightness. It should be noted that the recovery of the electrostimulated luminescence intensity was not due to a thermal effect of the current. In fact, after the isochronal thermal annealing (for 30 min) of the irradiated LED's over the range from room temperature to 480 K the increase of the relative amplitude of the red luminescence peak (from 0.20 to 0.26) is the same as that observed when the forward bias current flows at 80 K. A similar effect has been observed earlier in [1].

Some of the non-irradiated diodes also showed an increase of luminescence brightness. In diodes being switched on for the first time under forward bias ($j = 0.5$ A/cm^2), the brightness increased during 10 min by 30 to 40% and did not change further. The magnitude of this effect was different for diodes from different charges and sometimes there was no effect at all.

Presumably, defects appeared in these diodes during the growth process which were similar to those, introduced during γ-irradiation and were recovered afterwards under the effect of electrostimulated annealing.

4. Discussion

γ-irradiation of LED's results in point defect creation (Frenkel pairs) and may also lead to a dissociation of Zn–O complexes. Besides, the irradiation may result in leakage through the sides of the diodes due to the creation of defects at the surface. However, etching of the side surfaces of the irradiated diode to a depth of about 2 μm showed that brightness degradation was not surface controlled.

The fact that after the irradiation the intensities of red and green bands decrease in the same proportion may suggest that the degradation is due to the creation of a competitive non-radiative recombination channel, which successfully competes with all radiative channels. This is also supported by the fact that the current at a given forward bias increases after the irradiation. There were no new noticeable radiative bands in the photo- and electroluminescence spectra with energy $h\nu > 1$ eV which also supports the non-radiative character of the created defects.

Defects of Frenkel type or/and associates of these defects are probably such non-radiative centres and they are responsible for the LED degradation. The annealing energy of radiation defects for LED's, obtained in our experiment (0.6 eV) approximates the activation energy for interstitial migration in the III–V compound semiconductors.

The forward bias may induce electrotransport which leads to defect migration to a more stable site. Another explanation may be the recombination enhanced annealing of radiation defects due to the energy released at the act of electron–hole pair recombination when a forward current flows [4]. At present the experimental data available do not afford unambiguously conclusions on the mechanism of radiation defect annealing by forward current in GaP diodes.

Acknowledgements

The authors are grateful to Prof. B. I. Boltaks and I. E. Maronchuk for discussion and permanent interest in this work.

References

[1] R. A. Logan, H. G. White, and R. M. Mikulyak, Appl. Phys. Letters 5, 41 (1964).
[2] A. S. Epstein, S. Share, R. A. Polimadei, and A. H. Herzog, IEEE Trans. Nuclear Sci. 19, 386 (1972).
[3] A. Onton, Phys. Rev. 186, 786 (1969).
[4] D. V. Lang and L. C. Kimerling, Appl. Phys. Letters 28, 248 (1976).

(Received April 27, 1978)

phys. stat. sol. (a) **49**, 217 (1978)

Subject classification: 14.4; 22

Institute of Semiconductor Physics, Academy of Sciences of the USSR, Siberian Branch, Novosibirsk

Electron and Hole Conduction in Silicon Nitride at Moderate Electric Fields

By

V. M. EFIMOV, V. A. KOLOSANOV, and S. P. SINITSA

The method of isothermal discharge of MNOS-structures with the thickness of SiO_2 of 80 to 100 Å is proposed for investigating electron and hole transport in fields of 10^5 to 10^6 V/cm through Si_3N_4 films. The hole transport is investigated on structures with molybdenum "grains" deposited on the SiO_2–Si_3N_4 interface. It is shown that transport of both electrons and holes occurs according to the Poole-Frenkel effect.

Предлагается метод изотермической разрядки МНОП-структур с толшиной окисла 80 до 100 Å для исследования переноса электронов и дырок через плёнки нитрида кремния в полях 10^5 до 10^6 V/cm. Перенос дырок исследовался на структурах с „зёрнами" молибдена, нанесёнными на границу раздела SiO_2–Si_3N_4. Показано, что перенос как электронов, так и дырок осуществляется по механизму Пула-Френкеля.

1. Introduction

The method of current–voltage characteristics is widely used at high electric fields $E > 10^6$ V/cm. It has been found [1 to 4] that in fields $E \times (3$ to $4) \times 10^6$ V/cm and at temperatures above 300 K, charge is transported in bulk silicon nitride films by the Poole-Frenkel mechanism and in the nearcontact region by the Fowler-Nordheim one [5]. Between 10^6 and 3×10^6 V/cm there is a section of current–voltage characteristics with weaker dependence on the average electric field, which is interpreted in [1, 2] as hopping conduction. In [5] it is assumed that the current in this region is due to the decay of charge trapped in the near-contact region. In fields less than 10^6 V/cm this method is not applicable due to the extremely low current passing through the nitride. There are no experimental results on conduction in silicon nitride in fields of 10^5 to 10^6 V/cm, and the mechanism of conduction in this range of fields remains unknown.

A variety of views is available concerning the nature of conduction in silicon nitride in the investigated range of high electric fields. In [1, 6] experimental data are interpreted on the basis of a one-band model and it is assumed that current is transferred by electrons. In [7, 9] it is asserted that both electron and hole conduction take place (or conductivity via two types of trapping centres). The results reported in [10 to 12] indicate prevalence of hole conductivity.

We studied the mechanism of electron and hole transport in silicon nitride in electric fields of 10^5 to 10^6 V/cm by using data on the decay of charge stored in a MNOS structure with a relatively thick SiO_2 layer (80 to 100 Å). The experimental results are interpreted on the basis of electron and hole participation in conduction.

2. Experimental Procedure

All experiments were performed with MNOS capacitors fabricated on 7.5 Ω cm, (100) oriented, n-type Si substrates. The oxide layer was grown in dry oxygen at 1000 °C. Then molybdenum "grains" were deposited on the SiO_2 layer by thermal sublimation in a vacuum of about 5×10^{-6} Torr. Silicon nitride was deposited by means of monosilane ammonolysis in hydrogen flow at 850 to 900 °C. The film thicknesses of SiO_2 and Si_3N_4 measured by ellipsometry were 80 to 100 Å and 600 to 800 Å, respectively. Together with this we made control samples with no molybdenum "grains at the SiO_2–Si_3N_4 interface. As well as in [13], deposition of molybdenum "grains" to the interface of the two dielectrics permits storage of both negative and positive charges in these structures (MN(Mo)OS), while control samples can store only negative charge. Both positive and negative charges were stored by applying a voltage of 40 to 50 V with negative ($\tau_p = 1$ s) and positive ($\tau_p = 1$ ms) polarity, respectively, to an Al electrode (the field in the oxide is $\approx 10^7$ V/cm) (Fig. 1a). The decay of the trapped charge was studied by measuring the flat band voltage U_{FB} at various times from 10^1 to 10^5 s after the writing pulse. The change of U_{FB} with time was studied in the temperature range from 300 to 600 K at various voltages U_g from 0 to 15 V applied to the Al electrode. In this case the voltage sign at Al was such that the external field was added to the field of the trapped charge in the silicon nitride bulk and had an opposite sign in SiO_2. Thus, the field in SiO_2 was 10^4 to 10^6 V/cm, in Si_3N_4 2×10^5 to 2×10^6 V/cm (Fig. 1b, c).

3. Experimental Results and Discussion

The use of a relatively thick SiO_2 layer changes the nature of current in the MNOS structure. Injection of carriers from silicon into the dielectric is replaced by that from the dielectric into silicon [8]. Naturally, the sign of the charge stored at the SiO_2–Si_3N_4 interface changes. With a SiO_2 film thickness of 80 to 100 Å, injection and storage of electrons from silicon are still proceeding, but hole injection is already blocked as the barrier for holes is about 1 eV larger than that for electrons. This explains the peculiarity of charging of the control samples. In addition, the use of an oxide layer of 80 to 100 Å thickness excludes charge carrier exchange between silicon and silicon nitride in fields of 10^5 to 10^6 V/cm during charge decay. This permits investigations of the transport of the stored charge (both negative and positive) through the silicon nitride bulk to the metal under different environmental conditions.

3.1 Decay of the negative charge

Fig. 2 shows the dependence of U_{FB} on t at fixed temperature and various voltages at the gate which increases the rate of decay of the charge (+Al) trapped in the

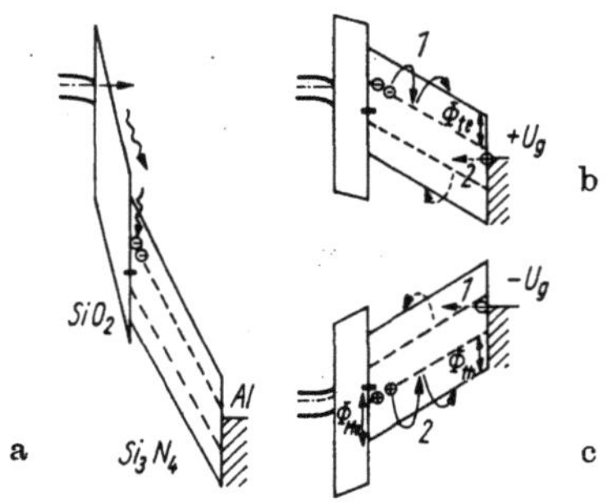

Fig. 1. The band diagram of a MN(Mo)OS structure. a) Negative charge storage, b) negative charge decay, c) positive charge decay

MN(Mo)OS structure. The decay curves of the negative charge are the same as for the control MNOS structures. A peculiarity of the field dependences of decay is that an increase in positive bias at the gate enhances the negative charge decay.

The temperature dependences of the negative charge decay given in Fig. 3 are similar. In this case a constant voltage $U_g = 10$ V was applied to the gate. The obtained results carry the necessary information on charge carrier transport in silicon nitride in the field range from 10^5 to 10^6 V/cm. In these fields the time of thermal excitation of charge carriers τ from the deep trapping centres is extremely long, and the charge transport will be determined by the value of τ rather than by the time of transport of carriers in the conduction band. It can be assumed in the first approximation that $\tau = \tau_0 \exp(\Phi/kT)$, where Φ is the barrier height for thermal excitation of carriers from the trapping centre (it can depend on the electric field); $\tau_0 \sim q\,\hbar/\Phi$, $\hbar$ is Planck's constant. It is evident that a study of the temperature and field dependences of τ can provide information on the temperature and field dependences of Φ and, consequently, on the mechanism of charge carrier transport in the dielectric. Thus, at low fields where the current–voltage method is not applicable (to silicon nitride), it is possible to explore the process of decay of the charge trapped in a dielectric, that is, the $U_{FB}(t)$ dependence.

If X_c is the distance in nitride from the SiO_2–Si_3N_4 interface at which the charge is trapped during a writing pulse with an initial value of Q_0, then the shift of the flat-band voltage caused by this charge is

$$\Delta U_{FB} = \frac{Q_0}{c} \frac{d_N - X_c}{d_N^*}, \tag{1}$$

where c is capacitance per unit area of the MNOS structure, d_N is the thickness of silicon nitride, $d_N^* = d_N + \varepsilon_N/\varepsilon_{ox}\, d_{ox}$; d_{ox} is the oxide thickness, ε_{ox}, ε_N are relative permittivities of oxide and nitride, respectively.

The charge Q and, hence, ΔU_{FB} will decrease with time due to charge carrier transport through the silicon nitride to the metal. From the very beginning we shall

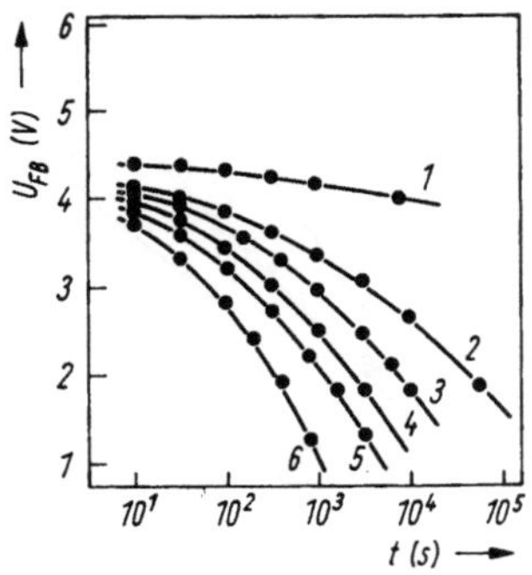

Fig. 2

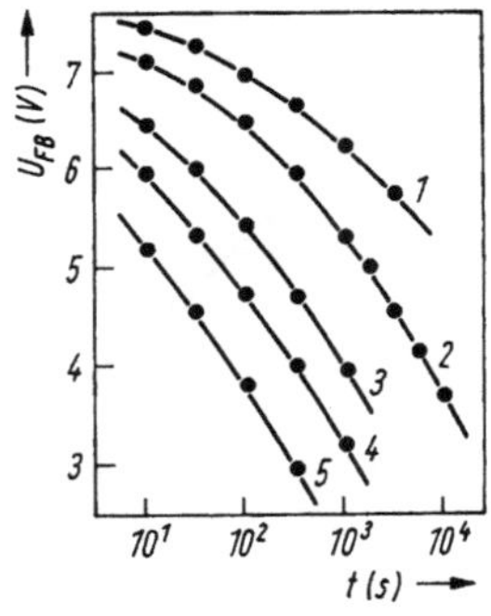

Fig. 3

Fig. 2. The change of U_{FB} with time at various voltages at the gate (+Al) increasing the rate of decay of a stored negative charge at a fixed temperature ($T = 433$ K). (1) $U_g = 0$, (2) 8, (3) 10, (4) 12, (5) 14, (6) 16 V

Fig. 3. The change of U_{FB} with time at an increase of the temperature of decay of a stored negative charge and the fixed voltage at the gate ($U_g = 10$ V). (1) $T = 445$, (2) 475, (3) 526, (4) 550, (5) 575 K assume that the Poole-Frenkel mechanism also operates at fairly low fields. Then the probability of escape of the charge trapped in the nitride is proportional to $\exp(-\Phi_t - \beta\sqrt{E_N}/kT)$, where Φ_t is the trap depth, k is Boltzmann's constant, T is the absolute temperature, E_N is the electric field in the nitride in the X_c–Al region The dependence of charge on time is described by the expression

$$\frac{\partial Q}{\partial t} = -\frac{Q(t)}{\tau_1}\exp\left(-\frac{\Phi_t - \beta\sqrt{E_N}}{kT}\right), \tag{2}$$

where τ_1 is a constant independent of T and E_N, $E_N(t) = E_g + E_Q(t)$, $E_g = V_g/d_N^*$ is the external field, $E_Q = U_{FB}/d_N^*[(d_N^*/d_N - X_c) - 1]$ is the intrinsic field of the trapped charge in the silicon nitride bulk.

Taking into account that in the experiment the change of the intrinsic field of the trapped charge is always much less than the external one, we obtain with an accuracy up to terms of the order of E_Q/E_N from (2)

$$2.3\lg\frac{t}{\tau_2(U_{FB})} \approx \frac{\Phi_t - \beta\sqrt{E_N}}{kT}, \tag{3}$$

where $\tau_2 = \tau_1 \lg \Delta U_{FB}(t)/\Delta U_{FB}(0)$, $\Delta U_{FB}(0)$ is the initial voltage shift of flat bands. After differentiation of both parts of (3) with respect to $\sqrt{E_N}$ and $1/kT$ at $\Delta U_{FB} = {}$ $= \text{const}$ we obtain

$$\beta^* \equiv -2.3\frac{\partial \lg t}{\partial\sqrt{E_N}}kT \approx \beta; \qquad T = \text{const}, \tag{4}$$

$$\varphi \equiv 2.3\frac{\partial \lg t}{\partial\, 1/kT} \approx \Phi_t - \beta\sqrt{E_N}; \qquad E_N = \text{const}. \tag{5}$$

It can be seen that the slope of the dependence $\lg t = f(1/T)$ at $U_{FB} = \text{const}$ determines the activation energy from the Frenkel centre, and the dependence $\lg t = {}$ $= f(\sqrt{E_N})$ permits to find the Frenkel constant β. If the trapped charge is near the

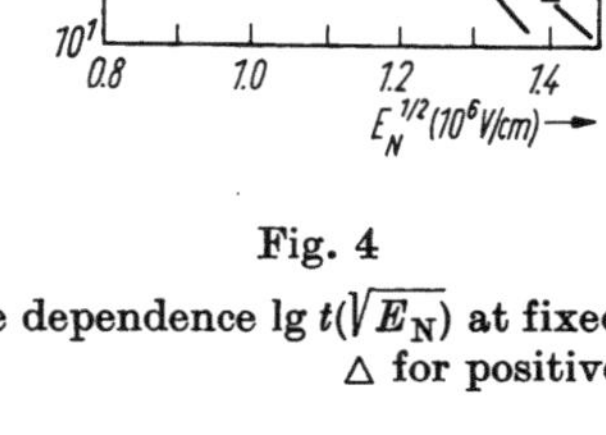

Fig. 4

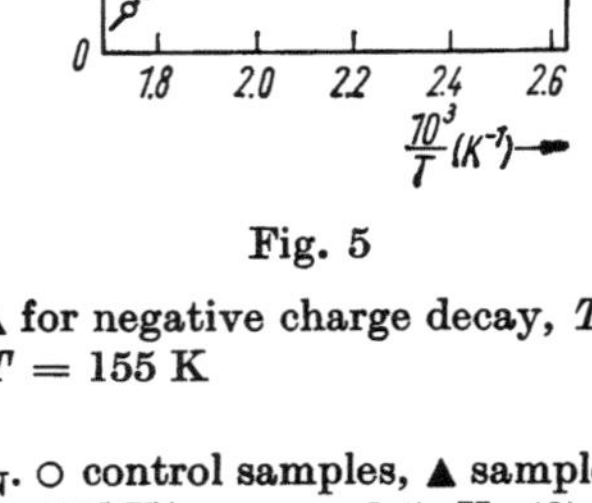

Fig. 5

Fig. 4. The dependence $\lg t(\sqrt{E_N})$ at fixed temperature; ▲ for negative charge decay, $T = 433$ K △ for positive charge decay, $T = 155$ K

Fig. 5. The dependence $\lg t(10^3/T)$ at various values of E_N. ○ control samples, ▲ samples containing Mo grains at the SiO_2–Si_3N_4 interface. (1) $E_N = 1.5 \times 10^5$ V/cm, $\varphi = 1.5$ eV; (2) 1.2×10^6, 1.1; (3) 2.9×10^6, 0.85

SiO_2–Si_3N_4 interface (that is, $x_c \approx 0$), in most cases, when $U_{FB} < U_g$, one can neglect the intrinsic field of the trapped charge and assume that the field in the nitride is $E_N \approx (U_g + \Delta U_{FBO}\, d_N^*/d_N)/d_N^*$.

Experimental dependences of lg $t(\sqrt{E_N})$ and lg $t(1/T)$ plotted according to the data of Fig. 2 and 3 are given in Fig. 4 and 5, respectively. It is seen that experimental points lie on straight lines in the corresponding coordinates. We obtained $\beta^* = (4.0 \pm 0.6) \times 10^{-4}$ eV/V$^{1/2}$ cm$^{-1/2}$ from the slope of the straight line in Fig. 4 which approaches the theoretical value of the Frenkel constant $\beta = 3.8 \times 10^{-4}$ eV/V$^{1/2}$ cm$^{-1/2}$.

The slope of the experimental straight line lg $t(1/T)$ at U_{FB} = const that determines the activation energy, decreases in accordance with (5) with increase of the field in nitride, as shown in Fig. 5. In addition, it can be seen from Fig. 5 that the activation energies for the decay of the negative charge trapped in MN(Mo)OS structures and control MNOS samples are the same. At $E_N = 1.2 \times 10^6$ V/cm the activation energy $\varphi_e = 1.1$ eV for both groups of samples. Extrapolation of the dependence of $\varphi(\sqrt{E_N})$ to zero field gives for the trap depth $\Phi_{te} = (1.6 \mp 0.05)$ eV.

3.2 Decay of the positive charge

The decay curves of a stored positive charge at various voltages on the gate and at fixed temperature are shown in Fig. 6. The temperature dependences of the decay at a fixed voltage on the gate are given in Fig. 7. It is seen that the field and temperature dependences which define the decay of a stored positive charge are similar to the dependence for a negative charge given in Fig. 2 and 3. The lg $t(\sqrt{E_N})$ and lg $t(1/T)$ dependences plotted from the data of Fig. 6 and 7 are given in 4 and 8, respectively. The activation energy of the positive charge decay in an external field $E_N = 1.2 \times$

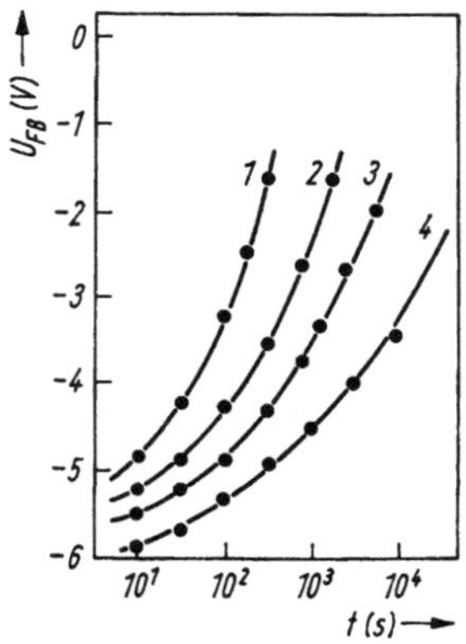

Fig. 6

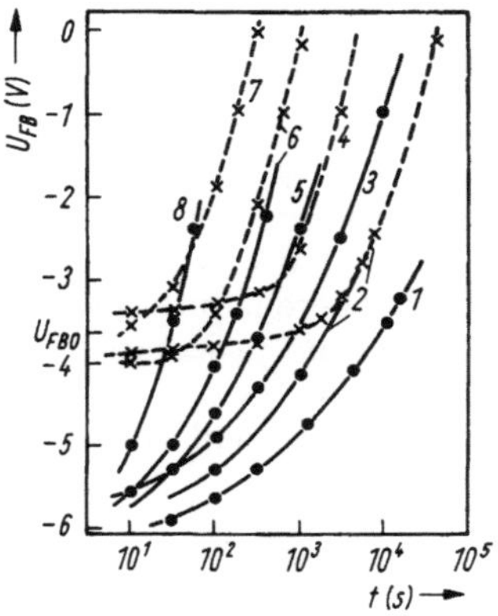

Fig. 7

Fig. 6. The change of U_{FB} with time at various voltages at the gate (—Al) increasing the rate of decay of a stored positive charge at a fixed temperature (T = 515 K). (1) $U_g = -15$, (2) -12, (3) -10, (4) -8 V

Fig. 7. The change of U_{FB} with time at an increase of the temperature of decay of a stored positive charge and fixed voltage at the gate. The dashed curves correspond to the change of an initial value of U_{FBO} (no stored charge) with time for the same sample at $U_g = -10$ V. (1) T = 476, (2) 500, (3) 515, (4) 532, (5) 535, (6) 556, (7) 580, (8) 590 K

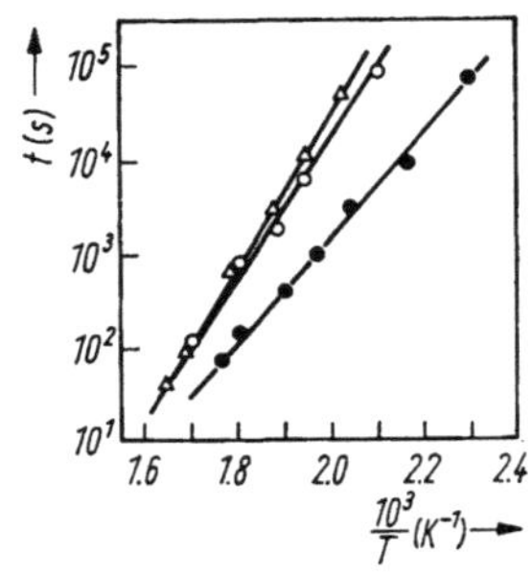

Fig. 8. The dependences lg $t(10^3/T)$ for two groups of samples with different time of molybdenum deposition; $E_N = 1.2 \times 10^6$ V/cm. For stored positive charge decay $\triangle$ $t_{ev} = 60$ s and $\circ$ 120 s. For stored negative charge decay: $\bullet$ $t_{ev} = 120$ s

$\times 10^6$ V/cm estimated from the curve slopes in Fig. 8 is (1.55 ± 0.1) eV (for two groups of samples with different times of molybdenum deposition). The results given for comparison were obtained from the temperature dependence of the negative charge decay in the field $E_N = 1.2 \times 10^6$ V/cm. As can be seen, the activation energies of decay of positive and negative ($\varphi_e = 1.1$ eV) charges differ significantly.

Let us consider the way the charge carrier transport occurs in the silicon nitride bulk at different polarities of the voltage applied to the gate. For a negative stored charge a positive voltage on the gate can cause either electron transport in the nitride conduction band to the metal (process 1 in Fig. 1b) or hole injection from Al (process 2 in Fig. 1b) and hole transport in the valence band with subsequent recombination with a stored negative charge. Similarly, for a positive stored charge a negative voltage on the gate causes hole transport through the nitride bulk to the metal (process 2 in Fig. 1c) and electron injection from Al and transport in the conduction band with subsequent recombination with a stored positive charge. The hole transport should be defined by the same activation energy as the electron transport at positive gate bias, if the stored charge is negative.

In control samples with no stored charge electron injection from Al at negative bias must have resulted in storage of a negative charge near the SiO_2–Si_3N_4 interface, hole injection from Al at positive bias in storage of a positive charge, that is, in variation of the initial value of U_{FBO}.

Fig. 9 shows the variation of U_{FB} from an initial value of U_{FBO} with time at $T = 580$ K and $U_g = \pm 10$ V for MNOS and MN(Mo)OS structures, when there is no stored charge initially. It is seen that for control samples the variation of U_{FB} does not exceed 0.2 to 0. 5 V while for the MN(Mo)OS structures $\Delta U_{FB} = 6.4$ V at $U_g = -10$ V. Since there is practically no change of U_{FB} in the control samples, one can say that there are no processes 2 (Fig. 1b) and 1 (Fig. 1c) in the samples under investigation in fields $< 2 \times 10^6$ V/cm and there is no generation of charge carriers in Si_3N_4 with their subsequent transport. The consequence is that the observed variation

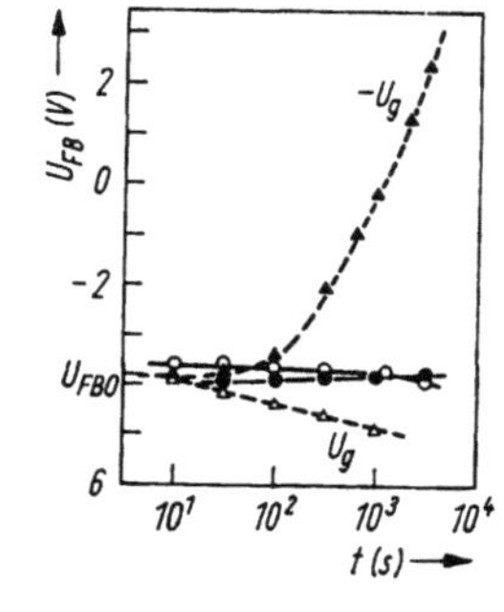

Fig. 9. The change of the initial value of U_{FB} with time for two groups of samples at $\pm$Al. Con- tinuous curves: $\circ$ (+Al), $\bullet$ (−Al). Dashed curves: $\triangle$ (+Al), $\blacktriangle$ (−Al); $T = 556$ K. The continuous curves correspond to control samples, the dashed ones to samples containing Mo grains at the SiO_2–Si_3N_4 interface ($t_{ev} = 120$ s; $U_g = \pm 10$ V)

of U_{FB} for molybdenum-containing samples may be associated only with hole generation and transport through the silicon nitride bulk over the valence band (process 2 in Fig. 1c). Note that after fabrication of a MN(Mo)OS structure the centres associated with molybdenum have no charge, which is indicated by the equality of U_{FB} values in these structures and control samples. However, if there is an electric field in the sample and the temperature rises, the centres generate free carriers, their acceptor properties being more well-defined than the donor ones. The activation energy of U_{FB} variation obtained from the slope of $\lg t(1/T)$ dependences for a set of dotted $U_{FB}(t)$ curves given in Fig. 7 agrees with that of decay of the positive charge preliminarily stored in the same structure, but differs from that of electron decay by 0.5 eV.

These results indicate that the decay of the stored negative charge is associated with electron transport in the conduction band. An activation energy of the positive charge decay may be defined by both hole transport via bulk traps and hole generation from centres connected with molybdenum. In the former case the observed value of the activation energy of (1.55 ± 0.1) eV will be determined by the trap depth Φ_{th} in the silicon nitride bulk, in the latter by the generation centre depth Φ_{Mo} (Fig. 1c).

4. Conclusions

1. A method was proposed for invertigating the mechanism of charge transport in silicon nitride in relatively low fields of 10^5 to 10^6 V/cm which uses data on trapped charge decay in MNOS structures with thick oxide (80 to 100 Å).

2. It has been shown that conduction in silicon nitride in the fields under investigation occurs according to the Poole-Frenkel mechanism; it is a two-band one.

3. The depth of electron traps in the silicon nitride bulk was estimated to $\Phi_{te} = (1.6 \pm 0.05)$ eV.

References

[1] S. M. Sze, J. appl. Phys. **38**, 2951 (1967).
[2] E. J. M. Kendall, Canad. J. Phys. **46**, 2509 (1968).
[3] D. Frohman-Bentchkowsky and M. Lenzlinger, J. appl. Phys. **40**, 3307 (1969).
[4] J. R. Yeargan and H. L. Taylor, J. Electrochem. Soc. **115**, 273 (1968).
[5] P. C. Arnett and D. J. Di Maria, J. appl. Phys. **47**, 2092 (1976).
[6] K. Kobayashi and K. Ohta, Japan. J. appl. Phys. **11**, 538 (1972).
[7] A. S. Ginovker, V. A. Gritsenko, and S. P. Sinitsa, Mikroelektronika **2**, 283 (1973).
[8] L. Ye. Kurdov, V. V. Pospelov, and B. I. Tsilibin, Mikroelektronika **2**, 263 (1973).
[9] A. S. Ginovker, V. A. Gritsenko, and S. P. Sinitsa, phys. stat. sol. (a) **26**, 489 (1974).
[10] S. M. Hu, D. R. Kerr, and L. V. Gregor, Appl. Phys. Letters **10**, 1 (1967).
[11] Z. A. Weinberg and R. A. Pollak, Appl. Phys. Letters **27**, 254 (1975).
[12] D. J. Di Maria and P. C. Arnett, IBM J. Res. Developm. **21**, 227 (1977).
[13] D. Kahng, W. J. Sundburg, D. M. Boulin, and J. R. Ligenza, Bell Syst. tech. J. **53**, 1723 (1974).

(Received June 6, 1978)

phys. stat. sol. (a) 49, 225 (1978)

Subject classification: 13.5.2 and 14.3.3; 22.1.2

Physico-Energetical Institute, Academy of Sciences of the Latvian SSR, Riga

Plasmon Mechanism for Secondary Breakdown

By

T. PURITIS

Secondary breakdown studies in silicon p–n junctions reveal the existence of different types of mesoplasma, viz., warm mesoplasma and hot mesoplasma. To explain the behaviour of these mesoplasmas, a plasmon mechanism of ionization and pinch is suggested. The conditions of plasmon excitation are considered. The possibility of ionization by a plasmon having sufficient energy is discussed. It is found that the presence of plasmons may be responsible for the form of the spectral distribution of the infrared and visible radiation of mesoplasma. A qualitative explanation is presented for the basic relationships of hot and warm mesoplasma in terms of the proposed secondary breakdown mechanism.

При исследовании вторичного пробоя кремниевых p–n переходов отмечено существование разных видов мезоплазмы: теплая мезоплазма и горячая мезоплазма. Для объяснения свойств разных видов мезоплазмы в данной работе предложен плазмонный механизм ионизации и пинч-эффекта. Обсуждены условия возбуждения плазмонов. Анализирована возможность плазмонов с достаточно большой энергией ионизировать. Установлено, что присутствие плазмонов влияет на вид спектрального распределения эмиттированного инфракрасного и видимого излучения мезоплазмы. При помощи предложенного механизма вторичного пробоя дано качественное объяснение основным свойствам горячей и теплой мезоплазмы.

1. Introduction

The secondary breakdown of silicon p–n junctions is linked with the appearance of a negative dynamic resistance and of current localization occurring in a small-diameter channel emitting visible light [1]. The localized plasma observable in the case of secondary breakdown of a p–n junction is referred to as mesoplasma [2]. Experimentally, two types of mesoplasma are observable. In the first case, mesoplasma is seen to emit dark-red light appearing with some delay ($\approx 30\ \mu s$) after the onset of secondary breakdown current [3]. This type of mesoplasma will be referred to as warm mesoplasma. In the second case, the mesoplasma is observed to emit high-intensity red-yellow light, which is visible before the onset of the secondary breakdown current [4 to 13]. This type of mesoplasma will be referred to as hot mesoplasma. The diameter of the hot mesoplasma is approximately equal to $10\ \mu m$, whereas for warm mesoplasma it is larger almost by one order of magnitude. In order to have a warm mesoplasma, it is necessary to attain a critical value of local temperature in the p–n junction. The value is known to be close to the intrinsic conductivity onset temperature [14]. The same applies to the hot mesoplasma [10].

The microscopic examination of the mesoplasma reveals bands of light occurring on either side of the p–n junction, which is observed in the form of a dark (non-light-emitting) line [15]. This pattern is opposite to that of the case of avalanche breakdown when the p–n junction itself is emitting light.

The spectral distribution of mesoplasma emission studied in silicon p–n diodes [15] and in sapphire-based epitaxial silicon diodes [16] has its maxima in the infrared region. The spectral distribution of mesoplasma emission is shown in Fig. 1, where curves 1 and 2 correspond to warm mesoplasma and curve 3 to hot mesoplasma [15].

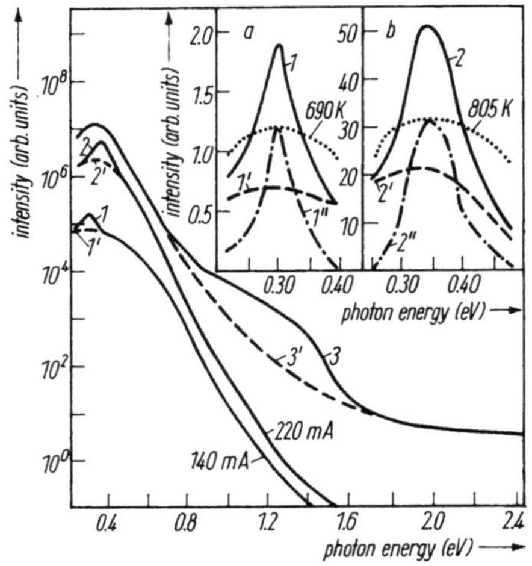

Fig. 1. Spectral distribution of energy intensities (on a logarithmic scale) for the warm mesoplasma: (1) and (2) in the case of two different currents and (3) for the not mesoplasma. Curve (3) is arbitrarily shifted with respect to curves 1 and 2. Fig. 1 a and b show the region of maximum for curves 1 and 2 (on a linear scale). ——— Experimental curve; — — — bulk of emission; ·—·—· the hump; .... the black-body emission

These spectra may be treated as a superposition of not less than three radiations. As in the case of microplasma radiation [17], the bulk of infrared emission may be attributed to the direct transitions of holes between the subbands of the valence band, whereas the tail of the distribution in the visible spectral region is associated with indirect recombination transitions of hot electrons from the absolute minima of the conduction band to the maximum of the valence band in silicon as in the case of visible microplasma radiation [18]. In addition to these basic components of radiation which form smooth spectral distribution curves (curves 1′, 2′, and 3′ in Fig. 1), there are some additional components which are responsible for the appearance of the humps on the spectral distribution curves as singled out in Fig. 1 by dash-dotted lines (curves 1″ and 2″ in Fig. 1a and 1b). In the case of warm mesoplasma, an increase in the bias current causes higher intensities of emission and a certain shift of the emission maximum towards higher photon energies (see curves 1 and 2 in Fig. 1), which is typical for emission of thermal origin. However, it is substantially different from the black-body radiation. This is shown in Fig. 1a and 1b. The main difference between the warm mesoplasma and the hot mesoplasma spectra is that the former does not emit high-energy photons, and the location of the humps is also substantially different.

The occurrence of secondary breakdown is usually attributed to thermal effects [19], i.e. to thermal ionization due mainly to the phonon mechanism. However, the phonon mechanism of recombination is of low efficiency, since recombination occurs through local levels which, at high carrier densities, are intensively screened [20]. For the same reason, the phonon mechanism of ionization at high carrier densities (high temperatures) is also of low efficiency. The nature of the hot mesoplasma as well as some characteristics of the warm mesoplasma cannot be explained in terms of the phonon mechanism of ionization.

The present work is an attempt to describe the nature of hot mesoplasma and warm mesoplasma in terms of a plasmon mechanism of secondary breakdown, comprising the plasmon mechanism of the pinch effect and the plasmon mechanism of ionization.

2. The Plasmon Excitation Conditions

The condition for the excitation of a plasmon with a frequency ω_p, i.e. the plasmon of energy [21 to 23]

$$E_p = \hbar\omega_p\,, \tag{1}$$

may be written as [23]

$$\nu < 2\omega_p\,, \tag{2}$$

where ν is the frequency of charge carrier scattering. However, in the case of a semiconductor material with a great effective mass of charge carriers, such as silicon, relation (2) does not hold at least for the region of impurity conduction [24]. Consequently, plasmons can appear in silicon only under the condition that the temperature of intrinsic conduction is attained, or at high densities of nonequilibrium charge carriers. Let m_e and m_h be the effective masses of electrons and holes, respectively, and the carrier densities $n_e = n_h = n$. The plasma frequency may then be expressed as

$$\omega_p = \left[\frac{4\pi n e^2}{\varepsilon}\left(\frac{1}{m_e} + \frac{1}{m_h}\right)\right]^{1/2}, \tag{3}$$

where ε is the lattice component of the dielectric constant for the semiconductor material [21 to 23].

The slightly damped plasma oscillation must, furthermore, obey the condition [21 to 23]

$$L_D < \frac{1}{k}\,, \tag{4}$$

where k is the plasmon wave number and L_D the Debye screening length,

$$L_D = \left[\frac{k_B T}{2\omega_p^2}\left(\frac{1}{m_e} + \frac{1}{m_h}\right)\right]^{1/2}, \tag{5}$$

where k_B is the Boltzmann constant and T the absolute temperature. The value of k and the compliance with condition (4) depend on the mode of plasmon excitation.

In the case of an electron passing through a p–n junction, the electron may lose energy corresponding to four quasi-discrete excitations of the plasma [25]. We are concerned with the excitation of a plasmon having the maximum energy, viz., an optical bulk plasmon. According to the laws of energy and momentum conservation, a hot electron (hole) with an effective mass m can excite a plasmon if

$$\frac{\hbar^2 p^2}{2m} = \frac{\hbar^2 (p - k)^2}{2m} + \hbar\omega_p\,, \tag{6}$$

where $\hbar p$ is the initial momentum of the exciting electron (hole) and $\hbar(p - k)$ is the final electron (hole) momentum [21, 22].

Let a hot charge carrier give all its energy and momentum to a plasmon, i.e. the condition that $\hbar(p - k) = 0$. We then obtain from (6)

$$\frac{1}{k} = \left(\frac{\hbar}{2m\omega_p}\right)^{1/2}. \tag{7}$$

Substituting L_D and $1/k$ from (5) and (7) into (4) and taking into account (1), we may write

$$E_p > k_B T m \left(\frac{1}{m_e} + \frac{1}{m_h}\right). \tag{8}$$

Using $m_e = 0.26 m_0$, $m_h = 0.39 m_0$, and $T = 750$ K, we obtain from (8) that electrons ($m = m_e$) under the condition that $\hbar(p - k) = 0$ can excite plasmons when $E_p > 0.11$ eV or, in the case of a hole ($m = m_h$), $E_p > 0.16$ eV.

The occurrence of plasmons is accompanied by the appearance of plasmon exchange forces [21] which, to the author's opinion, are responsible for current constriction, i.e., causing upon the mesoplasma the plasmon pinch effect.

3. The Possibility of Ionization by Plasmons

According to (1) and (3), ω_p and E_p increase with n. At sufficiently high values of n, when

$$E_p \geqq E_G\,, \tag{9}$$

where E_G is the value of the forbidden gap in the semiconductor, the effective recombination mechanism, i.e. the plasmon mechanism, may occur [20, 26]. Due to this mechanism the abrupt quenching of luminescence occurs when a certain doping ratio is exceeded [26 to 29]. If there exists a plasmon mechanism of recombination, then, in accordance with the detailed balancing principle, a reverse mechanism, i.e. a plasmon mechanism of ionization, should also be available. In this case the plasmon energy is spent for creating an electron–hole pair. The mechanism of ionization should be available if condition (9) is obeyed.

Assuming $m_e = 0.26m_0$, $m_h = 0.39m_0$, $\varepsilon = 11.7$, and $E_G = 0.90$ eV (at $T = 750$ K), we obtain the minimum electron density required for the ionization by the plasmon, $n = 1.1 \times 10^{21}$ cm^{-3}. The value exceeds the usual electron density by several orders of magnitude.

However, formula (3) implies plasma oscillation of free non-drifting carriers. In the case of a drifting plasma, ω_p tends to increase [21 to 23]. The second mechanism of the enhancement of n and ω_p is the current constriction caused by plasmon pinch effect. The third mechanism of the enhancement of ω_p is associated with virtual electron transitions from the valence band to the conduction band, which are responsible for a significant increase in the effective charge density n_{eff}. Studies of the luminescence quenching in various semiconductors suggest $n_{\text{eff}} \gg n$ [26 to 29]. This implies that relation (9) may hold also for silicon. It has been found out that a less effective ionization by plasmons, occurring through local levels, is possible already at lower E_p and, consequently, at lower n.

4. Plasmon Decay

A plasmon may decay in radiative and non-radiative way [30].

In the case of non-radiative decay the energy and momentum of the plasmon is transferred to a separate electron (hole) — this is a case of the so-called Landau damping [21 to 23]. Let us assume that $\hbar(p_1 - k)$ is the initial charge carrier momentum and $\hbar p_1$ the final charge carrier momentum after receiving energy from a plasmon, its effective mass being m_1. The laws of energy and momentum conservation in the given case lead to the expression

$$\hbar\omega_p + \frac{\hbar^2(p_1 - k)^2}{2m_1} = \frac{\hbar^2 p_1^2}{2m_1}. \tag{10}$$

Solving (6) together with (10), we obtain

$$p_1 = \frac{m_1}{m}\,p + \frac{1}{2}\left(1 - \frac{m_1}{m}\right)k\,. \tag{11}$$

If the same type of carriers (e.g., electrons) participates in plasmon excitation and decay, then $m_1 = m$ and, as follows from (11), $p_1 = p$. This implies that the momentum and energy of a hot electron (hole) are completely taken away by a plasmon from the region of the p–n junction and transferred to an adjacent region with n-(p-)type conductivity. As a result, hot electrons (holes) appear outside the region of the p–n junction, which due to recombination with holes (electrons) in a state of thermodynamic equilibrium emit photons of energy

$$E_{ph} = E_G + E_p + E_e + E_h \tag{12}$$

or

$$E_{ph} > E_G + E_p , \tag{12a}$$

where E_e and E_h are the initial energy of electrons and holes, respectively.

The emission maximum may be expected at E_p where, in the case of radiative decay, the energy of a plasmon is changed into photon energy. In general, such energy transfer is forbidden since the plasmon and photon represent two different modes of oscillation (longitudinal and transverse). The dielectric constant of a semiconductor in a longitudinal and a transverse field is different: $\varepsilon_{\parallel}(k, \omega) \neq \varepsilon_{\perp}(k, \omega)$. — According to one view point [21, 22], in an extreme case, when $k \to 0$, electrons do not "differentiate" between transverse and longitudinal waves: $\varepsilon_{\parallel}(0, \omega) = \varepsilon_{\perp}(0, \omega)$. In this case photons may excite plasmons and vice versa. — According to another view point [30], photons can interact only with a special kind of plasmons — radiative surface plasmons. The energy of a radiative surface plasmon approximately coincides with the energy of a bulk plasmon.

Despite all the theoretical intricacies, however, a good deal of experimental evidence [30, 31] suggests that metallic films and wafers under slow electron bombardment emit photons of an energy corresponding to the energy of bulk plasmons. If, in the case of secondary breakdown, hot electrons excite plasmons, then, similarly, part of them must decay into photons having an energy which corresponds to the energy of bulk plasmons, E_p.

5. Explanation of the Basic Relations of the Mesoplasma

The occurrence of the mesoplasma and a number of its properties and types may be accounted for in terms of the plasmon mechanism of ionization and pinch in the following way.

In the case of high currents of primary breakdown of a reverse-biased p–n junction, a separate region is heated up to temperatures sufficient for intrinsic conductivity. The hot electrons (holes), which occur upon crossing the p–n junction, can in this case excite plasmons in a region of n-type (p-type) conductivity, adjacent to the depletion layer of the p–n junction. The plasmon energy is still insufficient to effect band–band transitions. However, in the hot spots of the p–n junction, the current may rise due to ionization by plasmons through local levels. The plasmon exchange forces may pinch the current, which, in turn, causes the carrier density n, the plasmon energy E_p, the ionization efficiency, and the magnitude of the current passing through the p–n junction to rise. Negative dynamic resistance and the warm mesoplasma may occur. — At high local heating, when the area of increased conductivity is small, the increase of the current passing through the p–n junction is negligible. The electron density may proceed to increase. As a result, a critical value of n is reached for the onset of plasmon ionization, i.e., negative dynamic resistance and hot mesoplasma occur. — However, further heating and increase in n bring about an increase in E_p, so that hot electrons of energy E_k appear to be insufficiently "hot" to excite plasmons and, in consequence, the carrier density in the hot mesoplasma is stabilized.

Taking into account emission due to plasmons, the experimental spectral distributions [15, 16] of mesoplasma emission may be explained. The humps in the spectral distribution (which appear at 0.3 to 0.4 eV in the warm mesoplasma and at 0.9 eV in the case of the hot mesoplasma) may be attributed to plasmon emission, implying that the photon energies in the humps correspond to the plasmon energies E_p. — The visible emission (the high-energy tail in the spectral distribution) may be regarded as being due to electron heating in the Landau damping process and subsequent recombination of hot electrons. The length of the tail in this case is closely associated with energies in the humps according to (12) and (12a). In the case of a warm mesoplasma, $E_p = 0.3$ to 0.4 eV and, in accordance with (12a), $E_{ph} > (1.2$ to $1.3)$ eV, whereas in the case of a hot mesoplasma, $E_p \approx 0.9$ eV and, accordingly, $E_{ph} > 1.8$ eV. Such a differentiation in the visible spectra of warm and hot mesoplasma is in good agreement with experimental data.

References

[1] J. TAUC and A. ABRAHAM, Phys. Rev. **108**, 936 (1957).
[2] A. C. ENGLISH, Solid State Electronics **6**, 511 (1963).
[3] A. C. ENGLISH, IEEE Trans. Electron Devices **13**, 662 (1970).
[4] J. BALODIS, E. PENTYUSH, T. PURITIS, E. FONAVS, and V. EGLITIS, in: Poluprovodniki i ikh primenenie v elektrotekhnike, Vol. V, Zinatne, Riga 1971 (p. 139).
[5] E. PENTYUSH, E. FONAVS, T. PURITIS, and V. EGLITIS, Izv. Akad. Nauk LatvSSR, Ser. fiz. tekh. Nauk, No. 5, 35 (1970).
[6] E. PENTYUSH, E. FONAVS, T. PURITIS, and V. EGLITIS, Izv. Akad. Nauk LatvSSR, Ser. fiz. tekh. Nauk, No. 5, 42 (1970).
[7] E. PENTYUSH, E. DEKENA, T. PURITIS, E. FONAVS, V. EGLITIS, J. BALODIS, and B. DZELME, Izv. Akad. Nauk LatvSSR, Ser. fiz. tekh. Nauk, No. 3, 43 (1973).
[8] J. BALODIS and T. PURITIS, Izv. Akad. Nauk LatvSSR, Ser. fiz. tekh. Nauk, No. 3, 52 (1974).
[9] V. EGLITIS, T. PURITIS, and I. TAMANE, Izv. Akad. Nauk LatvSSR, Ser. fiz. tekh. Nauk, No. 4, 55 (1974).
[10] E. PENTYUSH, E. DEKENA, and T. PURITIS, Izv. Akad. Nauk LatvSSR, Ser. fiz. tekh. Nauk, No. 4, 17 (1974).
[11] E. PENTYUSH, E. DEKENA, and T. PURITIS, Izv. Akad. Nauk LatvSSR, Ser. fiz. tekh. Nauk, No. 5, 48 (1974).
[12] V. EGLITIS, J. LIBERIS, V. BAREIKIS, and T. PURITIS, Izv. Akad. Nauk LatvSSR, Ser. fiz. tekh. Nauk, No. 1, 29 (1976).
[13] V. EGLITIS, I. TAMANE, and T. PURITIS, Izv. Akad. Nauk LatvSSR, Ser. fiz. tekh. Nauk, No. 3, 27 (1976).
[14] H. MELCHIOR and M. J. STRUTT, Proc. IEEE **52**, 439 (1964).
[15] J. BALODIS and T. PURITIS, Izv. Akad. Nauk LatvSSR, Ser. fiz. tekh. Nauk, No. 5, 9 (1977).
[16] R. A. SUNSHINE and M. A. LAMPERT, IEEE Trans. Electron Devices **19**, 873 (1972).
[17] W. HAECKER, phys. stat. sol. (a) **25**, 301 (1974).
[18] A. G. CHYNOWETH and K. G. MCKAY, Phys. Rev. **102**, 369 (1956).
[19] H. A. SCHAFFT, Proc. IEEE **55**, 1272 (1967).
[20] V. L. BONCH-BRUEVICH and E. G. LANDSBERG, phys. stat. sol. **29**, 9 (1968).
[21] D. PINES, Elementary Excitations in Solids, W. A. Benjamin, Inc. New York/Amsterdam 1963.
[22] P. M. PLATZMAN and P. A. WOLFF, Waves and Interactions in Solid State Plasmas, Academic Press, New York/London 1973.
[23] M. C. STEELE and B. VURAL, Wave Interactions in Solid State Plasmas, McGraw-Hill Publ. Co., New York 1969.
[24] A. S. DAVIDOV, Teoriya tverdovo tela, Izd. Nauka (Gl. red. fiz. mat. lit.), Moskva 1976 (p. 100).
[25] Yu. A. ROMANOV, Fiz. tverd. tela **5**, 2988 (1963).
[26] P. TUSSING, W. ROSENTHAL, and A. HAUG, phys. stat. sol. (b) **53**, 451 (1972).
[27] H. J. QUEISSER and M. B. PANISH, J. Phys. Chem. Solids **28**, 1177 (1967).
[28] J. C. TSANG, P. J. DEAN, and P. T. LANDSBERG, Phys. Rev. **173**, 814 (1968).
[29] C. J. HWANG, J. appl. Phys. **42**, 4408 (1971).
[30] W. STEINMANN, phys. stat. sol. **28**, 473 (1968).
[31] P. G. BORZYAK, I. P. ZAPESOCHNII, I. A. KONOVALOV, V. A. KRITSKII, Yu. A. KULYUPIN, K. N. PILIPCHAK, S. S. POP, and P. M. TOMCHUK, Izv. Akad. Nauk SSSR, Ser. fiz. **40**, 1621 (1976).

(Received June 12, 1978)

phys. stat. sol. (a) 49, 231 (1978)

Subject classification: 13.4 and 20.3; 2; 19; 22

Research Laboratory of Semiconductor Physics, Latvian State University, Riga[1])

X-Ray Induced Paramagnetic States and Luminescence in As_2S_3 and As_2Se_3

By

A. V. MOSKALONOV and J. R. ZAKIS

Optical absorption and electron spin resonance (ESR) attributed to X-ray induced localized paramagnetic states are studied in glassy As_2S_3 and As_2Se_3 for $T \geqq 77$ K. The production of induced paramagnetic states is not detected in crystalline As_2S_3. X-ray irradiation of a crystal at 77 K excites luminescence with the same spectrum as that for photoluminescence. No X-ray excited luminescence is observed in glasses at 77 K. It is shown that a unified explanation for the data can be given on the basis of the Mott-Davis-Street model.

Исследовались оптическое поглощение и электронный парамагнитный резонанс (ЭПР) в стеклообразных As_2S_3 и As_2Se_3 при $T \geqq 77$ K, связанные с локализованными парамагнитными состояниями, индуцированными рентгеновским излучением. В кристаллическом As_2S_3 образование индуцированных парамагнитных состояний не обнаружено. Рентгеновское излучение при 77 K возбуждает в кристалле люминесценцию, спектр которой совпадает со спектром фотолюминесценции. В стеклах рентгенолюминесценция при 77 K не обнаружена. Показано, что единое объяснение полученных результатов может быть дано на основе модели Мотта, Дэвиса и Стрита.

1. Introduction

It is well known that light-induced absorption [1], ESR signals [1], and photoluminescence [2 to 10] of glassy As_2S_3 and As_2Se_3 are due to deep localized states in the gap. According to the MDS model [11, 12] these states are at dangling bonds which in the ground state are either empty or occupied by two electrons (so-called D^+ or D^- centres, respectively). It has been suggested [7 to 10] that photoluminescence can be excited by photons absorbed in the nearest vicinity of the D^- centres, the following energy-transfer processes taking place: (i) photon generates on electron–hole pair (or exciton), (ii) a D^- centre captures the hole and becomes a neutral D^0 centre, (iii) the D° centre subsequently captures the electron, a transition accompanied by light emission. This overall process is monomolecular and does not give the rise to the creation of a D^0 centre concentration. On the other hand, there is a certain probability for electrons and holes to remain captured in dangling bond centres. The evidence for this are light-induced absorption and ESR observed in chalcogenide glasses at 6 K [1]. According to [10 to 12] the presence of metastable D^0 centres also accounts for the fatiguing of photoluminescence [5, 6].

A previous article [13] described X-ray induced paramagnetic states in glassy As_2S_3 and As_2Se_3 observed at 77 K. From these experiments it seems likely that the efficiency of D^0 centre generation will be higher in the case of higher excitation energy. To provide further evidence for this, investigations are reported here of the

[1]) Riga 226100, USSR.

X-ray induced paramagnetic states and luminescence in glassy As_2S_3, As_2Se_3, and crystalline As_2S_3. It is shown that a unified explanation for the data can be given on the basis of the MDS model.

2. Experimental

We have used commercial glasses As_2S_3 and As_2Se_3 as well as high purity samples synthesized in the A. F. Ioffe Physico-Technical Institute of the Academy of Sciences of the USSR. We have not noticed any difference between the investigated properties of these samples. Crystalline samples of As_2S_3 were cut from the natural mineral.

For the optical measurements the glasses were polished and had a thicknesses of about 0.2 to 0.5 mm. These measurements were performed in a conventional liquid nitrogen cryostat at temperatures from 80 to 300 K. The induced absorption in the glasses was measured using low-level light intensity. ESR spectra were measured at 77 K on samples with thicknesses of about 1.0 to 1.5 mm.

The X-ray source had the following parameters: tungsten anode 55 kV, 16 mA.

3. Results

Virgin samples do not exhibit any ESR signal. After exposure to X-rays for 5 min at 77 K an ESR signal appears (Fig. 1). This signal does not show any significant decay in the dark during 1 h.

The X-ray induced ESR signal is always accompanied by an additional absorption in the spectral range below the band-edge absorption (Fig. 2 and 3). Fig. 4 shows the growth kinetics of the induced absorption in the glasses investigated; the ESR signal has the same growth kinetics. By using a calibrated ESR standard sample we estimate the paramagnetic centre concentration to be about 10^{17} cm^{-3}.

We have not observed any X-ray excited luminescence in glassy As_2S_3 and As_2Se at 77 K. In the case of crystalline As_2Se_3 there is not any X-ray induced ESR signal

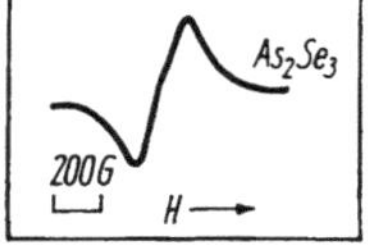

Fig. 1

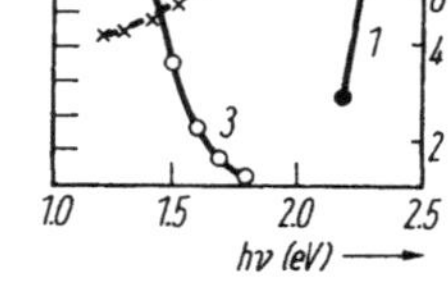

Fig. 2

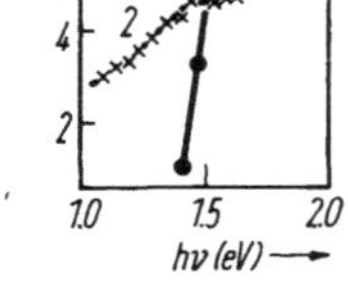

Fig. 3

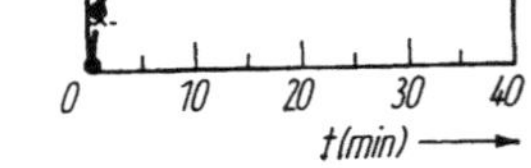

Fig. 4

Fig. 1. X-ray induced ESR spectra of chalcogenide glasses at 77 K

Fig. 2. (1) Band-edge absorption before irradiation for glassy As_2S_3, (2) X-ray induced absorption for glassy As_2S_3 and (3) X-ray induced luminescence spectrum for crystalline As_2S_3 at 85K

Fig. 3. (1) Band-edge absorption before irradiation and (2) X-ray induced absorption for glassy As_2Se_3 at 85 K

Fig. 4. X-ray induced absorption growth curve at 85 K for glassy (1) As_2S_3 and (2) As_2Se_3

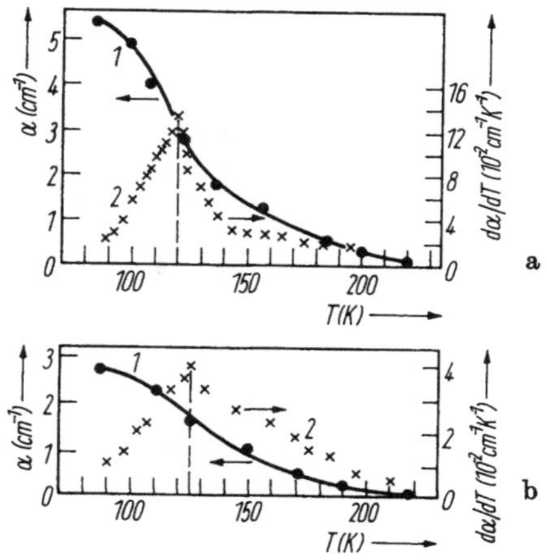

Fig. 5. Temperature dependences of the (1) X-ray induced absorption and (2) its derivative for a) glassy As_2S_3 at photon energy 1.6 eV; b) glassy As_2Se_3 at photon energy 1.2 eV

or absorption, but the samples give X-ray excited luminescence (Fig. 2) with the same spectrum as that for photoluminescence. No fatiguing is observed in the X-ray excited luminescence for crystalline As_2S_3.

X-ray induced absorption can be bleached by light with photon energies within this absorption band. This absorption can be bleached also by heating. Fig. 5 shows the temperature dependences of the X-ray induced absorption and its derivative for glassy As_2S_3 and As_2Se_3. The heating rate is (4.0 ± 0.1) K/min.

4. Discussion

Absorbed X-rays generate high-energy electrons which can transfer their energy to nuclei or can ionize atoms. Because of the large difference in masses the electron cannot transfer to a nucleus more than 0.015% of its initial energy in each collision (in the case of As and Se nuclei). We have used 55 keV X-rays and therefore the maximum energy transferred from the generated electron to a nucleus is less than the energy required to create a defect. It seems likely therefore that the difference between the action of X-rays and visible light is due to the different mechanisms of carrier production.

Our experiments are in good agreement with previous work [1, 2] in which light was used for excitation showing that we are dealing with the same kind of luminescence and paramagnetic states. The only difference is the absence of X-ray excited luminescence in glasses. It seems to be a significant feature and consistent with the following model:

1. As proposed by the MDS model the luminescence is due to electron and hole recombination at a dangling bond centre (Fig. 6). This process can be represented by the following reactions:

$$D^- + h \rightarrow D^0 , \tag{1}$$

$$D^0 + e \rightarrow D^- + h\nu^- , \tag{2}$$

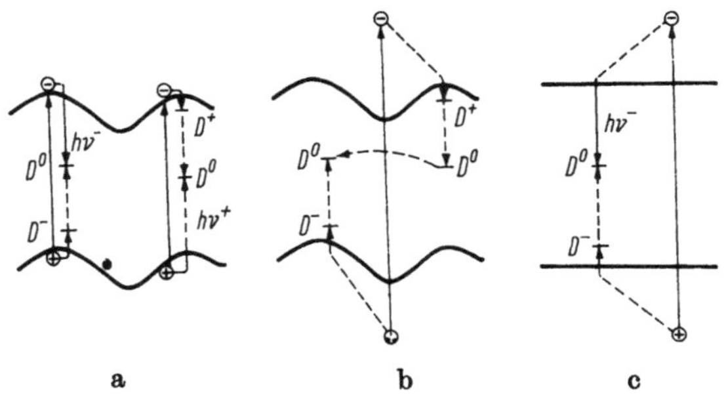

Fig. 6. Electron and hole recombination mechanisms. a) Glass at low excitation energies (according to the MDS model [11, 12]), b) glass at large excitation energies, c) crystal at large excitation energies

or possibly

$$D^+ + e \rightarrow D^0 , \tag{3}$$

$$D^0 + h \rightarrow D^+ + h\nu^+ , \tag{4}$$

where $h\nu^-$ and $h\nu^+$ are the energies of the luminescence radiation corresponding to the two mechanisms. Because there is only one emission band only one of these processes gives light emission. According to [7, 10] the light emission comes from a D^- centre. Therefore, D^+ may be a non-radiative recombination centre.

2. Glasses have a large concentration of dangling bonds. However, the equilibrium concentration of D^0 centres during photoexcitation is small when electron–hole pairs are generated in the near vicinity of D^- or D^+ centres and each pair recombines in the same centre (Fig. 6a) (monomolecular process). This is due to the low mobility of carriers in a disordered structure.

3. At high excitation energies (X-ray excitation) electrons and holes can separate and after thermalization their recombination takes place via different centres (Fig. 6b). This gives rise to steady-state concentration of D^0 centres which are responsible for the ESR signal and induced absorption. Because of the large D^0 centre concentration, tunneling between two D^0 centres becomes possible [11, 12],

$$2D^0 \rightarrow D^+ + D^- . \tag{5}$$

The probability of this process can be enhanced by a non-uniform distribution of D^0 centres in the sample.

The large concentration of D^0 centres is responsible for the fatiguing effect and quenches the luminescence [10 to 12]. This explains the absence of X-ray excited luminescence in glassy As_2S_3 and As_2Se_3 as well as the difference between photoluminescence excitation spectra for glasses and crystals (Fig. 7). Furthermore, the large concentration of D^0 centres increases the probability of non-radiative Auger recombination which may occur due to the presence of the trapped electrons and holes.

4. In crystalline samples electrons and holes recombine in the same way as in glassy samples. However, the steady-state concentration of D^0 centres is small because the carrier diffusion lengths are much longer in the crystals than in the glasses and it is not necessary for a photon to be absorbed near a centre to excite luminescence [10]. We consider in crystalline As_2S_3 the probability of reaction (3) to be small and the excited electron to always recombine with the trapped hole, as shown in Fig. 6c.

5. The thermal bleaching curve of the induced centres (Fig. 5) can be explained by assuming only one bleaching mechanism. The activation energy for the thermal bleaching is (0.10 ± 0.03) eV for As_2S_3 and (0.06 ± 0.03) eV for As_2Se_3. These energies are characteristic of the phonon-assisted tunneling mechanism which is consistent

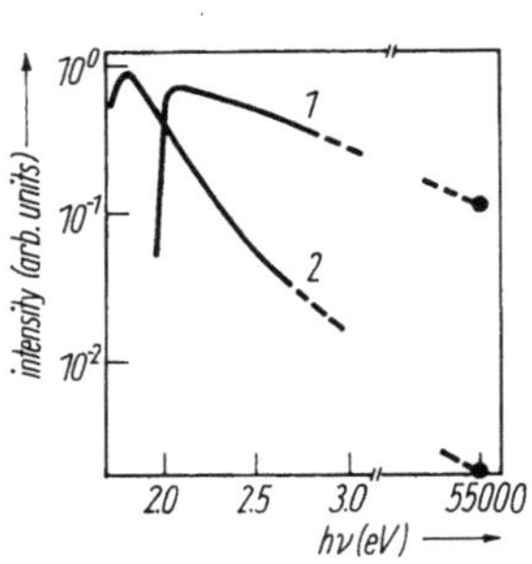

Fig. 7. Possible shape of the luminescence excitation spectrum for (1) chalcogenide crystals and (2) glasses in the visible light and X-ray regions (full lines represent data from [10])

with the MDS model [11, 12]. Therefore, the low-temperature annealing of the induced centres is due to electron tunneling between two D^0 centres according to reaction (5).

The high-temperature annealing (above 200 K) may be due to thermal excitation of trapped carriers from the D^0 centre to the band states. For As_2Se_3 the minimum energy required is approximately 0.45 eV for holes and 0.65 eV for electrons [9].

References

[1] S. G. Bishop, U. Strom, and P. C. Taylor, Phys. Rev. Letters **34**, 1346 (1975); Solid State Commun. **18**, 573 (1976); Proc. VI. Internat. Conf. Amorphous and Liquid Semiconductors, Vol. 2, Izd. Nauka, Leningrad 1976 (p. 241); Phys. Rev. B **15**, 2278 (1977).

[2] B. T. Kolomiets, T. N. Mamontova, and A. A. Babaev, J. non-crystall. Solids **4**, 289 (1970).

[3] B. T. Kolomiets, T. N. Mamontova, E. A. Smorgonskaya, and A. A. Babaev, phys. stat. sol. (a) **11**, 441 (1972).

[4] Yu. N. Ivashchenko, B. T. Kolomiets, and T. N. Mamontova, phys. stat. sol. (a) **24**, 401 (1974).

[5] J. Cernogora, F. Mollot, and C. Benoit à la Guillaume, phys. stat. sol. (a) **15**, 401 (1973).

[6] R. A. Street, T. M. Searle, and I. G. Austin, J. Phys. C **6**, 1830 (1973).

[7] R. A. Street, T. M. Searle, and I. G. Austin, Phil. Mag. **29**, 1157 (1974); **30**, 1181 (1974); **32**, 431 (1975).

[8] R. A. Street, I. G. Austin, and T. M. Searle, Proc. VI. Internat. Conf. Amorphous and Liquid Semiconductors, Vol. 2, Izd. Nauka, Leningrad 1976 (p. 232).

[10] R. A. Street, Adv. Phys. **25**, 397 (1976).

[11] R. A. Street and N. F. Mott, Phys. Rev. Letters **35**, 1293 (1975).

[12] N. F. Mott, E. A. Davis, and R. A. Street, Phil. Mag. **32**, 961 (1975).

[13] A. V. Moskalonov, Fiz. tverd. Tela **19**, 1440 (1977).

(Received April 27, 1978)

phys. stat. sol. (a) **49**, 237 (1978)

Subject classification: 10.2 and 12.2; 2

Laboratory of Physical Metallurgy, Faculty of Applied Sciences, Catholic University of Louvian[1])

Experimental Study of Crack Propagation in Brittle Amorphous Solids

Refraction of a Propagating Crack by Stress Waves[2])

By

N. TERAO

When pine-resin is melted by heating on a metallic tray and solidified rapidly by cooling from the metallic plate side, a thin layer of resin under strong tensile stress is obtained. The interaction of propagation cracks and especially the influence of a stress wave on the crack propagation in this thermally strained layer are examined. The cracks are initiated from two or three points in the pine-resin plate by electric spark shocks in water. The propagating cracks are observed to refract by the stress waves produced by the shocks. The velocity of crack propagation is estimated from the relation between the distance travelled by the crack and the stress wave.

En refroidissant énergiquement la base d'un plateau métallique sur lequel nous avons coulé de la résine de pin, nous obtenons une couche de résine sous tension. Nous avons, ensuite, en créant des décharges électriques dans l'eau, produit des chocs qui ont initié en deux ou trois endroits des cassures. Les interactions entre ces cassures qui se propagent, et surtout l'influence de l'onde de tension sur la propagation de la cassure de la couche de résine de pin ont été examinées. Nous avons ainsi observé que les directions de propagation des cassures sont modifiées par les ondes de tension produites par les chocs. Enfin, nous avons estimé la vitesse de propagation de la cassure en reliant la distance traversée par l'onde de tension et celle par la cassure.

1. Introduction

Hirata has observed, in some of his earlier work, crack propagation in a glass plate caused by an electric spark under water [1, 2]. Fig. 1 shows the interference of two cracks of a thin glass plate (about 0.5 mm thick) produced by two shocks almost at the same moment. Fig. 2 was obtained under the same conditions except that the two shocks were applied with a small time interval [3]. The cracks produced by the first shock were refracted by the second one. The mechanical shocks used in this experiment were obtained by strong electrical discharges under water. We observed a similar phenomenon on a pine-resin layer cooled rapidly after melting it on a metallic plate [4].

A thin layer of pine-resin under strong tensile stress is obtained if one melts pine-resin by heating on a metallic tray and cools it rapidly from the metallic plate side. When one touches a point on this superstressed resin plate by, for example, the tip of a pencil or tweezers, radial cracks start from this point and propagate instantaneously across the entire surface of the plate. Taguchi has examined this phenomenon and published many beautiful photographs of this type of cracks. He has also measured the propagation velocity of these cracks. and obtained the values 610 to 910 m/s as mean values during a fairly large propagation distance (10 to 20 cm) [5]. It may be possible that the real propagation velocity just near the "initial crack" is higher than these mean values [1, 5].

[1]) B-1348 Louvain-la-Neuve, Belgium.

[2]) This paper is based on one part of a presentation made at the Euromech Colloquium 91 on the Dynamic Crack Propagation, held in Jablonna, Poland, on August 17 to 20, 1977.

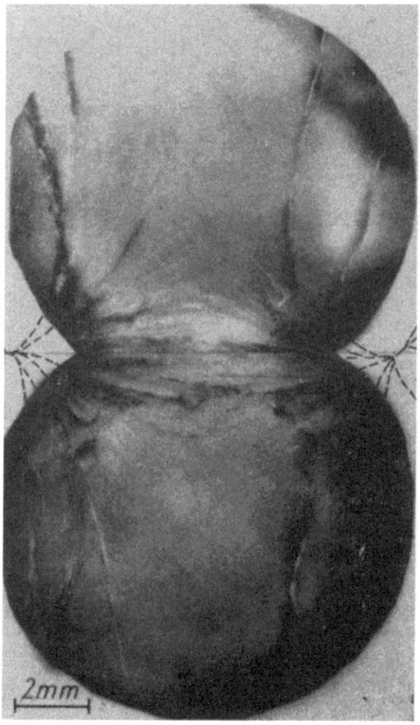

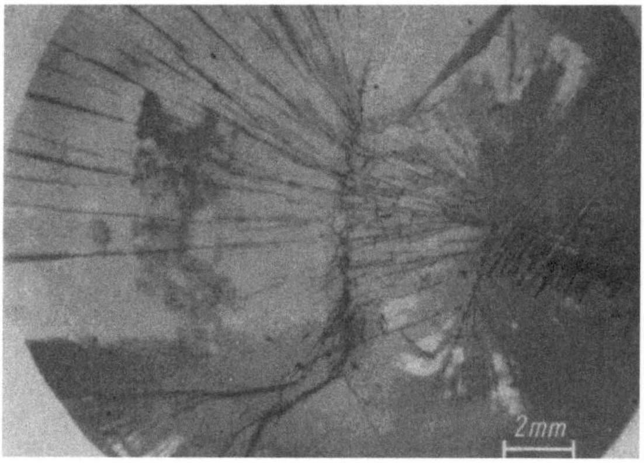

Fig. 2. Cracks in a thin glass plate produced by two shocks which were applied successively with a small time interval (by Hirata)

Fig. 1. Two cracks in a thin glass plate produced by two shocks almost at the same moment (by Hirata)

The crack propagation in thin thermally strained pine-resin layer was examined in this work by using strong mechanical shocks.

2. Experimental Procedures

We used circular trays of aluminium (25 and 30 cm diameter) or stainless steel (30 cm diameter) on which the pine-resin is melted by heating. The rapid solidification of the melted pine-resin was carried out by putting the metallic tray on the surface of water. The mechanical shocks produced by electrical discharges under water were used, as in the experiments by Hirata, in order to start the cracks in the superstressed resin plate from two or three points simultaneously or with a very small time interval. Fig. 3 shows the discharge electrode which consists of a platinum wire cathode (C) and a cylindrical brass anode (A). The metallic tray with a superstressed resin layer is put on two or three electrodes in water as shown in Fig. 4. The used electrical discharge circuit, shown in Fig. 5, consists of two parts: the main discharge circuit (M) and the discharge induction circuit (D.I.).

The experimental processes used in this work are as follows: The pine-resin on the

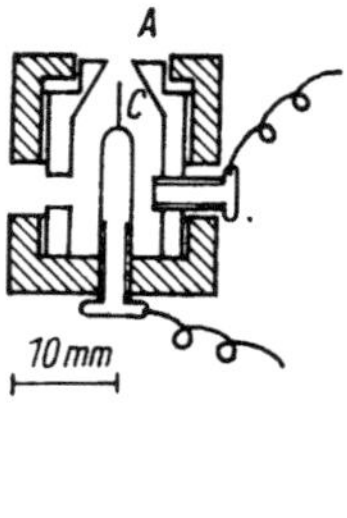

Fig. 3. Electrical discharge electrode. Shadowed part–bakelite, blanck part–brass; A anode with a small hole of discharge gap, C cathode (platinum wire)

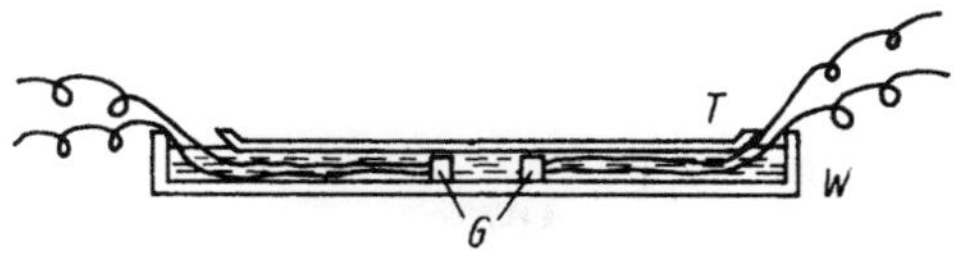

Fig. 4. Metallic tray T with pine-resin plate placed on discharge electrodes G in water. W is a water vessel

metallic tray is melted in an electric furnace at about 200 °C and then the melted resin layer of homogeneous thickness is cooled rapidly by placing the tray on the discharge electrodes submerged in water as shown in Fig. 4. Thethickness of the resin plate is about 0.2 to 1.0 mm. Then, the high voltages (30 to 40 kV to the main discharge circuit and about 7 kV to the discharge induction circuit) are applied to the electric circuit of Fig. 5 and the condensers are charged during the time in which the thin resin layer is solidified gradually by cooling. The cracks are started by switching S. The mechanical stress conditions in the resin plate depend on heat treatment, cooling time, and plate thickness. If the thin resin layer is solidified rapidly on a thin aluminium tray, the tensile stress becomes higher and the cracks propagate easily over the whole of the resin plate. A thick stainless steel tray, however, does not give a high stress state for a thick resin layer because of its low thermal conductivity and so the cracks do not propagate over a large region.

The cracks start with some time lag after the mechanical shock is applied. The fluctuation of this time lag becomes large if the shock is small. It is necessary, consequently, to use strong sparks in order to start the cracks simultaneously from two or three different positions. The circuit shown in Fig. 5 permits induction of one or several strong mechanical shocks simultaneously or within a very small time interval. Besides, the time duration of each shock is also very small, because the electrical spark duration is extremely small. In reality, as it will be seen afterwards, it is possible to start the cracks from two or three points within less than 10^{-6} s time interval. In comparison with this time interval of crack initiation, the time interval of electrical discharges is further negligibly small in our case. By using a trigger circuit, Hyodo could realize and control electronically the time interval between several sparks within the range

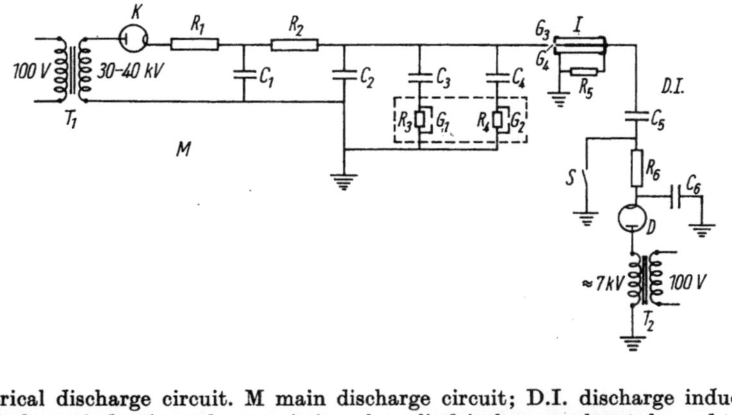

Fig. 5. Electrical discharge circuit. M main discharge circuit; D.I. discharge induction circuit; I electrical discharge induction tube, consisting of a cylindrical pyrex glass tube and two electrodes (one is a platinum wire and the other a brass cylinder with a small circular hole of discharge gap G_4. Dashed-line part indicates the metallic tray with pine-resin plate and discharge electrodes in water, shown in Fig. 4. G_1 and G_2 are the discharge electrodes shown in Fig. 3. C_1, C_2, ... , C_6 Leyden jars; R_1, R_2, ... R_6 water resistances

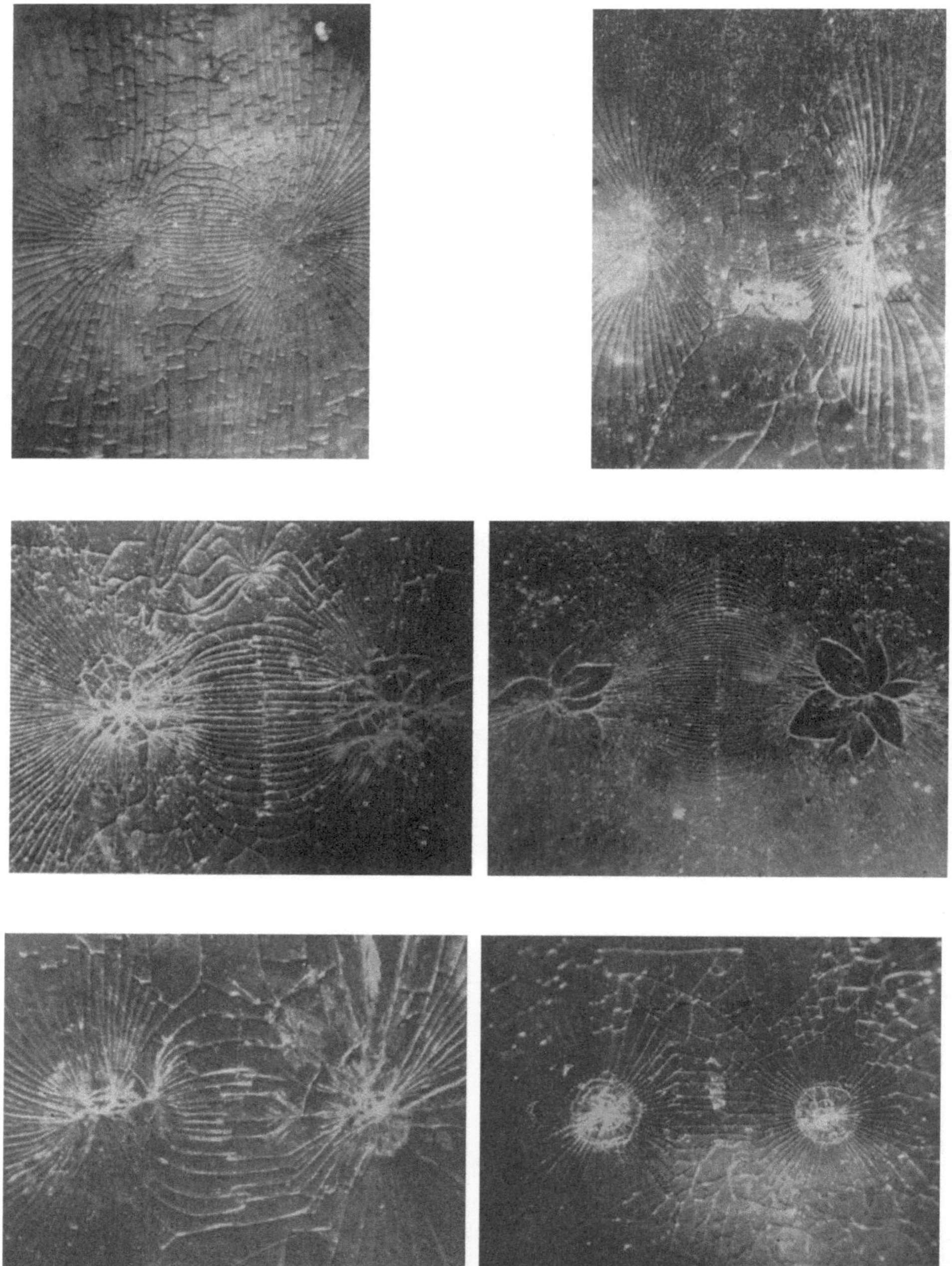

a

c

e

Fig. 6. Two groups of cracks produced by two simultaneous shocks in pine-resin plate. a) 1.00 b) 0.90 ×, c) 1.38 ×, d) 1.19 ×, e) 1.26 ×, f) 0.57 ×

of 2.5×10^{-7} to 8×10^{-5} s [6]. The time lag for crack propagation also depends on the distance between the discharge electrode and the metallic disc. This distance was in general about 1 mm in our experiments.

3. Results and Discussions

3.1 Case in which two groups of cracks start simultaneously

Fig. 6 a to f show some examples of cracks which started simultaneously from two points. The distance between two shock points is about 15 to 40 mm. The observed results indicate that the cracks starting from points A_1 and A_2 propagate along smooth curves if the stress in the resin plate is high, as shown in Fig. 7a. In this case, the cracks formed in the angles $\sphericalangle P_1A_1P_2$ and $\sphericalangle P_1A_2P_2$ propagate as if they are attracting each other, while the cracks out of these angles propagate as if they are rejecting each other. The angles $\sphericalangle A_1P_1A_2$ and $\sphericalangle A_1P_2A_2$ are nearly 70°. Fig. 6a shows one of these types of cracks. If the stress in the resin plate is not so high and the shocks are strong, the direction change of crack propagation occurs with a sharp angle as shown in Fig. 7 b and c and the angles $\sphericalangle A_1P_1A_2$ and $\sphericalangle A_1P_2A_2$ become smaller, about 60°. Fig. 6 b, e, and f are these types of cracks. In the case of Fig. 6f, the stress in the resin layer was relatively small, and consequently the cracks did not propagate over the whole region. In the case of Fig. 1, the angles $\sphericalangle A_1P_1A_2$ and $\sphericalangle A_1P_2A_2$ are nearly 90°. Concerning the critical points B_1, B_2, B_3, and B_4, shown in Fig. 7, which divide the attractive and repulsive regions of crack propagation, it is observed that the angles $\sphericalangle A_1B_1A_2$, $\sphericalangle A_1B_2A_2$, $\sphericalangle A_1B_3A_2$, and $\sphericalangle A_1B_4A_2$ are always 90°. The small cracks produced in the middle upper region in Fig. 6c are induced accidentally, without a direct mechanical shock, from a small air bubble in the resin layer. Fig. 6d shows flower-type cracks around the shock points, which seem to be formed by shear stress.

3.2 Case in which the formation of one group of cracks is delayed

Fig. 8 a to d show several examples of crack propagation when two shocks are given with some time interval or one shock is weaker than the other. This type of crack is often obtained by controlling the spark gaps of G_1 and G_2 in Fig. 5 or the distance be-

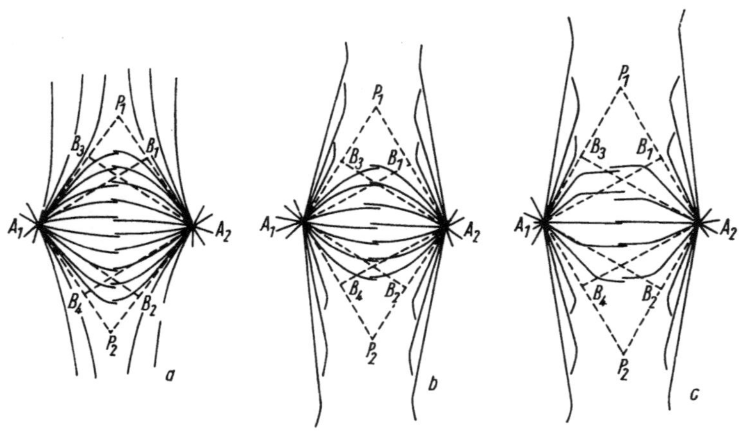

Fig. 7. Various aspects of propagating cracks refracted by two simultaneous shocks, which depend on the stress state in the resin plate. a) High stress, b) medium stress, and c) low stress

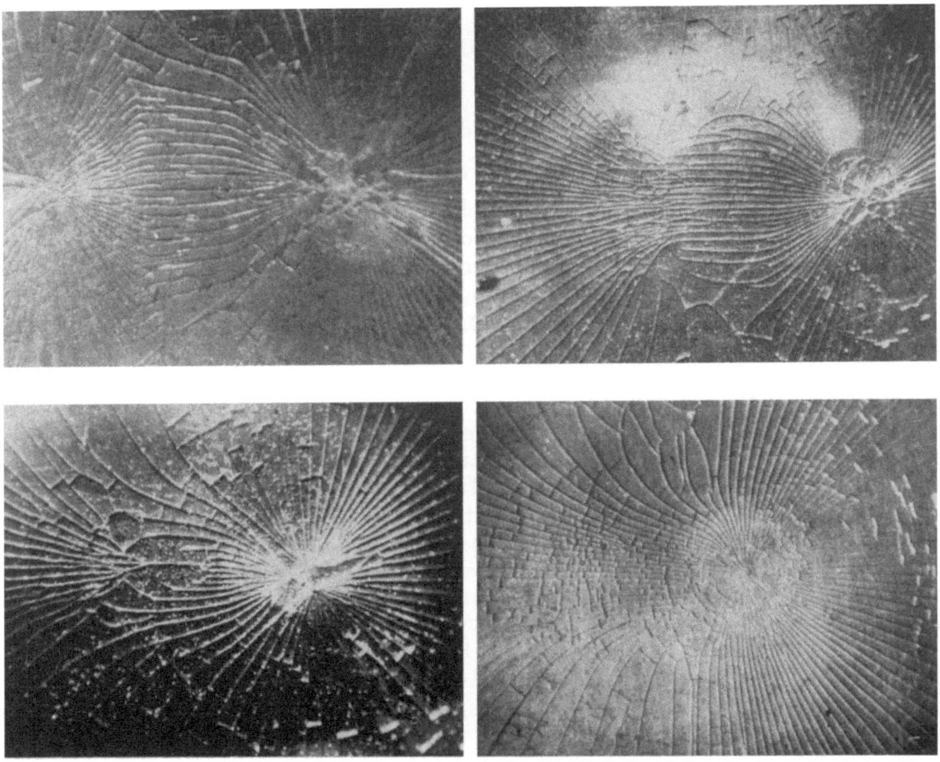

a b c d

Fig. 8. Crack propagation in a pine-resin plate. One group of cracks has a time lag. a) 1.25 ×, b) 0.98 ×, c) 0.97 ×, d) 0.97 ×

tween the metallic disc and the spark electrodes. Thus, two groups of cracks interfere as shown in Fig. 8a. The time interval of crack initiation from two points is in the order of 10^{-5} s, as discussed afterwards in Section 3.4. If the lag increases, the first group of cracks which has started from A_2 propagates as shown in Fig. 9a. The cracks formed in angle $\sphericalangle B_1A_2B_2$ approach the second shock point A_1 and then they diverge. Fig. 8 b and c show these types of cracks. When the second shock lags much more or is not sufficiently strong, the first group continues to propagate as shown in Fig. 8d and Fig. 9b.

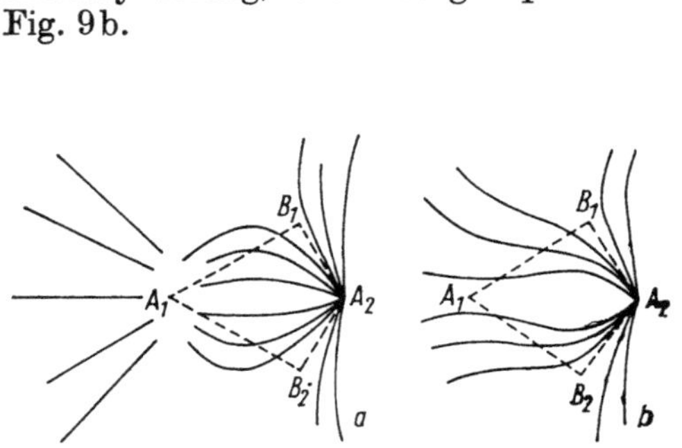

Fig. 9. Typical aspects of crack propagation when one group of cracks is started with some time delay or the second mechanical shock is smaller

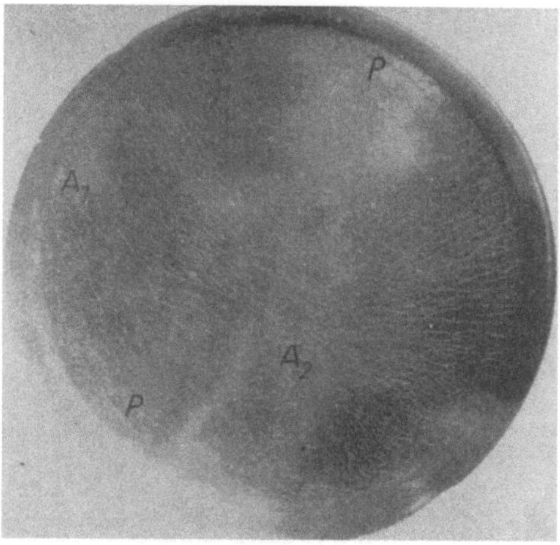

Fig. 10. Two groups of cracks, produced spontaneously from two points, A_1 and A_2, during cooling of the pine-resin plate (0.22 ×)

3.3 Refraction of propagating cracks

The following two effects can be considered as possible origins of the refraction of propagating cracks described above. The first is due to the propagating cracks themselves and the second to the applied mechanical shocks.

Fig. 10 shows two groups of cracks, produced spontaneously from points, A_1 and A_2, without mechanical shocks. These cracks of a similar type as those observed by Taguchi, are formed abruptly during cooling of the pine-resin layer. There often appear reflected wave-like patterns near the boundary of the tray, as shown in the region P of Fig. 10 when the stress in the resin layer is sufficiently high. The interval of these periodic patterns is about 10 mm. This type of crack figure is probably due to the elastic wave which is produced by the crack formation and propagates through the resin plate with a velocity higher than that of propagating cracks. This elastic wave is reflected at the tray boundary and interferes with cracks which are propagating towards the plate boundary after the elastic wave. The intersecting position of two groups of cracks in Fig. 10 shows that both of these cracks propagate with almost the same constant velocity. In this case, however, the propagating cracks interfere mutually only when they approach to a distance of about 5 mm and their interaction manner is very different from that produced by a mechanical shock.

Fig. 11 and 12 show two types of cracks: one is produced by strong mechanical shocks

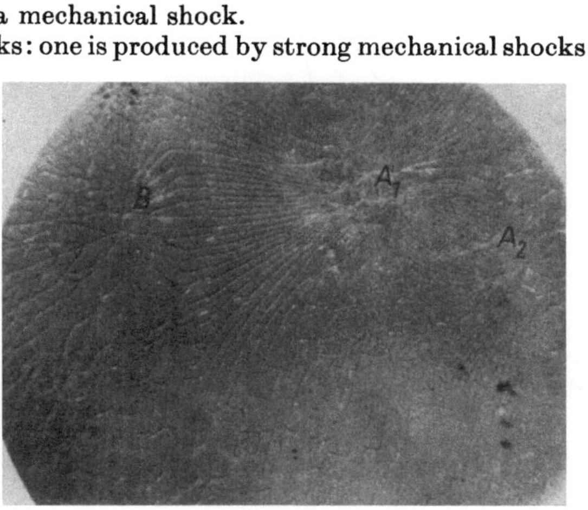

Fig. 11. Two groups of cracks produced by mechanical shocks at A_1 and A_2 and one group starting from B without mechanical shock in a pine-resin plate (0.90 ×)

Fig. 12. Many cracks formed by small shocks around two main groups of cracks wich are produced by strong mechanical shocks in a pine-resin plate (1.35 ×)

and the other by weak shocks or without mechanical shocks. These aspects clearly indicate the influence of mechanical shocks on the propagation of cracks. The cracks starting from A_2 without a shock or with only a small shock are refracted remarkably by a strong shock coming from A_1, while the cracks starting from A_1 propagate in straight lines as shown in Fig. 13 and the angle $\sphericalangle A_1BA_2$ is 90°.

From the above described observations, it is concluded that the refraction of a propagating crack is due to the stress wave which is produced by mechanical shocks and propagates through the glass or pine-resin plate. Fig. 14 show the refraction law: The stress wave arrives at B from A_1 while the crack propagates from A_2 to B. If the angle $\alpha > 90°$, the crack refracts towards the shock point A_1 as shown in Fig. 14 (a). If the angle $\alpha < 90°$, the crack refracts as if it goes away from the shock point A_1 (Fig. 14b). If the stress wave propagates with a constant velocity of the longitudinal elastic wave in resin, about 1600 m/s [7], the distance l indicates the time interval in which the stress wave propagates from A_1 to B and the crack propagates the distance s from A_2 to B. In reality, the shock is applied, not at a point but over some extended region and the crack path is not always straight. Consequently, there are some ambiguities on the measurement of s- and l-values. Nevertheless, the relation between these two distances is expected to give some information on the crack propagation. We shall call the curve showing the relation between s and l the travel-time curve.

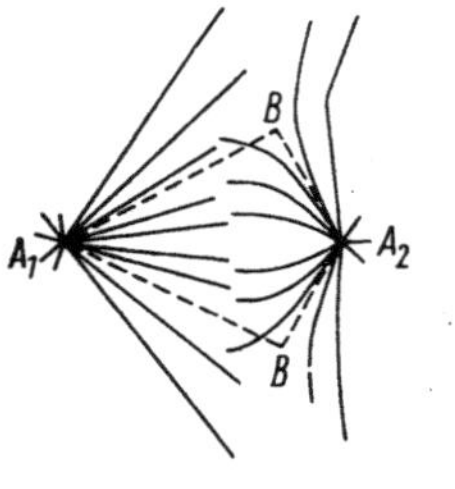

Fig. 13

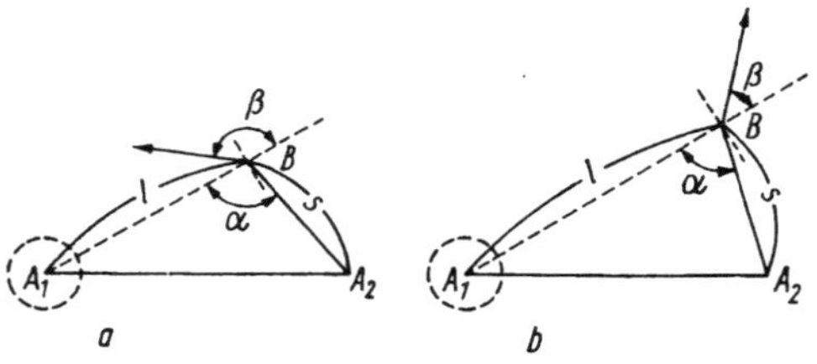

Fig. 14

Fig. 13. Scheme of crack propagation. One group of cracks starts from A_1 with strong mechanical shock and the other from A_2 without shock or with very weak shock. The cracks A_2 are refracted by stress wave coming from A_1

Fig. 14. Refraction of propagating crack. a) $\alpha > 90°$, b) $< 90°$; A_1 shock point, A_2 start point of crack, B refraction point

3.4 *Travel-time curves*

Fig. 15 shows several types of travel-time curve obtained in cases in which two groups of cracks start almost simultaneously by two mechanical shocks. (From the symmetrical form of cracks, the time interval of the initiation of these two groups of cracks is estimated to be less than 10^{-6} s.) In general, this curve is composed of two straight lines when the cracks change their propagation path abruptly by the stress wave due to a strong mechanical shock. The tangents of these straight lines are different in two regions, $\alpha > 90°$ and $< 90°$. The most ordinary examples are shown in Fig. 15 a and b. The tangent of the straight line in the region $\alpha > 90°$ is about 0.5 to 0.9 and, in general, higher than that in the region $\alpha < 90°$. If one takes the velocity 1600 m/s as the propagation velocity of the stress wave as mentioned above, the propagation velocity of the crack becomes about 800 to 1400 m/s. These values are nearly the same as or higher to some extent than those obtained by Taguchi [6] and seem to be fairly reasonable.

Fig. 1 suggests that the cracks propagate with almost the same velocity as that of the stress wave in the glass plate. Hyodo and Okuda [8] have measured the propagation velocity of cracks, produced in a thin glass plate by an under-water electric spark in a similar way as that in our work. They have observed, using the trigger delay circuit mentioned previously [6] and a high speed framing camera, that the maximum velocity could reach about 2.4 km/s, much higher than the usually accepted limiting value. The relation between this result and that of Fig. 1 should be clarified by future studies.

In some cases, however, anomalous types of travel-time curves are obtained as shown in Fig. 15c and d even though their occurrence is much less frequent than the previous ones: the tangent of the s–l straight line in these figures becomes higher than 1.00 or the tangent in the region $\alpha < 90°$ appears higher than in the region $\alpha > 90°$. Fig. 15c is obtained from the cracks formed in the lower-right region of Fig. 6b and 15d from the left-upper region of the same picture. Fig. 15e is obtained from the right region of

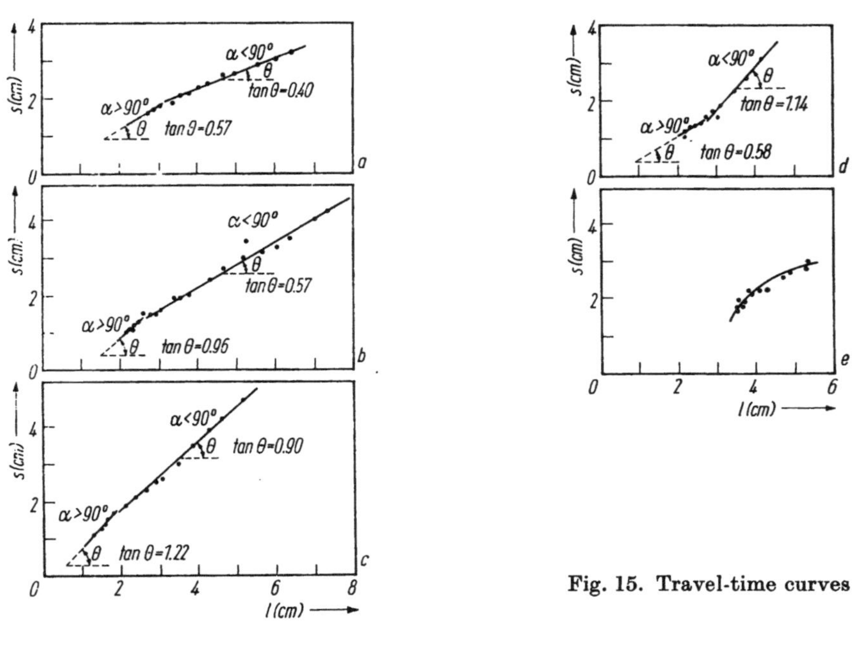

Fig. 15. Travel-time curves

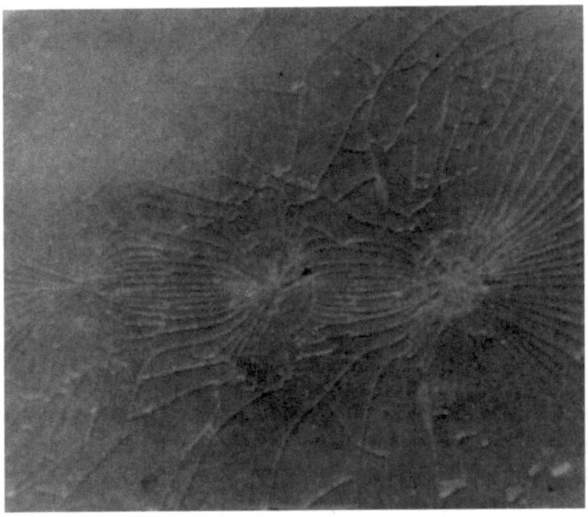

Fig. 16. Three groups of cracks produced by three simultaneous shocks in a pine-resin plate (0.72 ×)

the cracks shown in Fig. 6f. In this case, the thermal stress applied to the resin plate before crack propagation is small and so the cracks do not propagate in a large region. The travel-time curve suggests that the velocity of crack propagation in this case decreases gradually.

It is to be noted that no travel-time curve ever passes through the origin of the s–l diagram in the region $\alpha > 90°$ but its extension cuts the l-axis at the position of about $l = 5$ to 20 mm. This indicates that the cracks start with some time lag, in the order of 10^{-5} s, after the application of strong mechanical shocks. This time lag appears larger in the case of Fig. 15e where the thermal stress in the resin plate is small. When the thermal stress in the resin plate is high, the tangent of the s–l curve becomes higher corresponding to the higher velocity of shock propagation.

The reason why the tangents of the travel-time curve are different in the two regions, $\alpha > 90°$ and $< 90°$, is not clear. The observed results suggest that the mechanism of refraction of the propagating crack may be much more complicated and the propagation velocity of cracks seems to change, in general it decreases, by the stress wave. In order to check this latter point, we tried to examine crack propagation applying three mechanical shocks to the resin plate.

3.5 Refraction of propagating cracks by two successive shocks

Fig. 16 shows an example of cracks produced by three mechanical shocks applied almost simultaneously to the pine-resin plate. The cracks starting from A_3 are refracted at points 1, 2, 3, ..., as shown in Fig. 17, by the first stress wave coming from A_2. Then,

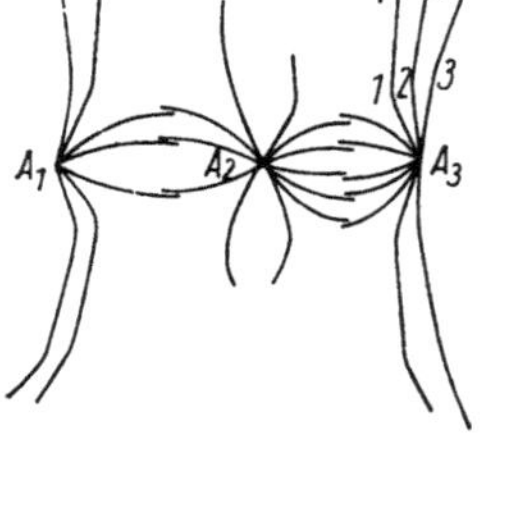

Fig. 17. Scheme of refraction of propagation cracks by two stress waves

Fig. 18. Travel-time curves of propagating cracks refracted by two successive stress waves

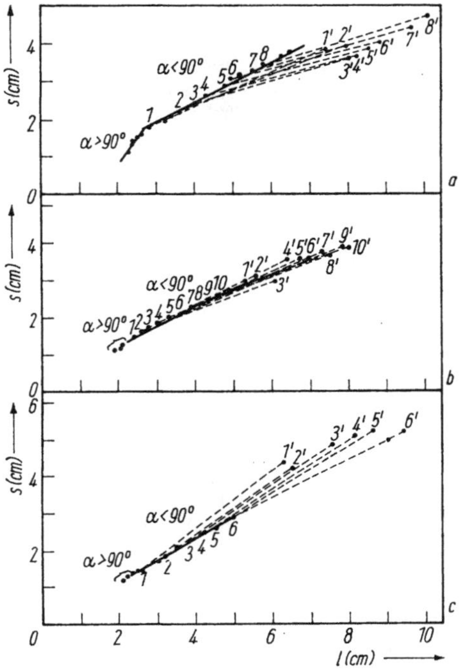

they are refracted further at points 1′, 2′, 3′, ..., respectively, by the second stress wave coming from A_1. Some of the observed travel-time curves for these types of cracks are shown in Fig. 18. The mean velocities after the first refraction points 1, 2, 3, ... are proportional to the inclination of the lines 1–1′, 2–2′, 3–3′, ..., if the velocity of the stress wave which produces the refraction of propagating cracks is constant. In general, these mean values decrease after the refraction as shown in Fig. 18a and b. In some cases, however, these apparent mean velocities appear as if they increase after the first refraction as shown in Fig. 18c. Nevertheless, these mean velocities show in all cases a decreasing tendency as the propagation distances increase. Besides, it seems that the propagation velocity of the stress wave responsible for the crack refraction is variable depending on the intensity of the mechanical shock.

The index of refraction is defined by the following formula as in the case of geometrical optics:

$$n = \frac{\sin \alpha}{\sin \beta}.$$

Here, α and β are the angles of incidence and refraction, respectively, as shown in Fig. 14b. We observed that this index of refraction increases as the angle α decreases in the region $\alpha < 90°$.

Fig. 19 shows one of our results on the refraction index of propagating cracks as a function of the incident angle. The small points and the cross marks indicate the results obtained at the first refraction points in the upper and the lower sides of cracks, respectively, in relation to the A_1–A_2 line in Fig. 14b. These observed values of refraction index are on a straight line passing through the point $n = 1.00$ at $\alpha = 90°$. Practically, the reliable measurement of the refraction index in the region $\alpha > 90°$ is difficult but our results show that the refraction index becomes higher as the angle of incidence $(180° - \alpha)$ becomes smaller. The results obtained at the secondary refraction

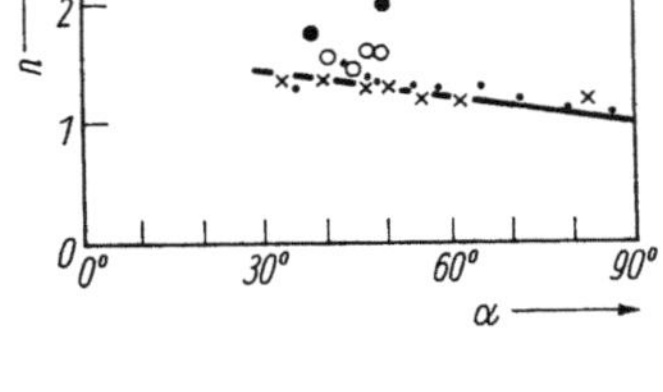

Fig. 19. Refractive index of propagating cracks as a function of incident angle α

points, such as 1′, 2′, 3′, ... shown in Fig. 17, are also indicated in Fig. 19 by open and full circles corresponding to both sides of the crack. The precision of these measurements is not so high because the crack paths lose their rectilinear nature. Nevertheless, the observed results show that the index of refraction in secondary refraction becomes higher than that of the primary refraction.

4. Conclusions

The refraction of propagating cracks occurs by a stress wave like that of a light beam at the boundary of two different media. If the angle $\alpha > 90°$ (α, angle between the propagation direction of the crack and that of the stress wave), the crack refracts towards the source of shock, while if $\alpha < 90°$, the crack refracts as if it goes away from the shock point.

The travel-time curves, relations between the distance traversed by the crack and that traversed by the stress wave, give some informations on the crack propagation. However, the apparent anomalous behaviour of the travel-time curves raises several questions about the nature of the stress wave produced by mechanical shock and the mechanism of crack refraction by the stress wave. The propagation velocity of the stress wave, responsible for the crack refraction, in a thin layer of pine-resin or in a thin glass plate is probably variable depending on the intensity of the mechanical shock.

Acknowledgement

The author would like to take this opportunity to show his gratitude to the late Prof. M. Hirata for much advice and encouragement during the progress of this work.

References

[1] M. HIRATA, Ôyô-Butsuri (Appl. Phys.) 5, 386, 437, and 482 (1936).
[2] M. HIRATA, Kagaku (Science) 7, 585 (1937).
[3] M. HIRATA, Conf. Appl. Phys. Soc. Japan, Tokyo, June 29, 1947.
[4] N. TERAO, J. Phys. Soc. Japan 7, 631 (1952).
[5] R. TAGUCHI, Riken-Ihô (Bull. Inst. Phys. chem. Res.) 10, 110 (1931).
[6] S. HYODO, Ôyô-Butsuri (Appl. Phys.) 29, 544 (1960).
[7] E. J. IRONS, J. sci. Instrum. 7, 323 (1930).
[8] S. HYODO and F. OKUDA, J. Phys. Soc. Japan 15, 2093 (1960).

(Received April 17, 1978)

phys. stat. sol. (a) **49**, 249 (1978)

Subject classification: 9; 10.2; 22.2.2

Zentralinstitut für Elektronenphysik der Akademie der Wissenschaften der DDR, Berlin

A Comparative Study of Zinc Tracer Diffusion Profiles in LEC, SSD, and CVD GaP

By

E. NEBAUER

Zinc tracer concentration profiles $C(x)$ in LEC, CVD, and SSD GaP samples are measured at 850 °C. In agreement with published investigations it is found that the single-crystalline Czochralski material having about 10^5 dislocations/cm^2 shows only the bulk part in $C(x)$ down to some 10^{17} Zn atoms/cm^3. The C profiles of well-grown CVD layers agree with those of the LEC crystals, but those of layers having high dislocation densities exhibit exponential tail regions. Similarly in polycrystalline SSD material the C distributions consist of two parts, the fast component being due to grain-boundary diffusion.

Es werden bei 850 °C Zink-Tracer-Konzentrationsprofile $C(x)$ in LEC, CVD und SSD GaP-Proben vergleichsweise gemessen. In Übereinstimmung mit anderweitig veröffentlichten Arbeiten wird gefunden, daß das Czochralski-Material (mit größenordnungsmäßig 10^5 Versetzungen/cm^2) bis zu einigen 10^{17} Zn-Atomen/cm^3 herab nur die Volumendiffusionskomponente in $C(x)$ aufweist. Die $C(x)$-Profile von CVD-Schichten passabler Perfektion stimmen mit denen von LEC-Kristallen überein, hingegen zeigen Schichten mit hohen Versetzungsdichten exponentielle Schwanzregionen. In ähnlicher Weise besitzen die $C(x)$-Verteilungen von polykristallinem Material zwei Abschnitte, wobei die schnelle Komponente durch Korngrenzendiffusion zustandekommt.

1. Introduction

A lot of experimental papers in the last few years has clearly demonstrated that one-dimensional crystal defects often strongly influence the electrical and optical properties of elemental as well as of compound semiconductors. Especially the efficiency of III–V optoelectronic devices can be substantially reduced by high dislocation densities n_d; for the connection between luminescence and n_d in GaP see e.g. [1]. Within this field it is also of interest to investigate in which way the diffusion profiles of essential dopants and the structure of p–n junctions are modified by the variation of the perfection of the GaP material used. To our knowledge up to now for no semiconductor the concentration profile $C(x)$ of an impurity has been measured (at constant annealing conditions) as a function of n_d. In contrast to the experimental situation the influence of n_d on $C(x)$ was investigated theoretically already some years ago in a series of papers (see Section 4). It was found that at appropriate parameter combinations of volume and dislocation diffusion coefficients D and D', annealing time t, and n_d a tail region in $C(x)$ can be found if some orders of magnitude in C are detectable. But such tails were seldom measured in semiconductors; a good example is the system CVD Si:As [2]. Concerning III–V compounds, the tails appearing sometimes in $C(x)$ were discussed as "short-circuit diffusion paths" up to now only on the systems AlSb:Zn and AlSb:Cd [3]. With respect to the practically very important GaP:Zn system corresponding investigations are wanting. However, an inspection of all published Zn tracer profiles in GaP shows that sometimes tail-like regions appear [4] and we feel that these tails may originate in dislocation pipe diffusion.

In this study we consider the ("chemical") diffusion behaviour of Zn in single-crystalline GaP samples of different origin and perfection and, for comparison, polycrystalline SSD samples at 850 °C by means of the tracer technique. Experimental details are given in the following section.

2. Experimental

For our Zn profile measurements three types of GaP material were available: firstly, undoped and Te-doped ($n \approx 2 \times 10^{18}$ cm^{-3}) LEC (liquid encapsulated Czochralski) single crystals with the usual average n_d values of (some) 10^5 cm^{-2}; secondly, a sample collection of single-crystalline CVD (chemical vapour deposition) layers, oriented 3° to 4° off ⟨111⟩ towards ⟨110⟩, deposited on lapped and Br_2–CH_2OH polished substrates by the HCl method in our institute; and — for comparison — thirdly, SSD (synthesis, solute diffusion) polycrystals also grown in our chemical laboratories. All samples were parallelepipeds with main faces of about 5 × 7 mm², the thickness of the LEC and SSD crystals amounted to 0.5 to 2 mm, the (optically determined) thicknesses of the CVD layers was in the range of 30 to 120 μm (deposited on 0.4 mm thick substrates). All samples were premicrotomed on a Jung microtome type K with a diamond knife. After that the Zn diffusion firings were made at a "standard" temperature of 850 °C and times of $t = 0.8$ to 4 h in evacuated quartz glass ampoules with appropriate amounts of elemental zinc powder ($\gtrapprox$ 0.3 mg Zn/cm³) containing the tracer Zn-65.[1])

In this way (nearly) saturated Zn pressures were maintained (one-temperature-zone procedure). During these firings the Zn sources acted as stationary ones. After the annealings the four side faces of the samples — disturbing the measurement of "one-dimensional" $C(x)$-curves — were reduced by grinding. Then the samples were again placed in the pneumatic holder of the microtome and the usual sectioning technique of layer removing and layer activity determination combined with appropriate weighing took place. Mostly 2 to 5 μm per measuring point were microtomed.

3. Experimental Results

Fig. 1 shows some typical Zn profiles for undoped and Te-doped LEC GaP crystals taken under our standard annealing conditions, besides — for comparison — a SIMS profile[2]) for a Te-doped sample. At equal times the two kinds of $C(x)$-curves exhibit no

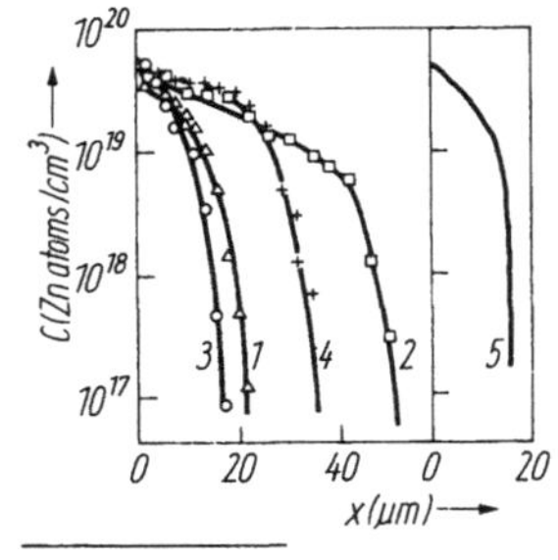

Fig. 1. Zn concentration profiles at $T = 850$ °C in LEC GaP crystals; undoped material: (1) 3.6×10^3, (2) 1.44×10^4 s; Te-doped samples: (3) 3.6×10^3, (4) 1.44×10^4, (5) 3.6×10^3 s. Curves (1) to (4): tracer technique, curve (5): SIMS technique

---

1) Such (or very similar) (T, p_{Zn}, t) combinations were frequently used by other workers [4, 5] and allow some comparisons of the own and the published (bulk) profiles. — From the point of view of "a priori" equilibrium conditions such a system with a "non-ternary" source is somewhat problematic but in our opinion the following results are not affected essentially by this circumstance (the surface erosions were insignificant).

Further on, test experiments with ampoules consisting of a double chamber (the zinc atmosphere interacts with the GaP only after equilibration of the sample) demonstrate that the $C(x)$ structure is the same as after annealings in the simple conventional ampoules.

2) Many thanks are due to Dr. R. Schimko and Mr. C.-E. Richter (VEB Werk für Fernsehelektronik Berlin) for carrying out this measurement.

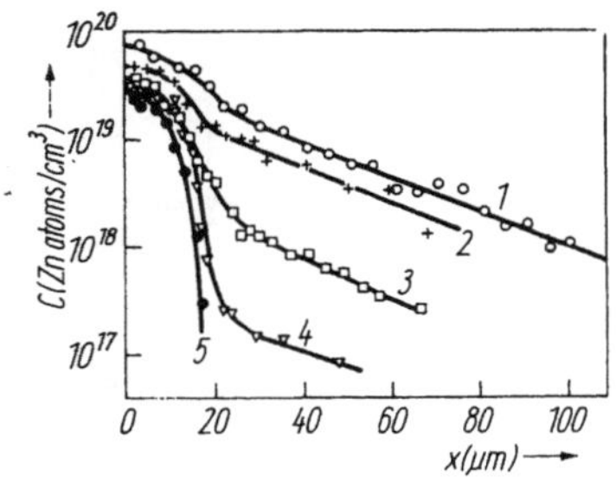

Fig. 2. Zn tracer profiles of some CVD GaP layers of different crystallographic perfection: curves (1) and (2) very high, curve (5) very low density of short-circuit diffusion paths. Parameters: $T = 850$ °C, (1) $t = 1.08 \times 10^4$ s; (2) to (5) $t = 3 \times 10^3$ s

strong differences in the solubility and (C-dependent) diffusivity values and show — at these relatively short annealing times — also approximately the regular time behaviour (cf. e.g. [4, 5]). The comparison of the "front" and "back" side profiles yields always good reproducibility. All profiles have no tail regions above some 10^{17} Zn atoms per cm³. As follows from SEM EBIC investigations the p–n-transition is, as expected, placed in the region of the steep C decrease and the planarity of the diffusion front is relatively good for $t = 1$ h, but already bad for $t = 4$ h. These curves of Fig. 1 may serve as "reference" curves for the following CVD and SSD profiles. Fig. 2 presents a series of $C(x)$ distributions of CVD GaP layers which were diffused under (nearly) the same conditions, the only varying relevant parameter being, in our opinion, the perfection of the samples. In Fig. 2 the curves (1) to (5) belong to samples with increasing perfection. Within our detection limit, many CVD samples — curve (5) is only one example — had tailfree $C(x)$ profiles which are very similar to the relevant curves of the bulk material (see Fig. 1 and [4, 5]), whereas layers deposited (partly intentionally) on rough or perhaps not sufficiently cleaned substrate surfaces have tails with such a high tail level that the bulk region was almost "drowned". The very high tail level of the $C(x)$ curve (2) was provoked by microtoming (instead of lapping) the GaP substrate before the CVD procedure.

The analytical structure of most of the measured tails could not be determined with sufficient accuracy because of the low counting rates and/or the limited layer thickness but it seems that an exponential law is effective. The slope of the tails is nearly the same.

In Fig. 3a a direct comparison is given of the profiles of the (bad) CVD layer (1) (cf. Fig. 2) and of the LEC GaP:Te substrate region (reference curve) of a layer–

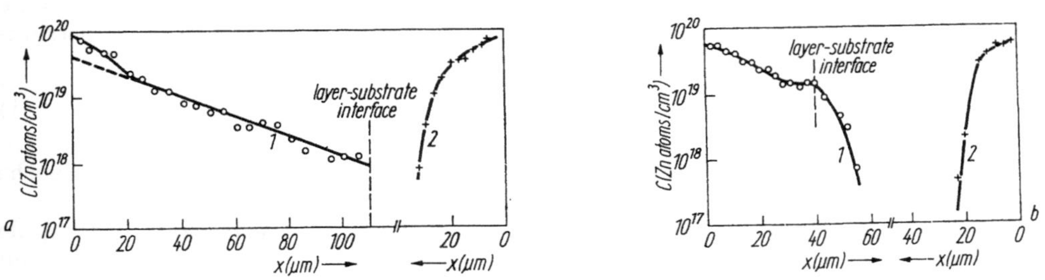

Fig. 3. a) Direct comparison between (1) CVD and (2) LEC $C(x)$ profile of a CVD layer–LEC substrate sample ($T = 850$ °C, $t = 1.08 \times 10^4$ s). b) Zn tracer profile of a low-perfection CVD GaP layer demonstrating the modification of the tail region by the layer–substrate boundary (1). For comparison, the "substrate curve" (2) is also given. Parameters: $T = 850$ °C, $t = 5.4 \times 10^3$ s

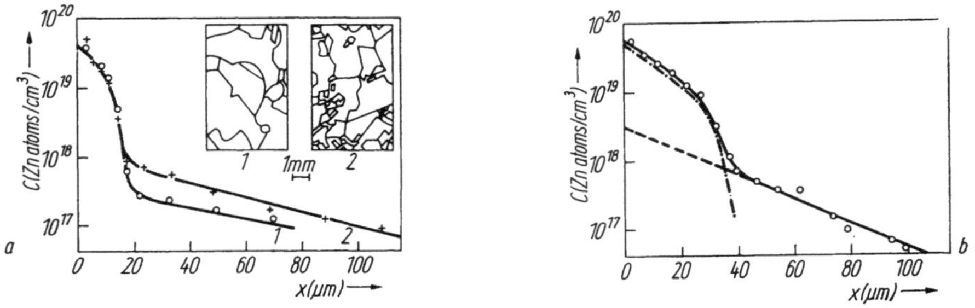

Fig. 4. a) Zn concentration distributions found in the (1) "front" and (2) "back" side regions of a SSD polycrystal having somewhat different lengths of grain boundaries (see the insets). Parameters: $T = 850$ °C, $t = 3.6 \times 10^3$ s. b) SSD Zn profile at $T = 850$ °C, $t = 1.44 \times 10^4$ s with the separated volume part (dash–dotted curve) found from a Fisher analysis

substrate sample. Further on, the following experiment served as a test: A CVD GaP layer of very low perfection was annealed until the tails intersected the layer–substrate interface region at relatively high C's, and the $C(x)$ measurement was extended to the substrate region. One expects that the strong reduction of the "short-circuit paths" on the boundary causes also a considerable decrease of the tail C level. This was really found, an example is given in Fig. 3b.

For comparison Zn profiles were taken and analysed on SSD samples having some amount of grain boundaries (but low n_d values in the grains). Fig. 4a demonstrates that the $C(x)$ distributions — like the profiles of bad CVD material — have two parts: an exponential tail C_{gb} follows the bulk region. Further, clearly the level of the fast component correlates with the total length l_{gb} of the grain boundaries/cm² in the areas investigates (l_{gb} is nearly constant within the measured x-region). The value $C_{gb}(1)/C_{gb}(2) = 1/2$ approaches that of $l_{gb}(1)/l_{gb}(2)$. Lastly in Fig. 4b a 4h curve is shown.

4. Discussions

From the experimental results we conclude that the $C(x)$ tails of the CVD samples are not caused by microtome-induced or diffusion-induced dislocations[3]); nonplanar diffusion fronts [6] or by a fast volume diffusion process, but by an "a priori dislocation pipe" diffusion process.

Already in 1951 Fisher [7] gave the first theoretical treatment of diffusion processes in solids including "short-circuit paths". He used a two-zone model with a (C-independent) D in the undisturbed volume regions and a $D' \gg D$ in the disturbed areas (in his model: grain boundaries with width δ). Specializing in the sectioning technique Fisher found in addition to the volume diffusion component a concentration distribution of the following type:

$$C \sim \exp\left(\frac{-Ax}{t^{1/4}}\right) \tag{1}$$

(A constant including D, D', δ), i.e. exponential tails at higher penetration depths appear.

Fisher's idea has been refined mathematically and extended, especially by Whipple [8], Suzuoka [8], and others for grain boundaries and by Mimkes and Wuttig [9] and

[3]) Of course, the creation of high diffusion-induced dislocation densities is probable under our conditions (cf. the TEM n_d investigations in Zn diffused CVD GaP undertaken recently by Kitada [11]) but it seems that this dislocation type produces no $C(x)$ tails.

others to dislocations. But the corresponding exact expressions are complicated, e.g the equation for the complete $C(x)$ profile of Mimkes including simultaneous diffusion via lattice sites and dislocation pipes (assuming constant surface concentration and the usual continuity conditions for C and the flux at the boundary between the D and D' region):

$$C(x, t) = C_0 \operatorname{erfc}(\xi) + \int_1^{\Delta}\int_{\varepsilon}^{1} \left(\frac{\varepsilon\varrho}{\pi\sigma^3}\right)^{1/2} \xi \exp\left(-\frac{\xi^2}{4\sigma} - \left(\frac{\sigma-1}{\Delta-1}\right)\left(\frac{k}{\varepsilon}\right)^2\right) \times$$
$$\times \operatorname{erfc}\left[\left(\frac{\Delta-1}{\Delta-\sigma}\right)^{1/2}\left(\frac{\varrho-\varepsilon}{2k} + \frac{k}{\varepsilon}\frac{\sigma-1}{\Delta-1}\right)\right] \mathrm{d}\varrho\, \mathrm{d}\sigma \qquad (2)$$

with $\xi = x/2(Dt)^{1/2}$, $k = (\pi D t n_\mathrm{d})^{1/2}$, $\Delta = D'/D$, $\varepsilon = r_\mathrm{d}(\pi n_\mathrm{d})^{1/2}$ (r_d pipe radius). It can be evaluated only by computers. But such calculations show that the basic structure of the C profile according to (1) is essentially preserved [9]. Further it was shown in [9] that the interesting parameters Δ, r_d, and n_d can be well determined by fitting the whole C profile.

With respect to our system, at present a direct application of this conception is not possible because the effect of concentration-dependent bulk diffusivity generally has not been considered in the above-sketched theoretical model up to now. But our experimental results on GaP:Zn and also the curves presented in [2] (where a $D_{\mathrm{As}}(C)$ dependence exists which is qualitatively similar to $D_{\mathrm{Zn}}(C)$ in GaP) show that the exponential dependence of the tail is at least approximately maintained. Carrying out a Fisher analysis (i.e. extrapolation of the tail to $x = 0$ and subtraction from the first part of C) it could be found that the bulk parts of curves (1) to (4) of Fig. 2 and of the curves of Fig. 4 a, b agree quite well with those of the tail-less LEC and CVD $C(x)$ curves, respectively. This indicates that 1. no significant differences in the Zn bulk diffusivity/solubility properties (or relevant point defect concentrations) between LEC and CVD material exist and 2. the tails of some Zn profiles were caused by the very high n_d values in the corresponding GaP layers. The order of magnitude of the dislocation density in the worse CVD layers can be estimated to $n_\mathrm{d} \approx 10^7\ \mathrm{cm}^{-2}$. This follows from the fact that the C values in the tails of such layers lie at least two orders of magnitude above that of the LEC samples (cf. also the levels of the tails found in [4]). Further, this can be corroborated by a comparison of the tail levels of CVD and SSD samples (of course, at equivalent diffusion conditions). From this n_d value a medium dislocation distance $d = 3\ \mu\mathrm{m}$ results, on the other side a penetration depth $(D_{\mathrm{Zn}}t)^{1/2}$ of some μm can be calculated, i.e. according to Hart's formula [10] the bulk diffusivity can already be somewhat increased.

Summarizing it has been demonstrated that on the basis of tracer concentration profile measurements of appropriate diffusants (well detectable tail levels) a (at least qualitative) distinction of the samples with respect to their crystallographic perfection is possible. Quantitative applications of the sketched method would require some improvements (raising of the limit of measurement in C, computer-fitting of the whole C curve).

Acknowledgements

I am very indebted to Dr. J. Noack and H. Fabig for the growth of the CVD and SSD samples, respectively, further Dr. J. Noack and Dr. H. Raidt for some discussions. The experimental help by J. Maege is gratefully acknowledged. Last not least I am much obliged to R. Zierach for his permanent cooperation in the tracer measurements.

References

[1] W. A. BRANTLEY, O. G. LORIMOR, P. D. DAPKUS, S. E. HASZKO, and R. H. SAUL, J. appl. Phys. **46**, 2629 (1975).
G. B. STRINGFELLOW, P. F. LUNDQUIST, T. R. CASS, and R. A. BURMEISTER, J. Electr. Mater. **3**, 497 (1974).
[2] D. R. CAMPBELL, K. N. TU, and R. O. SCHWENKER, Thin Solid Films **25**, 213 (1975).
[3] S. R. SHOWAN and D. SHAW, phys. stat. sol. **32**, 97 (1969).
D. SHAW and S. R. SHOWAN, phys. stat. sol. **34**, 475 (1969).
[4] L. C. LUTHER and K. B. WOLFSTIRN, J. Electr. Mater. **2**, 375 (1973).
B. TUCK and P. R. JAY, J. Phys. D **9**, 2225 (1976).
B. TUCK and P. R. JAY, J. Phys. D **10**, 1315 (1977).
B. TUCK and A. HOOPER, J. Phys. D **8**, 1806 (1975).
[5] L. L. CHANG and G. L. PEARSON, J. appl. Phys. **35**, 374 (1963).
L. L. CHANG and G. L. PEARSON, J. Phys. Chem. Solids **25**, 23 (1964).
[6] S. F. NYGREN and G. L. PEARSON, J. Electrochem. Soc. **116**, 648 (1969).
[7] J. C. FISHER, J. appl. Phys. **22**, 74 (1951).
[8] R. T. P. WHIPPLE, Phil. Mag. **45**, 1225 (1954).
T. SUZUOKA, J. Phys. Soc. Japan **19**, 839 (1964).
[9] J. MIMKES and M. WUTTIG, Phys. Rev. B **2**, 1619 (1970).
J. MIMKES, phys. stat. sol. (b) **58**, K31 (1973).
J. MIMKES, Thin Solid Films **25**, 221 (1975).
[10] E. W. HART, Acta metall. **5**, 597 (1957).
[11] M. KITADA, J. Mater. Sci. **12**, 207 (1977).

(Received June 13, 1978)

phys. stat. sol. (a) **49**, 255 (1978)

Subject classification: 1.2 and 12.2; 1.5; 1.6; 2; 22.1.1

Institut B für Physik der Technischen Universität Braunschweig

Surface Influences of Glass Substrates on Lattice Parameters and Phase Transition Temperature of Evaporated Ge Thin Films[1])

By

K. DETTMER, H. MÜLLER[2]), and F. R. KESSLER

The substrate temperature T_S during evaporation of Ge thin films highly determines the preparation of amorphous or polycrystalline films. In between these "phases" there exist characteristically sharp and different amorphous–polycrystalline transition temperatures T_{ST} on Duran 50 and Menzel HW glass substrates which are 135 and 235 °C, respectively. It is demonstrated that these transition temperatures of Ge films on glass depend on the kind of glass and its surface properties. The results are deduced by X-ray analysis of the crystallites in the polycrystalline films and are checked out by electron microscopic studies of the glass and film surfaces.

Die Substrattemperatur T_S während des Aufdampfens dünner Ge-Filme bestimmt im wesentlichen die Herstellung amorpher oder polykristalliner Schichten. Zwischen diesen „Phasen" existieren charakteristisch scharfe und unterschiedliche amorph–polykristalline Übergangstemperaturen T_{ST} von 135 auf Duran 50 und 235 °C auf Menzel HW Glassubstraten. Es wird gezeigt, daß diese Übergangstemperaturen von der Glassorte und dessen Oberflächeneigenschaften abhängen. Die Ergebnisse basieren auf Röntgenreflexanalysen der Kristallite in den polykristallinen Filmen und Elektronen-mikroskopischen Aufnahmen der Glas- und Filmoberflächen.

1. Introduction

By measurements of resistivity, refractive index, and X-ray analysis of Ge thin films on Duran 50 Dettmer et al. [2, 3] were able to prove films grown at substrate temperatures below 130 °C to be amorphous and those grown at 140 °C and higher temperatures to exist in a polycrystalline phase. Between these two temperatures a continuous but very sharp "phase change" takes place around the so-called transition temperature $T_{ST} = 135$ °C. In 1974 Johannessen [4] reported of a transition temperature of 235 °C for Ge thin films on Corning glass 7059 substrates. Now we want to show that this discrepancy is not contradicting. Especially there is no reason to assume systematical errors in temperature measurements.

In the literature we find a lot of different transition temperatures for Ge thin films on various substrates. Most of them are not comparable with one another on account of the different preparation and evaporation conditions on different substrate materials used (for instance CaF_2, Ge, Si, GaAs, C, mica, NaCl, glass). It is only correct to compare transition temperatures of Ge on different substrates under otherwise equal conditions. We have done this for two different kinds of glass, Menzel HW (Fa. Menzel, Braunschweig) and Duran 50 (Fa. Schott & Gen., Mainz).

In this paper we shall show by various experimental data analysis that the transition temperature between the amorphous and polycrystalline phases in the films then

---

[1]) Most of these results were presented at the Spring conferences of the DPG, Freudenstadt 1978 [1].

[2]) Now at Institut für Werkstoffkunde und Herstellungsverfahren, Technische Universität Braunschweig.

mainly depends on the type of glass substrate on account of its different surface properties, a fact on which was not spent much importance hitherto. The experimental verifications of these results are based on measurements of anisotropic lattice parameters, grain sizes, different nucleation by variable substrate surface roughnesses, internal stresses, and anisotropic strain by thermal mismatch.

Furthermore we shall discuss the influence of the glass substrates on the lattice constants of polycrystalline grains in Ge thin films. The question of anisotropic stress and strain will be dealt with in other papers.

2. Experimental

The glass substrates were all treated identically: washed in acetone, trichlore ethylene, distilled water and ultrasonically cleaned and heat stream dried. The substrates were then pressed on to a hot plate which heated them in a high vacuum bell jar by heat conduction to the controlled substrate temperature T_S which was determined by a Ni/CrNi thermocouple pressed on to the surface to be covered. Another thermocouple placed between substrate and hot plate indicated no temperature difference under equilibrium condition before the evaporation was started. During the evaporation we observed at most a temperature raise of about 5 degrees due to the source radiation absorbed and due to the heat of condensation of the adsorbed Ge atoms.

The Ge films were evaporated from a resistance heater at a vacuum pressure of 1.2×10^{-8} bar to thicknesses of about 1 μm, at high rates of between 100 and 700 Å/s. Most of the films discussed here are about 4000 Å thick.

The lattice constants were determined by X-ray analysis with a Siemens goniometer, Kristalloflex 4. The X-ray reflections were digitally recorded during stepwise scanning (step size: 1/100°). The measured results were modificated by a computer program correcting for the angle depending background, the true centre of peak gravity and broadening effects due to geometrical effects of the X-ray goniometer.

The beginning of detectable Bragg reflections in dependence on the substrate temperature allowed the exact determination of the transition temperature T_{ST} because it corresponded with the first appearance of a polycrystalline phase. In amorphous films no X-ray reflexes could be detected. If peaks grew up the (220) reflex always raised up at first. With increasing substrate temperature the number of peaks and the reflexes intensities increased. In all, we were able to record peaks of the (111) up to the (531) reflexes in the polycrystalline films.

Each reflection of a single film led to an adequate lattice plane distance. All these lattice plane distances for the reflections of one film were recorded against the well-known Nelson-Riley function [5].

$$NR = \frac{1}{2}\left(\frac{\cos^2\theta}{\sin\theta} + \frac{\cos^2\theta}{\theta}\right)$$

(2θ is the Bragg angle of the relevant reflex).
The lattice constant was then found by extrapolation of the least-squares fit straight line to $2\theta = 180°$. This lattice constant represents the mean value of the relevant constant a_z which is the lattice constant perpendicular to the substrate plane. The lattice constant a_{xy} in the film plane is different from a_z.[3]) a_{xy}, too, could be determined by a special X-ray stress analysis. The details of that analysis and the results will be given in a subsequent paper.

[3]) In the literature only a_z-values for thin films had been determined.

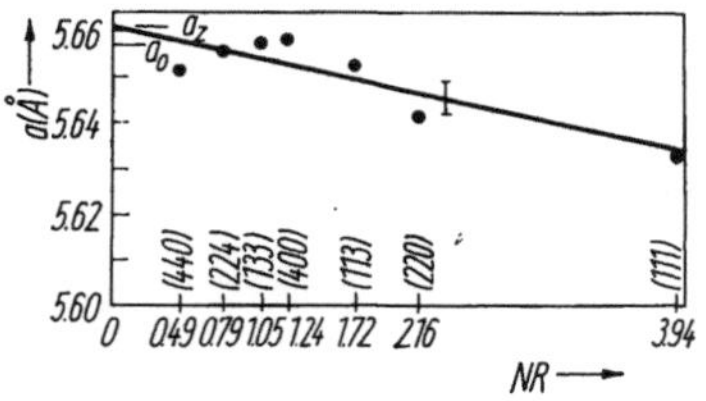

Fig. 1

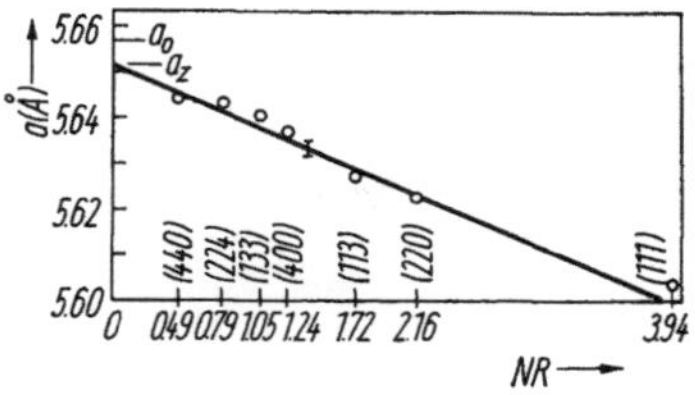

Fig. 2

Fig. 1. Example of an extrapolation to the lattice constant a_z of a 3410 Å thick germanium film on a Menzel HW substrate. The film was prepared at a substrate temperature of $T_S = 400$ °C. The exptrapolation to the Nelson-Riley (NR) function value 0, i.e. the Bragg angle $2\theta = 180°$, leads to $a_z = (5.6605 \pm 0.0033)$ Å in this case. a_0 is the bulk crystal value

Fig. 2. Extrapolation to the lattice constant a_z of a 2650 Å thick Ge film on a Duran substrate; $T_S = 400$ °C. Here a_z gives (5.6504 ± 0.0014) Å

3. Results

Two of the characteristic extrapolations to the lattice constants a_z are shown in Fig. 1 for Menzel HW glass and in Fig. 2 for Duran 50. We see the following remarkable results: At first, the reflex points for Ge on Menzel HW (Fig. 1) have a larger deviation from the least-squares fit straight line than those for Ge on Duran glass (Fig. 2). Secondly, the lattice constant a_z for films on Menzel HW is larger for the film on Menzel glass than for the film on Duran glass, which will be shown to be a general result.

Polycrystalline Ge films on Menzel HW substrates can be produced only in the temperature range of 235 and 445 °C, where the upper limit is given by reaching the softening region of this glass material. For films on these Menzel HW substrates the lattice constant a_z is shown in Fig. 3 in dependence on the substrate temperature.

We notice that a_z increases with increasing T_S and becomes larger than the bulk crystal value for $T_S \geqq 340$ °C. The latter fact is in contrast to results of Wissmann [6] and Paul et al. [7] who found that the lattice constants for thin films are lower than the bulk crystal value.

An increasing lattice constant a_z with T_S is also seen for Ge on Duran glass in Fig. 4 between 135 and 550 °C, which is the upper limit used by us. On Duran 50, however, a_z is always lower than the bulk value.

The comparison of Fig. 3 and 4 shows that the lattice constant a_z has the same value of 5.6485 Å for Ge at both substrates at the relevant transition temperatures (T_{ST} is 135 °C for Duran 50 and 235 °C for Menzel HW). We thus deduce the very

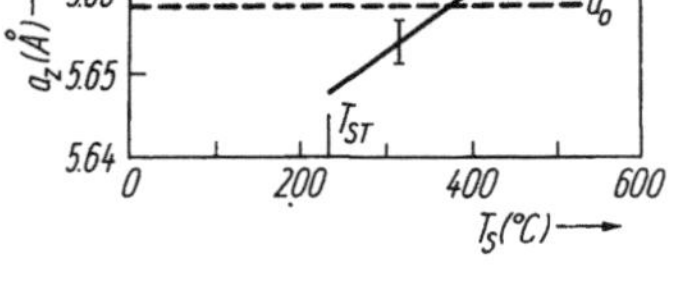

Fig. 3. Dependence of the lattice constant a_z on the suebstrat temperature T_S for Ge films on Menzel HW substrates. The mean deviation of the mean value is $\pm$ 0.0029 Å

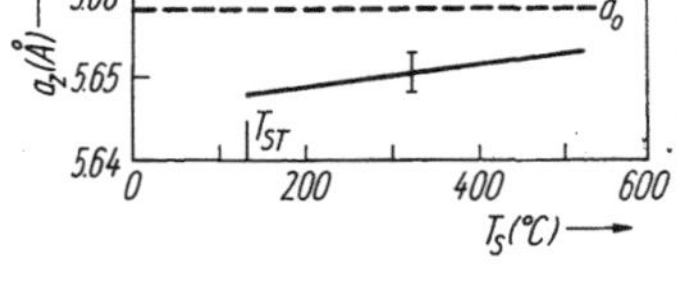

Fig. 4. Dependence of the lattice constant a on T_S for Ge films on Duran 50 substrates with a mean deviation of $\pm$ 0.0024 Å

Fig. 5 Fig. 6

Fig. 5. Typical electron micrograph of the surface of Menzel HW

Fig. 6. Electron micrograph of the surface of Duran 50

interesting statement that the appearance of a polycrystalline phase within the Ge films on both substrates at different T_{ST} starts with a lattice constant a_z which is equal and just 0.1% lower than the bulk crystal value.

The most important background for the explanation of the different behaviour of Ge films on Menzel HW and Duran 50 glass substrates was given by electron microscopic pictures of the two types of substrates. Such a picture (obtained by carbon replica technique) of a Menzel HW substrate in Fig. 5 shows a very rough surface structured by a lot of valleys and small craters.

The picture of the Duran 50 substrate in Fig. 6, however, gives a much smoother surface. There is another type of crater which is larger, but much less frequent. It is evident in Fig. 6 that the Duran 50 surface is not structured by valleys of considerably

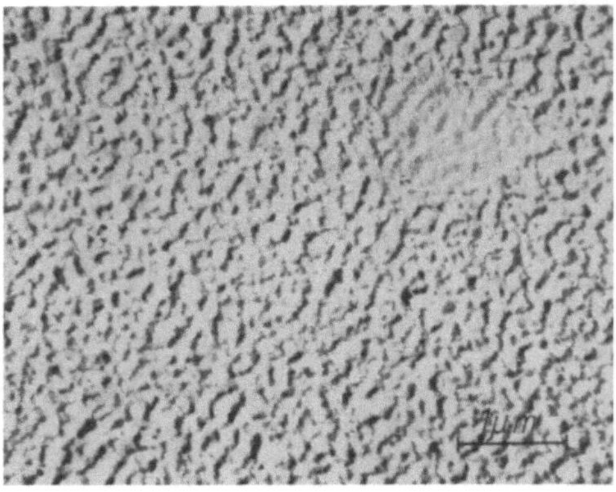

Fig. 7 Fig. 8

Fig. 7. Electron micrograph of the surface of a 5800 Å thick Ge film deposited at $T_S = 430$ °C on Menzel HW

Fig. 8. Electron micrograph of the Ge surface of a 5800 Å thick Ge film evaporated at $T_S = 500$ °C on a Duran 50 substrate bigness and depth. The very fine structure which is seen on the Duran surface must be attributed to the resolving power used for the replica technique.

At any case, all the glasses of one quality (Menzel HW and Duran 50) proved to show the same surface reliefs. Evidently Menzel HW glass possesses a rather rough surface structure whereas the surface of Duran 50 is relatively flat and undestroyed.

Characteristic differences, too, are seen in electron microscopic pictures of Ge films at the different substrate types. Surface pictures of 5800 Å Ge films are shown in Fig. 7 and 8 for evaporation on Menzel HW and on Duran 50, respectively. At the evaporated substrates the surface differences are not as pronounced as before. By evaporation with these 5800 Å thick films the coarse surface structures of the substrates are evidently flattened, and the small structures disappeared.

Our results are supported by a preparative observation: The period for getting the same vacuum pressure before evaporating on to Duran substrates takes about half the time than using Menzel HW substrates. This reflects the rougher and therefore larger surface of the Menzel HW glass for which the degasing time evidently is longer. This observation agrees with that of Zagar and Krause [8] who investigated the adsorption on glass materials. Mono- or bimolecular layers of gas molecules are accumulated on to the surface. The larger the surface the higher the gas adsorption. By that it is confirmed that Duran 50 has a lower surface roughness than Menzel HW.

4. Discussion

The evaporated Ge thin films on Menzel HW and Duran 50 substrates show qualitatively and quantitatively different results concerning the lattice constants a_z perpendicular to the surface plane and concerning the transition temperatures between amorphous and polycrystalline structure in dependence on the preparation temperature. Naturally, the different coefficients of expansion of substrate and film lead to different stresses and strains within the films after being cooled down from preparation to room temperature, and this will influence the crystalline grain sizes. But in this paper we are only concerned with the effects of different glass surface qualities on the general polycrystalline film growth and on the transition temperature between the amorphous and polycrystalline phase.

The differences between the substrates are mainly the differences in the surface quality which leads to different growth conditions to get polycrystalline films. A rough surface influences the surface mobility of the arriving atoms at the substrate surface. Crystalline grains can hardly be performed at low T_S, and amorphous films grow up. The rougher the surface is the higher the surface temperature must be in order to overcome a limit for the amorphous-polycrystalline transition. From this viewpoint it can be understood that the T_{ST} for Duran 50 lies 100 °C beneath that of the Menzel HW substrate on account of a smoother surface which initiates a quicker development of the critical grain size in dependence on the substrate temperature during evaporation. It is a very interesting fact that the glass surfaces of Corning 7059 used by Johannessen [4] showed similar surface properties in electron microscopic pictures as the rough Menzel HW substrates. The high transition temperature of $T_{ST} = 235$ °C found in these both cases is now well explained by the influence of the surface roughnesses.

In consequence of the higher surface roughness the nucleation on Menzel HW substrates will take place at capture islands at the surface which act as capture centers for the adatoms. With improved surface diffusion by higher substrate temperatures the crystallites will grow up with larger lattice constants perpendicular to the surface. This mechanism depends on the influence of internal strains which will be discussed in the following paper.

Acknowledgement

The authors are indebted to Mrs. Grusewski, Institut für Werkstoffkunde und Herstellungsverfahren, TU Braunschweig, who made the carbon replicas from the substrate surfaces and took the electron micrographs.

References

[1] K. Dettmer, H. Müller, and F. R. Kessler, Verh. DPG (VI) **13**, 44 (1978).
[2] H. Goebel, K. Dettmer, and F. R. Kessler, phys. stat. sol. (a) **16**, 61 (1973).
[3] K. Dettmer, H. Goebel, and F. R. Kessler, phys. stat. sol. (a) **27**, 393 (1975).
[4] J. S. Johannessen, phys. stat. sol. (a) **26**, 571 (1974).
[5] J. B. Nelson and D. P. Riley, Proc. Phys. Soc. 57, 160 (1945).
[6] P. Wissmann, Thesis, Hannover 1967.
[7] W. Paul, G. A. N. Connell, and R. J. Temkin, Adv. Phys. **22**, 531 (1973).
[8] L. Zagar and G. Krause, Forschungsber. d. Landes Nordrhein-Westfalen, Nr. 1718 (1966).

(Received May 8, 1978)

phys. stat. sol. (a) 49, 261 (1978)

Subject classification: 11 and 14.3.2; 22.1.2

Zentralinstitut für Kernforschung der Akademie der Wissenschaften der DDR, Rossendorf, [1])

A Very Simple Method for Profiling the Ion-Implanted Si-Surface

By

M. T. PHAM

The corrosion behaviour of the ion implanted Si in electrolyte solution containing redox agent is reported. The results show that the corrosion depends on the implantation parameters. Examples demonstrating the use of the corrosion for profiling of implanted surfaces are given.

Das Korrosionsverhalten von ionenimplantiertem Si in einer elektrolytischen Lösung mit Redoxmittel wird untersucht. Die Ergebnisse zeigen, daß die Korrosion von den Implantationsparametern abhängt. Beispiele werden angegeben, die die Benutzung der Korrosion für die Profilierung der implantierten Oberflächen zeigen.

1. Introduction

Recent studies [1, 2] of the contact silicon–electrolyte have demonstrated that the implanted layer on Si changes profoundly its electrochemical properties. The present paper describes the behaviour of the implanted Si-surface in contact with an electrolyte containing a redox component. A simple technique is reported which can be used to characterize the ion implanted layer on Si.

2. Experimental Methods

⟨111⟩ Si slices of both n- and p-type (18 mm in diameter and 0.3 to 0.4 mm thick) were used. After polishing with 0.25 μm diamond past the samples were etched for 3 min in a $HF:HNO_3:CH_3COOH$-mixture. One face of the sample was then exposed to a high dose implantation (n^+ on n-Si and p^+ on p-Si) and evaporated with Al to make the ohmic contact. The back face to be exposed to the solution was implanted with desired ions in a dose range from 10^{12} to 10^{17} cm^{-2}. The implantation energy was 30 and 12 keV. A misorientation of the sample provided a random implantation.

Fig. 1 shows the electrochemical cell and the measuring circuit. The rest potential was measured as a function of time. A saturated calomel electrode was used as the reference electrode. The dissolved Si-amount was determined radiochemically using the ^{31}Si formed by neutron activation. The electrolyte used was $HF–Na_2SO_4$ solution (0.1 N HF and 0.5 M Na_2SO_4) containing $K_3Fe(CN_8)$ in a appropriate concentration. All solutions were prepared with bidistilled water and all chemicals were reagent grade.

3. Results and Discussion

The change of the rest potential with time is shown in Fig. 2 to 7 for various implantation ions and doses. All curves indicate similar course. Immediately after immersion into the solution the rest potential drifts to negative values. The magnitude of this shift depends on the implantation dose. The rising rate is seen to be dose dependent, too. A slowing down of the drift follows this first phase resulting in a saturation period which is more obvious at high implantation doses. After this a reverse in the potential change occurs leading to a stable terminal value characteristic of the substrate state. During the potential change period an efficient dissolution of the Si was found. The thicknesses of the dissolved layer at the potential reversal are summarized in Table 1.

[1]) Postfach 19, DDR-8051 Dresden, DDR.

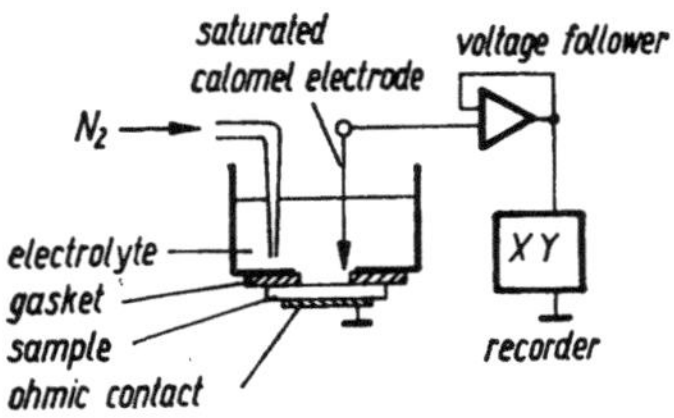

Fig. 1. Cell and circuit for measuring of rest potential

Table 1

The depth at which the potential reversal occurs represents approximately the thickness of the implanted layer

implanted ion	implantation parameters (cm^{-2})	(keV)	depth of potential reversal (Å)
Si	10^{15}	30	300
P	10^{16}	30	500
B	10^{15}	30	1500
Ar	10^{17}	30	480
	10^{16}	30	1100
In	10^{15}	12.2	180
	10^{14}	12.2	100
As	10^{15}	30	1100
	10^{14}	30	300
	10^{13}	30	150

It must be noted that the thickness values bear only qualitative character because the density of the implanted layer, particularly of the highly implanted one, is not known. The only reliably known quantity is the dissolved Si-amount following from the counting of the ^{31}Si in the electrolyte. The data in Table 1 are obtained using a value of 2.33 for the Si density. These data imply that the depth at which the reversal of the

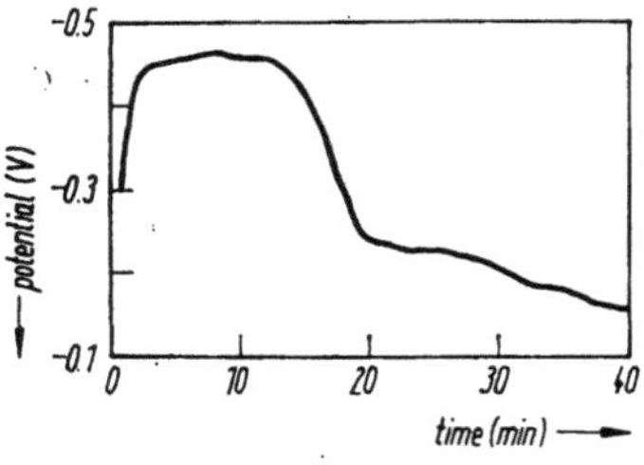

Fig. 2

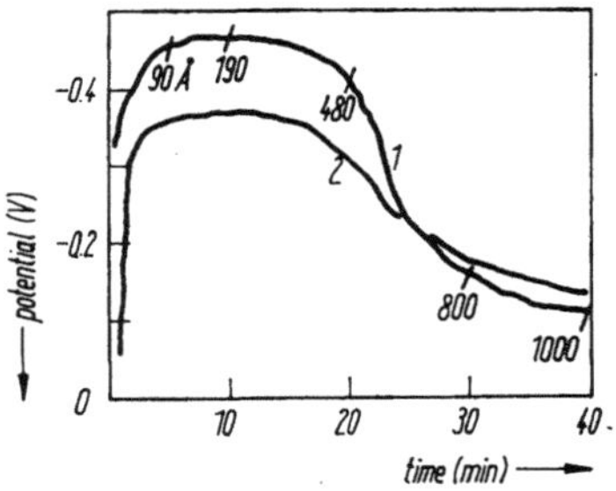

Fig. 3

Fig. 2. Rest potential for a n-Si (270 Ω cm) implanted with 10^{15} Si/cm^2 at 30 keV. Redox agent: 5×10^{-3} M $K_3Fe(CN)_6$

Fig. 3. Rest potential of n-Si (10 Ω cm) implanted with 10^{16} P/cm^2 of 30 keV. (1) Nonannealed; the numbers (in Å) denote the depth scale. (2) Annealed at 700 °C in N_2 atmosphere for 20 min. Redox agent: 5×10^{-3} M $K_3Fe(CN)_6$

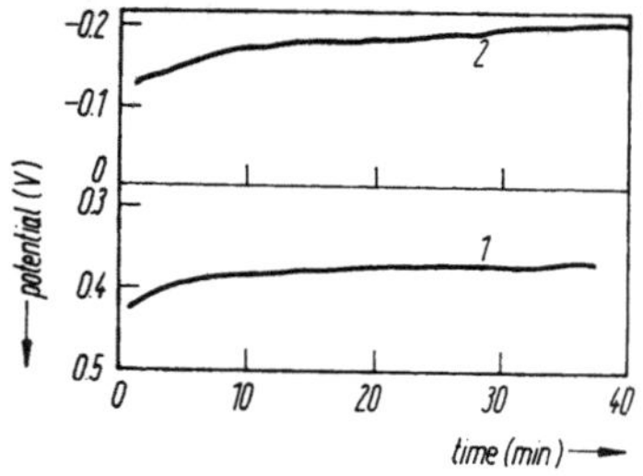

Fig. 4

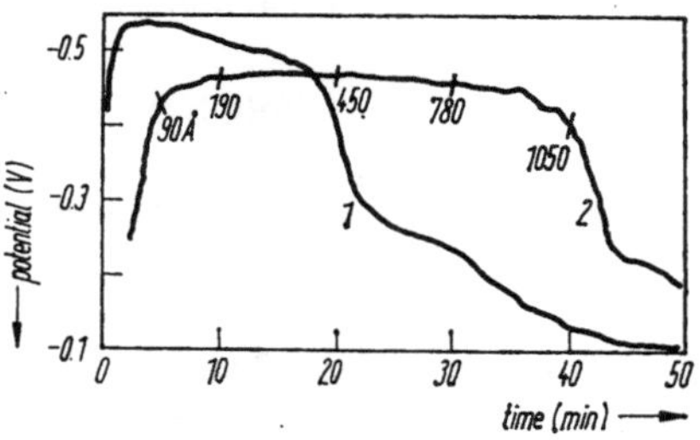

Fig. 5

Fig. 4. The rest potential behaviour of the B-implanted Si (7 Ω cm). Redox agent: 5×10^{-3} M $K_3Fe(CN)_6$. (1) 10^{17} B/cm² at 30 keV, (2) 10^{15} B/cm² at 30 keV

Fig. 5. The rest potential behaviour of the Ar-implanted Si (p-type, 10 Ω cm). Redox agent: 5×10^{-3} M $K_3Fe(CN)_5$. (1) 10^{17} Ar/cm² at 30 keV; (2) 10^{16} Ar/cm² at 30 keV. The numbers (in Å) denote the dehpt scale

potential occurs is dependent on the implantation dose, the ion energy, and the ion kind. Among the ions used the boron, Fig. 4, presents an exception the rest potential showing a time dependence different from that of the other ions.

It has been shown [3, 4] that introduction into the electrolyte of species that can be oxidized and reduced by simple electron transfer reactions shifts the stationary electrode potential (rest potential) according to the conducting type of the semiconductor. For a given redox component concentration the rest potential is a function of the semiconductor conductivity. In the absence of the redox substance the rest potential (i.e. the open-circuit potential) is determined by two opposite reactions: the anodic dissolution of the Si and the cathodic reduction of the H^+-ion (or dissolved O_2). No noticeable corrosion can occur in this case since the H^+-reduction extracts the conduction band electron, whereas the dissolution of Si is a process consuming holes. In the presence of a redox substance (our case: the oxidizing $K_3Fe(CN)_6$) having a positive

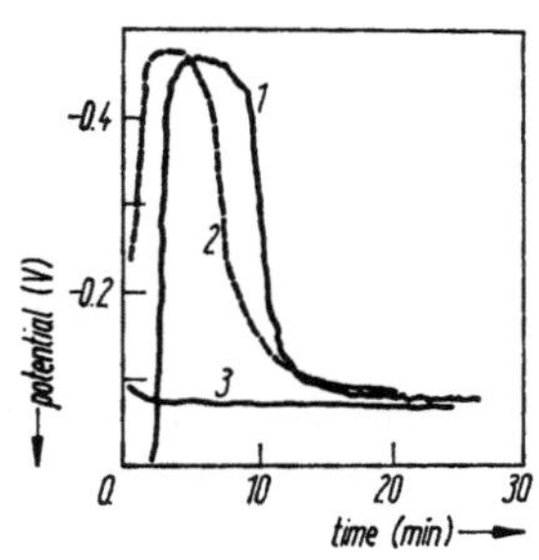

Fig. 6

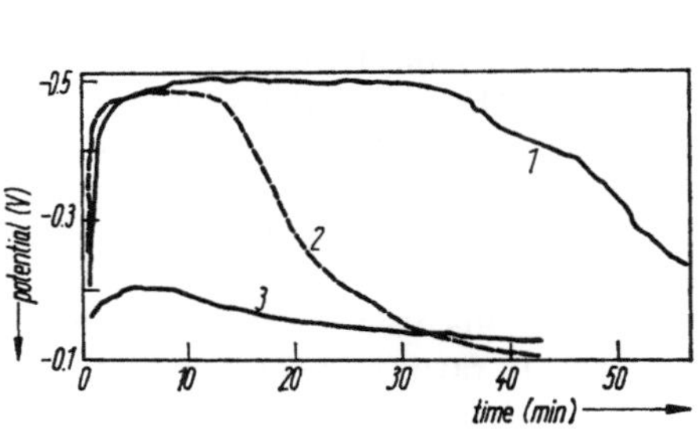

Fig. 7

Fig. 6. The rest potential behaviour of the In-implanted Si (p-type, 10 Ω cm). Redox agent: 5×10^{-3} M $K_3Fe(CN)_5$. (1) 10^{15} In/cm², at 12.2 keV, (2) 10^{14} In/cm², at 12.2 keV, (3) 10^{12} In/cm² at 12.2 keV

Fig. 7. The rest potential behaviour of the As-implanted Si (n-type, 270 Ω cm). Redox agent: 5×10^{-3} M $K_3Fe(CN)_6$. (1) 10^{15} As/cm², at 30 keV, (2) 10^{14} As/cm², at 30 keV, (3) 10^{13} As/cm², at 30 keV standard potential the cathodic reaction proceeds via the valence band resulting in a hole injection

$$Fe(CN)_6^{3-} \rightarrow Fe(CN)_6^{4-} + p\text{: cathodic}\,. \tag{1}$$

The anodic reaction is increased due to this hole supply

$$Si + 2p \rightarrow Si^{2+}\text{: anodic} \tag{2}$$

similar to the increase of the anodic dissolution by illumination. The rest potential adjusts to fulfil the condition of the zero net current. The consecutive character of the corrosion reactions (1) and (2) is more obvious with increasing n-type conductivity. For a given redox agent concentration the rest potential shifts therefore to positive values with increasing p-type conductivity of the Si. This dependence can be observed for p-Si. For n-Si the rest potential as a function of the conductivity indicates significant scattering [3, 5]. This conception is, obvious applicable only to the substrate material since dopants such as Si, Ar which are electrically inactive exhibit the same features in the rest potential behaviour as the electrically active dopants P, As, In. This fact suggests that the change in the anodic partial reaction determines the rest potential of the implanted Si surface. The crystal damage introduced by ion implantarion presents an efficient hole supply [1, 2] and thus increases the anodic partial curtent. Indeed, it is characteristic of all ions used that the corrosion of the implanted layer is always greater than that of the substrate. To compensate this increased anodic current component the rest potential must drift to the negative direction. By simply assuming that the anodic current is a function of the damage density the time dependence of the rest potential may be considered as an indication for the damage distribution. During the course of the corrosion the implanted layer is stripped successively and the rest potential falls to the value of the substrate. The time scale can be converted into the depth scale as demonstrated in Fig. 3 and 5 by measuring simultaneously the time dependence of the corrosion.

Fig. 7 shows the dose dependence of the rest potential. Saturation is clearly seen to cause a widening of the high dose curves. The lower dose limit appears to be around 10^{13} cm^{-2}. No significant visible change of the rest potential occurs at doses smaller than this value. This may be due to the different contributions of the hole supply to the anodic reaction by damage and corrosion. The contribution of the corrosion increases with decreasing dose (decreasing damage density) and at sufficiently low doses the anodic reaction is completely controlled by corrosion injected holes.

The influence of the thermal treatment after implantation is illustrated in Fig. 3 for samples annealed in a N_2-atmosphere for 20 min. The reduction of the rest potential effect suggests strongly, in addition to the similar feature of the rest potential curves for various ions, that the measuring effect correlates with the implantation damage.

In contrast to all ions used the B-implanted layer shows increased passivity. A reduction in the corrosion rate (in reference to the substrate) was found. This result is consistent with the rest potential which shows considerably positive values as indicated in Fig. 4. The widening of the curve is due to the low stripping rate. In addition, the penetration depth of the boron is larger than that of the heavy ions (at the same energy). The example of boron implantation demonstrates clearly that beside the crystal damage the kind of ion must be taken into consideration, at least in the high dose region. Two effects may be responsible for the behaviour of B-implanted layers: the compound formation and the precipitation of B excess. The latter seems to be significant due to the observation [5] that a platinum substrate after receiving a high implantation dose of B shows also increasing passivity.

4. Summary

It is demonstrated that the ion implantation changes considerably the corrosion of the Si-surface. The effect is attributed to the bombardment-induced damage and shows dependence on dose and kind of the ions, the implantation energy, and the annealing. The measuring of the rest potential appears to be useful in both the process control, for example in profiling of the implanted layer, and the investigation of effects related with ion bombardment. The method is simple and requires no more equipment than that indicated in Fig. 1. The exact knowledge of the depth scale requires, however, an independent measurement of the corrosion as a function of the layer etching.[2]) For qualitative monitoring purposes the rest potential–time curve offers a quick information. The preliminary results suggest a lower limit of the measurable ion dose of about 10^{13} cm^{-2}.

Acknowledgements

The author thanks Mr. Schoenreich and Mr. Schneider for performing the ion implantation.

References

[1] M. T. Pham and J. Hüller, J appl. Electrochem. 7, 531 (1977).
[2] J. Hüller and M. T. Pham, phys. stat. sol. (a) **43**, 505 (1977).
[3] D. R. Turner, J. electrochem. Soc. **103**, 252 (1956).
[4] H. Gerischer and F. Beck, Z. phys. Chem., NF **23**, 113 (1960).
[5] M. T. Pham, Laboratory report, 1977 (unpublished).
[6] M. T. Pham, Laboratory report, to be published.

(Received June 9, 1978)

[2]) The most reliable method appears to be the radiochemical one. The electrochemical dissolution by anodic polarization is, probably, not suitable since no unambiguous correlation between the anodic electricity and the resulting dissolved Si-amount seems to exist in the high dose range [6].

phys. stat. sol. (a) 49, 267 (1978)

Subject classification: 1.4; 1.5; 21.6

Institut d'Optique, Laboratoire associé au C.N.R.S., Centre Universitaire d'Orsay[1])

Study by Electron Microscopy of Some Twinned Gold Crystals with Variations in Thickness

By

K. Abdelmoula, B. Pardo, and D. Renard

Some large gold crystals obtained by epitaxy on alkali halide crystals (KCl) are studied by electron microscopy. Diffraction patterns as well as bright and dark field micrographs interpreted by dynamical calculations reveal thickness faults and some well known twins. But stacking faults appear to be randomly distributed in the thin film.

On a étudié par microscopie électronique en transmission à 100 kV, des grands monocristaux d'or obtenus par épitaxie sur un halogénure alcalin (KCl). L'interprétation dynamique des images obtenues en fond clair ou en fond noir, et des diagrammes de diffraction a montré d'une part la présence de défauts d'épaisseur, et d'autre part a confirmé la présence bien connue de macles; mais elle semble indiquer que les fautes d'empilement pouvant accompagner ces macles, sont relativement peu nombreuses.

1. Presentation of Samples

1.1 Preparation

The thin gold film studied here was obtained by evaporation in vacuum and by condensation on (100) faces of a KCl monocrystal heated to 200 °C. By this process, large gold monocrystals are prepared which have lateral dimensions greater than 100 microns, they are slightly curved and [001] axis is perpendicular to the surface.

1.2 Thickness

The mean thickness of the film is determined by the frequency variation of a piezoelectric quartz whose error of measurement is less than 5%. For the monocrystals studied in this paper the mean thickness is 450 Å. All the same it should be noted that, although the X-ray diagrams clearly showed the expected structure, it has not been possible to obtain fringes on the main "lines" (rays). This leaves us to suppose that variations of thickness of at least 50 Å exist inside the area radiated by the X-rays.

1.3 Presence of twins

The study by electronic diffraction of f.c.c. metals deposited on alkali halide monocrystals has already been carried out by numerous authors [1 to 4]. In addition to the normal spots of diffraction due to the 001 orientation of the metal, extra spots exist. Some of these spots are easily explained by the presence of microtwins inside the crystal. These microtwins are produced according to the {111} planes of the crystal and are therefore of four types (twins $1\bar{1}1$, twins $\bar{1}11$, twins $11\bar{1}$, and twins 111). The reciprocal lattice of each of these twins can be obtained from the reciprocal lattice of the normal crystal by a rotation of 180° round the $\langle 111 \rangle$ axis of the twin. Thus

[1]) Bât. 503, B.P. n° 43, 91406 Orsay Cedex, France.

a point (p, q, r) of the normal lattice becomes

$$\left(\frac{1}{3h}[-hp + 2kq + 2lr], \frac{1}{3k}[2kp - kq + 2lr], \frac{1}{3l}[2hp \pm 2qk - rl]\right),$$

where (hkl) is the $\{111\}$ plane of twinning. For example, for the 111 type twin, the spot $(\bar{1}11)$ becomes $\frac{1}{3}$ $(5\bar{1}\bar{1})$ and the spot (200) becomes $\frac{1}{3}$ $(\bar{2}44)$.

The twins are thus directly responsible for the extra spots, some having fractional indices. Furthermore, other spots present in the diagrams do not correspond either to the normal lattice or to the lattice of the twin. They can be explained by the effects of double diffraction when a diffracted beam of the normal crystal undergoes a further diffraction in the twin (or vice versa) [3, 4].

2. Study of the Local Variations of Thickness in the Non-Twinned Zones of the Crystal

2.1 Orientation of a crystal during studies in bright field and in dark field

In a previous article [5] it was shown that the study of images carried out with the beams $+g$ and $-g$ was productive for the orientations where the crystal should be in Bragg position for a "node" ng; as long as n was large enough. There is then interference between two fields of waves corresponding to two branches of the surface of dispersion which are reasonably far from each other; the result is pulsations for a small period in real space, from which we get, in the picture, the presence of very dense fringes of "isothickness". This study can be particularly well applied to crystals whose geometric form is close to the shape of a gutter, furthermore it should be noted that a zone of crystal in the Bragg position for a "node" ng of the reciprocal lattice is found in the Bragg position for a "node" $(n \pm 2)\, g$ provided that one passes from an image in a bright field to an image in a dark field with the $\pm g$ beam. In effect the movement to the dark field $+g$ or $-g$ is made by tilting the electronic beam by an angle of $2\theta g$ in one direction or another. The crystal studied below can be compared to a gutter with a very large curve. For the image in a bright field the photographed zone is found in the Bragg position for $3g$. It will therefore be in the Bragg position for g on Fig. 1a taken with the $-g$ beam; and it will be in Bragg position for $5g$ on Fig. 1b taken with the beam $+g$. The diffraction presented on Fig. 1c corresponds to Fig. 1a. One can see on the systematic line two spots evidently of equal importance which correspond to the incident beam and to the $+g$ beam; then one can see weaker spots which correspond to beams $-g$ and $2g$ and, finally, lateral spots of weak intensity which are caused by the twins. In this study g is the "node" of the reciprocal lattice indexed 200.

2.2 Theoretical study of the three cases above

We have used the notation which corresponds to the matrix relation of Fujimoto [6].

$$\{M\}\,\{C\} = 2K_z\,\Delta K_z\{C\} = \gamma\{C\}\,,$$

where $\{M\}$ is the matrix of diffusion, $\{C\}$ the collection of wave fields, and $2K_z\,\Delta K_z$ the eigenvalues being looked for. The calculations of dynamic theory are carried out with 15 waves. The matrix $\{M\}$ has a real part made up by the twinning terms U_g and an imaginary part made up by the terms U_g' which take the inelastic diffusions into account. The twinning terms are obtained from calculations of atomic scattering factors carried out by Doyle and Turner [7]. The imaginary terms have been determined by experiment [8]. The characteristics of the principle wave fields are indicated in Table 1 below for situations corresponding to our photos. A_g^γ is the amplitude of

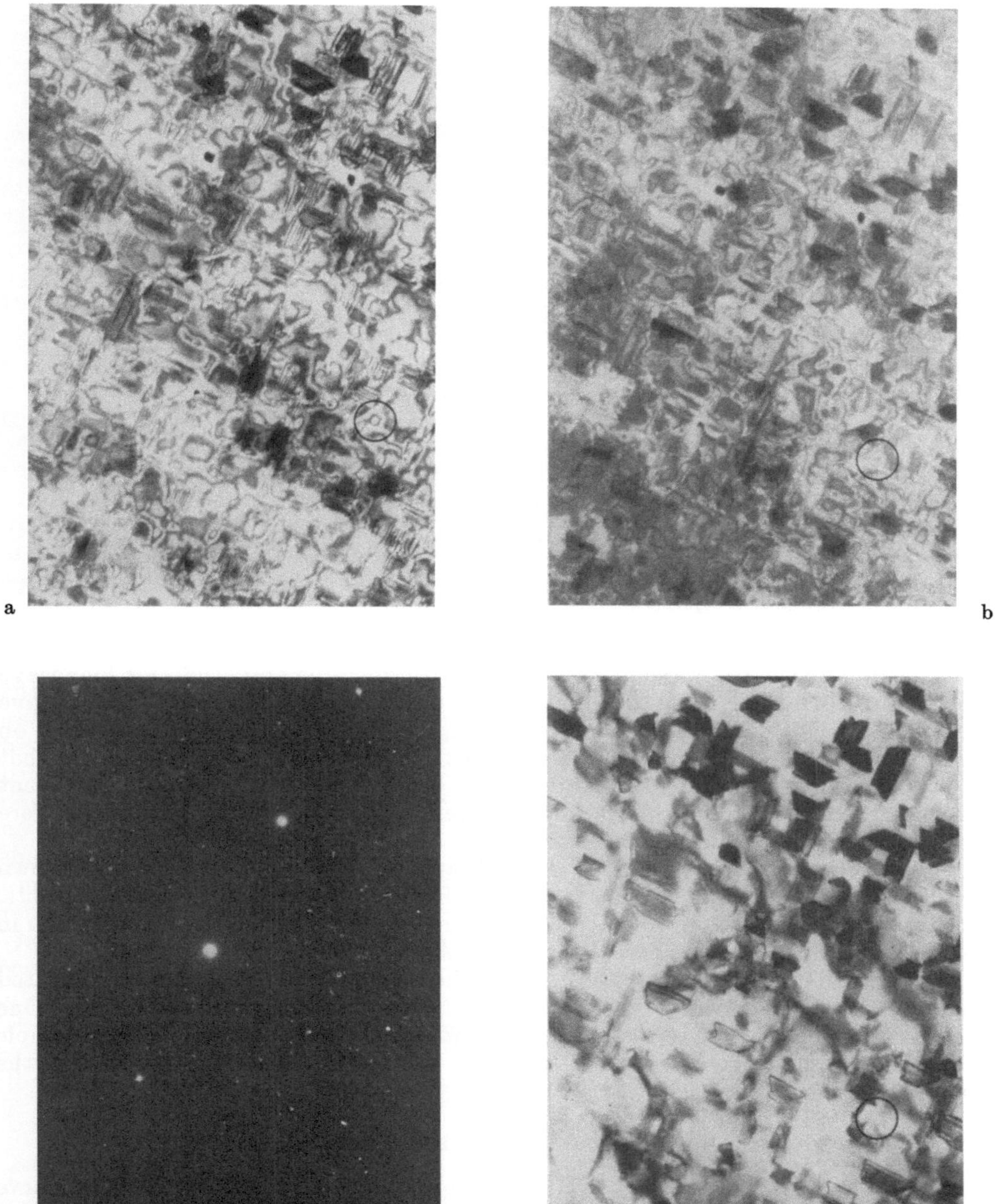

Fig. 1. a) Image in dark field $-g$ when g is in Bragg position, b) image in dark field $+g$ when $5g$ is in Bragg position, c) diffraction corresponding to Fig. 1a, d) image in bright field

Table 1

g in Bragg position dark field on the wave $-g$
ΔK is expressed as 10^{-3} A^{-1} and μ as Å^{-1}

$A^1_{-g} = 0.137$	$\Delta K^1 = 3.54$	$\mu^1 = 0.015$
$A^2_{-g} = 0.091$	$\Delta K^2 = -2.11$	$\mu^2 = 0.001$
$A^5_{-g} = -0.139$	$\Delta K^5 = -9.31$	$\mu^5 = 0.004$
$A^6_{-g} = -0.092$	$\Delta K^6 = -9.61$	$\mu^6 = 0.003$

$3g$ in Bragg position
incident wave

$A^3_1 = 0.4$	$\Delta K^3 = -0.2$	$\mu^3 = 0.004$
$A^4_1 = 0.53$	$\Delta K^4 = -0.5$	$\mu^4 = 0.003$

$5g$ in Bragg position
dark field on the wave $+g$

$A^3_g = 0.054$	$\Delta K^3 = 17.9$	$\mu^3 = 0.005$
$A^4_g = 0.054$	$\Delta K^4 = 17.6$	$\mu^4 = 0.004$
$A^6_g = -0.060$	$\Delta K^6 \cong 0$	$\mu^6 = 0.005$
$A^7_g = -0.063$	$\Delta K^7 \simeq -2.2$	$\mu^7 = 0.004$

the wave g in the wave field γ, ΔK^γ is the real eigenvalue which corresponds to this wave field, and μ^γ is its absorption coefficient.

One can see that the wave g (or $-g$) has a considerable amplitude for four wave fields. In reality these four wave fields group themselves in two wave fields to give interferences. For example, when $5g$ is in Bragg position the wave fields 3 and 4 on the one hand and 6 and 7 on the other have extremely similar characteristics, so that one can consider that there will be interference between two wave fields which present between them a difference of ΔK of the order of 20×10^{-3} Å. One must therefore expect, in real space, pulsations with a period of approximately $1/\Delta K$ or 50 Å.

For the incident beam when $3g$ is in Bragg position, we have the well-known results: two wave fields are both excited with slightly different absorption coefficients. Correspondingly their ΔK are very close to one another so that the interferences in real space will have a large period (here 3000 Å).

These results are summarised on Fig. 2 where the correlation of the transmitted intensity to the incident intensity is plotted as a function of the thickness. One clearly finds the expected results simply by studying the characteristics of the principle wave fields: the curve in a dark field $+g$, when $5g$ is in Bragg position, is the most interesting one.

2.3 Comparison with the experimental results

The curve 2a is associated to Fig. 1a, the curve 2b to Fig. 1b. On the first curve the period of the oscillations is about 140 Å or more than twice as large as the period of the oscillations on the second curve. This result can be found on the photos where the examination of a given detail shows twice as many fringes in one case than in the other. Thus the comparison between curves and photographs clearly shows the local variations in the thickness of the crystal of gold in the non-twinned zones. These variations can be evaluated from the curves. On Fig. 1b, one frequently sees two

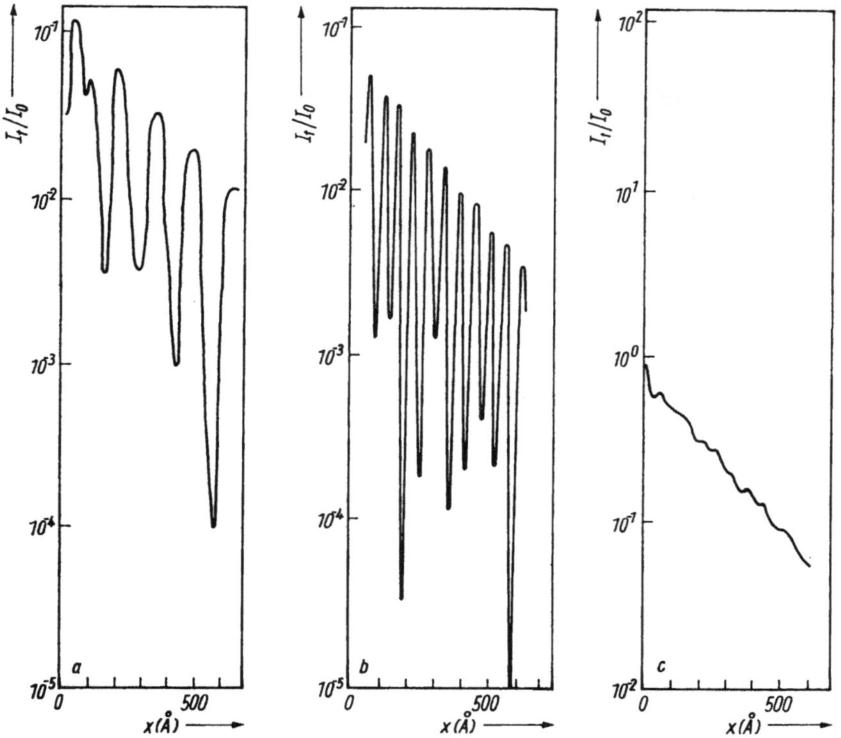

Fig. 2. Relation of the transmitted intensity to the indicent intensity as a function of the thickness of the gold crystal for various orientations (ordered logarithmic scale). a) Image in dark field $-g$ when g is in Bragg position; b) in dark field $+g$ when $5g$ is in Bragg condition; c) in bright field when $3g$ is in Bragg condition

white fringes and two black fringes inside a zone whose lateral dimension is not greater than 500 Å (see diagram). This corresponds to a variation in thickness of at least 100 Å. Under these conditions the absence of fringes on the X-ray diffraction pattern is entirely natural. Fig. 2c corresponds to the case of a bright field and shows a regular decrease of the intensity transmitted with very weak oscillations due to the presence of the most adsorbed wave field. The image in a bright field cannot, therefore, give precise information about the thickness of the crystal (Fig. 1d).

2.3.1 Remark

The quantitative study of the intensity transmitted has not been attempted here because it implies that the incident flux remains constant when one passes from one dark field to another. This has not been verified on our microscope. All the same, we have been able to verity that the orders of size found were correct. In the case of Fig. 1b the intensity transmitted for a thickness close to 450 Å is in the order of some thousandths of the incident intensity and in the order of one or two hundredths for Fig. 1a. Table 1 indicates clearly ($A_g^\gamma \approx 0.06$ and $A_{-g}^\gamma \approx 0.13$). The experimental study in a dark field necessitates, therefore, important exposure time and this certainly limits the precision of this method. In fact, the fringes of equal thickness obtained are all the more dense and the method all the more precise when the order n of the "node" in Bragg position is higher; but then the intensities transmitted decrease rapidly when n increases. We are thus led to make a compromise, experimentally it has not been possible for us to obtain correct photos with high indices "nodes" in Bragg position.

3. Study of the Contrast in the Twinned Areas

The contrast in the twinned areas depends, among other things, on the thickness of the non-twinned zones of the crystal, on the thickness of the zones in the twin position, and on the presence or absence of a stacking fault.

3.1 Characteristics of the twins observed

The angle between the planes {100} and the planes {111} being $\alpha = 54.73$ (Fig. 3) the size x of the activated zone in the image must be

$$x = (z + z_B)\,\gamma/\mathrm{tg}\,\alpha\,.$$

In this expression z is the total thickness of the crystal (450 Å), z_B is the thickness of the twin counted parallel to the beam, γ is the enlargement of the photo.

The observation of our photos shows then the presence of three characteristic thicknesses. Firstly the very fine twins for which the determination of thickness is very imprecise, in the calculations we have chosen to represent them by a thickness of 20 Å which is roughly equal to $\xi_{200}/10$. There are then the twins whose thickness is in the region of 100 Å ($\xi_{200}/2$) and then the twins of about 200 Å (ξ_{200}). We have no samples with very thick twins. Under these conditions the electronic beam crosses three different zones.

In zones I and III the beam crosses two crystals of different orientation. In zone II it crosses, in succession, the crystal in position 100, the crystal in twin position, and again the crystal in position 100. The twinned part may or may not be accompanied by a stacking fault which introduces a phase difference. For the twins of 20 Å, we have neglected parts I and III in the calculations.

3.2 The dynamic calculation

To deal with the crystal and its twin in the same dynamic calculation, it suffices to take as a lattice common to both the crystal and the twin, the lattice constructed on the three base vectors 300, 030, 003. This allows the diffractions by the planes $\frac{1}{3}(h, k, l)$ with the following selection rules:

(i) For the non-twinned crystal, h, k and l, are multiples of 3 and have the same parity.

(ii) For the twinned crystal, h, k and l have, similarly, the same parity and their sum is a multiple of 3.

When a stacking fault is present one must multiply the coupling terms $U_{g-g'}$ of the perfect crystal by $\exp\,[2\pi i\boldsymbol{R}(\boldsymbol{g} - \boldsymbol{g}')]$ where $\boldsymbol{R}$ is the displacement caused by the

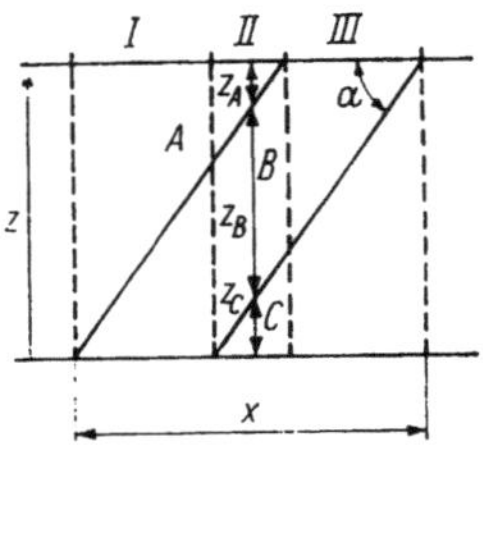

Fig. 3. Scheme of the characteristic of the twins observed (for further explanations see the text)

stacking fault [9 to 11]. The calculation is then continued in the following way: one determines the matrices of diffusion M_A, M_B, M_C, relative to the parts A, B, C of the crystal, then calculates their eigenvalues, γ_A, γ_B, γ_C, and constructs the diagonal matrices D_A, D_B, D_C so that the term i of the matrix D_α is written:

$$(D_\alpha)_i = \exp 2\pi i\left(K + \frac{\gamma_{\alpha i}}{2K}\right) z_\alpha$$

with α = A, B, or C.

One then chooses the conditions of incidence represented by a column vector $\boldsymbol{V}_0$ whose the first component is equal to one, the others being zero. The amplitude of the N waves transmitted or diffracted are the components of a vector $\boldsymbol{V}$ which is denoted by the following expression:

$$\boldsymbol{V} = C_C D_C \tilde{C}_C C_B D_B \tilde{C}_B C_A D_A \tilde{C}_A V_0 \,.$$

If there is no stacking fault, C_C is identical to C_A.

3.3 Results of the dynamic calculation and comparison with experiment

The number of waves used in the calculation was suggested by us from the diagrams of diffraction corresponding to the photos. It turned out that ten waves were enough to deal with all the experimental cases studied. These are the waves 400, $\bar{2}00$, the incident wave, the waves 200, 400, 600, 800, 10.000 for the non-twinned crystal and the waves $\pm\frac{1}{3}(151)$ for a twin $11\bar{1}$.

Two different orientations are presented here; wave (200) in Bragg position and wave (10 0 0) in Bragg position. According to the cases, we are interested in the directly transmitted beams, the beam $+g$ (g = 200), and the beam $-g$.

The results of the calculations are presented in Fig. 4, 5, and 6 in the form of curves.

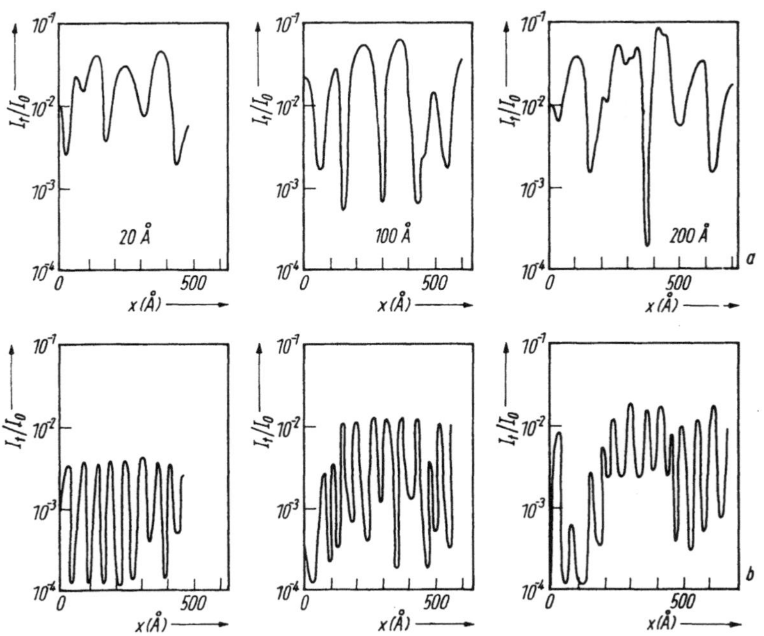

Fig. 4. Relation of the transmitted intensity to the incident intensity as a function of the thickness in the case of twins for various orientations. a) Dark field $-g$ when g is in Bragg position, b) dark field $+g$ when $5g$ is in Bragg position

The ordinate is logarithmic and gives the ratio I_t/I_0 of the transmitted intensity to the incident intensity, the abscissa indicates the value of x previously introduced.

Fig. 4a concerns the dark field obtained from beam $-g$ when the reflexion g is in Bragg position. Fig. 4b concerns the beam g when $5g$ is in Bragg position. These curves correspond to the twins in Fig. 1a and 1b. They are obtained without introduction of a stacking fault; a stacking fault hardly modifies the shape of the curves. It should also be noted here that there is a double interfringe between 4a and 4b which, furthermore, is identical to the one already obtained on curves 2a and b. It is thus clearly a question of fringes with equal depth of the defect. On Fig. 1a and 1b there are, of course, twins of all thicknesses (up to about 200 Å) but the number of fringes is clearly twice as large in one photo as in the other.

Fig. 5 provides the results for the beam directly transmitted (in a bright field) when g is in Bragg position. The results here are much varied and depend strongly on the

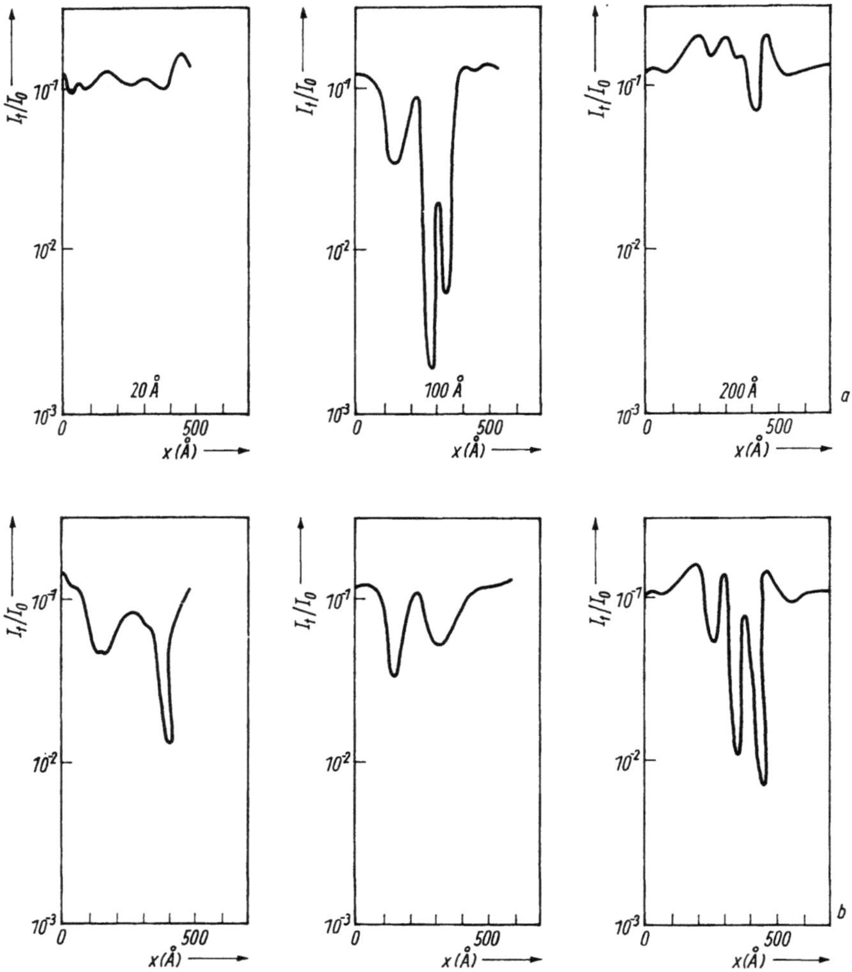

Fig. 5. Relation of I_t/I_0 as a function of the thickness in the case of twins for various orientations (bright field when g is in Bragg position). a) No introduction of stacking fault, b) introduction of stacking fault

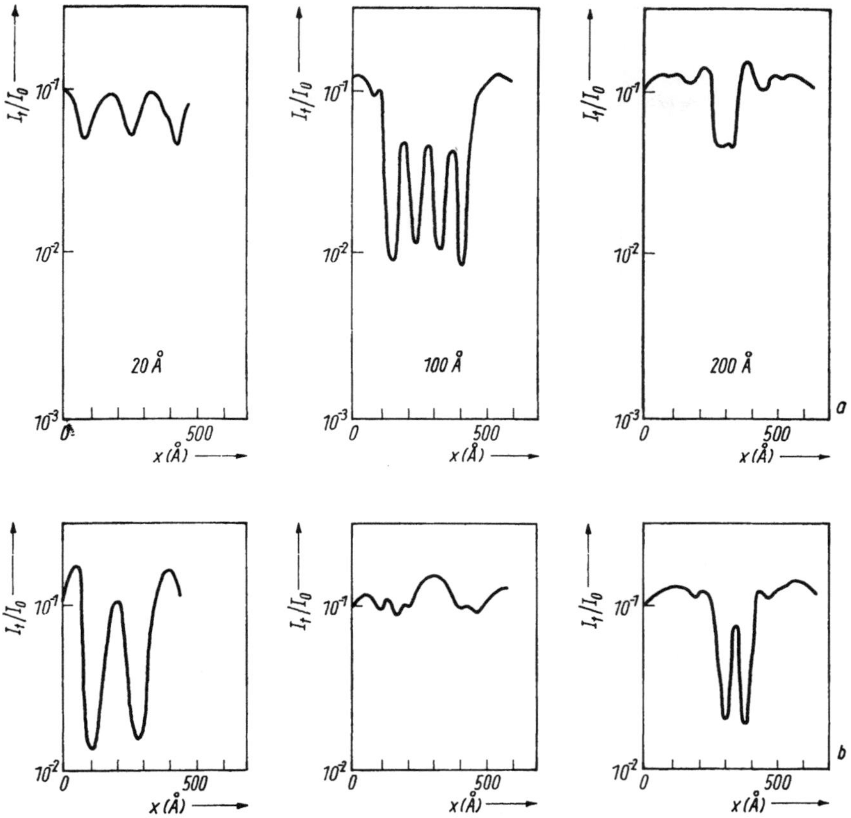

Fig. 6. Relation of I_t/I_0 as a function of the thickness in the case of twins for various orientations (dark field $+g$ when g is in Bragg position). a) No introduction of stacking fault, b) introduction of stacking fault

thickness of the twin. When the stacking fault is not introduced one notices three contrasting fringes for the fine twin, four contrasting fringes for the medium twin, and a single black fringe for the thick twin. The introduction of the stacking fault leads to the presence of two fringes of medium contrast for the twins 20 and 200 Å. On the other hand, it also vastly diminishes the contrast in the fringes in the case of the twins of 100 Å. Thus for the three cases studied (20, 100, and 200 Å), the presence of a stacking fault is made clear by the presence of two dark fringes or by the absence of a fringe.

The comparison with the photos is somewhat delicate because we do not know the thickness of the twins to a sufficient degree of exactness. The size measured on the photos allows us to distinguish fine, medium, or thick twins, but nothing more. Since the values chosen for the calculations are arbitrary, there certainly exist in the crystals twins of intermediate thickness which do not correspond to the theoretical cases studied. The observation of numerous twins on the photos in bright field has nonetheless frequently shown the presence of three or four fringes. Sometimes, scarcely contrasting twins or twins with two fringes do occur, but these are fairly rare cases as one can see on Fig. 7. This suggests that in most cases the twin is formed without stacking fault.

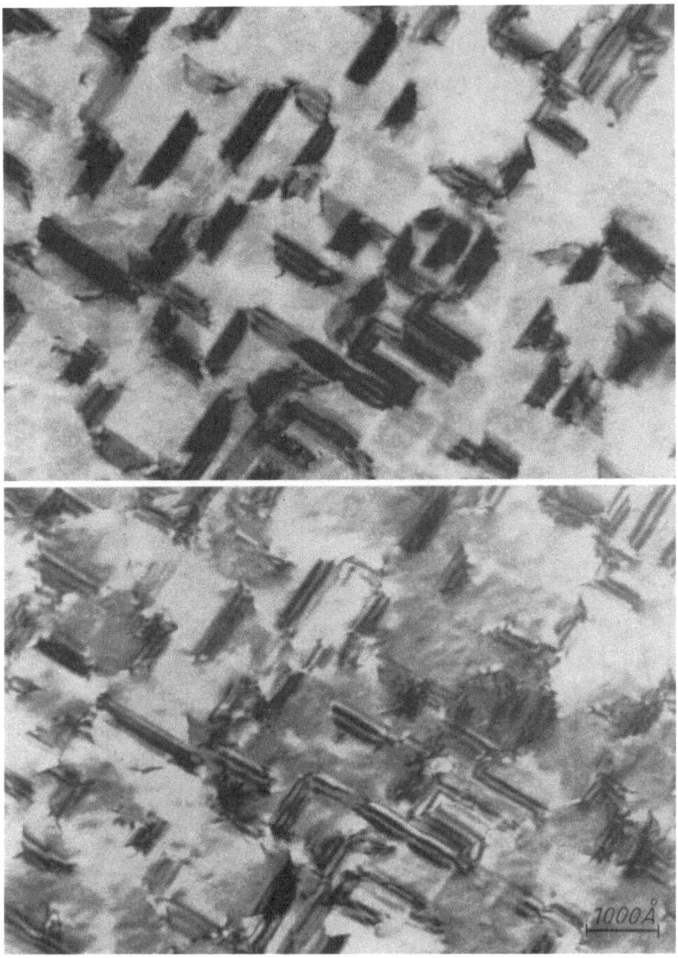

Fig. 7 Appearance of the twins in a) bright field and b) dark field (g is in Bragg position)

Fig. 6 relates to the image in a dark field obtained with the beam g when the corresponding reflection (here 200) is in Bragg position. Here the system of fringes is asymmetric in relation to the centre of the defect. The effect of the stacking feault is to reduce still further the contrast in the fringes for the medium twin and to create one for the twins 20 and 200 Å. The corresponding photo (7b) very often reveals the profile corresponding to a twin of 100 Å without stacking fault or the profile which corresponds to a twin of 200 Å with a stacking fault. The dark fields do not, therefore, supply supplementary information in this precise case.

It becomes evident from this study that the determination of the thickness of the twins is reasonably easy from the images in dark field presented on Fig. 1 because the interfringe is sufficiently small. The images in bright field obtained when a beam is in Bragg position, allow us to suppose that there are hardly any stacking faults. The study in a "classic" dark field would demand a quantitative comparison whether it can provide any sort of information.

4. Fine Structure of the Diffraction Patterns

The study carried out in the preceding section has shown that, when a crystal presents a microtwin, the images in dark field and bright field admit the presence of parallel fringes on the edge of the twin. This results from the fact that the amplitudes Ag($\boldsymbol{r}$) of the different electronic waves depend on the point of emergence. Fourier transformations of these amplitudes, $A_g(\boldsymbol{K})$ do not have the $\delta(\boldsymbol{K} - \boldsymbol{g})$ shape which is explained by the appearance of a fine structure in the diffraction spots. Each of these spots is split into one intense spot elongated in the direction normal to the fringes and into secondary spots which are also situated in this direction. Gevers and his collaborators have studied this fine structure in the case of a stacking fault by using two waves [12, 13]. Their study allows a qualitative explanation of the appearance of our diagrams.

4.1 The geometric aspect of the problem

In the case of gold crystals the faces of entrance and exit are parallel to (001). If $(\alpha\beta\gamma)$ is the plane of the twin or that of the stacking fault, its intersection with the plane of the faces is parallel to the direction $\langle\beta\bar{\alpha}0\rangle$ which is the direction of the fringes in the twins or the flat faults. The elongation of the spots will therefore be perpendicular to the direction $\langle\beta\bar{\alpha}0\rangle$. When the crystal is not tilted in the microscope, the diffraction pattern is a plane of the reciprocal lattice almost perpendicular to the axis [001]. It follows that the elongation of the spots of diffraction is made in the direction $\langle\alpha\beta0\rangle^*$ of the reciprocal lattice. Thus for the twins 111 and $11\bar{1}$, the elongation is parallel to the direction $\langle110\rangle^*$ and for the twins $1\bar{1}1$ and $1\bar{1}\bar{1}$, the elongation of the spots is parallel to the direction $\langle1\bar{1}0\rangle^*$. This is illustrated by Fig. 8a.

4.2 Appearance of the spots of diffraction

Gevers and his collaborators have shown that when the orientation of the crystal is close to the Bragg orientation, the satellites are very close to the normal spot and have the same intensity. When at a large distance from the Bragg orientation one of

Fig. 8 Appearance of the diagrams of diffraction, a) in the case of two important reflections, b) in the case of an orientation of high symmetry

the satellites becomes more intense at the expense of the other, the two satellites move away from the position of the normal spot whose intensity decreases, of course, as one moves the Bragg reflection further and further away.

Fig. 8b corresponds to a selected area centred on a zone of the crystal for which the axis [001] is strictly parallel to the beam. The orientation of each of the spots surrounding the central spot is very close to the Bragg orientation thus making the satellites very close to one another. Taking into account the exposition time the satellites are not separated and the spot has the appearance of a square. The diagnoal of this square is evidently in the region of $2 \operatorname{tg} \alpha/t$ where α is the angle of the twin plane, here equal to 54.73°, and t is the extinction distance in a two-wave theory. On Fig. 8a the beam 200 is almost in Bragg position, so that divergence to Bragg position for the other beams is relatively important. The spots are now separated and the elongation can be seen clearly. The normal spot is much less intense. Thus the appearence of the patterns can be qualitatively interpreted from the results obtained by Gevers and his colleagues. Nonetheless, certain difficulties have still to be resolved and a quantitative study will be undertaken. One must take into account the real number of waves in each case and here the faults are twins and not simples stacking faults. The dynamic calculation which we used in the preceding section is suitable for this study since Fourier transformation of diffracted amplitudes leads to the intensities of the spots of diffraction. The result of these calculations will be published later.

5. Conclusion

The epitaxy of gold on a monocrystal of KCl allows us to obtain very large monocrystals of gold with an interesting orientation. Nonetheless the study in electron microscopy has shown a certain number of faults:

1. The crystal presents numerous twins; but, from the comparison of the images (in dark and bright fields) and of the results of the calculations, we can conclude that there are few stacking faults. Furthermore, the thickness of the twins is always weak. The fringes observed in the twins are interpreted by dynamic calculation; they are directly responsible for the fine structure of the spots of diffraction.

2. The thickness of the monocrystal is not regular. This is shown by the contrast in the dark field images using a weak beam. There again, dynamic calculation shows the importance of the variation in thickness: 100 Å for a crystal with a mean thickness of 450 Å. This type of fault may depend on the technique of cleavage used. In this case it could without doubt be eliminated.

References

[1] H. LASSEN, Phys. Z., **35**, 172 (1934).
[2] H. LASSEN and L. BRUCK, Ann. Phys. (Lpz.) **22**, 65 (1935).
[3] H. R. THIRSK and E. J. WHITMORE, Trans. Faraday Soc. **36**, 565 (1940).
[4] D. W. PASHLEY and M. J. STOWELL, Phil. Mag. 8, 1605 (1963).
[5] B. PARDO and D. RENARD, phys. stat. sol. (a) **35**, 329 (1976).
[6] F. FUJIMOTO, J. Phys. Soc. Japan **14**, 1558 (1959).
[7] P. A. DOYLE and P. S. TURNER, Acta cryst. **A 24**, 390 (1968).
[8] D. RENARD, Thesis Orsay n° AO 6877 (1973).
[9] P. B. HIRSCH, A. HOWIE, R. B. NICHOLSON, D. W. PASHLEK, and M. J. WHELAN, Electron Microscopy of Thin Crystals, London 1965.
[10] J. VAN LANDUYT, R. GEVERS, and S. AMELINCKX, phys. stat. sol. **9**, 135 (1965).
[11] G. REMAUT, R. GEVERS, A. LAGASSE, and S. AMELINCKX, phys. stat. sol. **10**, 121 (1965).
[12] R. GEVERS, J. VAN LANDUYT, and S. AMELINCKX, phys. stat. sol. **18**, 343 (1966).
[13] J. VAN LANDUYT, R. GEVERS, and S. AMELINCKX, phys. stat. sol. **18**, 363 (1966).

(Received March 9, 1978)

phys. stat. sol. (a) **49**, 279 (1978)

Subject classification: 9 and 11; 22.2.4

VEB Werk für Fernsehelektronik, Berlin

Investigations on the Diffusion of Implanted Zinc in $GaAs_{1-x}P_x$ by Ion Microprobe

By

G. MÜLLER, M. HAUBOLD, R. SCHIMKO, M. TRAPP, and G. SCHWARZ

Zinc is implanted with an energy of 100 keV and a dose of 4×10^{15} cm^{-2} in $GaAs_{1-x}P_x$ VPE layers. The zinc concentration distribution after annealing at temperatures between 650 and 850 °C is measured by the ion microprobe. Zinc diffuses from the implanted source redistributed at the beginning of the annealing with a maximum concentration essentially below the maximum solubility of zinc in GaAsP. In the temperature range of 650 to 750 °C the implanted layer acts as an infinite source. The diffusion behaviour at 850 °C is analogous to the diffusion from a limited source. The diffusion coefficient D depends on concentration and location.

Zink wird mit einer Energie von 100 keV und einer Dosis von 4×10^{15} cm^{-2} in $GaAs_{1-x}P_x$-VPE-Schichten implantiert. Der Zink-Konzentrationsverlauf wird nach der Temperung bei Temperaturen von 650 bis 850 °C mit der Ionensonde gemessen. Zink diffundiert aus der zu Beginn des Temperprozesses umverteilten implantierten Quelle mit einerMaximalkonzentration wesentlich unterhalb der maximalen Löslichkeit von Zink in GaAsP. Im Temperaturbereich von 650 bis 750 °C wirkt die implantierte Schicht als unbegrenzte Quelle. Das Diffusionsverhalten bei 850 °C ist analog zur Diffusion aus einer begrenzten Quelle. Der Diffusionskoeffizient D ist konzentrations- und ortsabhängig.

1. Introduction

$GaAs_{1-x}P_x$ vapour phase epitaxial layers are widely used for the production of light-emitting diodes.

Zinc ion implantation in $GaAs_{1-x}P_x$ has been demonstrated to produce p-type layers with good electrical activation and efficient luminescence activity after proper annealing [1, 2].

In the last few years this subject was dealt with in a number of works which predominantly were concentrated on the annealing behaviour and the production of efficient LEDs [3, 4]. The results available for the annealing behaviour of implanted zinc in GaAsP were mostly obtained by means of electrical and optical measurements [1, 4, 5].

The mechanism of the diffusion from implanted layers in $GaAs_{1-x}P_x$, the causes for the variations in the results of the various authors, and the cause for the generation of the "double profiles" up to now could not be discovered completely.

The present work deals with the measurement of implanted zinc profiles in GaAsP after annealing at temperatures between 650 and 850 °C by ion microprobe.

The profile distribution is discussed and explanations on the influence of the surface-near $GaAs_{1-x}P_x$ layer on the diffusion process are given.

2. Experimental

$GaAs_{1-x}P_x$ vapour phase epitaxial layers on ⟨100⟩ GaAs substrates were used for the investigations.

Zinc ions with an energy of 100 keV and a dose of 4×10^{15} cm^{-2} were implanted at room temperature (RT) at an angle of 7° to the ⟨100⟩ direction.

Prior to annealing the $GaAs_{1-x}P_x$ surface was coated with a 3000 Å thick r.f.

sputtered layer of SiO_2 in order to prevent out-diffusion of implanted zinc and thermal dissociation of the $GaAs_{1-x}P_x$ [6, 7]. The annealing of the zinc-implanted GaAsP layer was carried out at temperatures between 650 and 850 °C during a period of 10 to 300 min in a hydrogen atmosphere. The zinc concentration profiles were measured by a secondary ion beam microanalyzer (SMI 300, Cameca) [8].

3. Experimental Results

Fig. 1, 2, and 3 show the zinc concentration profiles after implantation with 100 keV and a dose of 4×10^{15} cm^{-2} in $GaAs_{1-x}P_x$ at RT and after annealing at temperatures between 650 and 850 °C, for 10 to 300 min. For the sake of a clearer sight the surface-near increase of the zinc concentration in the diffused profiles is not shown.

The measured chemical concentration profiles show good agreement with the charge carrier concentration profiles determined by Hall measurements [1, 5].

As we have shown in [9] during the first minutes of annealing a strong redistribution of the implanted zinc occurs. The zinc is attracted by the SiO_2–GaAsP interface, where it is partly bound to defects and gallium vacancies and diffused by further annealing from this layer into the bulk. The shape of the profiles generated after annealing is analogous to the "double profile" described by Tuck and Kadhim [10] consisting of a shallow profile with high surface concentration superimposed on a deeper profile with a lower surface concentration. The two regions can be approximated by a Gaussian region and an exponential tail.

The profiles annealed of various temperatures are mainly different with regard to the exponential tail. While at 850 °C this tail already appears after an annealing of 10 min, at 750 °C 30 min are necessary for its appearance. At 650 °C an annealing time greater than 300 min is necessary to produce an exponential tail.

Differences also can be seen regarding the surface concentration of the profiles and the zinc quantity Q_v in the bulk. At 650 and 750 °C the surface concentration is

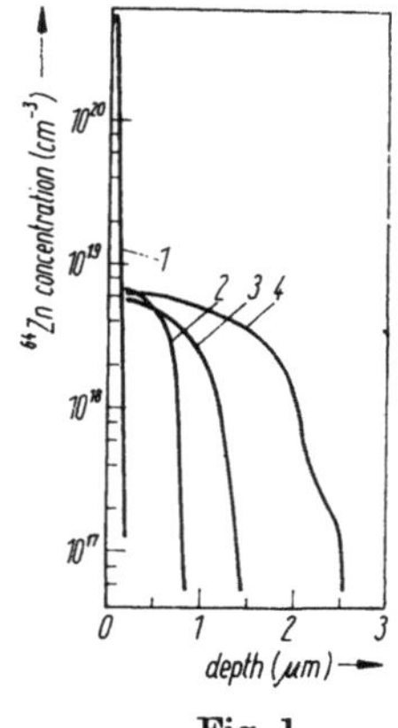

Fig. 1

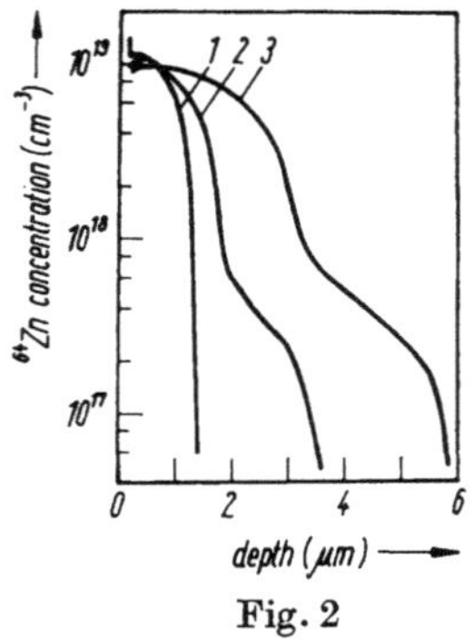

Fig. 2

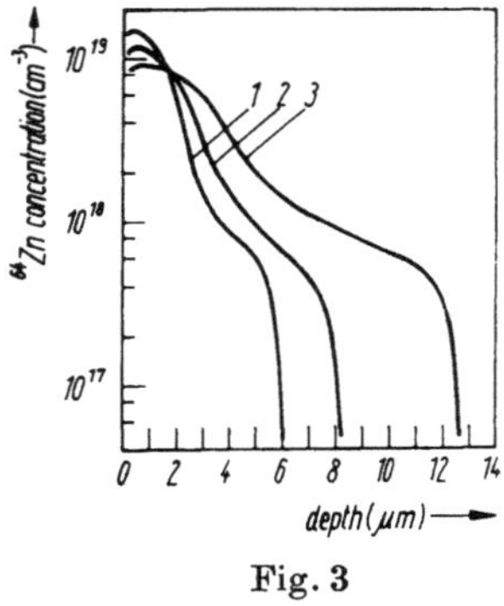

Fig. 3

Fig. 1. Concentration profiles of implanted zinc (100 keV, 4×10^{15} ions/cm^2) in $GaAs_{1-x}P_x$ at room temperature (1) and after annealing at 650 °C for (2) 30, (3) 100, and (4) 300 min

Fig. 2. Concentration profiles of implanted zinc (100 keV, 4×10^{15} ions/cm^2) in $GaAs_{1-x}P_x$ after annealing at 750 °C for (1) 10, (2) 30, and (3) 100 min

Fig. 3. Concentration profiles of implanted zinc (100 keV, 4×10^{15} ions/cm^2) in $GaAs_{1-x}P_x$ after annealing at 850 °C for (1) 10, (2) 30, and (3) 100 min nearly constant. The longer the time of annealing the greater the quantity of zinc in the bulk. At 850 °C the surface concentration decreases with increasing time of annealing, while the quantity of zinc in the bulk slightly increases.

After annealing for 100 min the total implanted zinc quantity is found in the bulk.

Table 1 shows the experimental values for N_s, Q_s, and the diffusion coefficient D_2 determined from the p–n junction depth.

Table 1

Surface concentratoin N_s, zinc quantity in the surface-near layer (Q_s), zinc quantity in the bulk (Q_V), and diffusion coefficient D_2 determined from the p–n junction depth after annealing (either erfc- or Gaussian-distribution is assumed)

T (°C)	t (min)	N_s (10^{19} cm^{-3})	D_2 for erfc (cm^2/s)	D_2 for Gaussian (cm^2/s)	Q_V (10^{15} cm^{-2})	Q_s (10^{15} cm^{-2})
650	30	0.7	3.7×10^{-13}	2.3×10^{-13}	0.27	3.73
	100	0.59	2.7×10^{-13}	1.9×10^{-13}	0.42	3.58
	300	0.64	2.9×10^{-13}	2.0×10^{-13}	0.96	3.04
750	10	1.14	2.5×10^{-12}	1.6×10^{-12}	0.96	3.04
	30	1.07	4.7×10^{-12}	3.1×10^{-12}	1.3	2.7
	100	0.97	4.1×10^{-12}	2.9×10^{-12}	2.21	1.79
850	10	1.5	4.0×10^{-11}	3.2×10^{-11}	2.8	1.2
	30	1.2	2.6×10^{-11}	1.9×10^{-11}	3.07	0.93
	100	0.92	2.0×10^{-11}	1.5×10^{-11}	3.85	0.15

4. Discussion

The analysis of the profiles, above all regarding the zinc quantity (Q_s) included in the surface-near layer at $-x_s \leqq x \leqq 0$ in the behaviour of the temperature–time programme, results in the following simplified model of the zinc diffusion from the implanted source (Fig. 4).

The implanted source and the bulk of the $GaAs_{1-x}P_x$ layer are regarded as different regions with the zinc distributions $N_1(x, t)$ and $N_2(x, t)$ and the diffusion coefficients D_1 and D_2. The out-diffusion into the SiO_2 layer is neglected [9]. In the temperature range of 650 to 750 °C at $t < 300$ min the drain characteristics of the surface-near layer are predominant. Zinc is kept in the implanted layer, it is released only gradually and thus supplied continuously. In the considered temperature and time intervals the implanted source can be regarded as infinite. The drive-in diffusion is carried out with a constant surface layer concentration N_s below the maximum solubility.

The width of the surface-near layer is dependent on temperature. For this reason in the following the entire zinc-implanted region with $-x_s \leqq x \leqq 0$, $x_s = 150$ nm, is regarded as surface-near layer. The implanted Gaussian profile is strongly redistributed during the first minutes of the annealing process. The initial distribution of the implanted zinc may by well described

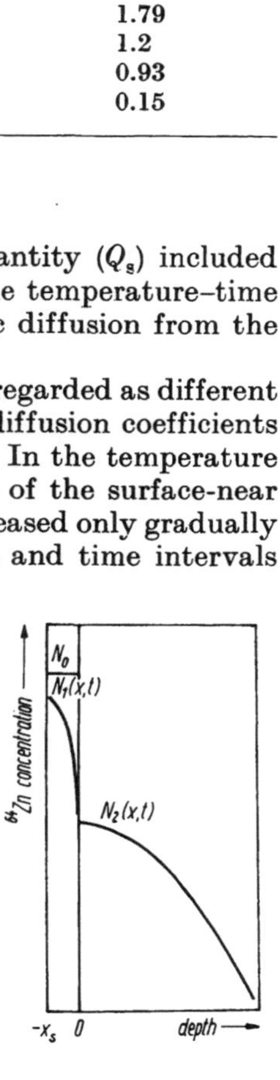

Fig. 4. Simplified model for the zinc diffusion in $GaAs_{1-x}P_x$ from the implanted source

in the following by a constant concentration $N_0 = N_D/x_s$ in the implanted layer with $-x_s \leqq x \leqq 0$. To solve this problem it is necessary to compute the following diffusion equations:

$$\frac{\partial N_1(x, t)}{\partial t} = D_1 \frac{\partial^2 N_1(x, t)}{\partial x^2}, \tag{1}$$

$$\frac{\partial N_2(x, t)}{\partial t} = D_2 \frac{\partial^2 N_2(x, t)}{\partial x^2} \tag{2}$$

with the initial conditions (3), (4)

$$N_1(x, t) = N_0 \quad \text{at} \quad -x_s \leqq x \leqq 0\,, \qquad t = 0\,, \tag{3}$$

$$N_2(x, t) = 0 \quad \text{at} \quad x > 0\,, \qquad t = 0\,, \tag{4}$$

and with the boundary conditions (5) to (8):

$$\frac{\partial N_1(x, t)}{\partial x} = 0 \quad \text{at} \quad x = -x_s\,, \tag{5}$$

$$N_1(x, t) = m N_2(x, t) \quad \text{at} \quad x = 0\,, \tag{6}$$

$$N_2(x, t) = 0 \quad \text{as} \quad x \to \infty\,, \tag{7}$$

$$D_1 \frac{\partial N_1(x, t)}{\partial x} = D_2 \frac{\partial N_2(x, t)}{\partial x} \quad \text{at} \quad x = 0\,. \tag{8}$$

These diffusion equations have been solved by Owen and Schmidt [11] for the case of diffusion from doped oxides into silicon. In the region 2 the diffusion equation has the following solution with respect to the initial and boundary conditions:

$$N_2(x, t) = \frac{N_0}{2} \frac{1-\alpha}{m} \left[\sum_0^\infty (-1)^n \alpha^n \operatorname{erfc} \frac{n 2 R x_s + x}{2\sqrt{D_2 t}} - \sum_0^\infty (-1)^n \alpha^n \operatorname{erfc} \frac{(n+1)\, 2 R x_s + x}{2\sqrt{D_2 t}} \right], \tag{9}$$

where N_0, x_s, D_2 have been defined already, t is the diffusion time, x the depth, m the segregation coefficient.
The parameters R and α are given by

$$R = \sqrt{D_2/D_1}\,, \qquad \alpha = \frac{R/m - 1}{R/m + 1}.$$

The analysis of the solution shows that the shape of the profile in the bulk (region 2) depends, in fact, only on the diffusion time and the characteristics of the implanted layer.

When the diffusion length of zinc in the implanted layer is much less than the thickness x_s, $\sqrt{D_1 t} < x_s$, the implanted layer tends to behave as an infinite source (erfc), but when the diffusion length is much greater than the thickness x_s, $\sqrt{D_1 t} > x_s$, it tends to behave as a limited source (Gaussian). In the temperature range of 650 to 850 °C and for $R/m = 20$ the diffusion of zinc from an implanted source can be well described by (9). The model shows also the explanation for the qualitative change of the diffusion process at 850 °C. The comparison of the experimental results with the model proposed here leads to the conclusion, that the diffusion coefficient is appreciably smaller in the implanted region than in the undisturbed $GaAs_{1-x}P_x$ in the bulk.

As we have already detected in the past, the generation of the double profiles is not due to the recrystallization of the zinc-implanted layer only, but above all to the influence of the surface-near layer with its source and drain characteristics in dependence on temperature and time [9]. Since the formation of the Gaussian section and of the exponential tail appear only at higher temperatures or after longer annealing times one may suppose that the formation of these two regions furthermore is connected with the depletion of the implanted source. In Fig. 5 the dependence of the zinc quantity in the surface layer (Q_s) is shown as the difference of the known implanted dose N_D and the zinc quantity (Q_v) contained in the bulk at $x > 0$ on the annealing time between 650 and 850 °C.

For the further determination of the characteristics of the surface-near layer the temperature dependence of the diffusion coefficient D (Fig. 6) was investigated:

— The diffusion depth of the exponential tail agrees well with the diffusion depths obtained by diffusion from the vapour phase. The temperature dependence of D agrees well with the results of Kobayashi et al. [12].

— The activation energy of the diffusion process of zinc from the implanted source is (2.2 ± 0.2) eV and agrees well with the value of 2.33 eV obtained by Kobayashi et al. [12] for zinc diffusion from the vapour phase.

The diffusion coefficient and the concentration distribution computed on the basis of the models presented above well describe the quantitative processes, though it is only an approximation using a constant diffusion coefficient for the complex experimental profile behaviours. The dependence of the diffusion coefficient on depth computed according to Titov [13] is shown in Fig. 7. In the range of $0 \leqq x \leqq 2{,}5$ μm

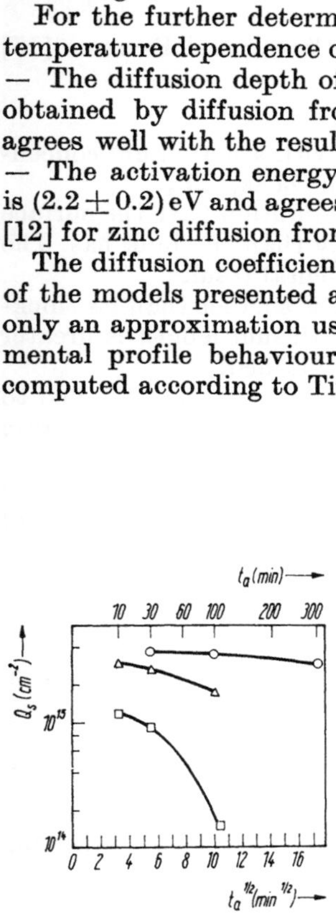

Fig. 5

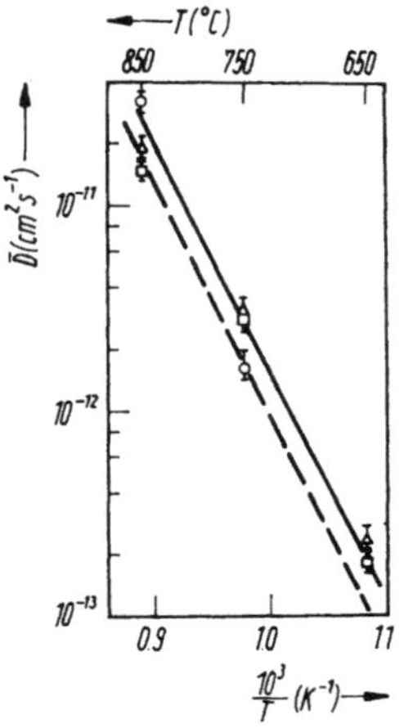

Fig. 6

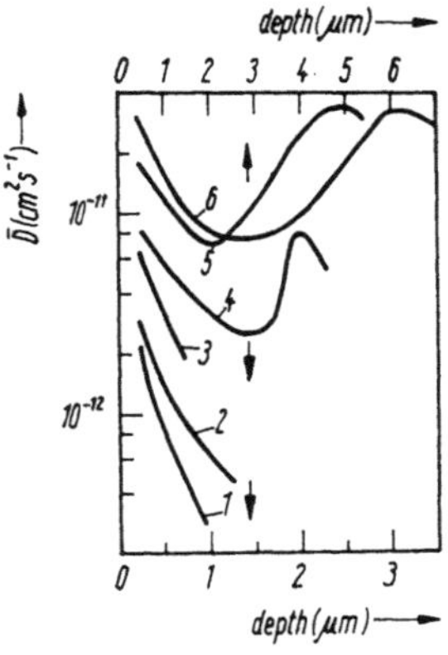

Fig. 7

Fig. 5. Zinc quantity in the surface layer (Q_s) at ○ $T = 650$, △ 750, and □ 850 °C in dependence on annealing time

Fig. 6. Temperature dependence of the effective diffusion coeffcient at different annealing times. ○ $t_a = 10$, △ 30, □ 100, and × 300 min, for comparison the values of Kobayashi are given (dashed line)

Fig. 7. Dependence on depth of the effective diffusion coefficient of zinc in the doped layer after annealing at 650 °C: (1) $t_a = 30 \ldots 100$ min and (2) $t_a = 100 \ldots 300$ min; 750 °C: (3) $t_a = 10$ to 30 min and (4) $t_a = 30 \ldots 100$ min; 850 °C: (5) $t_a = 10 \ldots 30$ min and (6) $t_a = 30 \ldots 100$ min. Curves 1 to 4: lower depth scale, curves 5 and 6: upper depth scale this dependence on depth can be interpreted by a concentration dependence of the diffusion coefficient.

The increase of the effective diffusion coefficient in larger depths indicates a greater influence of the faster interstitial diffusion mode in the diffusion in the exponential profile section, obviously due to smaller $[V_{Ga}]$ concentrations and thus to smaller interaction of $[V_{Ga}]$ with zinc.

The position of the local minimum in the location dependence of D ($x = 1.5$ to 2.5 μm) appearing at 850 °C allows to draw conclusions on the mean diffusion lengths of $[V_{Ga}]$ in the time interval observed.

The diffusion coefficients $D_{V_{Ga}}$ estimated from this are $(2 \pm 1) \times 10^{-12}$ cm²/s at 850 °C.

These values agree well with the results of Molnar [14], Chiang and Pearson, [15] and Blaschku et al. [16].

5. Summary

1. Zinc diffuses from the implanted source redistributed with a maximum concentration appreciably below the maximum solubility of zinc in $GaAs_{1-x}P_x$.

2. The diffusion behaviour can by described with the assumption that the surface layer has a smaller diffusion coefficient and higher solubility than the bulk has. The model proposed here is in good agreement with the experimental results.

3. If the diffusion length of zinc in the surface-near layer is smaller than its thickness, an erfc-distribution is obtained in the bulk. If the diffusion length is greater than the thickness of this layer, a Gaussian distribution is obtained.

4. In the temperature range 650 to 750 °C the diffusion source can be assumed to be infinite. At 850 °C the diffusion can be described by the model of a diffusion from a limited source.

5. The diffusion coefficient D is dependent on depth. From the local minimum of the diffusion coefficient at 850 °C a diffusion coefficient for the gallium vacancies of $(2 \pm 1) \times 10^{-12}$ cm²/s can be estimated.

References

[1] T. Itoh and Y. Oana, Appl. Phys. Letters **24**, 320 (1974).
[2] Y. Shiraki, T. Shimada, Y. Ono, and K. F. Komatsubara, J. appl. Phys. **14**, 1495 (1975).
[3] Y. Ono, K. Saito, Y. Shiraki, and T. Shimada, Japan. J. appl. Phys. **14**, 1489 (1975).
[4] T. Itch and Y. Oana, J. appl. Phys. **44**, 4982 (1973).
[5] H. Okabayashi, Proc. Internat. Conf. Ion Implantation in Semiconductors and Other Ma terials, Osaka 1974, Ed. S. Namba, Plenum Press, New York 1974 (p. 95).
[6] R. G. Hunsperger and O. J. Marsh, J. Elektrochem. Soc. **116**, 488 (1969).
[7] S. T. Picreaux, Proc. Internat. Conf. Ion Implantation in Semiconductors and Other Materials, Yorktown Heights 1972, Ed. B. L. Crowder, Plenum Press, New York 1973 (p. 641).
[8] G. Schwarz, M. Trapp, R. Schimko, G. Butzke, and K. Rogge, phys. stat. sol. (a) **17**, 653 (1973).
[9] G. Müller, M. Haubold, R. Schimko, C. E. Richter, and G. Schwarz, phys. stat. sol. (a) **42**, 579 (1977).
[10] B. Tuck and M. A. H. Kadhim, J. Mater. Sci. **7**, 585 (1972).
[11] A. E. Owen and P. F. Schmidt, J. Electrochem. Soc. **115**, 548 (1968).
[12] S. Kobayashi, M. Ohta, S. Shirose, E. Adachi, and I. Yamamoto, Hitachi Hyoron **53**, 896 (1971).
[13] V. V. Titov, phys. stat. sol. (a) **2**, 203 (1970).
[14] B. Molnar, J. Electrochem. Soc. **123**, 767 (1976).
[15] S. Y. Chiang and G. L. Pearson, J. appl. Phys. **46**, 2986 (1975).
[16] A. I. Blaszhku, B. I. Boltaks, and T. D. Dzhafarov, Fiz. Tekh. Poluprov. **6**, 1791 (1972).

(Received June 8, 1978)

phys. stat. sol. (a) **49**, 285 (1978)

Subject classification: 14.3; 11; 15; 22.2.3

School of Science, Griffith University, Nathan[1])

Defect Annealing and Transport Properties in High-Purity InSb Irradiated at 80 K

By

S. MYHRA

Galvanomagnetic and thermoelectric effects are measured at 80 K as functions of annealing temperature for "pure" n-type InSb irradiated with 1-MeV electrons. The effects of defects on both carrier concentrations and carrier scattering are discussed. It is found that ionized impurity scattering and a high degree of compensation play dominant roles. Hence donor and acceptor defects are injected and removed in a pair-wise fashion.

Galvanomagnetische und thermoelektrische Effekte werden bei 80 K in Abhängigkeit von der Temperungstemperatur für „reines" n-leitendes InSb gemessen, das mit 1-MeV-Elektronen bestrahlt wurde. Der Einfluß von Defekten auf die Ladungsträgerkonzentrationen und die Ladungsträgerstreuung wird diskutiert. Es wird gefunden, daß Streuung an ionisierten Störstellen und ein hoher Kompensationsgrad eine dominierende Rolle spielen. Somit werden Donator- und Akzeptordefekte paarweise eingeführt und beseitigt.

1. Introduction

Radiation effects in the III–V compound semiconductors have received considerable attention. The early work has been reviewed by Aukerman [1] and Eisen [2] while more recent developments have been discussed by Lang [3]. Relatively few radiation damage results on InSb have been published in recent years even through considerable progress has taken place in the basic understanding of the "intrinsic" properties of this material. This, no doubt, is at least partially due to the lack of commercial applications for InSb. In particular, there has been virtually no results forthcoming since 1970 on the recovery of irradiated InSb [3]. The present data are part of a systematic study of the damage production and defect annealing processes in high-purity and lightly doped specimens. Some of the earlier results are in the press [4]; the recovery results in the previous paper (henceforth denoted SMI) are for annealing temperatures up to 220 K and are based on measurements of carrier transport parameters at 10 K. The present data extend the annealing range to 330 K and use 80 K as the base temperature for isochronal measurements.

2. Experimental

The following transport effects were monitored: electrical resistivity ϱ (Ωm), relative transverse magnetoresistance $(\varrho_B/\varrho_0) - 1$, Hall constant R_H (m^3/C), and thermoelectric power Q (μV/K). All post-irradiation measurements were taken at a base temperature of 80 K. The experimental details regarding specimen characteristics, configuration and preparation have been described in SMI. Likewise, the cryostat, electronic instrumentation, and the experimental techniques have been discussed therein.

---

[1]) Nathan, Queensland 4111, Australia.

The high-purity InSb specimens were supplied by M.C.P. Electronics Limited; the as-received characteristics at 77 K were stated as $n_e = 5.49 \times 10^{13}$ (5.1×10^{13}) cm^{-3}, $\mu_e = 319000$ (273000) cm^2/Vs, and $\varrho = 0.356 \times 10^{-2}$ (0.38×10^{-2}) Ωm. The pre-irradiation characteristics actually measured at 80 K in the present experiment are quoted in brackets above.

3. Results

Pre-irradiation data as functions of $1/T$ for ϱ, R_H, and $(\varrho_B/\varrho_0) - 1$ are plotted in Fig. 1 for the temperature range 80 to 300 K. The 1 MeV electron fluence was approximately 5×10^{17} electrons/cm^2 with a beam current density of about 4 $\mu A/cm^2$.

Recovery results are shown in Fig. 2 (ϱ, R_H, and $(\varrho_B/\varrho_0) - 1$) and Fig. 3 ($Q$) following isochronal anneals between 85 and 330 K. The annealing pulses were of five minutes duration and $\Delta T/T$ was approximately 0.075 for successive annealing steps.

Statistical error bars are less than the size of the symbols in the graphs except for $|R_H|$ in the region of the sign change (Fig. 2) where the signal-to-noise ratio is poor and for Q (Fig. 3) where it was difficult to control ΔT along the specimen with adequate long term stability. Systematic errors arose mainly from the lack of accurate determination of specimen dimensions ($\pm 10\%$), the finite width of Hall probe contact points ($\pm 15\%$) and not being able to directly determine the actual ΔT along the specimen itself ($\pm 30\%$). However, relative features in the annealing data are not affected by these uncertainties.

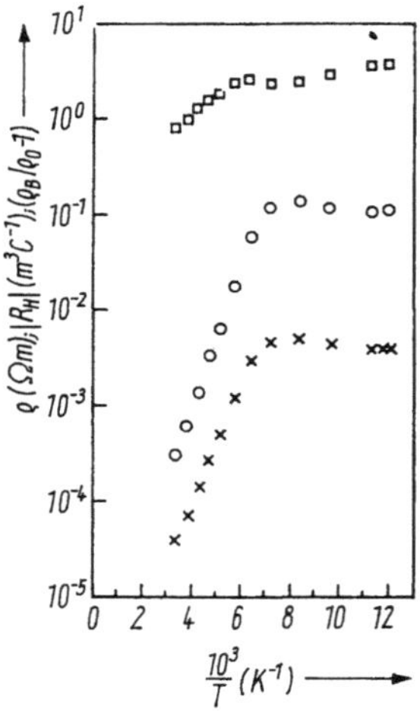

Fig. 1

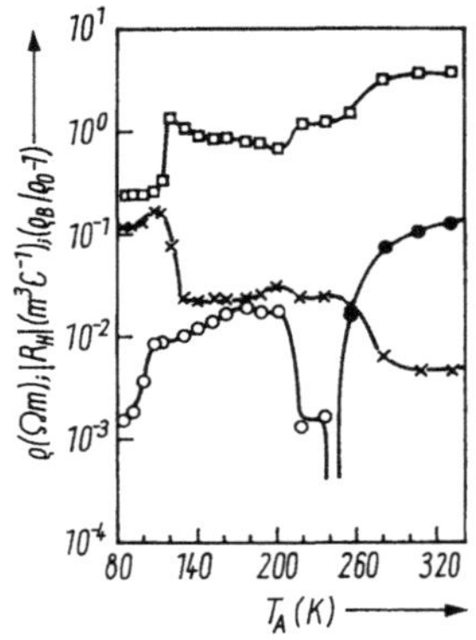

Fig. 2

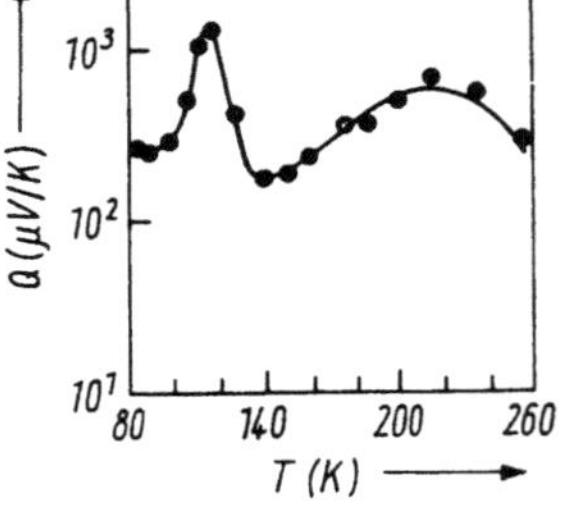

Fig. 3

Fig. 1. Pre-irradiation measurements for resistivity (×), Hall constant (○), and magnetoresistance (□) versus $1/T$; $B = 0.74$ T

Fig. 2. Recovery results for ϱ (×), R_H (○ R_H positive, ● R_H negative), and $(\varrho_B/\varrho_0) - 1$ (□) versus annealing temperature. Basis temperature was 80 K

Fig. 3. Recovery results for Q versus annealing temperature at a basis temperature of 80 K

4. Discussion

4.1 Formalism

The configuration of the specimen and the applied fields is a standard Hall arrangement with $\boldsymbol{E}$ equal to E_x and $\boldsymbol{B}$ equal to B_z such that the Hall field $\boldsymbol{E}_{\mathrm{H}}$ is generated along the y-axis. The ratio of specimen length L and width W was nearly 6:1. Therefore the geometrical effects due to partial shorting of the Hall field are small [5] ($\leqq 20\%$) as compared to the expected results for R_{H} and $(\varrho_{\mathrm{B}}/\varrho_0) - 1$ when $(L/W) \to \infty$. Hence the discussion below will be based on expressions where geometrical effects are ignored.

The choice of relevant transport formalism for InSb will depend on the following criteria:

a) Region of classical weak fields ($\mu^2 B^2 \ll 1$).

b) Region of classical strong fields ($\mu^2 B^2 \gg 1$, but $\hbar\omega^* < k_{\mathrm{B}}T$).

c) Region of quantizing fields ($\hbar\omega^* \geqq k_{\mathrm{B}}T$) where ω^* is the cyclotron frequency.

There also are intermediate field ranges when, for example, $\mu_{\mathrm{h}}^2 B^2 \ll 1 \ll \mu_{\mathrm{e}}^2 B^2$ and also when $\mu^2 B^2 \approx 1$ and $\hbar\omega^* \approx k_{\mathrm{B}}T$.

The following general expressions provide a convenient starting point [5, 6]:

$$\varrho = [e(n_{\mathrm{e}}\mu_{\mathrm{e}} + n_{\mathrm{h}}\mu_{\mathrm{h}})]^{-1}\,, \tag{1}$$

$$R_{\mathrm{H}} = \frac{n_{\mathrm{h}}\,\langle\mu_{\mathrm{hB}}^2\rangle - n_{\mathrm{e}}\,\langle\mu_{\mathrm{eB}}^2\rangle}{e(n_{\mathrm{h}}\,\langle\mu_{\mathrm{hB}}\rangle + n_{\mathrm{e}}\langle\mu_{\mathrm{eB}}\rangle)^2}\,, \tag{2}$$

where

$$\langle\mu_{i\mathrm{B}}^2\rangle = \left\langle\frac{\mu^2(E)}{1 + \mu^2(E)\,B^2}\right\rangle, \tag{3}$$

$$\langle\mu_{i\mathrm{B}}\rangle = \left\langle\frac{\mu(E)}{1 + \mu^2(E)\,B^2}\right\rangle, \tag{4}$$

and

$$\langle A(E)\rangle = \int_0^\infty \left(-\frac{\partial f_0}{\partial E}\right) A(E)\,k^3(E)\,\mathrm{d}E\,, \tag{5}$$

$$(\varrho_{\mathrm{B}}/\varrho_0) - 1 = \frac{n_{\mathrm{e}}\langle\mu_{\mathrm{e0}}\rangle + n_{\mathrm{h}}\langle\mu_{\mathrm{h0}}\rangle}{n_{\mathrm{e}}\langle\mu_{\mathrm{eB}}\rangle + n_{\mathrm{h}}\langle\mu_{\mathrm{hB}}\rangle} \times \left[1 + B^2\left(\frac{n_{\mathrm{h}}\langle\mu_{\mathrm{hB}}^2\rangle - n_{\mathrm{e}}\langle\mu_{\mathrm{eB}}^2\rangle}{n_{\mathrm{h}}\langle\mu_{\mathrm{hB}}\rangle + n_{\mathrm{e}}\langle\mu_{\mathrm{eB}}\rangle}\right)^2\right]^{-1} - 1\,, \tag{6}$$

$$Q = \varrho\sum_i \frac{k_{\mathrm{B}}}{e\varrho_i}\left[\frac{\langle E_i^*\mu_{i0}\rangle}{\langle\mu_{i0}\rangle} - E_{\mathrm{F}i}^*\right], \tag{7}$$

where the subscript i is a band index and the asterisk refers to $E^* = E/k_{\mathrm{B}}T$.

Further development of these expressions depend on assumptions with regard to criteria a), b), c) and the scattering mechanisms.

4.2 Background results

The characteristics of the results in Fig. 1 are intrinsic for $T \geqq 170$ K, intermediate for $170 \geqq T \geqq 100$ K and extrinsic for $T \leqq 100$ K. For electrons, we have $\mu_{\mathrm{e}} \approx 7.8\ \mathrm{m^2/Vs}$ at room temperature; since $\mu_{\mathrm{e}}^2 B^2 \approx 25$ for $B = 0.74$ T it is apparent that the strong field theory must be used. At 300 K we have $h\omega^* \approx 0.2\, k_B T$ for $B = 0.74$ T and $m_e \approx 0.014 m_0$ such that quantum effects should be unimportant; even at 80 K there are no significant deviations from the classical theory. The present experimental results at room temperature of $\varrho \approx 4 \times 10^{-5}\ \Omega$m and $(\varrho_B/\varrho_0) - 1 \approx 0.8$ are consistent with those reported elsewhere [5, 7]. Since the specimen length-to-width ratio was $\approx 6:1$ there should be negligible "end" effects. The values at 80 K for n_e and μ_e, calculated from the experimental data, are 5.1×10^{19} m^{-3} and 27 m^2/Vs, respectively; these are, within the experimental uncertainty, identical to the characteristics quoted by the supplier. The relatively low value for the mobility implies that the specimen is heavily compensated with a total ionized impurity concentration estimated as $\approx 1 \times 10^{21}$ m^{-3} in accord with the discussion of Nag and Dutta [8]. The dominant scattering mechanism at 80 K for the present n-type specimen is that due to ionized impurities; in the limit of $N_i^* \approx n_e \approx 1 \times 10^{19}$ m^{-3} the mobility has an upper bound due to lattice scattering which has been estimated to be ≈ 150 m^2/Vs [8]. Since radiation increases the concentration of ionized "impurities" (actually impurities plus defects), all transport properties in the present experiment are controlled by ionized impurity scattering as long as the specimen is n-type.

The hole mobility of compensated p-type InSb has been studied by Nasledov et al. at temperatures below 100 K [9]. They find that ionized impurity scattering for impurity concentrations as high as $\approx 1 \times 10^{21}$ m^{-3} has little effect above 30 K. However, it is found that the hole mobilities for highly compensated material at higher temperatures are much less than those predicted from lattice scattering; this has been ascribed to large-scale fluctuations of charged impurities.

In the present experiment we find that the irradiation induces p-type conduction at 80 K with $n_h \approx 1.3 \times 10^{21}$ m^{-3} before recovery. This gives $\mu_h \approx 0.015$ m^2/Vs which is a factor of ≈ 50 less than that predicted from lattice scattering. Hence it follows that the post-irradiated specimen is heavily compensated.

4.3 Recovery results

Annealing in InSb following electron irradiation have previously been reported by Eisen [2, 10] who found that five substages are resolvable above 80 K and that the major substage at ≈ 150 K is composed of two first-order stages. The features in the present annealing data are more numerous and a labelling system different from that of Eisen has been adopted. The variations in ϱ, R_H, $(\varrho_B/\varrho_0) - 1$, and Q for

Table 1

Summary of recovery effects in nominally pure InSb. Changes in transport parameters as functions of annealing temperature are indicated by arrows (↑ and ↓ for increasing and decreasing numerical values)

label	temp. range (K)	ϱ	R_H	$(\varrho_B/\varrho_0) - 1$	Q
A	$T \lesssim 115$	↑	+ ↑	↑	↑
B	$115 < T \lesssim 138$	↓	+ ↑	↓	↓
C	$138 < T \lesssim 160$	no change	+ ↑	no change	↑
D	$160 < T \lesssim 178$	↑	+ ↑	↓	↑
E	$178 < T \lesssim 200$	↑	+ ↓	↓	↑
F	$200 < T \lesssim 220$	↓	+ ↓	↑	no change
G	$220 < T \lesssim 240$	no change	no change	no change	↓
H	$240 < T \lesssim 300$	↓	− ↑	↑	no data
I	$300 < T \lesssim 330$	no change	− ↑	no change	no data

ight selected temperature ranges have been summarized in Table 1. In addition it s seen from Fig. 2 and 3 that R_H changes sign at ≈ 240 K and that Q reaches maxima t ≈ 106 and ≈ 220 K.

In highly extrinsic semiconductor materials we may have $|n_e - n_h| \gg n_i$ (n_i is the ntrinsic carrier concentration) and also have $|n_e - n_h| \ll N_A$ or N_D such that the naterial is simultaneously highly compensated and dominated by a majority carrier onduction process. If this is the case, we can, as a rough guide, set

$$R_H = [qn]^{-1}$$

nd

$$\varrho = [q\mu n]^{-1}$$

uch that the Hall mobility is independent of carrier concentration if scattering due to arriers can be neglected and when $N_I^* \gg n$ applies.

Carrier concentrations and Hall mobilities have been calculated from the recovery ata (Fig. 4); the more complicated transition region, where $n_h\langle\mu_{hB}^2\rangle \approx n_e\langle\mu_{eB}^2\rangle$, has eliberately been left out. It is seen that the specimen is p-type for $T_A \leqq 200$ K and hat n_h decreases in such a way that three substages can be resolved below this temerature; the annealing temperatures (T_A) corresponding to these sub-stages are $T_A \approx 106$, $110 < T_A < 144$ K and $148 < T_A < 170$ K. There may be a further ubstage resolvable between 180 and 200 K where n_h increases while μ_h goes down. Corresponding substages can be resolved from the curve for μ_h versus T_A. These ubstages are in good agreement with features observed by Vook [11] for the recovery f thermal conductivity in irradiated InSb. At 250 K, above the n ↔ p transition region, further substage is observed for n_e and μ_e versus annealing temperature. At the nd of this stage, at 330 K, both the free carrier concentration and the electron nobility have nearly returned to the pre-irradiation values.

The following points can now be made:

i) Changes in carrier mobilities are often more prominent than carrier removal ffects in irradiated semiconductors; for the present data we have that n_h decreases y ≈ 12 while μ_h increases by ≈ 80 during annealing to 200 K. This point has often ot been fully appreciated when carrier transport measurements are used to monitor he effects of irradiation. Clearly scattering may outweigh carrier removal effects if a igh degree of compensation is maintained during defect production or annealing; his will occur when ionized donor and acceptor defects are injected or removed in qual numbers. It is quite plausible that this will happen for electron irradiation where most of the damage is introduced as interstitial–vacancy pairs. In high-purity

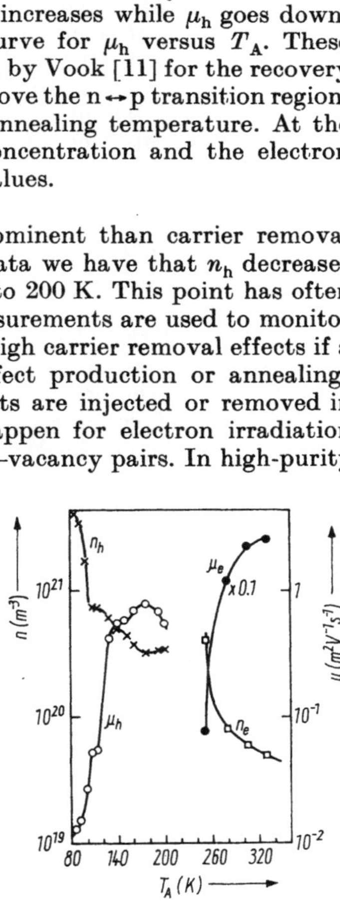

Fig. 4. Carrier concentration and Hall mobility versus annealing temperature

materials there will be little chance of defect–impurity complex formation during irradiation. Likewise, during recovery the lowest-lying substages are generally associated with annihilation of close pairs; only free migration can alter the relative proportions of interstitials and vacancies when defect–defect and defect–impurity complexes are formed. Hence compensation is maintained if the two basic defect species act as donor and acceptor centers, respectively, and if these are ionized at the base temperature where measurements are made. For high doses, when the defect concentration is much higher than the impurity dopant concentration, no further changes in free carrier concentration will therefore take place; however, the scattering relaxation time will continue to decrease with increasing concentration of ionized defects.

ii) The Hall mobility for the holes changes from 1.4×10^{-2} to $0.8\ m^2/V\,s$ during recovery up to 180 K. The latter value is close to the accepted hole mobility at 77 K when the limiting scattering mechanism is due to lattice vibrations. Ionized impurity scattering of holes at low temperature limits their mobility, μ_I, to $\approx 6\ m^2/V\,s$ at 20 K for a total concentration of ionized impurities, N^*, of $\approx 7 \times 10^{20}\ m^{-3}$. We have that

$$\mu_I = (A/N_I^*)\ T^{3/2}\ ,$$

where A is a constant. Using this expression it can be inferred that the post-irradiation concentration of ionized impurities at 80 K must be $\approx 2.5 \times 10^{24}\ m^{-3}$ in order to account for $\mu_h = 1.4 \times 10^{-2}\ m^2/V\,s$. Since we simultaneously have $n_h = 4 \times 10^{21}\ m^{-3}$ as well as $n_h = |N_A - N_D|$, the compensation is $N_D/N_A = 0.999$. The value for N_I^*, at which $\mu_I = \mu_l = 0.8\ m^2/V\,s$ at 80 K, is $5.6 \times 10^{22}\ m^{-3}$, where μ_l is the mobility limited by lattice scattering. Hence N_I^* must have decreased to about this value after recovery to 145 K. This is a change of ≈ 500. Simultanously, n_h decreases by about a factor of 10; this implies that the high degree of compensation is substantially maintained during recovery over the temperature range 85 to 145 K. Therefore, one can conclude that 1 MeV electron irradiation introduces donor and acceptor defect pairs and that the recovery up to 145 K consists of annihilation of close interstitial–vacancy pairs.

The requirement that N_I^* be $\approx 2.5 \times 10^{24}\ m^{-3}$ following irradiation means that a defect concentration of ≈ 30 ppm or more has been introduced. Considering the electron fluence in the present experiment, the maximum energy transfer of $T_M \approx \approx 75$ eV, the threshold energy for displacement $T_D = 6$ to 7 eV, the effect of secondary displacements and the McKinley-Feshbach displacement cross-section, we find that 30 ppm is quite a reasonable value.

iii) The recovery taking place between 250 and 350 K exhibits a decreasing free electron concentration while μ_e goes from ≈ 0.8 to $\approx 26\ m^2/V\,s$. The effect of ionized impurity scattering is again predominant. The lowest value for the mobility corresponds to an ionized impurity concentration of $\approx 3.8 \times 10^{22}\ m^{-3}$ while the carrier concentration, n_e, is $4 \times 10^{20}\ m^{-3}$. Therefore, we can deduce that between 180 and 250 K, where intermediate conduction takes place ($\mu_e n_e \approx \mu_h n_h$), there has been a total change in carrier concentration of $\approx 7.5 \times 10^{20}\ m^{-3}$, from $n_h \approx 3.5 \times 10^{20}$ at 180 K to $n_e \approx 4 \times 10^{20}$ at 250 K. Also, over this interval, the ionized impurity concentration has changed by $\approx 18 \times 10^{20}\ m^{-3}$. This means that there is rather less pair-wise removal of acceptor–donor defect pairs in comparison to a net decrease in acceptor defect concentration. If we accept that the interstitial is still the migrating defect in this temperature range (180 to 250 K), then these features correspond to the case where some proportion of the defects are removed by Frenkel-pair annihilations while the remainder is lost (loss of acceptors) at other trapping centers (interstitial complex formation and defect–impurity agglomeration). By implication it is required that interstitials act as acceptor sites.

We can now return to the temperature range 250 to 350 K. Here we find a return of μ_e to the pre-irradiation value of ≈ 30 m²/V s; hence the ionized impurity concentration must approach $\approx 1 \times 10^{21}$ m^{-3} from above by the end of the stage. There is also a substantial decrease in free electron concentration, from 4×10^{20} at 250 K to 4×10^{19} at 350 K. We therefore have a net change in N_I^* of $\approx 3.7 \times 10^{22}$ m^{-3} while n_e is altered by $\approx 3.6 \times 10^{20}$ m^{-3}. These numbers again require the pair-wise removal of donors and acceptors. Following the one-interstitial model, we must invoke a vacancy– migration mechanism such that vacancies are removed at trapped interstitial sites or other sinks; vacancies must accordingly be interpreted as donor sites.

4.4 *Correlation with magnetoresistance and thermopower measurements*

Hole conduction dominates for annealing temperatures below ≈ 200 K; the expressions (6) and (7) can therefore be simplified considerably:

$$(\varrho_B/\varrho_0) - 1 = \left[\frac{\langle\mu_{h0}\rangle}{\langle\mu_{hB}\rangle}\left(1 + B^2\,\frac{\langle\mu_{hB}^2\rangle^2}{\langle\mu_{hB}\rangle^2}\right)^{-1}\right] - 1\,. \tag{8}$$

We now derive two further expressions for the low and high-field limits, respectively:

$$(\varrho_B/\varrho_0) - 1 = B^2\,\frac{\langle\mu^3\rangle\,\langle\mu\rangle - \langle\mu^2\rangle^2}{\langle\mu\rangle^2}\,; \qquad \mu B \ll 1\,, \tag{9}$$

where the subscripts (h0, hB) have been dropped,

$$(\varrho_B/\varrho_0) - 1 = \frac{\langle\mu\rangle\,\langle\mu^{-1}\rangle}{1 + (\langle\mu^{-1}\rangle\,B^{-1})^2} - 1\,; \qquad \mu B \gg 1. \tag{10}$$

Since $\mu = |e|\,\tau/m^*$ and τ can be written in the form $\tau = \tau_0(E/k_BT)^r$ where $r = 3/2$ for ionized impurity scattering, the evaluation of the integrals is straightforward.

Initially, the pre-recovery value for μ_h is $\approx 1.5 \times 10^{-2}$ m²/Vs. Since $\mu B \ll 1$ applies, we can use (9) to calculate the theoretical magnetoresistance. Agreement is found for $\mu_0 = 0.1$ m²/V s where $\mu_0 = |e|\,\tau_0/m^*$. After annealing to 120 K the magnetoresistance has increased to 1.4 such that a value of $\mu_0 = 0.24$ is required for theoretical agreement. The standard expression for the relaxation time due to ionized impurity scattering is in the case of non-degenerate statistics and parabolic bands [12]

$$\tau_0 = \frac{64\pi^{1/2}\,\varepsilon^2\,(2k_BT)^{3/2}\,m^{*\,1/2}}{N_I^* Z^2 e^4}\,, \tag{11}$$

where ε is the dielectric permittivity. Even though the change in the value of μ_0 is in the right direction with decreasing values of N_I^*, the effect is only a factor of 2.5 as compared to a change of ≈ 15 in the Hall mobility for recovery up to 120 K. It is not clear how this discrepancy can be reconciled.

At the completion of the recovery, when the conduction has reverted to n-type, the specimen has a relative magnetoresistance approaching four. The high-field expression must now be used since $\mu B \gg 1$ and a value of $\mu_0 \approx 1.2$ m²/Vs is obtained. This gives a relaxation time, τ_0, for electrons after recovery to 330 K which is slightly lower than that for holes after recovery to 200 K when differences in the effective masses are considered. This is reasonable since even though N_I^* has decreased substantially, $(m_h^*)^{1/2}$ must be replaced by $(m_e^*)^{1/2}$ in (11).

The thermopower is positive for recovery up to 260 K; experimental results for Q were not obtained above this temperature. For a p-type specimen we have

$$Q = \frac{k_B}{|e|}\left[\frac{\langle E^* \mu_h\rangle}{\langle \mu_h\rangle} - E_F^*\right]. \tag{12}$$

Using $r = 3/2$ for ionized impurity scattering, we find that $Q \approx 4k_B/|e| \approx 350\ \mu V/K$ for $E_F^* \ll k_B T$. Therefore, the Fermi level is near the edge of the valence band before the recovery temperature, T_A, exceeds 100 K. For $T_A > 100$ K we find that Q rises to a maximum at 230 K of $1.2 \times 10^3\ \mu V/K$. The Fermi level is therefore rising above valence band edge to a position of about 0.07 eV.

Further recovery brings lattice scattering into play such that $r = -\frac{1}{2}$. The Fermi level must therefore again approach the valence band edge until in the limit $Q \approx \approx 2k_B/|e| \approx 175\ \mu V/K$ when once more $E_F^* \ll k_B T$ at $T_A = 140$ K. From 140 to 200 K the Fermi level finally rises to a position of ≈ 0.02 eV before the contribution of free electrons start to play a role.

Acknowledgements

The research was supported in part by the Australian Institute of Nuclear Science and Engineering. The experimental work was carried out at the Australian Atomic Energy Commission's Research Establishment.

References

[1] L. W. Aukerman, in: Semiconductors and Semimetals, Vol. 4, Ed. R. K. Willardson and A. C. Beer, Academic Press, New York 1962 (p. 343).

[2] F. H. Eisen, Radiat. Eff. 9, 273 (1971).

[3] D. V. Lang, Proc. I. Internat. Conf. on Radiation Effects in Semiconductors, Ed. N. B. Urli and J. W. Corbett, Dubrovnik 1976.

[4] S. Myhra, Radiat. Eff., in the press.

[5] S. Lakeou, S. Cristoloveanu, and A. Chovet, phys. stat. sol. (a) **43**, 213 (1977).

[6] L. Sosnowski, Soc. Ital. di Fisica, Scuola Internazionale di Fisica "Enrico Fermi", Rendiconti 22, 426 (1963).

[7] J. E. Drabble, M. M. Kaila, and H. J. Goldsmid, J. Phys. D 8, 790 (1975).

[8] B. R. Nag and G. M. Dutta, J. appl. Phys. **48**, 362 (1977).

[9] D. N. Nasledov, Yu. S. Smetannikova, and T. K. Tashkhodzaev, phys. stat. sol. (a) **20**, 101 (1973).

[10] F. H. Eisen, Phys. Rev. **148**, 828 (1966).

[11] F. L. Vook, Phys. Rev. **135**, A1750 (1964).

[12] B. K. Ridley, J. Phys. C **10**, 1589 (1977).

(Received May 10, 1978)

phys. stat. sol. (a) **49**, 293 (1978)

Subject classification: 14.2 and 14.3.3; 22.6

Sektion Physik der Friedrich-Schiller-Universität Jena

Properties of Quantum Interference Devices with V–VO_x–Pb Tunnel Junctions

By

F. DETTMANN

The properties of double contact quantum interference devices (SQUIDs) with V–VO_x–Pb tunnel junctions are studied experimentally. With decreasing temperature an increasing hysteresis in the (I–U) characteristics is observed. At lower temperatures there are multiple critical currents. At sufficiently low inductivities of the SQUID loop and small critical currents single valued critical currents can be reached. The dependences of the quantum interferences from magnetic field and from the temperature pointed out that small tunnel areas and a symmetric construction of the SQUID loop are advantageous.

Die Eigenschaften von Zwei-Kontakt-Quanteninterferometern (SQUIDs) mit V–VO_x–Pb Tunnelverbindungen werden experimentell untersucht. Mit fallender Temperatur wird eine zunehmende Hysterese in den I–U-Kennlinien beobachtet. Bei tieferen Temperaturen treten Vieldeutigkeiten des kritischen Stromes auf. Bei genügend geringen Induktivitäten der SQUID-Schleifen und bei kleinen kritischen Strömen können eindeutige kritische Ströme erreicht werden. Die Abhängigkeiten der Quanteninterferenzen vom Magnetfeld und von der Temperatur zeigen, daß kleine Tunnelflächen und ein symmetrischer Aufbau der SQUID-Schleife vorteilhaft sind.

1. Introduction

It is well known that V–VO_x–Pb Josephson tunnel junctions prepared in a suitable manner are characterized by high stability of their properties during storage at normal conditions [1 to 3]. Therefore, such junctions may be used successfully in superconducting quantum interference devices (SQUIDs). Yanson and Rybalchenko [4] informed about the application of V–VO_x–Pb junctions in SQUIDs, but they report only the remarkable peculiarity that their tunnel SQUIDs in the current–voltage characteristics do not show hysteresis. We have prepared and tested a series of SQUIDs with V–VO_x–Pb tunnel junctions and will inform about their properties in more detail.

2. Construction of the SQUIDs, Preparation, and Experimental Techniques

In preparation of the SQUIDs with V–VO_x–Pb junctions we have used the technology described previously [5]. A cross-section of our structure is shown in Fig. 1. First, on a clean glass substrate a vanadium film, about 200 nm thick, is evaporated through a mask. The surface of the vanadium film is oxidized in a dc glow discharge in an oxygen atmosphere. Then the upper lead electrodes of the tunnel junctions were evaporated through a mask on the edges of the vanadium stripes. After that an insulating film with a thickness between 1 and 3 μm is formed by evaporation of SiO. Last, the SQUID loop was closed by a second lead film.

On one substrate twelve SQUIDs were prepared simultaneously. The length of the vanadium stripes shown in Fig. 1 were between 2.5 and 4.5 mm. The tunnel junctions had areas of about 300 × 500 μm². The width of the upper lead film was between

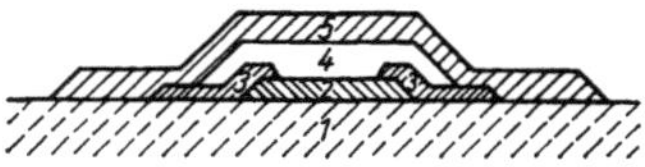

Fig. 1. Cross-section of the tunnel SQUID. 1 glass substrate, 2 vanadium filmwith oxidized surface, 3 lead electrodes of the tunnel junctions, 4 SiO insulation, 5 upper lead film

0.1 and 1.5 mm. The inductivity of the SQUIDs determined from their geometrical dimensions was in the range from 10^{-12} to 10^{-10} H.

The vanadium film and the upper lead film were contacted for four-point measurements with the SQUID. The magnetic flux through the SQUID loop has been generated either by an external magnetic field directed perpendicular to the cross-section or by a current flowing through the upper lead film. The method to measure the current–voltage characteristics and the critical currents of the SQUIDs was that used by Seifarth and Rentsch [1].

3. Current–Voltage Characteristics and Multiple Critical Currents

The critical temperatures of the vanadium films used in our SQUIDs were larger than 4.4 K. The current–voltage characteristics in the case of quasi-particle tunneling was as expected from measurements with single tunnel junctions [2, 5]. Near the critical temperatures of the vanadium films the critical currents were small and no hysteresis was observed. In the case of parallel junctions with smaller resistances even at 4.2 K no hysteresis appears in the characteristics. In general the hysteresis was found to increase with decreasing temperature [6].

A typical current–voltage characteristic at a lower temperature is shown in Fig. 2. The SQUID used in this measurement is characterized by the following data: length of the vanadium film 2.5 mm, width of the upper lead stripe 1.5 mm, total inductivity calculated from the dimensions of the sample 4.0×10^{-12} H. As shown in Fig. 2 this SQUID has a maximum critical current of 13.3 mA. The critical currents I_{c1} and I_{c2} of the two tunnel junctions of the SQUID are nearly equal, therefore

$$LI_{c1,2} \gg \Phi_0\,, \tag{1}$$

where L is the total inductivity of the SQUID. As Fulton et al. [7] have shown in this case multiple critical currents are possible.

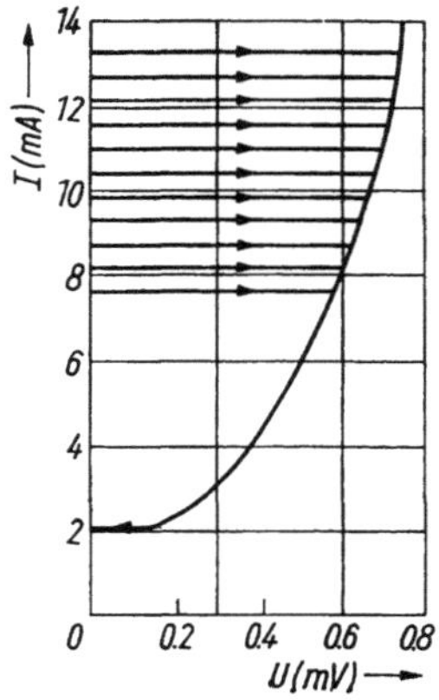

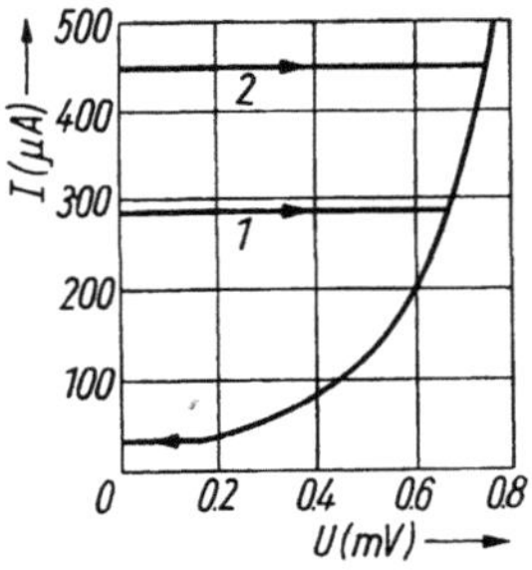

Fig. 3. I–U characteristic of a SQUID with $LI_c \approx \Phi_0$ measured at 2.1 K. The critical current is single-valued and varies within one period between (1) and (2)

Fig. 2. I–U characteristic of a tunnel SQUID with $LI_c \gg \Phi_0$, measured at 2.2 K with a frequency of 5.5 kHz. The various critical currents appear randomly one after another

The current–voltage characteristic shown in Fig. 2 is measured with a frequency of 5.5 kHz and is taken from the screen of the oscilloscope. Eleven critical currents we could observe at the same time. The different critical currents appear with various probabilities. The third critical current from top was observable with the highest probability. The probabilities of the other critical currents decreased with increasing distance from this third critical current. The distances δI_c between the different critical currents were equal. We have found from the characteristic $\delta I_c = 570\ \mu A$.

With increasing magnetic field acting on the area of the SQUID all critical currents are shifted upward until the upper critical current reaches its maximum value. At further increase of the field the upper critical current decreases rapidly below the smallest critical current. If the magnetic field increases to a value at which the flux in the SQUID loop is enhanced by one flux quantum we observe the same critical currents as before. The gradual increase and the rapid decrease of the critical currents with increasing magnetic field are due to the asymmetry of the inductances of the branches of the SQUID ($L_1 \gg L_2$). As shown in [7] the variation of the critical currents with magnetic field in the case realized with our SQUID is given by

$$\frac{dI_{c1}}{d\Phi} = \frac{1}{L_1}, \qquad \frac{dI_{c2}}{d\Phi} = \frac{1}{L_2} \tag{2}$$

with $\Phi = \mu_0 HA$ the flux through the SQUID loop and A the SQUID area. Following (2) the distance of the critical currents may be calculated from

$$\delta I_c = \frac{\Phi_0}{L_1}. \tag{3}$$

With our measured δI_c we obtain $L_1 = 3.63 \times 10^{-12}$ H. This value is in good agreement with the inductivity calculated from the dimensions of the SQUID.

For application of SQUIDs in measurements of magnetic fluxes multiple critical currents are disadvantageous. Therefore, we have prepared SQUIDs with sufficiently low critical currents. The current–voltage characteristic of such a sample is shown in Fig. 3. The inductivity of the SQUID is calculated to $L = 1.16 \times 10^{-11}$ H from the equation

$$L = \frac{\Phi_0}{\Delta H b} \tag{4}$$

with ΔH the period of the quantum interferences and b the width of the upper lead film. As shown in Fig. 3 the critical current varies near $H = 0$ within one period between 280 and 450 μA. These values are characterized by 1 and 2, respectively. The current at which the voltage at decreasing current jumps to zero does not depend on the critical current within one period.

4. Magnetic Field Dependence of the Quantum Interferences

We have measured quantum interferences of the SQUID whose current–voltage characteristic is shown in Fig. 3 starting at different magnetic fields. The results of these measurements are given in Fig. 4. The four curves are shifted in the direction of the H_1-axis in such a way that the minima are at the same values of H_1. As we see in the figure the amplitudes and the shapes and slopes of the quantum interferences within one period are different at various bias fields. This is due to the dependence of the critical currents of the two tunnel junctions on magnetic field. The length of the period of the quantum interferences is not affected by the magnetic bias field.

Fig. 5 gives an example of the dependence of the amplitude of the quantum interferences on magnetic field in a larger range of field H. The measured curve deviates

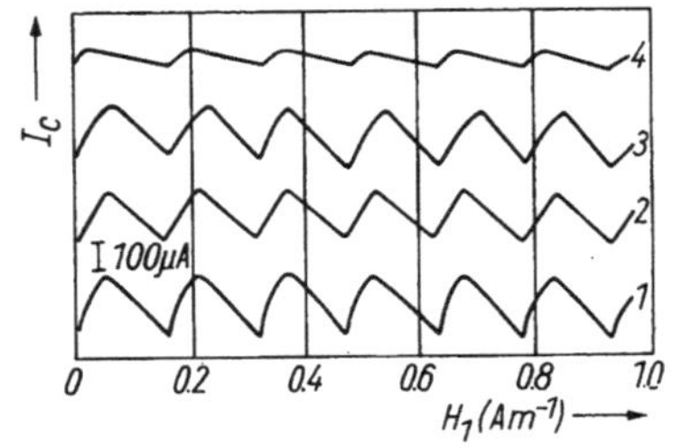

Fig. 4. Quantum interferences of a SQUID measured at various constant bias fields: (1) $H = 0$, (2) 109, (3) 364, (4) 729 A/m; temperature $T = 2.28$ K, H_1 is the variable magnetic field

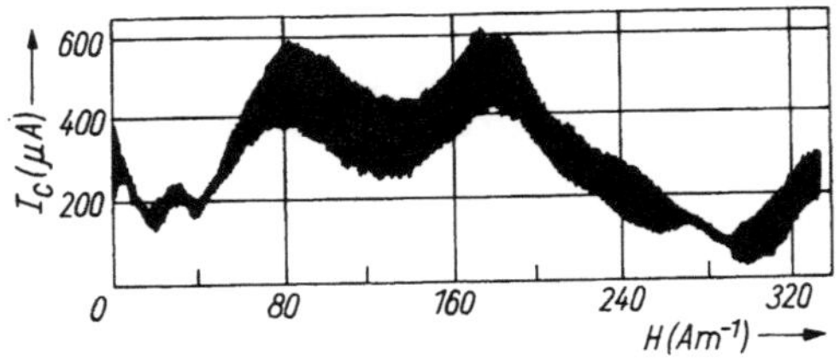

Fig. 5. Dependence of the amplitude of quantum interferences on magnetic field influenced by trapped flux in the tunnel junctions

strongly from the expected Fraunhofer pattern for a single tunnel junction modulated by quantum interferences. The reason for the deviations probably are trapped fluxes in the ranges of the V–VO_x–Pb tunnel junctions. This assumption is supported by the fact that after heating up to temperatures above the critical temperature of lead and repeated cooling in each case another dependence of the amplitude of quantum interferences on magnetic field was measured. Trapped flux in the range of the tunnel junctions may be avoided by using junctions with smaller dimensions, as our studies of single V–VO_x–Pb junctions have shown.

5. Temperature Dependence of the Quantum Interferences

At low magnetic fields we have measured the quantum interferences of our SQUIDs at different temperatures. We will discuss in detail the results obtained with a SQUID with a length of the vanadium stripe of 3.0 mm and a width of the upper lead film of 1.0 mm. The inductivity of this SQUID determined from its dimensions was 2.6×10^{-12} H. As mentioned above in the current–voltage characteristic near the critical temperature of vanadium there is no hysteresis. At lower temperatures hysteresis occurs. The quantum interferences of this SQUID between 2.2 and 3.6 K are shown in Fig. 6. At all temperatures the measurement was started at the magnetic field $H = 0$. Because of the different inductivities in the two branches of the SQUID loop and of the increasing critical currents with decreasing temperature, the effect of the magnetic fluxes of the currents in the SQUID branches is larger at lower temperatures. Therefore, the quantum interferences at lower temperatures are shifted in the direction of lower fields. The problem of the influence of the self-field on the critical current of SQUIDs was solved theoretically by Halse and Salleh [8] in the case of $LI_{c1,2} \gg \Phi_0$. In contrast to this, with our SQUIDs, the case $LI_{c1,2} \approx \Phi_0$ is realized and the calculation of the influence of the self-field is more difficult. We have to expect, however, that our results will agree qualitatively with those of Halse and Salleh.

For comparison we have plotted various critical currents of our SQUID versus temperature in Fig. 7. The upper dashed curve gives the maximum critical current within one period of the quantum interferences, the lower dashed line the minimum within the period. The triangles characterize the critical current curve I_c $(H = 0)$ and the open circles the critical current curve at a constant field of $H = 0.39$ A/m, respectively. At higher temperatures the curve of the maximum critical current corresponds to the value of I_c $(H = 0)$. With decreasing temperature an increasing difference between these critical currents occurs and at 2.25 K I_c $(H = 0)$ agrees with

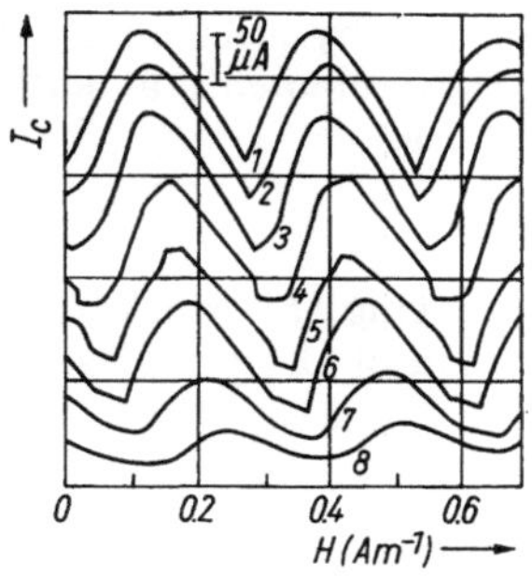

Fig. 6

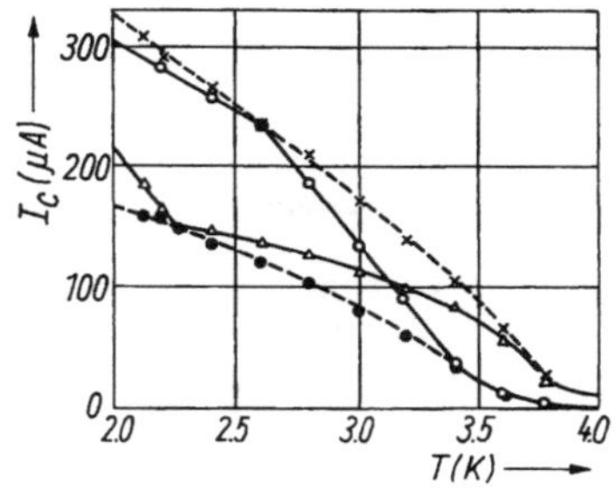

Fig. 7

Fig. 6. Quantum interferences at various temperatures: (1) 2.2, (2) 2.4, (3) 2.6, (4) 2.8, (5) 3.0, (6) 3.2, (7) 3.4, (8) 3.6 K. The zero point for each curve is shifted arbitrarily along the ordinate axis. Each curve starts at $H = 0$

Fig. 7. Critical currents of a SQUID as a function of temperature: × maximum critical current within one period near $H = 0$, ● minimum critical current within one period near $H = 0$, △ critical current at $H = 0$, ○ critical current at $H = 0.39$ A/m

the minimum critical current. At further decrease of temperature we observe a rapid growth of I_c ($H = 0$). The critical current I_c ($H = 0.39$ A/m) corresponds to the critical current minimum at higher temperatures, moves with decreasing temperature to the maximum critical current, and becomes only a little smaller than this with further decrease of temperature.

This behaviour, especially the step-like structure of $I_c(T)$, is qualitatively the same as given by Halse and Salleh [8] for different critical currents of the superconducting weak links and different inductivities of the two branches of the SQUID loop. The deviation from linearity within the steps of the critical current predicted theoretically, is due to the fact that we have not realized the case $LI_{c1,2} \gg \Phi_0$, and that the critical currents of the single tunnel junctions are not linear functions of temperature.

At lower temperatures where the self-fields of the currents in the branches of the SQUID loop have an essential influence on the flux in the SQUID loop, the shift of the quantum interferences is considerable. Because the critical current increases with decreasing temperature, the shift itself is also a function of temperature. This is demonstrated in Fig. 8. Here the ratio of the shift of the period to the total length of the period of the quantum interferences is plotted versus temperature. All shifts are compared with the position of the period at 3.6 K. At a temperature of 3.0 K, for instance, the slope of the curve is given by 0.30 periods/K. This means that for a measurement with this SQUID with a resolution of better than $10^{-4}\,\Phi_0$ the temperature variation must be lower than 3.3×10^{-4} K. Larger fluctuations of temperature are permissible in the case of SQUIDs with symmetric inductivities and critical currents, because there is no shift of the period. A better method, however, to reach high re-

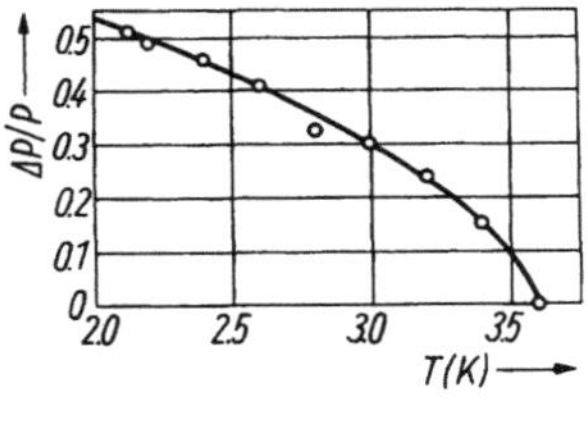

Fig. 8. Relative shift of the period of a SQUID as a function of temperature. The point $\Delta P = 0$ at 3.6 K was chosen arbitrarily

solution with SQUIDs, is the use of tunnel junctions formed by electrodes with high critical temperatures, because the critical currents of those elements at sufficiently low temperatures depend only very weakly on temperature.

Acknowledgements

I would like to thank Prof. Dr. G. Albrecht for stimulating this work and helpful discussions and Dr. K. Blüthner for the careful preparation of the samples.

References

[1] H. SEIFARTH and W. RENTSCH, phys. stat. sol. (a) **18**, 135 (1973).
[2] G. ALBRECHT, K. BLÜTHNER, F. DETTMANN, and P. PERTSCH, Proc. V. Internat. Cryogenic Engng. Conf. and Exhibition, Kyoto **1974** (p. 121).
[3] I. K. YANSON and L. F. RYBALCHENKO, phys. stat. sol. (a) **28**, 663 (1975).
[4] I. K. YANSON and L. F. RYBALCHENKO, Fiz. nizkikh temperatur **3**, 44 (1977).
[5] F. DETTMANN, K. BLÜTHNER, P. PERTSCH, P. WEBER, and G. ALBRECHT, phys. stat. sol. (a) **44**, 577 (1977).
[6] F. DETTMANN, phys. stat. sol. (a) **28**, K 21 (1975).
[7] T. A. FULTON, L. N. DUNKLEBERGER, and R. C. DYNES, Phys. Rev. **B 6**, 855 (1972).
[8] M. R. HALSE and K. M. SALLEH, Solid State Phys. **5**, 1643 (1972).

(Received May 9, 1978)

phys. stat. sol. (a) 49, 299 (1978)

Subject classification: 10.1; 1.1; 21; 21.1

Sektion Mathematik/Physik der Pädagogischen Hochschule Halle (Saale)[1])

About a Relation between Microhardness and the Average Radius of the Guinier-Preston Zones in Al–Zn Alloys

By

H. LÖFFLER, O. KABISCH, and R. KROGGEL

The shape of the Guinier-Preston zones, in dependence on their size, are analyzed in Al–Zn(12 to 18 at%) alloys in the ageing temperature range between 20 and 90 °C by means of X-ray small angle scattering investigations in order to obtain more information about effects which influence mainly the hardness of these alloys and at which mean size of the zones the transition from spherical to elliptical zones is going on. The main results are: a) The deviation from the spherical shape of the zones does not increase monotonously by growth of the zones, but minima and maxima appear. b) Some of these extrema coincide well with the minima and maxima observed in the microhardness plotted versus r_G within the range of experimental error, i.e. in the r_G-range $r_G = 22$ up to 48 Å, the course of the hardness is mainly caused by the degree of deviation from the spherical shape. c) By increase of the strains around the GP zones the rise of the hardness caused by the properties of the zones to act as obstacles for dislocations is much larger than the decrease of the hardness influenced by the increase of the distorsions around the zones.

Die Gestalt der Guinier-Preston-Zonen wird in Al–Zn(12 bis 18 At%)-Legierungen im Temperaturbereich von 20 bis 90 °C in Abhängigkeit von der Zonengröße mit Hilfe der Röntgenkleinwinkelstreuung gemessen, um Informationen zu erhalten, welche Effekte die Härte dieser Legierungen hauptsächlich beeinflussen und bei welcher Zonengröße der Übergang von den sphärischen zu den elliptischen Zonen geschieht. Die Hauptergebnisse sind folgende: a) Die Abweichungen von der Kugelform nehmen mit wachsendem r_G nicht monoton zu, sondern es erscheinen Minima und Maxima. b) Einige dieser Extrema stimmen im Rahmen des Fehlerbereiches mit entsprechenden Extrema der Härte überein. Das bedeutet, daß im r_G-Bereich von 22 bis 48 Å der Härteverlauf wesentlich durch den Grad der Abweichungen der Zonen von der Kugelform bedingt ist. c) Beim Anwachsen der Spannungen um die GP-Zonen ist der Härteanstieg, verursacht durch die Eigenschaft der Zonen als Hindernisse für die Versetzungsbewegung zu wirken, viel größer als die Härteabnahme, die durch die Zunahme der Verzerrungen des Gitters bedingt ist.

1. Introduction

Only little information is present concerning the influence of the average value of the radius r_G of the Guinier-Preston (GP) zones on the mechanical properties of Al–Zn alloys. Merz and Gerold [1] pointed out that in Al–Zn alloys, comparatively independent of the concentration of zinc, the annealing temperature T_q and the quenching rate, a minimum of Young's modulus appears during room temperature ageing at an average value of the GP zones of about 33 Å. In the opinion of these scientists this minimum of Young's modulus is caused by the transition of the GP zones from spherical to elliptical shape, i.e. this transition starts at a mean radius of the zones of about 33 Å. This variation of the shape of the GP zones is accompanied by a change of the coherence strain field around the zones from radial symmetrically to rhombohedrically distorted type and simultaneously by a decrease of the strains around the clusters. This decrease introduces an increase of Young's modulus, since a loss of strain reduce

[1]) DDR-402 Halle (Saale), DDR.

the distortions around the GP zones and (due to Zener's theory [2]) local distortions diminish Young's modulus.

In our opinion these arguments could not be quite sure since, on the other hand, a reduction of the strains around the clusters will lessen the property of the zones to act as obstacles for the movement of dislocations, in this way causing a decrease of Young's modulus. Furthermore Merz and Gerold [1] have measured the ratio of the smallest axis c and the longest axis a of the zones analyzing the X-ray small-angle scattering (XSAS) intensities and found at $r_G = 20$ Å, $c/a = 1$ (e.g. spherical GP zones) and at $r_G = 30$ Å, $c/a = 0.6$, this means the transition from spherical to elliptical GP zones starts (probably) before the minimum of Young's modulus appears.

It is the main aim of this work to obtain information on the shape of the GP zones in dependence on their size, in order to get a better knowledge of the effects which influence the mechanical properties of the Al–Zn alloys and at which mean sizes of the GP zones the transition from spherical to elliptical zones is going on.

2. Experimental Procedure

Four alloys of composition 12, 15, 16, and 18 at% zinc were prepared from 4N pure aluminium and 4N pure zinc. After homogenization the samples were annealed for 1 h at 490 °C and then quenched in water of the ageing temperature T_a ($T_a = 20$ to 90 °C). The XSAS measurements were done by a self-constructed camera of the Kratky type "in situ", i.e. at T_a in order to avoid additional quenching treatments. The camera is described elsewhere [3]. The hardness measurements are done as given in [4]. The shape of the clusters expressed by the ratio $c/a = v$ is determined by the method developed by Kawano et al. [5], analyzing the XSAS intensities in both the Guinier and the Porod range (method 1). The ratio $c/a = v$ can be found by determining the ratio $V_s = I(\beta_R)/I(2\beta_R)$, where $\beta_R = (175\ \text{Å}/6r_G)\ 2\pi\lambda^{-1}$ and $I(\beta_i)$ are the XSAS intensities at $\beta_i = 2\pi\lambda^{-1} \sin \varepsilon/2$ (λ wavelength of the X-rays). β_R must be in the Guinier range and $\beta 2_R$ in the range of the asymptotic approximation (Porod range). The relation between $v = c/a$ and V_s was taken from Fig. 1 of [5]. These results are controlled partially by use of the formulae given in [1] (method 2),

$$v = \frac{3\sqrt{\frac{\pi}{5}}\, I(0)}{r_G^2 \int\limits_0^\infty I(\beta)\, \beta\, d\beta},$$

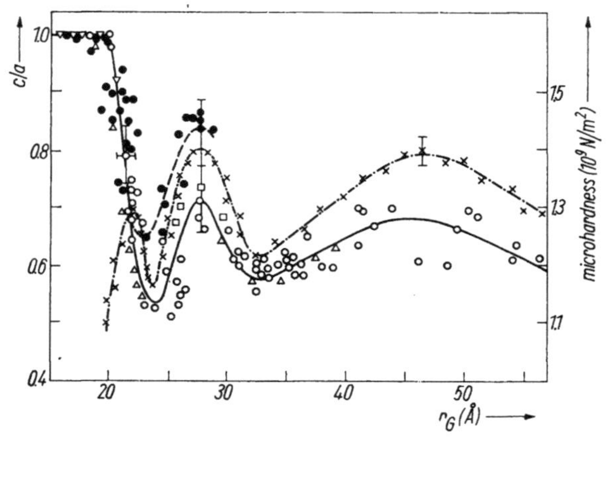

Fig. 1. The ratio $v = c/a$ of Al–Zn(Y) (Y = 12 (○), 15 (□), 16 (△) and 18 (▽) at%) using method 1 (empty symbols) and method 2 (full symbols) and the Vicker's hardness number of Al–Zn (12)(×) versus the Guinier radius r_G

where $I(0)$ is the scattering intensity at the scattering angle $\varepsilon = 0$ (obtained by extrapolating the Guinier straight line to $\varepsilon = 0$) and the integral is the integral intensity. The application of both methods supposes that in the range of $r_G > 22$ Å during the ageing the "normal" size distribution (defined by the mean oscillation amplitude of r_G, neglecting the deviations from the spherical shape) is not essentially changed.

We would like to remark that in both methods used the range of experimental error is rather large, namely about $\Delta v = \pm 0.06$, mainly caused by the uncertainty in the background of the XSAS intensities, the error in the slope of the Guinier straight line, and that one in the integral intensity. Therefore, in order to increase the reliability of this investigation more than 100 XSAS scattering curves are analyzed.

3. Results and Conclusions

In Fig. 1 the course of the c/a values of the Al-Zn(12 to 18 at%) alloys and that of the Vickers hardness number of the microhardness of the Al–Zn (12) alloy is plotted versus the Guinier radius. Only some points are drawn. Obviously, in the r_G range $r_G > 22$ Å the extrema of the $v = c/a$ value coincide well with the microhardness number.

We will discuss the relation between the size, the shape of the GP zones, and the microhardness number mostly with the help of the following premises.

P1) The microhardness is raised by an increasing amount of Zn atoms precipitated in the GP zones. This effect yields its maximum at the saturation of the integral intensity of the XSAS.

P2) The strain field around the spherical GP zones is increased by the growth of the zones causing a rise of the hardness.

P3) The deviation from the spherical shape of the GP zones (transition to elliptical GP zones) is accompanied by a decrease of the microhardness induced by the transition of the radial symmetrical strain field to a rhombohedrically distorted one.

P4) The increase of the mean distance d between the GP zones during the coarsening of the zones at constant volume fraction, i.e. at $r_G \gtrapprox 22$ Å, causes monotonous decrease of the hardness (e.g. $\Delta HV \sim d^{-1/2}$).

In order to check whether the r_G–d relation shows anomalous fluctuations in the r_G range considered, we plotted the mean distance d between the particles (calculated by the method of Synecek [11]) versus r_G for $c_{Zn} = (12$ to $16)$ at% (in this concentration range well-formed maxima in the XSAS curve appear). These plots show [12] a straight r_G–d line for each alloy, this means, no remarkable oscillations of d are observable during the coarsening process. Therefore, the oscillations of HV with r_G can not be caused by the dependence of d on r_G.

Now we will discuss the course of the $v = c/a$ and microhardness curves in dependence on the Guinier radius in detail.

R1) r_G range $\leqq 22$ Å, increase of HV.
The increase of the microhardness in this range should be mainly caused by the growth of the amount of Zn atoms precipitated in the GP zones. This interpretation is supported by the measurements of the integral intensity which yields its saturation value at about $r_G \approx 22$ Å.

The value of $v = c/a$ up to $r_G \approx 20$ Å is approximately unity, announcing that in the early part of the decomposition process spherical GP zones are growing and that the size of the zones is relatively unique. This interpretation is supported by the measurements of Harkness et al. [6]. These authors stated in Al–Zn(3.5 and 7.0 at%) alloys rather sharp maxima of the curve of the size distribution in the initial stage of the ageing process ($r_G = 20.9$ Å in Al–Zn(7 at%) alloy) which are growing wider at longer ageing times, i.e. for r_G values larger than 21 Å. Due to the investigation of Gould [7], this conclusion can be extrapolated to alloys richer in zinc content.

Due to these facts the strong decrease of the v–r_G curve after reaching $r_G \approx 21$ Å should be influenced remarkably by the increase of the size distribution of the GP zones. Probably, before reaching the saturation of the integral intensity the process of the Ostwald ripening is influencing essentially the decomposition of these alloys. Kawano et al. [5] stated likewise a decrease of the v-value during the process of the Ostwald ripening in an Al–Zn(2.8 at%)–Mg(1.4 at%) alloy.

R2) r_G range 22 Å $\leqq r_G \leqq$ 24 Å, decrease of HV and v.
As investigated by many authors (see, for example, [1, 6]) the amount of Zn atoms inserted in the GP zones remains constant during the growth of the GP zones after reaching the saturation of the integral intensity. Therefore, the decrease of the microhardness is caused by a change of the shape of the GP zones and/or by a variation of the size distribution of the zones. As investigated by Harkness et al. [6] and Gould [7] the size distribution of the GP zones is growing monotonously with increasing ageing time. Therefore, the "oscillations" of the microhardness and of the v-value plotted versus r_G should be mainly caused by the variation of the shape of the GP zones. Therefore, we discuss the curves of HV and v drawn versus r_G under the assumption that in the r_G range between 22 and 48 Å these courses should be mainly caused by the variation of the shape of the GP zones. For this reason, in our opinion, the decrease of HV and $v = c/a$ in this r_G range should mainly be caused by the transition of the GP zones from spherical to elliptical shape.

R3) r_G range 24 Å $\leqq r_G \leqq$ 28 Å, increase of HV and v
Due to the arguments given in R2 the increase of HV and $v = c/a$ in this r_G range should announce a decrease of the deviation of the GP zones from the spherical shape. The reasons of the increase of the ratio c/a in the r_G ranges from 24 to 28 Å and from 33 to 48 Å are not clear. Perhaps a further decrease of the ratio $c/a = v$ in these r_G ranges could go on only by a break-down of the full coherence of the GP zones to the matrix. The partial loss of coherence of the zones for comparatively small zones should be connected with a rather large increase of the surface energy of the zones, since a "dislocation" must be formed inside the zones by a transition from fully coherent to partially coherent zones. In other words: The increase of $v = c/a$ is a consequence of the thermodynamic system to find out the "most favourable" sum of the surface and volume energies of the zones. The decrease of the deviation from the spherical shape of the zones is accompanied by an increase of the strain field around the zones influencing the rise of the microhardness.

R4) r_G range 28 Å $\leqq r_G \leqq$ 32 Å, decrease of HV and v.
As discussed in R2 the decrease of HV should be mainly caused by the increase of the deviation of the zones from the spherical shape. Obviously after reaching a mean radius of about 28 Å the difficulties to continue the deviation from the spherical shape of the zones overcome. Merz and Gerold [1] and Morinaga et al. [8] found also a minimum of the mechanical properties at $r_G \approx 32$ Å.

R5) r_G range 32 Å $\leqq r_G \leqq$ 48 Å, increase of HV and v.
The reason for the increase of both HV and $v = c/a$ should be given by the arguments in R3.

R6) r_G range 48 Å, decrease of HV and r_G.
The maximum of HV at about 48 Å coincides well with the one detected by Graf [9] in an Al–Zn(4.5 at%) alloy (49 Å). We interpret the decrease after the maximum by the increase of the deviation from the spherical shape of the zones and by the transition of the fully coherent elliptical zones to the partially coherent α'_r-phase. Electron-microscopic and X-ray investigations confirmed that the transition from elliptical

GP zones to the partially coherent α'_r phase is going on at a mean radius of the zones of about 50 Å [10]. Obviously at this mean size the misfit between the matrix and the zones is growing so large that the formation of a "dislocation" inside the zones will be possible in the alloys.

We would like to remark that, in our opinion, the influence of the transition from spherical to elliptical GP zones on the mechanical properties is opposite to the one of Merz and Gerold [1], i.e., in our line of thinking, by an increase of the strains around the GP zones the rise of the hardness caused by the property of the zones to act as obstacles for dislocations is much larger than the decrease of the hardness influenced by the increase of the distortions around the zones. In our opinion, the comparatively good coincidence between the course of the c/a–r_G curve and the HV–r_G curve should support our conception.

Acknowledgements

We would like to express our thanks to Doc. Dr. V. Synecek and Dr. M. Simerska for valuable discussions and for producing the alloys with 12, 15, and 18 at% zinc, and Mrs. Grant, Krause, Gueffroy, Fabian, and Hartleib for their careful measurements of the XSAS and microhardness.

References

[1] W. Merz and V. Gerold, Z. Metallk. 57, 607 (1966).
[2] C. Zener, Acta cryst. 2, 163 (1949).
[3] R. Kroggel, H. Löffler, and O. Kabisch, phys. stat. sol. (a) **40**, K163 (1977).
[4] R. Kroggel, O. Kabisch, and H. Löppler, phys. stat. sol. (a) **37**, K85 (1976).
[5] O. Kawano, Y. Murakami, and T. Nakazawa, Trans. Japan. Inst. Metals **11**, 12 (1970).
[6] S. D. Harkness, R. W. Gould, and J. J. Hren, Phil. Mag. **19**, 115 (1969).
[7] R. W. Gould, J. Inst. Metals **99**, 1 (1971).
[8] M. Morinaga, K.-S. Liu, Y. Tomita, and Y. Murakami, Trans. Japan. Inst. Metals **15**, 62 (1974).
[9] R. Graf, C. R. Acad. Sci. (France) **246**, 1544 (1958).
[10] P. Bartuska, R. Kroggel, H. Löffler, M. Simerska, O. Simmich, and V. Synecek, Wiss. Z. Pädagogische Hochschule Halle **15**, 24 (1977).
[11] V. Synecek, J. Phys. Radium **23**, 828 (1962).
[12] H. Löffler, R. Kroggel, O. Kabisch, and B. Gueffroy, to be published.

(Received July 13, 1978)

phys. stat. sol. (a) 49, 305 (1978)

Subject classification: 20.2 and 20.3; 3; 20.1; 22.8.2

A. V. Shubnikov Institute of Crystallography, Academy of Sciences of the USSR, Moscow (a), P. N. Lebedev Physical Institute, Academy of Sciences of the USSR, Moscow (b), and Central Institute for Optics and Spectroscopy, Academy of Sciences of the GDR, Berlin (c)

Growth, Spectroscopic Investigations, and Some New Stimulated Emission Data of $Gd_3Ga_5O_{12}:Nd^{3+}$ Single Crystals

By

A. A. KAMINSKII (a), V. V. OSIKO (b), S. E. SARKISOV (a), M. I. TIMOSHECHKIN (b), E. V. ZHARIKOV (b), J. BOHM (c), P. REICHE (c), and D. SCHULTZE (c)

The spectral and laser properties of $Gd_3Ga_5O_{12}:Nd^{3+}$ and $Nd_3Ga_5O_{12}$ crystals are investigated. Highly perfect crystals are grown by a sophisticated Czochralski technique. Absorption, luminescence, and stimulated emission spectra are studied at 77 and 300 K. The luminescence lifetime of the metastable $^4F_{3/2}$ state in crystals with high Nd content and the dependence of the spectroscopic quality parameter X on the luminescence intensity parameter Y are determined accurately, using new methods. Further inter-manifold and "Stark" luminescence branching ratios, spontaneous transition probabilities, luminescence linewidths, and peak and effective transition cross-sections are obtained. Stimulated emission is observed at 300 K in the cw regime due to the basic $^4F_{3/2} \rightarrow {}^4I_{11/2}$ transition, and in the long pulse regime due to the additional $^4F_{3/2} \rightarrow {}^4I_{13/2}$ transition.

An Kristallen von $Gd_3Ga_5O_{12}:Nd^{3+}$ und $Nd_3Ga_5O_{12}$, die mit hoher Perfektion nach einer speziell ausgearbeiteten Czochralski-Technik gezüchtet wurden, werden die spektroskopischen und Laser-Eigenschaften untersucht. Absorptions-, Lumineszenz- und stimulierte Emissions-Spektren werden bei 77 und 300 K aufgenommen. Die Lumineszenz-Lebensdauer des metastabilen $^4F_{3/2}$-Zustandes in Kristallen mit hohem Nd-Gehalt sowie die Abhängigkeit des spektroskopischen Qualitätsparameters X vom Parameter der Lumineszenzintensität Y werden unter Anwendung neuer Methoden exakt bestimmt. Ferner werden die Lumineszenz-Verzweigungsverhältnisse zwischen den aufgespaltenen und „Stark"-Niveaus, die spontanen Übergangswahrscheinlichkeiten, die Lumineszenz-Linienbreiten und die maximalen und effektiven Übergangsquerschnitte ermittelt. Stimulierte Emission wird bei 300 K im Dauerstrich-Regime ($^4F_{3/2} \rightarrow {}^4I_{11/2}$-Grundübergang) sowie zusätzlich im Langpuls-Regime beim $^4F_{3/2} \rightarrow {}^4I_{13/2}$-Übergang erhalten.

1. Introduction

Stimulated emission (SE) of Nd^{3+} ions in gadolinium gallium garnet, $Gd_3Ga_5O_{12}:Nd$, was first mentioned in [1]. The SE, due to the basic $^4F_{3/2} \rightarrow {}^4I_{11/2}$ transition occurred at a high threshold energy, $E_{thr} = 250$ J. In recent investigations [2 to 4], the laser wavelength of this basic transition was determined accurately at 77 and 300 K, and SE was observed in the additional $^4F_{3/2} \rightarrow {}^4I_{13/2}$ transition at 77 K. The value of E_{thr} was diminished two orders of magnitude. The data obtained so far [1 to 4] are not sufficient to give a comprehensive characterization of the $Gd_3Ga_5O_{12}:Nd^{3+}$ system with respect to the various solid state laser materials [5].

The present work deals with the growth of $Gd_3Ga_5O_{12}:Nd^{3+}$ and $Nd_3Ga_5O_{12}$ crystals of high optical quality and sufficient dimensions. Further we report detailed investigations of the spectral properties and the SE parameters under different conditions of excitation at 77 and 300 K in the transitions $^4F_{3/2} \rightarrow {}^4I_{11/2;13/2}$.

Analysing the experimental data, problems arose from the determination of the Stark level positions of the $^4I_{13/2}$ manifold and the identification of the induced transition in the $^4F_{3/2} \rightarrow {}^4I_{13/2}$ channel. Furthermore it was necessary to define the parameter of spectroscopic quality X [4] more accurately, to determine the intermanifold and "Stark" liminescence branching ratios $\beta_{JJ'}$ and β_{ij}, the spontaneous transition probabilities $A_{JJ'}$ and A_{ij}, and to measure peak and effective cross-sections $\sigma^p_{e,ij}$ and σ_e^{eff} [6] of all registered laser transitions.

Accordingly, it was of interest to determine the radiative lifetime τ_{rad} of the metastable state of Nd^{3+} in self-activated neodymium crystals by using the known values of $\beta_{JJ'}$ and $A_{JJ'}$ obtained from absorption in the resonant $^4I_{9/2} \leftrightarrow {}^4F_{3/2}$ transition, using $Nd_3Ga_5O_{12}$ and $Gd_3Ga_5O_{12}:Nd^{3+}$ crystals. This question is important in connection with the actual search for high-concentrated and self-activated neodymium crystals for laser miniaturization.

2. Crystal Growth

Since the first report [7], remarkable progress has been achieved in the melt growth of $Gd_3Ga_5O_{12}$ crystals by a highly sophisticated Czochralski technology [8 to 13]. Most of these works aimed at the preparation of defect-free substrate crystals for the epitaxy of magnetic bubble domain layers. Concerning the growth of $Gd_3Ga_5O_{12}:Nd^{3+}$ and $Nd_3Ga_5O_{12}$ crystals, this work follows the procedure of [8]. Starting materials Gd_2O_3, Nd_2O_3, and Ga_2O_3, all of 99.999% purity[1]), were intimately mixed according to the stoichiometric ratio or to the congruent melting composition ($Gd_{3.05}Ga_{4.95}O_{12}$ after [14]) and pressed isostatically. In some cases, the mixture was pre-reacted at 1400 °C before densifying. Crystals ($Gd_3Ga_5O_{12}$, melting point 1725 °C [14]; $Nd_3Ga_5O_{12}$, melting point 1510 °C [8]) are pulled from rf heated iridium crucibles at a rate of 2 to 6 mm/h. The gas ambient was a dried mixture of nitrogen with a controlled content of 1.8 or 20 vol% oxygen for $Gd_3Ga_5O_{12}:Nd^{3+}$ or $Nd_3Ga_5O_{12}$, respectively, which was blown on the melt surface through iridium tubes to prevent formation and volatilisation of gallium suboxides [15, 16]. Inclusions and decorated dislocations which are characteristic of Czochralski grown $Gd_3Ga_5O_{12}:Nd^{3+}$ crystals (cf. [17] and references cited there), have been overcome by a special pulling procedure [8, 9]: Using defect-free $\langle 111 \rangle$ oriented seed crystals, the first part of the crystal is pulled at a rotation rate of 10 min^{-1} to form a conically shaped growth interface. During this, due to the formation of $\{211\}$ and $\{110\}$ facets at the interface, dislocations grow out of the crystal. However, the facetted growth results in considerable mechanical stresses. After the grow-out of dislocations, the rotation rate is increased to 80 min^{-1}. Thus, the convection pattern in the melt is changed resulting in a plane interface without facets. Then, supposing all conditions to be carefully controlled, crystals grew free from structural defects and stress. Usually, crystals 25 mm in diameter and 100 mm in length are grown, from which laser rods were fabricated with plane-parallel faces to within 5″. By the same technology, crystals of diameters up to 55 mm were grown.

3. Spectral and Laser Measurements

The experimental setup of SE excitation in $Gd_3Ga_5O_{12}:Nd^{3+}$ crystals and measurement of SE parameters is described elsewhere [18, 19]. Laser experiments were performed in a confocal optical resonator ($R = 576$ mm at $\lambda = 1.06$ μm; $R = 600$ mm at $\lambda = 1.35$ μm; transparency of dielectric multilayer interference mirrors about 1% at λ_g; illumination with Xe flash tubes ISP-250, IFP-400, and IFP-800). In the CW regime the active element and the pumping lamp have been cooled by flowing water. Experimental results are listed in Table 1.

[1]) Ga_2O_3 "OC4", Gd_2O_3 "ГД0-1", and Nd_2O_3 "НД0-1" (soviet trade marks).

Table 1
Spectral and laser characteristics of $Gd_3Ga_5O_{12}:Nd^{3+}$ crystal

laser transition	λ_g (Å)	T (K)	laser regime	excitation threshold J (W)	$\Delta\nu_{lum}$ (cm^{-1})	β_{ij}	A_{ij} (s^{-1})	$\sigma^{p}_{e,ij}$ (10^{-19} cm^2)	σ^{eff}_{e} (10^{-19} cm^2)
$^4F_{3/2} \rightarrow {}^4I_{11/2}$	10584 B	77	pulse	1*)	$\approx$ 2.3	0.0913	$\approx$ 490	5.5	5.5
	10599 Б	77	pulse	2*)	$\approx$ 2.3	0.0946	$\approx$ 530	5.7	5.7
	10591 B	300	pulse	—**)	6.7	0.0666	$\approx$ 490	1.8	1.8
	10606 Б	300	pulse	—**)	7.0	0.081	$\approx$ 530	1.9	2.0
	10621 A	300	pulse	1.3	7.2	0.097	810	2.8	2.9
		300	CW	1000					
$^4F_{3/2} \rightarrow {}^4I_{13/2}$	13307	77	pulse	7	2.5	0.016	90	1.4	1.5
	13315	300	pulse	5.3	7***)	0.025***)	170***)	—	$\approx$ 1.55

*) Data of [2].
**) Excited in laser with combined active media [2].
***) Data correspond to two closely lying transitions (see Fig. 2).

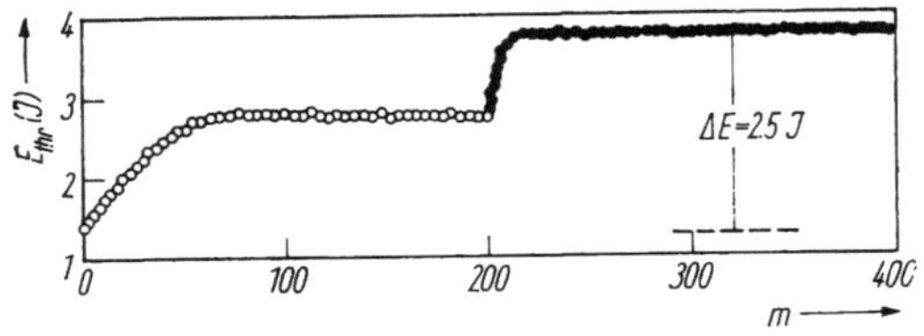

Fig. 1. "Ageing" curves of $Gd_3Ga_5O_{12}:Nd^{3+}$ crystals in laser experiments at 300 K. (Open dots: $E_{exc} = 100$ J and $\tau_{pul} = 400$ μs; full dots: $E_{exc} = 100$ J and $\tau_{pul} = 100$ μs)

The short-wavelength part of the excitation radiation leads to an "ageing" effect, resulting in an increase of the excitation threshold energy E_{thr} as dependent on the number of pulses m. This effect (Fig. 1) is estimated according to [20, 21]. It can be prevented by cutting the short wavelength part of the pumping light using a yellow filter (ZhS-17 glass) which is transparent above 0.5 μm. Our experiments have shown that crystals with Nd^{3+} concentration up to about 3 at% will exhibit laser energetical parameters sufficient for laser applications.

To determine the radiative lifetime τ_{rad} of the metastable state of Nd^{3+} ions in self-activated neodymium laser crystals, we use the known values of the branching ratio $\beta_{JJ'}$ and transition probability $A_{JJ'}$. The value of τ_{rad} of the $^4F_{3/2}$ state of Nd^{3+} ions is determined by the summary probability of the inter-manifold spontaneous transitions $^4F_{3/2} \to {}^4I_{9/2 \text{ to } 15/2}$:

$$\tau_{rad} = 1/\sum_{J'} A_{JJ'} \tag{1}$$

or

$$\tau_{rad} = \beta_{JJ'}/A_{JJ'} \,. \tag{2}$$

Using the connection of $A_{JJ'}$ with the probabilities of induced absorption within the same manifold system, usually one can determine $A_{JJ'}$ by

$$A_{JJ'} = \sum_{i,j} A_{ij} \,, \tag{3}$$

where

$$A_{ij} = \frac{8\pi n^2}{N_j \lambda_{ij}^2} \int k(\nu)\, d\nu \tag{4}$$

(n refractive index, λ_{ij} wavelength of Stark transition, $k(\nu)$ absorption coefficient at frequency ν, and N_j population of the j-th Stark level belonging to the $^4I_{9/2}$ ground state). To apply this method of τ_{rad} determination to $Nd_3Ga_5O_{12}$ crystals, we preliminarily appointed those lines from its absorption spectra ($^4I_{9/2} \to {}^4F_{3/2}$), belonging to the basic activator centre (Nd^{3+} ions in structural "c" positions). Using formulae (2) to (4) and the parameter $X = 0.38$ measured according to [4] by absorption into two manifolds $^2P_{1/2}$ and $^4I_{15/2}$ we find the value of $\tau_{rad} = 0.268$ ms. This value corresponds to $\beta_{J(9/2)} = 0.306$; $A_{J(9/2)} = 1148\ s^{-1}$; and $\sum_{J'} A_{JJ'} = 3752\ s^{-1}$. In the calculations the following values were used: $n = 1.95$ [22] and $N_0 = 1.225 \times 10^{22}\ cm^{-3}$ for Nd^{3+} ions in "c" positions. This calculated τ_{rad} value practically coincides with the experimentally measured $\tau_{rad} = 0.27$ ms of the isostructural and very similar $Gd_3Ga_5O_{12}:Nd^{3+}$ (0.3 at%) crystals with $X \approx 0.41$ [4]. Such a correspondence between calculated and measured values indicates the reliability of the method suggested above for the determination of τ_{rad} in crystals highly concentrated and self-activated with Nd^{3+} ions. This method was recently applied to self-activated $K_5Nd(MoO_4)_4$ laser crystals [23].

The second way of determining the spectroscopic quality parameter X of $Gd_3Ga_5O_{12}:Nd^{3+}$ crystals is based on measurements of the ratio of intermanifold luminescence intensities $I_{JJ'}$ of two non-resonant $^4F_{3/2} \to {}^4I_{11/2}$ and $^4F_{3/2} \to {}^4I_{13/2}$ transitions and on the dependences $\beta_{JJ'}(X)$, known from [4],

$$\frac{I_{J(11/2)}}{I_{J(13/2)}} = \frac{\beta_{J(11/2)}}{\beta_{J(13/2)}} = Y .$$

Here, $I_{JJ'}$ is the ratio of luminescence intensity to the number of photons. This relation, denoted as luminescence intensity parameter Y, is connected with the spectroscopic quality parameter X by the simple formula

$$X = 0.765Y - 2.96 . \qquad (5)$$

The measurements at 300 K yield $Y = 4.4$, corresponding to $X = 0.4$ to 0.5. The same agreement in the X parameter, derived from absorption and luminescence measurements, one of us has obtained for isotropic ($Y_3Al_5O_{12}$ and $Bi_4Ge_3O_{12}$) and anisotropic ($YAlO_3$, $KGd(WO_4)_2$, $LuAlO_3$, and $K_5Nd(MoO_4)_4$) laser crystals with Nd^{3+} ions [24].

Though the luminescence method is very simple, its application without considering absorption data, demands high care. The behaviour of the $\beta_{JJ'}(X)$ dependence is determined [4] by squares of matrix elements $\langle||U^{(t)}||\rangle$, which, in turn, depend on the wavefunctions of the Nd^{3+} states. The choice of the wavefunctions (intermediate coupled wavefunctions or truncated intermediate coupled wavefunctions) [4, 25] as well as their absolute vales [26], cannot be considered as well established[2]) at present. The same consideration holds for the X determination from absorption. However, since the coefficients $\beta_{JJ'}$ slightly depend on X, the choice of the wavefunctions does not lead to a significant accuracy in the values of $\beta_{JJ'}$. On the other hand, due to the strong dependence of the parameter X on Y the luminescence method yields the essential definition of $\beta_{JJ'}$ if one takes different wavefunctions.

Thus, the parameter $X = 0.41$ of $Gd_3Ga_5O_{12}:Nd^{3+}$ crystal corresponds to the following coefficients $\beta_{JJ'}$ [4]: $\beta_{J(9/2)} = 0.3144$, $\beta_{J(11/2)} = 0.5541$, $\beta_{J(13/2)} = 0.1255$, and $\beta_{J(15/2)} = 0.0598$. Using these data and the spectroscopic characteristics obtained from the analysis of luminescence spectra for the ${}^4F_{3/2} \to {}^4I_{11/2}$ and ${}^4F_{3/2} \to {}^4I_{13/2}$ transitions, values of β_{ij}, A_{ij}, and $\sigma^{p}_{e,ij}$ were determined by the formulae [5, 24]

$$A_{ij} = \frac{\beta_{ij}}{\tau_{rad}} \frac{\sum_i b_i}{b_i} \qquad (6)$$

and

$$\sigma^{p}_{e,ij} = \frac{\lambda^2_{ij}\beta_{ij}}{4\pi^2 n^2 \, \Delta\nu_{lum}\, \tau_{rad}} . \qquad (7)$$

($b_i = \exp[-\Delta E({}^4F_{3/2})/kT]$ is the i-th Stark level Boltzmann factor of the metastable ${}^4F_{3/2}$ state relative to its first low level ($i = 1$) and $\Delta\nu_{lum}$ is the luminescence linewidth). Determining σ^{eff}_e we have taken into account that in the luminescence spectra of $Gd_3Ga_5O_{12}:Nd^{3+}$ crystals in the ${}^4F_{3/2} \to {}^4I_{11/2}$ transition at 300 K intensive lines connected with SE partially mutually overlap (cf., Fig. 2 in [2]). This is necessary, because for the calculation of laser properties and measurements of spectroscopic characteristics by laser methods, one needs an effective cross-section σ^{eff}_e which directly determines the gain coefficient of the laser medium. For the SE line of the additional emission channel ${}^4F_{3/2} \to {}^4I_{13/2}$, σ^{eff}_e is given by $\sigma^{eff}_e = (\pi/2)\, \sigma^{p}_{e,ij}$ for an inhomogeneously broadened line. The account of the Boltzmann population of the ${}^4F_{3/2}$ levels in calculating σ^{eff}_e and $\sigma^{p}_{e,ij}$ gives the possibility of presenting the luminescence spectra for non-resonant transitions (partially for ${}^4F_{3/2} \to {}^4I_{11/2 \text{ to } 15/2}$) as frequency curves of laser gain (for example, in long pulse and CW regime). It is important to note that σ^{eff}_e of the most

[2]) Therefore, some difference may occur in the numeral values in (5) due to the choice of wavefunctions [4, 25].

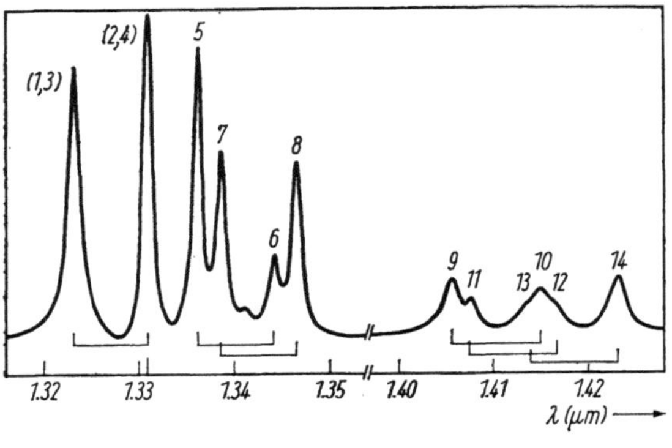

Fig. 2. Luminescence spectrum of Nd^{3+} in $Gd_3Ga_5O_{12}$ crystals at 300 K; ${}^4F_{3/2} \to {}^4I_{13/2}$ transition and crystal field splitting scheme for ${}^4F_{3/2}$ and ${}^4I_{13/2}$ manifolds at 77 and 300 K. Level positions in cm^{-1}; transitions among them in Å. The thick arrow indicates induced transition. Lines in the spectra and transitions in the schemes are connected by corresponding numeration. Splitting of the ${}^4F_{3/2}$ state is shown by brackets

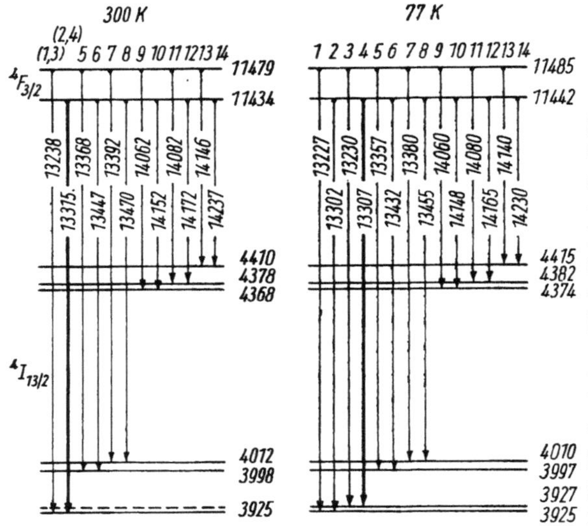

intensive line ($\lambda_g = 10621$ Å) of $Gd_3Ga_5O_{12}:Nd^{3+}$ crystals, connected with the 11479 cm^{-1} ${}^4F_{3/2} \to {}^4I_{11/2}$ 2064 cm^{-1} transition, is 12% smaller than that of the A-line ($\lambda_g = 10641.5$ Å) of $Y_3Al_5O_{12}:Nd^{3+}$ crystals ($\sigma_e^{eff} = 3.3 \times 10^{-19}$ cm^2) [5, 24]. This shows that $Gd_3Ga_5O_{12}:Nd^{3+}$ crystals may have some advantage in comparison with $Y_3Al_5O_{12}:Nd^{3+}$ crystals, for instance when using them in lasers operating in long pulse regime and Q-switched regime. The main results obtained from spectral and laser measurements are listed in Table 1.

Finally, Fig. 2 shows the Nd^{3+} luminescence spectrum in $Gd_3Ga_5O_{12}$ crystals at 300 K according to the additional ${}^4F_{3/2} \to {}^4I_{13/2}$ transition. Further, Fig. 2 shows the crystal fields splitting schemes of the ${}^4F_{3/2}$ and ${}^4I_{13/2}$ manifolds at 77 and 300 K with identification of the induced transitions. The stark level position of the ${}^4I_{13/2}$ manifold was also tested by the ${}^4I_{9/2} \to {}^4I_{13/2}$ absorption spectra.

Acknowledgements

The authors wish to express their sincere thanks to B. K. Vainshtein, A. M. Prokhorov, K. Junge, L. Wieczorek, and Kh. S. Bagdasarov for their interest in our work and to Tran-Ngoc, G. A. Denisenko, I. A. Sherbakov, L. Li for discussion and help.

References

[1] J. GEUSIC, H. M. MARCOS, and L. G. VAN UITERT, Appl. Phys. Letters **4**, 182 (1964).

[2] KH. S. BAGDASAROV, G. A. BOGOMOLOVA, M. M. GRITSENKO, A. A. KAMINSKII, A. M. KEVORKOV, A. M. PROKHOROV, and S. E. SARKISOV, Dokl. Akad. Nauk SSSR **216**, 2017 (1974).

[3] A. A. KAMINSKII, G. A. BOGOMOLOVA, and A. M. KEVORKOV, Izv. Akad. Nauk SSSR, Ser. neorg. Mater. **11**, 884 (1975).

[4] A. A. KAMINSKII and L. LI, phys. stat. sol. (a) **26**, 593 (1974).
[5] A. A. KAMINSKII, Laserniye kristally, Izd. Nauka, Moscow 1975.
[6] A. A. KAMINSKII, A. O. IVANOV, S. E. SARKISOV, I. V. MOCHALOV, V. A. FEDOROV, and L. LI, Zh. eksper. teor. Fiz. **71**, 984 (1976).
[7] R. C. LINARES, Solid State Commun. **2**, 229 (1964).
[8] C. D. BRANDLE and A. J. VALENTINO, J. Crystal Growth **12**, 3 (1972).
[9] B. COCKAYNE and J. M. ROSLINGTON, J. Mater. Sci. **8**, 601 (1973).
[10] D. F. O'KANE, V. SADAGOPAN, E. A. GIESS, and E. MENDEL, J. Electrochem. Soc. **120**, 1272 (1973).
[11] D. MATEIKA, J. HERRNRING, R. RATH, and C. RUSCHE, J. Crystal Growth **30**, 311 (1975).
[12] K. TAKAGI, T. IKEDA, T. FUKAZAWA, and M. ISHII, J. Crystal Growth **38**, 206 (1977).
[13] I. A. CHISTYI, V. F. KITAEVA, V. V. OSIKO, N. N. SOLOB, B. P. STARIKOVA, and M. I. TIMOSHECHKIN, Fiz. tverd. Tela **17**, 1434 (1975).
[14] M. ALLIBERT, C. CHATILLON, J. MARESCHAL, and F. LISSALDE, J. Crystal Growth **23**, 289 (1974).
[15] M. KH. ASHUROV, YU. K. VORONKO, V. V. OSIKO, A. A. SOBOL, P. P. STARIKOV, M. I. TIMOSHECHKIN, and A. YA. YABLONSKII, phys. stat. sol. (a) **35**, 645 (1976).
[16] G. V. MAKSIMOVA, V. V. OSIKO, A. A. SOBOL, and M. I. TIMOSHECHKIN, Izv. Akad. Nauk SSR, Ser. neorg. Mater. **9**, 1763 (1973).
[17] J. BOHM, H.-D. KÜRSTEN, P. REICHE, and R. SCHALGE, phys. stat. sol. (a) **39**, 517 (1977).
[18] A. A. KAMINSKII, D. SCHULTZE, B. HERMONEIT, S. E. SARKISOV, L. LI, J. BOHM, P. REICHE, R. EHLERT, A. A. MAYER, V. A. LOMONOV, and V. A. BALASHOV, phys. stat. sol. (a) **33**, 737 (1976).
[19] A. A. KAMINSKII, G. A. BOGOMOLOVA, KH. S. BAGDASAROV, and A. G. PETROSYAN, Optika i Spektroskopiya **39**, 1119 (1975).
[20] A. A. KAMINSKII, V. V. OSIKO, and M. M. FURSIKOV, Zh. eksper. teor. Fiz., Pisma **4**, 92 (1966).
[21] A. A. KAMINSKII, phys. stat. sol. (a) **20**, K51 (1976).
[22] J. BOHM, P. REICHE, D. SCHULTZE, and H. ZSCHAECK, Kristall und Technik **13**, K10 (1978).
[23] A. A. KAMINSKII, S. E. SARKISOV, J. BOHM, P. REICHE, D. SCHULTZE, and R. UECKER, phys. stat. sol. (a) **43**, 71 (1977).
[24] A. A. KAMINSKII, Laser Crystals, Springer-Verlag, Berlin/Heidelberg/New York 1978.
[25] T. S. LOMHEIM and L. G. DESHAZER, Optics Commun. **24**, 89 (1978).
[26] H. M. CROSSWHITE, Coll. internat. C.N.R.S. N° 255, 1977 (p. 65).

(Received June 12, 1978)

phys. stat. sol. (a) 49, 313 (1978)

Subject classification: 1.3 and 14.3.2; 1.5; 21.6; 22.1.2

Advanced Research and Applications Corporation, Sunnyvale[1])

Investigation of Solid–Solid Reactions of Au Films on Silicon[2])

By

T. J. MAGEE and J. PENG

Reactions between evaporated Au films deposited on Si wafers at substrate temperatures in the range 25 to 200 °C and subsequently annealed in vacuum or air at temperatures up to 400 °C are investigated using transmission electron microscopy/diffraction, Auger electron spectroscopy, and current–voltage measurements. Using newly developed radioisotopic image enhancement techniques, gold silicide reactants in thin ($\leqq$ 50 Å) surface or interfacial zones are identified directly from electron diffraction plates. At substrate deposition and/or anneal temperatures $\geqq$ 100 °C, the orthorhombic δ-gold silicide (Au_3Si) is detected at the Au–Si interface and within thin ($\approx$ 35 Å thick) zones at the surface of the Au film. At temperatures $\geqq$ 330 °C, the appearance of the δ(Au_3Si) phase and at least one or more additional intermetallic silicides are noted throughout the film. The development of prominent Au–Si mixing zones and Au–Si interdiffusion is correlated with reductions in Schottky barrier height and increases in forward current at deposition (or vacuum anneal) temperature $\geqq$ 100 °C. The formation of the mixed Au–Si phase within the film is also shown to reduce room temperature thermal aging effects and the formation of oxides at the surface of the film.

Mit Hilfe von Transmissionselektronenmikroskopie/Beugung, Auger-Elektronenspektroskopie und Strom–Spannungs-Messungen werden Reaktionen untersucht zwischen Goldbelägen, die bei Substratflächentemperaturen von 25 bis 200 °C auf Si-Scheiben aufgedampft und danach im Vakuum oder in Luft bei Temperaturen bis zu 400 °C wärmebehandelt wurden. Unter Verwendung einer kürzlich entwickelten Bildverstärkungstechnologie vermittels Radioisotopen ist es möglich, Goldsilizidreagenzien in dünnen ($\leqq$ 50 Å) Oberflächen- oder Grenzflächenzonen direkt auf den fotografischen Elektronenbeugungsaufnahmen zu identifizieren. Bei einer Substratflächenbedampfungs- und/oder Wärmebehandlungstemperatur $\geqq$ 100 °C kann das rhombische δ-Goldsilizid (Au_3Si) an der Au–Si-Grenzfläche und innerhalb dünner ($\approx$ 35 Å) Zonen an der Oberfläche des Au-Belags festgestellt werden. Bei Temperaturen $\geqq$ 330 °C wird das Erscheinen der δ-(Au_3Si)-Phase und zumindest einer oder mehrerer zusätzlicher intermetallischer Silizide über den ganzen Belag beobachtet. Die Entwicklung hervortretender Au–Si-Mischzonen und Au–Si-Interdiffusion wird mit einer Verminderung der Schottky-Schwellenhöhe und dem Ansteigen des Durchlaßstromes bei Bedampfungs-(oder Vakuum-Wärmebehandlungs-)temperaturen $\geqq$ 100 °C in Verbindung gebracht. Ebenfalls wird nachgewiesen, daß die Bildung der Mischphase Au–Si innerhalb des Belags die thermischen Alterungseffekte bei Raumtemperatur und die Bildung von Oxiden an der Oberfläche des Belags vermindert.

1. Introduction

Metal–semiconductor contacts are usually formed by depositing a thin-metal film on the semiconductor substrate and then subjecting the structure to a heat treatment. Thin-metal films on single-crystal silicon have received the greatest attention because of the importance of these systems in silicon planar device technology where metal–silicon junctions are used to provide ohmic electrical contacts to device or to form Schottky barrier rectifying contacts.

[1]) Sunnyvale, California 94086, USA.

[2]) Research supported by Army Research Office Contract No. DAAG29-77-C-0024.

The Au–Si system has been studied by a number of investigators [1 to 5]. It has been established that at temperatures > 150 °C (below the eutectic temperature of 370 °C), silicon migrates through Au thin-film layers and accumulates on the surface in the form of silicon dioxide. The oxide layer is nonuniform in thickness and its growth rate is influenced by both the oxidizing ambient and the crystallographic orientation of the silicon substrate. Recent results [5] obtained using Auger electron spectroscopy profiling on the Au–Si structure subjected to extended vacuum anneals have suggested that at temperatures < 200 °C, Au reacts with Si to form thin-layer (30 Å) silicides at the surface. However, the composition of the silicides was only inferred from Auger spectral line splitting and amplitude ratio measurements. To date, no detailed studies have been undertaken of solid–solid reactions within the Au film or substrate mixing zone at temperatures significantly below the eutectic temperature. The purpose of this paper is to provide a more consistent evaluation of Au–Si reactions using correlated data obtained from transmission electron microscopy/diffraction (TEM/TED), Auger electron spectroscopy (AES), and current–voltage (I–U) measurements.

2. Experimental

Polished, single-crystal Si wafers of (100) or (111) orientation obtained from Semi Metals Corporation were used for this study. The samples were n-type with resistivities ranging from 2.7 to 3.8 Ω cm. The wafers were chemically cleaned and immersed in a dilute hydrofluoric acid solution to remove surface oxidation before depositing the Au thin films. Gold films ranging in thickness from 100 to 1000 Å were deposited on the silicon surfaces at a vacuum level of $\approx 10^{-7}$ Torr. During deposition, substrate temperatures were maintained in the range of 25 to 200 °C. Postdeposition thermal anneals were done under vacuum or in air at temperatures up to 400 °C for times varying from 5 min to 12 h.

Annealed or "as-deposited" Au films were removed from the Si by dissolving the substrate in a solution of five parts HF and one part HNO_3. The film was then lifted from the solution with a 3 mm diameter grid, and was subsequently rinsed and dried for TEM examination. In separate experiments, the Au film was stripped from the Si substrate using an aqua regia solution. The sample was then jet-thinned from the back surface for TEM/TED analysis of Si–Au interfacial and mixing regions of the substrate. Annealed samples were prepared for TEM by first immersing the specimen in a dilute HF solution and then exposing it to (KI + I_2) and aqua regia solutions to remove SiO_2, Au + Si mixtures, and residual Au before jet thinning.

Since TED patterns from thin (35 to 50 Å) layers within interfacial or mixing zones are often obscure or ill-defined, it was necessary to develop a technique for clearly and reproducibly imaging diffraction data. Using a radiotoning procedure adapted from a technique described by Thackray et al. [6, 7], we intensified electron diffraction data recorded on Ilford plates and obtained autoradiographs on secondary plates to reconstruct the intensified diffraction data. This procedure has proven effective in these experiments and other studies of interfacial transitional zones of heterojunctions and will be discussed in detail in a separate paper [8].

Samples used for electrical measurements were coated on the back surface with a 1000 Å thick Au–Sb film and alloyed at 450 °C for 30 min prior to deposition of Au films. Gold dots, ≈ 1 mm in diameter and 1000 Å thick, were vacuum-deposited on the front surface at substrate temperatures of 25, 100, and 160 °C, respectively. After deposition, a number of samples were vacuum-annealed for 30 min at temperatures ranging from 150 to 400 °C. Air anneals were done in a tube furnace in the same temperature range.

Auger profiling of as-deposited and annealed structures was conducted in the vacuum chamber of an Auger spectroscopy unit obtained from Physical Electronics Inc. Single and dual Ar-ion sputtering guns were used to provide depth removal rates up to a maximum of 100 Å/min.

3. Results

3.1 Transmission electron microscopy/diffraction

Transmission electron micrographs were obtained on Au thin films evaporated on Si wafers at substrate temperatures ranging from 25 to 190 °C. A representative bright-field electron micrograph is shown in Fig. 1 for a film evaporated at a substrate temperature of 100 °C. Average grain size measurements obtained from similar

Table 1

Average grain sizes in Au films evaporated on (100) Si wafers as a function of Si substrate deposition temperature

substrate temperature (°C)	grain size (Å)
20	50 to 500
50	1000
100	2000 to 5000
150 to 160	$<$10000
170 to 190	$\geqq$10000

electron micrographs are summarized in Table 1. The data in Table 1 indicate that Au grain size increases rapidly as a function of increasing substrate temperatures from

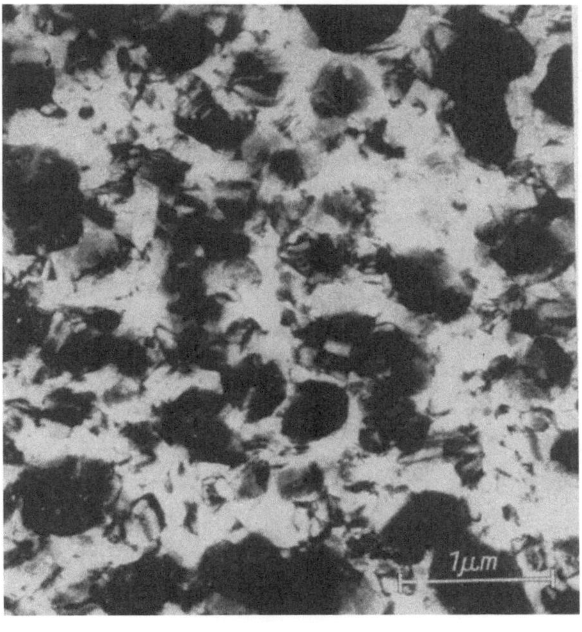

Fig. 1. Bright field electron micrograph of Au film evaporated on (100) silicon at substrate deposition temperature of 100 °C

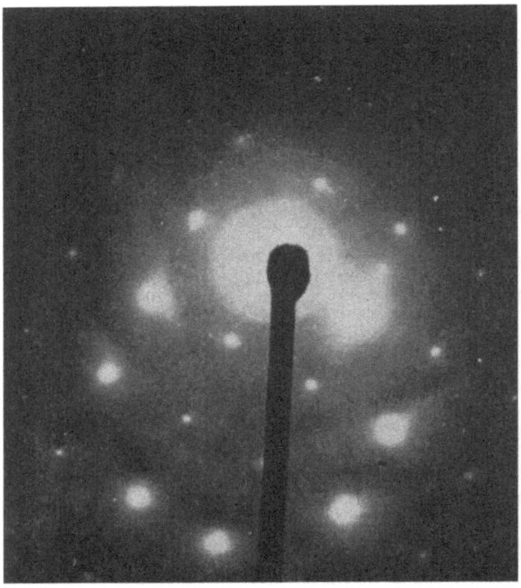

Fig. 2. Transmission electron diffraction pattern obtained from mixing zone of Au/Si structure. Substrate deposition temperature 150 °C; no post deposition anneal. Diffraction pattern enhanced using radioisotopic toning techniques

Fig. 3. Transmission electron diffraction pattern obtained from mixing zone of annealed Au/Si structure. Substrate deposition temperature 50 °C; annealed in air for 12 h at 400 °C

25 to 160 °C. At higher temperatures, the growth rate is substantially reduced, and apparent size saturation ($\approx 1 \mu m$) is obtained at temperatures below the Au–Si eutectic (370 °C).

After Au films were removed from the substrate with aqua regia, the as-prepared Si samples were thinned from the back surface for TEM analysis. Examination of interfacial and near-surface regions of the Si showed no detectable reactions or disordering of the lattice at substrate temperatures < 100 °C. However, at temperatures > 100 °C, diffraction patterns obtained from the mixing region exhibited a series of additional polycrystalline rings, as shown in Fig. 2. Table 2 presents an analysis of data from this diffraction pattern and comparison with published data [9]. These data indicate that the second-phase crystallites are orthorhombic δ-gold silicide

Table 2

Analysis of transmission electron diffraction pattern (Fig. 2)*)

measured "d" values (Å)	ASTM and published "d" values (Å)	
	Si	δ-phase (Au_3Si)
3.195		3.195 (210)
3.116	3.116 (111)	
2.558		2.543 (301)
2.284		2.254 (220)
1.920	1.920 (220)	
1.637	1.638 (311)	
1.518		1.519 (126)
1.359	1.357 (400)	

*) (*hkl*) values are given in parentheses.

Table 3
Analysis of transmission electron diffraction pattern (Fig. 3)*)

measured "d" values (Å)	ASTM published and calculated "d" values (Å)							
	Si		α-phase		γ-phase		δ-phase	
5.540					5.629	(222)		
3.185	3.138	(111)						
3.049					3.083	(260)		
2.877			2.772	(021)			2.759	(020)
2.283			2.263	(222)			2.254	(220)
2.038					1.990	(844)		
1.915	1.920	(220)						
1.763			1.753	(420)				
1.636	1.638	(311)						
1.579							1.561	(107)
1.506							1.484	(430)
1.360	1.357	(400)						
1.289							1.272	(602)
1.234	1.246	(331)						
1.185							1.193	(228)

*) (*hkl*) values are given in parentheses.

(Au_3Si) containing 17.9 at% Si. Selected area electron diffraction patterns showed the presence of Au_3Si crystallites distributed throughout the Au–Si mixing region.

Au–Si structures prepared at variable substrate temperatures were annealed in a vacuum system at 10^{-7} Torr to the eutectic point (≈ 370 °C) of the Au–Si system. TEM examination of the Au films revealed that grain growth occurs rapidly on structures annealed at temperatures < 200 °C. Above anneal temperatures of 200 °C, recordering is also observed at the top surface of the Au in the vicinity of grain boundaries. Mixing regions within the film and near-surface regions of the vacuum-annealed Au–Si structures were characterized by the presence of δ-phase gold silicide (Au_3Si) crystallites. As the temperature and anneal time were increased, the number of crystallites and sharpness of electron diffraction rings also increased. In particular, the development of Au_3Si crystallites was readily detected in samples annealed for long intervals slightly below the eutectic temperature.

In contrast to the results obtained for vacuum-annealed structures, a thick SiO_2 layer develops on the surface of gold films of Au–Si structures annealed in air. The thickness of this layer is proportional to the initial thickness of the Au thin film and the annealing time, in agreement with previous results [4].

Within the mixing region reactions are often complex resulting in the possible formation of more than one phase of gold silicide. The δ-phase was identitied, although a number of weak diffraction rings were observed that could not be correlated with either the δ-phase, Si, or Au. In an earlier paper, Anatharaman et al. [10] identified two face-centered cubic intermetallic phases of the Au–Si system, but unfortunately no detailed information was presented regarding indexing of the diffraction lines. We calculated "d" values of all possible reflections for the f.c.c. lattice, using the published values for the lattice parameters of the α-phase ($a = 7.84$ Å) and the γ-phase ($a = 19.50$ Å). These data were then used to determine if measured "d" values obtained from diffraction patterns could be correlated with calculated values for the α-phases and γ-phases.

Fig. 3 shows a representative diffraction pattern obtained within the mixing region of an Au–Si structure annealed in air at 400 °C for 12 h. Comparison of established and experimentally measured "d" values is shown in Table 3 and indicates good agreement for Si and δ-gold silicide lines. A number of additional lines can then be correlated with calculated d-values for the α- and γ-phases. However, the limited number of lines and the correlation with calculated rather than reported "d" values makes the identification of the α- and γ-phases somewhat tentative. Since the appearance of more than one phase is noted within the mixing regions of all Au–Si structures annealed in air, we can assume formation of the orthorhombic δ-phase and at least one or more additional intermetallic silicides, possibly similar to the α- and γ-phases.

3.2 Auger electron spectroscopy

AES profiling analyses were obtained on "as-deposited" and vacuum- or air-annealed Au/Si samples as a function of temperature. Fig. 4 shows the results obtained on unannealed structures formed by vacuum deposition of Au at Si substrate temperatures of 25, 100, and 160 °C, respectively. Fig. 4a shows that room temperature deposition of the Au thin film produces no significant outdiffusion of Si into the Au layer. As the substrate temperature is increased (Fig. 4b and 4c), Si diffuses rapidly into the Au film. At 100 °C, Si has migrated to the surface of the Au and a mixed Au–Si film is formed. Similar experiments on structures annealed in vacuum at temperatures up to 330 °C indicate that the amount of Si outdiffusion increases as the annealing temperature and annealing time increase. In all cases, there is a noticable diffusion of Au into the Si substrate as a function of increasing substrate deposition and/or annealing temperature.

Fig. 5 shows AES profiles for Au films that have been deposited on Si at substrate temperatures of 25, 100, and 160 °C, respectively, and annealed in air at 200 °C. As the substrate deposition temperature is increased, significant intermixing of Si and Au occurs within the film and at the interface. The maximum near-surface concentration of oxygen is also present on films deposited at 20 °C. At higher substrate deposition temperatures, the oxygen concentration is reduced, implying that the thickness of

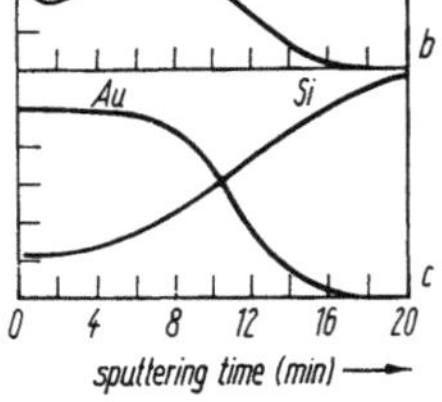

Fig. 4. AES depth profiles of Au/Si structures prepared at various substrate deposition temperatures (no post-anneal). a) $T_{dep} = 25$, b) 100, c) 160 °C

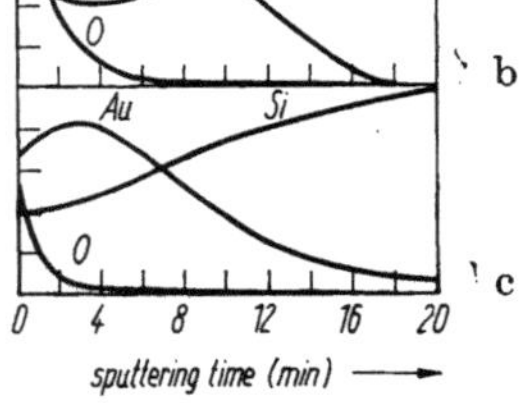

Fig. 5. AES depth profiles of 200 °C air annealed (30 min) Au/Si structures prepared at various substrate deposition temperatures. a) $T_{dep} = 25$, b) 100, c) 160 °C

SiO_2 overlayers on Au/Si structures at moderate (air) annealing temperatures is inversely proportional to the substrate temperature when the films were deposited.

We have also observed varying amounts of silicon and oxygen on the surfaces of films deposited at room temperature and allowed to age for several days in air. From AES profiles, the thicknesses of the oxide layers were determined to be ≈ 200 Å for samples deposited at room temperature, ≈ 50 Å for films deposited at 100 °C, and ≦ 20 Å for layers deposited at 160 °C.

For films deposited at room temperature, Si diffuses rapidly through the Au to accumulate at the surface. Annealing in air results in the formation of irregularly distributed oxides that reach a saturation thickness in proportion to the Au film thickness. Films deposited at substrate temperatures > 100 °C exhibit a pronounced mixing phase and relatively inhibited growth of oxides at the surface. In addition, the mixed Au + Si phase within the film reduces room temperature thermal aging effects and the formation of oxides.

We observed two low energy Auger peaks that can be associated with the formation of silicides in as-deposited and vacuum-annealed (200 °C, 30 min) films deposited at 25 °C. These were positioned at 85 and 88.5 eV and are attributed to the splitting of the Si-LVV peak at 92 eV. In agreement with Green and Bauer [5], we observed an apparent silicide thickness of ≈ 35 Å at the surface of the film. However, our data indicate that the probable composition of the silicide is Au_3Si, as confirmed in direct identification from electron diffraction plates using radioisotopic image enhancement techniques.

3.3 Electrical measurements

The forward current–voltage characteristics will be particularly sensitive to the development of interfacial reactants or transitional layers at the metal–semiconductor interface. For comparison and correlation with the results discussed in the previous section, we show in Fig. 6 to 8 the I–U characteristics for samples prepared under various substrate deposition and annealing conditions.

The thermionic emission theory for charge transport in Schottky barrier structures describes the barrier height, Φ_B, in the following equation [11]:

$$\Phi_B \approx \frac{kT}{q} \ln \frac{A^{**}T^2}{J_s}, \tag{1}$$

where k is Boltzmann's constant; A^{**} is the effective Richardson constant; and J_s is the saturation current density. If the forward current is extrapolated to the current axis on an $\ln I_f$ vs. U_f plot, the intersection on the axis defines the saturation current in the relation $I_s = J_s A$, where A is the metallurgical area of the contact. Using (1), the value of Φ_B can then be determined. However, this approach must be used cautiously since (1) is valid only in the pure thermionic emission region for moderately doped substrates when the ideality factor ≈ 1 and where the empirical approach yields the smallest percentage error in the value of Φ_B [12].

Under these restrictions, we determined approximate barrier heights from Fig. 6 for films deposited at 25 and 100 °C to compare with results reported earlier in similar experiments. The barrier height for the Au film deposited at 25 °C is ≈ 0.81 eV, in reasonable agreement with previously published results. At a deposition temperature of 100 °C, $\Phi_B \approx 0.77$ eV reflecting the initial development of the interdiffusion regions and thin interfacial layers on the substrate. For films deposited at 160 °C, the I–U characteristics show that the ideality factor departs significantly from 1.00 and the possible error in determining Φ_B using the extrapolated slope method can be large [12]. In all cases, we observe a significant displacement in the forward I–U charac-

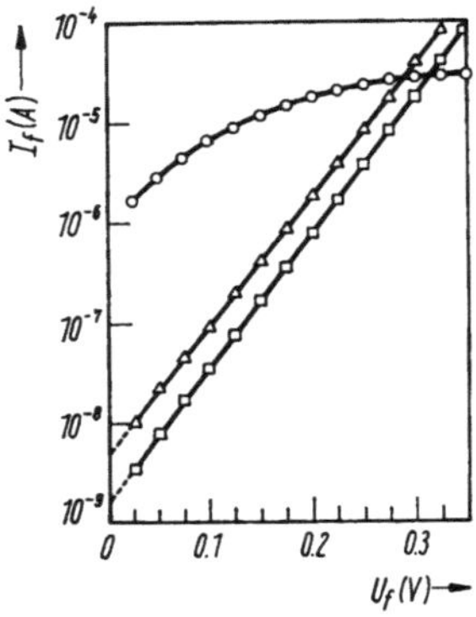

Fig. 6

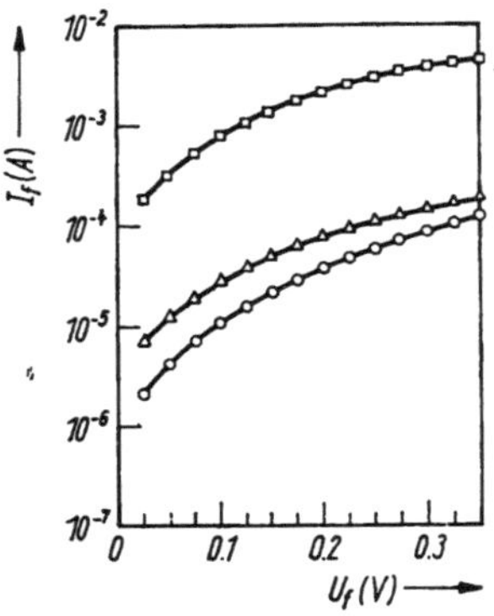

Fig. 7

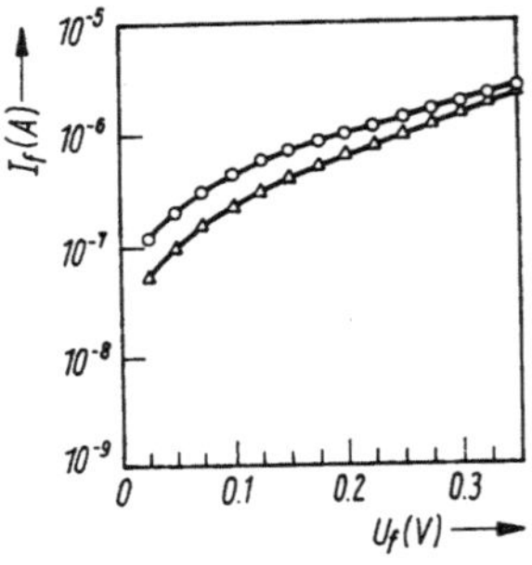

Fig. 8

Fig. 6. Forward current–voltage characteristics of Au films deposited on (100) silicon at variable substrate temperatures (no post-anneal). □ sample 1 AD, $T_{dep} = 25$ °C; △ sample 2 AD, $T_{dep} = 100$ °C; ○ sample 3 AD, $T_{dep} = 160$ °C

Fig. 7. Forward current–voltage characteristics of 200 °C vacuum annealed (30 min) Au films deposited on (100) silicon at variable substrate temperatures. □ sample 1 V, $T_{dep} = 25$ °C; △ sample 2 V, $T_{dep} = 100$ °C; ○ sample 3 V, $T_{dep} = 160$ °C

Fig. 8. Forward current–voltage characteristics of 200 °C air annealed (30 min) Au films deposited on (100) silicon at variable substrate temperatures. △ sample 1 A, $T_{dep} = 100$ °C; ○ sample 2 A, $T_{dep} = 160$ °C

teristics at lower voltages as the deposition temperature is increased from 25 to 160 °C, reflecting the development of prominent Au–Si mixing phases and the transition from thermionic emission toward field emission dominated conduction.

The effect of 200 °C vacuum annealing on charge transport properties of films deposited at variable substrate deposition temperatures is shown in Fig. 7. As the substrate temperature is increased from 25 to 160 °C, the forward current also exhibits a pronounced increase relative to unannealed samples deposited at temperatures $<$ 160 °C. The most pronounced variation, however, occurs for annealed samples deposited at 25 °C. On these samples, the forward current increases several orders of magnitude after annealing, corresponding to the development of a highly mixed region within the film and on apparent deep indiffusion of Au into the substrate. Correlated AES and TED data indicate that prominent mixing and silicide formation occur during deposition at substrate temperatures $>$ 100 °C. The I–U characteristics then suggest that the prior formation of Au–Si and subsequent motion of Au into the Si substrate during deposition will inhibit the outdiffusion of Si and subsequent motion of Au into the Si substrate during annealing. This conclusion is further supported by the similarity of I–U characteristics for as-deposited and annealed samples deposited at 160 °C.

In contrast to the results obtained for vacuum-annealed samples, structures annealed in air at 200 °C develop varying thicknesses of SiO_2 films at the surface. Samples deposited at 25 °C and annealed at 200 °C exhibit relatively thick discontinuous SiO_2 films on the surface, and the I–U measurements show a large amount of variability, depending upon the uniformity and lateral coverage of the oxide. At deposition temperatures $\geqq$ 100 °C, oxide layers on the ennealed samples are relatively thin, as indicated in AES and TED experiments. The I–U characteristics of these samples were relatively consistent and reproducible. Representative I–U plots are shown in

Fig. 8 for air-annealed samples deposited at 100 and 160 °C. In comparison to the results shown in Fig. 7, the forward currents are somewhat lower for air-annealed samples, implying that the thin oxide layers at the surface are dominating injection at the contact probe.

4. Conclusions

In this study, we have shown that the grain size of Au films formed at variable substrate deposition temperatures on Si substrates increases from 50 to 500 Å at room temperature to an apparent size saturation of $\approx 1\ \mu m$ at temperatures > 170 °C. At substrate deposition and/or anneal temperatures $\geqq 100$ °C, the orthorhombic δ-gold silicide (Au_3Si) is detected within the mixing region. The initial phase of growth is observed at the Au–Si interface and within thin (≈ 35 Å thick) zones at the surface of the Au film. Correlated TEM and AES results suggest that the growth of the silicide at the surface of the film is initiated preferentially at grain boundary sites. At higher anneal temperatures ($\geqq 330$ °C), the appearance of the δ-(Au_3Si) phase and at least one or more additional intermetallic silicides, possibly similar to the α- and γ-silicide phases, is noted throughout the film.

At substrate deposition or vacuum anneal temperatures > 100 °C, Si outdiffuses readily into the Au film, producing a heavily mixed Au–Si region and Au indiffusion. Annealing in air produces similar Au–Si mixing zones and a discontinuous SiO_2 surface layer that increases in thickness as the Au film thickness increases. Growth of the oxide layer in annealed films deposited at substrate temperatures $\geqq 100$ °C, however, is inhibited by the formation of mixing regions and silicides. In like manner, the development of oxide films during room temperature aging is more pronounced for films deposited at 25 °C and is negligible for films deposited at temperatures > 100 °C.

Correlated *I–U* measurements on samples deposited at variable substrate temperatures, and annealed in vacuum or air below the eutectic temperature, also reflect the metallurgical changes occurring within the Au film and at the Au–Si interface. At deposition (or vacuum anneal) temperatures $\geqq 100$ °C, the development of prominent Au–Si mixing zones yields a reduction in Schottky barrier height and a marked increase in forward current. Samples deposited at 25 °C and subsequently annealed in vacuum at 200 °C exhibit a larger increase in forward current than those deposited at higher temperatures. The results indicate that the formation of silicides and intermixing regions during deposition will be rate-limiting for additional metallurgical changes during post-deposition annealing. In contrast, samples annealed in air will exhibit variable *I–U* characteristics relative to the vacuum-annealed samples, reflecting the varying thickness and discontinuity of SiO_2 layers at the surface of the air-annealed Au-films.

Additional research is required to further identify changes in Schottky barrier height as a function of variable deposition and post-anneal temperature using the *C–U* method. In particular, precise *C–U* measurements would provide definitive information on the differences in room temperature aging effects for Au films deposited at variable substrate deposition temperatures. These experiments are currently in progress and will be described in a future publication.

References

[1] A. Hiraki, M. A. Nicolet, and J. W. Mayer, Appl. Phys. Letters **18**, 178 (1971).
[2] A. Hiraki and E. Lugujjo, J. Vacuum Sci. Technol. **9**, 155 (1972).
[3] A. Hiraki, E. Lugujjo, M. A. Nicolet, and J. W. Mayer, phys. stat. sol. (a) **7**, 401 (1971).
[4] A. Hiraki, E. Lugujjo, and J. W. Mayer, J. appl. Phys. **43**, 3643 (1972).
[5] A. K. Green and E. Bauer, J. appl. Phys. **47**, 1284 (1976).

[6] M. Thackray, Autoradiography of Photographic Images, Australian Atomic Energy Commission Tech. Rep. No. AAEC/E317, September 1974.
[7] M. Thackray, D. Roman, E. L. R. Hetherington, and H. R. Brian, Internat. J. appl. Radiation and Isotopes **23**, 79 (1972).
[8] T. J. Magee, J. Peng, M. Thackray, and R. Pettijohn, to be published (1978).
[9] G. A. Anderson, J. L. Bestel, A. A. Johnson, and B. Post, Mater. Sci. Engng. 7, 83 (1971).
[10] T. R. Anatharaman, H. L. Luo, and W. Klement, Jr., Nature **210**, 1040 (1966).
[11] S. M. Sze, Physics of Semiconductor Devices, John Wiley, New York 1969 (p. 393).
[12] J. D. van Otterloo and L. J. Gerritsen, J. appl. Phys. **49**, 723 (1978).

(Received June 19, 1978)

phys. stat. sol. (a) **49**, 323 (1978)

Subject classification: 1.2 and 10.1; 14.2; 21.1

Sektion Grundlagen des Maschinenwesens (a) and Sektion Physik (b) der Technischen Universität Dresden[1]) and Zentralinstitut für Festkörperphysik und Werkstofforschung der Akademie der Wissenschaften der DDR, Dresden[2]) (c)

On the Lattice Transformation in Plastically Deformed V_3Si Single Crystals Studied by the Kossel Technique

By

H.-J. ULLRICH (a), U. REINHOLD (b), S. DÄBRITZ (b), P. PAUFLER (b), K. KLEINSTÜCK (b), and B. PIETRASS (c)

Lattice parameters of specimens deformed plastically at high temperatures are determined between 7 and 293 K. Above and below the well known phase transition temperature of non-deformed crystals tetragonal and rhombic symmetry is detected after deformation, respectively. The results agree qualitatively with the temperature dependence of lattice parameters calculated for V_3Si single crystals under uniaxial pressure. Dislocations created by plastic deformation appear to have the same effect.

Gitterkonstanten an Proben, die bei hohen Temperaturen plastisch verformt wurden, werden zwischen 7 und 293 K gemessen. Oberhalb bzw. unterhalb der bekannten Phasenumwandlungstemperatur unverformter Kristalle wird tetragonale bzw. rhombische Symmetrie nach der Deformation nachgewiesen. Die Ergebnisse stimmen qualitativ mit dem berechneten Temperaturverlauf der Gitterkonstanten von V_3Si-Einkristallen unter einachsiger Druckspannung überein. Die durch die Verformung erzeugten Versetzungen rufen anscheinend die gleiche Wirkung hervor.

1. Introduction

As Batterman and Barrett [1, 2] pointed out, the intermetallic compound V_3Si (Cr_3Si type of structure) undergoes at $T_m \approx 21$ K a lattice transformation cubic–tetragonal that seems to be correlated to the high transition temperature, T_c, into the superconducting state. The lattice parameters a and c as well as T_c have been shown to depend on the chemical composition within the range of homogeneity of this compound [3 to 6]. In [7] a correlation of lattice transformation and T_c by means of the chemical potential is discussed. Paper [3] suggests an increase of T_c, provided the phase transition could be suppressed. The dependence of T_c on uniaxial pressure has been investigated from a theoretical point of view by one of the authors [8]. The predicted preferred orientation of tetragonal domains has been found experimentally in [9]. Even single-domain crystals may be produced. It is interesting to note a decrease of critical current density, j_c, with pressure [5].

No results of the influence of true plastic deformation on lattice transformation have become known hitherto. The following report is part of an investigation of the effects of irreversible deformation on superconducting properties.

2. Preparation and Characterization of the Specimens

Cylindrical single crystals of V_3Si were grown by Jurisch (for details. cf. [10]) and have been plastically deformed ($\approx 10\%$) at temperatures between 1340 and 1420 °C at stresses of about 1.8×10^7 Nm^{-2} [11]. Then the samples were sliced into wafers by a spark cutting technique. The wafers were thinned electrolytically and perforated

[1]) Mommsenstraße 13, DDR-8027 Dresden, DDR.

[2]) Postfach, DDR-8027 Dresden, DDR.

by jet polishing using Burck's device [12]. Fig. 1 shows a scanning micrograph and the profile of a specimen.

The total content of impurities amounted to less than 270 ppm. According to an autoradiographic study after neutron activation the impurities are mainly distributed homogeneously [10]. The chemical composition of the main components was obtained by a combined measurement of residual resistivity ratios $r_{20\,K}^{-1} \equiv R_{293\,K}/R_{20\,K}$ or $r_{17.5\,K}^{-1} \equiv R_{293\,K}/R_{17.5\,K}$ and of lattice parameters [10]. Table 1 summarizes the results.

Table 1

Chemical composition x, lattice parameter a, and resistivity ratios r^{-1} of specimens used

spec. No.	$r_{20\,K}^{-1}$	$r_{17.5\,K}^{-1}$	$a_{293\,K} \pm \Delta a$ (Å)	x (at% Si)
67/7	25.65	27.65	$4.7261 \pm 2 \times 10^{-4}$	24.6 ± 0.1
67/3	25.65	27.65	$4.7261 \pm 2 \times 10^{-4}$	24.6 ± 0.1
65/3	48.6	56.1	$4.7254 \pm 2 \times 10^{-4}$	24.9 ± 0.1
98/KS3		16	$4.7261 \pm 4 \times 10^{-4}$	24.6 ± 0.2
44/2			$4.7251 \pm 4 \times 10^{-4}$	25.0 ± 0.2
44/7			$4.7254 \pm 2 \times 10^{-4}$	24.9 ± 0.1

Thin foils have been investigated by TEM and HEM techniques at 150, 200, and 1000 kV, respectively [11, 13, 14]. After deformation contrary to the results of TEM investigation of as-grown crystals [15] the following typical defect configurations have been detected: curved single dislocations, loops, dipoles, grain boundaries, dislocations running along preferred directions.

The density of individual dislocations was estimated to $\varrho_d \approx (1 \text{ to } 3) \times 10^7\ cm^{-2}$, that of regions containing network and along grain boundaries to $(5 \text{ to } 6) \times 10^7$ and $(0.8 \text{ to } 3) \times 10^8\ cm^{-2}$, respectively [11, 13, 14]. The dominating glide system of dislocations is $\{110\}\ \langle 001 \rangle$ [14].

Kossel lines of some of the specimens were considerably broadened. Other specimens delivered no reflections at all. Only six out of fifteen specimens allowed to obtain reflections sharp enough for the determination of lattice parameters.

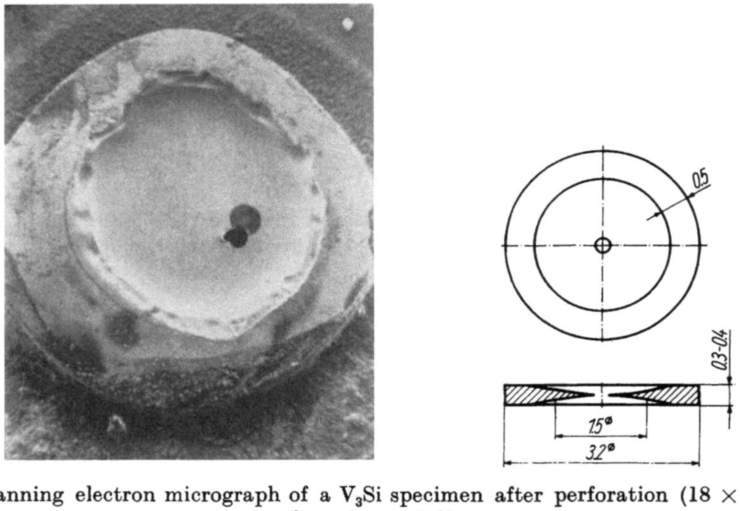

a b

Fig. 1. a) Scanning electron micrograph of a V_3Si specimen after perforation (18 ×); b) profile of specimens [13]

3. Kossel Equipment

The lattice transition has been detected by a Kossel technique [4, 6, 9]. In order to minimize the heating of the sample by electron beam its diameter was chosen to 50 μm. The acceleration voltage amounted to 40 kV, whereas the current through the sample varied between 0.15 and 0.5 μA, resulting in exposure times between 3 and 15 min.

The lattice parameters were investigated at temperatures from 7 to 25 K, at 80, and at 293 K. Temperatures were measured with the aid of an Au (+0.02% Fe)–Cu thermocouple in contact with the sample. Care has been taken for a good thermal contact between specimens and specimen holder (made of copper). The rise in temperature due to heating by the electron beam turned out to be as high as 1 to 6 K.

4. Determination of Lattice Parameters

To determine the lattice parameters a, b, and c, the system of Kossel lines near the [111] pole has been used (see Fig. 2 and 3). a, b, and $c = f(g_{hkl}, g_{klh}, g_{lhk})$ are calculated from the distances g_{hkl}, where $\{hkl\}$ are the $\{123\}$ and $\{210\}$ reflections on a photographic plate.

On the other hand, complete Kossel line patterns can be simulated with the aid of a computer provided the lattice parameters are known [16]. The error in drawing these patterns amounts to ± 0.1 mm and is of the order of magnitude of the error in the distance measurement. Fig. 3 shows a simulated pattern for cubic V_3Si.

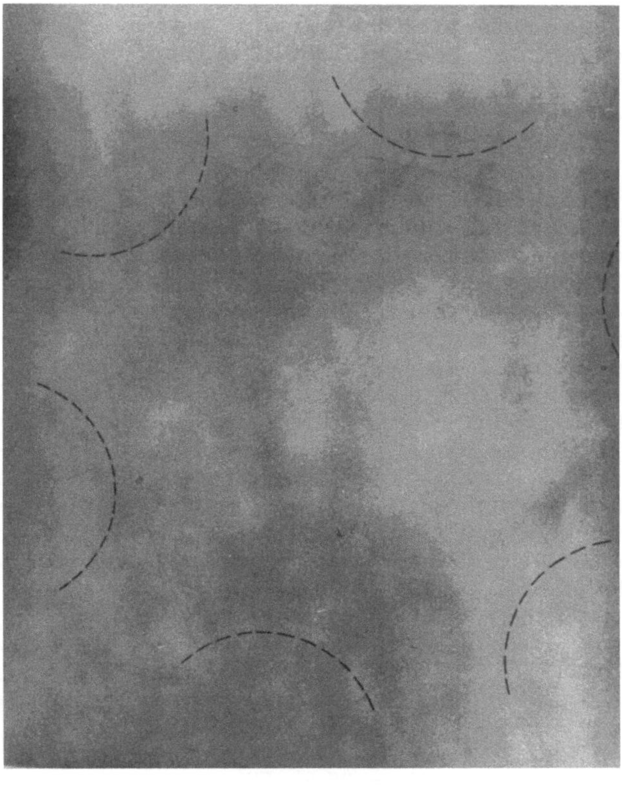

Fig. 2. Kossel pattern from a V_3Si crystal using V-K$\alpha_{1,2}$ radiation. Reflection lines are visible

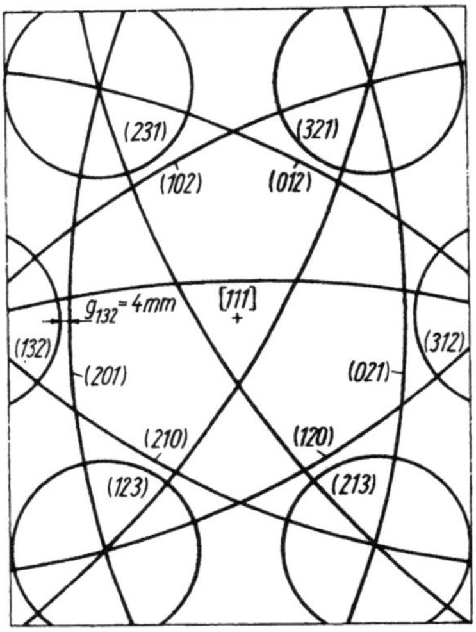

Fig. 3. Simulated Kossel pattern of cubic V_3Si. Lattice parameter $a = 4.7173 \times 10^{-10}$ m corresponding to 80 K. The distances g between all Kossel lines of the types {132} and {201} are equal

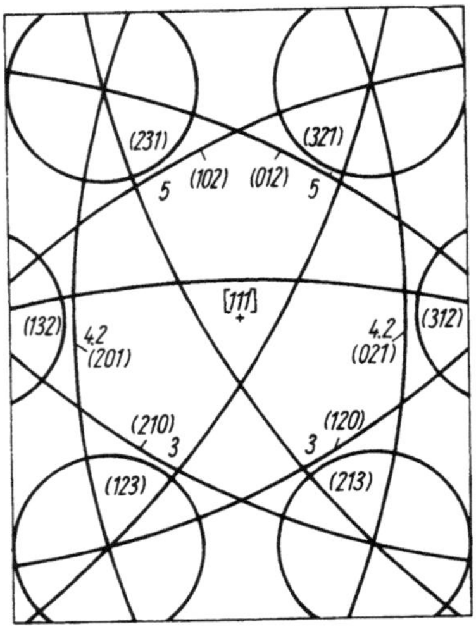

Fig. 4. Simulated Kossel pattern of tetragonal V_3Si. Lattice parameters $a = 4.7148$ and $c = 4.7223 \times 10^{-10}$ m. Numbers give the distances of g in mm

The angular aperture of the {123} cones of diffracted rays and hence the g_{hkl} are very sensitive to lattice distortions. The lattice parameter a may thus be determined with an error $\Delta a/a = 10^{-4}$ to 10^{-5} Fig. 4 contains a simulated Kossel line pattern for a tetragonal unit cell of V_3Si.

5. Results

Lattice parameters and orientations of tetragonal domains are summarized in Table 2. The domain orientations are given by the direction of the fourfold axis with respect to a cartesian coordinate system, where z coincides with the stress axis.

A plastic deformation less than 10% yields a lattice transition like that known with non-deformed crystals. Specimen No. 44/7 showed a splitting of cubic lattice parameters reaching its maximum value already at 25 K, i.e. its transition temperature T_m increased and should be higher than 25.5 K. Specimen No. 98/KS3 exhibits a tetragonal unit cell with $c/a < 1$ between 20 and 23.5 K and a cubic one at higher temperatures. It has also been deformed less than 10% but more than No 44/7. Moreover, tetragonal unit cells with $c/a < 1$ were observed even at room temperature with specimens 44/2 and 67/3 which had been deformed more than 10%. At low temperatures the picture changed completely. All distances g_{hkl} differ from one another and from the systems of tetragonal symmetry (see Fig. 5a). The symmetry of the pattern is less than tetragonal. We propose it to be orthorhombic. Stereographic projections on the basis of this proposal are in good quantitative agreement with the pattern observed (Fig. 5b).

In Fig. 6 lattice parameters are plotted versus temperature. The lattice transition at $T_m = 21.5$ K prior to deformation is characterized by a tetragonality $c/a > 1$. Below $T_c = 17$ K no change of lattice parameters takes place. The lattice appears to be "frozen in". In the absence of external stresses, different domain orientations could be observed within the same specimen.

After plastic deformation with strains $< 10\%$ a tetragonality $c/a < 1$ has been observed at $T < T_m \approx 25$ K. Higher strains stabilize the tetragonal unit cell even at

Table 2

Type of crystal lattice, type of domains, and lattice parameters of specimens used

spec. No.	pattern No.	T (K)	a (Å)	b (Å)	c (Å)	Δa (10^{-4} Å)	domain	crystal lattice	c/a
67/7	1	293 ± 2	4.7261			2		cub.	
	2	80 ± 1	4.7194			2		cub.	
	3	16 ± 0.5	4.7185		4.7200	4	x, y	tetr.	1.0003
44/7	1	293 ± 2	4.7254			2		cub.	
	2	80 ± 1	4.7189			2		cub.	
	3	25 ± 2	4.7173		4.7207	1	x, y	tetr.	1.0007
	4	7 ± 0.5	4.7169		4.7200	3	x, y	tetr.	1.00066
98/KS3	1	293 ± 2	4.7261		·	3		cub.	
	2	80 ± 1	4.7194			4		cub.	
	3	23.5 ± 0.5	4.7198		4.7180	4	z	tetr.	0.99962
	4	22.5 ± 0.5	4.7202		4.7167	3	z	tetr.	0.99926
	5	21 ± 0.5	4.7193		4.7166	5	z	tetr.	0.99943
	6	20 ± 0.5	4.7202		4.7168	4	z	tetr.	0.99928
67/3	1	293 ± 1	4.7262		4.7250	2	z	tetr.	0.99975
	2	80 ± 1	4.7196		4.7185	4	z	tetr.	0.99977
	3	23.5 ± 0.5	4.7165	4.7183	4.7207	4		rh.	
	4	22 ± 0.5	4.7156	4.7177	4.7217	4		rh.	
	5	21 ± 0.5	4.7158	4.7174	4.7214	4		rh.	
	6	20 ± 0.5	4.7157	4.7172	4.7213	4		rh.	
	7	19 ± 0.5	4.7159	4.7170	4.7212	4		rh.	
44/2	1	293 ± 1	4.7253		4.7245	3	z	tetr.	0.99983
	2	80 ± 1	4.7253		4.7177	3	z	tetr.	0.99953
	3	24 ± 0.5	4.7195		4.7165	3	z	tetr.	0.99935
	4	23.3 ± 0.5	4.7188		4.7170	2	z	tetr.	0.99962
	5	19.5 ± 0.5	4.7193		4.7167	3	z	tetr.	0.99945
	6	17.6 ± 0.5	4.7154	4.7195	4.7200	2		rh.	
	7	17 ± 0.5	4.7145	4.7198	4.7216	2		rh.	

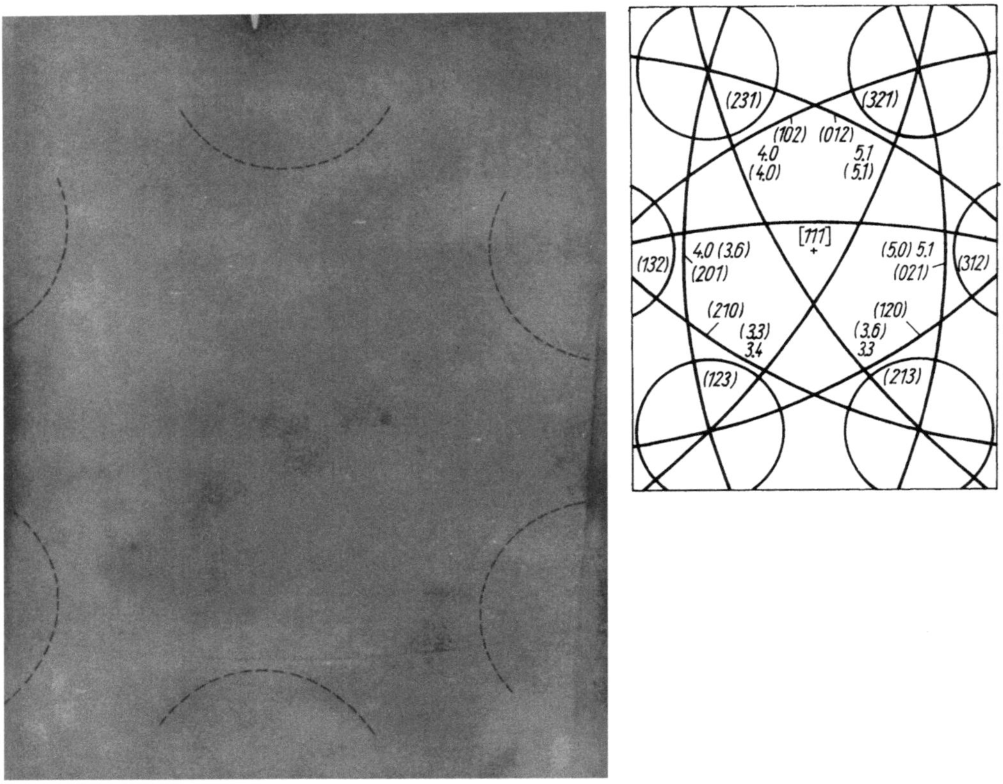

Fig. 5. a) Kossel pattern of V_3Si (No. 44/2) at 17 K after plastic deformation, showing symmetry less than tetragonal (cf. Fig. 4). b) Simulated Kossel pattern of orthorhombic V_3Si. Lattice parameters $a = 4.7145 \times 10^{-10}$, $b = 4.7198 \times 10^{-10}$, $c = 4.7215 \times 10^{-10}$ m. The measured distances on the pattern g_{hkl}^{exp} between the {123} and {012} Kossel lines are compared with those of the computer simulation g_{hkl}^{theor} (in parenthesis)

room temperature. Below T_m the symmetry is orthorhombic. Further decreasing temperature yields a unit cell that is approximately tetragonal with $c/a > 1$. Although neither the specimen holder nor the temperature gradient prefer one of the $\langle 100 \rangle$ axes, only single domain crystals were observed after plastic deformation.

6. Discussion

Most of the results mentioned above may be understood from a theoretical point of view by the assumption that dislocations created during plastic slip generate a (quasi-homogeneous) uniaxial compressional stress along a $\langle 100 \rangle$ direction. Recently, one of the present authors [8] studied in some detail the influence of uniaxial stress on the lattice transition of V_3Si using the phenomenological Landau theory. Fig. 7 shows the theoretical temperature dependence of lattice parameters near the lattice transition temperature calculated by using the results of [8].

At stresses below the tricritical stress ($\approx 0.8 \times 10^7$ Nm^{-2}) the phase transition is of first order, at higher stresses of second order.

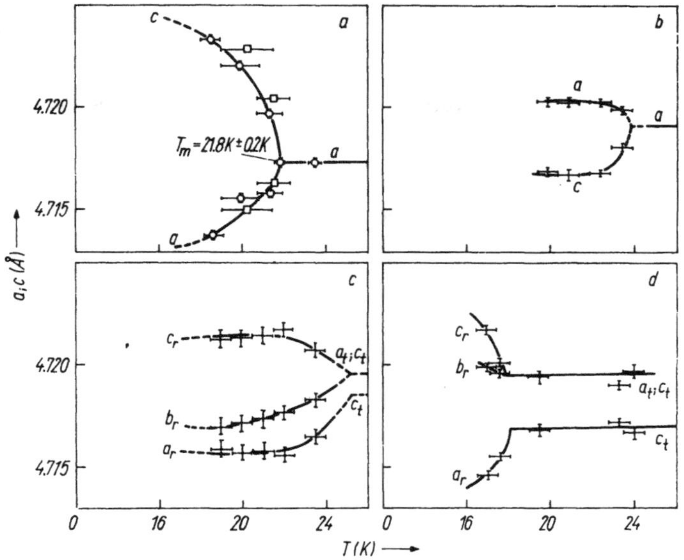

Fig. 6. Temperature dependence of lattice parameters of V_3Si single crystals after plastic deformation a) 0% ○ No. 27/2 L1, □ No. 45/2.4; b) 8%, No. 98/KS 3; c) 13%, No. 67/3; d) 12.9%, No. 44/2

Comparing experiment (Fig. 6c) and theory (Fig. 7) one comes to the conclusion that the effective internal stress should surmount the tricirtical stress and amount to (1 to 2) $\times$ 10^7 Nm^{-2}. In Fig. 6d the behaviour of a_r as well as of c_r compares well with theory. The temperature dependence of b_r might be due to an incomplete structural transition tetragonal → orthorhombic so that the curve labelled b_r corresponds to a metastable tetragonal z-domain.

On cooling specimen 98/KS3 (cf. Fig. 6b) down to 20 K no lattice transition occurred. The tetragonal distortion of the unit cell due to stress at 300 and 80 K is too small to be detected. It increases, however, due to the softening of the elastic moduli $c_{11} - c_{12}$ and at 25 K reaches a measurable size.

From Fig. 6d a lattice parameter ratio $c/a \approx 0.99947$ above the transition temperature can be derived. Interpolation of the theoretical values of Fig. 7 yields an uniaxial stress of $\sigma \approx 1.1 \times 10^7$ Nm^{-2}. Such stress corresponds to a dislocation density $\varrho_d \approx 36\sigma^2/G^2b^2 \approx 3 \times 10^8$ cm^{-2} (G shear modulus, $\boldsymbol{b}$ Burgers vector) that has really been observed. In view of the relatively large diameter of the electron beam, the question arises why local variations of dislocation stress fields can be observed. TEM work indicated, however, extended regions of parallel dislocations.

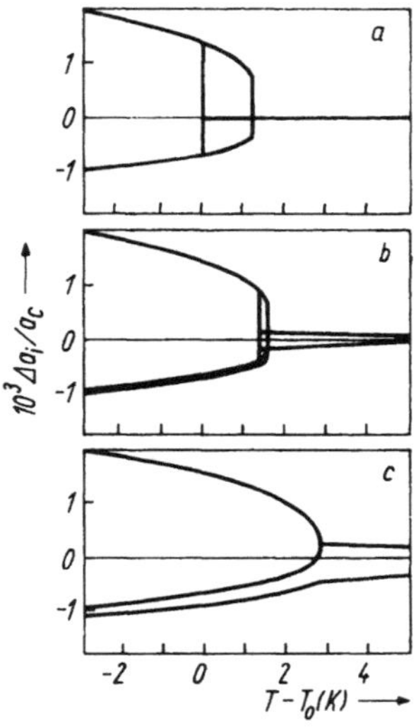

Fig. 7. Temperature dependence of the lattice parameters of V_3Si under a uniaxial compressional stress along [100] direction. T_0 is the temperature of the instability of the cubic lattice at zero stress. The curves were calculated from the phenomenological theory of phase transitions [8]. The maximum theoretical termal hysteresis is indicated in a) and b). a) zero stress; b) 0.3 $\times$ 10^7, c) 1.5 $\times$ 10^7 Nm^{-2}

It is interesting to note that the point symmetry of a cubic V_3Si crystal containing a straight screw dislocation $\boldsymbol{b} = \langle 100 \rangle$ becomes tetragonal (point group 422) [17], as observed at room temperature. The low-temperature phase at zero stress belongs to the point group 4/mmm [18]. Just this symmetry is lowered by dislocations $\boldsymbol{b} = [100]$ and [001] to orthorhombic (point group 222 for screws, point group mm2 for edges [19]). No effort has been made in the present work, however, to determine the point symmetry of the samples.

A number of authors have already reported on characteristic changes of the X-ray diffraction pattern and/or of the transition curve of powdered samples as compared to bulk material [20 to 23]. They occurred with intermetallic compounds with Cr_3Si-type structure and in some cases turned out to be reversible. The present results let us suppose that external and/or internal stresses generate a distortion which is enhanced due to elastic softening at low temperatures:

Acknowledgements

The authors express their gratitude to a number of colleagues for experimental support, especially to Dr. M. Jurisch and Mr. G. Behr for growing the single crystals, to Dr. K.-H. Berthel for measurement of resistivity and to Dr. U. Krämer and Dr. A. Luft, Dr. P. Burck, Mrs. C. Reinhold, and Miss K. Keßler for TEM work.

References

[1] B. W. BATTERMAN and C. S. BARRETT, Phys. Rev. Letters **13**, 390 (1964).
[2] B. W. BATTERMAN and C. S. BARRETT, Phys. Rev. **145**, 296 (1966).
[3] K.-H. BERTHEL, D. ECKERT, and A. HANDSTEIN, presented at the Abstracts VI. Internat. Symp. Reinststoffe in Wissenschaft und Technik, Dresden 1975 (p. 179).
[4] H.-J. ULLRICH, K-H. BERTHEL, M. JURISCH, S. DÄBRITZ, and K. KLEINSTÜCK, see [3] (p. 181).
[5] K.-H. BERTHEL, A. HANDSTEIN, N. E. ALEKSEEVSKII, M. JURISCH, and H.-J. ULLRICH, in: Problemy fiziki tverdogo tela i materialovedeniya, Izd. Nauka, Moskwa 1976 (p. 275).
[6] H.-J. ULLRICH, W. SCHATT, S. DÄBRITZ, and V. GEIST, Microchimica Acta (Wien) **II**, 167 (1977).
[7] L. P. GORKOV, Solid State Commun. **19**, 1107 (1976).
[8] B. PIETRASS, phys. stat. sol. (b) **68**, 553 (1975).
[9] H.-J. ULLRICH, G. ARTMANN, K.-H. BERTHEL, S. DÄBRITZ, M. JURISCH, K. KLEINSTÜCK, and H. THRAEN, in: III. Arbeitstagung Mikrosonde, Berlin 1975, Physikalische Gesellschaft der DDR, Berlin 1975 (p. 83).
[10] M. JURISCH, K.-H. BERTHEL, and H.-J. ULLRICH, phys. stat. sol. (a) **44**, 277 (1977).
[11] K. KESSLER, Diploma Work, TU Dresden (Sektion Physik), 1975.
[12] P. BURCK, Proc. V. European Congr. Electron Microscopy, Manchester 1972 (p. 308).
[13] C. REINHOLD, Diploma Work, TU Dresden (Sektion Physik), 1977.
[14] U. KRÄMER, unpublished work.
[15] P. PAUFLER, E. ZEDLER, H.-J. ULLRICH, K.-H. BERTHEL, U. KRÄMER, M. JURISCH, K. RICHTER, and K. EICHLER, phys. stat. sol. (a) **44**, 499 (1977).
[16] H.-J. ULLRICH, K. THIELE, S. DÄBRITZ, H. SCHREIBER, K. GÖTZE, and F. FELDHOFER, Kristall und Technik 7, 1153 (1972).
[17] P. PAUFLER, Kristall und Technik **11**, 607 (1976).
[18] YU. A. IZYUMOV, V. E. NAISH, and V. N. SYROMYATNIKOV, Kristallografiya **21**, 256 (1976).
[19] P. PAUFLER, Kristall und Technik **10**, 829 (1975).
[20] R. VISWANATHAN and H. LUO, Solid State Commun. **9**, 1733 (1971).
[21] T. TAKASHIMA, Japan. J. appl. Phys. **12**, 781 (1973).
[22] E. M. SAVITSKII, V. V. BARON, A. A. KOROSTELIN, A. R. KADYRBAEV, V. N. SUMAROKOV, and V. A. GUSEV, Dokl. Akad. Nauk SSSR **227**, 840 (1976).
[23] M. ROHR and P. MÜLLER, phys. stat. sol. (a) **42**, 477 (1977).

(Received May 18, 1978)

phys. stat. sol. (a) 49, 331 (1978)

Subject classification: 13.4 aud 20.3; 13.5.1; 22.2.2

Sektion Physik der Karl-Marx-Universität Leipzig, Arbeitsgemeinschaft A^{III}–B^{V}-Halbleiter[1])

The Spectral Dependence of the Photoneutralization Cross-Section of Oxygen in GaP Near the Threshold

By

R. BINDEMANN, H. FISCHER, and K. KREHER

Using the green S–C donor–acceptor pair luminescence as a probe the threshold $h\nu_O$ of the photoneutralization cross section of O in GaP is determined by photoluminescence excitation spectroscopy with high accuracy: $h\nu_O = (1.4515 \pm 0.001)$ eV. Combining this value with measurements of pair spectra and of donor conduction band transitions revised values of the gap $E_g = (2.350 \pm \pm 0.002)$ eV, the binding energy of the free exciton $E_x = (22 \pm 2)$ meV and of the binding energies of acceptors in GaP may be determined which are in good agreement with revised values reported recently in the literature.

Unter Benutzung der grünen S–C-Donator–Akzeptor-Paarlumineszenz als Sonde wird die Schwellenenergie $h\nu_O$ für die Photoneutralisation von O in GaP mittels Anregungslumineszenzspektroskopie mit hoher Genauigkeit bestimmt: $h\nu_O = (1{,}4515 \pm 0{,}001)$ eV. Kombiniert man diesen Wert mit Paarspektren und Donator–Leitungsband-Übergängen, so erhält man revidierte Werte für die Bandlücke $E_g = (2{,}350 \pm 0{,}002)$ eV, die Bindungsenergie des freien Excitons $E_x = (22 \pm 2)$ meV und die Akzeptorbindungsenergien in GaP, die in guter Übereinstimmung mit neueren Literaturwerten sind.

1. Introduction

The investigation of optical cross-sections of deep impurities is of general interest as it reveals information about purely electronic parameters (energy of the ground state, impurity potential, and wave functions) and about electron–phonon coupling connected with lattice deformation.

For deep impurities the electron–phonon coupling is strong, therefore the phonon assisted transitions must be separated from the purely electronic ones. In order to accomplish this separation measurements very close to the threshold at very low cross-sections are important. For O in GaP this was done in [1, 2] using the photoexcitation of the O–C donor–acceptor pair luminescence.

The photoneutralization threshold $h\nu_O$ of O in GaP is of fundamental importance for the determination of electronic energy level in this material: As the ground state energy of the electron bound to O relative to the minimum of the conduction band can be derived from new measurements [3] the value of the gap E_g follows from $h\nu_O$.

Combining E_g with the well-known exciton gap E_{gx} [4] gives the binding energy of the free exciton; from a combination of $h\nu_O$ with the results of donor–acceptor pair luminescence the values of the acceptor binding energies can be found.

Because of this key role of $h\nu_O$ in the present paper the photoexcitation spectrum of the green S–C donor–acceptor pair luminescence via the donor O was measured. This new method has high sensitivity and accuracy as it is selective for oxygen, works best at very low temperature, and is not disturbed by the overlap of emission and excitation spectra.

[1]) Linnéstr. 5, DDR-701 Leipzig, DDR.

2. Experimental Method

There are a lot of methods available to determine optical cross-sections of impurities. Straightforward transmission measurements require large optically perfect crystals and are not selective for a special impurity. All types of electrical measurements, especially those on p–n junctions or Schottky barriers, are difficult at low temperature and may be influenced by electrical fields. Photoconductivity measurements give good sensitivity but are also not selective for a particular impurity. Therefore, we used photoluminescence excitation of the S–C luminescence according to Fig. 1. In a first step electrons are excited from the valence band to the O donor (cross-section σ_{p1}), in a second step these electrons are excited from the O donor to the conduction band (cross-section σ_{n1}). Shallow donors (S) and acceptors (C) have high capture cross-sections, so at low temperature the electrons and holes excited to the bands are readily captured by these centres and finally recombine via pair emission.

As the energy required for the neutralization of O is larger than the energy required for photoionization, $\sigma_{p1} \ll \sigma_{n1}$ near the threshold of photoneutralization. Thus the photoneutralization is the rate determining process, σ_{n1} has no pronounced spectral dependence in the interesting region and the excitation spectrum reflects the spectral dependence of $\sigma_{p1}(h\nu)$. This method is selective for deep centres which allow a two-step excitation of luminescence [5]. In GaP no other impurity than O is known to show this effect, so the method is selective for O. The sensitivity is high due to the large capture cross-sections of the shallow centres and due to the large volume of the sample which is excited at the low absorption coefficients near the threshold. Whether the green S–C luminescence or the infrared O–C luminescence is a better probe depends on the samples available. p–doped GaP shows intense O–C luminescence, whereas in n-doped GaP the luminescence due to shallow donors dominates.

In principle two methods can be used to determine σ_{p1}:

i) The intensity of the exciting radiation I_{exc} is held constant. Then $\sigma_{p1}(h\nu)$ is proportional to the luminescence intensity I_L if I_L is proportional to $I_{exc}(h\nu)$. I_{exc} must be measured in photons $cm^{-2}\ s^{-1}$.

In medium doped p-GaP the intensity of the O–C luminescence depends linearly on I_{exc} whereas for the S–C luminescence $I_L \sim I_{exc}^{1.4}$ [1, 5]. So in p-GaP only the O–C luminescence can be used for the measurement of σ_{p1}.

In low-doped highly compensated n-GaP the S–C luminescence depends linearly on the excitation intensity, so this luminescence can be used.

ii) The intensity of the luminescence I_L is held constant by controlling the intensity of the exciting radiation $I_{exc}(h\nu)$. In this case $\sigma_{p1} \sim I_{exc}^{-1}(h\nu)$.

This method has the advantage that the occupation of all energy levels is held constant, so the nonlinearities cannot disturb. The disadvantage is that the stabilisation of I_L is not possible at very low intensities, so i) represents the more sensitive method.

The experimental arrangement is described thoroughly in [6, 7]. The sample was immersed in liquid He or was attached to a cold copper finger. In the latter case it

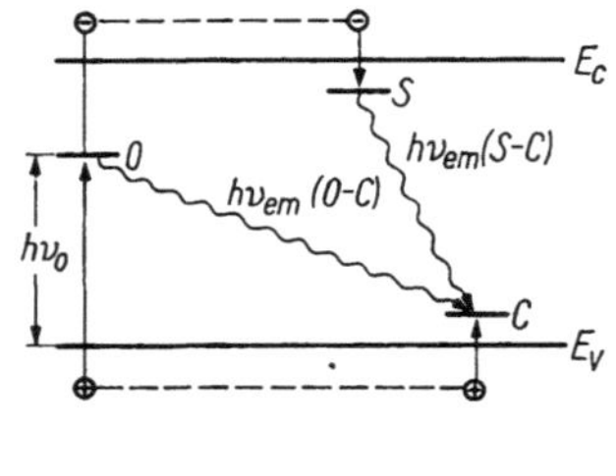

Fig. 1. Transitions which are connected with emission excited via the O level

was carefully screened from room temperature infrared radiation by a fused silica window cooled down to liquid He temperature. This was necessary because otherwise deviations of the linear relationship $I_L \sim I_{exc}$ occurred which are due to infrared stimulated carrier transfer processes [8].

The samples were grown by the synthesis solute diffusion (SSD) techniques. They were not intentionally doped and had a high degree of compensation ($N_A/N_D \geqq 0.5$). Residual impurities are C, S, O, and N. The concentrations of C and S are in the region of 1 to 3×10^{16} cm^{-3}.

3. Experimental Results

The spectral dependence of the cross-section $\sigma_{p1}(h\nu)$ was measured with both methods i) and ii) described in Section 2. The results were identical as was expected according to the linear dependence of luminescence intensity on excitation intensity.

We obtained five orders of magnitude of the rise of σ_{p1} above the threshold. In the spectral region between 1.45 and 1.6 eV pronounced fine structure can be seen. Fig. 2 shows a plot of σ_{p1}^2 versus $h\nu$ which consists of straight lines marking the thresholds of the first row of Table 1. The steep rise at 1.4515 eV is the threshold for zero-phonon transitions, the higher thresholds correspond to phonon assisted transitions with the phonon energies of Table 1. The small tail below 1.4515 eV is probably due to broadening effects discussed in [1].

The absorption coefficient in this spectral region was estimated to be ≈ 1 cm^{-1}; no threshold could be detected by absorption measurements. The comparison with the excitation spectrum shows clearly the high sensitivity and selectivity of the photoluminescence excitation method.

Table 1

Thresholds of the photoneutralization of O in GaP

threshold energy (eV)	difference to the threshold $h\nu_0$ (meV)	phonons involved	energy of the phonons (meV)
1.4515	—	no	—
1.471	19.5	$\hbar\omega_1$	19
1.490	38.5	$2\hbar\omega_1$	38
1.500	48.5	$\hbar\omega_2$	48
1.5175	66	$\hbar\omega_1 + \hbar\omega_2$	67
1.5398	88.3	$2\hbar\omega_1 + \hbar\omega_2$	86
1.5845	133	$2\hbar\omega_1 + 2\hbar\omega_2$	134

The error is ± 1 meV; a correction of -1.0 meV due to the spectral width of the exciting radiation (2.5 meV) is included. The last column gives the phonon energies due to [1, 2].

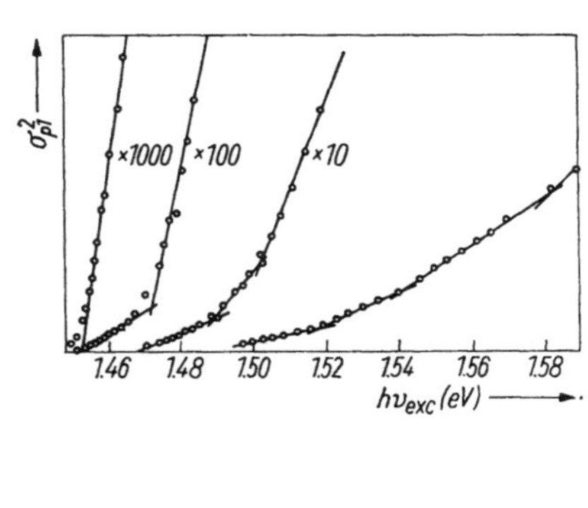

Fig. 2. σ_{p1}^2 vs. $h\nu$. Threshold values are summarized in Table 1

4. Theory and Discussion

According to [9] the optical cross-section $\sigma_j(h\nu)$ between an impurity level E_0 and an isotropic band $E_j(\boldsymbol{k})$ can be written in the form

$$\sigma_j(h\nu) \sim \frac{1}{h\nu} |A_0(\boldsymbol{k})|^2 \, |\boldsymbol{M}_{0j}(\boldsymbol{k})|^2 \, \varrho_j(\boldsymbol{k}) \, \delta(E_j(\boldsymbol{k}) - E_0 + h\nu) \, . \tag{1}$$

Deriving this equation the impurity wave function $\psi_0(\boldsymbol{r})$ was written

$$\psi_0(\boldsymbol{r}) = \sum_{\boldsymbol{k}} A_0(\boldsymbol{k}) \, |0, \boldsymbol{k}\rangle \, , \tag{2}$$

where $|0, \boldsymbol{k}\rangle$ is the Bloch function of the band $E_0(\boldsymbol{k})$ to which the impurity is connected. $\varrho_j(\boldsymbol{k})$ is the density of states in the band $E_j(\boldsymbol{k})$, and $\boldsymbol{M}_{0j}(\boldsymbol{k})$ the dipole matrix element between the bands $E_0(\boldsymbol{k})$ and $E_j(\boldsymbol{k})$.

In the case of photoneutralization $E_j(\boldsymbol{k})$ is the valence band with mainly p-character, whereas the final state wave function of the electron bound to O is mainly composed of s-like wave functions from the conduction band near the Γ- and the X-points, the Γ-like contributions $A_0(\boldsymbol{k})$ with small k-vectors are the strongest ones [10]. The valence band–oxygen transition thus contains an allowed zero-phonon transition with the matrix element $\boldsymbol{M}_{0j}(\boldsymbol{k}) = \text{const.}$

The expansion coefficients $A_0(\boldsymbol{k})$ depend on the potential which binds the electron; the potential can be approximated by a δ-function [11].

Near the Γ-point one gets

$$A_0(\boldsymbol{k}) \sim \frac{1}{E_i^{\mathrm{d}} + \dfrac{\hbar^2 k^2}{2 m_{\mathrm{c}}(\Gamma)}} \, , \tag{3}$$

where $m_{\mathrm{c}}(\Gamma)$ is the effective mass in the isotropic $\Gamma_{1\mathrm{c}}$-minimum and E_i^{d} the binding energy of the donor relative to the $\Gamma_{1\mathrm{c}}$-minimum. From (1) and (3) the photoneutralization cross-section for zero-phonon transitions is obtained:

$$\sigma_{\mathrm{p1}}(h\nu) \sim \frac{1}{h\nu} \frac{\sqrt{h\nu - E_0}}{\left[\dfrac{m_{\mathrm{c}}(\Gamma)}{m_{\mathrm{v}}} E_i^{\mathrm{d}} + h\nu - E_0\right]^2} \, , \tag{4}$$

where m_{v} is a properly averaged valence band mass. E_0 is the energy of the oxygen level relative to the valence band maximum and is equal to the photoneutralization threshold.

Near this threshold only the spectral dependence of the numerator is important, so one gets

$$\sigma_{\mathrm{p1}}^2 \sim h\nu - E_0 \tag{4a}$$

in agreement with experiment.

Phonon assisted transitions involving configuration coordinate (CC) phonons $\hbar\omega_{\mathrm{n}}$ of A_1-type have shown to be responsible for the higher thresholds. They have the same lineshape near their thresholds [1, 2], so

$$(\sigma_{\mathrm{p1}}^{\mathrm{n}})^2 \sim h\nu - E_0 - \hbar\omega_{\mathrm{n}} \tag{4b}$$

for these transitions.

According to formulae (4a) and (4b) all the fine structure of Fig. 2 could be explained using the two CC-phonons found in [2] (see Table 1). For higher energies the influence of the real band structure (nonparabolic behaviour, anisotropy, degeneracy between light and heavy holes, influence of the split-off band) must be taken into consideration. This is discussed thoroughly in [1, 2].

5. Consequences Concerning Binding Energies and the Gap

In the last time several papers appeared due to which the indirect gap and the binding energy of the free exciton in GaP are 11 meV larger than previously assumed [4, 12]. This follows from revised values of the binding energy of acceptors differing by 8 meV [13 to 15] and of donors by 3 meV [3, 16]. Contrary to these results photoluminescence excitation spectra of donor acceptor pairs [17] are consistent with the "old" binding energies [12].

Our interpretation of the band peaked at 2.241 eV in S-doped GaP at 77 K as being caused by free (hole) to bound (electron at the S donor) transition [18] gives the energy difference between the donor energy of S and the top of the valence band $h\nu_s = 2.2437$ eV (at 4.2 K). Combining this value with sulfur-acceptor pair transition energies extrapolated to infinite separation ($h\nu_\infty$) [12] results in the acceptor binding energies E_A of Table 2, column 3. Our own measurements of the S–C pair spectrum [19] gave $h\nu_\infty$ (S–C) = 2.1890 eV in very good agreement with [12].

Table 2

Revised acceptor energies derived from $h\nu_s$ (column 3) and $h\nu_O$ (column 4)

acceptor X	$h\nu_\infty$ (S–X) (eV)	$E_A = h\nu_s - h\nu_\infty$ (S–X) (meV)	E_A from $h\nu_O$ (meV)	E_A [15] (meV)	E_A [12] (meV)
C	2.1884 [12]	55.3	54.9	54.7	46.4
	2.1890 [19]	54.7			
Zn	2.1731 [12]	70.6	70.4	70.1	61.7
Cd	2.1405 [12]	103.2	102.9	102.6	94.3

Column 5 shows the results of [15], column 6 the "old" values [12] assuming $E_g = 2.339$ eV.

Combining the value $h\nu_O = 1.4515$ eV of this paper with the O–C pair transition energy $h\nu_\infty = 1.3966$ eV [12] gives a further independent value of the binding energy of carbon: $E_A(C) = 54.9$ meV. Taking the differences $E_A(Zn) - E_A(C) = 15.5$ meV and $E_A(Cd) - E_A(C) = 48.0$ meV from [20] one gets the other acceptor binding energies of the 4th column of Table 2.

These two rows of independently determined acceptor binding energies are in perfect agreement with values from measurements of the recombination of electrons bound to nitrogen pairs with holes on acceptors [15] shown in column 5 of Table 2, but not with the results of [17]. The difference of 792.2 meV between $h\nu_s$ and $h\nu_O$ is equal to the difference of the donor binding energies of O and S. It is in very good agreement with the value of 791.9 meV which can be derived from the values $h\nu_\infty$ for the O and S pair spectra with the same acceptor. It proves that not only the pair spectra involving O but also the spectra involving shallow donors show no deviations from the Coulomb law

$$h\nu = E_g - (E_A + E_D) + \frac{e^2}{\varepsilon R} \quad \text{for} \quad R > 40 \text{ Å} .$$

Further conclusions can be drawn accepting the donor binding energy of S $E_D(S) = (107 \pm 1)$ meV [3] which gives the donor binding energy of O $E_D(O) = E_D(S) + (h\nu_s - h\nu_O) = (899 \pm 2)$ meV and a gap energy $E_g = E_D(S) + h\nu_s = E_D(O) + h\nu_O = (2.350 \pm 2)$ meV. Using the well-known exciton gap $E_{gx} = 2.328$ eV [4] the binding energy of the free exciton amounts to $E_x = (22 \pm 2)$ meV consistent with the recent theoretical value of 20.1 meV [21] which takes into account the polaron shift due to electron-phonon coupling.

6. Summary

1. The photoneutralization cross-section of O in GaP was measured using the excitation of shallow donor acceptor pair luminescence via the two-step excitation.

2. The spectral dependence of the photoneutralization cross-section near the threshold $h\nu_0 = 1.4515$ eV corresponds to an allowed zero-phonon transition; fine structure due to phonon assisted transitions involving CC phonons $\hbar\omega_1 = 19$ meV and $\hbar\omega_2 = 48$ meV could be resolved in agreement with [1, 2].

3. Revised values for the acceptor binding energies of Zn, Cd, and C in GaP were derived combining the independently measured photoneutralization thresholds of S and O with data from donor acceptor pair spectra.

4. Using the donor binding energy of S from [3] revised values for the GaP gap energy $E_g = 2.350$ eV and the binding energy of the free exciton $E_x = 22$ meV were derived.

References

[1] L. I. SAMUELSON, Thesis, Lund Institute of Technology 1977.
[2] B. MONEMAR and L. I. SAMUELSON, J. Lum. **12/13**, 507 (1976).
[3] A. A. KOPYLOV and A. N. PIKHTIN, Fiz. Tekh. Poluprov. **11**, 867 (1977).
[4] P. J. DEAN and D. G. THOMAS, Phys. Rev. **150**, 690 (1966).
[5] A. SCHINDLER, R. BINDEMANN, and K. KREHER, phys. stat. sol. (b) **59**, 439 (1973).
[6] H. FISCHER and R. BINDEMANN, Jenaer Rundschau **22**, 6 (1977).
[7] H. FISCHER, Exper. Tech. Phys. **25**, 3 (1977).
[8] R. BINDEMANN, H. FISCHER, and K. KREHER, phys. stat. sol. (a) **34**, K 63 (1976).
[9] W. W. ANDERSON, Solid State Electronics **18**, 235 (1975).
[10] M. JAROS, J. Phys. C **8**, 2455 (1975).
[11] H. GRIMMEISS and L. A. LEDEBO, J. Phys. C **8**, 2615 (1975).
[12] A. T. VINCK, R. L. A. VAN DER HAYDEN, and I. A. W. VAN DER DOES DE BYE, J. Lum. **8**, 105 (1973).
[13] V. BERNDT, A. A. KOPYLOV, and A. N. PIKHTIN, Soviet Phys. — J. exper. theor. Phys., Letters **22**, 284 (1975).
[14] V. BERNDT, A. A. KOPYLOV, and A. N. PIKHTIN, Fiz. Tekh. Poluprov. **11**, 1782 (1977).
[15] M. D. STURGE, A. T. VINK, and F. P. KUIJPERS, Appl. Phys. Letters **32**, 49 (1978).
[16] A. C. CARTER, P. J. DEAN, M. S. SKOLNICK, and R. A. STRADLING, J. Phys. C **10**, 5111 (1977).
[17] R. A. STREET and W. SENSKE, Phys. Rev. Letters **37**, 1292 (1976).
[18] R. BINDEMANN, R. SCHWABE, and T. HÄNSEL, phys. stat. sol. (b) **87**, 169 (1978).
[19] R. BINDEMANN, E. HEMPEL, and K. KREHER, to be published.
[20] P. J. DEAN, C. H. HENRY, and C. F. FROSCH, Phys. Rev. **168**, 812 (1968).
[21] G. BENI and T. M. RICE, Solid State Commun. **23**, 871 (1977).

(Received May 12, 1978)

phys. stat. sol. (a) 49, 337 (1978)

Subject classification: 1.1 and 1.4; 21.1; 21.6

Rijksuniversitair Centrum Antwerpen[1])

A High Resolution Study of Ordering in Au_4Mn [2])

I. The Microstructure of Fully Ordered Au_4Mn

By

G. van Tendeloo and S. Amelinckx[3])

The ordered Au_4Mn alloy is studied by high resolution dark field electron microscopy. Fully ordered Au_4Mn is isostructural with Ni_4Mo and it presents features which make it particularly suitable for structure imaging especially if the crystal is in a [001] orientation and if the image is formed exclusively by superlattice reflections. By this technique detailed "atomic level" information can be obtained concerning the structure of the domain boundaries. At inclined twin boundaries evidence is given for the occurrence of a coincidence structure. At antiphase boundaries relaxation of the structure can be observed in the vicinity of the boundary region.

Die geordnete Au_4Mn-Legierung wird mit hochauflösender Dunkelfeld-Elektronenmikroskopie untersucht. Vollständig geordnetes Au_4Mn ist isostrukturell mit Ni_4Mo und zeigt Einzelheiten, die es für Strukturabbildungen besonders geeignet machen, insbesondere wenn sich der Kristall in einer [001]-Orientierung befindet und wenn das Diagramm ausschließlich durch Übergitterreflexionen gebildet wird. Mit dieser Technik wird eine detaillierte Information auf „atomarer Ebene" bezüglich der Struktur der Domänengrenzen erhalten. Bei geneigten Zwillingsgrenzen werden Hinweise für das Auftreten einer Koinzidenzstruktur gefunden. An Antiphasengrenzen kann eine Relaxation der Struktur in der Nähe des Grenzbereiches beobachtet werden.

1. Introduction

In the course of a systematic study of gold–manganese alloys with LPAPB structures, we have examined Au_4Mn using electron diffraction and high resolution imaging. Fully ordered Au_4Mn is isostructural with Ni_4Mo [1], of which the microstructure has been studied previously [2] using ordinary electron microscopy and of which a high resolution study is in progress [3]. Special attention is paid in Part II of this paper to the transition structure which results after quenching from 900 °C on annealing at 360 °C during less than 2 min. In this part we shall describe aspects of the microstructure of the Ni_4Mo type structure as revealed by the high resolution images. Electron diffraction studies of this alloy have been published previously [4].

2. Experimental

Specimens of nominal composition Au_4Mn are prepared by melting together the constituent elements in an arc furnace. Foils were made by rolling. After homogenization for 20 h at 900 °C the thick foil specimens were quenched and subsequently annealed for various periods of time at 360 °C. The foils were finally thinned by electropolishing using the method described by Wolf [5].

[1]) Groenenborgerlaan 171, B-2020 Antwerpen, Belgium.

[2]) Work performed under the auspices of the association R.U.C.A.-S.C.K. with the financial support of I.I.K.W.

[3]) Also at S.C.K./C.E.N., Boeretang, B-2400 Mol, Belgium.

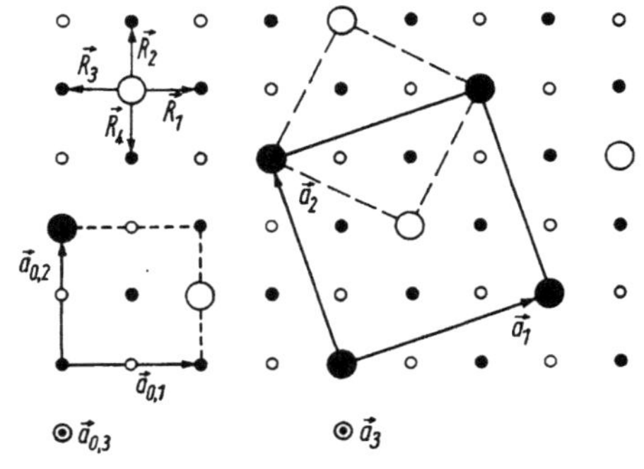

Fig. 1. Structure of Au_4Mn. The large circles represent Mn in two different levels (0, 1/2), whereas the small circles represent gold atoms at two levels (0, 1/2). The base vectors of the f.c.c. basic structure are $\boldsymbol{a}_{0,1}$, $\boldsymbol{a}_{0,2}$, and $\boldsymbol{a}_{0,3}$; the base vectors of the Ni_4Mo structure are $\boldsymbol{a}_1$, $\boldsymbol{a}_2$, and $\boldsymbol{a}_3$. Displacement vectors for anti-phase boundaries $\boldsymbol{R}_1$, $\boldsymbol{R}_2$, $\boldsymbol{R}_3$, and $\boldsymbol{R}_4$ are indicated as well

As quenched from 900 °C the alloy contains mainly the transition structure, to be described in Part II. After 2 min at 360 °C regions of Ni_4Mo structure are formed, which increase in surface area and in domain size after increasing annealing times.

3. Crystal Structure of Fully Ordered Au_4Mn

The structure of fully ordered Au_4Mn ($a = 6.45$ Å, $c = 4.03$ Å) is a centered tetragonal superstructure of the f.c.c. structure. It is shown in Fig. 1. It can be derived from the DO_{22} structure by crystallographic shear [6]. The reciprocal lattice is represented in Fig. 2a, together with the most relevant section $hk0$ for a single domain and for a twinned crystal (Fig. 2b, c).

4. Microstructure: Geometrical Considerations

4.1 Orientation variants

The point group of Au_4Mn is 4/m, i.e. of the order 8. The point group of the disordered f.c.c. structure being of order 48, the number of orientation variants is six. As a variant generating group [7] one can choose the point group 32 with its threefold axis oriented along the [111] direction of the f.c.c. structure. The different orientation variants are represented in Fig. 3 together with the variant generating group.

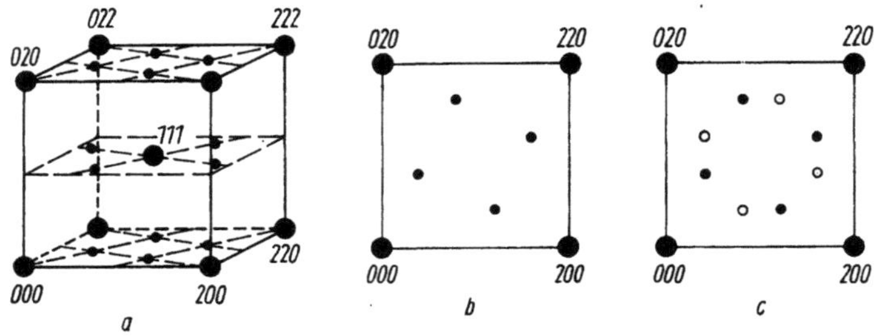

Fig. 2. Reciprocal lattice of Au_4Mn. a) in space, b) [001] section of a single domain, c) [001] section for two coaxial variants

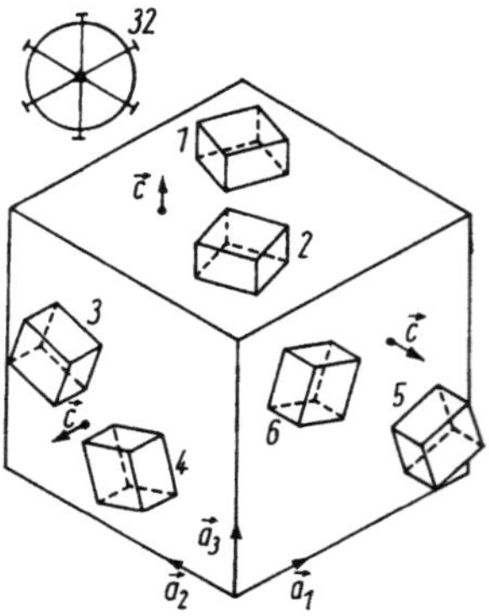

Fig. 3. Schematic representation of the six orientation variants of the Au_4Mn structure. The variant generating group is 32

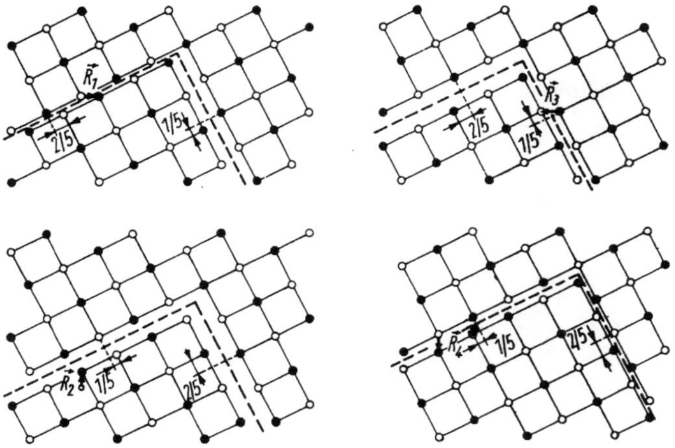

Fig. 4. Four different APB's; the contact planes are parallel with (120) and (2$\bar{1}$0). Note the offsets of the atom rows

There are only two essentially different interfaces: the anti-parallel and the perpendicular twin [2], the first consisting of two coaxial variants their common four-fold axis being oriented along a cube direction of the f.c.c. structure, the second having their c-axes mutually perpendicular in the two variants. Twins can easily be identified from the diffraction patterns. The twin formed by two coaxial variants produces the pattern of Fig. 2c.

4.2 Translation variants

The volume of the primitive unit cell of the Au_4Mn structure is 1.25 times the volume of an f.c.c. unit cell, which has itself a volume equal to four times the volume of the primitive unit cell of the f.c.c. disordered structure. The number of translation variants is thus five [7]. The four possible vectors of the antiphase boundaries (APB) are represented in Fig. 1. One can thus distinguish four different APB's which are represented in Fig. 4, assuming the boundary plane to be (120) or (2$\bar{1}$0), which is in fact the observed preferential plane. The boundaries are represented with the structures in the two domains continuing unmodified up to the boundary. In general the boundaries are thus non-conservative; by adding or removing manganese atoms along the boundary the composition can be changed of course.

5. Observations by High Resolution Microscopy

5.1 The Au_4Mn structure

In previous papers it was shown that under certain favourable circumstances the atomic structure of alloys can be imaged [6, 8]. The Au_4Mn structure has the required favourable features to achieve successful imaging. Along the c-axis the unit cell is only one atomic layer thick; atoms of the same chemical species form columns along this direction. The images are made in the dark field, collecting all superstructure beams in the aperture excluding the beams due to the basic f.c.c. structure. In this way redundant information is rejected. Moreover the atomic scattering amplitude of Au and Mn are rather different, which is also a favourable feature.

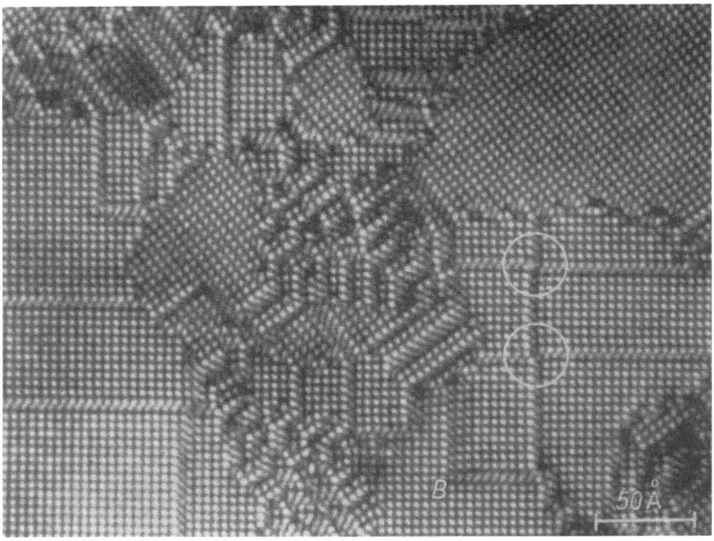

Fig. 5. Dark field structure image of multidomain structure of Au_4Mn APB's as well as twin interfaces can be recognized

The image of perfect parts of the structure can be observed in Fig. 5. Like in the previously studied gold–manganese alloys the pattern of bright dots has the configuration, orientation, and scale of the pattern of manganese columns in Au_4Mn. Their geometry is compared with a correctly scaled model of the structure in the inset of Fig. 12. One can conclude beyond any reasonable doubt that the bright dots image columns of single manganese atoms within a gold matrix. Under certain imaging conditions one can obtain dark dots as well. The pattern of dots is only slightly dependent on focus and foil thickness; this seems to be a feature characteristic of column structures.

5.2 *Microstructur*

Fig. 5 also shows an area of Au_4Mn exhibiting a pronounced microstructure. Since the atom columns in coaxial variants are parallel, structure resolution can be achieved in two variants simultaneously. The two orientation variants, with their common c-axis parallel to the beam direction, and related by a mirror operation with respect to a cube plane mirror, can easily be recognized in Fig. 5. The other variants, with their c-axis perpendicular to the beam direction, are imaged as dark areas since they do not contribute to the beams collected in making the image.

The APB's are revealed by the offset which they cause in the "atom rows"; they have highly preferential directions along the (120) planes of the f.c.c. structure (i.e. the (110) planes of the Au_4Mn structure). They usually have an angular shape consisting of mutually perpendicular segments. This shows that the APB energy is highly anisotropic.

In [6] we have shown how the displacement vectors of the APB's in Au_5Mn_2 can be determined from the lateral shift of rows of columns of manganese pairs along two different directions. The same method could of course also be applied to Au_4Mn, however in the present case the shifts along the two mutually perpendicular atom rows are always 1/5 and 2/5 as shown in Fig. 4 a differentiation between different

APB's is thus not possible on this basis alone. From the detailed atomic structure of the boundary such a distinction is possible, however. The boundary of Fig. 5, (B) can thus be compared directly with the model of Fig. 4c.

5.3 Coincidence structure at domain boundaries [9]

In a number of cases the contact plane separating coaxial orientation variants is not parallel with the [001] viewing direction; interesting dot patterns then result in the overlap region (Fig. 6). This region exhibits a square pattern of prominently bright dots with a mesh size equal to 2.5*a* (*a* lattice parameter of the f.c.c. structure) and with its edges parallel to the cube plane of the f.c.c. structure. This pattern can readily be interpreted in terms of the structure by means of a simple geometrical model.

Considering the pattern of manganese columns on both sides of the interface it turns out that the bright dots image in fact the coincidence lattice of the structure in both domains, as well in scale as in orientation. This can be concluded from the geometrical model of Fig. 7 where the atom columns in one domain have been represented as open dots and those in the other domain as full dots. Coincident open and closed dots clearly represent coincident columns. The manganese columns at coincidence sites are continuous in both variants; this is the reason why they are imaged as prominently bright dots. This observation is a strong indication that the bright dots image "columns" and not "atoms".

The coincidence lattice can also be calculated by the simple method the of Appendix.

In particular we want to consider the effect on the coincidence pattern of a parallel translation *R* of one of the domains. This effect can directly be revealed in an overlap region intersected by an antiphase boundary. A spatial view of such a situation is shown in Fig. 8. The geometrical model for the coincidence pattern on both sides of

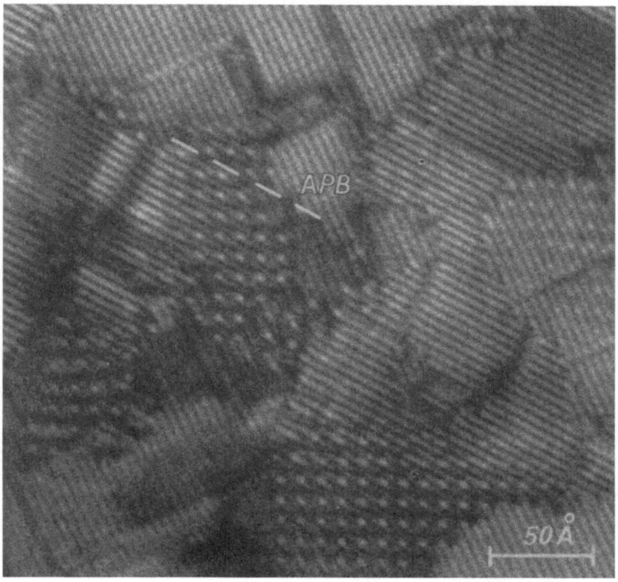

Fig. 6. Overlap pattern of two coaxial variants. Note the square pattern of extra white dots corresponding with the coincidence columns. Note also the shift of the coincidence pattern along the APB

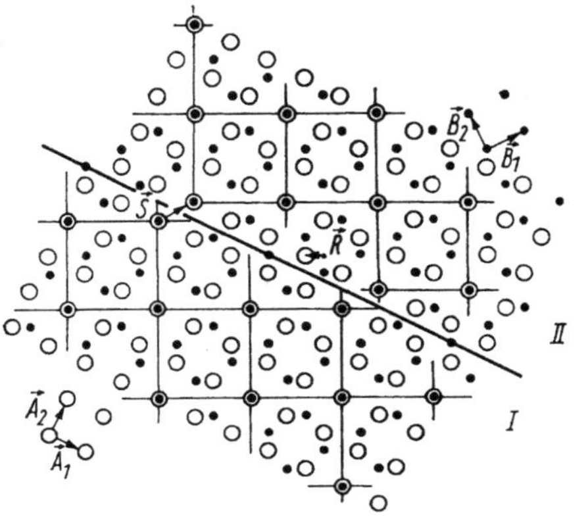

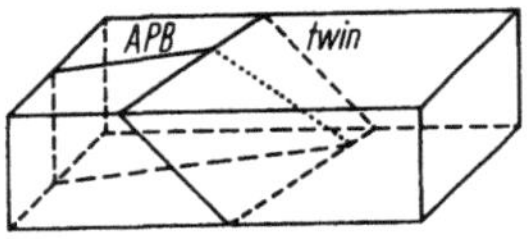

Fig. 8. Spatial view of the intersection of an APB with an inclined twin boundary

Fig. 7. Geometrical model for the coincidence structure which can be compared directly with Fig. 6

the APB is represented in Fig. 7. It is obvious that the coincidence patterns on either side of the APB have suffered a relative shift over a vector $\boldsymbol{S}$, which is different from $\boldsymbol{R}$, the displacement vector of the APB. (Also the shift vector can be calculated, see Appendix.) Such a shift is visible in Fig. 6 where the trace of an APB has been indicated. The correspondence between the geometrical model and the observed configuration is striking.

5.4 The structure of the antiphase boundary region

5.4.1 Streaks at APB's

Most of the antiphase boundaries produce striking images consisting of streaks connecting atom positions at the edge of one domain with atom positions at the edge of another domain. Such features are visible in the encircled areas of Fig. 5. The reproducibility of these features suggests that they presumably have some structural significance. In Fig. 9 we have represented these features schematically. We tentatively interpret these streaks as meaning that there is a stepwise shift of the manganese columns; neighbouring pieces of manganese columns would be too close to be resolved and hence streaks would be produced. The intensity of the image being smaller than that produced at a complete column we assume that the gradual shift occurs by short segments of columns as represented in Fig. 10.

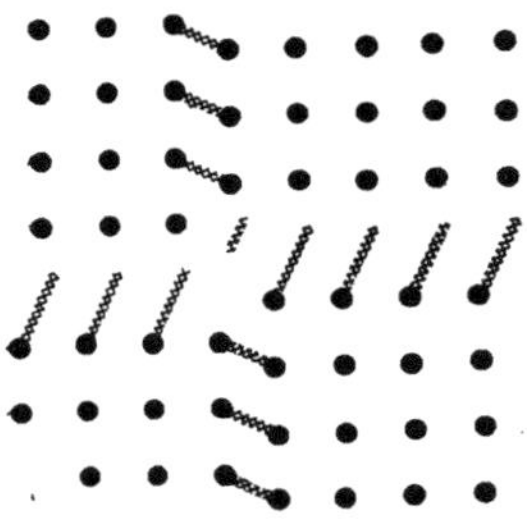

Fig. 9

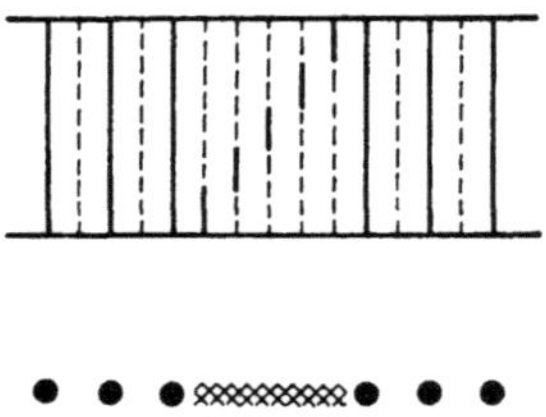

Fig. 10

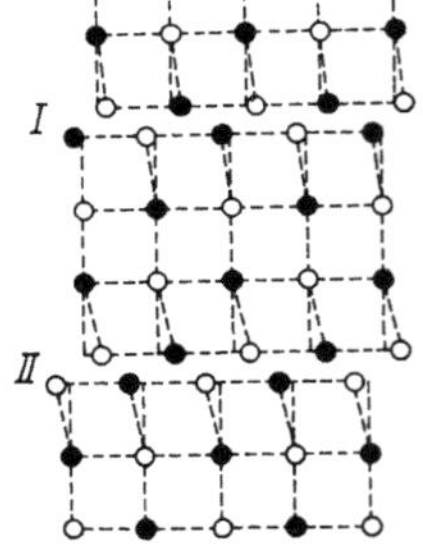

Fig. 11

Fig. 9. Schematic view of streaks at APB's as observed in the encircled parts of Fig. 5

Fig. 10. Structure of columns at the interface producing streaks in the image

Fig. 11. Schematic view of the bending of atom row in the vicinity of anti-phase boundaries

Fig. 12. Enlarged area of Fig. 5. Bending of lattice rows is visible in the vicinity of an APB. The inset shows a correctly scaled model of the structure where only Mn atoms are shown

5.4.2 Relaxation along APB's

It was shown previously that some relaxation occurs along certain types of antiphase boundaries in Ni_3Mo and Ni_4Mo [10]. Use was here made of the effect on the extinction conditions of the fringe pattern of a small additional displacement, superimposed on the crystallographic displacement associated with the interface. The present observations reveal that along many APB's those lattice rows which are normal to the boundary plane are slightly bent in the close vicinity of this boundary; this is shown schematically in a somewhat exaggerated way in Fig. 11. It is possible that this is in part an imaging effect due to the inclination of the interface. However, since the bending is systematic in such a way as to maximize the average distance between manganese columns, it is probable that this is a relaxation effect and a visual manifestation of the repulsion between manganese atoms. The bending is most pronounced along the boundary parts where the lateral shift of the perpendicular rows is 1/5 and where the distance between manganese columns would therefore become smallest if no relaxation would take place.

A difference in the magnitude of the relaxation is to be expected for boundaries of type I and II of Fig. 11. In type I the rows of manganese columns that meet with a 1/5 shift are at different levels in the structure whereas they are at the same level in the type II boundary. In the latter case the relaxation is therefore expected to be somewhat larger. An observed example is reproduced in Fig. 12.

5.5 Boundary regions between transition structure and Au_4Mn

In specimens which are not yet fully ordered in the Au_4Mn structure, one can observe small regions which have still the transition structure. Along the boundary regions between the transition structure and the Au_4Mn structure, one observes a dot pattern which is characteristic of the Au_4Mn structure but in which not all the spots have the same intensity (Fig. 5). It is suggested that this means that certain manganese columns are incomplete, i.e. certain manganese sites are occupied by gold.

6. Conclusions

The crystal structure of fully ordered Au_4Mn presents features which make it particularly suitable for structure imaging especially if the image is formed exclusively by superlattice spots. The structure images give direct evidence on an "atomic" level, for the domain structure. In particular detailed information can be obtained on the structure of the boundary regions. Evidence is given for the occurrence of a coincidence structure at inclined twin domain boundaries. At antiphase boundaries relaxation of the structure occurs as can be deduced from the bending of atom rows.

Acknowledgement

We wish to thank Dr. D. van Dyck for useful discussions concerning the coincidence lattice calculations.

Appendix

Coincidence lattice

The coincidence lattice can be calculated by means of the following method. The base vectors of the two-dimensional lattices $\mathcal{A}$ and $\mathcal{B}$ of the projected manganese columns are $(\boldsymbol{A}_1, \boldsymbol{A}_2)$ and $(\boldsymbol{B}_1, \boldsymbol{B}_2)$, respectively; they are related by the relations

$$\begin{pmatrix} \boldsymbol{B}_1 \\ \boldsymbol{B}_2 \end{pmatrix} = M \begin{pmatrix} \boldsymbol{A}_1 \\ \boldsymbol{A}_2 \end{pmatrix},$$

where M represents the operation relating the structures in the two domains.

$$M = \begin{pmatrix} 3/5 & 4/5 \\ -4/5 & 3/5 \end{pmatrix}, \qquad M^{-1} = \begin{pmatrix} 3/5 & -4/5 \\ 4/5 & 3/5 \end{pmatrix}.$$

In the present case it is a rotation over an angle $\delta = 2 \operatorname{arctg} (1/2)$ from $(\boldsymbol{A}_1, \boldsymbol{A}_2)$ towards $(\boldsymbol{B}_1, \boldsymbol{B}_2)$ about a common lattice point.

An arbitrary vector of lattice $\mathcal{A}$ can be represented by

$$\boldsymbol{r}_{\text{A}} = (l_1, l_2) \begin{pmatrix} \boldsymbol{A}_1 \\ \boldsymbol{A}_2 \end{pmatrix} = (l_1, l_2)\, M^{-1} \begin{pmatrix} \boldsymbol{B}_1 \\ \boldsymbol{B}_2 \end{pmatrix},$$

whereas an arbitrary vector of lattice $\mathcal{B}$ can be represented as

$$\boldsymbol{r}_{\text{B}} = (p_1, p_2) \begin{pmatrix} \boldsymbol{B}_1 \\ \boldsymbol{B}_2 \end{pmatrix} = (p_1, p_2)\, M \begin{pmatrix} \boldsymbol{A}_1 \\ \boldsymbol{A}_2 \end{pmatrix},$$

where l_1, l_2, p_1, and p_2 are integers. The conditions for coincidence is that $\boldsymbol{r}_{\text{B}}$ should also be a vector of lattice $\mathcal{A}$, this will be the case if

$$(p_1, p_2)\, M \begin{pmatrix} \boldsymbol{A}_1 \\ \boldsymbol{A}_2 \end{pmatrix} = 0 \quad \text{modulus a lattice vector of } \mathcal{A}.$$

With the expression for M this leads to a set of Diophantine equations

$$\left.\begin{aligned} (1/5)\,(3p_1 - 4p_2) &= 0 \\ (1/5)\,(4p_1 + 3p_2) &= 0 \end{aligned}\right\} \text{(mod 1)} \tag{A1}$$

with solutions (m and n integers)

$$\begin{cases} p_1' = n \\ p_2' = 2n \end{cases} \quad \text{and} \quad \begin{cases} p_1'' = 2m \\ p_2'' = -m \end{cases}. \tag{A2}$$

One solution is thus

$$(n, 2n) \begin{pmatrix} \boldsymbol{B}_1 \\ \boldsymbol{B}_2 \end{pmatrix} = n(\boldsymbol{B}_1 + 2\boldsymbol{B}_2) = n\boldsymbol{C}_1$$

and the other

$$(2m, -m) \begin{pmatrix} \boldsymbol{B}_1 \\ \boldsymbol{B}_2 \end{pmatrix} = m(2\boldsymbol{B}_1 - \boldsymbol{B}_2) = m\boldsymbol{C}_2 .$$

Since the system of equations is linear, any linear combination is evidently a solution as well. The most general solution is thus

$$\boldsymbol{C} = n\boldsymbol{C}_1 + m\boldsymbol{C}_2 \qquad (n, m \text{ integers}) \tag{A3}$$

which represents the coincidence lattice with base vectors

$$\boldsymbol{C}_1 = \boldsymbol{B}_1 + 2\boldsymbol{B}_2 \quad \text{and} \quad \boldsymbol{C}_2 = 2\boldsymbol{B}_1 - \boldsymbol{B}_2 .$$

Since the problem is symmetrical in A and B one can of course also reason as follows:

$$\boldsymbol{r}_A = (l_1, l_2)\, M^{-1} \begin{pmatrix} \boldsymbol{B}_1 \\ \boldsymbol{B}_2 \end{pmatrix} = 0 \quad \text{modulus a lattice vector of } \mathcal{B}.$$

This leads to the set of Diophantine equations

$$\left.\begin{aligned} (1/5)\,(3l_1 + 4l_2) &= 0 \\ (1/5)\,(-4l_1 + 3l_2) &= 0 \end{aligned}\right\} \pmod 1 . \tag{A4}$$

The solution of this system of equations is

$$\begin{cases} l_1' = 2m \\ l_2' = m \end{cases} \quad \text{and} \quad \begin{cases} l_1'' = -n \\ l_2'' = 2n \end{cases} \tag{A5}$$

or

$$m(2\boldsymbol{A}_1 + \boldsymbol{A}_2) \quad \text{and} \quad n(-\boldsymbol{A}_1 + 2\boldsymbol{A}_2)$$

or

$$\boldsymbol{C} = n\boldsymbol{C}_1 + m\boldsymbol{C}_2$$

with $\boldsymbol{C}_1 = -\boldsymbol{A}_1 + 2\boldsymbol{A}_2$ and $\boldsymbol{C}_2 = 2\boldsymbol{A}_1 + \boldsymbol{A}_2$. The vectors $\boldsymbol{C}_1$ and $\boldsymbol{C}_2$ are the same as those given above.

Effect of a translation in one domain

Let the translation be represented by

$$\boldsymbol{R} = (s_1 s_2) \begin{pmatrix} \boldsymbol{A}_1 \\ \boldsymbol{A}_2 \end{pmatrix} = (r_1 r_2) \begin{pmatrix} \boldsymbol{B}_1 \\ \boldsymbol{B}_2 \end{pmatrix} \tag{A6}$$

with in our particular case $s_1 = -2/5$, $s_2 = -1/5$ and $r_1 = -2/5$, $r_2 = 1/5$. The displaced coincidence lattice can be represented by

$$\boldsymbol{C} = n\boldsymbol{C}_1 + m\boldsymbol{C}_2 + \alpha\boldsymbol{C}_1 + \beta\boldsymbol{C}_2 = (n + \alpha)\,\boldsymbol{C}_1 + (m + \beta)\,\boldsymbol{C}_2 ,$$

where α and β are constants, $0 < \alpha < 1$, $0 < \beta < 1$, n and m are integers. We shall refer all vectors to the lattice ($\boldsymbol{B}_1$, $\boldsymbol{B}_2$) which is continuous in the two domains.

An arbitrary lattice vector $\boldsymbol{r}'_A$ of the displaced $\mathcal{A}$ lattice can then be represented by

$$\boldsymbol{r}'_A = (l_1 l_2)\, M^{-1} \begin{pmatrix} \boldsymbol{B}_1 \\ \boldsymbol{B}_2 \end{pmatrix} + (r_1 r_2) \begin{pmatrix} \boldsymbol{B}_1 \\ \boldsymbol{B}_2 \end{pmatrix}.$$

We express the condition for coincidence with lattice $\mathcal{B}$ as

$$(l_1 l_2)\, M^{-1} + (r_1, r_2) = 0 \quad (\text{mod } 1)$$

or explicitly

$$\left.\begin{aligned} 1/5\,(3l_1 + 4l_2) - 2/5 &= 0 \\ 1/5\,(-4l_1 + 3l_2) + 1/5 &= 0 \end{aligned}\right\} \quad (\text{mod } 1)\,. \tag{A7}$$

The general solution of the system of Diophantine equations (A7) consists of the general solution of the homogeneous set (A4) to which a particular solution of the complete set is added. As a particular solution one finds $(l_1, l_2) = (1, 1)$. The general solution is hence

$$\begin{pmatrix} l_1 \\ l_2 \end{pmatrix} = \begin{pmatrix} 2 \\ 1 \end{pmatrix} m + \begin{pmatrix} -1 \\ 2 \end{pmatrix} n + \begin{pmatrix} 1 \\ 1 \end{pmatrix},$$

i.e. explicitly

$$\begin{aligned} \boldsymbol{C} &= m(-\boldsymbol{A}_1 + 2\boldsymbol{A}_2) + m(2\boldsymbol{A}_1 + \boldsymbol{A}_2) + (\boldsymbol{A}_1 + \boldsymbol{A}_2) \\ &= n\boldsymbol{C}_1 + m\boldsymbol{C}_2 + \boldsymbol{S} \end{aligned}$$

with $\boldsymbol{S} = \boldsymbol{A}_1 + \boldsymbol{A}_2 = 1/5\,\boldsymbol{C}_1 + 2/5\,\boldsymbol{C}_2$ i.e. $\alpha = 1/5$ and $\beta = 2/5$.

References

[1] A. Watanabe, Acta cryst. **10**, 483 (1957).
[2] E. Ruedl, P. Delavignette, and S. Amelinckx, phys. stat. sol. **28**, 305 (1968).
[3] G. van Tendeloo and S. Amelinckx, to be published.
[4] D. Watanabe, J. Phys. Soc. Japan **15**, 1251 (1960).
S. K. Das, P. R. Okamoto, P. M. J. Fisher, and G. Thoams, Acta metall. **21**, 913 (1973).
H. Sato, R. S. Toth, G. Shirane, and D. E. Cox, J. Phys. Chem. Solids **27**, 413 (1966).
H. Sato, R. S. Toth, and G. Honjo, J. Phys. Chem. Solids **28**, 137 (1967).
[5] R. Wolf, Micron 8, 171 (1977).
[6] G. van Tendeloo, R. Wolf, D. van Dyck, and S. Amelinckx, phys. stat. sol. (a) **47**, 105 (1978).
[7] G. van Tendeloo and S. Amelinckx, Acta cryst. **A 30**, 432 (1974).
[8] G. van Tendeloo, R. Wolf, J. van Landuyt, and S. Amelinckx, phys. stat. sol. (a) **47**, 539 (1978).
[9] G. van Tendeloo and S. Amelinckx, phys. stat. sol. (a) **47**, 555 (1978).
[10] G. van Tendeloo and S. Amelinckx, phys. stat. sol. (a) **22**, 621 (1974); **27**, 93 (1975).

(Received April 14, 1978)

phys. stat. sol. (a) **49**, 347 (1978)

Subject classification: 3 and 11; 22.1.2

Istituto di Fisica dell'Università, Padova[1]) *(a) and Istituto di Fisica dell'Università Catania*[2]) *(b), Unità G.N.S.M.—C.N.R., and Laboratorio La.M.El. — C.N.R., Bologna*[3]) *(c)*

Two-Stage Laser Annealing of Lattice Disorder in Phosphorus Implanted Silicon[4])

By

G. BATTAGLIN (a), G. DELLA MEA (a), A. V. DRIGO (a), G. FOTI (b), G. G. BENTINI (c), and M. SERVIDORI (c)

Q-switched ruby laser is used to anneal phosphorus implanted ⟨100⟩ silicon wafers. Residual lattice disorder is analyzed by combined TEM and backscattering techniques. Different kinds and densities of residual damage can be observed as a function of the laser pulse energy density. Two annealing stages are observed for energy densities below and above 1.5 J/cm^2. Above this threshold a very good annealing takes place leading to the formation of an almost perfect single crystal.

Ein gütegeschalteter Rubinlaser wird benutzt, um phosphorimplantierte ⟨100⟩-Siliziumscheiben zu tempern. Der Reststrahlenschaden wird mit kombinierter TEM und Rückstreutechnik untersucht. Verschiedene Arten und Dichten des Restschadens können als Funktion der Energiedichte des Laserimpulses beobachtet werden. Zwei Temperungsstufen werden für Energiedichten unterhalb und oberhalb 1,5 J/cm^2 beobachtet. Oberhalb dieser Schwelle findet eine sehr gute Ausheilung statt, die zur Bildung eines fast perfekten Einkristalls führt.

1. Introduction

The implantation technique, extensively used in semiconductor device technology, requires thermal treatments to minimize the disorder induced by ions in the crystal lattice [1].

High fluence implantations [2], effect of substrate orientation [3], temperature of wafer during implantation [1], and light ion implantation [1] result in the generation of lattice damage that cannot be fully annealed by standard thermal treatments. Recent investigations [4] have shown that high power pulsed laser irradiation of implanted semiconductors can be used to solve this problem. In particular a very good annealing of self-ion implanted amorphous layer on ⟨100⟩ and ⟨111⟩ silicon single crystals has been observed after laser irradiation [5]. Electrical measurements of activation of implanted impurities give sharp profiles after laser annealing [6]. Moreover high substitutional fractions of boron [7], arsenic [8], and tellurium [9] have been detected by channeling techniques. Sheet conductivity of phosphorus implanted in ⟨100⟩ silicon has been measured as a function of the incident laser energy density: it is found comparable to observations at high temperature thermal annealing [10] and reveals that the electrical activation occurs above a well defined energy density threshold.

The aim of this work is to analyze the structure of the crystallized layer near the threshold energy density for Q-switched ruby laser irradiation on silicon implanted with phosphorus ions.

---

[1]) Via Marzolo, 8—35100 Padova, Italy.

[2]) Corso Italia, 95100 Catania, Italy.

[3]) Via dei Castagnoli, 1—40126 Bologna, Italy.

[4]) Work supported in part by G.N.S.M. — C.N.R..

2. Experiment

⟨100⟩ silicon wafers (1 Ω cm, p-type) were implanted with 100 keV P^{31} ions at room temperature. The total fluence was 1×10^{15} ions/cm² and an amorphous layer about 1800 Å thick was produced. To minimize channeling effects during the implantation the crystal was tilted 6° off the ⟨100⟩ direction avoiding any principal plane.

Laser irradiation was performed in air at room temperature with Q-switched ruby laser pulse of about 15 ns duration. The energy density ranged between 1.0 and 3.0 J/cm² with a spot of 1.8 cm diameter. The samples were analyzed with 1.6 MeV He^+ ions in backscattering geometry using the channeling technique. The energy resolution of the overall detection system was about 20 keV, which corresponds to about 400 Å depth resolution in silicon.

After the channeling measurements, the samples were thinned for the transmission electron microscopy (TEM) observations, which were performed with a 125 kV Siemens Elmiskop 102.

3. Results

3.1 Damage evolution during laser irradiation

Fig. 1 shows the experimental spectra obtained with 1.6 MeV He^+ ions incident along the ⟨100⟩ direction of silicon samples laser irradiated. For comparison the spectra of as-implanted and unimplanted samples are also reported.

At an energy density above 1.1 J/cm² the aligned yield decreases drastically compared with an as-implanted specimen. This result indicates that an amorphous–crystalline transition has taken place as a consequence of laser irradiation. However, high residual disorder is still present in the crystallized layer for energy densities below 1.5 J/cm². In fact the depth distribution of defects, as shown by the channeling spectra for 1.2 and 1.4 J/cm² irradiated samples, exhibits a residual high disorder peak near the substrate-to-crystallized layer interface. For high energy laser pulses, above

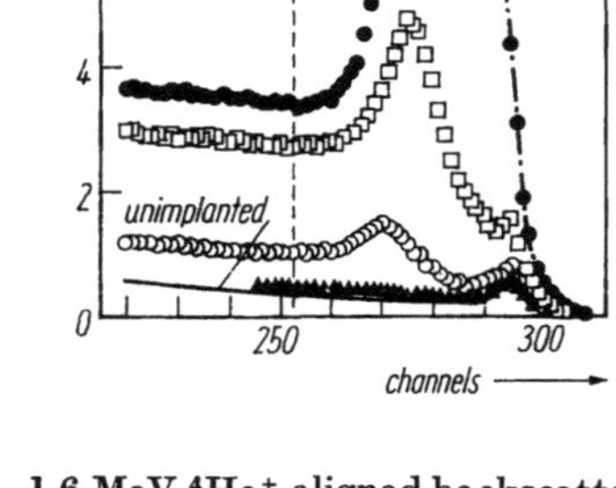

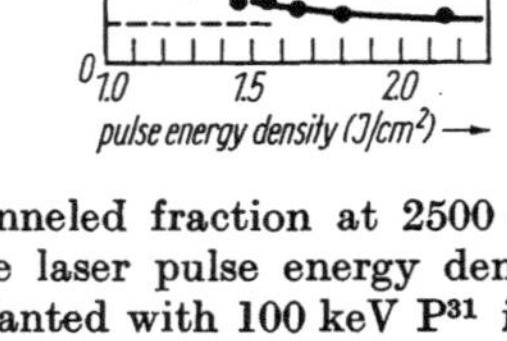

Fig. 2. Dechanneled fraction at 2500 Å depth as a function of the laser pulse energy density of ⟨100⟩ Si wafers implanted with 100 keV P^{31} ions, the total fluence was 1×10^{15} ions/cm²

Fig. 1. 1.6 MeV $^4He^+$ aligned backscattering spectra obtained from ⟨100⟩ silicon wafers after different laser pulse irradiations. For comparison the aligned unimplanted spectrum (continuous line) and the random spectrum (dash-dotted line) are reported. ● as-implanted, □ after 1.2 J/cm², ○ after 1.4 J/cm², ▲ after 1.6 J/cm²

1.5 J/cm^2, the channeling aligned yields are comparable with an unimplanted, defect-free silicon.

A parameter related to the amount of residual damage in the regrown layer is the dechanneled fraction measured at a depth larger than the damaged layer thickness. The effects of the laser annealing have been hence followed by plotting in Fig. 2 the dechanneled fraction measured at 2500 Å depth as a function of the laser pulse energy density. The experimental values clearly indicate the existence of two stages in the laser annealing process: in the first stage, between 1.1 and 1.5 J/cm^2 irradiation, the dechanneled fraction decreases steeply from the value corresponding to the as-implaned sample to about 5%; in the second stage, from 1.5 to 2.20 J/cm^2, the dechanneled fraction slowly approaches the unimplanted aligned yield of about 3%. The channeling measurements of the residual disorder have been correlated with TEM observations for a better description of the two-stage laser annealing.

The TEM micrographs and the selected area diffraction (SAD) patterns reported in Fig. 3 show the damage evolution following the laser spot energy. Fig. 3a reveals in an as-implanted sample the presence of an amorphous layer superimposed to the perfect substrate. After irradiation with energy density of 1.2 J/cm^2 an amorphous-to-crystalline transition takes place leading to the formation of a very disordered single crystal layer. This disorder is mainly associated to a high density of twins dispersed

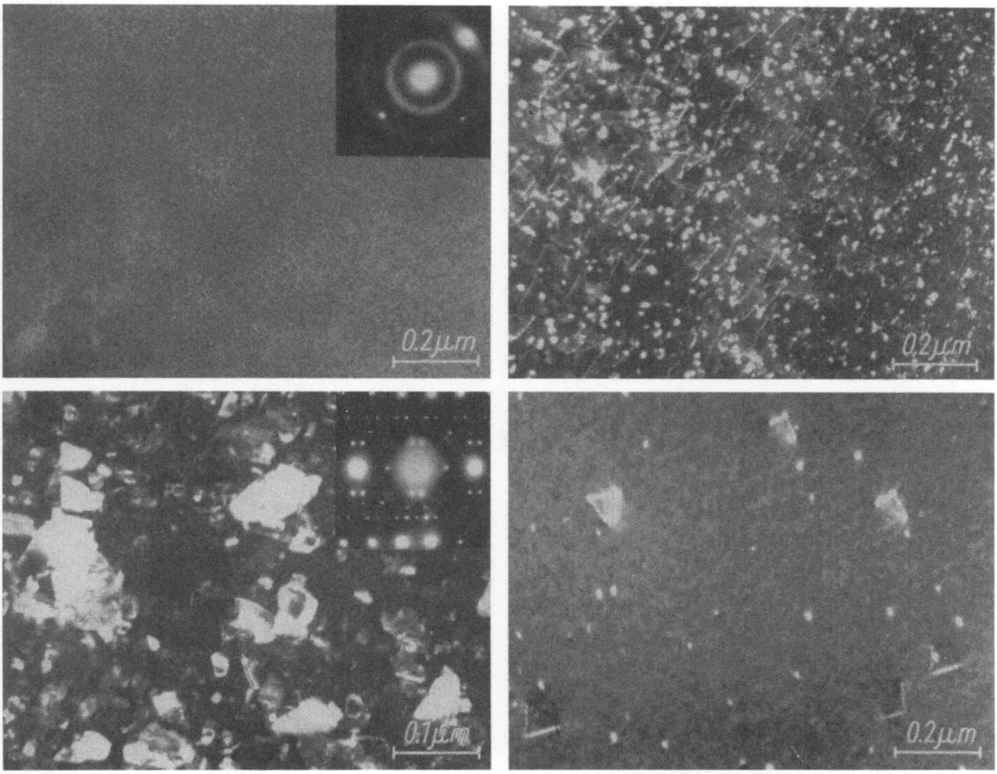

Fig. 3. TEM micrographs and SAD patterns showing damage evolution during laser annealing. a) First Debye ring dark field (DF) image, as implanted sample; b) 111 twin spot DF image, 1.2 J/cm^2; c) 220 dark field weak beam (DFWB) image, 1.4 J/cm^2; d) 220 DFWB image, 1.6 J/cm^2

in the matrix. Fig. 3b refers to the same sample and shows in fact a dark field image taken with the 111 twin spots, with the matrix exactly oriented in the (510) projection. For a laser pulse of 1.4 J/cm² a large amount of V-shaped dislocations, point defect clusters, and dislocation microloops is observed (Fig. 3c). The defect density decreases after 1.5 J/cm² irradiation and only a small region showing low density of V-dislocations and microloops (Fig. 3d) is present after 1.65 J/cm² irradiation. After 2.15 J/cm² annealing TEM images (not shown) reveal that a high quality, defect-free crystallized layer is obtained.

A depth analysis of the defect distribution was also performed by TEM. In an as-implanted sample and in samples irradiated up to 1.4 J/cm² a layer about 1800 Å thick was removed from the surface by anodic stripping. Fig. 4 shows the presence of a large amount of point defect clusters, whose concentration decreases after laser irradiation. These clusters were observed to be confined within a depth of about 1500 Å, that is 3300 Å below the unstripped surface. This value is in very good agreement with the junction depth observed by the staining technique.

The high quality of the crystallized layer obtained with laser pulses above 1.5 J/cm² with about 10^{-8} s duration, rules out the description of the growth of silicon layer as due to a solid phase reaction. The growth rate measured just below the melting temperature is orders of magnitude lower than that necessary to explain the crystallization of the implanted layer. The second annealing stage seems to be correlated with the existence of a melted layer during laser irradiation. Very simple calculations,

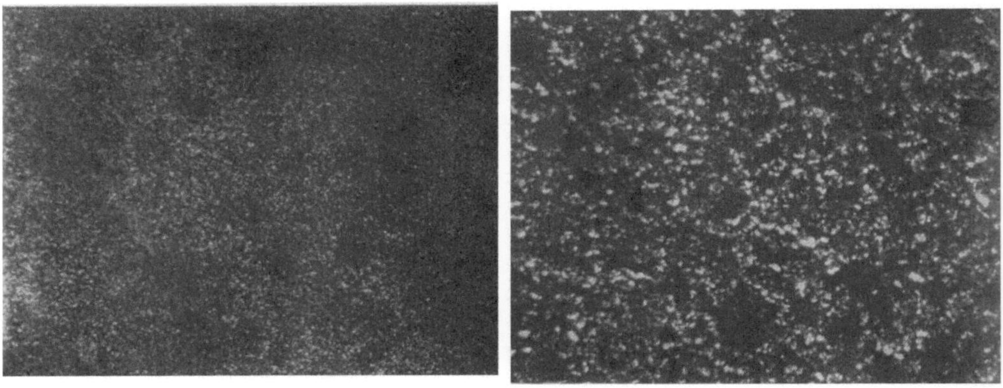

a

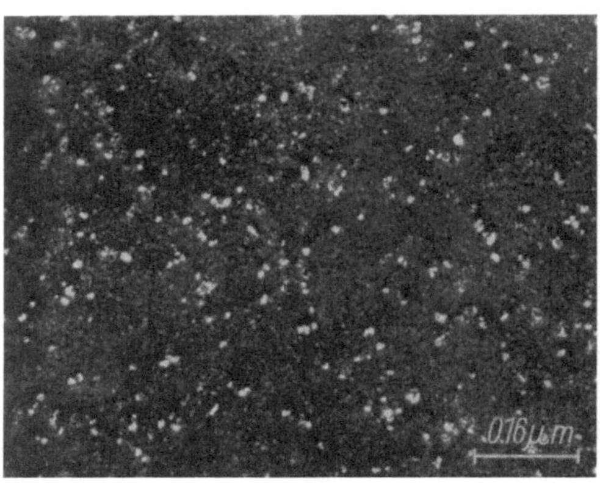

c

Fig. 4. TEM micrographs of the damage at 1800 Å below the silicon surface. a) As-implanted sample; after irradiation with 1.2 J/cm², b) and 1.4 J/cm², c) energy densities reported in [11], indicate that the melting might occur in silicon for laser energies higher than 1.5 J/cm². For a 20 ns ruby laser pulse, the relationship between incident energy density, E, and melted thickness, t, is approximately $E \approx E_0 + \alpha t$, where $E_0 \approx \approx 1.5$ J/cm² and $\alpha \approx 2.5 \times 10^4$ J/cm³. Because the thickness of the amorphous layer after implantation was 1800 Å, to melt this layer an energy of 1.9 J/cm² is required, which is in reasonable agreement with the experimental value observed for the beginning of the second annealing stage.

3.2 Homogeneity of laser spot

In view of the possible technological application of laser annealing the investigation is of relevance of both chemical effects on the surface and of spatial uniformity of the crystallized layer, whenever suitable filtering systems to homogenize the laser spot are not available. Because of the high temperature rise contamination of the silicon surface cannot be excluded during irradiation in air. Light impurities like carbon, oxygen, and nitrogen can be easily detected by measuring the backscattering yield in aligned conditions. Moreover channeling technique allows a lateral analysis by moving the incident beam on the irradiated spot.

Typical spectra for an energy density of 2.15 J/cm² are reported in Fig. 5, where in the upper portion the laser spot is sketched. In a inner circular region of about 1.3 cm diameter, labelle A, the aligned spectra are comparable with a good single crystal. The minimum yield is 3.5% and the measured amount of carbon and oxygen is about 10^{16} C/cm² and 3×10^{15} O/cm². In the region labelled B, the oxygen and carbon peaks increase of a factor three and a nitrogen component of about 10^{16} N/cm² is also present.

The different behaviour of the region A compared with the region B points out that the uniformity of laser spot and/or lateral temperature gradient effects must be investigated in order to use the laser annealing in technological processes.

For lower energy densities, near the threshold, the region B disappears but different amounts of residual damage have been evidenced by backscattering in different zones of the irradiated area some millimeters apart, due to the non-homogeneity of the laser spot.

In addition a non-uniformity on a smaller scale was observed by TEM analysis. Different amounts and kinds of defects are present within each TEM sample as sketched in Fig. 6. By increasing the energy density of the laser pulse the twinned area be-

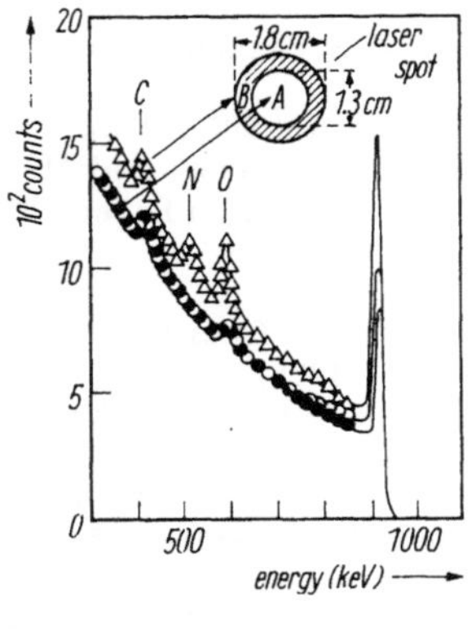

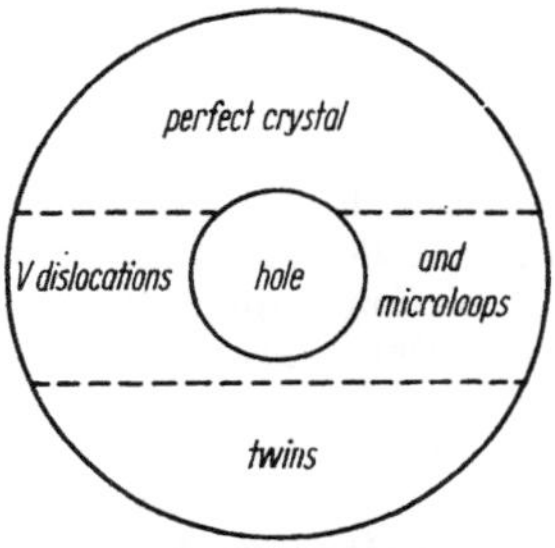

Fig. 6. Schematic view of the lateral defect distribution observed by TEM in the transparent area of samples irradiated with energies corresponding to the threshold region

Fig. 5. Backscattering aligned spectra taken in different regions of a sample irradiated at 2.15 J/cm². A sketch of the irradiated area is reported in the upper part of the figure. Open circles and triangles refer to spectra taken in region A and B of the irradiated samples, respectively; full circles refer to an unimplanted sample comes smaller and smaller until only a very small portion of the overall transparent area reveals a low dislocation density.

3.3 Conclusions

Analyses performed with the ion channeling technique and TEM observations indicate the existence of two stages in the laser annealing process of phosphorus implanted silicon.

The first one, between 1.1 and 1.5 J/cm^2 irradiations, is characterized by a high residual disorder in the crystallized layer. This damage, strongly dependent on the energy density of the laser pulse, is clearly evidenced by channeling. Moreover TEM observations indicate the prevalent presence of twins. In the second stage, between 1.5 and 2.2 J/cm^2 irradiation, a high quality crystallized layer is obtained. This effect seems to be correlated with the existence of a melted layer during laser irradiation. In this second stage the channeling aligned yields approach slowly the unimplanted silicon values and TEM micrographs show the presence of V-dislocations with decreasing density by increasing the energy density of the laser pulse.

To obtain the best crystallization it is convenient to operate between 1.8 and 2.2 J/cm^2. In this range of energy density the quality of the regrown layer is nearly independent of inhomogeneities of the laser spot and so a good recrystallization takes place over a very large area. An appreciable residual damage is found only at the boundary of the laser spot.

Work is in progress to determine the lattice location and the depth distribution profile of phosphorus after laser annealing. Preliminary results indicate that phosphorus is located nearly completely on substitutional sites.

Acknowledgements

The authors wish to thank E. Iannitti for the laser irradiations and L. Dori for his technical assistance in TEM work.

References

[1] J. W. MAYER, L. ERIKSSON, and J. A. DAVIES, Ion Implantation in Semiconductors, Academic Press, 1970.
[2] L. CSEPREGI, E. F. KENNEDY, S. S. LAU, J. W. MAYER, and T. W. SIGMON, Appl. Phys. Letters **29**, 645 (1976).
[3] L. CSEPREGI, J. W. MAYER, and T. W. SIGMON, Appl. Phys. Letters **29**, 92 (1976).
[4] G. A. KACHURIN, V. A. BOGATYRIOV, S. I. ROMANOV, and L. S. SMIRNOV, Ion Implantation in Semiconductors, Plenum Press, 1976.
[5] G. FOTI, E. RIMINI, W. TSENG, and J. W. MAYER, Appl. Phys. **15**, 365 (1978).
[6] I. B. KHAIBULLIN, E. I. SHTYRKOV, M. M. ZARIPOV, R. M. BAYAZITOV, and M. F. GALYATUDINOV, On some Peculiarities of Laser Annealing of Implanted Silicon Layers, First USSR-USA Seminar on Ion Implantation, Albany — New York 1977.
[7] G. FOTI and G. DELLA MEA, Lettere Nuovo Cimento **21**, 89 (1978).
R. T. YOUNG, C. W. WHITE, G. J. CLARK, J. NARAYAN, W. H. CHRISTIE, M. MURAKAMI, P. W. KING, and S. D. KRAMER, Appl. Phys. Letters **32**, 139 (1978).
[8] J. KRYMIEKI, J. SUSKI, S. UGMIEWSKI, R. GROTISCHEL, R. K. KLABES, U. KUISSIG, and J. RUDIGER, Phys. Letters **61 A**, 181 (1977).
[9] S. U. CAMPISANO, G. FOTI, E. RIMINI, and G. VITALI, J. appl. Phys. (1978), in the press.
[10] I. B. KHAIBULLIN, R. M. BAYAZITOV, E. I. SHTYRKOV, and M. M. ZARIBOV, Proc. Conf. Ion Implantation in Semiconductors, Budapest 1975 (p. 212).
[11] P. BAERI, S. U. CAMPISANO, G. FOTI, and E. RIMINI, Appl. Phys. Letters (1978), in the press.
G. FOTI, E. RIMINI, M. BERTOLOTTI, and G. VITALI, Phys. Letters **65 A**, 430 (1978).

(Received June 9, 1978)

phys. stat. sol. (a) 49, 353 (1978)

Subject classification: 10.2 and 14.4; 22.5.2

Department of Metallurgy and Materials Science, Imperial College, London[1])

The Role of Dimers and Trimers in the Aggregation of Impurity–Vacancy Dipoles in Alkali Halides

By

A. L. GUERRERO, S. C. JAIN[2]), and P. L. PRATT

Recent experimental data on the conductivity of NaCl crystals doped with Mn is reanalysed in terms of a model which includes the presence of stable dimers and trimers. The analysis shows that the main product of the initial aggregation of dipoles is in the form of stable dimers with a binding energy $E_{dm} = 0.38$ eV. However, the binding energy E_{dm} appears to be dependent on the concentration of dipoles and dimers. Possible causes for this dependence are discussed. A new analysis of the kinetics of the decay of Mn dipoles in NaCl, observed by measurement of dielectric loss, shows that they are consistent with a second order reaction when the back reaction is taken into account. The analysis yields a value of 0.90 eV for the migration energy of Mn dipoles in NaCl. The analysis of Unger and Perlman is extended to enable the binding energy E_{dm} of the dimers to be determined. In this way a value of $E_{dm} = 0.44$ eV is obtained for Ba dimers in KCl from the experimental data of Unger and Perlman.

Les resultats des experiences sur la conductivité des cristaux de NaCl dopés avec du Mn sont analysés de nouveau en termes d'un modèle qui tient compte de la stabilité des dimères et trimères. L'analyse montre que le produit de l'agglomeration initiale de dipoles prend la forme de dimères stables qui ont une énergie de liaison de $E_{dm} = 0{,}38$ eV. L'énergie de liaison E_{dm} parait dépendre de la concentration de dipoles et de dimères. Les causes possibles de cette dependence sont discutées. Une nouvelle analyse des cinétiques de la disparition des dipoles Mn dans le NaCl mesurées par perte diélectrique démontre qu'ils s'accordent à une réaction de deuxième ordre, quand la réaction rétroactive est prise en considération. L'analyse donne une valeur de 0,90 eV pour l'énergie de migration de dipoles Mn dans le NaCl. L'analyse de Unger and Perlman fut poursuivie pour permettre le calcul de l'énergie de liaison des dimères. Dans ces conditions, la valeur pour les dimères de Ba dans le KCl est $E_{dm} = 0{,}44$ eV a l'aide des resultats experimentaux de Unger et Perlman.

1. Introduction

At temperatures in the extrinsic range, the conductivity of ionic crystals is usually analysed in terms of the association of divalent impurity ions with cation vacancies to form dipoles and of the subsequent precipitation of the impurity (e.g. Jain and Dahake [1], Kirk and Pratt [2]). In view of the commonly-held belief that the dipoles aggregate first into stable trimers obeying third order kinetics, (Cook and Dryden [3, 4]), Chapman and Lilley [5] have presented a more detailed analysis of the conductivity which takes into account the formation of trimers as an intermediate stage between the formation of dipoles and of the precipitate. However, the recent work of Unger and Perlman [6 to 9] on the analysis of dielectric loss data, suggests that dimers are the main species into which the dipoles aggregate, at least until the plateau in the dipole concentration–time plot is reached. They have pointed out that the original data of Cook and Dryden [3, 10, 11], and other similar data including their own, are consistent with a second order decay of dipoles into dimers provided the back reaction term, neglected by Cook and Dryden, is taken into account. In support of this view

[1]) London S.W. 7, Great Britain.

[2]) Now at: Defence Research and Development Office, Delhi 110007, India.

point, Hor et al. [12], performing thermal depolarisation experiments on KCl:Sr^{2+}, have concluded that the main product at the first stage of dipole aggregation is in the form of dimers, and they have found one peak which they attribute to the relaxation of dimers with a configuration having a net dipole moment. In view of these considerations, in this paper we have reanalysed the conductivity data of Chapman and Lilley [5], and earlier data obtained in this laboratory (Kirk and Pratt [2], Kirk et al. [13]), in terms of a model which includes the presence of stable dimers and trimers. We have also reanalysed the data on the decay of Mn dipoles in NaCl obtained in this laboratory earlier (Bannaghan et al. [14], Bannaghan [15], according to the method suggested by Unger and Perlman.

In view of Unger and Perlman's results, the question of the binding energy of dimers acquires a new significance. The values of this binding energy calculated theoretically are summarised in Table 1:

Table 1
Calculated values of the binding energy of the most stable dimers

crystal	E_{dm} (eV)	reference
NaCl	0.377	Crawford [16]
KCl	0.413	Crawford [16]
KCl	0.41	Strutt and Lilley [17]

These values are so high that a large concentration of dimers is expected as a result of the aggregation of dipoles. The only experimental value available to compare with the calculations is that of Strutt and Lilley [18]. By a computer analysis of the dielectric loss data of Dryden [11] and of Symmons and Kemp [19], they obtained a binding energy $E_{dm} = 0.37$ eV for Mn dimers in NaCl. However, in their subsequent publication, Strutt and Lilley [17] have suggested that the rate of formation of these strongly bound dimers is so low that the aggregation of dipoles into trimers by the Crawford [16] mechanism is preferred. According to this mechanism, dipoles first form loosely bound dimers (with $E_{dm} \approx 0.02$ to 0.30 eV) and thereafter further dipoles and the loose dimers decay into trimers. Crawford [16] has shown how this process can lead to the apparent third order kinetics used by Cook and Dryden to interpret their experiments.

New experimental values for E_{dm} emerge from our reanalysis of the conductivity and dielectric loss data, and in the final section of this paper we show how the binding energy of dimers can be determined directly from the analysis of Unger and Perlman [6].

2. Analysis of the Conductivity Data

2.1 The method

We consider an NaCl crystal doped with Mn^{2+} ions which may contain free impurity, dipoles, dimers and trimers, but the temperature of the crystal is sufficiently high that higher aggregates or precipitates of the impurity do not occur. Since the total impurity concentration must be conserved,

$$C - x_v - x_{dp} - 2x_{dm} - 3x_{tm} = 0 . \quad (1)$$

C is the total divalent impurity concentration (molar fraction), x_v the concentration of free vacancies, x_{dp}, x_{dm}, x_{tm} the concentration of dipoles, of dimers, and of trimers, respectively. Using the law of mass action (Lidiard [20]), the following relations can be obtained:

$$x_{dp} = \left(x_v \, e^{-\frac{E_{DH}}{2kT}}\right)^2 \left(12e^{\frac{E_{dp}^{nn}}{kT}} + 6e^{\frac{E_{dp}^{nnn}}{kT}}\right). \tag{2}$$

Similarly,

$$x_{dm} = x_{dp}^2 z_{dm} \, e^{\frac{E_{dm}}{kT}} \, e^{-F_1(x_{dp}, T)} \tag{3}$$

and

$$x_{tm} = x_{dp}^3 z_{tm} \, e^{\frac{E_{tm}}{kT}} \, e^{-F_2(x_{dp}, T)} \,. \tag{4}$$

E_{DH} is the Debye-Hückel energy, E_{dp}^{nn}, E_{dp}^{nnn}, E_{dm}, and E_{tm} are the binding energies of nearest neighbour dipole, next-nearest neighbour dipole, dimer, and trimer, respectively, z_{dm} and z_{tm} are constants for dimers and trimers, T the temperature.

The factors e^{-F_1} and e^{-F_2} have been introduced to take into account any possible variation of the binding energies E_{dm} and E_{tm} with concentration of dipoles, dimers and trimers or with temperature (see later discussion in Section 2.3).

We now define a quantity δ such that the total impurity present in the form of dimers and trimers is δx_{dp}^2, where

$$\delta = \frac{C - x_v - x_{dp}}{x_{dp}^2} \,. \tag{5}$$

Combining (1) to (5) we obtain

$$\delta = 2z_{dm} \, e^{\frac{E_{dm}}{kT}} \, e^{-F_1} + 3z_{tm} x_{dp} \, e^{\frac{E_{tm}}{kT}} \, e^{-F_2} \tag{6}$$

and in the usual case of $F_1 = F_2 = 0$, (6) reduces to the simple form

$$\delta = 2z_{dm} \, e^{\frac{E_{dm}}{kT}} + 3z_{tm} \, e^{\frac{E_{tm}}{kT}} \, x_{dp} \,. \tag{7}$$

In the method we have used, δ and x_{dp} are calculated from the conductivity data (as discussed below) and a plot of δ against x_{dp} is made. Equation (6) predicts that the plot will be a straight line (provided $F_1 = F_2 = 0$) with zero slope if trimers are absent (Unger and Perlman model) and with slope $3z_{tm} \, e^{E_{tm}/kT}$ if only trimers, or if both dimers and trimers, are present. The relative concentration of dimers is given by the intercept $2z_{dm} \, e^{E_{dm}/kT}$ of the line at $x_{dp} = 0$. Knowing the relative concentrations of dimers and trimers at different temperatures, the binding energies E_{dm} and E_{tm} can be determined in principle.

2.2 Values of the parameters

The number of free vacancies x_v is calculated from the equation

$$\sigma T = x_v A_1 \, e^{-\frac{H_1}{kT}} \tag{8}$$

using the observed values of the conductivity. Here σ is the conductivity, A_1 the pre-exponential factor for mobility of vacancies $\left[A_1 = \frac{4a_0^2 q^2}{kV} \nu_D \exp\left(\frac{\Delta S_1}{K}\right)\right]$ with a_0 interionic distance, ν_D Debye frequency, ΔS_1 the entropy of migration for vacancies, H_1 the enthalpy of migration of vacancies, q the electronic charge, V the volume of 1 molecule of pure salt.

In the σT versus $1/T$ plots, region III represents the association of free Mn^{2+} ions and cation vacancies into dipoles, and region IV is that of precipitation (Kirk and Pratt [2]). In between these two regions, aggregates higher than dipoles, i.e. dimers, trimers etc., are expected to form and it is from this region that the conductivity data has been taken.

Thus, to calculate x_{dp} from (2) and x_v from (8) we need the values of the parameters E_{dp} in equation (2) and A_1 and H_1 in equation (8). The values of A_1 measured by different workers vary from 4×10^6 to 16×10^6 (Ω^{-1} cm^{-1} K) and those of H_1, from 0.65 to 0.75 eV (Allnatt and Pantelis [21], Kirk et al. [13], Allnatt et al. [22], Chapman and Lilley [5], and Nadler and Rossel [23]). In analysing the data of Chapman and Lilley [5] we have used the values $A_1 = 12g \times 10^6$ (Ω^{-1} cm^{-1} K) and $H_1 = 0.71$ eV given by these authors. The factor g is the Pitts drag mobility factor (Pitts [24], Lidiard [20]) which depends on the temperature and the free vacancy concentration and varies from 0.80 to 0.92 in the range of temperature and concentration analysed in this paper. Similarly in analysing the data of Kirk and Pratt [2], we have used the values given by these authors, i.e. $A_1 = 5.7 \times 10^6$ (Ω^{-1} cm^{-1} K) and $H_1 = 0.68$ eV.

The values of E_{dp}^{nn} vary from 0.3 to 0.42 eV (Watkins [25], Dreyfus and Nowick [26], Brown and Hoodless [27], Kirk et al. [13], Chapman and Lilley [5]). Again, we have used the values 0.386 eV (Chapman and Lilley) and 0.34 eV (Kirk et al.) in analysing the respective data of these two groups of authors. The effect of varying the values of the parameters within the observed ranges was found not to alter the results. The values of E_{dp}^{nnn} are obtained from:

$$E_{dp}^{nnn} = E_{dp}^{nn} - 0.039 \text{ (eV)}$$

(Symmons [28]).

2.3 The results

The values of the Mn^{2+} impurity concentration determined by chemical analysis have been given by both sets of authors. However in the case of Kirk and Pratt, it was possible to calculate the Mn^{2+} concentration actually in solid solution from their conductivity data in the temperature range 500 to 600 °C where association was negligible. The values are summarised in Table 2. The value of 505 (ppm) for sample 4 is taken from Fig. 2 of Chapman and Lilley [5]. However, the 'experimental solvus curve' in their Fig. 10 seems to indicate a lower value for this crystal ($\approx$ 475 ppm). The latter value gives better results and is the one we have used in our calculations. In the case of the data from Kirk and Pratt, the impurity concentration determined from the conductivity data, given in the last column, was used in the calculations.

Table 2

C (ppm)

sample	Chapman and Lilley quoted by authors	Kirk and Pratt quoted by authors	Kirk and Pratt from σ
1	42	130	125
2	99	186	177
3	180	325	308
4	505 (475)	686	770
5	700		

Typical plots of δ versus x_{dp} are shown in Fig. 1 and 2. The error bars are calculated assuming 5% error both in σ and C. There is no increase in the value of δ for large values of x_{dp}, as would be expected from (6) if trimers were present. We conclude that the concentration of trimers is negligible in the range of temperature and concentra-

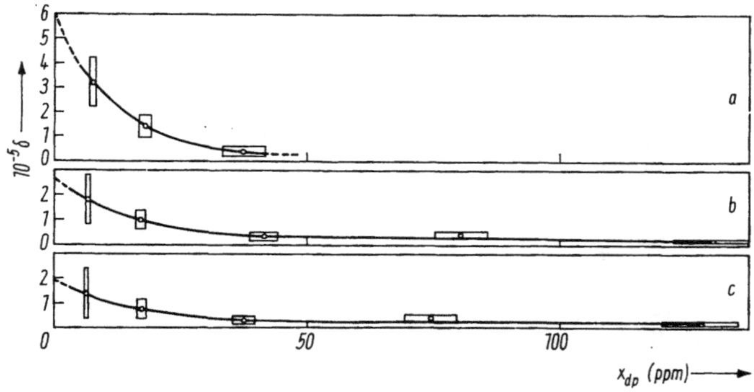

Fig. 1. Plots of δ against dipole concentration, x_{dp}. The experimental data is taken from Chapman and Lilley [5]. (a) $T = 264$, (b) 322, (c) 352 °C

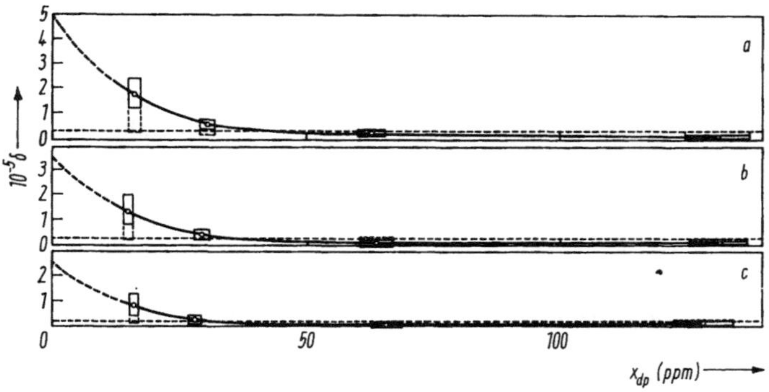

Fig. 2. The same as Fig. 1 but with experimental data from Kirk and Pratt [2]. (a) $T = 298$, (b) 322, (c) 352 °C

tion of impurity considered here. Dimers appear to be the main product of the aggregation of the dipoles. However, in both figures, there is a decrease in δ when the concentration of dipoles is increased and this decrease is outside the likely experimental error. The dashed error bars in Fig. 2 show the magnitude of the error necessary for the points to lie on a straigt line; in some cases the error would have to be up to 22% in σ; furthermore, the intercepts of the horizontal dashed lines with the δ-axis yield values of z_{dm} and E_{dm} much lower than expected. Were this an error of measurement, the deviation is unlikely to be towards decreasing δ in each case.

Thus, these results suggest that the curvature in the δ plots is genuine, and hence that the value of e^{-F_1} in (6) is not unity. A numerical calculation shows that

$$e^{-F_1} = e^{-\frac{1}{kT}\left(G\frac{x_{dp}}{T^{1/2}}\right)}, \tag{9}$$

where $G = 9 \times 10^{-2}$ ($\pm 6\%$) for Fig. 2 gives a satisfactory fit for the equation

$$\delta = 2z_{dm}\, e^{\left(\frac{E_{dm}}{kT} - F_1\right)}$$

within the limits of the experimental data. Here x_{dp} is expressed in (ppm) units.

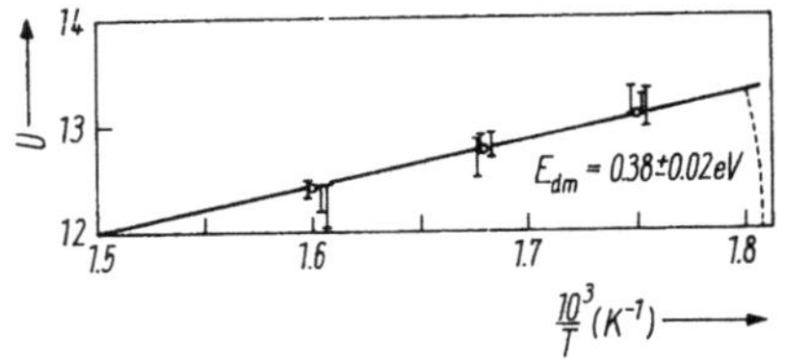

Fig. 3. Plot of $U = \ln \delta + F_1$ against $1/T$ for the data of Kirk and Pratt [2]. The vertical error bars shown are at $1/T = 1.6 \times 10^{-3}$, 1.68×10^{-3}, and 1.75×10^{-3} but have been shown slightly displaced from each other for clarity

The plot of $\ln \delta + F_1 \equiv U$ versus $1/T$ is shown in Fig. 3, which yields a value of $E_{dm} = 0.38$ eV for Fig. 2. The value of E_{dm} can also be obtained from the δ-axis intercepts using (7). This method yields $E_{dm} = (0.38 \pm 0.04)$ eV for Fig. 1 and $E_{dm} = (0.39 \pm 0.03)$ eV for Fig. 2. The values of E_{dm} obtained here are in reasonable agreement with the theoretical value given in Table 1 and with the experimental value of 0.37 eV for Mn in NaCl obtained by Strutt and Lilley [18]. Note that if the rise in δ for small values of x_{dp} is ignored as being due to unknown errors (dotted rectangles in Fig. 2) the values of E_{dm} obtained from the intercepts of the extrapolated horizontal portions are ≈ 0.25 eV, considerably lower than the expected values for the most stable dimers.

The analysis of the experimental data suggest that, provided there is no precipitate phase present, the apparent binding energy of dimers decreases as the concentration x_{dp} or C increases or as the temperature decreases. This behaviour is qualitatively similar to the reduction of binding energy of a dipole due to the Debye-Hückel effect (Lidiard [20]). The Debye-Hückel reduction in the binding energy of dipoles is $E_{DH} = q^2\varkappa/[\varepsilon(1 + \varkappa R)]$, $\varkappa$ being the Debye-Hückel screening factor $= (8\pi q^2 x_v / V \varepsilon k T)^{1/2}$, ε the dielectric constant, R the distance of closest approach between ions in solid solution. It is not clear whether the dipoles can also increase their stabilisation by interaction with charged free vacancies, free impurities, and other dipoles.

In view of these results, the data of Chapman and Lilley at lower temperatures and higher concentrations should be reanalysed since their analysis assumes the presence only of trimers in the range of C and T we have considered.

2.4 The influence of precipitation

To make sure that the previous behaviour of δ has nothing to do with the presence of a precipitate phase, the calculations were extended to values of temperature and

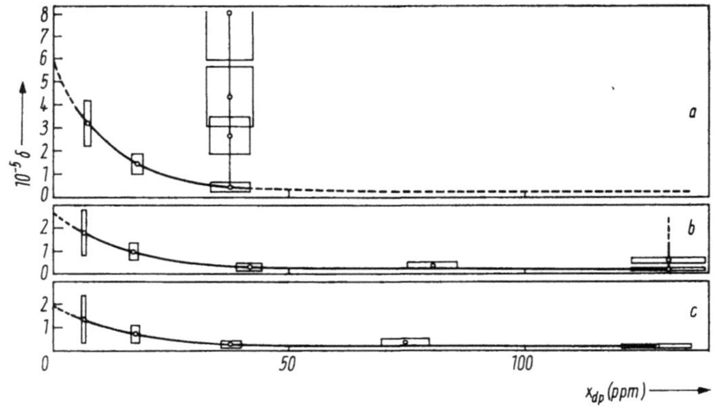

Fig. 4. The same as Fig. 1, but extended to higher total Mn concentrations showing the formation of the Suzuki-phase precipitate. (Temperatures as in Fig. 1)

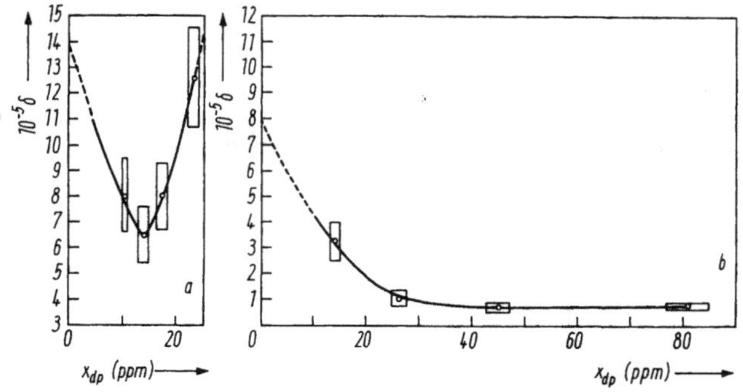

Fig. 5. The same as Fig. 2, but at lower temperatures showing the formation of the Suzuki-phase precipitate. (a) $T = 232$ and (b) 264 °C

total Mn concentration in which, according to the conductivity plots, precipitation does occur.

Fig. 4 shows a plot of δ versus dipole concentration for the same data source, and with the same values of the parameters, as in Fig. 1 but extended to higher concentrations. There is a sharp increase in δ when the Suzuki phase starts to form. This abrupt rise in δ is a direct consequence of the superimposition of the original conductivity curves (Chapman and Lilley [5], Fig. 2) in the precipitation region.

The conductivity curves of Kirk and Pratt do not superimpose in the precipitation region (Kirk and Pratt, [2], Fig. 9), so a rise in δ as sudden as the one in Fig. 4 cannot be expected. Fig. 5 shows δ versus x_{dp} plots for the same data source and parameter values as in Fig. 2, at two different temperatures 264 and 232 °C, representing the beginning and almost the end of the region of precipitation respectively in the original conductivity curves. At the beginning of this region ($T = 264$ °C) Fig. 5 shows that the values of δ start to increase at the highest concentrations considered. At lower temperatures ($T = 232$ °C) when the precipitation has occurred for a wider range of concentration, the plot of δ shows the expected steep rise. The difference in behaviour between the two sets of results at 264 °C does not reflect a difference in the methods of measuring ionic conductivity; in each case holding times of the order of one day were used at temperatures below 350 °C to ensure that equilibrium had been achieved. The difference probably is due to the different atmospheres in which the crystals were grown, those of Kirk and Pratt containing a higher concentration of OH^-. The effect that the presence of OH^- may have on the ionic conductivity in the precipitation region was first discussed by Dreyfus [29].

3. Analysis of Dipole Decay Data

3.1 Unger and Perlman's method of analysis

The rate equation for the reaction between dipoles and dimers is

$$\frac{dn}{dt} = -c_1 n^2 + c_2 p\,, \tag{10}$$

where c_1 and c_2 are the rate constants, and n and p are the concentration of dipoles and dimers, respectively.[3]) Equation (10) is valid when the concentration of trimers is negligible.

[3]) In the rest of this paper for convenience we will follow Unger and Perlman's notation [6, 7].

The first term corresponds to the formation of dimers from dipoles and is the one considered by earlier workers. The second term takes into account the back reaction when the dimers dissociate. Since experimentally observed decay curves show an approach to equilibrium, Unger and Perlman [6, 8] pointed out that the back reaction term c_2p is important and cannot be neglected. If this term is ignored misleading results will be obtained.

The solution of (10) can be written conveniently in the following form (Unger and Perlman [6]):

$$\ln Q = c_1'/t + \ln Q_0 , \tag{11}$$

where

$$Q = \frac{n-b}{n-a}, \tag{12a}$$

$$a = n_s , \quad b = -\left(\frac{n_0 n_s}{n_0 - n_s}\right), \tag{12b}$$

$$c_1' = (a - b)\, \nu_0\, e^{-E/kT} , \tag{12c}$$

$$Q = Q_0 , \quad \text{and} \quad n = n_0 \quad \text{when} \quad t = 0 . \tag{12d}$$

Here n_s is the concentration of dipoles at the plateau in the n versus t plots (i.e. when $dn/dt \approx 0$), E is the activation energy for the migration of dipoles and ν_0 is a frequency factor.

3.2 *The slope c_1' and the migration energy E. Reanalysis of Bannaghan's data*

We have calculated $Q = (n - b)/(n - a)$ for different values of t from the data taken from Bannaghan et al. [14] and found that plots of $\ln Q$ versus time obtained in this manner are straight lines, Fig. 6, in agreement with (11). The plot of $\ln c_1'$ versus $1/T$, Fig. 7, is also a straight line corresponding to an activation energy ≈ 0.90 eV, in reasonable agreement with the value 0.95 eV for the energy of Mn dipole migration in NaCl obtained from diffusion measurements (Stewart and Reed [30]). However

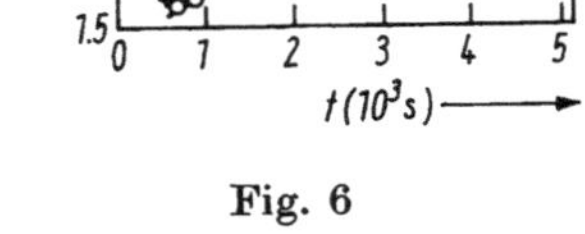

Fig. 6

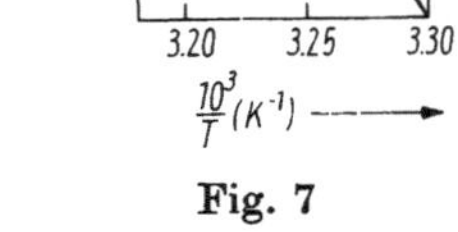

Fig. 7

Fig. 6. Plots of $\ln Q$ as a function of time (see equation (11) in the text). The data for the number of Mn dipoles in NaCl at different times required to calculate Q at these four temperatures is taken from the paper of Bannaghan et al. [14]. (a) $T = 37.7$, (b) 36.2, (c) 34.0, (d) 32.2 °C

Fig. 7. Plot of $\ln c_1'$ against $1/T$; c_1' is the slope of the straight lines in Fig. 6. The straight line shown is the least squares fit to the data points and its slope yields a value 0.90 eV for the energy of migration of Mn dipoles in NaCl

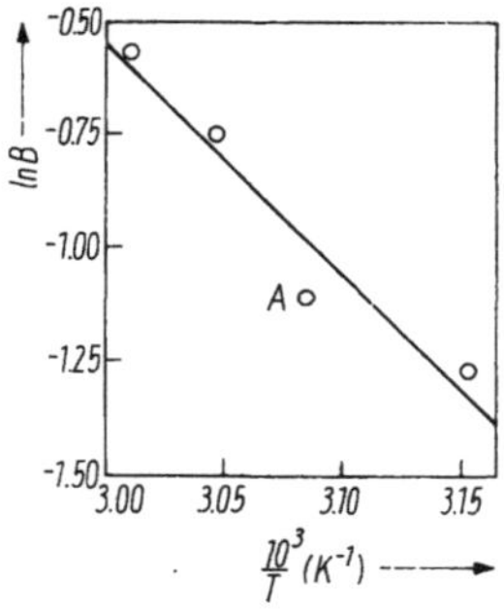

Fig. 8. Plot of ln B against $1/T$ (equation (14)) for Ba dipoles in KCl. The slope gives $E_{dm} = 0.44$ eV for the binding energy of Ba dimers in KCl

uncertainty in our value is large since the errors involved in reading the data are appreciable. Bannaghan and Pratt [31] are analysing their later as yet unpublished data to obtain more dependable values of these parameters.

3.3 The intercept Q_0 and the binding energy E_{dm}. Reanalysis of Unger and Perlman's data

It can be seen from equations (12) and the relation (see e.g. Crawford [16])

$$2n_s^2/(n_0 - n_s) = z^{-1}\, e^{-E_{dm}/kT}\,, \tag{13}$$

where z is a constant, that

$$B \equiv Q_0^{1/2} + Q_0^{-1/2} - 2 = \frac{1}{2n_0 z}\, e^{-E_{dm}/kT}\,. \tag{14}$$

Equation (14) shows that if Q_0 is read from Unger and Perlman plots at $t = 0$ and a plot of ln B against $1/T$ is made, we should get a straight line, whose slope should yield a value of E_{dm}.

We have read the values of the intercepts Q_0 from Fig. 2 of Unger and Perlman [7]. The value of n_s/n_0 determined from the observed data point on the y-axis, i.e. at $t = 0$, in Fig. 2 of Unger and Perlman [7] is slightly different from that determined from the plateau in their Fig. 1. However the difference is within the combined errors of reading the data in the two figures. ln B is plotted against $1/T$ in Fig. 8 and the straight line drawn is the least squares fit to the data points. This yields $E_{dm} = 0.44$ eV for Ba in KCl, which compares well with the theoretically calculated values given in Table 1 in the Introduction. Note that if the point marked A is ignored, the other three points lie almost exactly on a straight line with the same slope as the line shown. The point A corresponds to the intercept of the line for 51 °C in Fig. 2 of Unger and Perlman. This line contains only 3 experimental points and is likely to be in error.

The computer analysis of the unpublished data of Bannaghan and Pratt [31] should yield accurate values of the binding energies for Mn in NaCl. Extrapolation of the points in Fig. 6 of the present paper is unlikely to yield a reliable value.

Ackowledgements

We are grateful to the S.R.C. for a Senior Visiting Fellowship to support Prof. S. C. Jain, to Conicit Venezuela for a Research Scholarship to support Mr. A. L. Guerrero, and to Prof. J. G. Ball for the provision of research facilities at Imperial College.

References

[1] S. C. Jain and S. L. Dahake, Indian J. pure appl. Phys. **2**, 71 (1964).
[2] D. L. Kirk and P. L. Pratt, Proc. Brit. Ceram. Soc. No. **9**, 215 (1967).
[3] J. S. Cook and J. S. Dryden, Proc. Phys. Soc. **80**, 479 (1962).
[4] J. S. Cook and J. S. Dryden, Phys. Rev. B **12**, 5995 (1975).
[5] J. A. Chapman and E. Lilley, J. Physique **34**, C9–455 (1973).
[6] S. Unger and M. M. Perlman, Phys. Rev. B **10**, 3692 (1974).

[7] S. UNGER and M. M. PERLMAN, Phys. Rev. B 12, 809 (1975).
[8] S. UNGER and M. M. PERLMAN, Phys. Rev. B 12, 5997 (1975).
[9] S. UNGER and M. M. PERLMAN, Phys. Rev. B 15, 4105 (1977).
[10] J. S. COOK and J. S. DRYDEN, Austral. J. Phys. 13, 260 (1960).
[11] J. S. DRYDEN, J. Phys. Soc. Japan, Suppl. III, 18, 129 (1963).
[12] AH. MEE HOR, P. W. M. JACOBS, and K. S. MOODIE, phys. stat. sol. (a) 38, 293 (1976).
[13] D. L. KIRK, D. C. PHILLIPS, and P. L. PRATT, Proc. Brit. Ceram. Soc. No. 19, 105 (1970).
[14] A. J. BANNAGHAN, D. R. HAYMANN, and P. L. PRATT, Reactivity of Solids Conf. Rep., Bristol 1972 (published by Chapman and Hall).
[15] A. J. BANNAGHAN, Ph. D. Thesis, University of London, 1974.
[16] J. H. CRAWFORD, J. Phys. Chem. Solids 31, 399 (1970).
[17] J. E. STRUTT and E. LILLEY, phys. stat. sol. (a) 33, 229 (1976).
[18] J. E. STRUTT and E. LILLEY, Reactivity of Solids Conf. Rep., Bristol 1972 (published by Chapman and Hall).
[19] H. F. SYMMONS and R. C. KEMP, Brit. J. appl. Phys. 17, 607 (1966).
[20] A. B. LIDIARD, Hdb. Phys., Vol XX, Ed. S. FLÜGGE, Springer-Verlag, Berlin 1957 (p. 246).
[21] A. R. ALLNATT and P. PANTELIS, Solid State Commun. 6, 309 (1968).
[22] A. R. ALLNATT, P. PANTELIS, and S. J. SIME, J. Phys. C 4, 1778 (1971).
[23] C. NADLER and J. ROSSEL, phys. stat. sol. (a) 18, 711 (1973).
[24] E. PITTS, Proc. Roy. Soc. (London) A 217, 43 (1953).
[25] G. D. WATKINS, Phys. Rev. 113, 79 (1959).
[26] R. W. DREFUS and A. S. NOWICK, Phys. Rev. 126, 1367 (1962).
[27] N. BROWN and I. M. HOODLESS, Solid State Commun. 6, 597 (1968).
[28] H. F. SYMMONS, J. Phys. C 3, 1846 (1970).
[29] R. W. DREYFUS, Appl. Phys. Letters 3, 175 (1963).
[30] W. H. STEWARD and C. A. REED, J. chem. Phys. 43, 2890 (1965).
[31] A. J. BANNAGHAN and P. L. PRATT, to be published (1978).

(Received July 4, 1978)

phys. stat. sol. (a) **49**, 363 (1978)

Subject classification: 14.3.4; 22.2.3

School of Studies in Physics, Vikram University, Ujjain[1])

Effect of a Density Gradient on the Electromagnetic Wave Instability in Semiconductor Plasmas

By

S. GUHA and S. GHOSH

The effect of the density gradient on the instability of an e.m. wave propagating perpendicular to the crossed electric and magnetic fields is investigated. Using an hydrodynamic model of the plasma, the dispersion relation $D(\omega, \boldsymbol{k}) = 0$ is derived. The propagation characteristics and the possibility of instability of the wave in the collision dominated regime are analyzed in detail taking into account hot carrier effects. Calculations are made for the case of InSb at 77 K. It is seen that the growth rate and the propagation constant increase with the electric field but decrease with the increase of the magnetic field. The positive density gradient in the direction of the electric field enhances the growth rate.

Der Einfluß des Dichtegradienten auf die Instabilität einer elektromagnetischcn Welle, die sich senkrecht zu gekreuzten elektrischen und magnetischen Feldern ausbreitet, wird untersucht. Mit einem hydrodynamischen Plasmamodell werden die Dispersionsbeziehungen $D(\omega, \boldsymbol{k}) = 0$ abgeleitet. Die Ausbreitungscharakteristiken und die Möglichkeit von Welleninstabilitäten im stoßgesteuerten Bereich werden ausführlich unter Einbeziehung der Effekte heißer Träger analysiert. Rechnungen für den Fall des InSb bei 77 K werden durchgeführt. Es wird gezeigt, daß die Wachstumsgeschwindigkeit und die Ausbreitungskonstante mit dem elektrischen Feld zunehmen jedoch mit wachsendem Magnetfeld abnehmen. Der positive Dichtegradient in Richtung des elektrischen Feldes erhöht die Wachstumsgeschwindigkeit.

1. Introduction

During the past few years a number of theoretical papers have appeared which discuss the propagation and instabilities of electromagnetic waves in semiconductor plasmas subject to crossed electric and magnetic fields [1 to 4]. None among the earlier authors has analyzed the effect of a density gradient on the instabilities of the wave. However, the instabilities of the plasma density waves in the presence of a density gradient in thin plates of semiconductors have been studied theoretically and experimentally by a large number of workers [5 to 7]. The microwave emission from InSb in the presence of electric and magnetic fields has been accounted by the occurrence of the instability of the plasma density wave in the presence of a density gradient [5]. Miyazaki and Yamaguchi [6] have observed the onset conditions of the oscillation in a Ge slice in a large transverse magnetic field in the presence of a density gradient and interpreted it following Suzuki's work. Similarly, Zavyalov et al. [7] have studied experimentally the threshold electric field as a function of the magnetic field in n-Ge at room temperature for the growth of density fluctuations associated with the density gradient. Recently, Newton and Short [8] have analyzed the gradient-induced convective instability of a longitudinal wave in a configuration different from ours for a one-component plasma (electrons only). Thus it is seen that little attention has been given to the problem how the carrier density gradient influences the propagation characteristics and instability of a hybrid mode in two-component semiconductor plasmas. In this paper we report on a study of the propagation characteristics and the possibility

[1]) Ujjain (M.P.) 456010, India.

of an instability of an electromagnetic wave travalling along the z-axis which is subjected to a static electric field $\boldsymbol{E}_0$ (x-axis) and a dc magnetic field $\boldsymbol{B}_0$ (x-axis). The density gradient ∇n_0 is taken in the direction of the electric field $\boldsymbol{E}_0$. As a typical case, calculations have been made for the case of InSb for which the momentum and energy transfers of the carriers are assumed to be due to acoustic and polar optical mode scattering, respectively, at 77 K. The result shows that a positive density gradient in the direction of the static electric field increases the phase constant and the growth rate considerably, and in the absence of an electric field, the diamagnetic drift due to the density gradient can cause an instability.

2. Theoretical Formulation

2.1 Dispersion relation and related equations

The motion of carriers of type j is governed by the equation of momentum transfer as

$$\frac{\mathrm{d}\boldsymbol{v}_j}{\mathrm{d}t} = \frac{q_j}{m_j}(\boldsymbol{E} + \boldsymbol{v}_j \times \boldsymbol{B}) - \nu_j \boldsymbol{v}_j - \frac{v_{\mathrm{T}j}^2}{2}\frac{\nabla n_j}{n_j} \tag{1}$$

in which $\boldsymbol{v}_j$, q_j, m_j, $v_{\mathrm{T}j}$, and n_j denote the velocity, charge, mass, thermal velocity, and the carrier concentration, respectively. The subscript j takes e for electrons and h for holes.

The carrier concentration and carrier velocity are related by the continuity equation

$$\nabla(n_j \boldsymbol{v}_j) + \frac{\partial n_j}{\partial t} = 0\,. \tag{2}$$

Each of the variables, electric field intensity $\boldsymbol{E}$, magnetic flux density $\boldsymbol{B}$, velocity $\boldsymbol{v}$, and carrier concentration n, is assumed to be the sum of a time-invariant part (subscript 0) and a small time-dependent part (subscript 1), e.g.

$$n = n_0(r) + n_1(r) \exp\left[i(\omega t - \boldsymbol{k}\cdot\boldsymbol{r})\right].$$

Here ω and $\boldsymbol{k}$ denote the wave angular frequency and the propagation vector, respectively, $\boldsymbol{r}$ is the displacement vector.

The time-dependent (first-order) part of the momentum transfer equation in the collision-dominated regime ($\nu \gg \omega$, kv_0) is

$$\nu_j \boldsymbol{v}_{1j} + \boldsymbol{v}_{1j} \times \boldsymbol{\omega}_{\mathrm{c}j} = \frac{q_j}{m_j}(\boldsymbol{E}_1 + \boldsymbol{v}_{0j} \times \boldsymbol{B}_1) + v_{\mathrm{T}j}^2 \frac{n_{1j}}{n_{0j}} \boldsymbol{k}\,, \tag{3a}$$

where $\boldsymbol{\omega}_{\mathrm{c}j} = -q_j \boldsymbol{B}_0/m_j$ is the cyclotron frequency and ν_j the momentum transfer collision frequency of the carrier.

From (2) one gets

$$\frac{n_{1j}}{n_{0j}} = \frac{(\boldsymbol{k} + i\boldsymbol{\delta}_j)\cdot \boldsymbol{v}_{1j}}{\bar{\omega}_j + i\boldsymbol{\delta}_j \cdot \boldsymbol{v}_{0j}}, \tag{3b}$$

where $\bar{\omega}_j = \omega - \boldsymbol{k}\cdot\boldsymbol{v}_{0j}$ and $\boldsymbol{\delta}_j \equiv \nabla n_{0j}/n_{0j}$ and from (3a) and (3b),

$$\nu_j \boldsymbol{v}_{1j} + \boldsymbol{v}_{1j} \times \boldsymbol{\omega}_{\mathrm{c}j} = \frac{q_j}{m_j}(\boldsymbol{E}_1 + \boldsymbol{v}_{0j} \times \boldsymbol{B}_1) + i v_{\mathrm{T}j}^2 \frac{(\boldsymbol{k} + i\boldsymbol{\delta}_j)\cdot \boldsymbol{v}_{1j}}{\bar{\omega}_j + i\boldsymbol{\delta}_j \cdot \boldsymbol{v}_{0j}} \boldsymbol{k}\,. \tag{3c}$$

For the geometry of the fields under consideration the components of the perturbed part of the velocity $\boldsymbol{v}_{1j}$ are, using the approximation that the spatial rate of change of the static quantity is much smaller than the time-dependent quantity, i.e. $|\boldsymbol{\delta}_j| \ll |\boldsymbol{k}|$:

$$v_{1xj} = \frac{q_j}{m_j\nu_j(1+\eta_j^2)}(-E_{1x} + \chi_j\eta_j E_{1z}),$$

$$v_{1yj} = \frac{q_j}{m_j\nu_j}\frac{\omega_{cj}}{\omega}E_{1y},$$

and

$$v_{1zj} = \frac{q_j}{m_j\nu_j(1+\eta_j^2)}\left[\frac{1}{\omega}(kv_{0jx} - \eta_j\bar{\omega}_j)\,E_{1x} - \chi_j E_{1z}\right], \tag{4}$$

where $\eta_j = \omega_{cj}/\nu_j$, $\chi_j = 1 + k^2v_{Tj}^2/\omega_{pj}^2$ and $\omega_{pj}^2 = n_{0j}q_j^2/m_j\varepsilon_0\varepsilon_L$ is the plasma frequency of the carrier, ε_L and ε_0 being the lattice dielectric constant and the absolute permittivity, respectively.

The dc (zero-order) part of the equation of motion gives

$$\boldsymbol{v}_{0j} + \boldsymbol{v}_{0j} \times \boldsymbol{\eta}_j = \mu_j\boldsymbol{E}_0 - D_j\boldsymbol{\delta}_j,$$

where

$$\mu_j = \frac{q_j}{m_j\nu_j} \quad\text{and}\quad D_j = \frac{v_{Tj}^2}{\nu_j}.$$

Hence

$$v_{0xj} = \frac{1}{1+\eta_j^2}(\mu_jE_0 - D_j\delta_j)$$

and

$$v_{0zj} = -\eta_j v_{0xj}.$$

The momentum transfer collision frequency ν_j is assumed to be due to acoustic phonon scattering given as

$$\nu_j = \nu_{0j}\left(\frac{T_j}{T_0}\right)^{1/2},$$

where T_j is the effective carrier temperature, T_0 the lattice temperature, and ν_{0j} the collision frequency when $T_j = T_0$. T_j is determined from the energy balance equation. In the steady state we get

$$\frac{\boldsymbol{J}_{0Ej}\cdot\boldsymbol{E}_0}{n_{0j}} = -\left\langle\frac{\partial\varepsilon}{\partial t}\right\rangle_{\text{coll}}, \tag{5}$$

where $\boldsymbol{J}_{0Ej}$ is the dc current density due to the applied electric field only and $\langle\partial\varepsilon/\partial t\rangle_{\text{coll}}$ the energy loss assumed to be due to the polar optical phonon scattering which is given by equation (3.6.26) of [9].

The time-dependent current densities are of interest in the dispersion relation and are given by

$$\left.\begin{aligned} \boldsymbol{J}_{1j} &= n_{0j}q_j\boldsymbol{v}_{1j} + n_{1j}q_j\boldsymbol{v}_{0j} \\ \text{and}\quad \operatorname{div}\boldsymbol{J}_{1j} + i\omega n_{1j}q_j &= 0. \end{aligned}\right\} \tag{6}$$

The general wave equation is given as

$$\boldsymbol{k}\times\boldsymbol{k}\times\boldsymbol{E}_1 = i\omega\mu_0\boldsymbol{J}_{1j} - \frac{\omega^2}{c_L^2}\boldsymbol{E}_1, \tag{7}$$

where $c_L = (\mu_0\varepsilon_0\varepsilon_L)^{-1/2}$ is the velocity of the electromagnetic wave in the medium, μ_0 being the absolute permeability. From (6) and (7), one get the dispersion relation as

$$\begin{bmatrix} D_{xx} & 0 & D_{xz} \\ 0 & D_{yy} & 0 \\ D_{zx} & 0 & D_{zz} \end{bmatrix}\begin{bmatrix} E_{1x} \\ E_{1y} \\ E_{1z} \end{bmatrix} = 0, \tag{8}$$

where

$$D_{xx} = n^2 + 1 - \sum_j \frac{i\omega_{pj}^2}{\omega\nu_j(1+\eta_j^2)} \left[\frac{\overline{\omega}_j}{\omega} + \left(\frac{kv_{0xj}}{\omega}\right)^2\right],$$

$$D_{xz} = -\sum_j \frac{i\omega_{pj}^2 \chi_j \eta_j}{\omega\nu_j(1+\eta_j^2)} \left(1 + \frac{kv_{0xj}}{\overline{\omega}_j \eta_j}\right),$$

$$D_{yy} = -n^2 + 1 - \sum_j \frac{i\omega_{pj}^2}{\omega\nu_j} \frac{\overline{\omega}_j}{\omega},$$

$$D_{zx} = -\sum_j \frac{i\omega_{pj}^2 \eta_j}{\omega\nu_j(1+\eta_j^2)} \left(\frac{kv_{0xj}}{\overline{\omega}_j \eta_j} - 1\right),$$

$$D_{zz} = 1 - \sum_j \frac{i\omega_{pj}^2 \chi_j}{\overline{\omega}_j \nu_j(1+\eta_j^2)},$$

and

$$n^2 = \frac{k^2 c_L^2}{\omega^2}.$$

Equation (8) yields either

$$D_{yy} = 0 \tag{8a}$$

or

$$D_{zx} D_{xz} - D_{xx} D_{zz} = 0\,. \tag{8b}$$

Equation (8a) gives the dispersion relation for the linearly polarized mode (E_{1y}) in the direction of the magnetic field. It is easy to see that it does not show any spatial growth or instability and has not been considered. Assuming $\chi_j \approx 1$ (i.e. long-wavelength region), equation (8b) gives the dispersion relation of the hybrid mode (E_{1x}, E_{1z}) as

$$\left[1 - i\left(\frac{b_e \nu_e}{\overline{\omega}_e} + \frac{b_h \nu_h}{\overline{\omega}_h}\right)\right] \left\{n^2 + 1 - i\left[b_e \nu_e \overline{\omega}_e \left(1 + \frac{k^2 v_{0xe}^2}{\overline{\omega}_e^2}\right) + b_h \nu_h \overline{\omega}_h \left(1 + \frac{k^2 v_{0xh}^2}{\overline{\omega}_h^2}\right)\right]\right\} =$$

$$= (b_e^2 \omega_{ce}^2 + b_h^2 \omega_{ch}^2) + b_e^2 \nu_e^2 \left(\frac{kv_{0xe}}{\overline{\omega}_e}\right)^2 + b_h^2 \nu_h^2 \left(\frac{kv_{0xh}}{\overline{\omega}_h}\right)^2, \tag{9}$$

where

$$b_e = \frac{\omega_{pe}^2}{(1+\eta_e^2)\,\nu_e^2} \quad \text{and} \quad b_h = \frac{\omega_{ph}^2}{(1+\eta_h^2)\,\nu_h^2}.$$

One can infer from (9) that for $B_0 = v_{0x} = 0$ the system is described by two uncoupled modes given by

$$1 - i\frac{b_e \nu_e}{\overline{\omega}_e} - i\frac{b_h \nu_h}{\overline{\omega}_h} = 0$$

and

$$n^2 + 1 - i\left[b_e \nu_e \overline{\omega}_e \left(1 + \frac{k^2 v_{0xe}^2}{\overline{\omega}_e^2}\right) + b_h \nu_h \overline{\omega}_h \left(1 + \frac{k^2 v_{0xh}^2}{\overline{\omega}_h^2}\right)\right] = 0\,.$$

If $v_{0x} = 0$ but $B_0 \neq 0$, the modes are coupled by the magnetic field. On the other hand, if $v_{0x} \neq 0$ but $B_0 = 0$ they are coupled by the drift. In the present case, if the carrier drift due to the applied electric field is absent, the diamagnetic drift which is due to the density gradient and the magnetic field causes the coupling of the modes.

Using (8), we can write

$$D_{xx}E_{1x} + D_{xz}E_{1z} = 0$$

and

$$D_{zx}E_{1x} + D_{zz}E_{1z} = 0 \, .$$

Eliminating E_{1x} and assuming

$$n^2 \gg 1 - \sum_j \frac{i\omega_{pj}^2}{\omega\nu_j(1+\eta_j^2)} \left\{ \frac{\overline{\omega}_j}{\omega} + \left(\frac{kv_{0xj}}{\omega}\right)^2 \right\},$$

a approximation valid for slow waves [1], we obtain

$$(-n^2 D_{zz} + D_{zx}D_{xz})\, E_{1z} = 0 \, . \tag{10}$$

The wave represented by (10) is known as modified quasi-longitudinal wave [1]. If the magnetic field is absent, the wave will be a purely longitudinal wave. However, with $B_0 \neq 0$, the wave becomes a hybrid wave.

If we neglect the drift velocity of the holes and treat them as a stationary background conducting dielectric, equation (10) becomes for a two-component plasma (electrons and holes):

$$\omega^2(A_1) + \omega(A_2 + iA_3) + (A_4 - iA_5) = 0 \, , \tag{11}$$

where

$$A_1 = -k^2c_L^2 + Z_eZ_h\eta_e\eta_h(\chi_e + \chi_h) + Z_e^2\chi_e\eta_e^2 + Z_h^2\chi_h\eta_h^2 \, ,$$

$$A_2 = 2k^3c_L^2v_{0z} + Z_eZ_h\eta_hkv_{0x}(\chi_e - \chi_h) + Z_eZ_h\eta_e\eta_hkv_{0z}(\chi_e + \chi_h) - 2Z_e^2\chi_e\eta_e^2kv_{0x} \, ,$$

$$A_3 = k^2c_L^2(Z_c\chi_e + Z_h\chi_h) \, ,$$

$$A_4 = -k^4c_L^2v_{0z}^2 - Z_e^2\chi_ek^2v_{0x}^2 + Z_e^2\chi_e\eta_e^2k^2v_{0z}^2 \, ,$$

$$A_5 = Z_e\chi_ek^3c_L^2v_{0z}^2 \, ,$$

$$Z_e = \frac{\omega_{pe}^2}{\nu_e(1+\eta_e^2)} \quad \text{and} \quad Z_h = \frac{\omega_{ph}^2}{\nu_h(1+\eta_h^2)} \, .$$

(For simplicity $\boldsymbol{v}_{0e}$ is written as $\boldsymbol{v}_0$.) For real values of the wave number k, equation (11) is a quadratic equation in terms of the complex wave angular frequency $\omega = (\beta + i\alpha)$.

2.2 Instability of the wave

An instability of the wave arises when the imaginary part of the wave angular frequency α becomes negative for real values of k [1, 3, 10 to 12]. We may write the roots of (11) for real k as

$$\omega_{1,2} = \frac{1}{2A_1}\{(A_2 + iA_3) \pm [(A_2^2 - A_3^2 - 4A_1A_4) + 2i(A_2A_3 + 2A_1A_5)^{1/2}\} \, . \tag{12a}$$

For zero carrier drift velocity ($E_0 = \delta = 0$), we get

$$\left.\begin{aligned} \omega_1 &= i(A_3/A_1) \, , \\ \omega_2 &= 0 \, . \end{aligned}\right\} \tag{12b}$$

Then the mode is a damped aperiodic one. When the magnetic field is absent ($B_0 = 0$), the roots become

$$\omega_{1,2} = \frac{1}{2A_1}[iA_3 \pm i(A_3^2 + 4A_1A_4)^{1/2}] \, . \tag{12c}$$

Since $A_3^2 > 0$ and $4A_1A_4 > 0$, both modes are aperiodic. This is in agreement with the result of Gueret [1].

The general solution of (11) is

$$\beta = \frac{-A_2 \pm h}{2A_1} \quad \text{and} \quad \alpha = -\left(\frac{A_3 \mp p}{2A_1}\right), \tag{12d}$$

where

$$h = \left(\frac{r + f}{2}\right)^{1/2}, \qquad p = \left(\frac{r - f}{2}\right)^{1/2},$$

$$r = +(f^2 + g^2)^{1/2}, \qquad f = A_2^2 - A_3^2 - 4A_1A_4,$$

and

$$g = 2(A_2A_3 + 2A_1A_5).$$

To ensure the existence of a wave instability, we use the Briggs' criterion [13], i.e. a vawe instability exists if the imaginary part of ω is negative when k is replaced by the pure imaginary ik [14]. Hence (11) becomes

$$\omega^2(A_1) + \omega[i(A_2 - A_3)] + (A_4 - A_5) = 0.$$

The roots are

$$\omega_{1,2} = \frac{1}{2A_1}[A_2^2 \pm i(A_2^2 + 4A_1A_3)]$$

and

$$\beta_1 = \beta_2 = \frac{A_2^2}{2A_1}, \qquad \alpha_1 = -\alpha_2 = (A_2^2 + 4A_1A_3)^{1/2}.$$

This establishes the existence of the wave instability.

From (6) and (10), the real part of the effective conductivity (for $\eta_j^2 \gg 1$) comes out to be

$$\text{Re}\,\sigma_{||} = \frac{\varepsilon_L \omega_{pj}^2 \omega \chi_j}{\nu_j \eta_j^2 \overline{\omega}_j}. \tag{13}$$

Re $\sigma_{||}$ would be negative if

$$0 < \omega < \boldsymbol{k} \cdot \boldsymbol{v}_{0E} + \omega_e^*, \tag{14}$$

where $\omega_e^* = \delta_j v_{Tj}^2 k/\omega_{cj}$ is the electron drift wave frequency and v_{0E} the drift due to the dc electric field only. It may be noted that if the direction of the dc electric field is in the direction of the non-uniformity, i.e. $\boldsymbol{E}_0 \cdot \nabla n_0 > 0$ and $\boldsymbol{k} \cdot \boldsymbol{v}_{0E}$ and ω_e^* have the same sign, then the instability can be more easily excited. Further, for the instability to occur, the phase velocity ($v_\varphi = \beta/k$) should be less than the effective drift velocity (v_{0z}) in the direction of propagation. From (14), the threshold electric field comes out as

$$E_{th} = \frac{1}{\mu_e}(v_\varphi - \omega_{e1}^*),$$

where

$$\omega_{e1}^* = \frac{\delta_j v_{Tj}^2}{\omega_{cj}}.$$

Thus it is seen that the threshold electric field E_{th} for the instability is reduced in the presence of a positive density gradient.

3. Results and Discussion

As a typical case, equation (11) has been solved numerically with the help of the computer IBM 360/44 for the case of InSb using the following parameters: $\varepsilon_L = 17.54$,

$\varepsilon_\infty = 15.7$, $n_{0e} = 2.8 \times 10^{20}\ \mathrm{m^{-3}}$, $n_{0h} = 3.0 \times 10^{20}\ \mathrm{m^{-3}}$, $T_0 = 77\ \mathrm{K}$, $\nu_{0e} = 3.5 \times 10^{11}\ \mathrm{s^{-1}}$, $\nu_{0h} = 4.4 \times 10^{11}\ \mathrm{s^{-1}}$, $m_e = 0.014 m_0$, $m_h = 0.40 m_0$ and $\theta_D = 278\ \mathrm{K}$.

One of the two modes is unstable ($\alpha < 0$) and the other is damped ($\alpha > 0$), but both have the same value of $|\alpha|$. The results for the unstable mode are shown in Fig. 1 to 4. From Fig. 1 to 3, it is seen that the growth rate $|\alpha|$ and the phase constant β increase with the increase of static electric field E_0 and propagation constant k. But at high values of k, the phase constant becomes almost independent of k, whereas the growth rate continues to increase. This will lead to an aperiodic mode. As the magnetic field increases, the phase constant as well as the gowth rate decrease and at high magnetic fields, they are independent of it.

Fig. 4 shows the dependence of the phase constant and the growth rate on the density gradient when the electric field is absent. From equation (14) it is seen that there an instability is possible even in the absence of the dc electric field, provided the diamagnetic drift wave frequency ω_e^* is greater than the wave angular frequency. In the present case $1.7 \times 10^5 \leqq \omega_{e_j}^* \leqq 8.6 \times 10^6$ and $1.4 \times 10^5 \leqq \beta \leqq 7.1 \times 10^6$; thus $\omega_e^* > \beta$ in every case. As ω_e^* is governed by the restriction $|\boldsymbol{\delta}| \ll |\boldsymbol{k}|$, the excited wave frequency is small. To have a numerical appreciation of the magnitude of the growth rate, we consider a typical case: $B_0 = 0.3\ \mathrm{T}$, $E_0 = 12\ \mathrm{kV\ m^{-1}}$, $\delta = 10.0$, and $k = 5 \times 10^5\ \mathrm{m^{-1}}$; $|\alpha| = 2.4 \times 10^2$. Thus an appreciable magnitude of the growth rate can be obtained in a semiconductor for usual values of the relevant parameters.

It is found that if the direction of the dc electric field is in the direction of the nonuniformity, i.e. $\boldsymbol{E}_0 \cdot \nabla n_0 > 0$, the growth rate and the phase constant can be enhanced in the present field configuration.

Under crossed field configuration complete homogeneity of a plasma system cannot be achieved due to the Lorentz force. If the system has gradients of density, temperature, pressure, magnetic field etc. a plasma current or a particle drift exists, otherwise in the presence of dc electric field the gradients enhance the particle drift. The present investigation suggests that the wave may be coupled in and out of the material, provided that the direction of wave vector is property oriented with respect to the other system parameters.

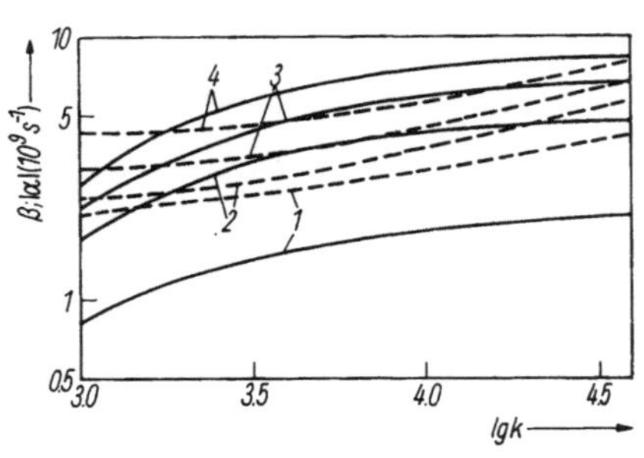

Fig. 1

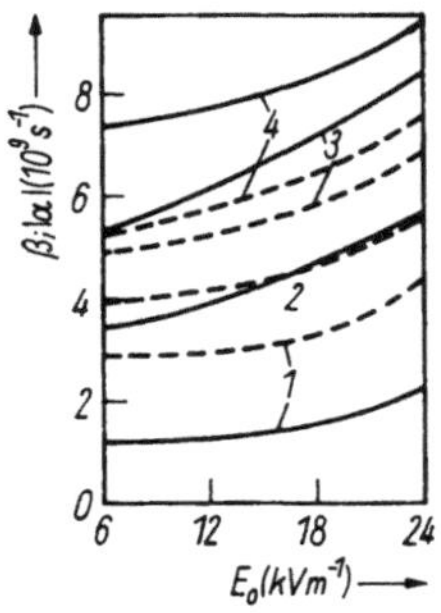

Fig. 2

Fig. 1. Variation of the phase constant β (solid curves) and the growth rate $|\alpha|$ (dashed curves) with the propagation constant k for different values of δ: (1) 0.0, (2) 10.0, (3) 50.0, (4) 100.0; $E_0 = 12\ \mathrm{kV\ m^{-1}}$ and $B_0 = 0.3\ \mathrm{T}$

Fig. 2. Variation of the phase constant β (solid curves) and the growth rate $|\alpha|$ (dashed curves) with the applied electric field E_0 for different values of δ: (1) 0.0, (2) 10.0, (3) 50.0, (4) 100.0; $k = 5 \times 10^5\ \mathrm{m^{-1}}$ and $B_0 = 0.3\ \mathrm{T}$

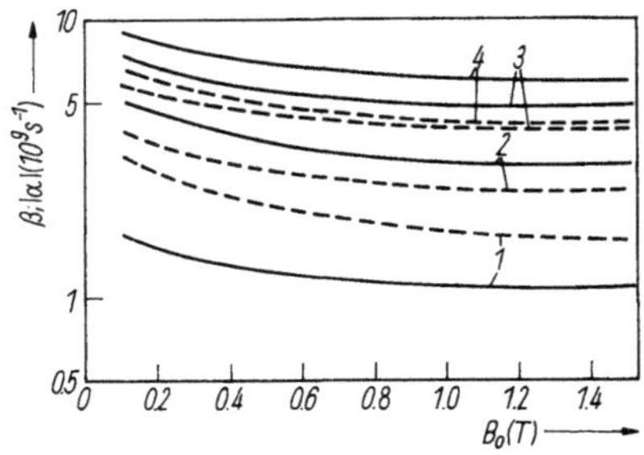

Fig. 3

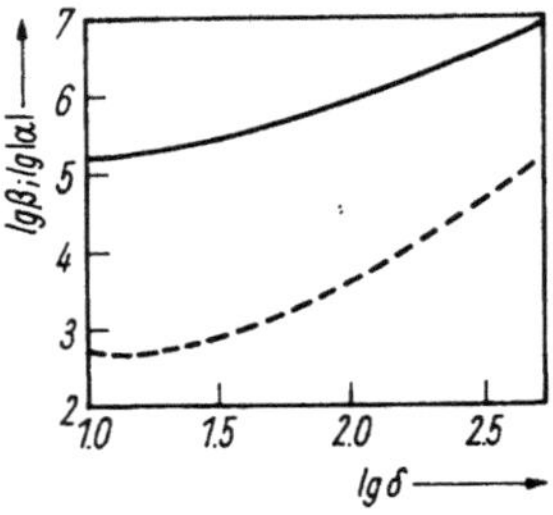

Fig. 4

Fig. 3. Variation of the phase constant β (solid curves) and the growth rate $|\alpha|$ (dashed curves) with the applied magnetic field B_0 for different values of δ: (1) 0.0, (2) 10.0, (3) 50.0, (4) 100.0, $k = 5 \times 10^5\ m^{-1}$ and $E_0 = 12\ kV\ m^{-1}$

Fig. 4. Dependence of the phase constant β (solid curve) and the growth rate $|\alpha|$ (dashed curve) on δ; $E_0 = 0.0$, $B_0 = 0.3$ T, and $k = 5 \times 10^5\ m^{-1}$

Acknowledgements

We gratefully acknowledge the financial support from U.G.C. India under the research project "Instabilities in Solid State Plasmas". One of the authors (S.Gh.) is indebted to P. K. Sen for many clarifying discussions. The computation work is done at the computer centre of Physical Research Laboratory, Ahmedabad.

References

[1] P. GUERET, J. appl. Phys. **39**, 2136 (1968).
[2] G. F. FREIRE, Int. J. Electronics **28**, 1 (1970).
[3] H. C. HSIEH, Phys. Rev. B **10**, 4297 (1974).
[4] S. GUHA and S. GHOSH, phys. stat. sol. (a) **41**, 249 (1977).
[5] T. SUZUKI, J. Phys. Soc. Japan **21**, 2000 (1966).
[6] K. MIYAZAKI and J. YAMAGUCHI, Japan. J. appl. Phys. **7**, 1059 (1968).
[7] A. V. ZAVYALOV, G. F. KARLOVA, and L. L. LYUZE, Soviet Phys. – Semicond. **5**, 504 (1971).
[8] R. H. C. NEWTON and D. M. SHORT, J. Phys. A **6**, 722 (1973).
[9] E. M. CONWELL, High-Field Transport in Semiconductors, Academic Press, New York 1967. (p. 159).
[10] R. R. KIKVIDZE, A. A. RUKHADZE, and E. P. FETISOV, Soviet Phys. – Solid State **9**, 1055 (1967).
[11] A. HASEGAWA, Plasma Instabilities and Nonlinear Effects, Springer-Verlag, New York 1975 (p. 17).
[12] P. A. ANDREEV, Soviet Phys. – J. techn. Phys. **19**, 1001 (1975).
[13] R. J. BRIGGS, Electron Stream Interaction with Plasmas, MIT Press, Cambridge 1964 (p. 8).
[14] A. SASAKI, J. appl. Phys. **41**, 3813 (1970).

(Received April 20, 1978)

phys. stat. sol. (a) 49, 371 (1978)

Subject classification: 1.5 and 14.3.1; 2; 13.1; 22.6

Department of Physics, Banaras Hindu University, Varanasi[1])

Effect of Nonstoichiometry on the Electrical Conductivity of Erbium Oxide Films

By

U. SAXENA and O. N. SRIVASTAVA

Thin films of erbium oxide show a thickness dependent amorphous-to-crystalline transformation. Both, the amorphous and crystalline phases are found to be nonstoichiometric with respect to oxygen content. Electrical measurements of *I–U* characteristics of these as-grown films in a sandwich configuration, Al–erbium oxide–Al/Au, exhibit a voltage controlled negative resistance region (VCNR). The nature of the characteristic is independent of electrode material, electrode polarity, and vacuum conditions. It is postulated that the inherent nonstoichiometry of the film gives rise to a 'built in' localized energy band in the forbidden gap of the oxide films which is responsible for the occurrence of VCNR.

Dünne Schichten von Erbiumoxid zeigen einen dickenabhängigen Übergang amorph–kristallin. Es wird gefunden, daß sowohl die amorphe als auch die kristalline Phase nichtstöchiometrisch ist bezüglich des Sauerstoffgehalts. Elektrische Messungen der *I–U*-Charakteristiken von diesen gewachsenen Schichten in einer Schichtstruktur, Al–Erbiumoxid–Al/Au, ergeben einen spannungsgesteuerten Bereich negativen Widerstands (VCNR). Die Art der Charakteristik ist unabhängig vom Elektrodenmaterial, der Elektrodenpolarität und den Vakuumbedingungen. Es wird angenommen, daß die innere Nichtstöchiometrie der Schichten zu einem „eingebauten" Energieband in der verbotenen Zone der Oxidschicht führt, das für das Auftreten der VCNR verantwortlich ist.

1. Introduction

Rare-earth oxides constitute a celebrated class of solid state materials which possess excellent insulating and semiconducting properties even at high temperatures. Due to this reason they are finding increasing application in industries and technology. Thus the electrical conductivity of these oxides, being one of the most important properties from the application standpoint, has been the subject of intensive studies during the recent years. A good deal of work is available regarding the dependence of electrical conductivity on pressure, temperature [1], neutron irradiation [2], etc. in bulk samples. However, the field relating to the study of the influence of stoichiometric variations in these oxides remains rather unexplored. Various investigations on stoichiometric and nonstoichiometric compounds have revealed that the stoichiometry plays a very important role in deciding the conduction behaviour of solids. To cite a few examples, nonstoichiometry significantly affects the electrical conductivity of CdS films [3]. In the EuO system [4], the oxygen rich crystals are found to be insulators, while Eu rich crystals are either metallic or show an insulator to metal transition. In the Ni–S system also the insulator → metal transition which appears at 30 kbar in NiS_2, disappears rapidly in $NiS_{2.15}$ [5].

In the case of amorphous materials (films or bulk) which cannot be doped easily [6], the role of stoichiometry is all the more important in the sense that this is the only parameter in hand which can be controlled to get the desired conductivity behaviour. Motivated by this idea and the fact that such studies have not been made previously, we studied electrical conductivity of the as-grown aluminium–erbium oxide–alumi-

[1]) Varanasi-221005, India.

nium/gold thin film sandwiches. Measurements of I–U characteristics revealed in these as-grown sandwiches the existence of voltage controlled negative resistance regions. This is rather strange in the sense that unlike the present results, VCNR region in MIM sandwiches usually occurs only in formed samples [7 to 16].

As has already been shown in previous communications [17, 18] the as-grown thin films of erbium oxide ($\leqq$ 700 Å) are amorphous. With increasing thickness, they slowly transform into a f.c.c. crystalline phase. Also it has been established through electron diffraction studies that the crystalline phase is nonstoichiometric and contains oxygen vacancies. In the present communication, the results regarding the structural phases and nonstoichiometry have been discussed in brief, the main emphasis being on the influence of nonstoichiometry on electrical conductivity. Attempts have been made to show that the voltage controlled negative resistance (VCNR) region in the I–U characteristics of these films is the consequence of the inherent nonstoichiometry of these films with respect to the oxygen content. It is proposed that the nonstoichiometry results in the creation of a broad band of closely spaced localized states which is responsible for the occurrence of VCNR.

2. Experimental Technique

To study the nonstoichiometry and the structural phases of erbium oxide, the films were deposited on formvar coated copper grids by thermal evaporation in a vacuum of about 10^{-6} Torr. The oxide film thickness was determined by mass-distance calibration which was checked beforehand by multiple beam interferometry.

The samples for the measurement of I–U characteristics were prepared in the form of parallel-plate capacitors of the type Al–erbium oxide–Al. The sandwich geometry was similar to that employed for the studies of the dielectric behaviour of erbium oxide films [17]. These samples were prepared on glass slides, precoated with formvar films, so that they could be easily transferred on copper grids for electron microscope examination at different stages of voltage treatment. Electrical connections employing copper wire and indium soldering were made on these slides before depositing the films, lest the formvar film should interfere with the measurements.

The I–U characteristics of the MIM sandwiches were measured by employing a conventional dc arrangement [7], using a voltage measuring device (Simpson AC/DC VTVM, model 221-I) with very high impedance. The measuring circuit was pretested and the initial linear I–U characteristic gave the same value of resistance of the oxide film as obtained by direct resistance measurement.

3. Results

3.1 Study of nonstoichiometry in erbium oxide films

In the present investigation films of thicknesses ranging between 200 and 1000 Å were employed. As has been discussed in a previous publication [17], the films in the said thickness range show a thickness-dependent amorphous → crystalline transformation. While thin films ($\leqq$ 700 Å) were amorphous as evidenced by broad halo-type diffraction patterns and rapid crystallization on short pulse annealing treatment (Fig. 1), thicker films showed a slow transformation from amorphous to f.c.c. crystalline phase. Fig. 2 shows a diffraction pattern from a mixed phase film which is predominantly amorphous. The discrete lines due to the f.c.c. phase which are superimposed on broad halo-type rings due to the amorphous phase, are clearly seen in the figure. The transmission micrograph of the crystalline phase (Fig. 3) reveals its microcrystalline nature. The lattice parameter of this phase, however, was found to vary significantly from film to film. This observation suggested that the crystalline phase might correspond to a nonstoichiometric variant. This was further verified by pulse

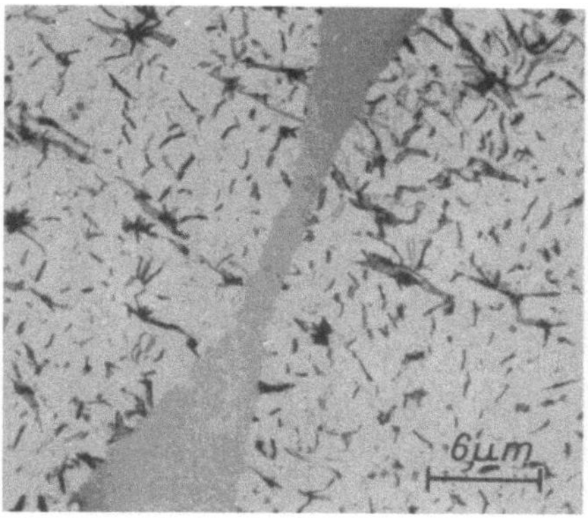

Fig. 1

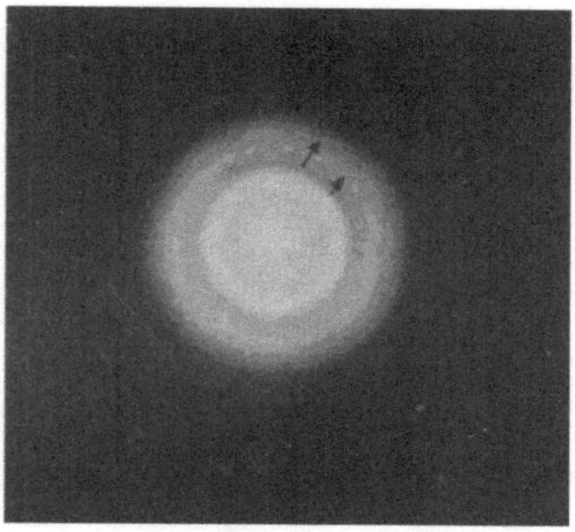

Fig. 2

Fig. 1. Transmission electron micrograph of an electron beam irradiated erbium oxide film showing crystallization of the amorphous phase

Fig. 2. Diffraction pattern of as-grown mixed phase (amorphous + crystalline) film. The discrete lines due to crystalline phase, superimposed on broad halos due to amorphous phase have been marked by arrows

annealing the sample at about 900 °C. As a result of this, a uniform grain growth in the sample occurred which is expected for a microcrystalline film. The diffraction patterns taken under normal beam illumination from these single crystal regions invariably revealed the presence of intense diffuse streaks along with the sharp spots due to parent f.c.c. lattice (Fig. 4), thus indicating that the phase was indeed a nonstoichiometric one embodying a short-range order arrangement of atoms or vacancies [19]. The intense diffuse streaks suggested the presence of a rather severe nonstoichiometry. In the present case, the nonstoichiometry presumably arises due to oxygen vacancies which are initially arranged in a short-range manner but finally achieve a long-range order on prolonged annealing treatment [19]. It should be mentioned here that there it is not easy to assess nonstoichiometry in the amorphous phase. But, since in the present case, the microcrystalline state which is very near to the amorphous one exhibits a significant amount of nonstoichiometry, it can be taken that the amorphous phase of erbium oxide films also corresponds to a nonstoichiometric variant [6]. It is opportune to mention that nonstoichiometry in amorphous films of Yb_2O_3 has been conjectured by Jawalkar [20] also.

Fig. 3. TEM picture of the crystalline phase

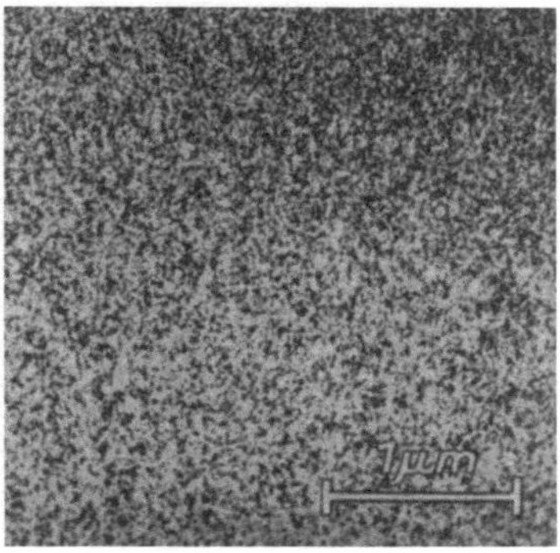

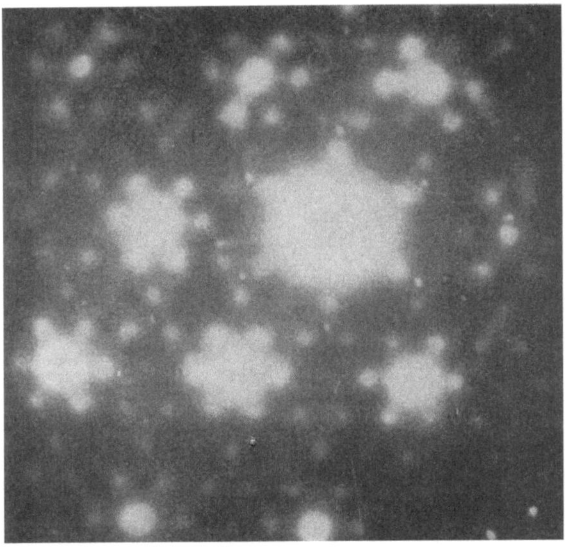

Fig. 4. Single-crystal diffraction pattern of erbium oxide film. The existence of intense diffuse streaks indicates the presence of severe nonstoichiometry and the accompanying short-range order

3.2 Study of electrical conductivity of erbium oxide films

The effect of nonstoichiometry on electrical properties was investigated by measuring the I–U characteristics of sandwich samples (Al–erbium oxide–Al) with different oxide film thicknesses in the range 200 to 1000 Å. The measurements were taken at room temperature and atmospheric pressure. A representative I–U characteristic obtained in the present case is shown in Fig. 5. As can be seen, the nature of the characteristic is different from that usually observed for dielectric films [21]. For the case of dielectric films, the current increases with voltage following various high-field conduction mechanisms, till at a particular voltage it shoots up resulting in the breakdown of the dielectric film, whereas in the present case, the current first increases with voltage and then after reaching a fixed voltage ($\approx$ 2.5 V) starts decreasing, thus showing the presence of VCNR region. Measurements of samples with different oxide film thicknesses showed that the nature of the characteristic was essentially the same in all cases and the position of the peak was independent of voltage. The only difference amongst the various characteristics was in the magnitude of the current peak. It was however observed that the best I–U characteristics in regard to the reproducibility and well developed VCNR region were obtained in films of thicknesses ranging between 600 and 800 Å. This is the same thickness range where the amorphous phase mixed with the f.c.c. crystalline phase embodying nonstoichiometry has been found to exist [17 to 19]. The same VCNR regions were traced if the polarity of the electrodes was changed. Thus the VCNR did not depend on the polarity of the electrodes.

To explore the possible causes which may be responsible for the appearance of VCNR region in the I–U characteristic, the effect of nature of electrode material and the ambient

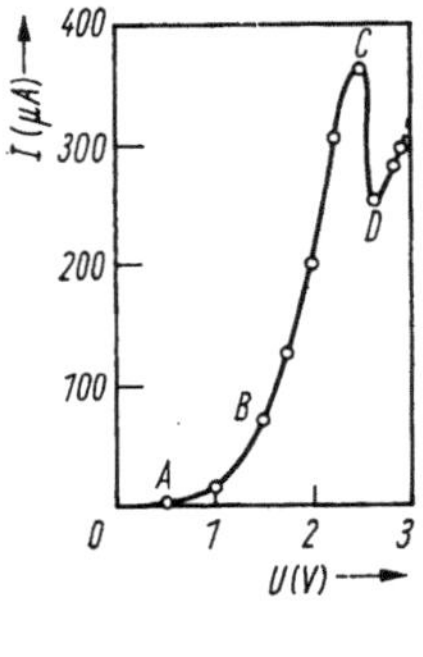

Fig. 5. I–U characteristic of Al–erbium oxide–Al sandwich with oxide film thickness 700 Å

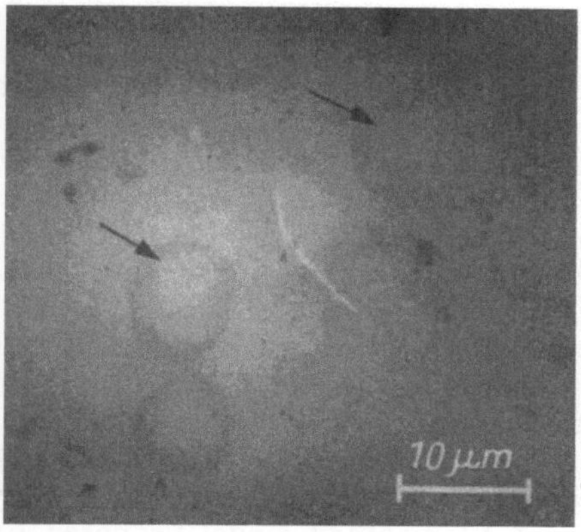

Fig. 6

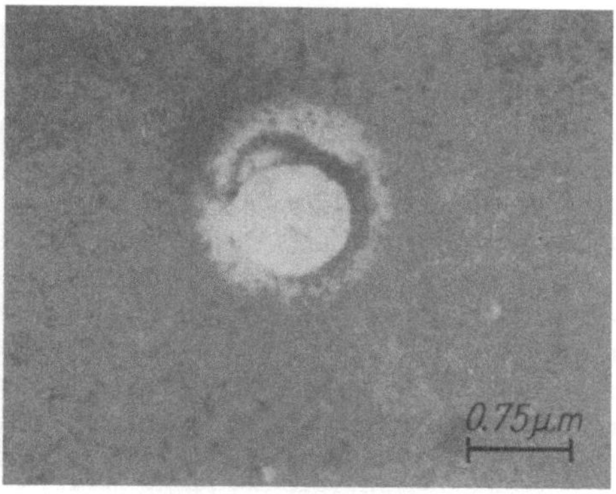

Fig. 7

Fig. 6. TEM picture of a sandwich film treated up to a voltage corresponding to point C of Fig. 5 showing the presence of blister regions. Some of them have been marked by arrows

Fig. 7. TEM picture of a sandwich film treated up to a voltage beyond the point D in Fig. 5 shows the presence of a circular localized centre over which the counter electrode has melted away. Usually several such centres are present in the worked films

atmosphere on the I–U characteristic were investigated. The measurements were carried out by incorporating two variations in the experimental conditions. In one of these, a gold counterelectrode was employed (sandwich geometry being Al–erbium oxide–Au) and in the other one the measurements were done in vacuum ($\approx 10^{-4}$ Torr). It was found that with the Au electrode the nature of the I–U characteristic was essentially the same except differences in position and height of the current peak associated with VCNR region. Measurements in vacuum again did not change the nature of the observed characteristic.

To investigate the structural changes taking place as a consequence of voltage application, the Al–erbium oxide–Al sandwiches were examined at different stages in an electron microscope. For samples which had been treated up to the voltage region where the current slowly rises (AB region in Fig. 5), the microstructure did not show any special feature and appeared to be that of the superposition of aluminium and erbium oxide films. The diffraction patterns from these films either corresponded to aluminium or else to aluminium together with faint lines coming from erbium oxide films. The films which had been treated up to or slightly beyond 2.5 V, invariably showed the presence of blister regions. Fig. 6 shows a micrograph of such a film. Some of the blister regions have been marked by arrows. The films which had been treated up to a voltage far beyond 2.5 V, exhibited well defined circular regions over which the top electrode had evaporated (Fig. 7).

The evaporation of the top electrode at localized regions was confirmed by retracing the I–U characteristic after the appearance of blister regions with decreasing and increasing voltages. In both the cases the characteristic was retraceable but the current was found to be decreased by one order or more. Also the position of the peaks varied slightly from that of the first cycle. The suppression of current in reverse and second cycles makes us believe that the blister regions correspond to the early stage of the local centres at which the top electrode later on gets evaporated (Fig. 7).

4. Discussion

In the recent past VCNR has been observed for a large variety of materials which include oxides, sulphides, nitrides, and organic compounds [7 to 16], and several theories have been put forward to explain the phenomenon. These observations, however, differ from the present one in one respect. This relates to the fact that in all the previous MIM configurations, unlike the present case, VCNR could not be observed in as-grown films, instead it ensued only after a so-called "forming treatment". This treatment involves the application of a certain voltage ($\approx$ 5 to 10 V), preferably in near vacuum conditions. In all the theories which have been proposed to explain the VCNR phenomenon the forming treatment is assumed to result in the creation of a localized band of closely spaced energy states in the forbidden gap of the dielectric material either by injection of electrode ions [8 to 10] or by the creation of defect centres [11]. According to Hickmott [7] these states are already present in the insulator and they merely get ionized on "forming". Alternately in a filamentary model proposed by Dearnaley [12], Greene et al. [13], and Sutherland [14], the VCNR is taken to arise as a result of growth and thermal rupture of many conducting filaments. In a more recent work the filaments are taken to correspond to broad impurity or defect bands [15] and VCNR and other related phenomena are then explained in terms of the electronic behaviour of these filaments.

As is evident from the above discussion, to date the consensus in regard to the occurrence of VCNR is in favour of the creation of an energy band consisting of localized levels within the forbidden band of the dielectric, the genesis of this band being ascribed to various reasons such as diffusion of electrode material into the insulator, inherent impurities, creation of defect centres etc. It seems that in the present case also the origin of the negative resistance is implicit in some way in the creation of localized band. In the present case, since no forming treatment was imparted to the film prior to measurement, we believe that the localized energy band here corresponds to "built-in" states which are inherent in the as-deposited film. These "built-in" states are apparently a property of the oxide and do not depend on electrode (material or polarity) or the atmospheric environment. This fact is strongly supported by our observation that the nature of the I–U characteristic is independent of electrode material, electrode polarity, and vacuum conditions. Since the as-grown oxide films do not contain any impurity, the most attractive alternative capable of creating localized states is the nonstoichiometry of the film with respect to oxygen content.

In the case of rare-earth oxides, the nonstoichiometry due ot the variation in oxygen content has been reported by several workers [22 to 25]. In the case of erbium oxide films we have also found that the nonstoichiometry is quite severe such that the chemical formula of the as-grown films corresponds to $R0_x$ with x varying between 1 and 2 [18; 19]. Due to their high concentration these oxygen vacancies are expected to lie in the perturbation range of the potential energy of each other. In amorphous as well as in microcrystalline structures with short-range order these oxygen vacancies will not have a uniquely defined environment and there will be no consistency in nearest and next-nearest neighbour configurations. Thus the potential energy as well as the perturbation of these vacancies will not be well defined within the matrix. In such a matrix the interactions amongst the vacancies would lead to the formation of a broad band instead of a discrete energy state in the forbidden gap of the insulator.

Keeping the foregoing facts in mind we have proposed an energy band model (Fig. 8) based on which the present observation on VCNR becomes intelligible. This model is in keeping with other similar models where creation of additional states in the forbidden band has been postulated [8 to 12]. However, in the present case the additional states are taken to result due to nonstoichiometry rather than due to impurities, elec-

Fig. 8. Schematic energy band diagram of the Al–erbium oxide–Al sandwich at $U = 0$ showing the localized energy bands due to nonstoichiometry in the forbidden band of the oxide

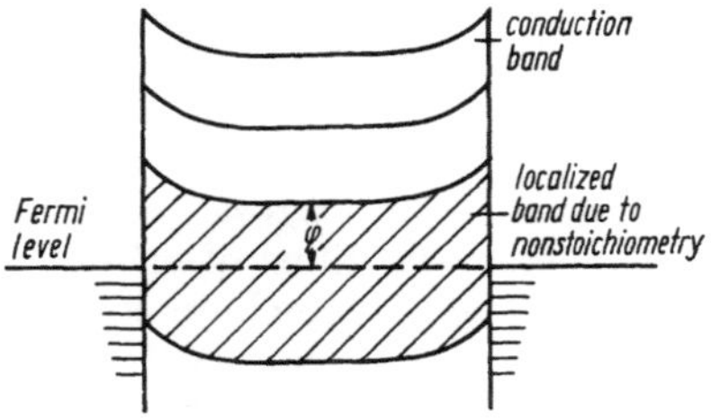

trode diffusion, or defect centres. It should be mentioned here that the existence of similar states due to nonstoichiometry in oxygen content has already been reported for the case of rutile [26].

It is well established that the band structure of a solid is a consequence of short-range rather than long-range order within the solid and hence all the essential features of the crystalline state are carried over to the microcrystalline and amorphous state [6]. Thus in the present model the band edges are assumed to be sharp and the band gap well defined as in the case of the crystalline state.

Still another assumption which has been made is in regard to the blocking contacts on both sides. This is a reasonable assumption in view of the fact that the energy gap of Er_2O_3 is about 3.28 eV [27]. Since the work functions of the two electrodes Al and Au used are 4.2 and 4.7 eV, respectively, and the electron affinity for insulators is generally $\lessapprox 1$ eV, ψ_m will always be greater than ψ_i, which is the condition for blocking contacts. The width of the localized band has been taken arbitrarily.

The localized band extends up to an energy φ above the electrode Fermi level. Once the localized band is taken to exist, the occurrence of VCNR can be explained in the same way as in the case of other MIM sandwiches where the localized energy band originates [8 to 12]. As is well known, electrical conduction in such a system takes place by indirect tunnelling of an electron from one electrode to another. The tunnelling takes place at the Fermi level of the electrode between adjacent sites located at the broad band. Theoretical calculations reveal that if the top of the localized band extends up to an energy φ above the electrode Fermi level, then under an applied voltage, the current will continue to rise up to $U < \varphi/e$. This is so since the easy passage of electrons which have been injected from the electrode from an energy level situated at the Fermi level of the electrode is still possible (see Fig. 9a). However, when $U > \varphi/e$ is applied, the electrons injected at the Fermi level can penetrate only up to the top of the localized band and hence cannot pass through. Consequently the current decreases and VCNR region appears (Fig. 9b). At a still higher voltage the current rises again and this results in a tail rising after point D. This rise is attributed to either the thermal excitation of electrons or else to a thinning of the barrier between the top of the localized band and the bottom of the conduction band. It may thus be noted that the basic difference in the conduction process relating to the present MIM system and other

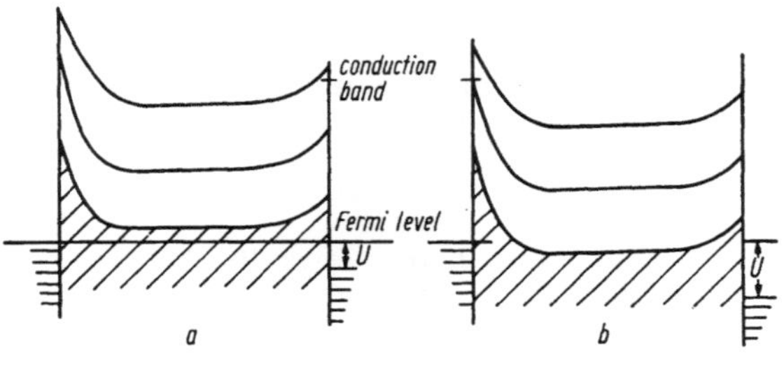

Fig. 9. Schematic energy band diagram showing the appearance of VCNR. a) $U < \varphi/e$, b) $U > \varphi/e$

similar systems lies in the origin of the band of the localized states. Unlike the other cases it is believed to arise due to the high oxygen vacancy concentration in the as-grown films.

5. Conclusion

The as-grown films of erbium oxide are nonstoichiometric with respect to oxygen content. The nonstoichiometry exists in both amorphous and crystalline phases which depend on film thickness. Its influence on electrical conductivity is manifested in the form of appearance of a voltage controlled negative resistance (VCNR) region in the I–U characteristics of as-grown Al–erbium oxide–Al sandwiches. It is proposed that the inherent nonstoichiometry results in the creation of a "built-in" band of localized states in the forbidden gap of the oxide which is responsible for the occurrence of VCNR.

Acknowledgements

The authors are thankful to Dr. A. K. Jonscher (Chelsea College, University of London), Dr. K. L. Chopra (Indian Institute of Technology, Delhi), and Dr. B. R. Nag (Institute of Radio Physics and Electronics, Calcutta) for useful discussions. One of us (U.S.) is grateful to C.S.I.R., India, for the award of senior research fellowship.

References

[1] L. EYRING, in: High Temperature Oxide. II, Oxides of Rare Earths, Titanium, Zirconium, Hafnium, Niobium, and Tantalum, Ed. A. M. ALPER, Academic Press, New York/London, 1970 (pp. 58–59).
[2] V. I. SPITSYN, G. N. PIRGOVA, and N. S. STELMAKH, Dokl. Akad. Nauk SSSR, **212**, 1389 (1973).
[3] A. YOUSHIKAWA and Y. SAKAI, Japan J. appl. Phys. **15**, 1861 (1976).
[4] M. W. SAFER, J. B. TORRANCE, and T. PENNEY, J. Phys. Chem. Solids **33**, 2251 (1972).
[5] F. GAUTIER, G. KRILL, and M. F. LAPIERRE, Solid State Commun. **11**, 1201 (1972).
[6] N. F. MOTT and E. A. DAVIS, Electronic Processes in Noncrystalline Materials, Clarendon Press, Oxford (1971).
[7] T. W. HICKMOTT, J. appl. Phys. **33**, 2669 (1962); **35**, 2118, 2679 (1964); **36**, 1885 (1965).
[8] J. G. SIMMONS and R. R. VERDERBER, Proc. Roy. Soc. **A301**, 77 (1967).
[9] R. R. VERDERBER and J. G. SIMMONS, Radio Electr. Engng. **33**, 347 (1967).
[10] J. PIVOT, M. BOUDEULLE, A. CACHARD, and C. H. S. DUPUY, phys. stat. sol. (a) **2**, 319 (1970).
[11] I. EMMER, Thin Solid Films **20**, 43 (1974).
[12] G. DEARNALEY, Phys. Letters A **25**, 760 (1967).
[13] P. D. GREENE, E. L. BUSH, and I. R. RAWLING, Proc. Symp. Deposited Thin Film Dielectric Materials, Montreal, Oct. 1968, The Electrochem. Soc., New York 1969.
[14] R. R. SUTHERLAND, J. Phys. D **4**, 468 (1971).
[15] J. E. RALPH and J. M. WOODCOCK, J. non-crystall. Solids **7**, 236 (1972).
[16] CHUN CHIANG, Appl. Phys. Letters **31**, 553 (1977).
[17] U. SAXENA and O. N. SRIVASTAVA, Thin Solid Films **33**, 185 (1976).
[18] U. SAXENA and O. N. SRIVASTAVA, Thin Solid Films **42**, L_1 (1977).
[19] V. K. KAUL and U. SAXENA, Acta cryst. **A33**, 992 (1977).
[20] S. R. JAWALKAR, Indian J. pure appl. Phys. **7**, 801 (1969).
[21] P. J. HARROP and D. S. CAMPBELL, in: Thin Film Technology, Ed. L. I. HAISSEL and R. GLANG, McGraw-Hill Publ. Co, 1970 (pp. 16–23).
[22] S. A. SEMILTOV, M. I. IMAMOV, N. A. REGIMLI, and L. I. MAN, Thin Solid Films **32**, 325 (1976).
[23] B. M. S. BIST, J. KUMAR, and O. N. SRIVASTAVA, phys. stat. sol. (a) **14**, 197 (1972).
[24] A. E. SOLOVEVA, Izv. Akad. Nauk SSSR, Ser. neorg. Mater. **10**, 1555 (1974).
[25] Z. S. VOLCHENKOVA and D. S. ZUBANKOVA, Izv. Akad. Nauk. SSSR, Ser. neorg. Mater. **12**, 1396 (1976).
[26] N. IZUMI, Japan J. appl. Phys. **16**, 273 (1977).
[27] H. B. LAL and N. DAR, Indian J. pure appl. Phys. **14**, 788 (1976).

(Received May 2, 1978)

phys. stat. sol. (a) **49**, 379 (1978)

Subject classification: 1.5; 1.4; 2; 22.1.1

Department of Physics, Indian Institute of Technology, Delhi[1])

Structural Properties of Polycrystalline Ge–Metal Films

By

V. Dutta, P. Nath, V. D. Vankar, and K. L. Chopra

Polycrystalline Ge–metal films with 0.5 to 10 at% concentration of Al, Ga, Ag, Au, Cu, and Fe are prepared by codepositing onto glass/rock salt substrates held at 723 K and also by crystallizing amorphous films of the same materials by heating at 723 K. The structural details of these films are established by transmission electron microscopy and electron diffraction. For metal concentrations $\lessapprox$ 5 at%, the electron diffraction patterns of these films are essentially similar to that of pure Ge. The grain size decreases rapidly with increasing concentration of the metal. Beyond 5 at%, segregation of the metal and compound formation determine the structural properties. In the case of Al which has a finite solubility in Ge, the lattice constant decreases and the grain size is nearly independent of metal concentration. In the case of Ag and Au which have negligible solubility in Ge, the films are heterogeneous having segregated metal particles, and the grain size decreases rapidly. The diffraction pattern corresponding to a mixture of Ge and Cu_3Ge and ordered phases/compounds of Fe with Ge are observed in the Ge–Cu and Ge–Fe films and the grain size increases with metal concentration.

Polykristalline Ge-Metallschichten mit 0,5 bis 10 At% Konzentration von Al, Ga, Ag, Au, Cu und Fe werden durch gleichzeitiges Aufbringen auf Glas/Steinsalzsubstrate mit einer Temperatur von 723 K hergestellt, sowie durch Kristallisation von amorphen Schichten desselben Materials durch Aufheizen auf 723 K. Die Sturktureinzelheiten dieser Schichten werden mittels Transmissionselektronenmikroskopie und Elektronenbeugung festgestellt. Für Metallkonzentrationen $\lessapprox$ 5 At% sind die Elektronenbeugungsmuster dieser Schichten im wesentlichen denen von reinem Ge ähnlich. Die Korngröße nimmt schnell mit steigender Metallkonzentration ab. Oberhalb 5 At% bestimmen Segregation des Metalls und Bildung von Verbindungen die Struktureigenschaften. Im Falle des Al, das eine endliche Löslichkeit in Ge hat, nimmt die Gitterkonstante ab und die Korngröße ist nahezu unabhängig von der Metallkonzentration. Im Falle des Ag und Au, die eine vernachlässigbare Löslichkeit in Ge besitzen, sind die Schichten heterogen und besitzen ausgefällte Metallteilchen, und die Krongröße nimmt schnell ab. Beugungsmuster, die einer Mischung von Ge und Cu_3Ge und geordneten Phasen/Verbindungen von Fe mit Ge entsprechen, werden in den Ge–Cu- und Ge–Fe-Schichten beobachtet; die Korngröße nimmt mit der Metallkonzentration zu.

1. Introduction

Electronic properties of semiconductors doped and alloyed with metals are of considerable technological and fundamental interest. Although detailed studies on several useful semiconductors have been done in bulk form, very little work has been done in thin films of such semiconductors. Because of the atom by atom condensation process during vapour deposition of a thin film, one expects the solubility condition of two or more materials to be relaxed considerably [1]. Therefore, it is of further interest to study solubility characteristics of metals and the formation of alloys and compounds in semiconductor thin films. Thus, we have undertaken a detailed study on the formation, and structural and electronic properties of Ge–metal alloys/compounds.

Germanium has been chosen as the host material because its electronic properties and impurity effects thereon are well known. The metals have been chosen on the basis of their solubility in Ge and compound forming tendency [2, 3] and these are Al, Ga,

[1]) New Delhi-110029, India.

Ag, Au, Cu, and Fe. Whereas Al and Ga have a finite solubility (3 to 4 at% for Al and 2 at% for Ga) in Ge, Ag, and Au have negligible solubility ($\lesssim 10^{-3}$ at%) and further, these metals do not form any compound with Ge. On the other hand, Cu and Fe have negligible solubility, but several compounds/ordered phases are known to occur in Ge–Cu and Ge–Fe systems. Further, such studies would allow a direct comparison with the properties of corresponding amorphous Ge-alloy films which have been studied in great details in our laboratory [4].

This paper reports on the structural properties of polycrystalline Ge–metal films. Results of other studies on these films will be reported separately.

2. Experimental Details

Ge-metal alloys or homogeneous mixtures of different compositions (0.5 to 10 at% metal) were prepared by melting high purity (99.999%) constituents in an alumina/graphite crucible for 1 h in a vacuum $\approx 10^{-6}$ Torr. Ge–metal films were obtained by thermal evaporation from a tungsten basket in a vacuum $\approx 10^{-6}$ Torr. The rate of evaporation was monitored using a precalibrated quartz crystal oscillator and was kept fixed at 10 Å/s in all cases. Polycrystalline films were obtained by evaporating the material onto freshly cleaved (in air) NaCl substrates kept at 723 K. A molybdenum strip heater was used for heating and the temperature was monitored using a chromel–alumel thermocouple. Crystallized films were obtained by heating amorphous films (prepared together with the crystalline films by depositing on NaCl substrates kept at room temperature) at 723 K for 1/2 h in a vacuum $\approx 10^{-6}$ Torr. The composition of the Ge–metal films was checked by an X-ray fluorescence method and it was found to be within 0.2 at% of the composition of the evaporated material. Structural order and microstructure of the films were established by electron diffraction and transmission electron microscopy. For these studies carbon ($\approx$ 100 Å thick) was deposited onto the films. The films were floated off the substrates by dissolving NaCl in distilled water and then transferred onto a copper grid. The diffraction pattern were analyzed using a Joyce Lowbell double beam recording densitometer. The linear grain size and its distribution were determined from the transmission electron micrographs.

3. Result and Discussion

3.1 Pure Ge films

Ge films deposited at 723 K and also films crystallized by heating amorphous Ge films to 723 K for 1/2 h exhibit polycrystalline structure as indicated by the ring-type electron diffraction patterns, shown in Fig. 1. An interesting feature in the diffraction pattern of crystalline films is the presence of an extra ring corresponding to (222) reflection which is normally forbidden in the elements having a diamond cubic structure. This ring is also seen in bulk Ge [5] and has been interpreted as arising due to nonsymmetry of the local charge density [6]. In the tetrahedral coordinated elements, instead of the outer electrons having spherically symmetric sp^3 orbitals, the charges are tetrahedrally positioned. This type of charge distribution modifies the structure factor (f_{hkl}) making f_{222} finite instead of zero as is the case with spherically symmetric charge distribution. Nonoccurrence of the (222) reflection ring in crystallized Ge films suggests that though there might still be a tetrahedral charge distribution around each atom, the total effect (because of small grain size) is due to a net spherically symmetric charge distribution.

Electron micrographs of the crystalline and crystallized Ge films are also shown in Fig. 1. As excepted, because of the differences in the growth process [1, 7], the linear

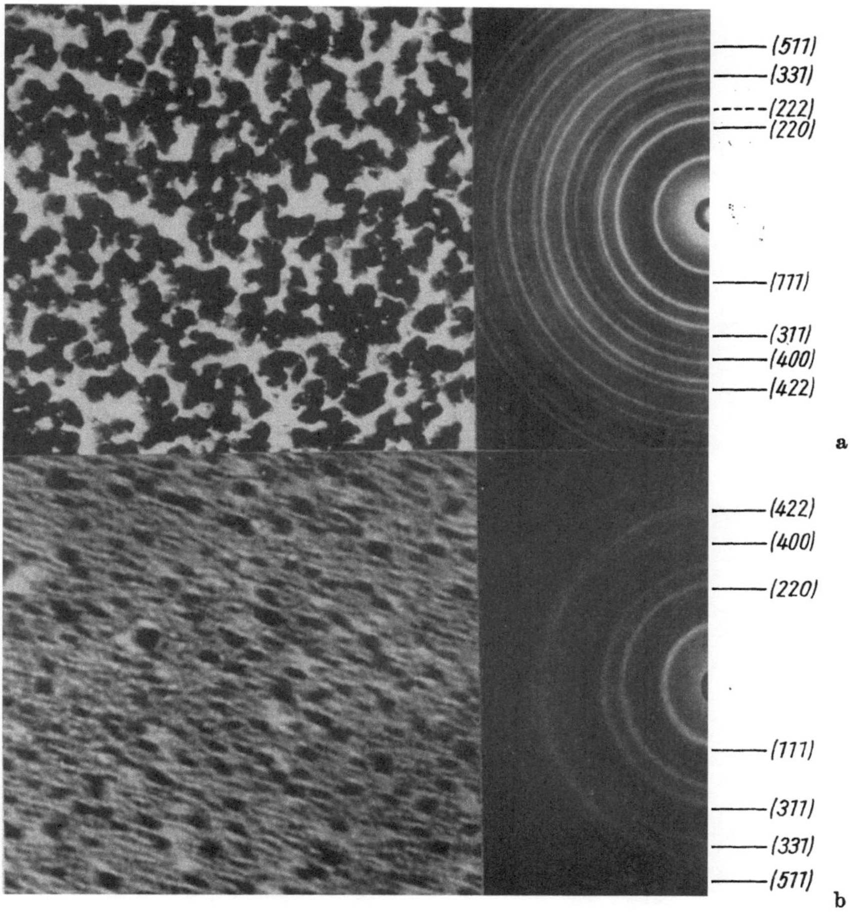

Fig. 1. Electron diffraction patterns and transmission electron micrographs of a) crystalline, b) crystallized Ge films

grain size for crystallized films ($\approx$ 100 Å) is very much smaller compared to that for pure Ge films ($\approx$ 680 Å). The distribution of linear grain size in the crystalline films is Gaussian.

3.2 Ge-metal films

3.2.1 Ge–Al and Ge–Ga films

Electron diffraction patterns of crystalline and crystallized Ge–Al and Ge–Ga films show sharp rings and are essentially similar to that for pure Ge films. Fig. 2 shows a typical diffraction pattern corresponding to crystalline Ge–Ga (10 at%) films. With increasing metal concentration spots appear on the rings suggesting an oriented growth. The (222) reflection ring is present in these films also and is fainter in crystallized films compared to that in crystalline films. Crystalline and crystallized Ge–Al and Ge–Ga films have lattice parameter equal to that of pure Ge up to $\approx$ 5 at% metal concentration. The films remain homogeneous and no separate rings corresponding to the metal are observed. Beyond 5 at% concentration of Al, faint rings corresponding to

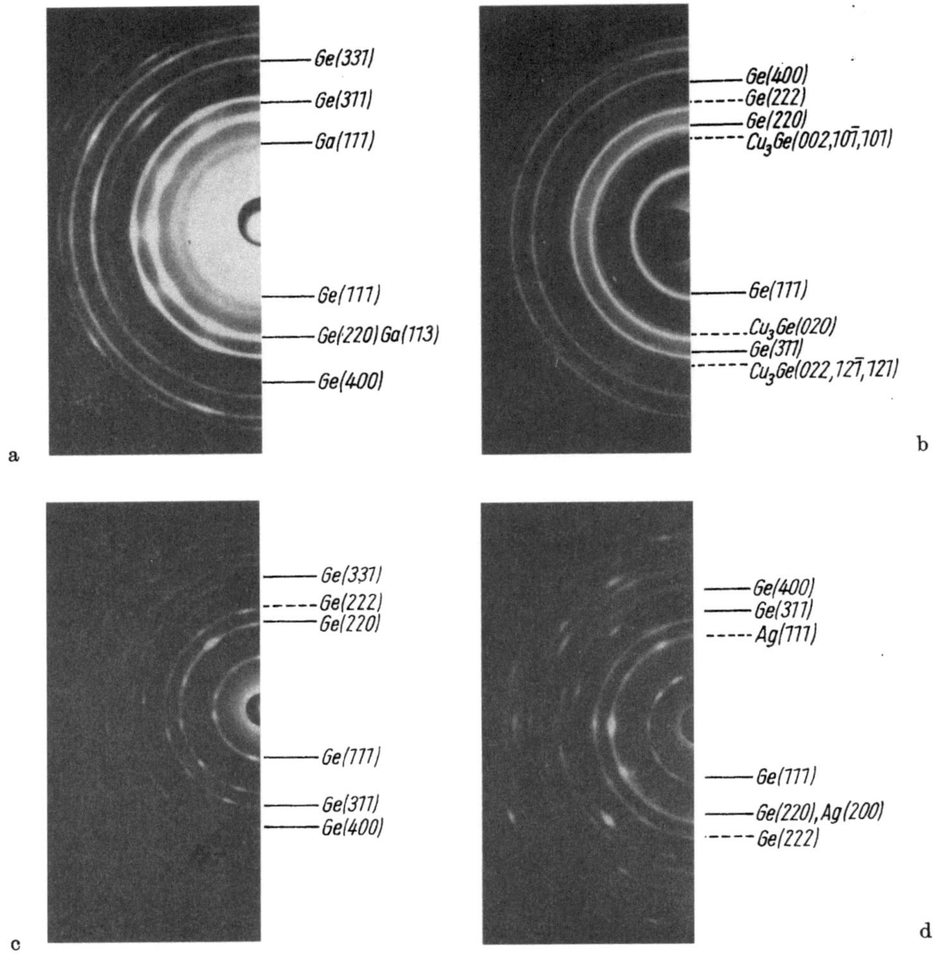

Fig. 2. Typical electron diffraction patterns of crystalline Ge–metal films. a) Ge–Ga (10 at%), b) Ge–Cu (10 at%), c) Ge–Ag (5 at%), d) Ge–Ag (10 at%)

the metal appear and a slight decrease ($\approx 2\%$) in the lattice parameter is observed. However, in Ge–Ga films there is no change in lattice parameter even up to 20 at% of Ga, although segregation of the metal is observed. Segregation of metal atoms is expected to occur beyond the solubility limit (3 to 4 at% for Al and 2 at% for Ga). Our observations show that the solubility of Al and Ga in Ge films is extended considerably beyond the bulk value.

As in the case of pure Ge films, crystallized Ge–metal films also have smaller grain size (nearly equal to that in crystallized Ge films) than crsytalline films. The grain size distribution in crystalline films as obtained from the transmission electron micrographs (Fig. 3) is Gaussian and the linear grain size (most probable as well as mean) decreases with increasing metal concentration (Fig. 4). Beyond 5 at%, saturation of the grain size occurs. The observed decrease in the grain size with increasing metal concentration can be understood on the basis of the growth process of thin films [1].

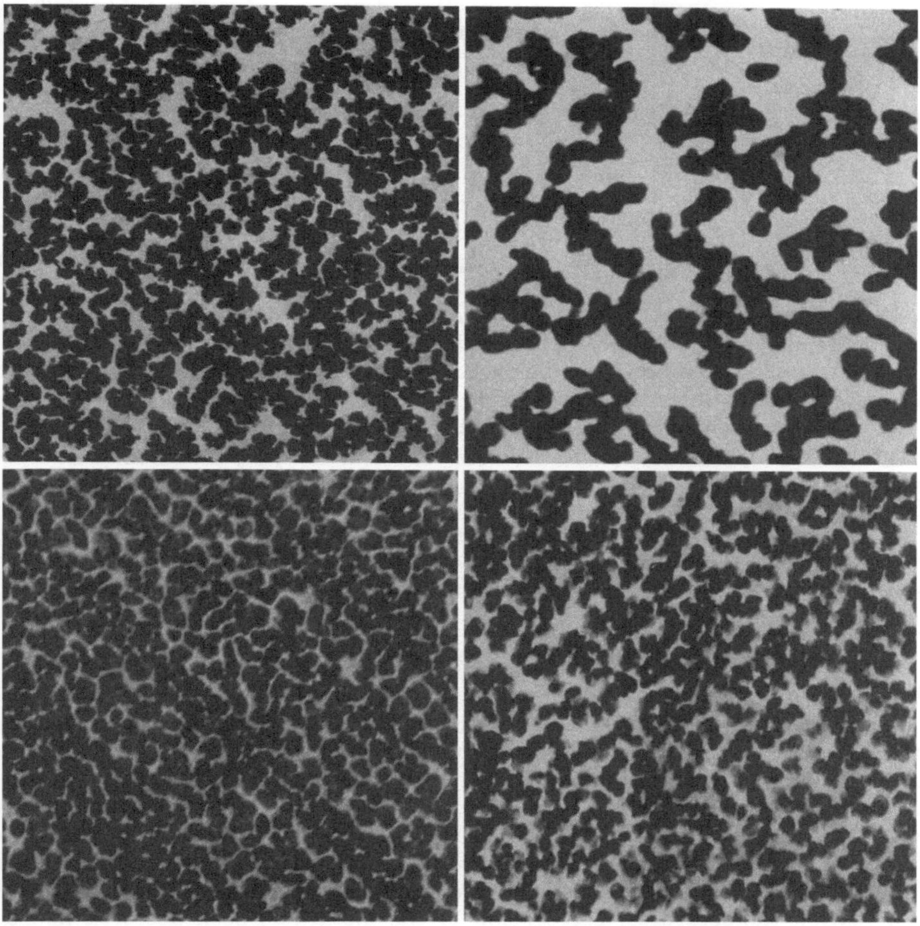

a d

c b

Fig. 3. Typical transmission electron micrographs of crystalline Ge–metal films. a) Ge–Ga (10 at%) b) Ge–Cu (10 at%), c) Ge–Ag (5 at%), d) Ge–Ag (10 at%)

When codeposited, the metal atoms inhibit the motion of the impinging Ge atoms and also provide extra nucleation sites for growth of the film. The net result is the formation of a larger number of nucleation centres and lower ad-atom mobility. This means that the number of supercritical nuclei is large, but their size is small.

3.2.2 Ge–Ag and Ge–Au films

Lattice parameter does not show any change in the crystalline and crystallized Ge–Ag and Ge–Au films even up to 10 at% metal concentration. Metal segregation is detected in crystalline Ge–Ag films with Ag concentration $\gtrsim$ 5 at% and in crystalline Ge–Au films with Au concentration $\gtrsim$ 10 at%. However, in crystallized Ge–Ag and Ge–Au films the segregation is observed for metal concentration $\gtrsim$ 10 at%.

Silver and gold, which have negligible ($\lesssim 10^{-3}$ at%) solubility in Ge, should segregate at very low concentrations. Our observations suggest that the metals are dissolved in the Ge matrix and are precipitated out only when the metal–metal interaction is sufficiently strong. Since Au has a smaller diffusion coefficient in Ge compared to that

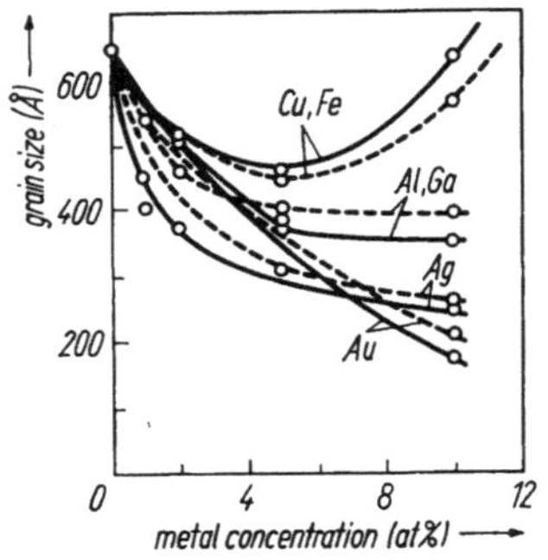

Fig. 4. Compositional dependence of grain size of crystalline Ge–metal films. ——— most probable, - - - - average

for Ag [8], segregation takes place at relatively lower concentrations in Ge–Ag films than in Ge–Au films. In order to further, confirm the observed extension of solubility, $Ge_{95}Ag_5$ and $Ge_{90}Ag_{10}$ films were deposited onto NaCl substrates at different temperatures ranging from room temperature to 723 K, simultaneously. It was found that films deposited at temperatures below 373 K were amorphous. No Ag was detected in films deposited at temperatures below 623 K. At higher temperatures, rings corresponding to Ag were observed. This clearly shows that, at temperatures below 623 K, there is solid solubility of Ag and Au in Ge films. Segregation at higher deposition temperatures is due to higher thermal energy available to the atoms, which facilitates segregation of the metal atoms.

The compositional dependence of linear grain size in crystalline Ge–Ag and Ge–Au films is shown in Fig. 4. The grain size decreases rapidly with metal concentration in both cases. However, the effect is larger in case of Au due to its smaller diffusion coefficient. Crystallized Ge–Ag and Ge–Au films have grain size nearly equal to that for crystallized Ge films.

3.2.3 Ge–Cu and Ge–Fe films

As indicated by the electron diffraction patterns (Fig. 2), Ge–Cu and Ge–Fe films are inhomogeneous for metal concentration $\gtrapprox 5$ at%. The diffraction pattern corresponds to that for Ge and Ge–metal compounds/ordered phases. The compound in case of Ge–Cu films is identified as Cu_3Ge. An unambiguous identification of the segregated phase present in Ge–Fe films is not possible, since Fe forms a number of compounds/ordered phases such as Fe_2Ge, Fe_3Ge, $Fe_{1.5}Ge$, etc. which yield a number of overlapping diffraction rings [9]. The compound/ordered phase is expected to be readily formed at high deposition/crystallization temperatures used in these films. As expected, at low crystallization temperatures, segregation of the metal has been detected in Ge–Cu films [10].

The compositional dependence of the grain size in crystalline Ge–Cu and Ge–Fe films is shown in Fig. 4. The grain size first decreases and then increases beyond 5 at%. This is possibly because at low concentrations metal atoms would be forming nuclei for growing thin films whereas at high concentrations the compound/ordered phase molecules would be forming the nuclei. In the latter case, a reduction in the number of nuclei and hence an increase in the grain size would occur.

4. Conclusions

Incorporation of different metals affects the microstructure of Ge–metal films. The effects depend on the solubility and/or compound forming tendency of the metal with Ge and are more pronounced in crystalline than in crystallized films. The solubility condition is relaxed and metal segregation/compound formation is detected only at metal concentration $\gtrapprox 5$ at%. The relaxation of solubility is higher in crystallized Ge–metal films than in crystalline Ge–metal films. The crystallized films tend to retain the homogeneous structure of amorphous Ge–metal films whereas in crystalline films higher ad-atom mobility helps in metal segregation/compound formation.

References

[1] K. L. CHOPRA, Thin Film Phenomena, McGraw-Hill Publ. Co., New York 1969.
[2] W. B. PEARSON, A Handbook of Lattice Spacings and Structures of Metals and Alloys, Pergamon Press, London 1958.
[3] A. E. VOL, Handbook of Binary Metallic Systems, Structure and Properties, Vol. II, Israel Program for Scientific Translation, Jerusalem 1967.
[4] K. L. CHOPRA, D. K. PANDYA, and P. NATH, Indian J. Phys. (Commemoration Volume) **1**, 9 (1976).
[5] J. J. DEMARCO and R. J. WEISS, Phys. Letters A **18**, 92 (1965).
[6] R. J. WEISS, X-Ray Determination of Electron Distributions, North Holland Publ. Co., Amsterdam 1966.
[7] A. BARNA, P. B. BARNA, and J. F. POCZA, J. non-crystall Solids **8–10**, 36 (1972).
[8] B. J. BOLTAKAS, Diffusion in Semiconductors, Infosearch, London 1963.
[9] Diffraction Data Index, Joint Committee on Powder Diffraction Standards, Pennsylvania 1972.
[10] H. S. RANDHAWA, L. K. MALHOTRA, H. K. SEHGAL, and K. L. CHOPRA, phys. stat. sol. (a) **37**, 313 (1976).

(Received March 29, 1978)

phys. stat. sol. (a) 49, 387 (1978)

Subject classification: 2; 21.1

Department of Metallurgy and Materials Science, University of Toronto[1]) *(a) and Bell Laboratories, Murray Hill*[2]) *(b)*

A Structural Study of Metallic Glasses Containing Boron (Fe–B, Co–B, and Ni–B)

By

Y. WASEDA (a) and H. S. CHEN (b)

The structure of metallic glasses containing boron (Fe–B, Co–B, and Ni–B) is investigated by X-ray diffraction. Using the common Fourier analysis, the atomic radial distribution function is evaluated from which interatomic distance and coordination number are obtained. The present results indicate that the atomic distribution of metallic glasses with boron content less than 20 at%, differs from that observed previously for a number of metallic glasses such as $Fe_{83}P_{17}$.

Die Struktur metallischer Gläser, die Bor (Fe–B, Co–B und Ni–B) enthalten, wird röntgenographisch untersucht. Die atomare Intensitätsverteilungsfunktion wird mit Hilfe einer Fourieranalyse bestimmt, und Atomabstände und Koordinationszahlen werden berechnet. Die Resultate zeigen, daß die Atomverteilung in metallischen Gläsern mit weniger als 20At% Bor nicht mit der bisher in Gläsern des $Fe_{83}P_{17}$-Typs gefundenen Verteilung übereinstimmt.

1. Introduction

The typical amorphous alloy systems are known to consist of a transition metal with about 20 to 25 at% of one or more kinds of metalloids such as B, C, P, and Si. One of the current interests in the field of metallic glasses focuses on the properties of amorphous Fe–B based alloys which show the excellent properties such as the large saturation magnetization and the Invar characteristics [1]. However, the structural data for this alloy systems are scarce, within the best knowledge of the authors. The main purpose of this work is to provide information about the structure of binary glasses containing boron (Fe–B, Co–B, and Ni–B), and to point out the distinct features exhibited in the atomic radial distribution of these alloy glasses as compared with that reported previously for other metallic glasses.

2. Experimental Procedures

The amorphous samples were prepared in the shape of ribbons (about 1.5 mm width and 0.05 mm thickness) by rapid quenching from the melt. The compositions of the samples investigated are $Fe_{84}B_{16}$, $Fe_{80}B_{20}$, $Fe_{75}B_{25}$, $Co_{83}B_{17}$, $Co_{77}B_{23}$, and $Ni_{82}B_{18}$. The basic arrangements for measurements (Mo radiation and lithium fluoride monochromator on incident beam) and operating procedures for X-ray scattering intensity and correction of the observed intensity data were almost identical to those described previously [2]. The measurements were carried out in the wave-vector (Q) range between 1.0 and 17.0 $Å^{-1}$. In this work, the observed intensity data at Q less than 1.0 $Å^{-1}$ have been smoothly extrapolated to zero at $Q = 0.0$ $Å^{-1}$, because the isothermal compressibility data of these metallic glasses are not available. The effect of the extrapolation and truncation up to $Q = 17.0$ $Å^{-1}$ is known to give no critical contribution

[1]) Toronto, M5S 1A4, Canada.

[2]) 600 Mountain Avenue, Murray Hill, New Jersey 07974, USA.

to the radial distribution function by using the Fourier transformation of (1), as shown in the previous works [3, 4]:

$$G(r) = 4\pi r[\varrho(r) - \varrho_0] = \frac{2}{\pi}\int\limits_0^\infty Q[S(Q) - 1] \sin (Qr) \, \mathrm{d}Q \, , \tag{1}$$

where $S(Q)$ is the structure factor derived from the measured intensity data, $\varrho(r)$ is the radial density function and ϱ_0 is the average number density of atoms.

3. Results and Discussion

The uncertainty in the structural information obtained in this work will now be discussed. The accumulated counts, varying from 5×10^4 at low angles to 1×10^5 at high angles, were chosen to hold the counting statistics approximately uniform. According to the usual method [5 to 7] for evaluating errors in normalization proposed by Rahman [5] and the detailed discussion for systematic errors in X-ray diffraction of non-crystalline materials by Greenfield et al. [7], the total error in the structure factor obtained in this work was estimated to be 1.9%. As a result, the uncertainty in the values of the reduced distribution function $G(r)$, which is the Fourier transform of $S(Q)$, is probably similar to that of $S(Q)$, because the procedures [8, 9] used in the Fourier transformation minimizes computational error.

The structure factor and the reduced radial distribution function for six samples measured are given in Fig. 1 to 4 and the relevant values such as density and peak positions are listed in Table 1. The density were measured by using the Archimedes method. In order to facilitate the clear indication of the characteristic features of metallic glasses with boron, the structural information of typical metal–metalloid glasses exemplified by an amorphous $Fe_{83}P_{17}$ [10] are also illustrated in the figures.

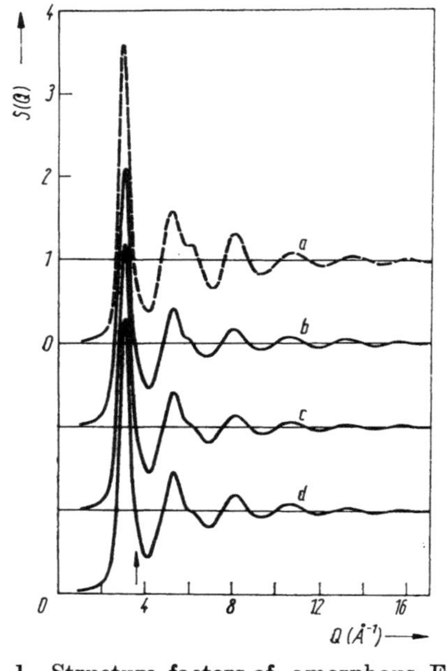

Fig. 1. Structure factors of amorphous Fe–B alloys. (a) $Fe_{83}P_{17}$, (b) $Fe_{75}B_{25}$, (c) $Fe_{80}B_{20}$, (d) $Fe_{84}B_{1}$

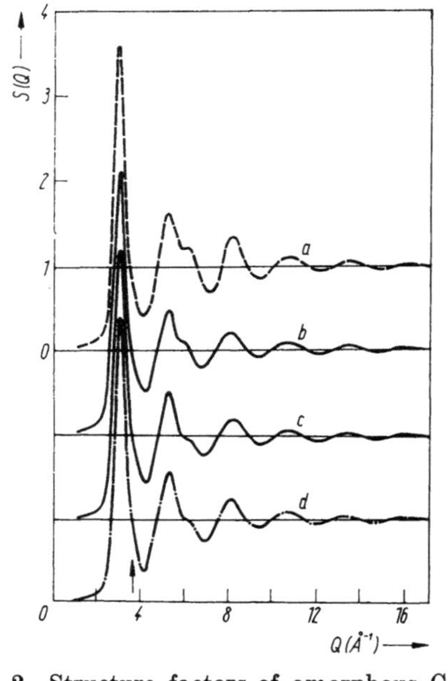

Fig. 2. Structure factors of amorphous Co–B and Ni–B alloys. (a) $Fe_{83}P_{17}$, (b) $Co_{77}B_{23}$, (c) $Co_{83}B_{176}$, (d) $Ni_{82}B_{18}$

Fig. 3. Radial distribution functions of amorphous Fe–B alloys. (a) $Fe_{83}P_{17}$, (b) $Fe_{75}B_{25}$, (c) $Fe_{80}B_{20}$, (d) $Fe_{84}B_{16}$

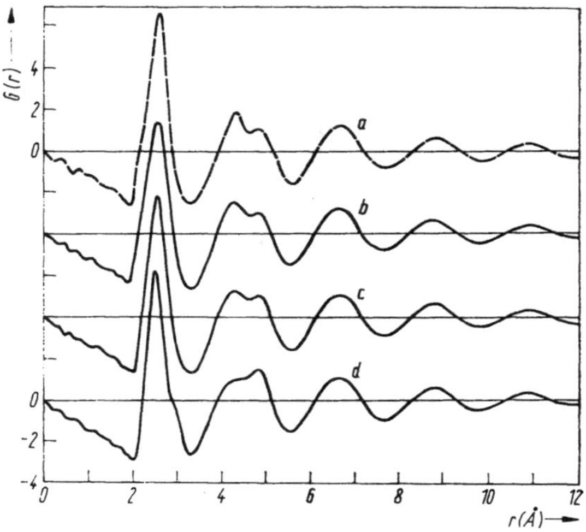

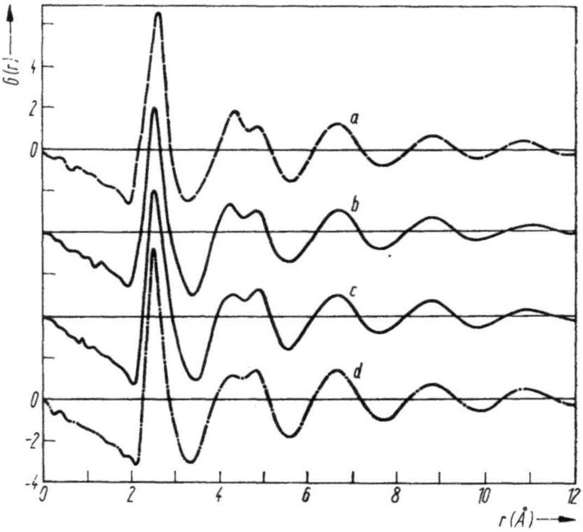

Fig. 4. Radial distribution functions of amorphous Co–B and Ni–B alloys. (a) $Fe_{83}P_{17}$, (b) $Co_{77}B_{23}$, (c) $Co_{83}B_{17}$, (d) $Ni_{82}B_{18}$

As shown in these figures, the basic profile of $S(Q)$ and $G(r)$ can be classified into the same group of other metallic glasses for metal–metalloid systems. Namely, the profile of all samples is composed of a relatively sharp first peak and subsequent smaller oscillations. We also found the second peak shoulder in $S(Q)$ and the second peak splitting in $G(r)$. Thus it appears that the general conclusion which has been obtained in various metallic glasses should be applicable to the structure of binary metallic glasses with boron. However, the following two points, emerging from the present results, are considered to be noteworthy:

1) The shoulder of the second peak in $S(Q)$ for all samples is not as intense as that of $Fe_{83}P_{17}$ whose profile is a typical one in amorphous metal–metalloid systems. With increasing boron content, the second peak shoulder becomes more pronounced, but still less than that of $Fe_{83}P_{17}$.

2) The samples containing boron less than 20 at% are found to show the subpeak height ratio of the second peak splitting in $G(r)$ reversed as compared with that of

Table 1

Density, peak positions and near-neighbour correlations in metallic glasses containing boron

	density (g/cm³)	peak positions in $S(Q)$ (Å^{-1}) first	second	shoulder	peak positions in $G(r)$ (Å) first	second	shoulder	n_1 (atoms)
$Fe_{84}B_{16}$	7.38	2.99	5.26	6.10	2.58	4.40	4.88	11.6
$Fe_{80}B_{20}$	7.31	3.01	5.26	6.10	2.57	4.35	4.93	11.9
$Fe_{75}B_{25}$	7.22	2.98	5.23	6.10	2.62	4.33	4.90	12.2
$Co_{83}B_{17}$	8.34	3.02	5.25	6.10	2.57	4.36	4.91	11.5
$Co_{77}B_{23}$	8.16	2.98	5.26	6.05	2.61	4.32	4.92	12.4
$Ni_{82}B_{18}$	8.36	3.04	5.28	6.01	2.57	4.33	4.87	11.4

$Fe_{83}P_{17}$, in which the first subpeak is more intense than the other. However, the first subpeak is enhanced with increasing boron content and the corresponding peak height ratio of the samples with boron content more than 20 at% becomes similar to that of $Fe_{83}P_{17}$.

These features imply that the atomic distribution of metallic glasses with boron, particularly with the boron content less than 20 at%, differs from that of other typical metal–metalloid glasses. It may be noted that the anomalous behaviour regarding the second peak profile in $G(r)$ for samples with low boron content resembles rather close to that of the original calculation by Bennett [11] based on the dense random packing of hard sphere model as shown in Fig. 5 using the result of $Co_{83}B_{17}$ as an example. This contrasts with the extended calculation by using the relaxed Bennett model [12] which gives the peak heights in $G(r)$ more like those shown in $Fe_{83}P_{17}$. The different peak profile for metallic glasses with boron content less than 20 at% is also reproduced by the single size hard sphere model of Percus-Yevick equation [13] being frequently applied to liquid metals. Here, we use the assumption that the observed structure factor mainly reflects the atomic distribution of Co atoms because of the contribution of B atoms to X-ray scattering intensity is considerably small. The different behaviour for metallic glasses with boron is of considerable interest and consistent with the recent model calculations by Boudreaux [14, 15]. However no definite reason on this anomalous behaviour of metallic glasses with boron could be drawn from the present experimental data.

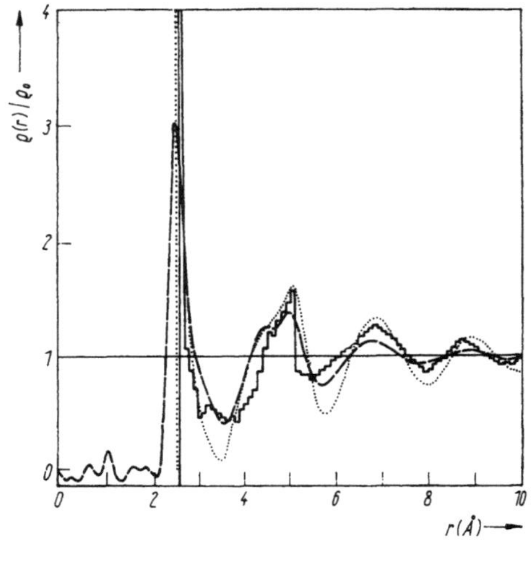

Fig. 5. A comparison of the structure of amorphous $Co_{83}B_{17}$ (dashed line) with model calculations of Bennett (histogram) (solid line) and of Hard Sphere Percus Yevick (dotted line) in the case of the packing density of 0.60. ($g(r) = \varrho(r)/\varrho_0$)

On the basis of the previous experimental data [16, 17], it is acceptable that the near neighbour correlation, which corresponds to the first peak in $G(r)$ of Fig. 3 and 4, is interpreted as the superposition of the correlations of M–M and M–B pairs, where M denotes Fe, Co, and Ni, although these correlations are not resolvable in Fig. 3 and 4. However, it was noted that the slight asymmetry of the first peak in the structure factor was found in the region of the high angle side of the first peak (see arrows in Fig. 1 and 2). This might be caused by the contribution of M–B pairs, although the estimated contribution of M–B pairs is only 7% and that of B–B pairs is negligibly small in these alloy systems. The estimated numbers of near neighbours within the distance of the first minimum on the right hand side of the principal peak of about 3.2 Å (n_1) are listed in Table 1. The experimental values are of the order of about 11.8 ± 0.5 within the samples investigated. These values which correspond to the M–M and M–B correlations for metallic glasses with boron is slightly smaller than those ($n_1 \approx 13$) obtained in metallic glasses with phosphorus [10, 17] and calculated by the dense random packing of hard sphere model [15]. The number of near neighbour correlations of n_1 is one of the useful information as a first step in the structural study of metallic glasses, but its interpretation is limited. For a detailed discussion, more information is required such as the partial structures of the respective pairs which are under investigation.

Based on the previous experimental data and model calculation for several kinds of metallic glasses [16, 17], it is probably to say that the atomic distribution of metallic glasses with boron depends mainly upon the dense random packing type structure of transition metal atoms (Fe, Co, and Ni) in which the boron atoms, of relatively small size (1.64 Å), occupy the vacant space formed by transition metal atoms (about 2.60 Å). This model leads to the relatively high density of glasses and is consistent with the measured densities for samples investigated (see Table 1) which are only 1 to 2% less than the crystalline values.

Acknowledgements

This work was carried out as a joint work during the session 1978 to 79 while Y. Waseda stayed at the Department of Metallurgy and Materials Science, University of Toronto. During this period, he wishes to acknowledge partial contractural support from the National Research Council of Canada.

References

[1] K. Fukamachi, N. Kikuchi, S. Esashi, and T. Masumoto, Solid State Commun. **23**, 955 (1977).
[2] Y. Waseda and T. Masumoto, Z. Phys. **B 21**, 235 (1975).
[3] K. Furukawa, Rep. Progr. Phys. **25**, 392 (1962).
[4] C. N. J. Wagner, H. Ocken, and M. L. Joshi, Z. Naturf. **20a**, 235 (1965).
[5] A. Rahman, J. chem. Phys. **42**, 3540 (1965).
[6] Y. Waseda and S. Tamaki, Phil. Mag. **32**, 273 (1975).
[7] A. J. Greenfield, J. Wellendorf, and N. Wiser, Phys. Rev. A **4**, 1607 (1971).
[8] R. Kaplow, S. L. Strong, and B. L. Averback, Phys. Rev. **138**, A 1336 (1965).
[9] A. H. Narten, D. M. Danford, and H. A. Levy, J. chem. Phys. **46**, 4875 (1967).
[10] Y. Waseda, H. Okazaki, and T. Masumoto, Proc. Internat. Symp. on the Structure of Non-Crystalline Materials, Cambridge Sept. 1976, Soc. of Glass Technology, London 1977 (p. 95).
[11] C. H. Bennett, J. appl. Phys. **43**, 272 (1972).
[12] L. V. Heimendahl, J. Phys. F **5**, L 141 (1975).
[13] G. J. Throop and P. J. Bearman, J. chem. Phys. **42**, 2408 (1965).

[14] D. S. BOUDREAX, Amorphous Magnetism II, Ed. R. A. LEVY and R. HASEGAWA, Plenum Press, New York 1977 (p. 463).
[15] D. S. BOUDREAX, Abstracts of AIP Topical Symp. on the Atomic Scale Structure of Amorphous Solids, Yorktown Heights (New York), April 1978 (L-8).
[16] G. S. CARGILL III, Solid State Phys. **30**, 227 (1975).
[17] Y. WASEDA, H. OKAZAKI, and T. MASUMOTO, J. Mater. Sci. **12**, 1927, (1977).

(Received May 23, 1978)

phys. stat. sol. (a) 49, 393 (1978)

Subject classification: 14.3.3 and 16; 22.2.3

Department of Electrical Engineering, Aino Shams University, Cairo

Photovoltaic Operation of Nonuniformly Doped Intrinsic Photoconductors

By

A. SOLIMAN[1]) and N. SALEH

A detailed analysis of the photovoltaic operation of nonuniformly doped intrinsic photon detectors is presented. The electric field produced by the impurity distribution assists the transport and collection of minority carriers created by photons absorbed in the detector. As a result of the analysis, a narrow band infrared detector of p-type (InSb), with exponentially distributed impurities is investigated, and the response of this detector is examined.

Eine ausführliche Analyse der Photospannung von nichtgleichförmig dotierten eigenleitenden Photodetektoren wird mitgeteilt. Das durch die Störstellenverteilung hervorgerufene Feld assistiert den Transport und die Sammlung von Minoritätsträgern, die durch im Detektor absorbierte Photonen erzeugt werden. Als Ergebnis der Analyse wird ein schmalbandiger p-leitender Infrarotdetektor (InSb) mit exponentiell verteilten Störstellen untersucht und die Response dieses Detektors bestimmt.

1. Introduction

The internal electric field resulting from an impurity gradient may be used to enhance the motion of minority carriers. In particular such aiding drift fields have been suggested for enhancement of the current collected in photovoltaic solar cells [1, 2]. The present study includes drift fields due to impurity gradients in a nonuniformly doped intrinsic photon detector and the design considerations for p-type (InSb) infrared photon detectors with exponentially distributed impurities. In the present analysis direct recombination and low level injection are considered.

2. Analysis

A nonuniformly doped semiconductor sample of width d is considered. The impurities are nonuniformly distributed in the z-direction, and the sample is uniformly illuminated over its surface as in Fig. 1. For intrinsic photoconductivity all the impurity atoms are assumed ionized, and from the neutrality equation we have

$$p(z) - n(z) = N_a(z) - N_d(z) .$$

If $N_d(z)$ and $n(z)$, the donor and electron densities are small, the hole density is given by

$$P(z) \approx N_a(z) .$$

Since the acceptor density is a function of z, the gradient of the hole density produces a current $D_p(dP(z)/dz)$ until the slight displacement of majority carriers which thus occurs, creates an electric field E_d and an opposing and equal current $\mu_p P(z) E_d$ (E_d electrostatic drift field produced by impurity gradient in V/cm, μ_p the hole mobility in cm²/Vs).

Hence in equilibrium, the majority carrier current is zero [3, 4]

$$qN_a(z) \mu_p E_d - qD_p \frac{d}{dz} N_a(z) = 0$$

[1]) Now at Department of Electrical Engineering, University of Alberta, Edmonton, Alberta, Canada T6G 2E1.

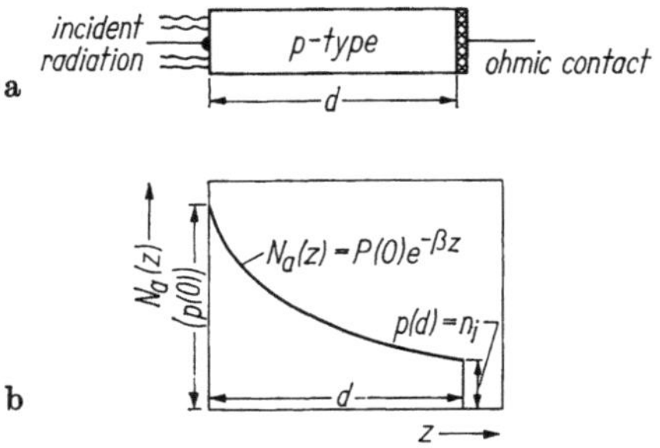

Fig. 1. a) Basic arrangement for the sample used, b) impurity distribution

and the drift field caused by the impurity gradient is given by

$$E_{\rm d} = \frac{D_{\rm p}}{\mu_{\rm p} N_{\rm a}(z)} \frac{\rm d}{{\rm d}z} N_{\rm a}(z) \tag{1}$$

which according to Einstein's relation may be written as

$$E_{\rm d} = \frac{kT}{q} \frac{1}{N_{\rm a}(z)} \frac{\rm d}{{\rm d}z} N_{\rm a}(z) \tag{2}$$

($N_{\rm a}(z)$ is the acceptor distribution in $\rm cm^{-3}$, $D_{\rm p}$ the hole diffusion coefficient in $\rm cm^2 s^{-1}$, q the electronic charge).

Equations (1) and (2) indicate that if the acceptor concentration decreases from the light exposed surface, its z change of variation is negative and the field $E_{\rm d}$ is negative pointing towards the $z = 0$ face, and helping the minority carriers (electrons) injected by light to travel in the positive z-direction towards the dark face.

2.1 Minority carriers distribution

The starting point of the derivation of the minority carrier distribution $\delta n(z)$ (density of minority carriers above the equilibrium density) is the solution of the continuity equation under steady state conditions, i.e.

$$\frac{1}{q} \nabla \cdot J - \frac{\delta n(z)}{\tau_{\rm n}} + g(z) = 0$$

($\delta n(z)$ photogenerated electrons/$\rm cm^3$, $\tau_{\rm n}$ electron lifetime in s, J total current density, $J = J_{\rm n} + J_{\rm p}$), where $g(z)$ depicts the number of minority carriers generated by electron-hole pair generation due to photon absorption in a layer of unit area and thickness dz, at a distance z from the light exposed surface

$$g(z) = \alpha Q_0 {\rm e}^{-\alpha z}\,, \tag{3}$$

where α is the absorption coefficient in $\rm cm^{-1}$ and Q_0 the number of phonons ($\rm cm^2 s^{-1}$). The one-dimensional drift plus diffusion current densities due to $\delta n(z)$ are given by

$$J_{\rm n} = q\mu_{\rm n}\, \delta n(z)\, E_{\rm d} + qD_{\rm n} \frac{\rm d}{{\rm d}z} \delta n(z) \qquad ({\rm Acm^{-2}}), \tag{4}$$

where $\mu_{\rm n}$ electron mobility, $D_{\rm n}$ electron diffusion coefficient. Substituting for $E_{\rm d}$ from (1) into (4)

$$J_{\rm n} = q\mu_{\rm n}\, \delta n(z) \frac{D_{\rm p}}{\mu_{\rm p} N_{\rm a}(z)} \frac{\rm d}{{\rm d}z} N_{\rm a}(z) + qD_{\rm n} \frac{\rm d}{{\rm d}z} \delta n(z)\,,$$

since

$$\frac{D_n}{D_p} = \frac{\mu_n}{\mu_p}; \qquad J_n = q\frac{D_n}{N_a(z)}\frac{d}{dz}N_a(z) + qD_n\frac{d}{dz}\delta n(z). \tag{5}$$

Equations (2), (3), and (5) are combined to give

$$D_n\frac{d}{dz}\left\{\frac{1}{N_a(z)}\frac{d}{dz}N_a(z) + \frac{d}{dz}\delta n(z)\right\} - \frac{\delta n(z)}{\tau_n} + \alpha Q_0\,e^{-\alpha z} = 0. \tag{6}$$

Equation (6) is a more general diffusion equation applicable for uniform as well as nonuniform distribution of impurities. Equation (6) may be put in a more simplified form when written in terms of $b'(z)$ instead of $\delta n(z)$ where $b'(z)$ is the relative excess density (Appendix A):

$$b'(z) = \frac{\delta n(z)}{n_0(z)}.$$

In the present analysis a solution for exponential distribution of acceptors in p-type semiconductors is considered, i.e.

$$N_a(z) = P(0)\,e^{-\beta z},$$

where $P(0)$ is the surface concentration of acceptors (cm^{-3}). For this case, $\frac{dN_a(z)}{dz} = -\beta N_a(z)$ and $\frac{d^2N_a(z)}{dz^2} = \beta^2 N_a(z)$. Substituting $N_a(z)$, $\frac{dN_a(z)}{dz}$ and $\frac{d^2N_a(z)}{dz^2}$ in (6) gives

$$\frac{d^2\delta n(z)}{dz^2} - \beta\frac{d}{dz}\delta n(z) - \frac{\delta n(z)}{L_n^2} + \frac{\alpha Q_0}{D_n}e^{-\alpha z} = 0; \qquad L_n = \sqrt{D_n\tau_n}. \tag{7}$$

A general solution for equation (7) is carried out in Appendix B to satisfy the following boundary conditions:

i) Since, for most materials α is high in the intrinsic absorption range, the distribution is carried out to satisfy the following three inequalities:

$$d \gg \frac{1}{\alpha}, \qquad d \gg \frac{1}{L_n}, \qquad \text{and} \qquad L_n \gg \frac{1}{\alpha}.$$

For this case $\delta n(z)$ is finite. If the contact at the dark face is perfectly ohmic, the surface recombination velocity at the dark face S_{2n} assumes a very high value and all the minority carriers are recombined [1].

ii) Surface recombination takes place at the light exposed surface with the recombination velocity at the front face S_{1n} expressed mathematically as

$$qS_{1n}\,\delta n(z)|_{z=0} = J_n|_{z=0}.$$

The solution of equation (7) is therefore (Appendix B):

$$\delta n(z) = \frac{\alpha Q_0}{D_n\left(\alpha^2 + \alpha\beta - \frac{1}{L_n^2}\right)}\left(\frac{S_{1n} + D_n(\alpha+\beta)}{S_{1n} - D_nM_1}\,e^{(\beta/2 - F_1)z} - e^{-\alpha z}\right), \tag{8}$$

where

$$M_1 = -\beta/2 - \sqrt{(\beta/2)^2 + \frac{1}{L_n^2}} \qquad \text{and} \qquad F_1 = \sqrt{(\beta/2)^2 + \frac{1}{L_n^2}}\ \ (\text{cm}^{-1}).$$

2.2 Opencircuit voltages

In order to get an expression for the potential difference between the dark and illuminated faces, the built-in total electric field E_z is first derived and then integrated over the sample width. Under open–curcuit conditions, the total electron plus hole current density components are zero, that is

$$(J_{\mathrm{ndrift}} + J_{\mathrm{pdrift}}) + (J_{\mathrm{ndiff}} + J_{\mathrm{pdiff}}) = 0 .$$

J_{ndrift}, J_{pdrift} are the electron and hole drift current densities respectively, and J_{ndiff}, J_{pdiff} are the corresponding diffusion current densities.

By substituting for the different current components we get

$$(n_0(z)\, q\mu_{\mathrm{n}} + p_0(z)\, q\mu_{\mathrm{p}})\, E_z - qD_{\mathrm{p}} \frac{\mathrm{d}}{\mathrm{d}z} p_0(z) + qD_{\mathrm{n}} \frac{\mathrm{d}}{\mathrm{d}z} n_0(z) + + qD_{\mathrm{n}} \frac{\mathrm{d}}{\mathrm{d}z} \delta n(z) - qD_{\mathrm{p}} \frac{\mathrm{d}}{\mathrm{d}z} \delta p(z) = 0 ,$$

where $n_0(z)$, $p_0(z)$ is the equilibrium concentration of electrons/holes (cm^{-3}), $\delta p(z)$ are the photogenerated holes.

For direct recombination, $\delta n(z) = \delta p(z)$ and therefore

$$E_z = \frac{D_{\mathrm{p}}}{\mu_{\mathrm{p}}[b n_0(z) + p_0(z)]} \frac{\mathrm{d}}{\mathrm{d}z} (P_0(z) - n_0(z)) + \frac{D_{\mathrm{p}}(1 - b)}{\mu_{\mathrm{p}}(b n_0(z) + p_0(z))} \frac{\mathrm{d}}{\mathrm{d}z} \delta n(z) , \tag{9}$$

where b is mobility ratio $\mu_{\mathrm{n}}/\mu_{\mathrm{p}}$. For a p-type sample $p_0(z) \gg n_0(z)$ and equation (9) is reduced to

$$E_z = \frac{D_{\mathrm{p}}}{\mu_{\mathrm{p}} p_0(z)} \frac{\mathrm{d}}{\mathrm{d}z} p_0(z) + \frac{D_{\mathrm{p}}}{\mu_{\mathrm{p}} p_0(z)} (1 - b) \frac{\mathrm{d}}{\mathrm{d}z} \delta n(z)$$

which for convenience may be written as

$$E_z = E_{\mathrm{d}} + E_{\mathrm{ph}} , \tag{10}$$

where

$$E_{\mathrm{d}} = \frac{D_{\mathrm{p}}}{\mu_{\mathrm{p}} p_0(z)} \frac{\mathrm{d}}{\mathrm{d}z} p_0(z)$$

is the same expression given by equation (1), and where

$$E_{\mathrm{ph}} = \frac{D_{\mathrm{p}}}{\mu_{\mathrm{p}} p_0(z)} (1 - b) \frac{\mathrm{d}}{\mathrm{d}z} \delta n(z) . \tag{11}$$

A point worth noting is that as a result of the analysis separation between the field due to impurity distribution E_{d} and that due to photogenerated carriers E_{ph} is possible.

For exponential distribution of impurities,

$$N_{\mathrm{a}}(z) = P(0)\, \mathrm{e}^{-\beta z} ,$$

the drift field E_{d} is therefore

$$E_{\mathrm{d}} = -\frac{D_{\mathrm{p}}}{\mu_{\mathrm{p}}} \beta = -\frac{kT}{q} \beta$$

which means the drift field E_{d} is a negative constant helping the negative charge carriers photogenerated to travel in the positive z direction. Integration of E_{d} gives

$$V_{\mathrm{d}} = + \int_0^d \frac{kT}{q} \beta = \frac{kT}{q} \beta d . \tag{12}$$

The drift voltage V_d is therefore a function of the product βd, and may be written as

$$V_d = m V_T ,$$

where

$$V_T = \frac{kT}{q}$$

is the thermal potential, and m is the drift factor which may be written as $m = V_d/V_T$ or $m = \beta d$.

The drift factor defined above, is therefore equal to the product βd and is inversely porportional to the temperature T.

In order to obtain an expression for the photogenerated voltage V_{ph}, integration of E_{ph} is carried out over the sample width d after substitution for $(d/dz)\, \delta n(z)$ from equation (8), where

$$\frac{d}{dz} \delta n(z) = \frac{Q_0 \alpha}{D_n \left(\alpha^2 + \alpha\beta - \frac{1}{L_n^2}\right)} \left(\frac{S_{1n} + D_n(\alpha + \beta)}{S_{1n} - M_1 D_n} \left(\frac{\beta}{2} - F_1\right) e^{\left(\frac{\beta}{2} - F_1\right) z} + \alpha\, e^{-\alpha z}\right).$$

Accordingly,

$$V_{ph} = - \int_0^d E_{ph}\, dz = \frac{D_p \alpha Q_0 (b - 1)}{D_n \mu_p P(0)} \int_0^d \frac{e^{\beta z}}{\left[\alpha^2 + \alpha\beta - \frac{1}{L_n^2}\right]} \times$$

$$\times \left(\frac{S_{1n} + D_n(\alpha + \beta)}{S_n - M_1 D_n} \left(\frac{\beta}{2} - F_1\right) e^{\left(\frac{\beta}{2} - F_1\right) z} + \alpha\, e^{-\alpha z}\right) dz =$$

$$= \frac{D_p \alpha Q_0 (b - 1)}{D_n \mu_p P(0) \left[\alpha^2 + \alpha\beta - \frac{1}{L_n^2}\right]} \frac{S_{1n} + D_n(\alpha + \beta)}{S_{1n} - M_1 D_n} \frac{\left(\frac{\beta}{2} - F_1\right)}{\left(\frac{3\beta}{2} - F_1\right)} \left(e^{\left(\frac{3\beta}{2} - F_1\right) d} - 1\right) +$$

$$+ \frac{D_p \alpha Q_0 (b - 1)}{D_n \mu_p P(0) \left[\alpha^2 + \alpha\beta - \frac{1}{L_n^2}\right]} \frac{\alpha}{(\beta - \alpha)} [e^{(\beta - \alpha) d} - 1] . \qquad (13)$$

Again equation (13) is a general solution valid for exponential distribution of impurities and for uniform distribution where β is set equal to zero.

Because the surface recombination velocity at the front face is a function of the surface conditions, S_{1n} may be controlled and we have to lower this velocity, otherwise a strong surface recombination for photogenerated carriers takes place. The photogenerated voltage for very small values of S_{1n} is reduced to

$$V_{ph} = \frac{D_p (b - 1)\, \alpha Q_0}{\mu_p P(0)\, D_n \left(\alpha^2 + \alpha\beta - \frac{1}{L_n^2}\right)} \frac{\alpha + \beta}{M_1} \frac{\left(\frac{\beta}{2} - F_1\right)}{\left(\frac{3\beta}{2} - F_1\right)} \left[e^{\left(\frac{3\beta}{2} - F_1\right) d} - 1\right] +$$

$$+ \frac{D_p (b - 1)\, \alpha Q_0}{\mu_p P(0)\, D_n \left(\alpha^2 + \alpha\beta - \frac{1}{L_n^2}\right)} \frac{\alpha}{(\beta - \alpha)} [e^{(\beta - \alpha) d} - 1] . \qquad (14)$$

Again equation (14) is a general expression for the photogenerated voltage. If β is set equal to zero, comparison between the photogenerated voltage for uniform distribution of impurities, with that of exponentially distributed impurities is possible.

3. Discussion

The impurity distribution function $N_a(z)$ may be written in terms of the drift factor $m = \beta d$ as

$$N_a(z) = P(0)\, e^{-\left(\frac{m}{d}\right) z}$$

or

$$m = \ln \frac{P(0)}{P(d)} \tag{15}$$

where $P(d)$ is the acceptor concentration at the dark face (cm^{-3}).

According to equation (15) the drift factor m and accordingly the drift field E_d can be increased by either an increase of $P(0)$, a decrease of $P(d)$, or both. Actually the increase of m is limited for the following reasons [5]:

i) The maximum impurity concentration is limited to the solid solubility of impurities in the semiconductor,

ii) the operating temperature of the detector,

iii) the minimum concentration at the dark face should be n_i.

iv) The increase of impurities to very high dopings may alter the main absorption edge of the sample.

v) Because the impurity gradients for high dopings may cause the optical energy gap to be a function of z, the problem should be treated in a completely different manner with all the different parameters being functions of z [3].

According to the previous limitations the maximum drift factor m is obtained when $P(d) = n_i$:

$$m_{\max} = \ln \frac{P(0)}{n_i}.$$

The surface concentration $P(0)$ is given by [6]

$$P(0) = n_i\, e^{(E_{F_i} - E_F)/kT}$$

and because the extrinsic Fermi level E_F shifts with doping, the optical energy gap also shifts, and E_F is bound to the edge of the valence band while the intrinsic Fermi level E_{F_i} may be taken in the middle of the energy gap

$$E_F - E_{F_i} \approx \frac{E_g}{2},$$

where E_g is the optical energy gap,

$$\frac{P(0)}{n_i} = e^{(E_{F_i} - E_F)/kT},$$

accordingly,

$$\ln \frac{P(0)}{n_i} = \frac{E_{F_i} - E_F}{kT},$$

giving

$$m = \frac{E_{F_i} - E_F}{kT} \approx \frac{E_g}{2kT}. \tag{16}$$

Application of the above analysis to InSb for which the optical energy gap lies in the infrared range of the spectrum, gives according to (16) a maximum drift factor

$m_{\max} \approx 4$ at room temperature. Since the drift factor m and accordingly β are design parameters subject to the above-mentioned constraints, for $\beta \gg 1/L_n$, the photogenerated voltage given by equation (14) is reduced to

$$U_{\text{ph}} = C \frac{\alpha}{\beta^2 - \alpha^2} (e^{(\beta-\alpha)d} - 1) , \tag{17}$$

where

$$C = \frac{D_p(b - 1) Q_0}{\mu_p P(0) D_n} .$$

Because the absorption coefficient α for InSb varies over a wide range of frequency, three operational ranges are of interest:

i) The frequency at which the numerical values of α and β are equal. At this frequency

$$V_{\text{ph}} = C \frac{d}{2};$$

which is proportional to the sample width d.

ii) The high-frequency range for which $\alpha \gg \beta$. In this range

$$V_{\text{ph}} = \frac{C}{\alpha} . \tag{18}$$

That is, the photogenerated voltage in the high-frequency range is inversely proportional to α and accordingly to the frequency.

iii) The low-frequency range

$$\alpha \ll \beta$$

$$V_{\text{ph}} = \frac{C\alpha}{\beta^2} e^{\beta d} , \tag{19}$$

where it is clear that the detector response drops to a very small value in a very narrow band of frequency around the main absorption edge.

According to equation (17) the photogenerated voltage U_{ph} is proportional to

$$\frac{\alpha}{\beta^2 - \alpha^2} [(e^{(\beta-\alpha)d} - 1)] \tag{20}$$

with C (V/cm) as a constant of proportionality. Computations for equation (20) were carried out for InSb subject to the constraint $\beta d = 4$. The absorption coefficient α for InSb was taken to vary over a wider range of frequency from small values up to 10^4 cm^{-1}.

Computations were carried out for two cases:

i) The effect of variation of the grading factor β, for a fixed sample width ($d = 0.04$ cm) on the detector performance is considered. The results of computations are plotted in Fig. 2. As shown in Fig. 2 the detector response increases with the increase of β and the band width decreases. The response has a peak at a frequency very close to the main absorption edge. In the high-frequency range the response is seen to be independent of β (equation (18)).

ii) The effect of variation of the sample width d for a fixed value of $\beta = 80$ cm^{-1} is considered and plotted in Fig. 3.

Comparison of Fig. 2 and 3 for the same values of βd shows that the detector performance is more affected by the variation of d than by the variation of β of the same product βd.

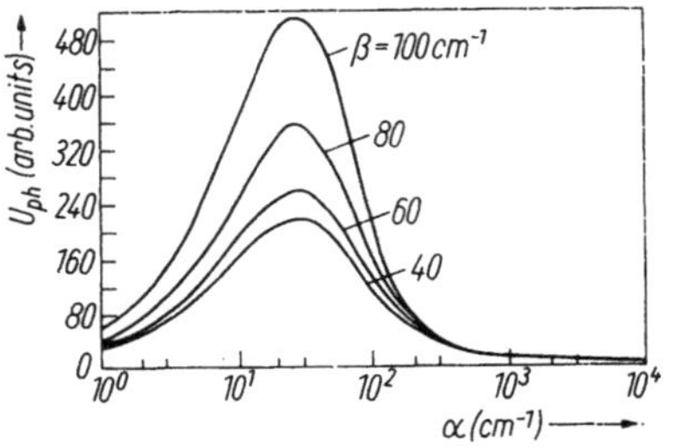

Fig. 2. Relative response of the InSb detector for different values of β, $d = 0.04$ cm

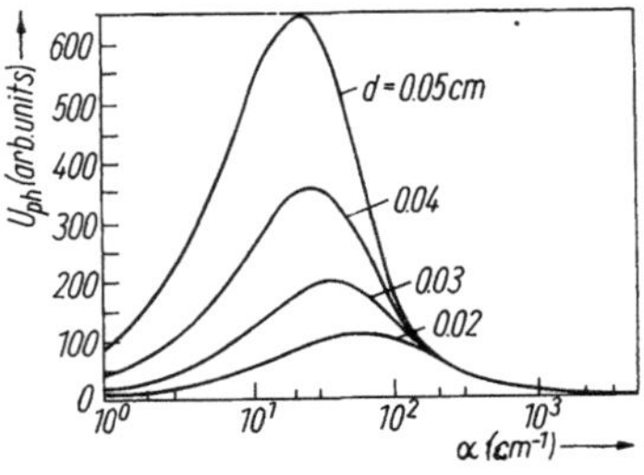

Fig. 3. Relative response of the InSb detector for different values of d, $\beta = 80$ cm^{-1}

4. Conclusion

A detailed analysis for nonuniformly doped photon detectors was performed and applied to InSb for the case of an exponential distribution. The response of the suggested photovoltaic infrared detector is quite different from the uniformly doped one, where the first shows a narrow band response. Experimentation is needed to verify the above theory and its applicability to other semiconductors. The effect of the impurity gradient for homo- and heterojunction photon detectors; the effect of the impurity gradient for degenerate (n-type InSb); the variation of the different parameters with variation of the optical energy gaps; and the effect of application of intense magnetic field on the performance, are all important points for further investigations.

Appendix A

The derivation of the continuity equation for the case of nonuniform distribution of impurities

The steady state continuity equation for minority carriers $\delta n(z)$ is given by

$$\frac{1}{q} \nabla \cdot J_\text{n} - \frac{\delta n(z)}{\tau_\text{n}} + \alpha Q_0 \, \text{e}^{-\alpha z} = 0 \,, \tag{A1}$$

defining $b'(z)$ as the relative excess density

$$b'(z) = \frac{\delta n(z)}{n_0(z)} = \frac{\delta n(z) \, N_\text{a}(z)}{n_0(z) \, N_\text{a}(z)} = \frac{\delta n(z) \, N_\text{a}(z)}{n_\text{i}^2},$$

this gives

$$\delta n(z) = \frac{b'(z) \, n_\text{i}^2}{N_\text{a}(z)},$$

n_i being the intrinsic carrier density.

The minority carrier current is given by

$$J_\text{n} = q\mu_\text{n} \, \delta n(z) \, E_\text{d} + qD_\text{n} \frac{\text{d}}{\text{d}z} \delta n(z) \,. \tag{A2}$$

Substituting for E_d from equation (1) and $\delta n(z)$ from equation (A1) into equation (A2) gives

$$J_n = q\mu_n \left(\frac{D_p}{\mu_p N_a(z)} \frac{d}{dz} N_a(z)\right) \frac{b'(z)\, n_i^2}{N_a(z)} + qD_n \frac{d}{dz} \frac{n_i^2 b'(z)}{N_a(z)} =$$

$$= q \frac{\mu_n}{\mu_p} D_p \frac{n_i^2 b'(z)}{N_a^2(z)} \frac{d}{dz} N_a(z) + qD_n n_i^2 \frac{db'(z)}{dz} - q \frac{D_n n_i^2 b'(z)}{N_a^2(z)} \frac{d}{dz} N_a(z) =$$

$$= q \left[\frac{\mu_n}{\mu_p} D_p - qD_n\right] n_i^2 b'(z) \frac{d}{dz} N_a(z) + qD_n \frac{n_i^2}{N_a(z)} \frac{d}{dz} b'(z)\,.$$

According to Einstein's relation, the quantity

$$\frac{\mu_n}{\mu_p} D_p - D_n = 0$$

and J_n is given by

$$J_n = qD_n \frac{n_i^2}{N_a(z)} \frac{d}{dz} b'(z)\,. \tag{A3}$$

Substituting from (A3) into (A1) the following diffusion equation is obtained:

$$\frac{d}{dz}\left(\frac{n_i^2}{N_a(z)} \frac{d}{dz} b'(z)\right) - \frac{b'(z)\, n_i^2}{N_a(z)\, L_n^2} + \frac{\alpha Q_0}{D_n} e^{-\alpha z} = 0\,. \tag{A4}$$

Equation (A4) is a more general diffusion equation, and is applied to uniform distribution when $N_a(z)$ is constant.

Appendix B

Minority carrier distribution for exponential distribution of impurities

In this case

$$N_a(z) = P(0)\, e^{-\beta z}\,.$$

Substituting into (A4) we have

$$\frac{d}{dz}\left(\frac{n_i^2}{P(0)\, e^{-\beta z}} \frac{d}{dz} b'(z)\right) - \frac{n_i^2 b'(z)}{P(0)\, e^{-\beta z}\, D_n} + \frac{\alpha Q_0\, e^{-\alpha z}}{D_n} = 0\,,$$

$$\frac{n_i^2}{P(0)} e^{+\beta z} \frac{d^2}{dz^2} b'(z) + \beta \frac{n_i^2}{P(0)} e^{\beta z} \frac{d}{dz} b'(z) - \frac{n_i^2 b'(z)}{L_n^2 P(0)} \cdot e^{+\beta z} + \frac{\alpha Q_0}{D_n} e^{-\alpha z} = 0$$

which may be written as

$$\frac{d^2 b'(z)}{dz^2} + \beta \frac{db'(z)}{dz} - \frac{b'(z)}{L_n^2} = -Y_1\, e^{-\gamma z}\,. \tag{B1}$$

Solution of (B1) is given by

$$b'(z) = C_1\, e^{M_1 z} + C_2\, e^{M_2 z} - \frac{Y_1\, e^{-\gamma z}}{\gamma^2 - \gamma\beta - \dfrac{1}{L_n^2}}\,, \tag{B2}$$

where

$$Y_1 = \frac{\alpha Q_0 P(0)}{D_n n_i^2} \quad (\text{cm}^{-2}) \tag{B3}$$

and

$$\gamma = (\alpha + \beta)\ (\text{cm}^{-1})\,,$$

$$M_1 = -\frac{\beta}{2} - \sqrt{\left(\frac{\beta}{2}\right)^2 + \frac{1}{L_n^2}} = -\frac{\beta}{2} - F_1\,,$$

$$M_2 = -\frac{\beta}{2} + F_1 \quad (\mathrm{cm}^{-1})\,,$$

$$F_1 = \sqrt{\left(\frac{\beta}{2}\right)^2 + \frac{1}{L_n^2}} \quad (\mathrm{cm}^{-1})\,.$$

Substituting for

$$\delta n(z) = \frac{n_i^2 b'(z)}{N_a(z)} = \frac{n_i^2 b'(z)}{P(0)\,\mathrm{e}^{-\beta z}}\,,$$

$$\delta n(z) = \frac{n_i^2}{P(0)}\left\{C_1\,\mathrm{e}^{\left(\frac{\beta}{2} - F_1\right)z} + C_2\,\mathrm{e}^{\left(\frac{\beta}{2} + F_1\right)z} - Y_2\,\mathrm{e}^{-\alpha z}\right\},$$

where

$$Y_2 = \frac{Y_1}{\left(\alpha^2 + \alpha\beta - \frac{1}{L_n^2}\right)}\,.$$

C_1 and C_2 are constants to be obtained from the following boundary conditions:
i) $\delta n(z)$ is finite. This gives $C_2 = 0$ and

$$\delta n(z) = \frac{n_i^2}{P(0)}\left\{C_1\,\mathrm{e}^{-\left(\frac{\beta}{2} + F_1\right)z} - Y_2\,\mathrm{e}^{-\alpha z}\right\}. \tag{B4}$$

ii) Surface recombination takes place at the light exposed surface with a surface recombination velocity S_{1n} which is a function of the surface conditions, mathematically this is expressed as

$$qS_{1n}\,\delta n(z)|_{z=0} = J_n|_{z=0}\,,$$

i.e.,

$$qS_{1n}\,\frac{n_i^2}{P(0)\,\mathrm{e}^{-\beta z}}\,b'(z)|_{z=0} = qD_n n_i^2 \frac{\mathrm{d}}{\mathrm{d}z}\,b'(z)|_{z=0}\,. \tag{B5}$$

From equation (B2) with $C_2 = 0$ and $\gamma = (\alpha + \beta)$,

$$b'(0) = C_1 - \frac{\alpha Q_0 P(0)}{D_n n_i^2\left(\alpha^2 + \alpha\beta - \frac{1}{L_n^2}\right)}\,, \tag{B6}$$

$$\left.\frac{\mathrm{d}b'(z)}{\mathrm{d}z}\right|_{z=0} = C_1 M_1 + \frac{(\alpha + \beta)}{\alpha^2 + \alpha\beta - \frac{1}{L_n^2}}\,. \tag{B7}$$

Substituting for $b'(0)$ and $(\mathrm{d}/\mathrm{d}z)\, b'(z)|_{z=0}$ from (B6) and (B7) into (B5), we get

$$\frac{n_i^2}{P(0)}\,[S_{1n} - D_n M_1]\,C_1 = \frac{n_i^2}{P(0)}\,Y_2[D_n(\alpha + \beta) + S_{1n}]$$

giving

$$C_1 = Y_2\left[\frac{D_n(\alpha + \beta) + S_{1n}}{S_{1n} - M_1 D_n}\right].$$

Substituting for C_1 into (B4) we get

$$\partial n(z) = \frac{n_i^2}{P(0)\,e^{-\beta z}} Y_2 \left[\frac{S_{1n} + D_n(\alpha + \beta)}{S_{1n} - M_1 D_n} e^{-\left(\frac{\beta}{2} + F_1\right) z} - e^{-(\alpha+\beta) z} \right]$$

which after substitution for Y_1 and Y_2 gives

$$\partial n(z) = \frac{\alpha Q_0}{D_n \left(\alpha^2 + \alpha\beta - \frac{1}{L_n^2}\right)} \left(\frac{S_{1n} + D_n(\alpha + \beta)}{S_{1n} - M_1 D_n} e^{\left(\frac{\beta}{2} - F_1\right) z} - e^{-\alpha z} \right). \tag{B8}$$

Acknowledgements

The authors would like to show their appreciation to Prof. Dr. Saad El Din Youssef for his helpful suggestions and encouragement and to G. French for her typing of the manuscript.

References

[1] A. G. Jordan and A. G. N. Millnes, IRE Trans. ED 7, 242 (1960).
[2] M. Wolf, Proc. IEEE 5, 674 (1963).
[3] W. Murray Bullis and W. R. Runyan, IEEE Trans. ED 14, 75 (1967).
[4] F. J. Blatt, Physics of Electronic Conduction in Solids, McGraw-Hill, Publ. Co. New York 1968.
[5] A. Ambroziak, Semiconductor Photoelectric Devices, Gordon & Breach, New York 1968.
[6] M. S. Sze, Physics of Semiconductor Devices, Academic Press, London 1969.

(Received May 12, 1978)

phys. stat. sol. (a) 49, 405 (1978)

Subject classification: 14.3.3; 11; 13.3; 22.1.2

Institute of Semiconductors, Academy of Sciences of the Ukrainian SSR, Kiev[1])

Effect of Laser Beam Irradiation on the Electrophysical Properties of the Interface in Boron Ion Implanted MIS Structures

By

V. S. LYSENKO and A. N. NAZAROV

The spectrum of traps formed in the thin silicium surface layer of Si–SiO_2 and Si–SiO_2–phosphosilicate glass structures after the high-power laser beam irradiation, boron ion implantation, and the combined effect of these two factors is studied by the methods of $C(U)$ characteristics and thermally stimulated charge release. Both the laser beam irradiation and the B^+ ion implantation through the dielectric layers are shown to cause a formation of traps with close parameters at the interface of MIS structures. The mechanism of trap formation at the treatments used is proposed.

Методами вольт-фарадных характеристик и термостимулированного освобождения заряда проведены исследования спектров ловушек, возникающих в тонких приповерхностных слоях Si структур Si–SiO_2 и Si–SiO_2-фосфоросиликатное стекло после мощного лазерного облучения, имплантации ионами B^+ средных энергий и совместного воздействия указанных факторов. Показано, что лазерное облучение и имплантация бора через слой диэлектрика формирует на границе разделе МДП-структур близкие по параметрам центры захвата. Высказаны предположения о механизме образования уровней при используемых обработках.

1. Introduction

As it has been shown previously [1 to 3], in Si–SiO_2 structures after boron ion implantation through the oxide layer, a trap spectrum appears on the interface, the nature of these traps being connected with the radiation damage.

The method of high-temperature annealing used to remove the radiation damage and to activate the implanted impurity does not lead to the full restoration of the electrophysical characteristics of the Si–SiO_2 interface to their initial values even at small implantation doses [4]. Recently a particular interest has been created in the investigations of the annealing processes caused by laser beam irradiation. It has been shown that this method may be effectively used for lattice restoration in thin surface layers of silicium previously implanted with high implantation doses [5].

In this work the effect of high-power laser beam irradiation on the properties of the SiO_2–Si interface is investigated and the possibilities of using such an irradiation for removing the radiation damage caused by the process of boron implantation at small implantation doses are discussed.

2. Experimental

n-Si substrates doped with phosphorus (10^{15} cm^{-3}) and cut along the crystallographic plane (111) were used. The samples were oxidized in dry oxygen atmosphere at 1150 °C; the oxide width was about 1500 Å. After oxidizing the part of samples was stabilized with a phosphorsilicate glass (PSG). Then B^+ ions of energy 40 keV and implantation dose 1.25×10^{12} ion/cm^2 were implanted. During the implantation process a part of each sample was covered with a screen for control measurements. As calculations showed [6, 7], for the indicated implantation energy the projection of the B^+ ion path

[1]) Prospekt Nauki 115, Kiev-28, USSR.

in SiO_2 was about 1800 Å; thus, the distribution maximum of the implanted impurity lay close to the SiO_2–Si interface.

The 0.5 cm^2 areas of the implanted and control samples were then irradiated by the non-focused beam of a Q-switched ruby laser with pulse duration of 50 ns. The intensity of the incident radiation varied from 0.9×10^7 to 1.8×10^7 W cm^{-2}. Aluminium contacts 0.3 mm^2 in area were evaporated both on the irradiated and non-irradiated surfaces of the samples. The change of the electrophysical properties of the Si–SiO_2 interface was controlled by the methods of high-frequency $C(U)$ characteristics and of thermally stimulated charge release (TSCR) [8]. The latter measurements were carried out in the temperature range from 80 to 300 K with constant heating rate of 0.2 K s^{-1}.

3. Experimental Results and Discussion

3.1 *C(U) characteristics*

It is found that laser irradiation affects the high-frequency $C(U)$ dependence of the non-implanted samples only if the laser power exceeds a "threshold" value of about 1.5×10^7 W cm^{-2}. The effects of such irradiation consist in the following: (1) the $C(U)$ curves shift to more positive voltages (from 1 to 2 V) at the capacitance level corresponding to the "flat"-band potential; (ii) the slope of the $C(U)$ dependence increases (Fig. 1).

In the case of boron implanted samples the effect of laser irradiation is similar but more considerable (Fig. 1).

As a comparison with our previous results [3] shows, the change of the $C(U)$ dependence indicates that the laser beam treatment results in the formation of new traps on the Si–SiO_2 interface and in the appearance of deep levels in the space charge region, which compensate the charge of the ionized donors in the depletion region of silicon. Besides the shape of $C(U)$ curves denotes that the laser beam treatment does not result in the annealing of radiation damage but, on the contrary, such a treatment leads to the increase of the trap density.

To determine the parameters of centres induced by the laser beam treatment and ion implantation, the method of TSCR has been used.

3.2 *TSCR spectra*

In the TSCR experiments the traps on the Si–SiO_2 interface are occuped in the temperature and electric field ranges at which the observed trap spectrum is "saturated". This takes place at potentials on the metal electrode which correspond to the formation of the accumulation layer in the surface region of the semiconductor. Fig. 2 shows the typical TSCR spectra obtained on PSG–SiO_2–Si structures after different treatments. Similar spectra are obtained on SiO_2–Si structures. As seen, both the laser beam treatment and ion implantation result in the formation of a trap spectrum which manifests itself in the same temperature region. But the laser irradiation of the implanted struc-

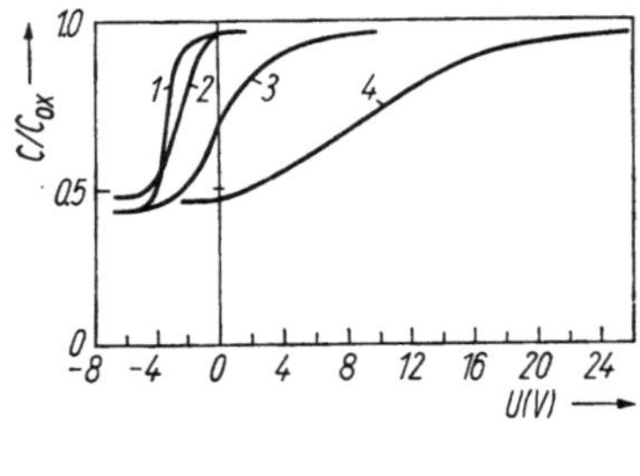

Fig. 1. $C(U)$ characteristics of Si–SiO_2–PSG structures after different treatments: (1) the non-treated sample; (2) laser beam irradiation ($I_0 = 1.5 \times 10^7$ W cm^{-2}); (3) boron ion implantation ($E = 40$ keV, $D = 1.25 \times 10^{12}$ ion/cm^2); (4) laser beam irradiation and boron ion implantation

tures leads to the expansion of the trap spectrum to higher temperatures. In these structures the largest amplitudes of the current peaks in the TSCR spectra are observed. Note that the relation between the peak amplitudes depends on the type of dielectric layer (SiO_2 or PSG–SiO_2).

To evaluate the parameters of centres responsible for TSCR spectra observed we used the methods of thermal [9], fractional [10], and field [11] cleaning of levels. The method of field cleaning consists in the successive decrease of the voltage applied to the field electrode at elevated temperature. It results in different occupations of traps with varying depth located on the dielectric–semiconductor interface. The method of the field cleaning is effective in the case of surface traps because it makes it possible to determine the trap distribution more exactly as compared with the thermal cleaning (Fig. 3). The trap level positions evaluated from the slope of the initial rise by the above-listed methods prove to be in a good agreement. Besides, it is found by the field cleaning that at zero potential on the metallic electrode only the $E_c - 0.40$ eV level is present in the TSCR spectra of boron implanted structures. This fact points to the appreciable depletion of the surface region after the implantation process. Note that the use of the field cleaning method does not lead to a separation of peaks which indicates that the traps distributed in the silicium volume are rather deep.

The above-listed methods of cleaning of TSCR spectra in boron implanted samples show the occurrence of four levels with close lying energies. These latter have been determined from the slope of the initial rise and prove to be $E_c - 0.11 \pm 0.02$, $E_c - 0.16 \pm 0.02$, $E_c - 0.21 \pm 0.03$, and $E_c - 0.40 \pm 0.03$ eV. Using these experimental values, a theoretical calculation of TSCR spectra has been made for the case of mono-energetic traps (i.e., traps occupy a single energy level) when the lifetime of the majority carriers in the conduction band is sufficiently short; $dn_c/dt \ll n_c/\tau$. Here τ and n_c are the lifetime and the density of the free electrons in the conduction band, respectively. Then, for random retrapping and random initial occupation of

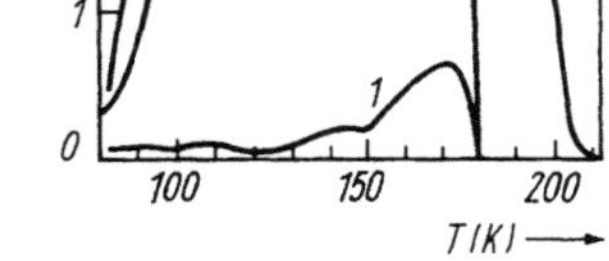

Fig. 2

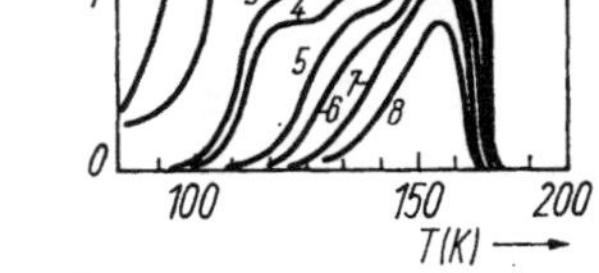

Fig. 3

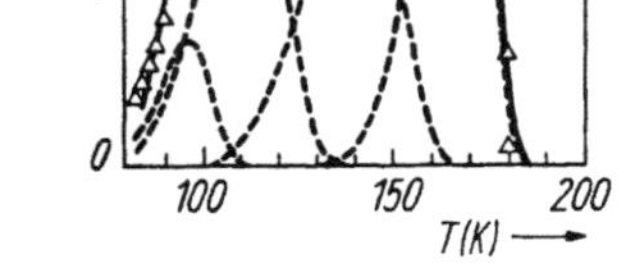

Fig. 4

Fig. 2. The TSCR current measured on Si–SiO_2–PSG structures after different treatments: (1) laser beam irradiation and boron ion implantation; (2) boron ion implantation; (3) laser beam irradiation

Fig. 3. Field cleaning of the trap spectrum: (1) $U = 15.5$; (2) 10; (3) 8; (4) 6; (5) 5; (6) 4; (7) 3; (8) 2 V

Fig. 4. Experimentally obtained (triangles) and calculated (dashed curve) TSCR spectrum of boron implanted structures. The solid line shows the resulting theoretical curve

Table 1

Comparison of the parameters of traps after the different treatments

$E_c - E_t$ (eV)	0.11			0.16			0.22			0.40			0.51		
treatment of $Si–SiO_2$–PSG structure	S_t (cm^2)	N_t (cm^{-2})	α	S_t (cm^2)	N_t (cm^{-2})	α	S_t (cm^2)	N_t (cm^{-2})	α	S_t (cm^2)	N_t (cm^{-2})	α	S_t (cm^2)	N_t (cm^{-2})	α
laser treatment ($I_0 = 1.5 \times 10^7$ W/cm^2)	10^{-22}	1.7×10^9	0	9×10^{-21}	4.3×10^9	10	4×10^{-20}	7.5×10^9	10	4×10^{-16}	1.9×10^{10}	4	–	–	–
boron implantation ($E = 40$ keV, $D = 1.25 \times 10^{12}$ cm^{-2})	10^{-22}	1.2×10^{10}	0	9×10^{-21}	5.4×10^{10}	10	4×10^{-20}	5.8×10^{10}	10	4×10^{-16}	7.5×10^{10}	4	–	–	–
boron implantation and laser treatment	10^{-22}	2×10^{10}	0	9×10^{-21}	5.4×10^{10}	10	4×10^{-20}	7.6×10^{10}	10	5×10^{-16}	1.1×10^{11}	10	10^{-14}	3×10^{10}	0

levels, the differential equation [12]

$$\frac{dn'_t}{dT} = -\frac{\nu}{\beta}\frac{n'_t \exp(-E_t/kT)}{1 + \alpha(1 - n'_t)}$$

has been solved by the Runga-Kutta method. Here $n'_t = n_t/N_t$; n_t is the electron density at the trap level, N_t the density of traps, β the heating rate of the sample, ν the frequency factor of the traps ($\nu = S_t N_c v$), S_t the cross-section of electron capture by traps, N_c the effective density of states at the conduction band edge, $\alpha = \tau S_t N_t v$ the factor of carrier retrapping.[2])

Using the parameters of levels listed in Table 1, a sufficiently good agreement between theoretical evaluations and experimental results on TSCR spectra measured after different treatments is obtained. As seen, the energy positions of levels are similar in the cases of TSCR spectra obtained both on boron implanted and laser beam treated structures. But, after a laser irradiation of implanted samples the density of shallow traps is increased and a new deep level at $E_c - 0.51 \pm 0.03$ eV appears. The presence of the $E_c - 0.51$ eV level in the TSCR spectrum causes the increase of retrapping by the $E_c - 0.40$ eV level. Fig. 4 shows the calculated and experimental TSCR spectra obtained on $Si-SiO_2$ structures after B^+ ion implantation.

3.3 The nature of the observed levels

As it has been shown above, B^+ ion implantation and laser beam treatment lead to the formation of a system of electrically active centres with the same parameters on the SiO_2-Si and $PSG-SiO_2-Si$ interfaces.

In the case of B^+ ion implantation the radiation damage formed is connected with the generation of point defects, partially vacancies, in the region of elastic collision of boron ions with the atoms of the host lattice and their subsequent interaction with different impurities, oxygen and other vacancies. Because the dielectric-semiconductor interface is a natural source of point defects (large surface strain, high electric fields), the process of formation of stable isolated defects is most active just in the surface region.

Hence, we may suppose that the creation of non-equilibrium point defects in the surface region of Si leads to formation of the same energy spectrum of levels at the interface as in the case of ion implantation. The formation of an excess of point defects is likely to take place at irradiation of the structures by a ruby laser beam. Indeed, our calculations according to [13] show that at a laser pulse power of about 1.5×10^7 W cm^{-2} the temperature in the surface layer of Si of 2.0 to 2.5 μm thickness reaches 1000 K.[3]) Under these conditions point defects are generated due to the thermal fluctuations. Then, when the sample is quickly cooled (quenching), such defects are non-equilibrium ones and may be involved in the formation of different complex defects. It must be noted that an excitation of the non-equilibrium electron-hole pairs under laser irradiation of Si may promote an appreciable increase of the density of point defects [14]. Besides, the plastic strain may serve as an additional source of point defects; such a strain is connected with the thermal stress appearing at rapid cooling of the surface region of Si after laser pulse irradiation.

But, in the case of laser beam treatment the point defects are formed to a depth of several microns, in distinction to the ion implantation process when, at a chosen im-

---

[2]) The heterogeneity of the surface parameters has not been taken into account; this problem requires an additional investigation.

[3]) An absorption in the SiO_2 layer has not been considered in this calculation; the reflectivity of Si has been taken as 0.3 [15].

plantation energy, the point defects are generated mainly on the interface. As a result, the complexes of defects are likely to be formed after the laser pulse irradiation in our experiments; the penetration depth of such complexes is greater than that in the case of ion implantation. The significant temperature gradient in the Si volume arising during laser irradiation promotes, in addition, the spatial distribution of the defects.

The increase of the trap density in the case of laser beam treatment of the implanted structures possibly is due to the fact that, because of the thermal fluctuations, the formation of interstitial atoms and vacancies proceeds more easily near complex defects [14].

As a comparison shows, the energy positions of traps obtained in the present work are close to those observed in Si at different particle irradiations. In particular, in the case of electron irradiation [16, 17] the levels at $E_c - 0.16$ and $E_c - 0.40$ eV have been detected, which are associated with vacancy–oxygen (A-centre) and vacancy–phosphorus (E-centre) complexes. The centres at $E_s - 0.11$ and $E_c - 0.21$ eV have also been observed in [18, 19].

The levels at $E_c - 0.16$ and $E_c - 0.40$ eV observed by us are likely to be associated with defects similar to the above-mentioned A- and E-centres. Indeed, in the course of high-temperature oxidation of Si the oxygen contamination at the interface and in the surface region is increased and it may lead to the formation of vacancy–oxygen complexes under the treatment used. In turn, the interaction between vacancies and the phosphorus impurity in the surface region of Si is responsible for the $E_c - 0.40$ eV level. This conclusion is supported by the fact that the density of levels at $E_c - 0.40$ eV in laser beam treated PSG–SiO_2–Si structures is higher than in SiO_2–Si structures. This may be explained by the higher phosphorus density at the SiO_2–Si interface.

The level at $E_c - 0.21$ eV may originate from the oxygen–bivacancy complexes, similar to the centres in bulk samples [20].

As to the levels at $E_c - 0.11$ and $E_c - 0.51$ eV, it is difficult to make now a conclusion about their nature. Note that the level at $E_c - 0.54$ eV has been observed, previous to this work, on SiO_2–Si structures implanted with gold [19], silicon [21], and high-energy boron [22] ions. Because of this, it may be supposed that this level is associated with intrinsic defects.

4. Conclusions

1. The annealing caused by the laser beam irradiation does not result in an appreciable activation of the implanted boron impurity.
2. The high-power laser beam irradiation leads to the formation of a trap spectrum due to radiation damage on the SiO_2–Si interface and in the surface region of silicon, which spectrum is similar to that formed at $^{11}B^+$ ion bombardment.
3. A mechanism for the formation of such traps is proposed.

Acknowledgements

The authors are greatly indebted to M. M. Lokshin for providing the samples and S. B. Kashieva for assistance with the experiments.

References

[1] T. TOKUYAMA, I. YOSHIDA, and T. WARABISAKO, Ion implantation in semiconductors and other materials, Ed. B. GROUDER, New York 1973 (p. 159).
[2] A. P. DUBCHAK, V. S. LYSENKO, V. T. ATAMANENKO, M. M. LOKSHIN, and YU. YA. POLISHUK, Ukr. fiz. Zh. **21**, 1197 (1976).
[3] V. S. LYSENKO, A. V. SACHENKO, A. P. DUBCHAK, A. N. NAZAROV, V. T. ATAMANENKO, M. M. LOKSHIN, and YU. YA. POLISHUK, phys. stat. sol. (a) **38**, 131 (1976).

[4] D. H. Lee and J. W. Mayer, Proc. IEEE 62, 70 (1974).
[5] Kazkaz Abdul-Ghaffar, These doct. phys. Unit. Sci. et mét, Grenoble 1973 (p. 91).
[6] J. W. Mayer, L. Eriksson, and J. A. Davis, Ion implantation in semiconductors, Academic Press, New York-London 1970.
[7] A. F. Akkerman and S. A. Akkerman, Fiz. Tekh. Poluprov. 8, 975 (1974).
[8] A. F. Saunders and G. T. Wright, Electronics Letters 6, 207 (1970).
[9] K. H. Nicholas and J. Woods, Brit. J. appl. Phys. 15, 783 (1964).
[10] H. Gobrecht and D. J. Hofmann, Phys. Chem. Solids 27, 509 (1966).
[11] V. S. Lysenko, A. N. Nazarov, M. M. Lokshin, and S. B. Kaschieva, Fiz. Tekh. Poluprov. 11, 2254 (1977).
[12] A. H. M. Shousha, Thin Solid Films 20, 33 (1974).
[13] J. F. Ready, Effects of high-power laser radiation, Academic Press, New York/London 1971.
[14] V. L. Vinetskii and G. A. Holodar, Statistic interaction of electrons and defects in semiconductors, Naukova Dumka, Kiev 1969 (in Russian).
[15] S. M. Ryvkin, Fiz. tverd. Tela 9, 1340 (1967).
[16] G. J. Bemski, J. appl. Phys. 30, 1195 (1959).
[17] J. Corbett and G. Watkins, Phys. Rev. Letters 7, 314 (1961).
[18] V. D. Tkachov, in: Actual physical problems of semiconductors and semiconductors devices, Ed. S. Kalvenas, Vilnus 1969 (p. 113) (in Russian).
[19] K. L. Wang and A. O. Evwaraye, J. appl. Phys. 47, 4574 (1976).
[20] I. P. Kozlov, T. A. Logvinenko, P. F. Lugakov, and V. D. Tkachov, Fiz. Tekh. Poluprov. 8, 1431 (1974).
[21] K. L. Wang, Appl. Phys. Letters 29, 700 (1976).
[22] V. S. Lysenko, A. N. Nazarov, S. B. Kaschieva, M. M. Lokshin, Yu. Ya. Polischuk, and V. E. Ustilko, Abstracts of the VI. All-Union Conf. Physics of the Surface Phenomena in Semiconductors, Vol. 2., Naukova Dumka, Kiev 1977 (p. 44) (in Russian).

(Received June 12, 1978)

phys. stat. sol. (a) **49**, 413 (1978)

Subject classification: 1.4 and 1.5; 3; 21.6

Polytechnical Institute, Voronezh

Interphase Coincidence Boundaries in Au–Si and Au–CaF_2 Systems

By

V. M. Ievlev, A. V. Bugakov, and V. A. Ammer

In the systems of Au–Si and Au–fluorite the possibility of realization of coincidence orientations on the interphase boundary in the oriented crystallization is experimentally proved. It is shown that the misfit, characterizing the interphase coincidence boundary, determines the additional energy of the boundary. Taking account of this energy, some peculiarities of realization of coincidence orientations may be explained.

На примере систем золото–кремний и золото–флюорит экспериментально доказана возможность реализации ориентаций совпадения на межфазной границе при ориентированной кристаллизации. Показано, что несоответствие, характеризующее межфазную границу совпадения, определяет добавочную энергию границы. Учет этой энергии позволяет объяснить некоторые особенности проявления ориентаций совпадения.

1. Introduction

In most cases of oriented crystallization the experimentally observed orientation relationships correspond to the principle of parallelism between the close-packed directions of the conjugated lattices of the substrate and the film. In [1] the possibility of the so-called "irregular epitaxy" was suggested with the formation of film–substrate interphase boundaries which have a structure similar to that of the grain boundaries with high misorientation angles of the crystal lattices. In the present paper this suggestion is experimentally confirmed for the first time.

The choice of the systems for investigation is governed by the following considerations.

(i) Between the interphase and grain boundaries an analogy may be drawn on the assumption that the former are usually characterized by the misfit of the lattice parameters of the conjugated phases.

(ii) On the interphase boundary as well as on the grain boundary a coincidence orientation with corresponding energy minima may exist. For systems with small misfit the parallel orientation is most favourable. If the value for misfit is high, the energy of the boundary at parallel orientation may exceed the energy of the boundary corresponding to the coincidence orientation, and during the oriented crystallization the interphase coincidence boundaries (ICB's) may appear. For the realization of such boundaries a sufficiently strong bonding on the interphase surface is required.

On this assumption, for the experimental test of the possibility of realization of ICB's in the oriented crystallization the Au–Si(III) and Au–fluorite (III) systems have been chosen. The choice of Si and CaF_2 substrates is governed by the fact that they have almost the same values of the lattice parameters which suggests the same coincidence orientations in the systems considered.

2. The Possible Coincidence Orientations in Au–Si and Au–CaF_2 Systems

To determine the coincidence orientations the methods of the 0-lattice theory may be applied [2]. The structural matrix of 0-lattice $X^{(0)}$ in the crystal coordinates of the first crystal is

$$X^{(0)} = [\mathsf{I} - (S^{(2)})^{-1} R^{-1} S^{(1)}]^{-1} , \tag{1}$$

where I is the unit matrix, $S^{(1)}$ and $S^{(2)}$ are the structural matrices of the first and the second crystal, respectively, R is the rotation matrix.

To get the coincidence site lattice (CSL) the transformation of $X^{(0)}$ into the matrix with integral coefficients Y is necessary. The column vectors Y are the CSL basis. The value of Σ, characterizing the degree of coincidence, is

$$\Sigma = \det |Y| . \tag{2}$$

Consider two plane hexagonal lattices ((III)-planes in f.c.c. crystals) with different parameters. Due to the favourable positions of condensed atoms over the substrate interstitials both the atoms and the interstitials have been taken into account as substrate sites.

Then

$$S^{(1)} = a_1 \begin{pmatrix} 1 & 1/2 \\ 0 & \sqrt{3}/2 \end{pmatrix}, \qquad S^{(2)} = \frac{1}{\alpha} S^{(1)} , \tag{3}$$

where a_1 is the Au-lattice parameter, $\alpha = \sqrt{3}\, a_1/a_2$, a_2 is the substrate lattice parameter and

$$X^{(0)} = \frac{1}{\Delta} \begin{pmatrix} 1 - \alpha \cos\theta + \dfrac{\alpha \sin\theta}{\sqrt{3}} & \dfrac{2\alpha \sin\theta}{\sqrt{3}} \\ -\dfrac{2\,\alpha \sin\theta}{\sqrt{3}} & 1 - \alpha\cos\theta - \dfrac{\alpha \sin\theta}{\sqrt{3}} \end{pmatrix}, \tag{4}$$

where $\Delta = 1 - 2\alpha \cos\theta + \alpha^2$, and θ is the rotation angle ($\theta = 0$ corresponds to $\langle 110 \rangle$ Au $\|$ $\langle 112 \rangle$ of the substrate).

For the existence of CSL it is necessary for θ and α to satisfy the following relations:

$$\operatorname{tg} \theta = \sqrt{3}\, r , \qquad \alpha^2 = p , \tag{5}$$

where r and p are rational numbers.

The coincidence orientations with $\Sigma \leqq 13$ for Au–Si and Au–CaF_2 systems are given in Table 1.

In contrast to the grain boundaries the ICB is characterized not only by the value of Σ but also by the deformation f, which is necessary for the exact coincidence of sites.

Table 1

Coincidence orientations on the (111) interphase boundaries of Au–Si and Au–CaF_2 systems. $\Sigma \leqq 13$

θ_c, (K)		0	3	5.2	10.9	19.1	24.8	27
Σ		9	13	7	7	4	13	7
f (%)	Au–Si	2.5	−2.3	4.7	0.7	1.7	−2.3	4.7
	Au–CaF_2	3.1	−1.8	5.3	1.2	2.2	−1.8	5.3
$\boldsymbol{b}$		$\frac{1}{8}\langle 1\bar{1}0 \rangle$	$\frac{1}{42}\langle \bar{5}41 \rangle$	$\frac{1}{26}\langle 1\bar{4}3 \rangle$	$\frac{1}{12}\langle 1\bar{2}1 \rangle$	$\frac{1}{14}\langle 2\bar{3}1 \rangle$	$\frac{1}{42}\langle \bar{5}41 \rangle$	$\frac{1}{26}\langle 1\bar{4}3 \rangle$

This deformation may be called the misfit of ICB. The misfit in ICB may be accommodated by the boundary dislocations (BD's) similar to the grain boundary dislocations. The Burgers vectors $\boldsymbol{b}$ of the BD's are defined by the DSC lattice which may be constructed using the relations that the DSC lattice is reciprocal to CSL constructed on the reciprocal lattices for the same rotation [3]. The vectors $\boldsymbol{b}$ (in coordinates of the Au-lattice) are given in Table 1. Besides the accommodation of misfit the BD's may compensate for the possible deviation from the exact coincidence orientation.

The existence of coincidence orientations with low Σ and f in Au–Si and Au–CaF_2 systems supports the realization of the corresponding ICB's at the oriented crystallization.

3. Methods and Experimental Results

Au was condensed in a high (6×10^{-7} Torr) and ultra-high (5×10^{-9} Torr) vacuum on to the preliminarily prepared (111) cleavage surfaces and the as-cleaved (i.e., formed as a result of cleavage of crystals directly in a condensed flow of atoms) surfaces at substrate temperatures T_s up to 180 °C for Si and up to 400 °C for CaF_2. The island films were backed with carbon. For the transmission electron microscope investigation the films were separated by etching of the substrates. The azimuthal orientation of the films with respect to the substrate was defined by the clear-cut edge of the free film. This edge is parallel to the ⟨110⟩ substrate direction.

The main results of the investigations are as follows.

1. Au–Si system. In the case of (111) Au || (111) Si the following azimuthal orientation relationships have been found:

(i) at $T_s = 25$ °C on the as-cleaved surfaces prepared in a high and ultrahigh vacuum the preferred orientation is

$$\langle 1\bar{1}0 \rangle \text{ Au} \parallel \langle 11\bar{2} \rangle \text{ Si} \tag{6}$$

which corresponds to $\theta_c = 0°$;

(ii) at $T_s = 120$ °C along with the orientation (6) twelve more exact orientations are formed (Fig. 1a)

$$\langle 1\bar{1}0 \rangle \text{ Au} \parallel \langle \bar{5}41 \rangle \text{ Si} \tag{7}$$

which are schematically represented in Fig. 1b. The relations (7) correspond to the coincidence orientation $\theta_c = 19.1°$ ($\Sigma = 4$).

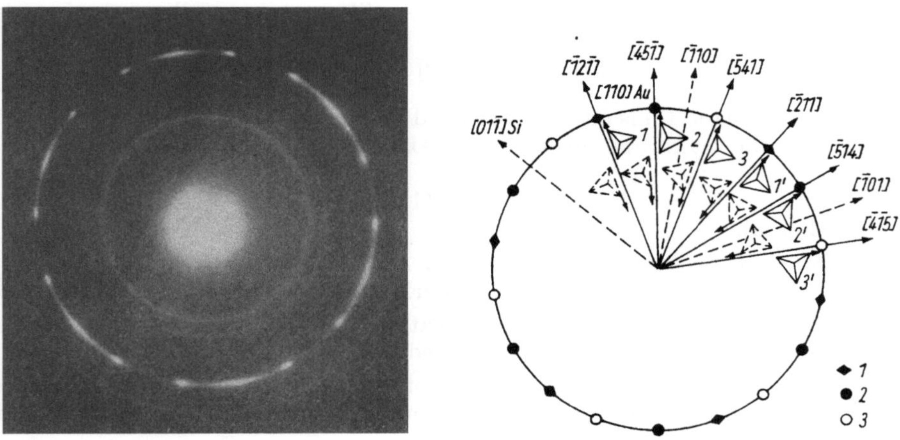

a b

Fig. 1. a) Electron diffraction pattern of Au-film ((111) Au || (111) Si) deposited on Si at $T_s = 120$ °C, b) scheme of orientations. 1: [$\bar{1}$10] Au || ⟨211⟩ Si, 2; 3: [$\bar{1}$10] Au || ⟨541⟩ Si

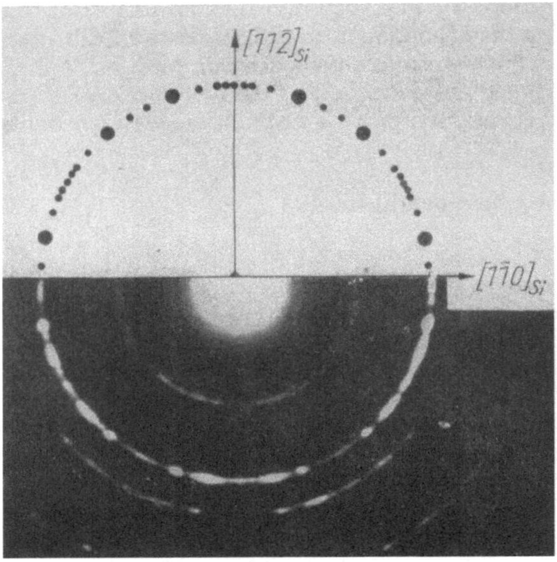

Fig. 2. Calculated and experimental electron diffraction patterns of Au-film on Si

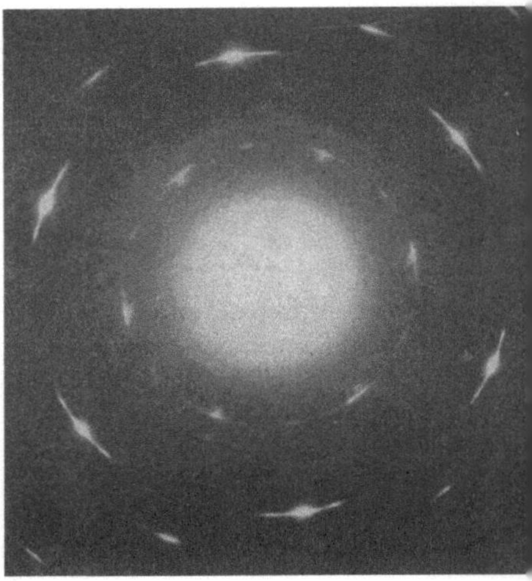

Fig. 3. Electron diffraction pattern of an island Au-film on the as-cleaved surface of CaF_2 ($T_s = 370$ °C)

(iii) With the increase of T_s the orientation (6) is no longer predominating; extra reflections appear (Fig. 2). The comparison with the calculated electron diffraction pattern (scheme in Fig. 2) which have been obtained in view of the realization of all the coincidence orientations of the table, shows that the smeared reflections in the angular gap of 5 to 10° may correspond to the orientations close to $\theta_c = 5.2$ and 10.9° ($\Sigma = 7$).

2. Au–CaF_2 system. In the temperature region for the oriented crystallization ($T_s > 300$ °C) with (111) Au || (111) CaF_2 the following regularities have been found:

a) On the non-as-cleaved surfaces the Au-films are crystallized in the orientation

$$[1\bar{1}0] \text{ Au} \parallel [1\bar{1}0] \quad \text{or} \quad [\bar{1}10] \text{ CaF}_2 . \tag{8}$$

b) For the island films on the as-cleaved surfaces in the high and ultra-high vacuum along with the main orientation (8) the extra azimuthal orientation

$$[1\bar{1}0] \text{ Au} \parallel \langle 1\bar{1}0\rangle \pm 9^\circ \text{ CaF}_2 \tag{9}$$

exists (Fig. 3).

c) The dark-field analysis of singular surface regions (free from cleavage steps) showed that the islands with the extra orientations are randomly distributed.

d) As a result of the secondary nucleation during the later stage of active coalescence the extra orientations are conserved (9).

e) Within the regions of high density of elementary steps which are oriented along ⟨110⟩ CaF_2 and which are the preferred nucleation places of the film, the islands in the extra orientation are absent (Fig. 4).

f) The continuous epitaxial films received under the same conditions are characterized by orientation (8) only.

a 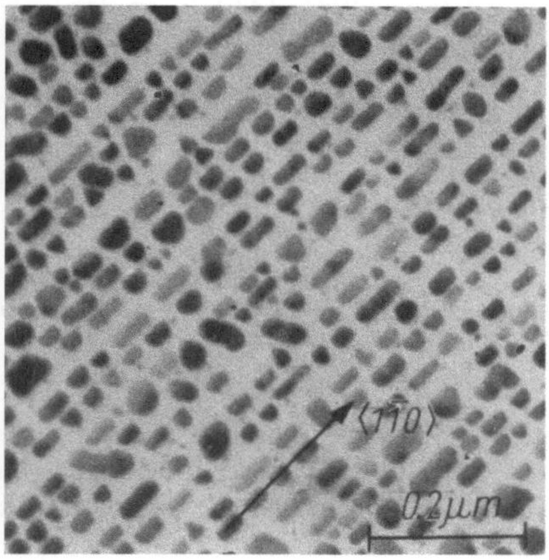  b

Fig. 4. a) Morphology and b) orientation of island Au-films in the CaF_2-area with high density of slip steps

4. Discussion

The results show that the formation of extra orientations is a pure interphase effect. For this effect to become possible, firstly, an as-cleaved surface (i.e. free from adsorbed gas and vapour) and sufficiently strong bonding on the interphase boundary are required. Secondly, the regularities of crystallization of Au on CaF_2 show the necessity of a defect-free surface. In particular, in [4] in island epitaxial films of Au on (111) matched faces of CaF_2 the effect of discrimination of the twin azimuthal orientations has been found. It has been shown that the orientation (8) is favoured by the nucleation of the condensate at point defects on the substrate surface (e.g. on F-vacancies). The preferred nucleation at point defects hinders the realization of coincidence orientations in the Au–CaF_2 system and their portion in the total bulk of the condensate is not large.[1])

Moreover, in contrast to the Au-Si system the orientations (9) of the Au–CaF_2 system have a great angular smearing and do not correspond to coincidence orientations with low Σ. This peculiarity may be explained by the following factors.

Firstly, for the second system the interphase interaction is weaker.[2]) This leads to the flattening of the energy minima corresponding to θ_c.

Secondly, these systems have different values for the misfit. Consider the role of the misfit of the interphase coincidence boundary in detail. The equilibrium deformation of the condensate ε will be defined by the same factors as those for the epitaxial boundaries [5], i.e., by the elastic strain energy of the film and the interphase energy. But in this case the interphase energy is dependent on the misorientation angle. Therefore, the boundary with minimum energy may not correspond to the exact coincidence orientation. The driving force of the deviation is connected with the interphase energy; the mechanism of this deviation is the introduction of misfit dislocations having twist components. This deviation will be favoured by the fact that near θ_c with low Σ orien-

[1]) If we assume that the main orientation (8) is connected with defects and the orientations (9) are connected with the defect-free surface regions, we may differentiate between the processes of nucleation on defects and the statistical nucleation on the defect-free surface.

[2]) This is true only for the defect-free surface regions because the local interaction in the point defect area is great.

tations with a higher Σ exist, but with a lower f. In particular, near $\theta_c = 19.1°$ the orientations $\theta_c = 22.9°$ ($\Sigma = 31$) and $21.85°$ ($\Sigma = 49$) exist with $f \approx 0$ for the Au–CaF_2 system and some larger f for the Au–Si system.

The dependence of the interphase boundary energy upon the deviation angle $\Delta\theta$ near θ_c may be understood on the assumption that the boundary energy is equal to the energy of the mixed boundary dislocation network which accomodates the misfit and brings about the deviation $\Delta\theta$. Then the film energy is

$$E = E_0 + \frac{Gb}{2\pi} \sin \varphi \left\{\text{ctg}\, \varphi\, \Delta\theta(I - \ln \Delta\theta) + \frac{f - \varepsilon}{1 - \nu} [1 - \ln (f - \varepsilon)]\right\} + \frac{2G(I + \nu)}{1 - \nu} \varepsilon^2 h\,. \tag{10}$$

Here E_0 is the boundary energy at exact coincidence, φ the angle between the Burgers vector $\boldsymbol{b}$ and the dislocation line $(\text{ctg}\, \varphi = \Delta\theta/(f - \varepsilon))$, G the shear modulus, ν the Poisson ratio, h the film thickness, and ε the equilibrium deformation corresponding to the energy minimum.

Let $f(\Delta\theta)$ be a function which becomes zero at a certain $\Delta\theta = \delta$. Fig. 5 shows the energy versus $\Delta\theta$ dependences for various h for the case that $f = f_0(1 - \Delta\theta/\delta)$ at $f_0 = 2.5\%$, $\delta = 1.5 \times 10^{-2}$ rad. The use of another function $f(\Delta\theta)$ leads to no qualitative changes of $E(\Delta\theta)$.

Fig. 5 shows that with the increase of h the energy minimum shifts from $\Delta\theta = 0$ at the initial stages of growth to $\Delta\theta \approx \delta$ at later ones. Therefore, during the film growth changes in its orientation may occur; as h defines the equilibrium deformation one may determine the influence of the film deformation on the location of the energy minimum, namely, the larger the film deformation the smaller the deviation from θ_c.

The location of the minimum depends on the relation between f_0 and δ. In Fig. 6 the plots for $f_0 = 1.5$ and 2.5% at the same $\delta = 3 \times 10^{-2}$ rad are given. It is clear that the deviation of the minimum from θ_c decreases with the decrease of f_0.

Thus, the results shown in Fig. 5 and 6 qualitatively explain the peculiarities of the realization of the coincidence orientations on the interphase boundary Au–CaF_2. The smeared reflections in the angular gap of 5 to 10° for the Au–Si system (Fig. 2) may be explained in a similar way. This suggests that the coincidence orientation may be realized as a set of boundaries in the angular gap the width of which has the same order of magnitude as the value for the misfit.

The application of the CSL concept to the interphase boundaries makes it possible to explain the regularity of some orientation relationships in the epitaxial growth of metal on metal. In particular, in [6] it has been stated that in case of condensation of

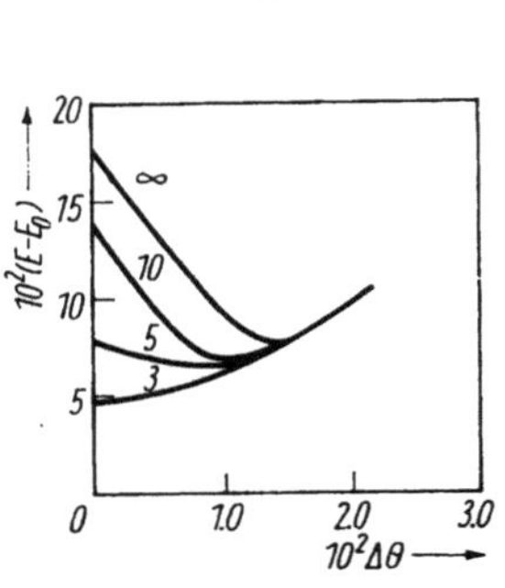

Fig. 5. Dependence of $E - E_0$ (in units of $Gb/2\pi$) on $\Delta\theta$ for various values of h/b (the figures at the curves are the values for h/b)

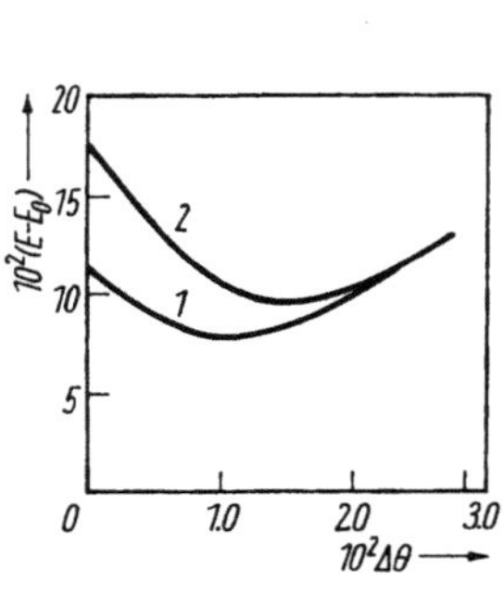

Fig. 6. Dependence of $E - E_0$ (in the units of $Gb/2\pi$) on $\Delta\theta$ for various values of f_0 at constant $\delta = 3 \times 10^{-2}$ rad. (1) $f_0 = 1.5$, (2) 2.5%

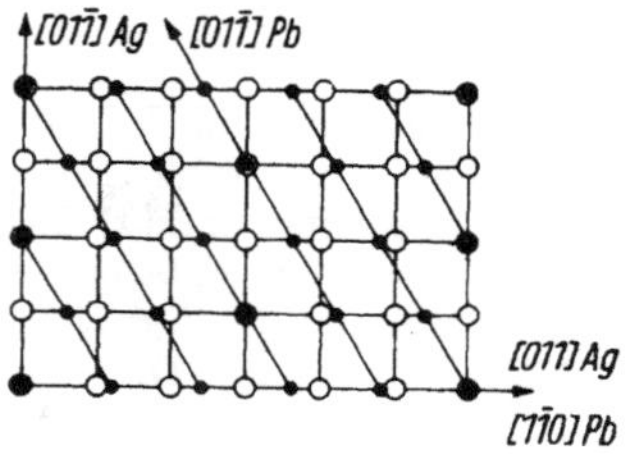

Fig. 7. Coincidence orientation in the interphase boundary (111) Pb || (100) Ag. ○ Ag, ● Pb

Pb on Ag the following relation is fulfilled:

$$(111), [1\bar{1}0]\ \text{Pb}\ \|\ (100), [011]\ \text{Ag}\,. \tag{11}$$

Orientation (11) co-exists with the parallel one for Au–Ni and for Ag–Ni systems too.

This experimental fact may be explained as follows. The lattice parameter ratio α is equal to 1.21 for Pb–Ag and 1.16 for Au–Ni and Ag–Ni systems. On the interphase boundary of orientation (11) the coincidence orientation with $\Sigma = 5$ for Pb ($\Sigma = 6$ for Ag) is possible (Fig. 7). This coincidence requires different α for two different directions, [011] and [0$\bar{1}$1] Ag (1.155 and 1.200), i.e., an anisotropic deformation is necessary. Thus, in this case the coincidence orientation is characterized by two values of misfit, f_1 and f_2.

The boundary energy at the exact coincidence E_0 may be estimated on the assumption that the co-existence of two orientations in the Au–Ni system supports the equality of their energies.

The energy for parallel orientation E_1 is (in units of $Gb/4\pi(1 - \nu)$)

$$E_1 = 2F(1 - \ln F)\,, \tag{12}$$

where F is the misfit for parallel orientation.

The energy for orientation (II) is

$$E_2 = E_0 + f_1(1 - \ln f_1) + f_2(1 - \ln f_2)\,. \tag{13}$$

For the Au–Ni system $E_1 = E_2$. This yields $E_0 = 0.72$. Using this value of E_0 for the Pb–Ag system we obtain $E_1 = 1.09$ and $E_2 = 0.98$. Thus, for the Pb–Ag system the orientation (11) is more preferable than the parallel one.

The analysis of the dependence of E_1 and E_2 upon α shows that at $\alpha > 1.27$ the parallel orientation is more preferable again. The transition to parallel orientation at high values of α is confirmed in the Pb–Ni system, where $\alpha = 1.41$.

Thus, this treatment of the interphase boundaries makes it possible to explain the experimental results and this confirms the validity of the proposition on the realization of coincidence orientations.

5. Conclusion

1. The results of our investigations support the fact that the principle of parallelism between the close-packed directions correspond to the particular case of oriented crystallization. In the general case, analyzing the orientation relationships in the interphase boundary the possibility of coincidence orientation formation should be taken into account.

2. The interphase coincidence boundary misfit defines the additional energy of the boundary. Taking into account this energy we may explain some peculiarities in the realization of coincidence orientations.

References

[1] L. S. PALATNIK, M. YA. FUKS, and V. M. KOSEVICH, Mekhanizm Obrazovaniya i Substruktura Kondensirovannykh Plenok, Nauka, Moskva (1972), (p. 75).
[2] W. BOLLMANN, Crystal Defects and Crystalline Interfaces, Springer-Verlag Berlin 1970.
[3] H. GRIMMER, Scripta metall. **8**, 1221 (1974).
[4] V. M. IEVLEV, E. V. SHVEDOV, and V. A. AMMER, Kristallografia **23**, 670 (1978).
[5] W. A. JESSER and D. KUHLMAN-WILSDORF, phys. stat. sol. **19**, 95 (1967).
[6] S. SCHIRAI, Y. FUKUDA, and M. NOMURA, J. Phys. Soc. Japan **16**, 1989 (1961).

(Received May 2, 1978)

See Short Note by Z. LILIENTAL and G. KÄSTNER

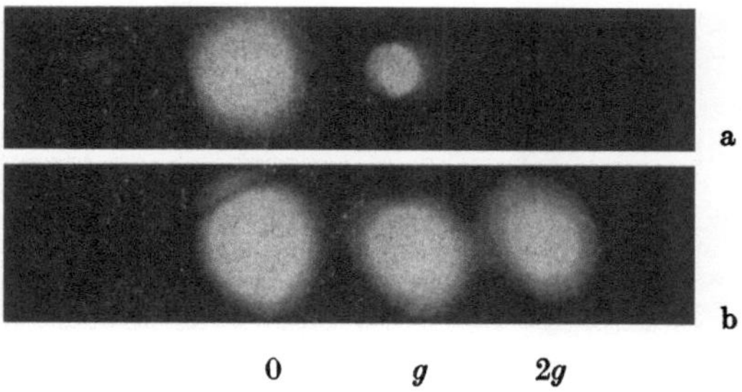

Fig. 1. Diffraction pattern showing ⟨220⟩ systematic reflections of a Cu-doped ZnTe crystal set at the second-order Bragg position. a) 600, b) 675 kV

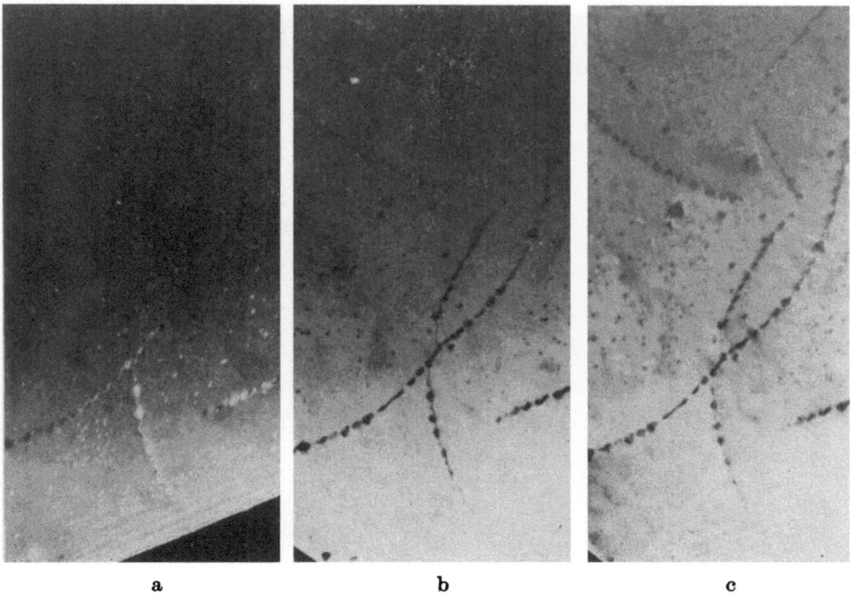

Fig. 2. Defects in a wedge-shaped Cu-doped ZnTe crystal imaged at second-order DF $(2g/2g)$, second-order and first-order BF, respectively; ⟨220⟩ systematic row, acceleration voltage 630 kV $\approx V_c$. a) $(2g/2g)$, b) $(0, 2g)$, c) $(0, g)$

See Short Note by YA. M. POGOSYAN et al.

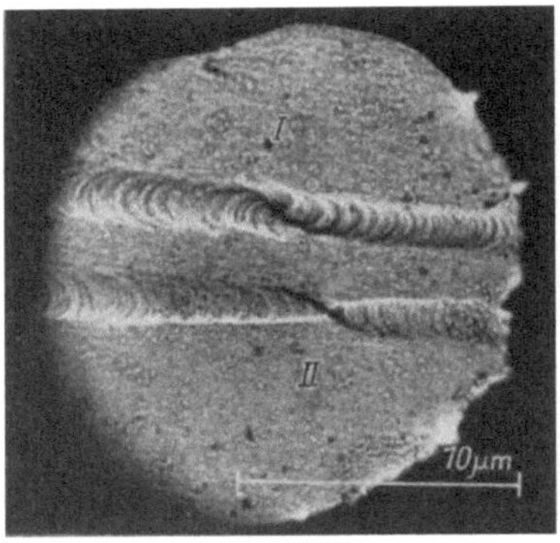

Fig. 1. Generated channel structure of magnetization (Lorentz-photo) with transition elements (type I and II) between channel sections with antiparallel "in-plane" magnetization components in a permalloy film under the influence of a "bias" field $H_{\perp}$ (100 to 200 Oe) normal to the film plane (the easy axis is lying parallel to the channels; $t = 600$ Å)

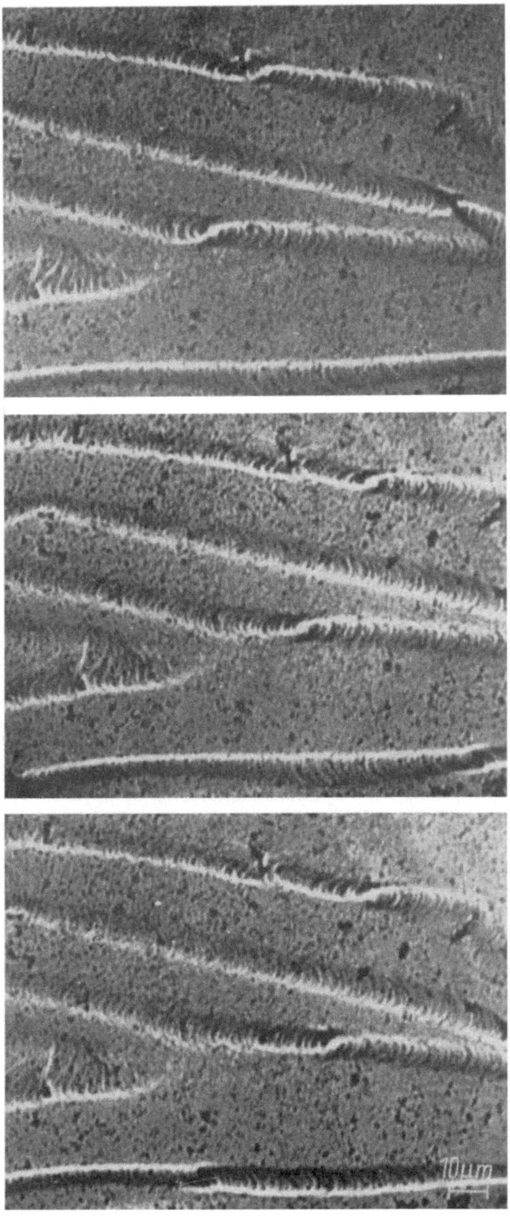

a

b

c

Fig. 3. Initial state a) and shifting b), c) of transition elements (type I) along their channels by an increasing in-plane field perpendicular to the channels

See Short Note bei YA. M. POGOSYAN et al.

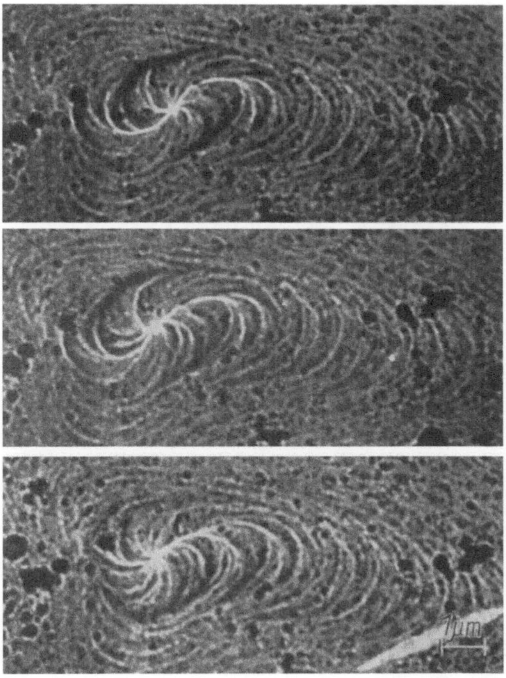

a

b

c

Fig. 4. Lorentz-photos of one transition element (type I) stopping by film irregularities under the influence of an increasing in-plane field

See Short Note by V. KOBZAREŇKO et al.

Fig. 1. SEM microphotograph of the surface of the one-packed polytype film of $ZnIn_2S_4$(I)

See Short Note by J. Llopis and J. Piqueras

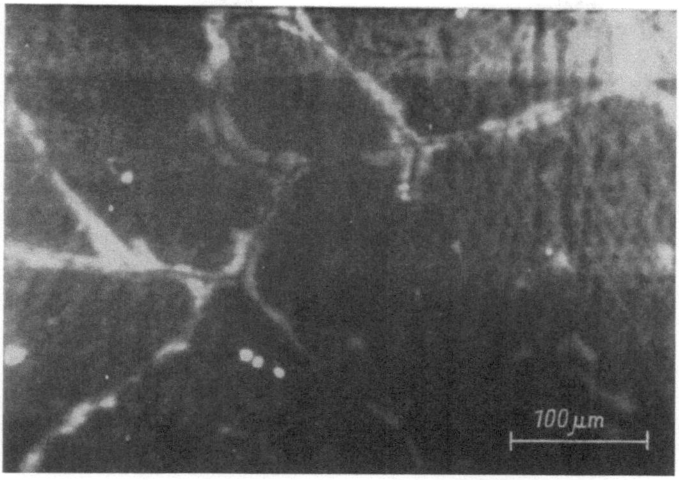

Fig. 1. Cathodoluminescence image of a MgO crystal showing lines of enhanced intensity

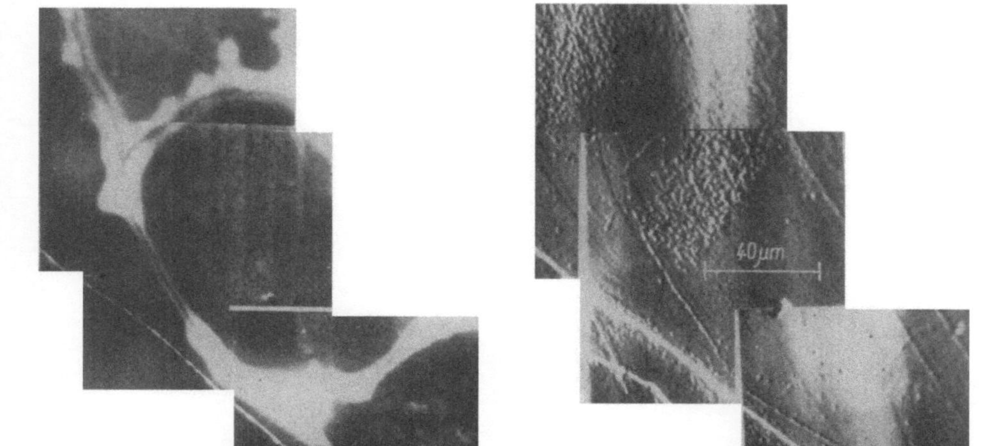

Fig. 2. Area of an etched MgO crystal observed in a) CL mode, b) secondary electron mode

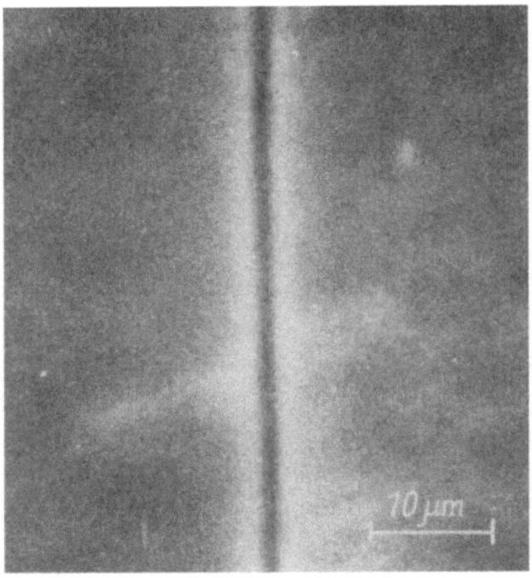

Fig. 3. CL image of a grain boundary in GaP

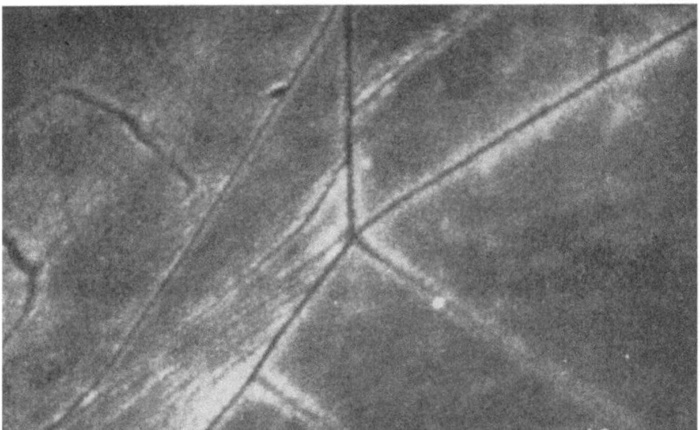

Fig. 4. CL image of several grains in GaP

See Short Note by H. NELKOWSKI and K. PETERMANN

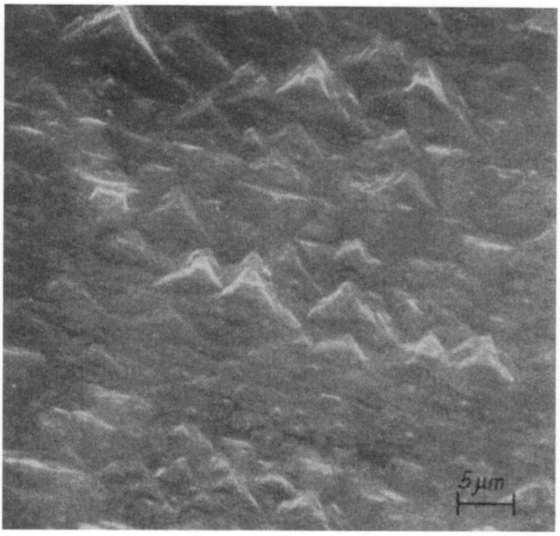

Fig. 2. Scanning electron micrograph of the surface of a ZnS–$CuGaS_2$ heterojunction with triangular growth hillocks and twins; 30 kV

phys. stat. sol. (a) 49, K1 (1978)

Subject classification: 10.2; 22.4.3

Institute of Physics, Polish Academy of Sciences, Warsaw (a)
and Institut für Festkörperphysik und Elektronenmikroskopie
der Akademie der Wissenschaften der DDR, Halle (Saale) (b)

Voltage Dependence on Precipitation Contrast

By

Z. LILIENTAL (a) and G. KÄSTNER (b)

In this note, the critical voltage for ZnTe was determined. The crystal was oriented in the {111} direction. Systematic reflections of the ⟨220⟩ type were excited in the second-order Bragg position. As commonly known, in the case of three beams two branches of the dispersion surface touch each other under critical voltage conditions. Although the ZnTe unit cell does not have a centre of symmetry, for the ⟨220⟩ reflection $U_g = U_{-g}$ and the following equation will also be satisfied /1/:

$$U_g^2 = U_{2g}\, g^2 + U_{2g}^2 \quad . \tag{1}$$

On the basis of condition (1), taking into consideration relativistic effects in the initial energy of the electrons incident on the crystal, a second-order critical voltage of $V_c = 611$ kV was calculated for the ⟨220⟩ systematic row in ZnTe. The Debye-Waller correction for the Debye temperature $\Theta_D = (176 \pm 8)$K /2/ was taken into account.

Experimentally the critical voltage was determined for ZnTe doped with Cu by a diffusion treatment at 900 °C during 48 h.

A second-order critical voltage of $V_c = (640 \pm 20)$kV was estimated from the reversal of the asymmetric intensity profile of the zero-order Kikuchi line crossing the $\vec{g} = [\bar{2}20]$ reflection (Fig. 1, see on the photo pages before the Short Notes part). A more precise determination is difficult since in this range of voltages the focusing property of the microscope does not allow accurate selected area diffraction work. Within this limitation the experimental and theoretical values of the critical voltage are in satisfactory agreement.

Fig. 2 (see on the photo pages before the Short Notes part) was taken near the critical voltage in order to compare the visibility of defects under different imaging conditions in dependence on the specimen thickness. The latter increases according to the wedge shape of the specimen the edge of which is seen in the lower part of the micrographs.

At a thickness up to $\approx 10\,\xi_g$, the second-order dark field (2g/2g) exhibits the typical bright contrast of defects as commonly observed near V_c /1, 3/. In thicker regions the contrast gradually reverses thus getting similar to the bright field which is presented by the micrographs taken at (0/2g) and (0/g). This contrast reversal should by caused by inelastic scattering as known from the weak-beam dark field /4, 5/. Additionally, the reversal of contrast should also depend on the depth position of a given defect, since it is observed that the precipitates decorating the dislocations exhibit the reversal of contrast at different thicknesses of the specimen.

The best overall visibility of defects is observed in the common BF (0/g) which yields a maximum penetration at large thickness in the upper part of the micrographs The BF (0/2g) has the advantage of better resolution and lower background modulation as typically observed also at other voltages.

References

/1/ J.S. LALLY et al., Phil. Mag. 25, 321 (1972).

/2/ R.J. BLATTNER et al., J. appl. Phys. 43, 935 (1972).

/3/ E.P. BUTLER, Proc. 3rd Internat. Conf. High Voltage Electron Microscopy, Ed. P.R. SWANN et al., Oxford 1974 (p. 22).

/4/ A. MELANDER and R. SANDSTRÖM, ibid. (p. 27).

/5/ G. KÄSTNER, ibid. (p. 172).

(Received May 22. 1978)

phys. stat. sol. (a) 49, K3 (1978)

Subject classification: 18.2.1; 21.1; 21.1.1

Erevan State University (a) and Zentralinstitut für Kybernetik und Informationsprozesse der Akademie der Wissenschaften der DDR, Berlin[1)] (b)

Submicroscopical Magnetization Structures in Permalloy Films Induced by Fields Normal to the Film Plane

By

YA.M. POGOSYAN (a), S.D. GEVORKYAN (a), and R. STRAUBEL (b)

Using Lorentz-microscopical methods magnetization structures in permalloy films with the thickness t = 300 to 1200 Å induced by a strong field $H_\perp$ normal to the film plane have been investigated. Fixing the samples at the usual position in the microscope the field was generated directly by the objective lens. The magnetic contrast for the photos was obtained by defocusing the pictures by means of the condensor and the intermediate lens alone. This procedure guaranteed that changing of the objective lens current had a small influence on the defocusing only.

A typical Lorentz-microscopical photo of a permalloy film is shown in Fig. 1 ($H_\perp$ = 100 to 200 Oe) (see on the photo pages before the Short Notes part). (The field strength was estimated by the current flowing through the objective lens under the assumption that in the state of highest resolution this field is about 3 kOe /1/.) We can see some channels - lying parallel to the easy axis of the film - with a magnetic fine structure caused by an "in-plane" component of the magnetization. Outside the channels there is a magnetization component aligned parallel to $H_\perp$. Increasing of $H_\perp$ leads to a decrease of the channel width K (see Fig. 2) until the channels vanish at a field $H_\perp$ of about 2 kOe. Furthermore from Fig. 1 it can be seen that there are two types of transition elements between channel sections with antiparallel in-plane magnetization components. The type I has a magnetization structure similar to a circle Bloch line /2/ in a crosstie wall (see also Fig. 4a) whereas the type II looks like a cross line. Both types can be shifted slightly by an in-plane field. (Until now we have generated the necessary field gradient by shifting the sample-holder inside the microscope.)

---

1) DDR-1199 Berlin, DDR.

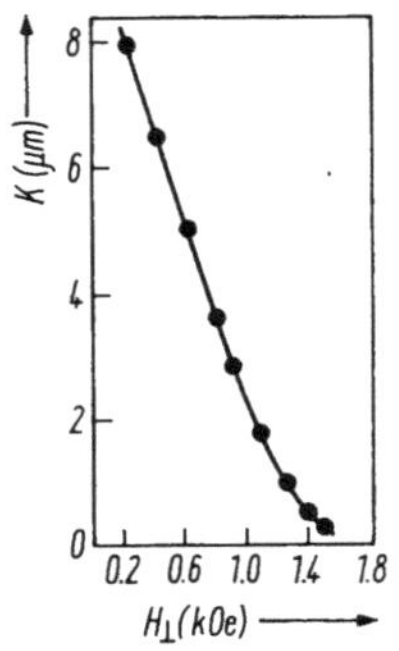

Fig. 2. The channel width K as function of $H_\perp$ (Ni 80/Fe 20 film; t = 600 Å

Beginning with the initial state (Fig. 3a) Fig. 3b and c (see on the photo pages before the Short Notes part) show the shifting of transition elements (type I) under the influence of an increasing field H_s adjusted perpendicular to the easy axis of the film. We took the photos after the shifting process was stopped by irregularities inside the film. For a given polarity of H_s the direction of the shifting depends alone on the magnetization directions in the neighbouring channel sections.

Fig. 4a (see on the photo pages before the Short Notes part) shows a Lorentz-photo of one transition element (with a high resolution). A symmetrical S-shaped magnetization structure (similar to a circle Bloch line) can be recognized. Fig. 4b and c (see on the photo pages before the Short Notes part) which show the same element under the influence of a field H_s - stopped by film irregularities - give an impression of what is happening inside a transition element during the shifting process. Increasing the field amplitude H_s leads to an increase of the asymmetry of the magnetization distribution around the element centre.

The described channel structure can also be observed starting from the "saturated" state of the film. During the decrease of the "bias" field $H_\perp$ germs for channels are generated at locations of the highest anisotropy deviations inside of the films to reduce the magnetic stray-field energy. Therefore starting from the saturated state of the film the formation of the channel structure in one film is a reproducible process and leads to the same pattern always.

More quantitative investigations and theoretical explanations of the possibilities to influence the transition elements are in progress.

References

/1/ A.N. PILYANKEVICH and A.N. KLIMOVICKII, Elektronye Mikroskopi, Tekhnika Kiev 1976.

/2/ S. METHFESSEL, S. MIDDELHOEK, and H. THOMAS, IBM J. Res. Developm. 4, 96 (1960).

(Received July 17, 1978)

phys. stat. sol. (a) 49, K5 (1978)
Subject classification: 1.5; 22.8

Institute of Applied Physics, Academy of Sciences of the Moldavian SSR, Kishinev

Thin Films of the One-Packed Polytype of $ZnIn_2S_4$(I)

By

V. KOBZARENKO, F. DONIKA, and S. RADAUTSAN

In the $ZnS-In_2S_3$ system the existence of the compounds $ZnIn_2S_4$, $Zn_2In_2S_5$, and $Zn_3In_2S_6$ has been established. According to X-ray study of single crystals /1/ these compounds have a large number of polytype modifications.

All the above-mentioned ternary compounds have good photoluminescent and photoelectric properties in the blue and near ultraviolet regions of the spectrum.

The physical characteristics of the compounds depend on composition and structure.

Thin films of various polytype modifications of these ternary phases essentially extend the possibility of their practical application.

In /2/ some properties of the $ZnIn_2S_4$ three-packed polytype, the technology of which was better worked out, both in the bulk and in the thin film form have been studied. That is why the measurements of the physical characteristics of this compound were made up to now mainly on the three-packed polytype /2, 4/, or on crystals and films of unknown structure /5/.

In order to obtain the other polytype forms of this compound in the form of thin films one must observe more strictly some conditions of the growth process. The thin films of the one-packed polytype of $ZnIn_2S_4$ (I) were prepared by the thermal evaporation of $ZnIn_2S_4$ crystals in vacuum. Applying a certain combination of temperatures of the evaporator and the glass substrate films of the one-packed polytype of $ZnIn_2S_4$ (I) were grown. The structure of the films was controlled by X-ray analysis and with an optical and scanning electron microscope (SEM).

The X-ray analysis showed that the films obtained consisted of crystals with hexagonal symmetry of the lattice, the parameters of which a = 3.85 Å; c = 12.34 Å correspond to the structure of the one-packed polytype of $ZnIn_2S_4$ (I) /3/. It is necessary to note that the layers of the one-packed polytype grew up to a certain critical thickness (d_{crit}). The growing rate was 1 to 2 Å/s. With increasing film thickness

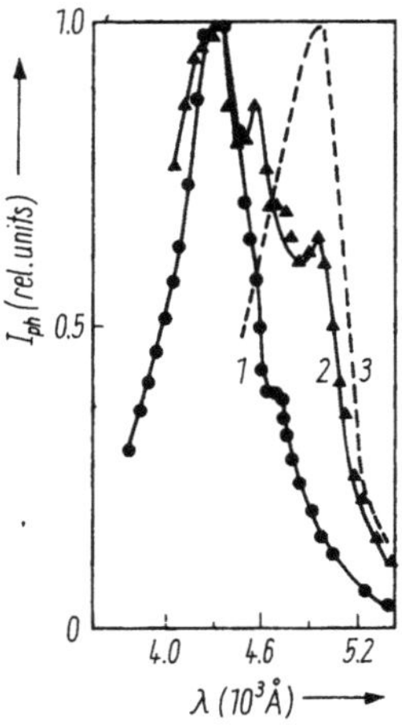

Fig. 2. Spectral distribution of the photoconductivity of the $ZnIn_2S_4$ layers; (1) one-packed polytype, (2) three-packed polytype, (3) $ZnIn_2S_4$ films of unknown structure

above d_{crit} a mixture of various polytype modifications of $ZnIn_2S_4(II)_a$, $ZnIn_2S_4(II)_b$, and $ZnIn_2S_4(III)_a$ was formed.

The $ZnIn_2S_4$(I) textured films consisted of crystallites in the form of lamellar hexagonal plates with the (0001) face parallel to the substrate plane. In Fig. 1 (see on the photo pages before the Short Notes part) a SEM microphoto of the surface of the one-packed polytype film of $ZnIn_2S_4$ grown on glass substrate is shown.

The structural investigation showed that quality and perfectness of the films depend on the temperatures of evaporation and deposition, and on the growing rate.

There are no published data about the physical properties both of bulk samples and thin films of the one-packed polytype of $ZnIn_2S_4$(I). The layers investigated by us had high resistance and photosensitivity, their resistivity varied from 10^3 to 10^8 Ω cm depending on the conditions of preparation. The spectral distribution of the photoconductivity of the one-packed polytype layers of $ZnIn_2S_4$ (I) are plotted in Fig. 2 (curve 1). The spectrum has a peak at a wavelength of 430 nm and a shoulder at 470 nm, which apparently correspond to intrinsic and impurity absorption. For samples with a thickness of less than 0.1 nm this peak is displaced at shorter wavelength due to the interaction of the incident wave with the layer of a certain thickness.

The absence of the other peaks in the photoconductivity spectrum as compared with those revealed in /2/ (curve 2), may be apparently explained by the fact that we had layers containing only the one-packed polytype of $ZnIn_2S_4$(I), while in /2/ the samples contained the three-packed polytype and some non-controllable quantity of other polytypes of this compound. The curve 3 is taken from /5/, where the measurements were carried out on $ZnIn_2S_4$ thin films of unknown structure.

Some differences in other characteristics (luminescence, absorption coefficient) of the one-packed polytype as compared with the more studied three-packed one, were also observed.

References

/1/ S. RADAUTSAN, F. DONIKA, G. KYOSSE, and I. MUSTYA, phys. stat. sol. 37, K123 (1970).

/2/ V. KOBZARENKO, F. DONIKA, V. ZHITAR, and S. RADAUTSAN, Dokl. Akad. Nauk SSSR 235, 1297 (1977).

/3/ F. DONIKA, Ș. RADAUTSAN, S. SEMILETOV, F. DONIKA, I. MUSTYA, and V. ZHITAR, Kristallografiya 15, 813 (1970).

/4/ S. RADAUTSAN, V. ZHITAR, and V. RAILEAN, Fiz. Tekh. Poluprov. 9, 2278 (1975).

/5/ F. ANDREANI and N. ROMEO, Thin Solid Films 31, 217 (1976).

(Received July 3, 1978)

phys. stat. sol. (a) 49, K9 (1978)

Subject classification: 10.2 and 20.3; 22.2.2; 22.6

Departamento de Física del Estado Sólido, Facultad de Ciencias Físicas, Universidad Complutense, Madrid

Cathodoluminescence from Grain Boundaries in MgO and GaP

By

J. LLOPIS and J. PIQUERAS

One of the factors which influence the cathodoluminescence (CL) emission of some ionic and semiconductor solids is the presence of lattice defects. Most of the investigations on this fact considered the dislocation CL in semiconductor crystals of technological interest. Often the scanning electron microscope (SEM) was used to observe the light emission localized in the defects.

The observations of CL from grain boundaries have been less numerous. Brümmer and Schreiber /1, 2/ studied the CL from dislocations and subgrain boundaries in CdS by means of the electron microprobe scanning technique and found that the luminescence from subgrain boundaries depends on the previous thermal treatment given to the crystal which modifies the decoration degree of the defect. The non-decorated subgrain boundaries showed dark contrast but after annealing, zones of increased luminescence appear besides the dark line whose intensity also changes. This effect was attributed to the point defects which decorate the boundary. More recently /3/, lines of reduced intensity identified as subgrain boundaries, were observed in $Ga_xIn_{1-x}P$ by means of the CL mode in the SEM.

In the present work, the CL of subgrain boundaries in MgO single crystals and of grain boundaries in GaP polycrystals was observed in the SEM.

The MgO single crystals used (W.C. Spicer and Co.) had a purity of 99.9% and a size of 10x10x5 mm^3 as received. The crystals were cleaved along (100) faces to their final dimensions of approximately 7x3x2 mm^3 and some of them were etch pits. Specimens were coated with a transparent conducting film and observed in the scanning electron microscope in the secondary and cathodoluminescence modes at 20 kV.

Fig. 1 (see on the photo pages before the Short Notes part) shows a CL micrograph of a MgO crystal, where a network of lines with different intensity as the background can be observed. Each line has a dark part in the middle and a bright band in each side with a width of about 10 μm and a peak at about 5 μm from the dark part. This intensity distribution is qualitatively similar to that observed /1, 2/ in some subgrain boundaries of CdS.

Observation of the same area of an etched crystal in the CL and secondary modes Fig. 2 (see on the photo pages before the Short Notes part), indicates that the lines observed in CL are subgrain boundaries. In general the width of the luminescent band besides the boundaries has the mentioned value of 10 μm, but this value is not constant along the band and in some boundary junctions and inflexion points in a boundary, the luminescent region extends up to about 30 μm from the boundary line, Fig. 2.

Besides the boundaries other parts of the crystal have a high number or dislocations which does not show luminescent properties. This fact could be explained if the presence of point defects and/or impurities in the immediate neighbourhood of the boundary determine its luminescence. Some works /4, 5/ on CL from plastica deformed MgO have already indicated that point defects or their clusters close to the dislocations generated during deformation can favour the light emission.

On the other hand, several investigations, reviewed by Kingery /6/, show that near the grain boundaries of ceramic materials, including MgO, there is an increase of impurity concentration extending in some cases a distance of several tens of μm from the boundary. Such distance is comparable to the width of the cathodoluminesce zone observed in the subgrain boundaries of MgO in this work.

One of the driving forces for solute migration to grain boundaries is the presence of stress fields associated to the boundaries. In some points, for instance, the junctions of several boundaries, longer range stresses can be originated /7/ which implies a wider zone of influence of the boundary on impurities or other point defects. If the luminescence is associated to an increased concentration of impurities, the long range stresses can contribute to the observed effect of a wider cathodoluminesce zone around some of these particular points of the boundary.

As described above, the CL image of subgrain boundaries in MgO is qualitatively the same as the previously observed image of decorated sub-boundaries in CdS /1/. However, the observation of sub-boundaries as dark lines in $Ga_xIn_{1-x}P$ /3/ shows that this kind of contrast is not a general feature of the subgrain boundaries. This fact has been explained by the influence of the decoration degree on CL. In /3/, composition deviations or segregation of another phase at the boundary were not detected, but due to the fact that the isolated dislocations do not show dark contrast, the boundary contrast was not attributed to dislocations and the possible contribution of point defects was mentioned.

Several observations in GaP, for instanoe /8, 9/, show that dislocations in undoped and nitrogen doped crystals appear as black dots in the CL image in SEM. Only occasionally the black contrast was surrounded by lighter haloes /10/.

In the present work polycrystalline GaP was observed in the CL mode in the SEM in order to compare the luminescence from the grain boundaries with that previously observed from dislocations and with the above discussed CL from grain boundaries in other materials.

The GaP used was from Metals Research and nominally undoped. From the original rods of about 15 mm diameter, discs approximately 3 mm thick were cut perpendicular to the axis of the rod. The specimens were then mechanically and chemically polished. The chlorine saturated methanol used in the chemical process etched some grains revealing dislocation etch pits and grain boundaries. The specimens were observed in the SEM at 20 kV in the secondary and cathodoluminescent modes.

Fig. 3 (see on the photo pages before the Short Notes part) shows the CL image of a grain boundary. The intensity distribution is qualitatively as in MgO, but in GaP the width of the bright band is of about 6 μm and does not vary much along the boundary. Some of the boundaries in GaP showed a central part darker than the background while in MgO, at 20 kV, the intensity in this part was about the same as in the background. This fact agrees with the observation of several authors that dislocations in GaP appear frequently as black dots in the CL image. Fig. 4 (see on the photo pages before the Short Notes part) shows several grains in the CL mode. Due to the chemical etching some of the boundaries can be also observed in the secondary mode, but the CL image of the grain boundaries is more complete.

In our case, the dislocations inside the grains were not visible in CL. Since in general, dislocations in GaP appear as black dots it seems that the luminescence at the boundary is not directly due to a higher dislocation density and is probably influenced by impurities or other point defects present with a concentration gradient near the boundary. Such variation in the number of impurities or other microdefects around dislocations in GaP have been previously reported /10, 11/.

It appears from the present observations that the "bright band" CL contrast from boundaries is a feature common to rather different cathodoluminescent materials (MgO, GaP). The spatial distribution of intensity, and comparison with previous works, indicates the influence of point defects on CL from boundaries.

References

/1/ O. BRÜMMER and J. SCHREIBER, Ann. Phys. 28, 105 (1972).

/2/ O. BRÜMMER and J. SCHREIBER, Kristall und Technik 9, 817 (1974).

/3/ H. RAIDT, G. KÄSTNER, H. MENNIGER, and G. VOIGT, phys. stat. sol. (a) 43, 167 (1977).

/4/ M.A. VELEDNITSKAYA, V.N. ROZHANSKII, L.F. COMOLOVA, G.V. SAPARIN, J. SCHREIBER, and O. BRÜMMER, phys. stat. sol. (a) 32, 123 (1975).

/5/ J. LLOPIS, J. PIQUERAS, and L. BRU, J. Mater. Sci. 13, 1361 (1978).

/6/ W.D. KINGERY, J. Amer. Ceram. Soc. 57, 74 (1974).

/7/ M.J. MARCINKOWSKY and E.S.P. DAS, Phil. Mag. 26, 1281 (1972).

/8/ J.M. TITCHMARSH, G.R. BOOKER, W. HARDING, and D.R. WIGHT, J. Mater. Sci. 12, 341 (1977).

/9/ D.B. DARBY and G.R. BOOKER, J. Mater. Sci. 12, 1827 (1977).

/10/ S.M. DAVIDSON, M.Z. IQBAL, and D.C. NORTHROP, phys. stat. sol. (a) 29, 571 (1975).

/11/ A.J.R. DE KOCK, W.M. VAN DE WIJGERT, J.H.T. HENGST, P.J. ROKSNOER and J.M.P.L. HUYBREGTS, J. Crystal Growth 41, 13 (1977).

(Received July 18, 1978)

phys. stat. sol. (a) 49, K13 (1978)

Subject classification: 14.3.3 and 20.3;1.5; 3; 22.4.1; 22.8

Institut für Festkörperphysik der Technischen Universität Berlin, Berlin (West)

Preparation and Investigation of $ZnS-CuGaS_2$ Heterojunctions

By

H. NELKOWSKI and K. PETERMANN

In honour of Prof. Dr. F. Stöckmann's 60th birthday

Introduction In recent years II-VI compounds have been often used for preparation of light emitting diodes because of their good luminescence properties /1, 2/. ZnS with its wide energy gap would be one of the best materials for LED's with short wavelength emission, but it is always n-type and cannot be changed into p-type by subsequent diffusion of acceptors. At present ZnS homojunctions cannot be prepared. On the other hand, the analogous I-III-VI_2 compounds like $CuAlS_2$ and $CuGaS_2$ are always p-type. Therefore it should be possible to obtain short wavelength light emitting diodes by preparation of II-VI-I-III-VI_2 heterojunctions.

In order to prepare good quality heterojunctions, the lattice constants of both compounds should be almost equal. On this reason we have chosen cubic ZnS (a = = 5.41 Å) as n-type semiconductor and tetragonal $CuGaS_2$ (a = 5.35 Å, c = 10.48 Å) as p-type semiconductor. The lattice mismatch between these two materials is only 1.1% in direction of the tetragonal axis.

Experimental The diodes were prepared by liquid phase epitaxy (LPE). This method is well established on III-V compounds and there should be no difficulties in growing II-VI or I-III-VI_2 layers, if one has found a suitable solvent for these materials. For our system of ZnS and $CuGaS_2$ we used molten tin as solvent because of the rather low "free energies of formation" of -20.6 kcal/mol for Cu_2S and -47.4 kcal/mol for ZnS. The values indicate that these compounds are more stable than the most stable tin-sulfur compound, i.e. SnS, whose formation energy is only -19.7 kcal/mol /3/. Thus the molten tin cannot reduce these compounds. Other usual solvents like indium or gallium are not suitable in this case; the values of In_2S_3 and Ga_2S_3 are not known, but they are probably higher than the value of Cu_2S, because the epitaxial growth with these solvents has been unsuccessful.

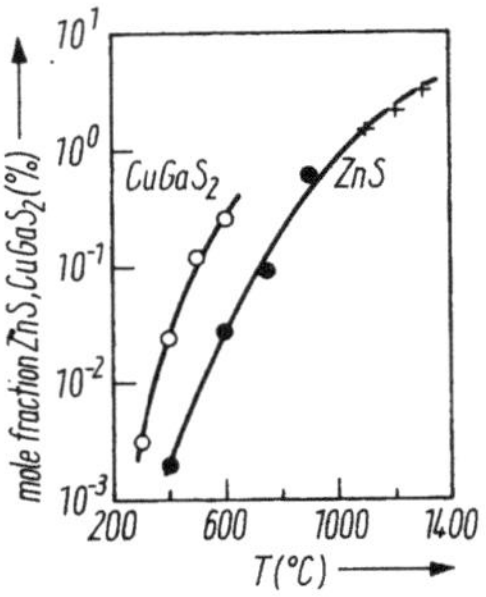

Fig. 1. Solubility of ZnS and $CuGaS_2$ in tin as a function of temperature. + values of Rubenstein /4/

Furthermore the solubility of ZnS and $CuGaS_2$ in tin must be examined. In the literature there is given information only on ZnS for temperatures between 1000 and 1300 °C /4/. But annealing experiments in tin with ZnS crystals of high conductivity revealed that the temperature for epitaxial growth must be less than 500 °C; otherwise the conductivity decreases very strongly - probably because of the outdiffusion of zinc. Therefore the solubility of ZnS and $CuGaS_2$ in tin has been determined experimentally in the temperature range between 300 and 800 °C. The results - shown in Fig. 1 - demonstrate that the solubility of $CuGaS_2$ is higher than the solubility of ZnS by a factor of about 10. Consequently ZnS single crystals were chosen as substrates and $CuGaS_2$ layers were grown epitaxially thereon.

The ZnS substrate crystals were grown in our institute by the halogen transport method with 5 N starting material. Subsequently, the crystals were cut into [111] oriented wafers of 4x6x1 mm^3 and annealed in liquid zinc at 950 °C. By this treatment the conductivity of the as-grown high resistivity samples increases to about 30 $\Omega^{-1} m^{-1}$. In order to prevent the decrease in conductivity during the epitaxial growth of the $CuGaS_2$ layer, a starting temperature of 450 °C was used. By cooling down to 420 °C with a rate of 10 K/h $CuGaS_2$ layers with a thickness of 1 to 5 μm were grown on the ZnS substrates. Fig. 2 shows the surface of a grown layer obtained by a scanning electron microscope (see on the photo pages before the Short Notes part).

Results A sample with a rather thick layer was beveled at an angle of about 3° in order to determine the composition in the junction region by electron microprobe analysis. The composition profile is shown in Fig. 3. The ZnS-$CuGaS_2$ junction

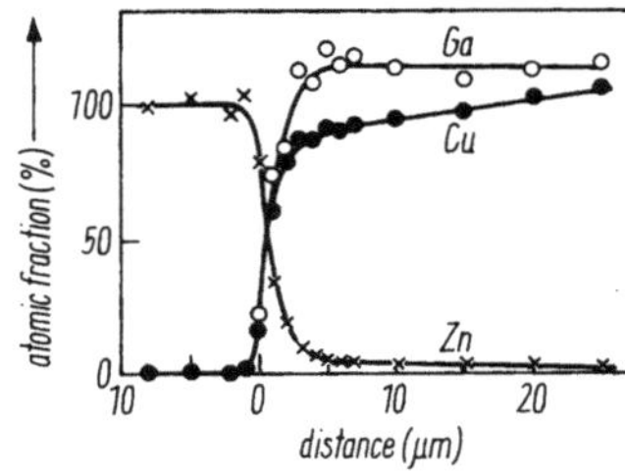

Fig. 3. Electron microprobe analysis of a ZnS-$CuGaS_2$ heterojunction prepared by liquid phase epitaxy. The values are related to a $CuGaS_2$ single crystal

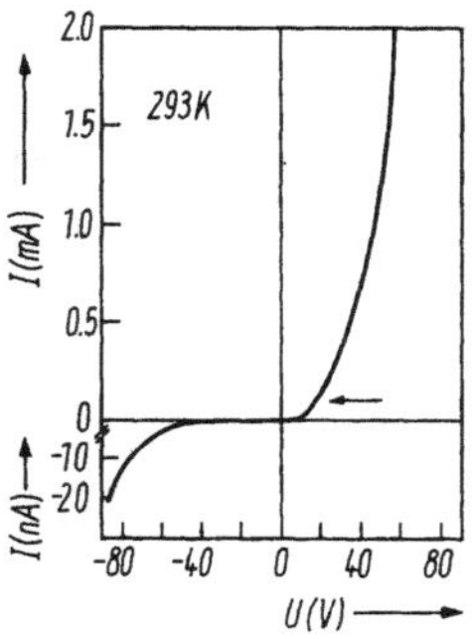

Fig. 4. Room temperature current-voltage characteristic of a ZnS-$CuGaS_2$ heterojunction. Note the different current scales in forward and reverse direction. The onset of forward conduction is at about 2 V, the onset of luminescence is marked by an arrow

is nearly abrupt, and in comparison to a $CuGaS_2$ single crystal one can see that at first a Ga-rich phase and later on almost stoichiometric $CuGaS_2$ is deposited. Till now this behaviour is not well understood and must be investigated further. The values greater than 100% are caused by the different morphology of the $CuGaS_2$ layer and the $CuGaS_2$ crystal.

The orientation of the grown layers was investigated by X-ray diffraction. As expected the layers are [112] oriented, i.e. parallel to the [111] direction of the ZnS substrate.

For electrical measurements gold contacts were evaporated on the $CuGaS_2$ layers and the substrates were contacted with indium. An analysis by the "hot point" and by the "rectifying" method revealed that the as-grown layers are p-type, but their conductivity is very poor. First attempts have shown that by annealing the diodes under maximum sulfur pressure at temperatures of about 700 ^{o}C the conductivity of the layers can be improved by a factor of about 10^4 /5/, but at the same time the surface conductivity of the ZnS substrates decreases very strongly. Therefore suitable coatings must be used or the top layer of the substrate must be removed. These experiments are in an early stage and further investigations are necessary.

If the grown layers are very thin, the conductivity of the diodes is sufficient for electrical measurements. The current-voltage characteristic of a preliminary device, shown in Fig. 4, demonstrates the good rectifying behaviour for voltages up to 70 V. In forward direction at currents above 0.1 mA injection luminescence was observed under the gold contact. The luminescence measurements are finished shortly; about the results we will report in detail at the International Conference on Luminescence 1978 in Paris.

Acknowledgements

The authors would like to acknowledge the support of this work by the Deutsche Forschungsgemeinschaft. We are grateful to M. Schlak for the collaboration on the solubility measurements.

References

/1/ S. WAGNER, J. appl. Phys. 45, 246 (1974).

/2/ YU.P. CHUKOVA and M.V. SHTILIKHA, Soviet Phys. - Semicond. 10, 1329 (1976).

/3/ Handbook of Chemistry and Physics, 57th ed., 1976/77.

/4/ M. RUBENSTEIN, Proc. Internat. Conf. Crystal Growth, 1968 (p. 309).

/5/ B. TELL and H.M. KASPER, J. appl. Phys. 44, 4988 (1973).

(Received June 9, 1978)

phys. stat. sol. (a) 49, K17 (1978)

Subject classification: 11 and 19; 14.3; 22.1.2

Physik-Department E13, Technische Universität München, Garching

EPR Study of the Annealing Behaviour of Neutron-Transmutation-Doped Silicon[1)]

By

G. STETTER[2)], H. COUFAL, and E. LÜSCHER

In honour of Prof. Dr. F. Stöckmann's 60th birthday

Semiconductors doped by neutron transmutation doping are of increasing industrial interest because of their homogeneous doping; most experiments performed up to now deal with silicon samples doped with phosphorus by this method /1, 2/. Neutron transmutation doping (NTD) means the neutron capture by statistically distributed nuclei of the irradiated sample with a subsequent β^--decay, thus increasing the atomic number of these atoms. Doping silicon in this method transmutates ^{30}Si (natural abundance 3.09%) by the process

$$^{30}\mathrm{Si}(n,\gamma)^{31}\mathrm{Si} \xrightarrow[2.6\ \mathrm{h}]{\beta^-} {}^{31}\mathrm{P}^+ + e^-\ (1.5\ \mathrm{MeV})$$

into ^{31}P atoms. This doping technique has the following advantages: Due to the statistical neutron captures the samples are statistically doped with P atoms and because of the small neutron capture cross section (σ = 0.13 barn) even large samples are doped homogeneously. This fact has been proved by spreading resistance and avalanche radiation measurements of doped silicon samples /3/.

Unfortunately not only transmutated atoms but also radiation damages are induced by this doping technique. Doping by reactor radiation causes radiation damages by two different kinds of processes:

1. Damage by recoil of the nuclear reactions; the recoil of the γ- and β^--decays creates defects in the vicinity of the transmutated atom.

2. Damage by fast neutrons; damage cascades are induced all over the sample by head-on collisions of these neutrons with lattice atoms.

These radiation damages must be carefully annealed for technical applications of NTD silicon. Otherwise high temperatures, used in the production of semiconductor

---

1) Research supported by BMFT under contract 03-45 C 06.

2) This work is part of a Ph.D. thesis at the Technische Universität München.

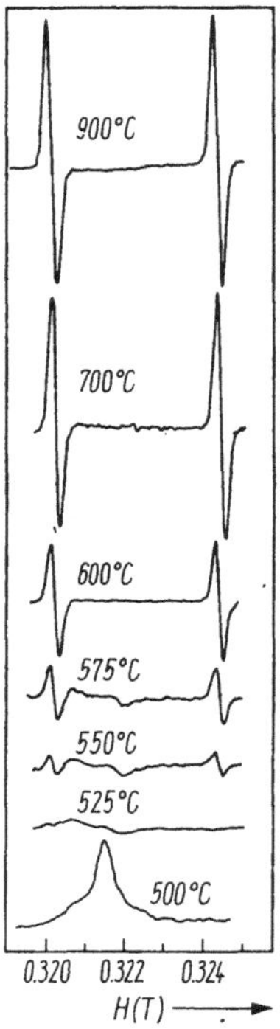

Fig. 1. ESR spectra of silicon samples doped by NTD for various annealing temperatures. Microwave frequency $9.015 \times 10^{9}\ s^{-1}$

devices, would change the resistivity of the NTD substrate. Usually, the annealing of radiation damages and the determination of the donor concentration of P atoms on substitutional lattice sites is monitored by measurements of the electrical resistivity /4 to 6/. ESR spectra of paramagnetic impurities in insulating samples show the surrounding of the impurity. Therefore, the number of P atoms on a lattice site in a silicon sample can be determined.

Dislocation free single crystals, grown by the floating zone technique, with a resistivity of 3800 Ω cm were used as an initial material. The samples were irradiated at the FRM Forschungsreaktor in Munich for 16.3 h with a thermal neutron flux of 10^{14} neutrons/cm^2 thus creating a phosphorus concentration of 10^{13} $^{31}P\ cm^{-3}$. By fast neutrons (neutron flux 1.4×10^{12} neutrons/c about $10^{18}\ cm^{-3}$ Frenkel defects are created simultaneously.

After irradiation the samples were etched with CP4 and were annealed under a vacuum of 3×10^{-4} Pa for 11 h at temperatures between 500 and 900 °C. The ESR spectra were recorded in the X-band using a Varian-V4502 spectrometer at a temperature of 5 K (Heli-Tran-Model LTD 3-110 of Air Products). The ⟨111⟩ direction of the samples was parallel to the external magnetic field.

Fig. 1 shows a summary of the original ESR spectra of samples with various annealing temperatures. The resonance of the substitutional ^{31}P atoms is split by hyperfine interaction (nuclear spin of ^{31}P $I = 1/2$) in a doublet. The g-factor of this doublet is 1.99865 ± 0.00015, its hyperfine-splitting is (117.98 ± 0.5) MHz in excellent agreement with Feher /7/.

These lines could be observed on samples with annealing temperatures above 500 °C, whereas at lower temperatures radiation damages caused an unsplit ESR line with a g-factor of 2.00315 ± 0.00015. The integral intensity of the doublet and thus the number of phosphorus atoms on substitutional lattice sites, as shown in Fig. 2, increases with raising annealing temperatures up to 900 °C.

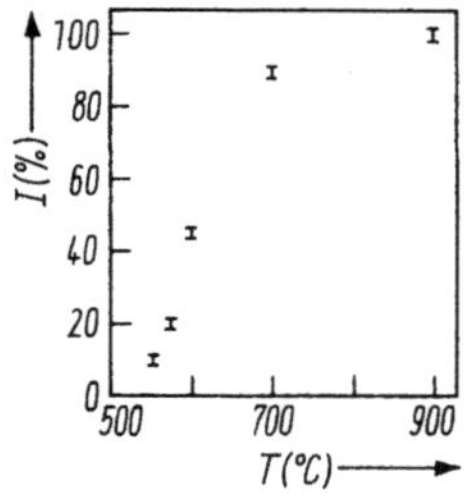

Fig. 2

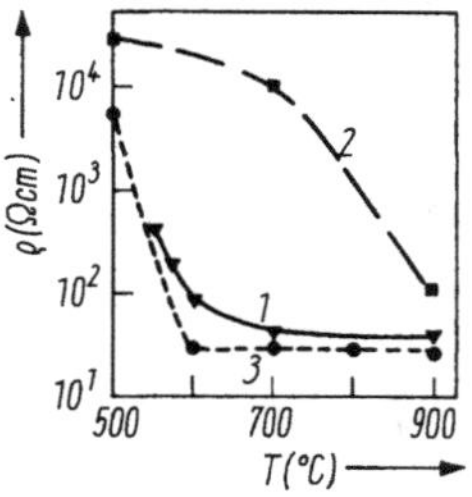

Fig. 3

Fig. 2. Integral intensity of the phosphorus doublet versus annealing temperature

Fig. 3. Annealing behaviour of the electrical resistivity of silicon samples irradiated at reactors with a Cd ratio of (1) 7, (2) 1, and (3) 10

The electrical resistivity of the samples was calculated from the number of phosphorus atoms on substitutional lattice sites. A comparison of these data with experimental results from resistivity measurements /1/ of samples irradiated at nuclear reactors with different Cd ratios is shown in Fig. 3. The annealing behaviour of a sample irradiated at a reactor with a Cd ratio of 1 (1 thermal neutron per 1 fast neutron) is dominated by defects caused by fast neutrons. Irradiating at a Cd ratio of 10 reduces the number of defect cascades accordingly.

Our samples irradiated at a Cd ratio of 7 show an annealing behaviour between these two limits, whereas the final resistivity is a function of the phosphorus concentration and thus of the thermal neutron dose. Resistivity measurements of silicon samples doped by ion implantation /6/ show an annealing behaviour similar to that of our samples.

References

/1/ H. HERZER, 3rd Internat. Symp. Silicon Material Science and Technology, Philadelphia 1977.

/2/ W.E. HAAS and M.S. SCHNÖLLER, J. Electronic Mater. 5, 57 (1976).

/3/ M.S. SCHNÖLLER, IEEE Trans. Electron Devices ED-21, 313 (1974).

/4/ M. TANENBAUM and A.D. MILLS, J. Electrochem. Soc. 108, 171 (1961).

/5/ J. MAYER, L. ERIKSON, and J.A. DAVIES, Ion Implantation in Semiconductors, Academic Press, 1970.

/6/ G. DEARNALEY, J.H. FREEMAN, R.S. NELSON, and J. STEPHEN, in: Defects in Crystalline Solids, Vol. 8, Ed. S. AMELINCKX, R. GEVERS, and J. NIHOUL, North-Holland Publ. Co., 1973.

/7/ G. FEHER, Phys. Rev. 114, 1219 (1959).

(Received April 10, 1978)

phys. stat. sol. (a) 49, K21 (1978)

Subject classification: 18.2; 21.1.1

Institut für Angewandte Physik der Universität Münster[1)]

On the Magnetostriction of Grain-Oriented Ferromagnetic Material

By

H. BITTEL, W. GROSSE-NOBIS, and TH. JANNEMANN

In honour of Prof. Dr. F. Stöckmann's 60th birthday

Introduction Polycrystalline material of silicon-iron alloys can be processed by special rolling and heat treatment to yield cubic grain-oriented strips with the crystallographic easy magnetic axis [100] parallel to the strip axis (cf. Fig. 1; Goss textured and cube-textured sheets). If the sample is magnetized by an external magnetic field directed exactly along the strip axis, the magnetic flux change is accomplished by shifting only 180° Bloch walls. In the demagnetized state the domains of parallel and antiparallel magnetization are of equal width in the mean and show a bar-like structure. The 180° processes do not change the sample length l by magnetic proceedings. Hence the magnetostriction $\lambda = \Delta l/l$ vanishes (Δl change of length).

Strong mechanic deformations (plastic deformation 0.1%) produce internal stresses in the originally grain-oriented sample. The induced position dependent uniaxial preferred axes due to the strain are often isotropically oriented in the mean. The strain can be so strong that the influence of the original grain orientation can be detected no longer. Since the local magnetization is now not parallel with the axis of the strip in the demagnetized state, approach to saturation changes the sample length considerably. Fig. 2 shows the different magnetostrictive behaviour of two strips. One

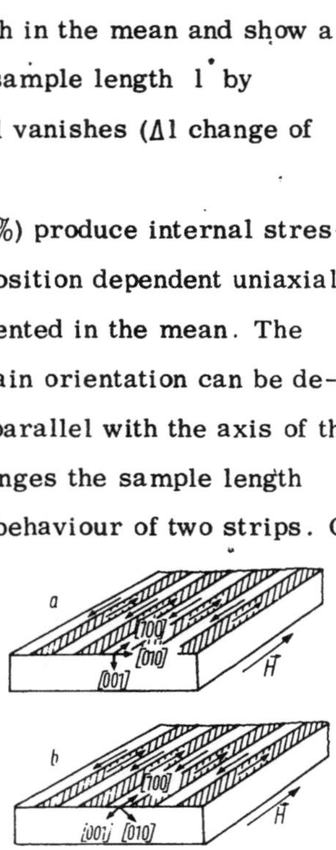

Fig. 1. Strips of oriented transformer steel with ideal grain orientation. 180° Bloch walls running quite through the strip forming a bar-like structure of domains with parallel and antiparallel magnetization with respect to the external field $\vec{H}$; a) cube texture, b) Goss texture

1) Roxeler Str. 70/72, D-4400 Münster, BRD.

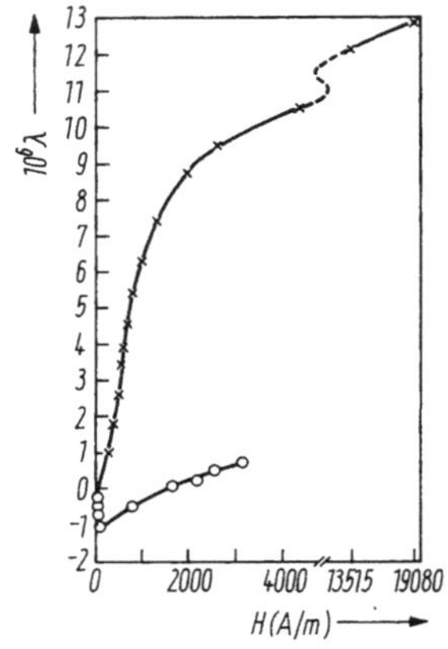

Fig. 2. Magnetostriction curve of grain oriented sample of good quality compared to that of a mechanically strongly deformed one. Initial curves, starting from the demagnetized state produced by demagnetization with 50 Hz measured with set-up shown in Fig. 3. × sample 6, heavily deformed, O sample 13, Goss; f = 50 Hz

having approximately isotropic preferred directions and the other having an undisturbed high-grade grain orientation. The figure also explains the fact that an ac transformer with a core of good grain oriented material has the advantage of producing much less mechanical vibrations owing to the minor excitation by magnetostrictive effects.

Occurrence of domains with transverse magnetization In practice, however, there remains a certain misorientation of the crystallites, and there is an influence of the grain boundaries, too. Especially the easy axis [100] of the crystallites often is not parallel to the sheet surface. As a consequence the spontaneous magnetization in [100]-direction leads to magnetic flux perpendicular to the sample surface (stray flux). This results in secondary domains which are transversely magnetized to the sample axis, and so contribute to a magnetostrictive shortening in the longitudinal direction in silicon-iron: Cube-textured material shows the well-known fir tree pattern. Goss material exhibits spikes at the surface which conduct the stray flux additionally by tube-like interior domains with [010] or [001] magnetization to the opposite surface /1/. Measurement of the magnetostriction of grain-oriented material yields useful information about the existence of transverse domains and consequently about details of the supplementary structure /2, 3/. This especially holds for the domain structure of the demagnetized state.

Device for measuring magnetostriction Strips of 3 wt% silicon-iron were investigated. Strain gauges and similar methods for measuring small changes in length cannot be applied. Therefore a Michelson interferometer with a He-Ne laser as the light source is used /3/, Fig. 3. The system produces circular interference rings on the screen. By subjective observation a measuring accuracy of $\lambda = \Delta l/l \approx 0.2 \times 10^{-6}$ for a sample length of l = 300 mm is achieved.

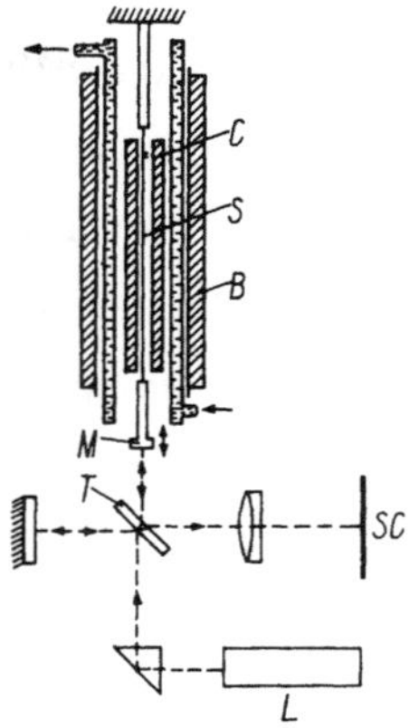

Fig. 3. Experimental set-up for interferometric magnetostriction measurements, schematic diagram. M mirror fixed to a brass cylinder; S vertically hanging specimen ($300x2x0.1\ mm^3$); L laser (wavelength 632.8 nm); B cylindrical coil for dc fields; C cylindrical coil producing ac fields; SC screen; T beam-splitter

Auxiliary windings of coil B are used to compensate the vertical component of the earth magnetic field. Furthermore an inner coil C impresses a magnetic ac field of a desired frequency whose peak value can be steadily decreased from 250 A/m to zero. It is used for the demagnetization of the samples. Special care was taken to avoid eddy current influence from parts other than the metallic sample, the complete device is located in a temperature stabilized room. The vessel between coils B and C is connected to the water circulation of a thermostat. All the measurements were made at room temperature. Fig. 2 and 4 show some examples. Preceeding the magnetostriction measurement the samples were annealed in vacuo at 750 ^{o}C for 1 h to get them free from internal stresses (coercive field $H_c \approx 24$ A/m). During the demagnetizing process and the measurement of the magnetostriction a small uniform tension of about $0.35x10^7\ N/m^2$ is applied to the sample. This is due to the mass of 70 g of the mirror holder. By this uniaxial stress the magnetization is additionally favoured to lie parallel to the longitudinal direction of the sample.

Results and discussion In the case of the annealed Goss strips the curves of Fig. 4 show with increasing field starting from the demagnetized state initially a magnetostrictive shortening. The minimum length is reached at about 150 A/m and thereafter the length of the sample increases. In connection with the above-mentioned transverse domains this run of the curves means that starting at the demagnetized state firstly

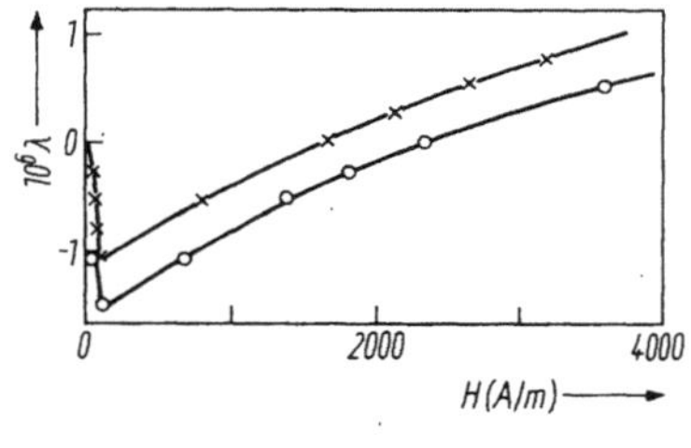

Fig. 4. Initial magnetostriction curves of a 3 wt% SiFe Goss sample after demagnetization with different frequencies. Sample 13, Goss, × 50 Hz, ○ 1 kHz

transversely magnetized secondary domains will form. Only for fields larger than 150 A/m the magnetization is preferably aligned parallel to the sample length. The transverse domains disappear with increasing field.

However, the curves are different depending on the way of demagnetization with an ac field of 50 Hz or of 1 kHz, respectively. Both curves of Fig. 4 as well as those of other samples show at least a uniform run in the field interval from the minimum up to the highest field strength: They merely are shifted in parallel by about $\delta\lambda = 0.5\text{x}10^{-6}$. Thus we suggest that the curves should be brought to coincide in that interval and that the mentioned difference δl should be attributed to the different demagnetized states since there will be no detectable influence of the demagnetization procedure at those high fields: The samples are shorter in length by relatively $0.5\text{x}10^{-6}$ when demagnetized with 50 Hz in comparison to the demagnetized state by 1 kHz.

The influence of the frequency of the demagnetizing ac field is explained as follows At the higher frequency also a bar-like primary domain structure with parallel and antiparallel domains formed, but the width of the domains is smaller than in the case of lower frequency /4, 5/. Besides other influences the finite propagation velocity of irreversible magnetizing processes owing to eddy current damping is responsible for this effect /6, 7/.

In the case of a weak misorientation of the grain-oriented material the component of the magnetization perpendicular to the sample surface causes a stray field outside the sample with a relative large stray field energy stored in it. If this energy exceeds a certain threshold the sample forms supplementary domains which collect the stray flux and direct it more economically back into the sample. If the primary bar-like domains are very narrow this threshold is not exceeded and the more economical arrangement is, that mainly the flux escapes into the exterior of the sample. However, in the case of wide domains produced by ac demagnetization with low frequency, secondary domains with transverse magnetization are formed more numerously. Henc this demagnetized state results in a shorter sample.

One concludes that depending on the demagnetizing method various demagnetized states can be obtained /8, 9/. There does not exist a unique demagnetized state as frequently is supposed erroneously. Besides the here differently chosen values of external parameters for the demagnetizing procedure additionally, owing to thermal activation in the considered solid, there should also be taken into account an influence on the domain structure by random forces /10/. Since these forces with strong likelihood merely cause changes on a much finer scale, there is no chance to detect these different demagnetized states by magnetostriction measurements.

Finally, the whole run of the curves of Fig. 4 may be interpreted by production and reduction of secondary domains: Starting at the demagnetized state increasing field diminishes unfavourably magnetized primary domains in favour of those primary domains which are parallel to the external field. The latter domains grow wider. Hence, owing to the misorientation present in a real sample the above-mentioned spikes and tube-like secondary domains form with the consequence of shortening of the sample. The volume of these transverse domains will attain a maximum at the minimum of the magnetostriction curve and will in larger fields decrease in favour of the longitudinal magnetization. Consequently, this now leads to an elongation of the sample. At the minimum of the magnetostriction curves the saturation value of the sample's gross magnetization is already nearly reached as a comparison with the hysteresis loop proves.

References

/1/ W.S. PAXTON and T.C. NILAN, J. appl. Phys. 26, 994 (1955).

/2/ G. KIETZMANN, Diploma work, University Münster, 1973.

/3/ H. BITTEL, W. GROSSE-NOBIS, and G. SCHÖNFELDER, Physica 80B, 83 (1975).

/4/ T.R. HALLER and J.J. KRAMER, J. appl. Phys. 41, 1034 (1970).

/5/ T.R. HALLER and J.J. KRAMER, J. appl. Phys. 41, 1036 (1970).

/6/ H. EIFRIG, W. GROSSE-NOBIS, and K. JANSEN, J. Magnetism magnetic Mater. 6, 73 (1977).

/7/ W. GROSSE-NOBIS and K. JANSEN, Proc. Conf. Soc. Magnetic Materials, Vol. 3, Bratislava 1977.

/8/ H. FAHLENBRACH and K. SIXTUS, Z. Metallk. 40, 187 (1949).

/9/ H. BITTEL, Forschungsbericht Nr. 1729 des Landes NRW, Westdeutscher Verlag Köln und Opladen, 1966.

/10/ J.L. PORTESEIL, Physica 93B, 201 (1978).

(Received May 16, 1978)

phys. stat. sol. (a) 49, K27 (1978)

Subject classification: 1.2; 22.9

Institute of Chemistry, Jagiellonian University, Cracow[1)]

Low-Temperature Phase-Transition in Urea

By

L. LEBIODA, S. HODOROWICZ, and K. LEWINSKI

Introduction The crystal structure of urea was investigated by several authors with the aid of X-ray and neutron diffraction. The most recent review is in the article by Pryor and Sanger /1/. These authors carried out very precise neutron measurements and combining their results with the previously collected X-ray data /2/ tried to get bonding features of the electron density on a difference X-N map. This attempt was not successful and they were convinced that the failure is a result of "a real discrepancy between the observed and calculated structure factors".

The displacive structural modulation in thiourea /3/ suggests that similar effects developing below room temperature can be responsible for these discrepancies.

Specific-heat measurements /4/ do not contradict some structural transition at this temperature. They revealed excess absorption of energy in the temperature range 230 to 265 K. Originally this was attributed to some presence of water in the investigated sample.

The low-temperature study of the crystal structure of urea /5/, carried out at 133 K, did not show substantial changes but it did not cover the discussed region.

Experimental X-ray powder diffraction data were collected with filtred CuK_α radiation on a DRON-1 (Burieviestnik) diffractometer using the previously described /6/ low-temperature attachment. The required temperatures, in the investigated range 293 to 93 K, were obtained by gradual decrease without thermal shock and were controlled with an accuracy of 0.1 K. The measured samples were kept at the given temperature for about 15 min before the recording of the diffraction pattern. We did not notice significant changes with longer exposures to low temperatures.

The photographs of the reciprocal lattice of urea crystals were taken using CuK_α radiation with an Enraf-Nonius precession camera equipped with a low-temperature attachment produced by the same company. The dimensions of the studied crystals

1) Krupnicza 41, 30-060 Cracow, Poland.

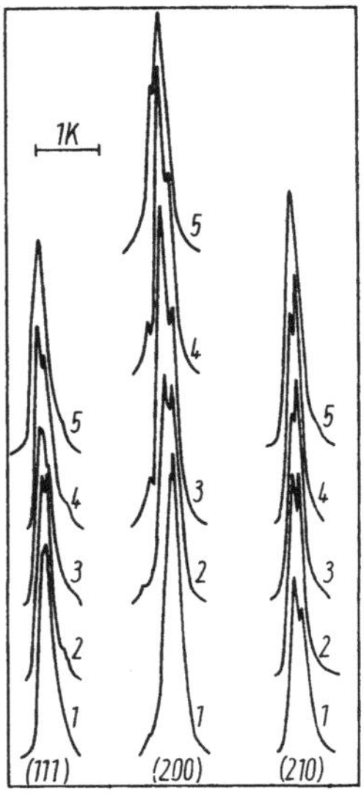

Fig. 1. Diffraction profiles of selected reflections at temperatures (1) 293, (2) 233, (3) 193, (1) 133, (5) 93 K

were about 0.1 mm. The crystals were somewhat out of the alignment to avoid superposition of the diffraction spots from the two Bragg positions. The superposition is never complete and might lead to some deformation of the reflection shape.

Lattice constants were obtained from the least-squares refinement based on θ angles of 15 reflections measured on a CAD-4 (Enraf-Nonius) single crystal diffractometer equipped with the same low-temperature device. The angles θ were about 50° and were derived from measurements on both sides of the primary beam.

Results The measurements were carried for two kinds of samples: commercially available urea (Ciech, Poland, pro analysis) and recrystallized by a slow evaporation of an aqueous solution.

Powder diffractograms, at room temperature, were essentially the same for both samples. The peaks were somewhat broader than for well crystalline etalons. At lower temperatures the diffraction patterns of recrystallized urea did not show significant changes. On the diffractograms of commercial urea, however, splitting of some peaks was observed as illustrated in Fig. 1. The relative intensities of maxima changed systematically with temperature. The obtained separability was much better than the accuracy of peak position measurements we could claim.

The photographs of reciprocal lattice layers $\{hk0\}$, $\{h0l\}$, and $\{hhl\}$ were taken for two different crystals from each sample. We did not observe satellite reflections neither for the crystals obtained by recrystallization nor for these selected from the commercial urea. Lattice constants obtained from these measurements are given in Fig. 2 together with values found in literature /5/. The systematic absences (h odd in $\{h00\}$), implied by the space group of urea, were preserved at low temperatures.

The last squares refinement of lattice parameters, based on single-crystal diffractometric measurements, was carried out for both the tetragonal and the orthorhombic system. The results showed that there is no reason to introduce lower symmetry.

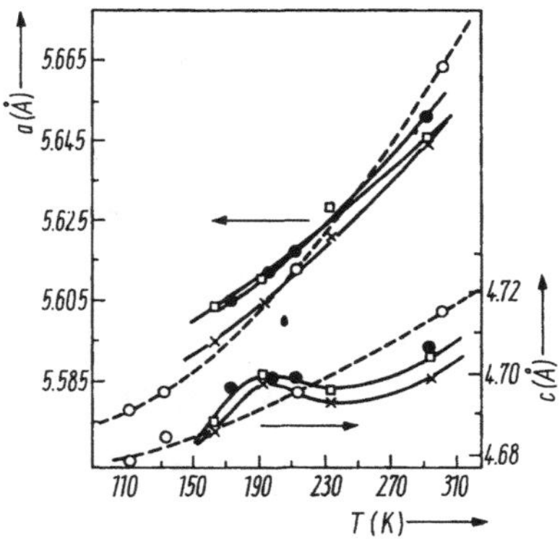

Fig. 2. Lattice constants versus temperature. O Sklar et al. /5/; recrystallized urea: □ from precession photographs, ● from single crystal diffractometer; rapidly crystallized samples: × from precession photographs.

Discussion The obtained results are dependent on the perfection of the investigated crystals. It is known from studies on the kinetics of phase transitions that nucleation of a new phase in crystalline solids is initiated in the environments of structural imperfections or inclusions /7/. In this study we observed diffraction effects inconsistent with the tetragonal urea structure for the samples which were crystallized rapidly and presumably were severly imperfect. The specimens suitable for single-crystal methods selected from the non-recrystallized urea contained certainly less defects than the rest of the sample and did not undergo transition. It is probable that in powder samples the density of active structural imperfections, being the potential centre of new-phase nucleation, is higher than in single crystals. The similar influence of the grain size on phase transition is described by Johanson /8/.

Basing on powder data only we could not conclude about the nature of the new phase. The systematic variation of the additional-maxima intensities with temperature is typical for the structure modulation, but sinusoidal displacements result in at least two additional maxima on both sides of the reflection while we observed only one. The hypothesis of a simple lowering of the urea symmetry is inconsistent with the number of additional maxima for particular reflections and with the observed intensity-temperature dependence.

The anomaly observed in the thermal expansion at about 190 K is probably connected with the effects observed for powder samples. Further measurement of lattice parameters is planned by the Bond method.

Conclusions We observed a phase transition in urea, which probably explains the discrepancies in the crystal structure described at room temperature /1/. The perfection of crystals is critical for the new-phase formation. The new phase cannot be described neither in a lower crystal symmetry nor as a sinusoidal periodic distortion incommensurable with the crystal lattice.

References

/1/ A.W. PRYOR and P.L. SANGER, Acta cryst. A 26, 543 (1970).

/2/ P. VAUGHAN and J. DONOHUE, Acta cryst. 5, 530 (1953).

/3/ Y. SHIOZAKI, Ferroelectrics 2, 245 (1971).

/4/ R.A. RUEHRWEIN and H.M. HUFFMAN, J. Amer. Chem. Soc. 68, 1759 (1

/5/ N. SKLAR, M.E. SENKO, and B. POST, Acta cryst. 14, 716 (1961).

/6/ S. HODOROWICZ, M. CIECHANOWICZ-RUTKOWSKA, J.M. JANIK, and J.A. JANIK, phys. stat. sol. (a) 43, 53 (1977).

/7/ D. TURNBULL, Primesi i defekty, Metallurgizdat, Moscow 1960 (p. 141).

/8/ G. JOHANSON, Arkiv Kemi (Sweden) 8, 33 (1955).

(Received May 12, 1978)

phys. stat. sol. (a) 49, K31 (1978)
Subject classification: 2; 21.6

Department of Metallurgy and Materials Science, University of Toronto[1)]

Structural Relaxation in Metallic Glasses

By
Y. WASEDA and W.A. MILLER

The metallic glasses produced by rapid quenching from the melt are not completely stable in the quenched condition, so that certain properties such as internal friction /1/ and specific heat /2/ show significant changes during annealing at temperatures below their crystallization temperatures. This behaviour has been interpreted in terms of structural relaxation /2/ in the amorphous state which is clearly separate from crystallization. Although structural relaxation has been frequently studied in metallic glasses by the indirect methods of internal friction /1, 3/, specific heat /2/, stress relief /4/, and creep rate /5/ measurements, very little work involving direct structural measurements has been reported. A slight change has been observed in the X-ray diffraction pattern of amorphous $Fe_{80}P_{13}C_7$ during low temperature annealing /6/. The accuracy of this measurements, however, was insufficient to allow definite interpretation of the data with respect to changes resulting from structural relaxation.

The main purpose of this note is to present some new X-ray measurements on structural changes which occur in the metallic glasses, $Pd_{80}Si_{20}$ and $Ni_{77}P_{23}$ during low temperature annealing.

Amorphous samples of $Pd_{80}Si_{20}$ were prepared in the shape of ribbons (about 3 mm in width x 0.04 mm in thickness) by rapid quenching from the melt. By contrast, an amorphous sample of $Ni_{77}P_{23}$ sheet (about 2x3x0.1 cm^3) was produced by electro-deposition on copper plate using a method previously described by Waseda and Tamaki /7/. Sample composition was determined by chemical analysis in each case. The experimental set-up, as well as the techniques for handling the X-ray scattering intensity, the correction of the observed intensity data and the Fourier analysis are essentially similar to the procedures employed previously /6/. The measurements were carried out in the Q-range between 0.75 and 17.0 $Å^{-1}$. According to the method

1) Toronto, M5S 1A4, Canada.

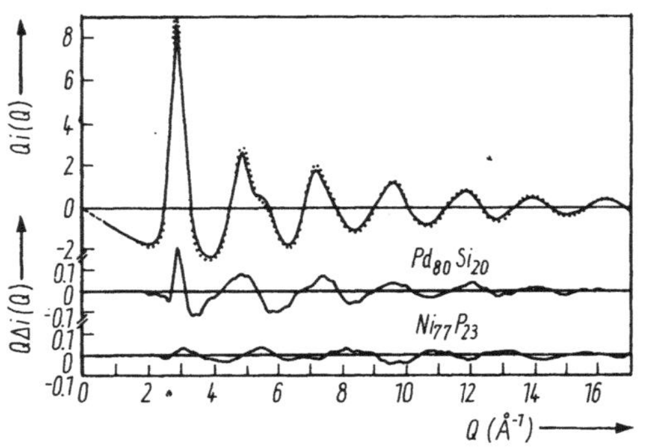

Fig. 1

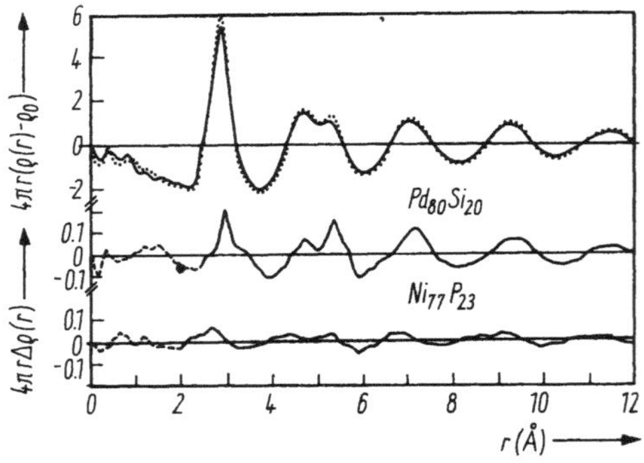

Fig. 2

Fig. 1. Effect of annealing on the function Qi(Q) and the relative change QΔi(Q) of rapid-cooled $Pd_{80}Si_{20}$ and electro-deposited $Ni_{77}P_{23}$. —— as-quenched; ··· annealed, 250 °C, 40 min

Fig. 2. Effect of annealing on the function $G(r) = 4\pi r\,[\rho(r) - \rho_o]$ and the relative change $4\pi r\Delta\rho(r)$ of rapid-cooled $Pd_{80}Si_{20}$ and electro-deposited $Ni_{77}P_{23}$. —— as-quenched; ··· annealed, 250 °C, 40 min

proposed by Rahman /8/ and the detailed discussion for systematic errors for X-ray diffraction by Greenfield et al. /9/, the total error in the Qi(Q) values obtained in this work were estimated to be about 1.9%, where i(Q) is the reduced interference function. However, the relative change QΔi(Q) produced by low temperature annealing is in error by less than 0.5%, i.e., the accuracy of the determination of the relative change QΔi(Q) is higher than that in the determination of Qi(Q) itself. Here the relative change QΔi(Q) is defined as $Q\Delta i(Q) = Q\,[i(Q)_{annealed} - i(Q)_{as\text{-}formed}]$. The radial distribution function $G(r) = 4\pi r\,[\rho(r) - \rho_o]$ is estimated from Qi(Q) by the usual Fourier transformation, where $\rho(r)$ is the radial density function and ρ_o is the average atomic number density. The uncertainties in the values of G(r) are probably similar to those for Qi(Q), because it has been confirmed that computational errors associated with the Fourier transformation are minimized by using well-known procedures /10, 11/. The relative change in G(r) is similarly defined as $4\pi r\Delta\rho(r) = 4\pi r\,[\rho(r)_{annealed} - \rho(r)_{as\text{-}formed}]$.

The experimental results for Qi(Q) and G(r) in amorphous $Pd_{80}Si_{20}$ in both the as-formed and the annealed conditions are shown in Fig. 1 and 2. Numerical values

Table 1

Density, peak positions in the functions Qi(Q) and G(r), and the area of the first peak in G(r) which corresponds to the average near-neighbour numbers of Pd-Pd and Pd-Si (or Ni-Ni and Ni-P) pairs, n_1

		density (g/cm^3)	peak position in Qi(Q) ($Å^{-1}$)			peak position in G(r)			n_1 (atoms)
			first	second	shoulder	first	second	shoulder	
$Pd_{80}Si_{20}$	as-quenched	10.62	2.82	4.85	5.62	2.81	4.69	5.29	11.9
	annealed at 250 °C for 40 min	10.66	2.82	4.82	5.60	2.81	4.68	5.33	12.1
$Ni_{77}P_{23}$	as-deposited	7.82	3.06	5.36	6.06	2.56	4.21	4.78	12.4
	annealed at 250 °C for 40 min	7.83	3.06	5.37	6.07	2.55	4.21	4.77	12.7

for density and peak positions are listed in Table 1. The densities were measured by using Archimedes' method in toluen. The small oscillations before the first peak in G(r) are spurious ones, which do not critically affect the estimation of interatomic distance and the area of the first peak in the radial distribution function /6, 9/.

Our results for the basic profile of Qi(Q) and G(r) are quite similar to the features which have been reported previously for other metallic glasses, i.e., the profiles are composed of a relatively sharp first peak and smaller subsequent peaks. In addition, we observed the familiar characteristic features of split second peaks in G(r) and the second peak shoulder in Qi(Q). These features are also found in the results of amorphous $Ni_{77}P_{23}$. The main effect of low temperature annealing on the structure of rapid-cooled $Pd_{80}Si_{20}$ was to increase the amplitude of the oscillations as shown in Fig. 1 and 2. Although the observed differences in Qi(Q) and G(r) are small, the profile of the relative changes indicate clearly the effect of low temperature annealing on the structure of rapid-cooled $Pd_{80}Si_{20}$. Our results may be summarized as follows;

(1) The heights of the first, second, and third peaks in Qi(Q) increase by about 2% but the height of the second peak shoulder decreases by about 3%. It should be noted that the opposite changes are observed in the heights of both the second peak and its shoulder in Qi(Q) when samples start to crystallize /6, 12/.

(2) The position of the first peak in G(r) and its area which corresponds to the average near-neighbour numbers of Pd-Pd and Pd-Si pairs are almost unchanged, although the density increases slightly by about 0.4% (see Table 1).

(3) Although the height of the first and second peaks in G(r) also increase by about 2%, the increase in the height of the second peak shoulder is more intense, i.e., an increase of about 6% is observed.

(4) The height of subsequent peaks in G(r) (the third, fourth, and fifth show increases of the order of 4 to 7%.

The values of percentage above-mentioned correspond to the change in amplitude the oscillations in the related functions. These observations agree well with those of Egami and Ichikawa /12/ on amorphous $Fe_{40}Ni_{40}P_{14}B_6$, although there are difference in detail such as the magnitude of the change in peak heights. The features observed in this work, especially the results for G(r), permit some comments to be made on the effect of low temperature annealing on the structure of rapid-cooled metallic glasses. In particular, point (2) above suggests that the basic structural arrangement of atoms (the so-called tetrahedral units of nearest neighbour atoms) is insensitive to low temperature annealing. However, points (3) and (4) clearly indicate that a rearrangement occurs in the relative positions of the tetrahedral units without crystallization, accompanied by a decrease in free volume.

It would appear that structural relaxation involves slight rearrangement leading to a somewhat more ordered atomic distribution without the long-range diffusion of atoms. Stress relief and the annihilation of the excess free volume accompany this process. The small observed increases in ordering and in density qualitatively correspond to some recent model calculations /13, 14/. It is also interesting to note that small density changes have been observed in studies of mechanical packings of equal spheres by Scott /15/. He showed that the density of mechanically stable dense random packings can vary continuously in a range from about 60% to 63.7% depending upon the amount of compaction. Thus structural relaxation in metallic glasses may correspond to changes which parallel these that occur during the compaction process in dense random packing of hard spheres.

In contrast to the results for rapid-cooled $Pd_{80}Si_{20}$, our results for electro-deposited $Ni_{77}P_{23}$ show only small corresponding changes due to low temperature annealing at 350 ^{o}C for 30 min as shown in the relative changes in Fig. 1 and 2. This is consistent with the previous results of Chi and Cargill /16/ for electro-deposited $Co_{77.8}P_{22.2}$. It would seem, therefore, that metallic glasses produced by the slower process of electro-deposition are quite stable prior to crystallization and thus structural relaxation is relatively small. By contrast, a highly disordered structure is frozen, in which metallic glasses are produced by rapid-quenching and this structure, being metastable, relaxes during low temperature annealing.

The authors express their gratitude to Prof. T. Egami for his useful comments and for providing us with a copy of his work prior to its publication. Financial support from the National Research Council of Canada is also gratefully acknowledged.

References

/1/ H.S. CHEN, H.J. LEAMY, and M. BARMATE, J. non-crystall. Solids 5, 444 (1970).

/2/ H.S. CHEN and E. COLEMAN, Appl. Phys. Letters 28, 245 (1976).

/3/ T. SOSHIRADA, M. KOIWA, and T. MASUMOTO, J. non-crystall. Solids 22, 173 (1976).

/4/ R.S. WILLIAMS and T. EGAMI, IEEE Trans. Magnetics MAG-12, 927 (1976).

/5/ R. MADDIN and T. MASUMOTO, Mater. Sci. Engng. 9, 153 (1972).

/6/ Y. WASEDA and T. MASUMOTO, Z. Phys. B 22, 121 (1975).

/7/ Y. WASEDA and S. TAMAKI, Z. Phys. B 23, 315 (1976).

/8/ A. RAHMAN, J. chem. Phys. 42, 3540 (1965).

/9/ A.J. GREENFIELD, J. WELLENDORF, and N. WISER, Phys. Rev. A 4, 1607 (1971).

/10/ R. KAPLOW, S.L. STRONG, and B.L. AVERBACK, Phys. Rev. 138, A1336 (1965).

/11/ A.H. NARTEN, D.M. DANFORD, and H.A. LEVY, J. chem. Phys. 46, 4875 (1967).

/12/ T. EGAMI and K. ICHIKAWA, Mater. Sci. Engng. 32, 293 (1978).

/13/ L.V. HEIMENDAHL, J. Phys. F 5, L141 (1975).

/14/ R. YAMAMOTO, H. MATSUOKA, and M. DOYAMA, J. Phys. F 7, L243 (1977)

/15/ G.D. SCOTT, Nature 188, 908 (1960).

/16/ G.C. CHI and G.S. CARGILL III, AIP Conf. Proc. 21, 359 (1976).

(Received May 18, 1978)

phys. stat. sol. (a) 49, K37 (1978)
Subject classification: 18.4; 8; 14.1; 21.1

The Research Institute for Iron, Steel and Other Metals, Tôhoku University, Sendai[1)]

Magnetic Phase Diagram of the Dilute Cr-Sb Alloy System

By
K. FUKAMICHI

It is well-known that the spin structure of antiferromagnetic Cr is modified by a small amount of additional element, and the magnetic phase diagrams have been determined for many kinds of Cr-based binary alloys containing transition metals /1 to 5/. The Néel temperature T_N is depressed by addition of V and Ta which reduce the electron-to-atom ratio, while T_N is increased by addition of Mn, Re, and other elements which increase the ratio /6/. On the other hand, the concentration dependence of the Néel temperature of Cr-based alloys containing non-transition metals does not indicate such a systematic tendency /7 to 11/.

In this note, the magnetic phase diagram determined from measurements of the electrical resistivity and thermal expansion for the dilute Cr-Sb alloy system is reported. Since Sb is soluble to about 5% in Cr /12/, the experiments were carried out in the solid solution range confirmed by X-ray analysis.

Cr-Sb alloys were prepared from 99.9% pure Cr and Sb by arc-melting in an argon atmosphere, and homogenized at 1150 $^{\circ}$C for 3 days in a silica ampoule. The specimens were cut into the form of square rods of about 2x2x30 mm^3. The composition of the specimens was determined by chemical analysis. The temperature dependence of the electrical resistivity was measured by a usual four-terminal method. The thermal expansion was measured by a differential transformer method, and a fused silica supplied by National Bureau of Standards was used as a standard material.

The temperature dependence of the relative change in electrical resistivity $\Delta\varrho/\varrho$ is shown in Fig. 1. In general, the resistivity-temperature curve of pure Cr and its alloys indicates a minimum below the Néel temperature due to the fractional truncation of the Fermi surface /13/. The Néel temperature, therefore, has been often determined from the ϱ-T curve /14, 15/. As seen in the figure, however, the curves

1) Sendai, 980, Japan.

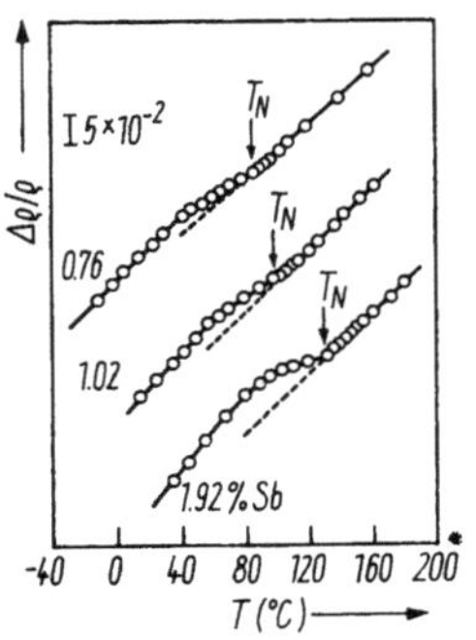

Fig. 1

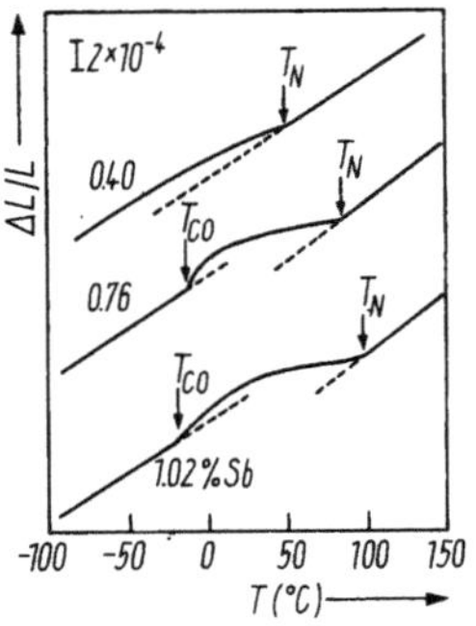

Fig. 2

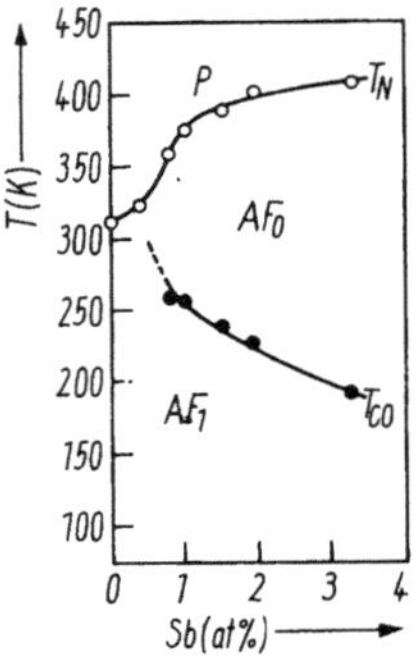

Fig. 3

Fig. 1. Relative change in the electrical resistivity as a function of temperature for Cr-Sb alloys

Fig. 2. Thermal expansion curves of the Cr-Sb alloys

Fig. 3. Magnetic phase diagram of the Cr-Sb alloy system

indicate no minimum. Therefore, the temperature, where the curve begins to deviate from the high temperature linear curve, was defined as the Néel temperature T_N.

Fig. 2 shows the thermal expansion curves of the Cr-Sb alloys. They are classified in two types. One is very similar to that of pure Cr and its thermal expansion coefficient α becomes slightly smaller below T_N as seen in the curve of the Cr-0.40% alloy. The other is similar to that of Cr-Sn alloys /9/ and α is reduced below T_N and becomes again large positive at lower temperatures, as seen in the curves of the Cr-0.76%Sb and Cr-1.02%Sb alloys.

Generally, Cr based alloys have three kinds of spin structures /3/. The magnetic transition from the paramagnetic state (P) to the commensurate antiferromagnetic state (AF_0) induces a decrease in the thermal expansion coefficient α /16/. On the other hand, the transition from P to the incommensurate antiferromagnetic state (AF_1) leads to a slightly smaller α /16/; α indicates a large positive value around the transition temperature from AF_0 into AF_1 /11, 17/. Therefore, the spin structure can be presumed from the thermal expansion characteristics. For example, the magnetic transition with decreasing temperature in the Cr-1.02 % Sb alloy will be:

$$P \xrightarrow[T_N]{} AF_0 \xrightarrow[T_{C0}]{} AF_1 ,$$

where T_{C0} indicates the AF_0 - AF_1 transition temperature, and hence the magnetic phase diagram is obtained.

Fig. 3 shows the magnetic phase diagram of the Cr-Sb alloy system. Here the temperature where the thermal expansion curve begins to deviate from the linear curve at low temperatures was defined as T_{C0}. With increasing Sb content, the Néel temperature T_N increases almost linearly with the Sb content while T_{C0} decreases. These results suggest that the triple point is below 1% Sb.

According to Fedders and Martin's model the relation between the magnetic moment and the Néel temperature is expressed as /6/

$$\frac{\mu}{\mu_{Cr}} = \frac{T_N}{T_{0N}} ,$$

where μ_{Cr} and T_{0N} are the magnetic moment and the Néel temperature of pure Cr, respectively. Therefore, the value of μ for Cr-1.92% Sb alloy is estimated as 0.54 μ_B by using the value of $\mu_{Cr} = 0.41\ \mu_B$ /7/.

Acknowledgements

The author wishes to thank Lecturer S. Takeyama and his co-workers at the Analytical Laboratory in our Institute for chemical analysis of specimens. Thanks are also due to Mr. Oki for X-ray analysis.

References

/1/ Y. ISHIKAWA, S. HOSHINO, and Y. ENDOH, J. Phys. Soc. Japan 22, 1221 (1967).

/2/ Y. ENDOH, Y. ISHIKAWA, and H. OHNO, J. Phys. Soc. Japan 24, 263 (1968).

/3/ B. LEBECH and MIKKE, J. Phys. Chem. Solids 33, 1651 (1972).

/4/ T. SUZUKI, J. Phys. Soc. Japan 41, 1187 (1976).

/5/ Y. TSUNODA and N. KUNITOMI, Solid State Commun. 18, 1103 (1976).

/6/ W.C. KOEHLER, R.M. MOON, A.L. TREGO, and A.R. MACHINTOSH, Phys. Rev. 151, 405 (1966).

/7/ A. KALLEL and F. DE BERGEVIN, Solid State Commun. 5, 955 (1967).

/8/ S. ARAJS, N.L. REEVES, and E.E. ANDERSON, J. appl. Phys. 42, 1691 (1971).

/9/ K. FUKAMICHI and H. SAITO, J. Phys. Soc. Japan 33, 1485 (1972).

/10/ J.W. CABLE, J. Magnetism magnetic Mater. 5, 112 (1977).

/11/ T. SUZUKI, J. Phys. Soc. Japan 43, 869 (1977).

/12/ H. NOWOTNY and O. ÅRSTAD, Z. phys. Chem. 38, 461 (1938).

/13/ M.B. SALANOM, K.S. SIMONS, and P.R. GARNIER, Solid State Commun. 7, 1035 (1969).

/14/ T. DE YOUNG, S. ARAJS, and E.E. ANDERSON, J. less-common Metals 32, 165 (1973).

/15/ S. ARAJS, K.V. RAO, and E.E. ANDERSON, Solid State Commun. 16, 331 (1975).

/16/ K. FUKAMICHI and H. SAITO, Sci. Rep. Res. Inst. Tôhoku Univ. 26, 299 (1977)

/17/ T.I. KOSTINA, V.YU. GALKIN, E.I. DONDORSKII, and A.V. KATERENCHUK, Fiz. Metallov i Metallovedenie 40, 851 (1975).

(Received May 18, 1978)

phys. stat. sol. (a) 49, K41 (1978)
Subject classification: 13.4 and 14.3.4; 22.9

Institute of Physics, Technical University of Łódź

SCL Currents and Trap Distribution in Polycrystalline Tetracene

By
W. MYCIELSKI and A. LIPIŃSKI

The space charge limited current (SCLC) technique is a well-known method of trap distribution investigation in high-resistivity materials /1/. In recent years new methods of analysis of SCLC curves were reported by several authors /2 to 4/. In particular, a very useful method which leads directly to the trap distribution (energy and density of localized states) was proposed by Manfredotti et al. /3/. This analysis is clearly valid in the regime of strong trapping, i.e. when $n_t \gg n_f$ (n_t is the density of trapped carriers, n_f is the free carrier density). In this case, the maximum of the dn_t/dE_F versus E_F plot (E_F is energy of the quasi-Fermi level) may be identified to a good approximation as a discrete or diffuse trapping level. In this letter the application of the above method to the analysis of SCLC curves for polycrystalline tetracene is presented. Fig. 1 shows an example of the j-U characteristic /5/. Plots of n_t versus E_F (curve a) and of dn_t/dE_F versus E_F (curve b) are shown in Fig. 2. These plots were deduced from the j-U characteristic using a simple version of the analysis described in /3/. Also, the data on the mobility in polycrystalline tetracene, reported by us elsewhere /6/, were used for the calculations. From Fig. 2 the energy depth of the traps, $E_t \approx 0.69$ eV, was obtained. The relatively small value of the full width at half maximum (FWHM = 0.02 eV) indicates the existence of a discrete trap level, with total density of states $N_t = 3.4\times10^{14}$ cm^{-3}. Previously, the depth of the traps,

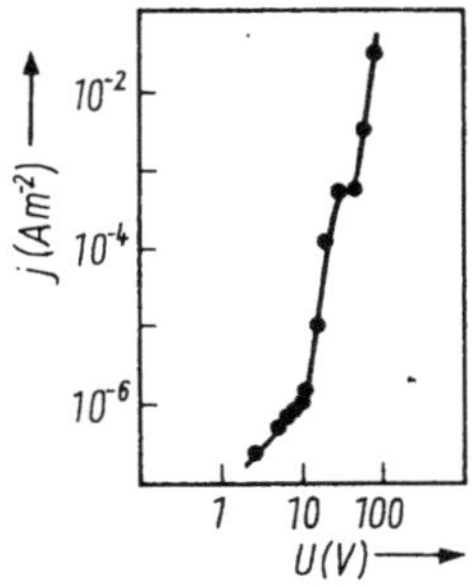

Fig. 1. The log-log plot of the current density (j) versus voltage (U) characteristic, measured at room temperature. The thickness of the polycrystalline tetracene sample was 7.5 µm (these results were extracted from /5/)

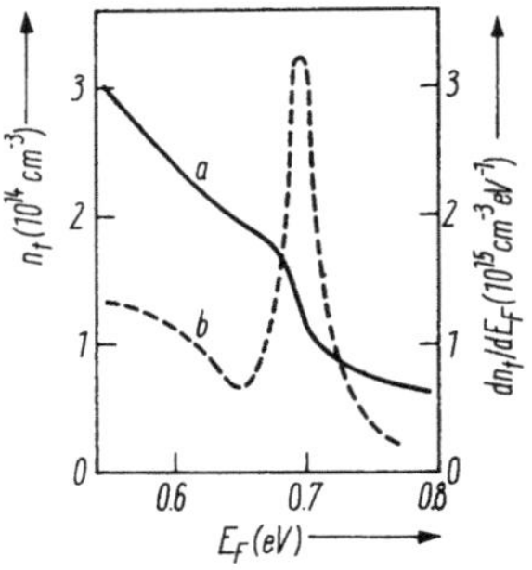

Fig. 2. Plots of n_t (curve a) and dn_t/dE_F (curve b) versus energy E_F of the quasi-Fermi level

E_t = 0.68 eV, was found in polycrystalline tetracene by Kasica and Kurczewska /7/ and Kasica /5/ using photocurrent decay measurements and thermally stimulated current technique, respectively.

A method of SCLC curve analysis reported by Nešpurek and Sworakowski /4/ was also used for the determination of the trap distribution in polycrystalline tetracene. This analysis gives two discrete trap levels: $E_t \approx 0.67$ eV and $E_t \approx 0.84$ eV. The second level was not previously found in polycrystalline tetracene.

Acknowledgement

The authors are greatly indebted to Dr. H. Kasica for the use of her results in this work.

References

/1/ M.A. LAMPERT and P. MARK, Current Injection in Solids, Academic Press, 1970.

/2/ J.C. PFISTER, phys. stat. sol. (a) 24, K15 (1974).

/3/ C. MANFREDOTTI, C. DE BLASI, S. GALASSINI, G. MICICCI, L. RUGGIERO, and A. TEPORE, phys. stat. sol. (a) 36, 569 (1976).

/4/ S. NEŠPUREK and J. SWORAKOWSKI, phys. stat. sol. (a) 41, 619 (1977).

/5/ H. KASICA, Ph.D. Thesis, Technical University of Łódź, 1975.

/6/ W. MYCIELSKI and A. LIPIŃSKI, Thin Solid Films 48, 133 (1978).

/7/ H. KASICA and H. KURCZEWSKA, Proc. Internat. Microsymp. Polarization and Conduction in Insulating Polymers, Bratislava-Harmonia, May 1972, Cables and Insulating Materials Research Institute, Bratislava 1972 (p. 137).

(Received May 19, 1978)

phys. stat. sol. (a) 49, K43 (1978)

Subject classification: 19; 22.8.1

Institute of Solid State Physics, Latvian State University, Riga

Analysis of Mn^{2+} EPR Spectra in Polycrystals Using Bir's Method to Account for the Transition Probability

By

J.G. KLIAVA and J.J. PURĀNS

Introduction Divalent manganese ions are often used as a paramagnetic probe in order to obtain information about the structure of crystals. However, in the case of polycrystals and glasses Mn^{2+} EPR spectra are so complicated that their interpretation requires comparison with the results of computer simulation.

In this work we confine ourselves to the case when the fine structure is much smaller than the Zeeman one ($|D| \ll g\beta H$). For this case, calculations of Mn^{2+} EPR spectra in polycrystals have been reported in a number of papers /1 to 4/. However, there is usually some discrepancy between experimental and calculated spectra arising mainly from the insufficient accuracy of the calculation of transition probabilities /5/.

Some years ago a method was developed by Bir /6, 7/ that permits to calculate the probabilities of allowed and forbidden hyperfine transitions for Mn^{2+} ions much more accurately. In the present work the Mn^{2+} EPR spectra in polycrystals are calculated using this method. The influence of the values of the crystal field parameter D on the spectra is considered. Computer simulated spectra are compared with the experimental Mn^{2+} EPR spectrum in polycrystalline $CaWO_4$. Good agreement between the experimental spectrum and the theoretical one is obtained.

Procedure of calculation A program for calculating the Mn^{2+} EPR spectrum in polycrystals was written for the following spin Hamiltonian:

$$\hat{\mathcal{H}}_S = g\beta \vec{H}\vec{S} + D[S_z^2 - \frac{1}{3}S(S+1)] + A\vec{S}\vec{I}\,, \tag{1}$$

where the symbols have their usual meanings, S = 5/2, I = 5/2, and z is the symmetry axis of the paramagnetic centre. Only the fine structure transitions $1/2 \leftrightarrow -1/2$ were considered. Lines corresponding to the other fine structure transitions must have a much smaller amplitude in polycrystals because of the much greater angular dependence of line positions. The allowed $\Delta m = 0$ and forbidden $\Delta m = \pm 1$ hyperfine transitions were taken into account. The spectra were calculated using

$$\mathcal{P}(H) = \sum_{i=-1}^{1} \sum_{\substack{m=-5/2 \\ |m+i| \leq 5/2}}^{5/2} \int_0^{\pi/2} W_{1/2,m;-1/2,m+i}(\vartheta) F\left(\frac{H-H_{1/2,m;-1/2,m+i}(\vartheta)}{\Delta H_{pp}}\right) \sin\vartheta \, d\vartheta \qquad (2)$$

where $W_{1/2,m;-1/2,m+i}(\vartheta)$ and $H_{1/2,m;-1/2,m+i}(\vartheta)$ are the transition probability and the resonance magnetic field for the transition $1/2, m \leftrightarrow -1/2, m+i$, respectively, ϑ being the angle between the external magnetic field $\vec{H}$ and the z axis, $F((H-H_{1/2,m;-1/2,m+i}(\vartheta))/\Delta H_{pp})$ is the lineshape which was taken to be either a Gaussian or a Lorentzian first derivative with the linewidth ΔH_{pp}.

Following Bir's method the transition probability may be written as a product of the probabilities of the electronic transition $|\omega^{(e)}_{1/2;-1/2}(\vartheta)|^2$ and of the nuclear one $|P^{(I)}_{m;m+i}(\mu_{1/2;-1/2})|^2$, where $\mu_{1/2;-1/2}$ is the cosine of the angle between the quantization axes of the nuclear spin for the electronic states 1/2 and -1/2. The functions $P^{(I)}_{m;m+i}$ are transformed like the irreducible representations of the rotation group. It can be shown /7/ that if $|D| \ll g\beta H$, the probability of the electronic transition is practically independent of ϑ, $|\omega^{(e)}_{1/2;-1/2}(\vartheta)|^2 \approx \text{const}$. So the transition probabilities in our calculations were described simply with the functions $|P^{(I)}_{m;m+i}(\mu_{1/2;-1/2})|^2$. The values of $\mu_{1/2;-1/2}$ were obtained in the second order of the perturbation theory /6/:

$$\mu_{1/2;-1/2} \approx 1 - 128 \frac{D^2}{(h\nu)^2} \sin^2 2\vartheta , \qquad (3)$$

where $h\nu$ is a microwave quantum. The perturbation theory up to the third order was used to calculate the resonance magnetic fields $H_{1/2,m;-1/2,m+i}(\vartheta)$ /3/.

Calculated spectra The calculated Mn^{2+} X-band (ν = 9030 MHz) EPR spectra in polycrystals for different values of D, g = 2.000, and $A/g\beta$ = -95G are shown in Fig. 1. The lineshape is the Gaussian first derivative with ΔH_{pp} = 7 G. Peaks in the spectra are attributed to six allowed $\Delta m = 0$ and ten forbidden $\Delta m = \pm 1$ hyperfine transitions. The most intense peaks correspond to the values of ϑ for which

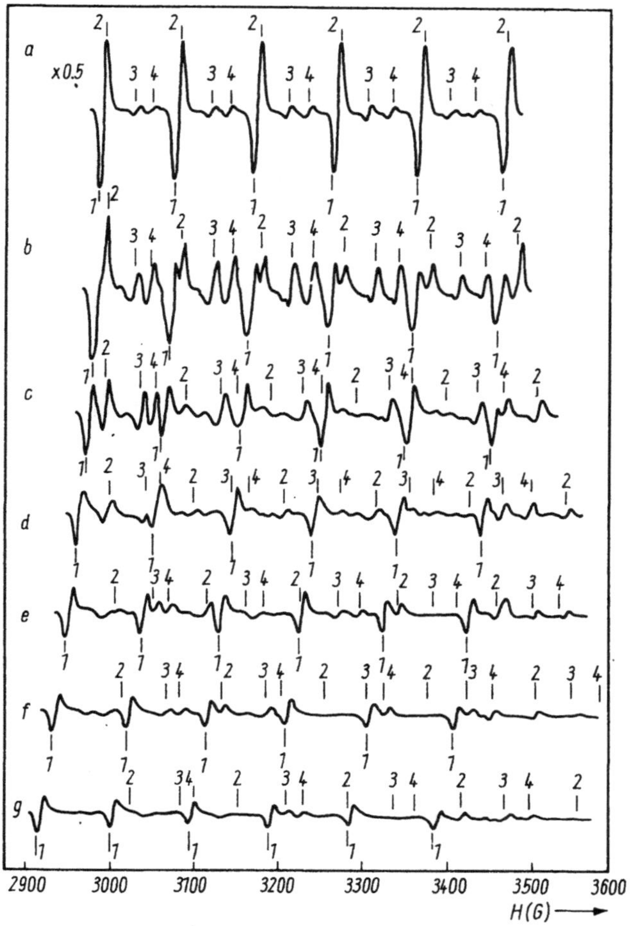

Fig. 1. Calculated Mn^{2+} EPR spectra in polycrystals with $g = 2.000$, $A/g\beta = -95$ G, $\nu = 9030$ MHz, $\Delta H_{pp} = 7$ G, and different values of $D/g\beta$: 50 G (a), 100 G (b), 150 G (c), 200 G (d), 250 G (e), 300 G (f), 350 G (g)

$$\frac{\partial H_{1/2,m;-1/2,m+i}(\vartheta)}{\partial \vartheta} = 0; \quad \sin\vartheta \neq 0; \quad W_{1/2,m;-1/2,m+i}(\vartheta) \neq 0 \ . \tag{4}$$

The above conditions lead to the values $\vartheta = 90^{\circ}$ and $\vartheta \approx 40^{\circ}$ for the allowed and $\vartheta \approx 40^{\circ}$ for the forbidden transitions. The calculated values of the corresponding resonance magnetic fields are indicated with the following numbers:

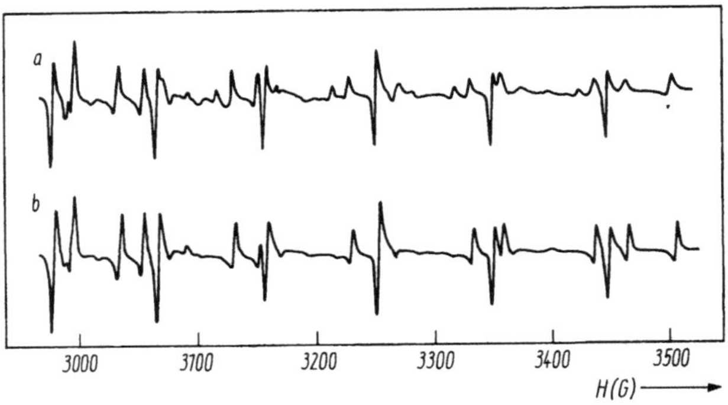

Fig. 2. (a) Experimental Mn^{2+} EPR spectrum in $CaWO_4$ at 110 K. ν = 9030 MHz. (b) Calculated spectrum with g = 2.000, $A/g\beta$ = -95 G, $D/g\beta$ = -150 G, ν = 9030 MHz, and ΔH_{pp} = 3.5 G

1 $(1/2, m \leftrightarrow -1/2, m, \vartheta = 90^{o})$,

2 $(1/2, m \leftrightarrow -1/2, m, \vartheta \approx 40^{o})$,

3 $(1/2, m \leftrightarrow -1/2, m-1, \vartheta \approx 40^{o})$,

4 $(1/2, m-1 \leftrightarrow -1/2, m, \vartheta \approx 40^{o})$.

One can see in Fig. 1 that the most intense peaks correspond to the allowed transitions at $\vartheta = 90^{o}$. However, for $D/g\beta > 100$ G the shape of these peaks is considerably distorted by a superposition of the peaks corresponding to the allowed and forbidden transitions at $\vartheta \approx 40^{o}$. The amplitudes of all the peaks depend strongly on the value of D. At different D the peaks corresponding to the transitions at $\vartheta = 90^{o}$ are either much greater or comparable with those at 40° (see Fig. 1).

Comparison with the experiment We have investigated the Mn^{2+} EPR spectrum in polycrystalline $CaWO_4$. Measurements have been carried out at 110 K by means of a RE 1301 X-band spectrometer. The experimental spectrum is shown in Fig. 2 in comparison with a computer simulated spectrum for g = 2.000, $A/g\beta$ = -95 G, $D/g\beta$ = -150 G. The lineshape is the Gaussian first derivative with ΔH_{pp} = 3.5 G. There is good agreement between the experimental and calculated spectra. It is possible to make this agreement still better by assuming a small distribution of D values: $\delta D/g\beta \approx 5$ G.

The parameters of the Mn^{2+} spin Hamiltonian in the monocrystalline $CaWO_4$ have been reported as follows /8/:

$g_{\parallel} = 2.00006$, $g_{\perp} = 2.00079$, $A_{\parallel}/g\beta = -95.2$ G, $A_{\perp}/g\beta = -95.9$ G, $D/g\beta = 147.3$ G
These values are very close to those obtained in the present work.

Acknowledgements

We wish to express our thanks to J.J. Kuzmin for numerous advices in writing the computer program and to I. Millere for providing us with the sample used in our experiment.

References

/1/ N.N. TIKHOMIROVA, S.N. DOBRYAKOV, and I.V. NIKOLAEVA, phys. stat. sol. (a) 10, 593 (1972).

/2/ N.J. STEINSCHNEIDER and G.M. ZHIDOMIROV, Zh. eksper. teor. Khim. 9, 414 (1973).

/3/ H.W. DE WIJN and R.F. VAN BALDEREN, J. chem. Phys. 46, 1381 (1967).

/4/ R.D. DOWSING, J. magnetic Resonance 2, 42 (1970).

/5/ P.T. DOBNEY, J. Phys. D 8, 1595 (1975).

/6/ G.L. BIR, Fiz. tverd. Tela 5, 2236 (1963).

/7/ G.L. BIR and L.S SOCHAVA, Fiz. tverd. Tela 5, 3594 (1963).

/8/ D H. LYONS and R.W. KEDZIE, Phys. Rev. 145, 148 (1966).

(Received May 25, 1978)

phys. stat. sol. (a) 49, K49 (1978)
Subject classification:14.3.3 and 14.3.4; 22.9

Department of Physics, Higher Institute of Civil Engineering and Architecture, Sofia[1)]

Thermionic-Field Emission from a Metal into Organic Polymer Semiconductor Films

By
M. VODENICHAROVA

The charge transfer phenomena in thin polymer films are not enough investigated in spite of the doubtless interest which they present from the theoretical and experimental point of view. Usually the studies of currents in thin film systems of the type metal-organic insulator (semiconductor)-metal are performed in a restricted temperature range and at relatively small electric fields in order to avoid breakdown phenomena and accompanying irreversible changes in the electrical conductivity of the samples. The interpretation of the experimental current-voltage characteristics very often "raises the problem to what extent the non-linear dependence j(v) is related to limitations of space charge or to field emission from the electrode" /1/. The conclusion itself about the dominating mechanism of charge transfer (Schottky emission, Poole-Frenkel effect, SCLC, and others) in systems with organic films is normally done on the basis of qualitative phenomenological considerations /2 to 7/, but not by means of severe quantitative criteria.

It was reported in previous investigations /8. 9/ that the currents in thin film systems with polyquinolinecarboxylic acid were due to thermally activated electron transitions above and through the top part of the effective potential barrier at the metal-polymer film interface. It was shown also that these currents were accurately enough described by $j(\mathcal{E}, T)$ relationships following from the theory of Christov /10, 11/ for injected currents in insulators in the whole extended thermionic emission region.

In the present paper for the first time experimental data for thermionic-field (T-F) emission in organic polymer semiconductors are reported which cover a great part of the intermediate emission region up to values of the ratio $T_k/T = 2.95$[2)].

1) 42 Sofia, Bulgaria.

2) T_k and T_c are the characteristic temperatures of Christov /10/ which can be used for delimitation of the different emission regions.

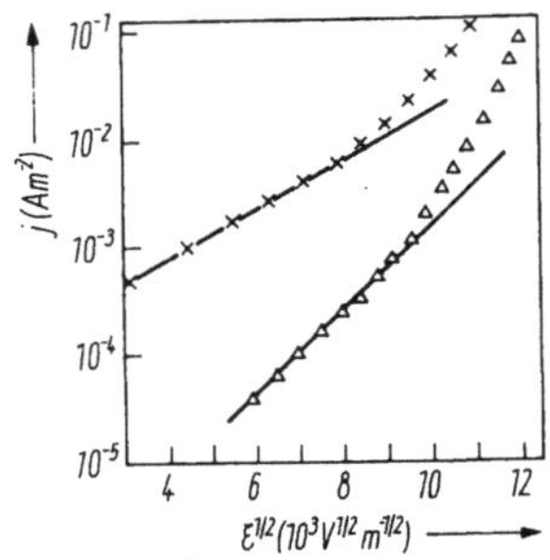

Fig. 1. Schottky characteristics ln j versus $\sqrt{\mathcal{E}}$ of sandwich samples Al-organic polymer semiconductor-Al; △ polymer I d = 285 nm, T = 245 K; × polymer II, d = 194 nm, T = 295 K

Experimental Thin films of two organic semiconductors denoted conditionally as polymer I and polymer II were used the polymer $(C_{32}H_{18}N_2O_4)_n$ with a quinoline ring in the conjugation chain, investigate /8, 9/, and the recently synthesized /12/ polymer $(C_{24}H_{14}N_4O_8)_n$ of the acenaphthene group. Sandwich samples with area of 1×10^{-6} m^2 and vacuum evaporated Al(99.999%) electrodes were used, the polymer films being deposited from a solution. The film thickness was measured by means of an interferometer (MII-10, USSR) and varied between 72 and 400 nm.

The current-voltage characteristics of the samples were studied under pulse conditions and recorded on an X-Y recorder (Endim 620.01, DDR) at 100 s duration of the pulses. The measurements were performed in air at different temperatures in the range 240 to 295 K.

Results and discussion In Fig. 1 the current-voltage characteristics at 245 and 295 K for two samples with the polymers I and II, respectively, are presented in Schottky coordinates ln j versus $\sqrt{\mathcal{E}}$. In the range of small fields the characteristics are linear and correspond to electron transitions over and through the upper part of the potential barrier at the metal-polymer film interface. The condition $T_k/T \leq 3/2$ defines the range of extended thermionic emission, according to the criterion of Christov.

The deviation of the experimental points from the linear ln j versus $\sqrt{\mathcal{E}}$ dependence at higher fields suggests tunneling of electrons mainly through the middle part of the barrier (above the Fermi level of the emitting electrode), therefore these points lie in the intermediate emission region. For the values of the ratio T_k/T reached in the measurements presented in Fig. 1 we get $z_1 > 1$ and $z_2 > 1$[3)], hence the experimental data in the range of higher fields refer really to the extended T-F emission region covering a great part of the "pure" T-F region (the region of Murphy and Good /15/).

3) z_1 and z_2 are arguments of two error integrals and are used in the theory of Christ for delimitation of the different emission regions like the characteristic temperatures T_k and T_c /10, 11/.

Fig. 2. T-F characteristics $\ln(j/\mathcal{E})$ versus $\mathcal{E}^2$ of the samples from Fig. 1

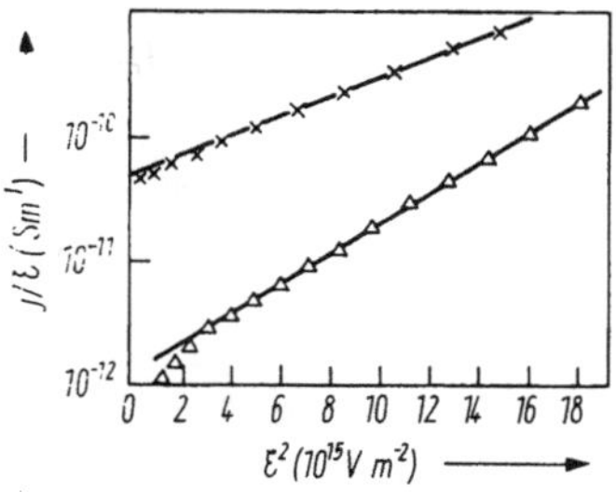

From the break point of the linear dependence $\ln j$ versus $\sqrt{\mathcal{E}}$ which corresponds to $T_k = 1.76\ T$ /13/ one could evaluate the effective mass of the charge carriers in the polymer films. Using this method the mean values $m_i = 0.92\ m_o \pm 6.5\%$ and $m_i = 0.64\ m_o \pm 5\%$ were obtained for the polymers I and II, respectively on the basis of the experimental data from many samples; m_o being the free electron mass.

The presence of a T-F emission region can be established also by representing the experimental current-voltage characteristics in the form $\ln(j/\mathcal{E}) = f(\mathcal{E}^2)$ /14/. For this region the above representation gives a straight line (Fig. 2) from the intercept of which with the $\ln(j/\mathcal{E})$ axis the value of the electron work function at the metal-polymer interface is determined approximately by the formula /16/

$$\chi_{T-F} = -kT \ln\left[\frac{h\ j}{e^2\sqrt{2\eta m_i kT}\ t(z)\ \mathcal{E}}\right];$$

h and k are the Planck and Boltzmann constants, respectively, e the electron charge, t(z) a function defined by Murphy and Good /15/ changing slightly with the field.

For the samples presented in Fig. 3 the values $\chi_{T-F} = 0.86$ eV and $\chi_{T-F} = 0.83$ eV are obtained while from the linear part of the characteristics $\ln j$ versus $\sqrt{\mathcal{E}}$ of the same ones (Fig. 1) according to the Richardson-Schottky equation the values $\chi_s = 0.86$ eV and $\chi_s = 0.87$ eV are found. The agreement between the values of the work function calculated from the theoretical $j(\mathcal{E}, T)$ relations for the current density in the different emission regions is quite acceptable and tells for inner coordination of the experimental data. This agreement justifies the model of an effective Schottky potential barrier at the metal-organic polymer semiconductor interface assumed in the considerations.

In Fig. 3 the experimental data for six samples with the polymer I are shown in a dimensionless coordinate system $\lg(j/j_{RS}) = f(T_k/T)$.[4)] The experimental conditions

4) Used in the study of electron emission from metals in vacuum and insulators /14, 16/. Such a system gives the possibility for a unique representation of the data for the field dependence $j(\mathcal{E})$ at T = const or the data for the temperature dependence $j(T)$ at $\mathcal{E}$ = = const from many investigated samples.

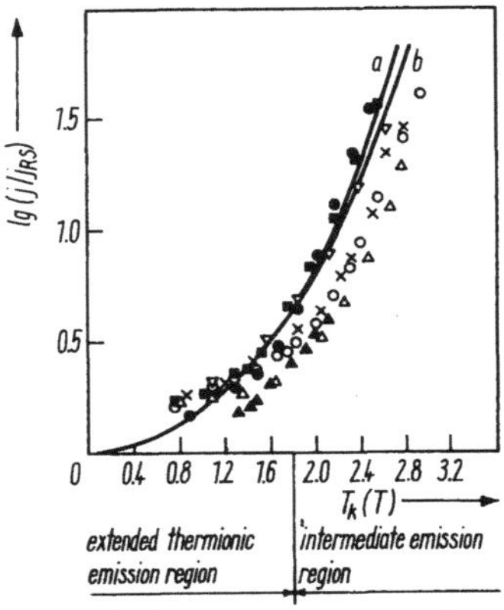

Fig. 3. Comparison of the experimental data from six samples of polymer I with the theoretical curves $\lg(j/j_{RS}) = f(T_k/T)$ calculated after Christov for $m_i = 0.9\ m_o$, $\chi = 0.88$ eV, and T = 250 K (curve a) and 295 K (curve b). The parameters of the samples are as follows: △ d = 285 nm, $\chi = 0.86$ eV, T = 245 K; ▲ d = 285 nm, $\chi = 0.86$ eV, $\mathcal{E}_1 = 7.05\times10^7$ V/m and $\mathcal{E}_2 = 1.06\times10^8$ V/m; ȸ d = 350 nm, $\chi = 0.87$ eV, T = 258 K; ▽ d = 200 nm, $\chi = 0.88$ eV, T = 272 K; ■ d = 130 nm, $\chi = 0.87$ eV, T = 273 K; ● d = 73 nm, $\chi = 0.88$ eV, T = 295 K; × d = 310 nm, $\chi = 0.88$ eV, T = 253 K

at which the measurements were carried out as well as the values of the work function and the polymer film thickness of each of the samples are also pointed out. On the same figure the theoretical curves $\lg(j/j_{RS})$ versus T_k/T are drawn by solid lines; j_{RS} being the current density according to the Richardson-Schottky equation, and j the current density after Christov (equation 34 of /10/, where the components j'' and j_2' corresponding to the thermionic and the thermionic-field emission regions are taken into account). The theoretical curves are calculated for two temperatures, 250 and 295 K at $m_i = 0.9\ m_o$ and $\chi = 0.88$ eV. The deviations of the experimental points from the theoretical curves in the extended T-F emission region do not exceed 80% at the greatest values of the ratio T_k/T reached in the present studies ($T_k/T = 2.95$).

In the same dimensionless coordinate system the experimental data for the temperature dependence of the current through one of the samples shown in Fig. 1 (d = $= 2.85\times10^{-7}$ m) are presented with ▲. The measurements have been carried out with two values of the field strength - $\mathcal{E}_1 = 7.05\times10^7$ V/m and $\mathcal{E}_2 = 1.06\times10^8$ V/m rising the temperature by 10 K from 245 to 295 K. In this temperature range at the fields mentioned above the ratio T_k/T changes from 1.32 to 2.16. The deviations between the theoretical curves and the experimental data are not beyond 40 to 60%.

As far as we know, experimental data for the thermionic-field emission in metal-organic polymer semiconductor-metal systems have not been published. The present paper shows that such a type of emission could be observed in polymer semiconductor films and it is not astonishing when taking into account the high breakdown field of the polymer films under investigations /17/ and the similarity of their electrical propertie with those of inorganic insulator films. T-F emission in the above systems occurs mainly in samples with good quality of the polymer films and usually preceeds the pre-breakdown conduction state.

References

/1/ M. KRYSZEWSKI and A. SZYMANSKI, J. Polymer. Sci. D 4, 246 (1970).

/2/ G. LENYEL, J. appl. Phys. 37, 807 (1966).

/3/ H. MANN, J. appl. Phys. 35, 2173 (1964).

/4/ R. CHRISTY, J. appl. Phys. 35, 2179 (1964).

/5/ P. EMTAGE and W. TANTRAPORN, Phys. Rev. Letters 8, 267 (1962).

/6/ A. BLAGODAROV and E. LUTSENKO, phys. stat. sol. (a) 21, K115 (1974).

/7/ D. BARBE and C. WESTGATE, J. chem. Phys. 52, 4046 (1970).

/8/ C. VODENICHAROV, M. VODENICHAROVA, and I. SCHOPOV, C.R. Acad. Bulg. Sci. 24, 1309 (1971).

/9/ M. VODENICHAROVA, phys. stat. sol. (a) 28, 263 (1975).

/10/ S. CHRISTOV, phys. stat. sol. 32, 509 (1969); 42, 583 (1970); (a) 7, 371 (1971).

/11/ S. CHRISTOV, Contemp. Phys. 13, 199 (1972).

/12/ I. SCHOPOV and P. KIROWA, Makromol. Chem. 176, 511 (1975).

/13/ C.M. VODENICHAROV, phys. stat. sol. (a) 42, 785 (1977).

/14/ C. VODENICHAROV and S. CHRISTOV, Solid State Electronics 15, 933 (1972).

/15/ E. MURPHY and R. GOOD, Phys. Rev. 102, 1464 (1956).

/16/ C. VODENICHAROV and S. CHRISTOV, phys. stat. sol. (a) 25, 387 (1974).

/17/ M. VODENICHAROVA and I. AMOV, Bulg. J. Phys. 2, 135 (1975).

(Received May 25, 1978)

phys. stat. sol. (a) 49, K55 (1978)

Subject classification: 19; 22

Department of Physical Chemistry, University of Geneva

Paramagnetic Resonance of Mn^{2+} in Na_2S Single Crystals

By

J.-M. MORET and R. LACROIX

We report the observation by electron paramagnetic resonance of Mn^{2+} in Na_2S single crystals. Measurements were made at 77 and 296 K on a Varian X-band spectrometer. The crystals have been grown from the melt by the Bridgman-Stockbarger method using a graphite crucible under an argon atmosphere. Experimental details on Na_2S single crystal preparation have been already reported /1/. Doping was done with manganese sulfide directly in the melt at initial concentrations of 10^{-3} to 10^{-2}.

The host lattice has an antifluorite cubic symmetry (O_h^5). The observed EPR spectrum with well resolved fine and hyperfine structure exhibits a trigonal symmetry. It is described by the usual spin Hamiltonian /2/:

$$\mathcal{H} = g_{\parallel}\beta B_z S_z + g_{\perp}\beta (B_x S_x + B_y S_y) + D\left[S_z^2 - (1/3)S(S+1)\right] +$$
$$+ (1/180)F\left[35\, S_z^4 - 30\, S(S+1)S_z^2 + 25\, S_z^2 + 3\, S(S+1)(S^2 + S - 2)\right] + A_{\parallel} S_z I_z +$$
$$+ A_{\perp} (I_x S_x + I_y S_y) .$$

We found the following values of the parameters:

$g_{\parallel} = 2.0026 \pm 0.0002$, $D = (-343.9 \pm 2)\ 10^{-4}\ cm^{-1}$, $F = (8.2 \pm 0.5)\ 10^{-4}\ cm^{-1}$,
$A_{\parallel} = (-65.7 \pm 0.5)\ 10^{-4}\ cm^{-1}$, $A_{\perp} = (-66.5 \pm 1.0)\ 10^{-4}\ cm^{-1}$ at 77 K and
$g_{\parallel} = 2.0026 \pm 0.0005$, $D = (-327.9 \pm 4)\ 10^{-4}\ cm^{-1}$, $F = (7.9 \pm 1)\ 10^{-4}\ cm^{-1}$,
$A_{\parallel} = (-65.2 \pm 1)\ 10^{-4}\ cm^{-1}$ at 296 K.

The hyperfine constant A has been shown to be related to the degree of covalent bonding between the 4 s state of Mn^{2+} ion and the ligands of particular cluster containing the impurity. Simanek and Müller /3/ have plotted A as function of Pauling's covalency parameter c/n. The observed value in Na_2S agrees well with those obtained with four sulfur ligands. Consequently we propose the following model for the center: the Mn^{2+} ion substitutes a Na^+ cation, the charge in excess being compensated by a near neighbour cation vacancy along a [111] trigonal axis. Such a charge compensation by cation vacancies occurs readily in alkali halides with Mn^{2+} ion impurities and has been already observed /1/ in sodium sulfide with Gd^{3+} ion impurities. The center produces light emission under X-ray excitation. At room temperature we observed a band at 515 nm with a halfwidth of 37 nm.

Results on Mn^{2+} in polycrystalline Na_2S have been recently published by Yanagisawa and Huzimura /4/. These authors attribute the observed spectrum to a substitutional Mn^{2+} ion associated with the nearest neighbour cation vacancy along [100] as charge compensator. Reported D, A and emission wavelength are very close to our values and suggest that the same center has been observed in polycrystalline and in single crystal Na_2S samples. Yanagisawa and Huzimura's speculation on the nature of the center is nevertheless not consistent with the symmetry of EPR spectra observed in single crystals.

References

/1/ J.-M. MORET and H. BILL, phys. stat. sol. (a) 41, 163 (1977).

/2/ A. ABRAGAM and B. BLEANEY, Electron Paramagnetic Resonance of Transition Ions, Oxford University Press, 1970 (p. 437).

/3/ E. SIMANEK and K.A. MÜLLER, IBM Res. Rep. RZ 320 (1969).

/4/ Y. YANAGISAWA and R. HUZIMURA, Japan. J. appl. Phys. 17, 253 (1978).

(Received May 26, 1978)

phys. stat. sol. (a) 49, K57 (1978)

Subject classification: 8 and 10; 12

Department of Physics, University of Athens[1)]

On the Pressure Variation of the Excessive Specific Heat Associated with Vacancies

By

P.A. VAROTSOS[2)]

Introduction We have recently proposed /1 to 3/ that the Gibbs formation energy g, the formation enthalpy h, and the (thermal) formation entropy s (per vacancy) can be expressed as

$$g = C\,B\,\Omega, \tag{1}$$

$$h = C\left(B - T\beta B - \frac{dB}{dT}\Big|_P\right)\Omega, \tag{2}$$

$$s = -C\left(\beta B + \frac{dB}{dT}\Big|_P\right)\Omega, \tag{3}$$

where B is the isothermal bulk modulus, β the thermal volume expansion coefficient, Ω the mean volume per atom, and C is a constant which is assumed to be temperature and pressure independent. The constant C can be found from the relation

$$C = \frac{h_o}{B_o\Omega_o} \tag{4a}$$

or alternately

$$C = \frac{h_{exp}}{B_o^{SL}\Omega_o}, \tag{4b}$$

where the subscript o denotes the corresponding values at absolute zero, h_{exp} is the experimental value of the formation enthalpy deduced from the linear part of the Arrhenius plot and B_o^{SL} the bulk modulus of the "harmonic crystal", i.e. the intercept of the linear part of the graph B = f(T) with the vertical axis.

The specific heat δC_p of a single vacancy is defined as /4, 5/

$$\delta C_p \equiv \frac{dh}{dT}\Big|_P \tag{5a}$$

---

1) Solonos Str. 104, Athens 144, Greece.

2) To the memory of Antonios Varotsos.

or alternately

$$\delta C_p \equiv T \left.\frac{ds}{dT}\right|_P . \tag{5b}$$

The contribution ΔC_p of the vacancies to the constant pressure specific heat C_p (per mole) of the solid is not simply equal to $n\delta C_p$, where n denotes the number of vacancies per mole (of the solid); it can be found from the relation /6, 7/

$$\Delta C_p = \left.\frac{d(nh)}{dT}\right|_P . \tag{6}$$

Taking into account that

$$n = N \exp\left(-\frac{g}{kT}\right) , \tag{7}$$

where N is the Avogadro number, (6) becomes

$$\Delta C_p = n\left(\frac{h^2}{kT^2} + \left.\frac{dh}{dT}\right|_P\right). \tag{8}$$

It is an experimental fact /8/ that the excessive specific heat of a solid depends on pressure. To the best of our knowledge no quantitative evaluation of the pressure dependence of ΔC_p has been proposed to date. It is the aim of the present paper to determine the pressure variation of ΔC_p but on macroscopic grounds. A microscopic calculation of $d\Delta C_p/dP|_T$ is an extremely difficult problem; this is obvious if one takes into account that even the calculation of the perfect lattice specific heat C_p^o is yet an unsolved problem whenever the solid shows higher order anharmonicity /9, 10/ i.e. near to the melting point.

Theory It is evident from (8) that the calculation of $d\Delta C_p/dP|_T$ is equivalent to the determination of the quantities n, h, dh/dT and of their pressure dependences as well. We proceed now to their determination by bearing in mind that the constant C is already known[3)] from (4a) (or from 4b).

Determination of n and dn/dP A combination of (1) and (7) leads to

$$n = N \exp\left(-\frac{CB\Omega}{kT}\right). \tag{9}$$

3) The microscopic calculation of h_o has been carried out by various authors with great success in a large variety of solids.

By differentiating (9) with respect to pressure we get

$$\frac{dn}{dP} = - \frac{n}{kT} \left.\frac{dg}{dP}\right|_T \quad \text{or} \quad \frac{dn}{dP} = - \frac{n}{kT} v^f , \tag{10}$$

where v^f is the vacancy formation volume. We have recently shown that the volume v^f is given by /11, 13/

$$v^f = C\left(\left.\frac{dB}{dP}\right|_T - 1\right)\Omega . \tag{11}$$

By combining (9), (10), and (11) we get

$$\frac{dn}{dP} = - \frac{CN\Omega}{kT}\left(\left.\frac{dB}{dP}\right|_T - 1\right) \exp\left(- \frac{CB\Omega}{kT}\right). \tag{12}$$

Determination of h and $dh/dP|_T$ The definition of the isothermal bulk modulus implies that

$$B = - \Omega \left.\frac{dP}{d\Omega}\right|_T \text{ and hence } \Omega = \Omega(0) \exp \int_0^P - \frac{dP}{B} , \tag{13}$$

where $\Omega(0)$ denotes the mean volume per atom at zero pressure.

Therefore (2) becomes

$$h = C\Omega(0) \left(B - T\beta B - \left.\frac{dB}{dT}\right|_P\right) \exp \int_0^P - \frac{dP}{B} . \tag{14}$$

The value of $dh/dP|_T$ can be determined from a direct differentiation of (2). However, in order to avoid complicated expressions we work as follows: By differentiating with respect to pressure the equation $g = h - T s$ and bearing in mind that $v^f = dg^f/dP|_T$ we get

$$\left.\frac{dh}{dP}\right|_T = v^f + T \left.\frac{ds}{dP}\right|_T . \tag{15}$$

According to the Maxwell relations we have $ds/dP|_T = - dv^f/dT|_P$, and hence (15) becomes

$$\left.\frac{dh}{dP}\right|_T = v^f - T \left.\frac{dv^f}{dT}\right|_P . \tag{16}$$

The definition of the thermal expansion coefficient β^f of the formation volume /11, 12/ implies that $dv^f/dT = \beta^f v^f$. Therefore (16) becomes

$$\left.\frac{dh}{dP}\right|_T = v^f(1 - T\beta^f) \quad . \tag{17}$$

But we have recently shown /11, 13/ that β^f can explicitly expressed through properties of the bulk material:

$$\beta^f = \beta + \frac{d}{dT}\left\{\ln\left(\frac{dB}{dP} - 1\right)\right\}. \tag{18}$$

By inserting in (17) the values of v^f and β^f given by (11) and (18) we get

$$\left.\frac{dh}{dP}\right|_T = C\Omega\left(\left.\frac{dB}{dP}\right|_T - 1\right)\left\{1 - T\left\{\beta + \frac{d}{dT}\ln\left(\frac{dB}{dP} - 1\right)\right\}\right\}. \tag{19}$$

Determination of dh/dT and its pressure dependence By differentiating (2) with respect to temperature we get /14/

$$\delta C_p \equiv \frac{dh}{dT} = -C\Omega T\left(\beta^2 B + 2\beta\left.\frac{dB}{dT}\right|_P + B\left.\frac{d\beta}{dT}\right|_P + \left.\frac{d^2B}{dT^2}\right|_P\right). \tag{20}$$

It is obvious that the pressure dependence of δC_p can now be directly obtained by differentiating (20) with respect to pressure. If one follows an analogous procedure with that indicated in /4/ one easily gets that the value of δC_p may be of some k, but at higher pressures it becomes appreciably smaller.

Conclusion The relations (9), (12), (14), (19), and (20) in connection with (8) show that whenever the functions B = f(T, P) and β = f(T, P) are known an explicit calculation of the pressure variation of ΔC_p can be done. Remember that in the case of an harmonic solid the bulk modulus does not depend on temperature and pressure and furthermore the coefficient β is zero /9, 10/. Therefore we conclude that the pressure variation of ΔC_p is governed by the anharmonicity of the solid.

Remark: If one is dealing with Schottky defects in alkali halides or with Frenkel defects in silver halides one can follows the same procedure as that indicated above; in these cases (7) should be replaced by n = N exp (-g/2kT), where now g correspon to a Schottky or Frenkel pair.

References

/1/ P. VAROTSOS and K. ALEXOPOULOS, J. Phys. Chem. Solids 38, 997 (1977).
/2/ P. VAROTSOS and K. ALEXOPOULOS, Phys. Rev. B 15, 4111 (1977).

/3/ P. VAROTSOS, J. Phys. F, in the press.

/4/ K. ALEXOPOULOS and P. VAROTSOS, J. Phys. F, in the press.

/5/ H.M. GILDER and D. LAZARUS, Phys. Rev. B11, 4916 (1975).

/6/ A.B. LIDIARD, in: Hdb. Phys., Ed. S. FLÜGGE, Vol. XX, Springer, Berlin (p. 246).

/7/ A.V. CHADWICK and H.R. GLYDE, in: Rare Gas Solids, Vol. II, Ed. M.L. KLEIN and J.A. VENABLES, Academic Press, 1977 (p. 1159).

/8/ F. BREMER and J. NÖLTING, Ber. Bunsenges. phys. Chem. 80, 1, 12 (1976).

/9/ W. LUDWIG, Recent Developments in Lattice Theory, Springer Tracts in Modern Physics, Springer, Berlin 1967.

/10/ G. LEIBFRIED and W. LUDWIG, Solid State Phys. 12, 275 (1961).

/11/ P. VAROTSOS, J. Physique, Lettres 38, L455 (1977).

/12/ P. VAROTSOS and W. LUDWIG, J. Phys. C 11, L305 (1978).

/13/ P. VAROTSOS, W. LUDWIG, and K. ALEXOPOULOS, Phys. Rev. B, in the press.

/14/ P. VAROTSOS, J. Physique, Lettres, to be published.

(Received May 26, 1978)

phys. stat. sol. (a) 49, K63 (1978)

Subject classification: 2 and 18.3; 21.1; 21.4

Zentralinstitut für Festkörperphysik und Werkstofforschung der Akademie der Wissenschaften der DDR, Institutsteil für magnetische Werkstoffe, Jena[1)]

Influence of Anisotropic Short-Range Order on Magnetic Properties of the Amorphous Ferrimagnet Gd-Co

By

H.-R. MÜLLER

It is generally acknowledged now that several mechanisms contribute to the phenomenon of magnetic anisotropy in sputtered amorphous $Gd_{1-x}Co_x$ films /1, 2/. We have shown recently that one mechanism, namely anisotropic short-range order, can easily be explained for preparation conditions involving resputtering processes /3/. In this communication, we will show that this model can explain other magnetic properties of Gd-Co films which have been observed experimentally. According to our model, the relative number n_{ij} of atomic pairs is different for the directions parallel and perpendicular to the film plane (i, j = Gd, Co; $n_{GdGd} + n_{CoGd} + n_{CoCo} = 1$). In Fig. 1 the number of Co-Co pairs is plotted versus Co content x in the film for the two directions and a target composition $Gd_{0.25}Co_{0.75}$. Here, a closed packed hexagonal lattice was used as a model for the amorphous structure, the binding energies of the pairs were assumed to increase in the sequence Gd-Gd, Gd-Co, Co-Co and only adatoms are expected to be resputtered. The difference between target and film composition is due to the preferred resputtering of Gd-atoms and increases with increasing bias voltage at the film as observed experimentally. The relative number of Co-Co pairs in plane, $n_{CoCo,\parallel}$, corresponds to the statistical distribution $n_{CoCo,s} = x^2$ for this example. In all cases considered in /3/, it turned out that there is not only an anisotropic pair distribution, but also a changed average number $\tilde{n}_{ij} = \frac{1}{3}(n_{ij,\perp} + 2\, n_{ij,\parallel})$ of pairs compared to the statistical distribution $n_{ij,s}$. This should lead to changes in the general magnetic behaviour of Gd-Co and we tried to estimate such effects.

The temperature and composition dependence of the magnetization for Gd-Co and similar rare-earth transition metal compounds can be described by a two-sublattice

1) Helmholtzweg 4, DDR-69 Jena, DDR.

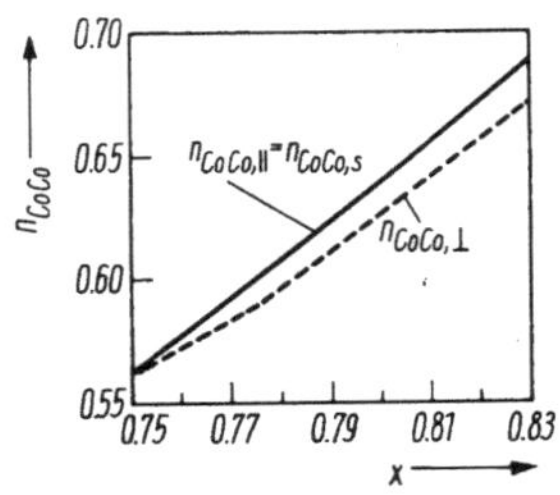

Fig. 1. Calculated relative number of Co-Co pairs in plane ($n_{CoCo,\parallel}$) and perpendicular to it ($n_{CoCo,\perp}$) for a bias sputtered film of composition $Gd_{1-x}Co_x$ versus x (target $Gd_{0.25}Co_{0.75}$, h.c.p. lattice, only adatoms backsputtered)

molecular field model firstly used by Hasegawa et al. /4/. The amorphous structure is approximated here, too, by a close packed hexagonal lattice with a number of z = : nearest neighbours for each atom. The subnetwork magnetization $\bar{S}_{Gd}$ is written as

$$\bar{S}_{Gd} = S_{Gd}\, B_{S_{Gd}}\left[\frac{2S_{Gd}}{kT}\,(z_{GdGd} I_{GdGd} \bar{S}_{Gd} + z_{GdCo} I_{GdCo} \bar{S}_{Co})\right].$$

Here S_{Gd}, S_{Co} are the spin values, z_{ij} and I_{ij} are the number of nearest neighbours and the exchange constants for the different atoms, respectively, $B_{S_{Gd}}$ is the Brillou function. The Co-subnetwork magnetization is described by a corresponding expression. In /4/, the z_{ij} values were approximated by their average values for a pure statistical distribution, namely $z_{GdGd} = z_{CoGd} = (1 - x)z$, $z_{CoCo} = z_{GdCo} = xz$. In our case of a non-statistical distribution, we have to use for the number z_{ij} of nearest neighbours the values

$$\tilde{z}_{GdGd} = \frac{1}{1-x}\,\tilde{n}_{GdGd}\, z, \qquad \tilde{z}_{CoGd} = \frac{1}{x}\,\tilde{n}_{GdCo}\,\frac{z}{2},$$

$$\tilde{z}_{GdCo} = \frac{1}{1-x}\,\tilde{n}_{GdCo}\,\frac{z}{2}, \qquad \tilde{z}_{CoCo} = \frac{1}{x}\,\tilde{n}_{CoCo}\, z .$$

If the $\tilde{n}_{ij}$ are written as $\tilde{n}_{ij} = n_{ij,s} + \Delta n_{ij}$, then in our lattice model

$$\Delta\tilde{n}_{GdGd} = \Delta\tilde{n}_{CoCo} = -\frac{1}{2}\,\Delta\tilde{n}_{GdCo} = \Delta\tilde{n}$$

is always valid. We calculated the subnetwork magnetizations for the statistical distribution and for $\Delta\tilde{n} = -0.01$, i.e. a 1% decrease of Gd-Gd and Co-Co pairs. For this the I_{ij} and S_i values from /5/ were used. It turned out that due to the weak coupling in the Gd subnetwork, there is a distinct effect for this magnetic subsystem. In Fig. the subnetwork magnetizations and the reduced total magnetization

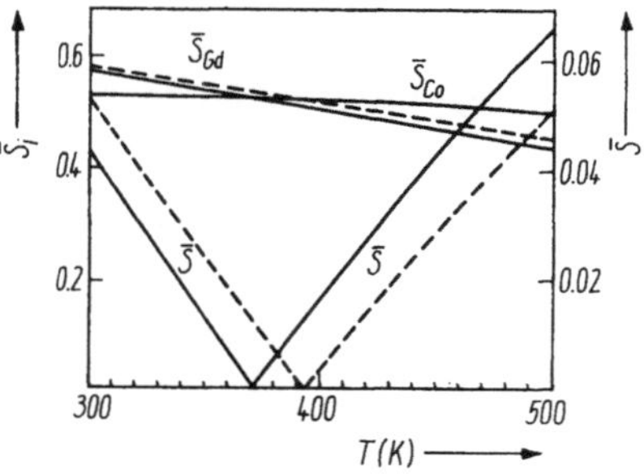

Fig. 2. Reduced subnetwork magnetization $\bar{S}_{Gd}$, $\bar{S}_{Co}$, and total magnetization $\bar{S}$ of a $Gd_{20}Co_{80}$ film in the region near T_{comp} for statistical ($\Delta\tilde{n} = 0$) and slightly non-statistical distribution ($\Delta\tilde{n} = -0.01$) of the atomic pairs. Note enlarged scale for $\bar{S}$ on the right. ——— $\Delta\tilde{n} = 0$, - - - $\Delta\tilde{n} = -0.01$

$$\bar{S} = |(1-x)\bar{S}_{Gd} - x\bar{S}_{Co}|$$

are plotted versus temperature in the region near magnetic compensation for the composition x = 0.8. Although there is only a small change in the Gd subnetwork, a rather big difference of about 20 K results for the magnetic compensation temperatures T_{comp}. In Fig. 3, the change of T_{comp} is plotted versus composition for the chosen example. It becomes clear from Fig. 2 and 3, that Gd-Co films of equal composition can show different magnetization values, when the atomic short-range order is different. Indeed, several authors observed different M(T) curves for seemingly equi-compositional films if prepared with different bias voltages (see, for instance /6/). But in most cases, it is not clear if the changed properties are due to different impurity inclusion which also strongly influences the magnetic properties.

Another possibility to test our picture of short-range order is to measure the changes of the magnetization after annealing. Decreased anisotropy means in general changed $\Delta\tilde{n}$ in our model and should therefore result in a shift $\Delta T_{comp} = T_{comp}$(annealed) - T_{comp}(as-prepared). Such shifts of T_{comp} have been reported in several papers /1, 7/. An own attempt to investigate the T_{comp} behaviour in annealing experiments of Gd-Co films resulted in rather large shifts to smaller T_{comp} values which were obviously due to preferential Gd oxidation. But the positive values of ΔT_{comp} observed in /1, 7/ for Gd-Co, Gd-Co-Mo, and Gd-Fe films annealed in

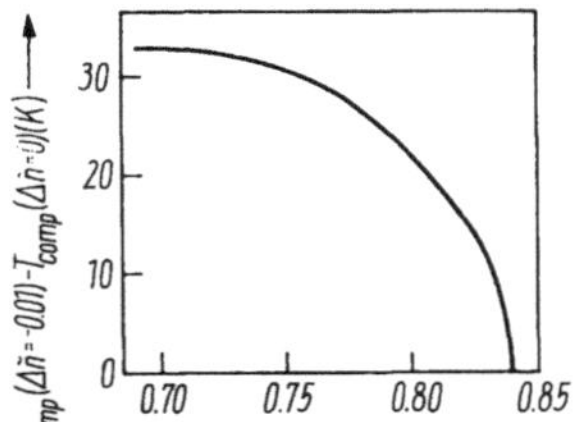

Fig. 3. Difference between the compensation temperatures of Gd-Co films with statistical ($T_{comp}(\Delta\tilde{n} = 0)$) and non-statistical distributions ($T_{comp}(\Delta\tilde{n} = -0.01)$) of the atomic pairs

UHV cannot be explained by an oxidation effect and seem to support our model of short-range order anisotropy in such films.

References

/1/ N. HEIMAN, N KAZAMA, D.F. KYSER, and V.J. MINKIEWICZ, J. appl. Phys. 49, 366 (1978).

/2/ W. KEILIG, H.-R. MÜLLER, and R. PERTHEL, 3rd Conf. Soft Magnetic Materials, Bratislava (ČSSR), Sept. 14 to 16, 1977.

/3/ H.-R. MÜLLER and R. PERTHEL, phys. stat. sol. (b) 87, 203 (1978).

/4/ R. HASEGAWA, B.E. ARGYLE, and L.J. TAO, AIP Conf. Proc. 24, 110 (1975).

/5/ H.-R. MÜLLER, E. JÄGER, and M. BITTERLICH, phys. stat. sol. (b) 80, K85 (1977).

/6/ N. HEIMAN, A. ONTON, D.F. KYSER, K. LEE, and C.R. GUARNIERI, AIP Conf. Proc. 24, 573 (1975).

/7/ T. KATAYAMA, K. HASEGAWA, K. KAWANISHI, and T. TSUSHIMA, 23rd Conf. Magnetism and Magnetic Materials, Minneapolis (USA) Nov. 8 to 11, 1977 (paper 2 P-10).

(Received June 12, 1978)

phys. stat. sol. (a) 49, K67 (1978)

Subject classification: 2 and 18.3; 11; 21.1; 21.4

Zentralinstitut für Festkörperphysik und Werkstofforschung der Akademie der Wissenschaften der DDR, Institutsteil für magnetische Werkstoffe, Jena[1)]

The Influence of Ion Bombardment on the Magnetic Anisotropy in Evaporated Amorphous Gd-Co Films

By

R. KOSICIK, P. DUDA, W. KEILIG, and R. MATTHEIS

It is well known from literature /1 to 4/ that for amorphous Gd-Co films the magnetic anisotropy is changing in dependence on the preparation technique. Whereas in the case of bias sputtering an easy axis perpendicular to the film plane may be obtained, in the evaporation process an in-plane anisotropy occurs. Bias sputtering differs from the evaporation mainly by a higher gaseous pressure (an additional inert gas) and by the occurrence of ion bombardment. Therefore both or one of these facts must be responsible for the different magnetic anisotropic behaviour. Therefore in a few first experiments we tried to simulate the ion bombardment in the evaporation process. In order to get this property, the substrate was held at a potential of -6 kV and in this way the vapour ions which are produced by electron bombardment in the e-gun, were accelerated to the substrate. An ionization rate of about 2 to 5 ‰ was obtained, determined by measuring the substrate current.

The samples were prepared by co-evaporation of Gd and Co from two e-guns in a background pressure of about 4×10^{-8} Torr. All samples were overcoated by silicon oxide. The film composition was adjusted by individual regulation of the Gd and Co evaporation rates which were measured by quartz crystal monitoring systems. The exact Gd-Co ratio was determined by electron microprobe analysis. Additionally, we found by this method always some atomic percent oxygen in our films.

Because of the finite distances between the two e-guns and between the e-guns and the substrate (5 cm and 35 cm, respectively) we got a composition profile at our glass substrates (4×4 cm^2). The substrates were not cooled during evaporation. All films had a thickness of about 5000 Å, the evaporation rate was 10 Å s^{-1}. The amorphous state was checked by electron diffraction patterns. Magnetic measurements

1) Helmholtzweg 4, DDR-69 Jena, DDR.

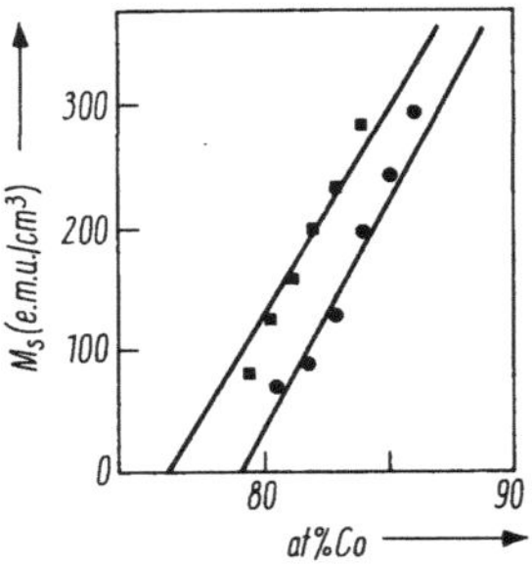

Fig. 1

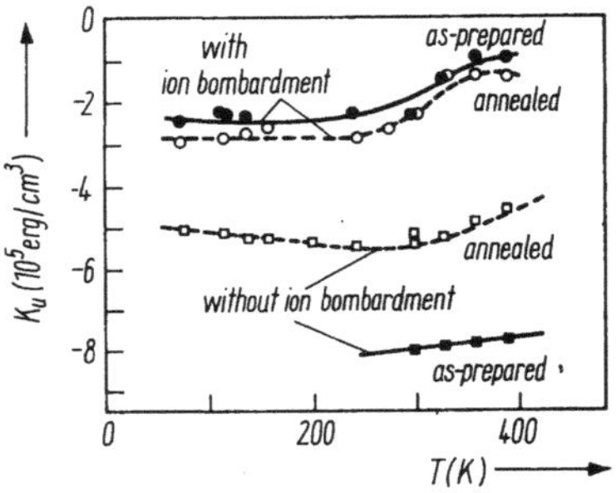

Fig. 2

Fig. 1. Saturation magnetization of samples evaporated with and without ion bombardment as a function of composition; ● without ion bombardment (4 at% O), ■ with ion bombardment (7 at% O)

Fig. 2. Magnetic anisotropy constant of with and without ion bombardment evaporated samples of composition $Gd_{0.18}Co_{0.82}$ versus temperature

were made on 5x5 mm^2 samples by a vibrating sample magnetometer and a torque equipment.

For the saturation magnetization we found values which are in agreement with the calculation of Müller et al. concerning the impurity influence on the magnetic properties in Gd-Co /5/. Fig. 1 gives an example for M_s values obtained with and without ion bombardment. The solid lines are the calculated values for oxygen impurities of 7 and 4 at%, respectively.

The anisotropy constant K_u of our samples is negative. Without ion bombardment evaporated films have room temperature values of $K_u = -(5 \text{ to } 8)\times10^5$ erg cm^{-3} in dependence of the composition; samples produced with ion bombardment show clearly changed values of $K_u = -(0.8 \text{ to } 2)\times10^5$ erg cm^{-3}. Fig. 2 represents the results for the composition $Gd_{0.18}Co_{0.82}$. The anisotropy constant K_u before and after annealing (9 h; 250 °C, 10^{-5} Torr) is plotted versus temperature. The absolute value of K_u decreases distinctly during annealing for the sample evaporated without ion bombardment, whereas it remains nearly constant for the ion bombardment sample.

The experimental results show that in the presence of ion bombardment the magnetic anisotropy changes to a less negative value, which cannot be explained by the different oxygen content, because after Brunsch and Schneider /6/ just the reverse behaviour should be obtained. It supports the picture that ion bombardment during preparation creates in GdCo a positive part in K_u, in agreement with the observed behaviour in sputtering experiments.

References

/1/ P. CHAUDHARI, J.J. CUOMO, and R.J. GAMBINO, IBM J. Res. Developm. 17, 66 (1973).

/2/ N. HEIMAN, A. ONTON, D.F. KYSER, K. LEE, and C.R. GUANIERI, AIP Conf. Proc. 24, 573 (1975).

/3/ R. PERTHEL, W. KEILIG, R. KOSICIK, U. RÖPKE, and C.-G. D'AMBLY, phys. stat. sol. (a) 40, K135 (1977).

/4/ R.C. TAYLOR and A. GANGULEE, J. appl. Phys. 47, 4666 (1976).

/5/ H.-R. MÜLLER, E. JÄGER, and M. BITTERLICH, phys. stat. sol. (b) 80, K85 (1977).

/6/ A. BRUNSCH and J. SCHNEIDER, J. appl. Phys. 48, 2641 (1977).

(Received June 20, 1978)

phys. stat. sol. (a) 49, K71 (1978)

Subject classification: 18.2 and 19; 18.3; 22.8.2

Institute of Solid State Physics, Czechoslovak Academy of Sciences, Prague

Note on FMR Absorption in a Low Resistivity Ferrite Cylinder

By

M. MARYŠKO

Recently, a theoretical study of the FMR has been made for a low resistivity ferrite cylinder positioned in the cavity at zero rf electric component (and maximum rf magnetic field) /1/. The calculated absorption curve has shown only one maximum which has been apparently in disagreement with the results of Watanabe et al. /2/. For cylindrical samples of manganese-zinc ferrites these authors have observed two distinct absorption maxima, one large broad main mode and a second small peak on the high field side of the main resonance. They ascribed them to the volume and surface mode, respectively. In this note we report the experimental results showing that the occurrence of the high field mode is connected with a non-zero amplitude of the rf electric field at the place where the sample is located. Let us note that similar effects have been already observed and discussed for the case of metal whiskers /3, 4/.

The FMR experiments in the rectangular cavity TE_{102} were performed at room temperature and frequency $f \approx 9.1$ GHz. The cylindrical sample with diameter 2a = = 0.25 mm and length 4.5 mm was prepared from a monocrystal of manganese-iron ferrite $Mn_{0.9}Fe_{2.1}O_4$ ($4\pi M_s \approx 5200$ G, $\sigma = 7\ \Omega^{-1}cm^{-1}$, for more information see e.g. /5/). For the ratio of the radius to the skin depth ($\mu = 1$) we get the value 0.2. The axis of the cylindrical sample was parallel to the static magnetic field and perpendicular to the wider wall of the cavity. The sketch of the cavity and the distribution of the rf fields are given in the inset of Fig. 1. In the symmetric plane σ_o ($\delta = 0.5\ \lambda_g$) the rf magnetic component h_x has a maximum while the rf electric field e_z equals zero. With decreasing δ the amplitude h_x decreases and e_z increases so that at $\delta = 0.25\ \lambda_g$, i.e. in the plane σ_1, h_x is zero and the rf electric field is at maximum. The sample has been first placed into the plane σ_o ($\delta = 0.5\ \lambda_g$), which is the usual position in measuring the magnetic insulators, and then into several positions specified by δ. Finally we have located the cylinder into the plane σ_1 (δ =

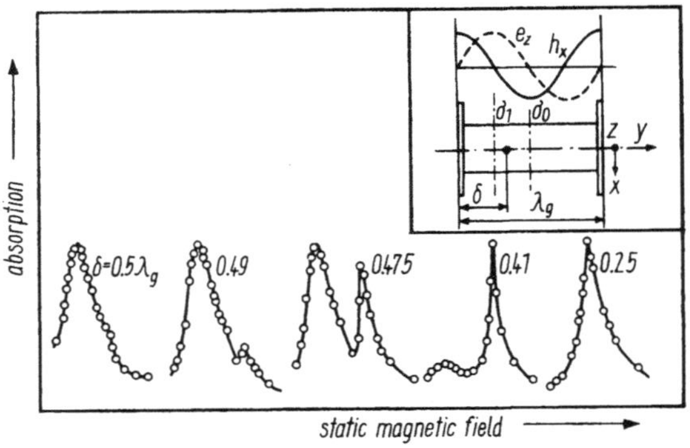

Fig. 1. The relative absorption for different positions of the sample. The inset shows the geometry of the cavity

$= 0.25\ \lambda_g$). The corresponding absorption curves have been detected and plotted in the relative scale in Fig. 1. (The absolute value of absorption has been strongly influenced by the changes of the quality factor of the cavity for the different positions of the sample.)

The absorption curves for the central position $\delta = 0.5\ \lambda_g$ as well as the low field maxima for $\delta = 0.49\ \lambda_g$, $0.475\ \lambda_g$, and $0.415\ \lambda_g$ obviously correspond to the nearly homogeneous mode excited by the rf component h_x. Really, the measured resonance field $H_1 \approx 550$ Oe has been found to be in rough agreement with the value calculated using the resonance condition

$$H_{res} = \frac{\omega}{\gamma} - 2\pi M_s$$

for an infinite cylinder.

From Fig. 1 it can be seen that there is also a high field resonance maximum at the field H_2, whose relative intensity increases when going to the positions with a larger rf electric field. For $\delta = 0.49\ \lambda_g$ the amplitude e_z, which is about 6% of the maximum rf electric field, is sufficient to produce a visible peak on the absorption curve. Even in the plane d_o, where $e_z = 0$ should hold, there is a certain indication of the second maximum which might be due to some disturbance of the rf electric field near the sample holder. We believe that a similar effect observed in /2/ may be explained in the same way. For $\delta = 0.25\ \lambda_g$ we observe the pure high field mode. The resonance field $H_2 \approx 1420$ Oe for this mode is approximately equal to the resonan field $H_{\perp}$ for parallel resonance in metals. It follows that the high field mode is probably excited by the rf tangential magnetic field inside the cylinder (the curl component /3/). This field has its origin in the rf electric current i_z inside the cylinder and thus primarily in the electric field e_z at the place of the sample.

The more detailed study concerning the possibility of calculating the corresponding absorption curves is in progress.

Thanks are due to Dr. S. Krupička, Dr. Z. Frait, and Dr. L. Kraus for valuable discussions.

References

/1/ M. MARYŠKO, phys. stat. sol. (a) 47, 277 (1978).

/2/ Y. WATANABE, S. SAITO, and S. TAKEMOTO, J. Phys. Soc. Japan 33, 1308 (1972).

/3/ D.S. RODBELL, J. appl. Phys. 30, 187S (1959).

/4/ Z. FRAIT, Czech. J. Phys. B10, 546 (1960).

/5/ M. MARYŠKO, Czech. J. Phys. B24, 1379 (1974).

(Received June 30, 1978)

phys. stat. sol. (a) 49, K75 (1978)

Subject classification: 14.2; 1.2

Zentralinstitut für Festkörperphysik und Werkstofforschung der Akademie der Wissenschaften der DDR, Dresden

Preparation of Superconducting Nb_3Au Diffusion Layers

By

P. MÜLLER

The phase Nb_3Au having the A15-structure shows a transition temperature into the superconducting state of 10.6 K /1 to 3/ up to 11.5 K /4/; the highest transition temperature is attainable only with exact stoichiometric composition /5/. Owing to this height of transition temperature this phase will be interesting for technical use. Therefore the aim of the present paper was the production of wires as they are required for the construction of superconducting magnets according to different diffusion methods. Studies by Röschel et al. /6/ yielded that the phase in the state diagram Nb-Au has a homogeneity range of 7 at% with nearly the same extent from stoichiometric composition to both the Au and the Nb side. Between the shape of the state diagram of a binary system and the layer thicknesses forming on diffusion couples there are relations originating from the concentration dependence of the free energies of the phases of a binary system /7/. The relative thickness of the diffusion layer of a A15-phase is the larger the wider its homogeneity range and the smaller the difference between the location of this range and the stoichiometric composition is, or the homogeneity range includes the stoichiometric composition. Thus the assumption of the formation of relatively thick Nb_3Au layers with nearly stoichiometric composition on Nb-Au diffusion couples is justified.

Experimental Two methods were applied to prepare diffusion layers: a) by direct diffusion of Nb into Au, using a Nb-plated Au wire, b) by selective diffusion of Au from a CuAu wire into a Nb-plating layer.

For the production of the wires Nb tubes having an outer diameter of 10 mm and a wall thickness of 1.5 mm were filled with a gold wire having a diameter of 6.9 mm (case a) or a CuAu alloy containing 18 at% Au molten in a vacuum arc furnace (case b), and then drawn together with an Fe sheat to wires of a diameter of 0.2 mm. The heat treatment of the wires was carried out in a cleaned argon-inert gas atmosphere at

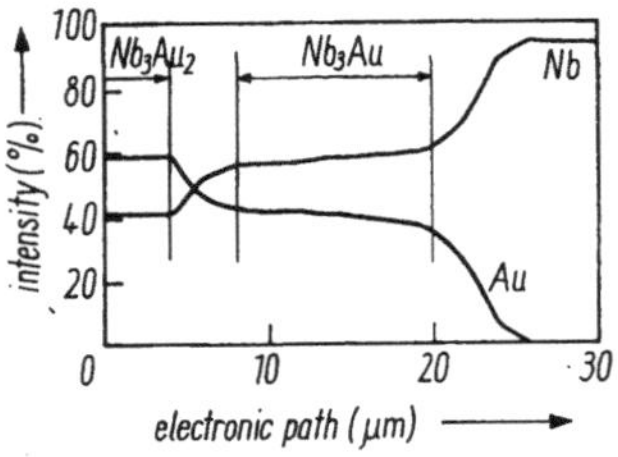

Fig. 1. Result of the electron microprobe analysis of a Nb-Au sample heat-treated at 1100 °C/100 h

temperatures of 800 to 1300 °C for 2 to 100 h; the samples were rapidly cooled in the furnace. Before measuring the superconducting properties the Fe sheat of the wires was removed by chemical etching. The transition curve to superconductivity was determined by measuring the relative change of the electrical resistivity. The critical current in dependence on an external magnetic field was determined at 4.2 K, a deviation of $U = 10\mu V$ from the zero potential being evaluated as the occurrence of normal conductivity. Moreover the wires were investigated by optical microscopy and electron microprobe studies.

Results All the samples prepared according to the method of direct diffusion showed the existence of NbAu layers. With larger thicknesses the layer thickness in dependence on the time t does not correspond any longer to the normal dependence $d^2 = D\,2t$, at 1100 °C and a diffusion period of 100 h markedly lower values were found:

t	2	20	100 h
d D	3.2×10^{-6} m 7.0×10^{-16} m^2/s	11.0×10^{-6} m 8.0×10^{-16} m^2/s	15.0×10^{-6} m 2.7×10^{-16} m^2/s

This points to a diffusion retardation resulting from an exhaustion of the Au source: By means of electron microprobe studies it was stated that towards the core still another phase, the Nb_3Au_2 phase, arises (Fig. 1); pure gold could not be found in the core of this sample. Furtheron from the electron microprobe analysis it is found that the tetragonal phase Nb_3Au_2 has an exact stoichiometric composition, whereas the A15 phase Nb_3Au has a phase range of 21 to 26.3 at%, that means a phase width of 5.3 at% (Fig. 1). This value is only a little smaller than that of 7 at% found by Röschel et al. /6/.

By means of the method of selective diffusion no A15-layers and also no Nb_5Au_3 phase are detectable by optical microscopy; obviously their thickness lies below the resolution limit. With the two diffusion methods nearly the same transition temper-

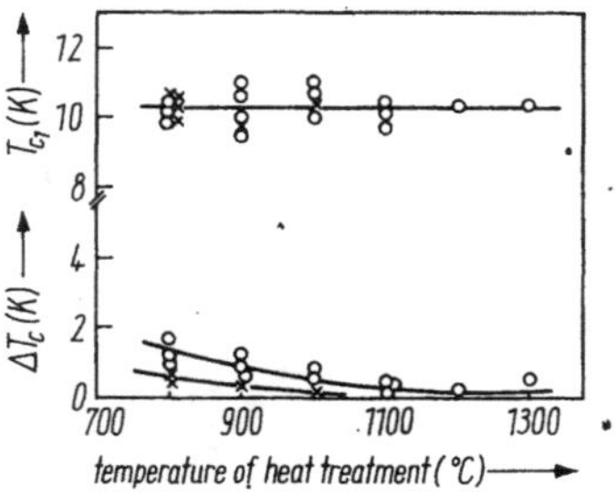

Fig. 2. Transition temperatures and transition widths of Nb-Au diffusion wires produced according to the methods of direct ($Au/Nb \rightarrow Nb_3Au$) and selective ($CuAu/Nb \rightarrow Nb_3Au$) diffusion

atures T_{c1} were obtained, merely for the case of selective diffusion the transition width to the complete superconductivity (ΔT_c) is a little smaller (Fig. 2). The measured transition temperatures of about 10.3 K are comparable with those ones usually obtained with samples produced by melting or sputtering and having stoichiometric composition. Thus also from the T_c measurements it results that by diffusion processes A15-layers with nearly stoichiometric composition are obtained.

The measured critical current densities in fields up to 12 T are represented for some typical examples in Fig. 3. The critical current densities for the samples produced by direct diffusion are still very low ($j_c \approx 10^4$ A/cm^2 in a field of 4 T) compared with Nb_3Sn strips, for example. The current densities for the samples produced according to the method of selective diffusion are expected to be essentially higher; since, however, no layer thickness could be measured, in Fig. 3 only the critical current strengths I_c are indicated. Here the I_c values are higher in any case than those of samples produced by direct diffusion under comparable conditions. By optimizing the heat treatment some essentially higher current strengths should be attainable. The curves in Fig. 3 already point to high critical magnetic field strengths H_{c2}. Measurements of the upper critical field at 4.2 K yielded the following values for samples produced by selective diffusion:

heat treatment	H_{c2} (T)
800 °C/20 h	15.3
800 °C/100 h	17.0
900 °C/20 h	13.4
900 °C/100 h	13.8

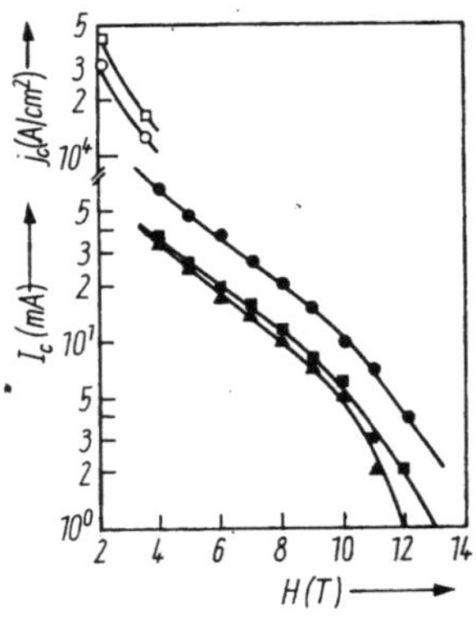

Fig. 3. Critical current strength and critical current density of Nb_3Au diffusion wires at 4.2 K; direct diffusion: ○ 900 °C/100 h, □ 1000 °C/2 h; selective diffusion: ■ 800 °C/20 h, ● 800 °C/100 h, ▲ 900 °C/20

By means of the dependence formulated by Karkin et al. /8/

$$H_{c2}(0) = H_{c2}(T)/[1 + (T/T_c)^2]/[1 - (T/T_c)^2]$$

the upper critical field at 0 K may be extrapolated to $H_{c2} \geqq 22$ T.

The special advantage of the Nb_3Au diffusion wires may be seen in the fact that with appropriate heat treatment besides high critical current strengths and high H_{c2} values the good conductivity of the residual Au or AuCu layers /9/ causes a stabilization of the superconducting current flow, if the formation of Nb_5Au_3 interfaces could be avoided by appropriate heat treatment. As to further diffusion couples, e.g. $Nb_3Sn/CuSn$ or $V_3Ga/CuGa$, complicated and expensive technologies are necessary to add copper for stabilization.

Acknowledgements

The author thanks V. Matthes and Dr. Vinzelberg for valuable help in performing the experiments and low temperature tests as well as Mr. Bitterlich for the electron microprobe analyses. Discussions with Dr. H.D. Wiesinger about stoichiometry of phases are acknowledged with pleasure.

References

/1/ R.D. BLAUGHER, R.E. HEIN, J.E. COX, and R.M. WATERSTRAT, J. low-Temp. Phys. 1, 539 (1969).

/2/ H.R. KHAN, E. RÖSCHEL, and CH.J. RAUB, Z. Phys. 262, 279 (1973).

/3/ E. BUCHER, F. LAVES, J. MÜLLER, and H. VON PHILIPSBORN, Phys. Letters 8, 27 (1964).

/4/ B.T. MATTHIAS, E.A. WOOD, E. CORENZWIT, and V.B. BALA, J. Phys. Chem. Solids 1, 188 (1956).

/5/ H.D. WIESINGER, phys. stat. sol. (a) 36, 61 (1976).

/6/ E. RÖSCHEL, O. LOEBICH, Jr., and C.J. RAUB, Z. Metallk. 64, 359 (1973).

/7/ P. MÜLLER and K. FISCHER, Freiberger Forschungshefte, to be published.

/8/ A.E. KARKIN, V.E. ARKHIPOV, B.N. GOSHCHITSKII, E.P. ROMANOV, and S.K. SIDOROV, phys. stat. sol. (a) 38, 433 (1976).

/9/ G.E.R. SCHULZE, Metallphysik, Akademie-Verlag, Berlin 1967.

(Received July 3, 1978)

phys. stat. sol. (a) 49, K81 (1978)

Subject classification: 18.2.1; 2; 21

Department of Physics K 203, Czechoslovak Technical University, Prague (a), Department of Microelectronics, Slovak Technical University, Bratislava (b), and Institute of Physics, Czechoslovak Academy of Sciences, Prague (c)

Contribution to the Technology of Gd-Co Films

By

J. ČERMÁK (a), V. TVAROŽEK (b), R. HARMAN (b), and J. KACZÉR (c)

Fabrication technique Amorphous films of $Gd_{1-x}Co_x$ (x = 0.75 to 0.81) were rf sputtered from a Gd_3Co_7 target. This target 152 mm in diameter was made from arc-melted alloy of the same composition, ground down to 10 μm particles and hydrostatically cold-pressed to the given shape. The resulting porosity of the target was 30%.

The sputtering apparatus used was a Perkin Elmer Ultek Randex 2400/8L equipped with a Veeco automatic pressure controller which assured a constant working pressure p = 1 Pa of argon during all experiments. The ultimate pressure of the system before letting in argon was approximately $2x10^{-5}$ Pa. The bias voltage was changed from -40 to -150 V in steps of 10 V, the target voltage 1 kV was constant. The power on the target was controlled by the power stabilizer and the reflected power never exceeded 1% of this value. The substrate-target distance was 40 mm. Under these conditions the deposition rate was between 540 and 600 nm/s.

Three types of substrates were used, Corning glass, Eutal, and Si wafers. Their cleaning before the deposition was carried out mechanically using a detergent solution and water and then they were ion-etched in the apparatus for 5 min.

The substrate was cooled through the water-cooled substrate holder but we were not able to measure the exact substrate temperature during film preparation. The deposition was carried out without substrate rotation.

After Gd-Co deposition a 0.2 μm thick layer of SiO_2 was deposited by rf sputtering without breaking the vacuum to passivate the surface of the magnetic films.

Measurement The film thickness was measured by a Varian Tolanski interferometer on a step made by photolitography and the density of the films was calculated from their weight. Fig. 1 shows the dependence between the measured film density

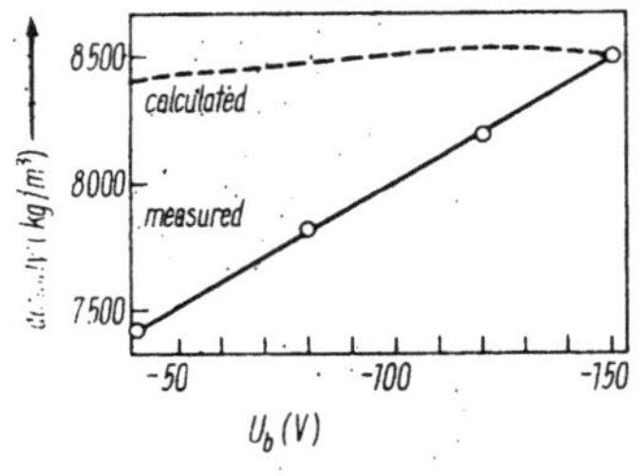

Fig. 1. Dependence of the film density on the applied bias voltage

and the applied bias voltage. The results of the density estimation were compared with the values calculated according to Tao et al. /1/ from the film composition and considerable discrepancies were obtained. The composition of the films was evaluated from electron microprobe measurements calibrated by magnetic momentun values. The saturation magnetization ($4\pi M_s$) was measured by a vibrating sample magnetometer. The results of $4\pi M_s$ versus the bias voltage and $4\pi M_s$ versus the film composition are summarized in Fig. 2. The film structure was observed by X-ray diffraction and no evidence of crystallization was detected.

Discussion The observed dependence of the film composition versus the applied bias voltage, expressed in Fig. 3, cannot be explained in terms of the expression suggested by Chaudhari et al. /2/,

$$(Co/Gd)_{film} = \frac{(Co/Gd)_{target}}{1 - K_1 U_b/U_t} .$$

A non-monotonous behaviour observed at the highest bias voltages U_b can be obtained by adding a second-order term in the denominator of the above expression (U_t is the target voltage):

$$(Co/Gd)_{film} = \frac{(Co/Gd)_{target}}{1 - K_1 U_b/U_t + K_2(U_b/U_t)^2} ,$$

where K_1 and K_2 are constants of the given system (in our case they reach the values 8.0 and 34, respectively). This cannot be explained by a constant sputtering yield of the individual components for the resputtering from the substrate. There are a few mechanisms which can affect the sputtering yields:

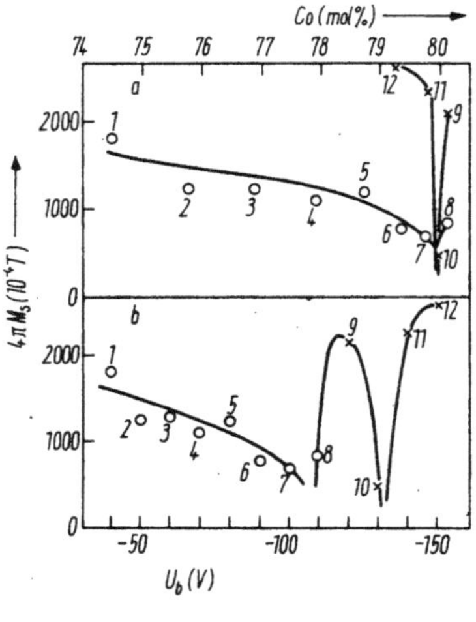

Fig. 2. Saturation magnetization $4\pi M$ of the films versus a) film composition and b) applied bias voltage

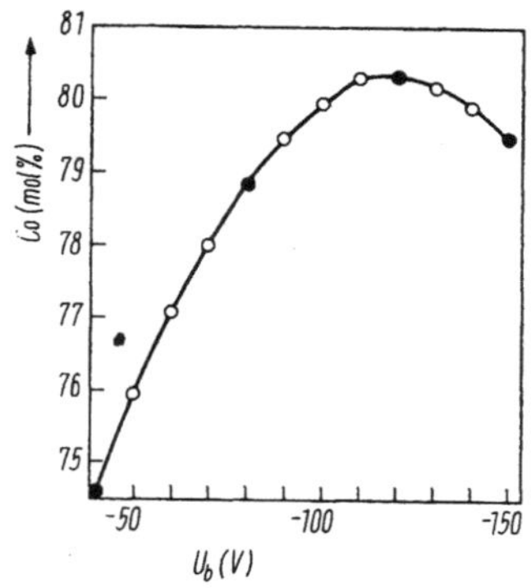

Fig. 3. The film composition as a function of the applied bias voltage. ● measured, ○ calculated with quadratic correction

a) The increase in the substrate temperature (increase of Gd vapour pressure) with increasing bias voltage cannot account for the observed dependence in Fig. 3 which shows an increase in Gd content in the film.

b) Oxidation effects appertain in the first place to the gadolinium oxide formation due to higher gadolinium reactivity in comparison with cobalt. The dependences of the film density on the bias (Fig. 1) and the saturation magnetization versus the bias (Fig. 2) confirm the fact that the gadolinium oxide content in the film decreases with increasing bias owing to a higher oxygen resputtering. Therefore the influence of the gadolinium oxides is minimum at the highest bias voltage and thus oxidation effects cannot explain the anomalous behaviour of the film composition (Fig. 3).

c) The results of saturation magnetization measurements displayed in Fig. 2 present two types of their dependences on the film composition, the first for the lower bias and the second for the higher ones, i.e. for bias voltages higher than -115 V. The fact that twodifferent dependences of the magnetic moment are observed for the same composition, demonstrates the existence of two types of structures, which grow under different sputtering conditions. These structure differences, caused by differing bias voltages, could affect the changes of resputtering rates of the individual components. This hypothesis, however, must be verified by further investigations of the film structure. Our X-ray diffraction did not distinguish differences in the films.

References

/1/ L.J. TAO, R.J. GAMBINO, S. KIRKPATRICK, J.J. CUOMO, and H. LILIENTHAL, 19th 3M Conf., Boston, December 1973.

/2/ P. CHAUDHARI, J.J. CUOMO, R.J. GAMBINO, S. KIRKPATRICK, and L.J. TAO, 20th 3M Conf., San Francisco, December 1974.

(Received July 4, 1978)

phys. stat. sol. (a) 49, K85 (1978)

Subject classification: 10.1; 21

Department of Metallurgy and Materials Science, Imperial College, London

Activation Analysis of Plastic Deformation of Ni_3Al-10 at% Fe

By

J.R. NICHOLLS and R.D. RAWLINGS

The flow stress σ of most materials decreases with increasing temperature. The flow stress may be considered to consist of two components, i.e. $\sigma = \sigma^* + \sigma_\mu$, where σ^* and σ_μ are the stresses required for dislocations to surmount short-range and long-range obstacles, respectively. σ_μ normally only varies with temperature through the elastic constants, consequently the temperature dependence of the flow stress is mainly due to changes in the effective stress σ^*.

In contrast to the usual behaviour, the flow stress of the $L1_2$ intermetallic compounds, e.g. $Ni_3Al(\gamma')$ /1/, Ni_3Ga /2/, and Ni_3Ge /3/, increases with increasing temperature reaching a peak value at temperature T_p which depends on stoichiometry, alloying additions, and orientation. Electron microscopy and mechanical properties measurements have revealed that five different modes of deformation occur in γ', each mode baving characteristic dislocation substructures, work-hardening rates, etc. /4/. Three of the modes (1, 2, and 3) occur at temperatures below T_p, one (4) in the region of T_p, and one (5) above T_p.

The temperature dependence of the flow stress of the γ'alloy Ni_3Al-10 at% Fe used in this investigation is given in Fig. 1. Note that even though ingots A and B were cast from the same master alloy their flow stresses differed in the temperature range 500 to 1000 K. This paper reports the results of applying the technique of activation analysis to Ni_3Al-10 at% Fe in order to determine σ^*, σ_μ, and the activation volume V^* over the temperature range 77 K to nearly T_p. Strain-rate cycling and stress relaxation tests were performed on polycrystalline compression specimens cut from the two ingots of nominally the same composition. Full experimental details are given in /5/.

The two most commonly used equations for the time (t) dependent decay of the applied stress during stress relaxation are

$$\sigma - \sigma_0 = \lambda \ln a + \lambda \ln(t + a) \text{ (Feltham /6/)}$$

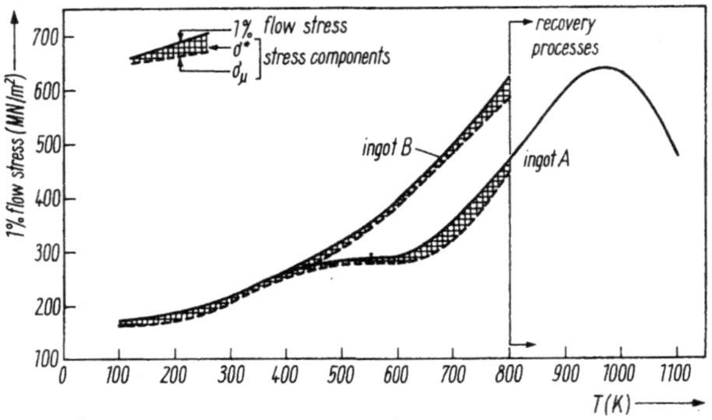

Fig. 1. The temperature dependence of the 1% flow stress, and the components σ_μ and σ^*, for two ingots (designated A and B) of Ni_3Al-10 at% Fe

and

$$\sigma - \sigma_\mu = K(t + a)^{-n}, \text{ (Li /7/)} ,$$

where σ_o is the flow stress at t = 0 and K, a, and n are constants. $\lambda = 2\,kT/V^*$ where k is Boltzmann's constant and T is temperature. It can be shown /5/ that a plot of $-d\sigma/d\ln(t + a)$ against σ should have a slope of n if Li's equation is obeyed and a slope of zero if Feltham's equation is applicable. The function $-d\sigma/d\ln(t + a)$ was found to be independent of stress at all temperatures and strains, thus it may be concluded that stress relaxations in γ' are best described by the Feltham equation.

At long times the applied stress relaxes to σ_μ. A linear regression computer program was used to extrapolate, according to Feltham's equation, the experimental data for times up to 1.2 ks to 100 ks and hence to obtain estimates for σ_μ. Long time relaxation tests at room temperature showed this procedure to be satisfactory in that good agreement between the value of σ_μ by extrapolation (345.8 MN/m^2) and the fully relaxed (346.2 MN/m^2) were obtained at 5% strain. The results of the analyses of the relaxations are presented in Fig. 2 and 3; values for the activation parameters could not be obtained at temperatures in excess of 800 K due to recovery during the relaxation. In these figures the effects of strain are represented by 1, 3, and 5% strain contours and the different response of ingot B, at the highest temperatures, as broken lines. The form of the stress-strain curves and the strain/temperature dependence and magnitude of the parameters σ^*, σ_μ, and V^* indicated that a number

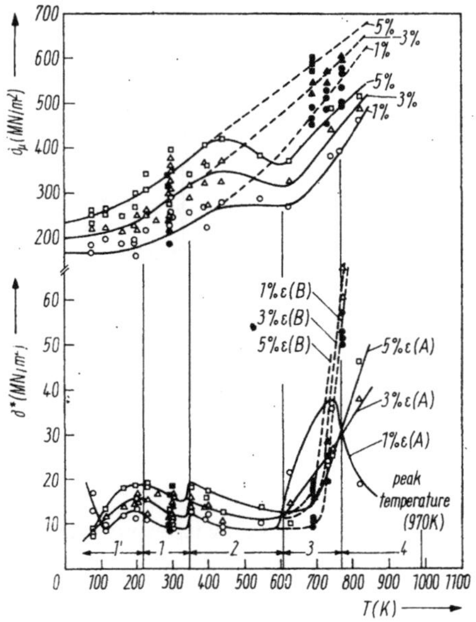

Fig. 2. Graph showing the temperature and strain dependence of σ_μ and σ^*. 0 to 2% strain: ○ ingot A, ● ingot B; 2 to 4% strain: △ ingot A, ▲ ingot B, 4 to 6% strain: □ ingot A, ■ ingot B

of modes of deformation marked 1' to 4 on Fig. 2 and 3, were operative over the temperature range of the investigation. Modes 1, 2, 3, and 4 correspond to the similarly numbered modes previously reported in nickel-rich Ni_3AlTi /4/ whereas mode 1' is considered to be a new low temperature deformation mode.

Previous workers /8, 9/ have reported different types of strain-rate response in γ', in the present work six responses were observed which could be correlated with the various deformation modes (Fig. 4). A number of methods for analysing strain-rate cycling data to obtain σ^* and σ_μ were considered but were found to be invalid due to various critical assumptions not being applicable to γ', e.g. the analysis proposed by Cheng /10/ requires σ_μ to be independent of strain. Consequently only V^* values were determined from the strain-rate cycling results and, as shown in Fig. 3, these were in good agreement with those determined by stress relaxation.

V^*, σ^*, and σ_μ are strain and temperature dependent and each mode of deformation, designated 1' to 4 in Fig. 2 and 3 has characteristic dependences and magnitudes of these parameters. The temperature dependence of V^* is similar to that which might be expected if there were one thermally activated mechanism below 400 K (corresponding approximately to modes 1' and 1), a region of athermal behaviour between 400 and 600 K (700 K for ingot B) and the onset of another thermally activated mechanism at higher temperatures (mode 3). However, it is apparent from Fig. 2 that this is too simple an explanation as over the complete temperature range there is a significant σ^*. Furthermore, only modes 1 and 2 follow the usual behaviour with σ^* decreasing with increasing temperature and even for these modes the trends

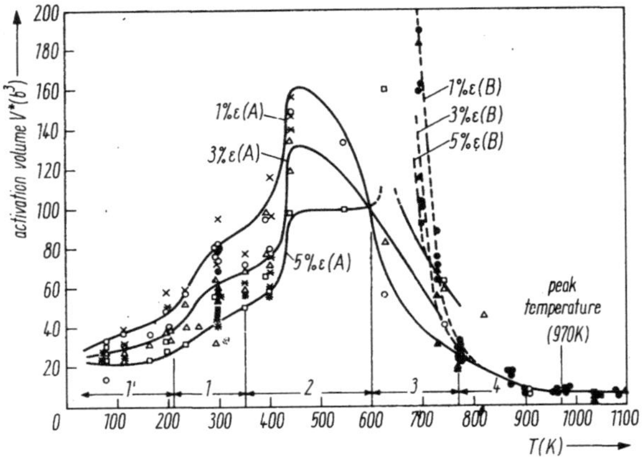

Fig. 3

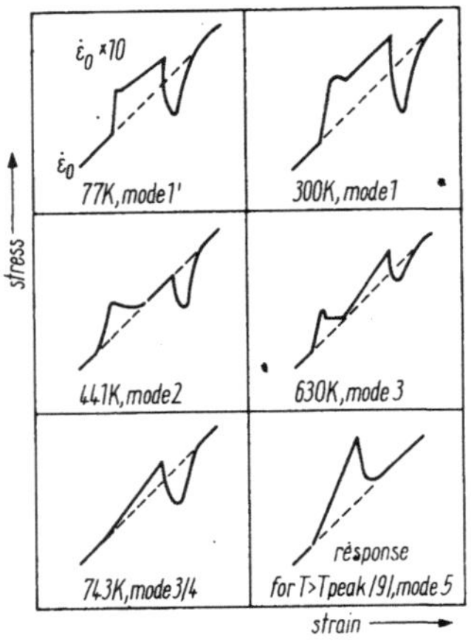

Fig. 4

Fig. 3. Plot of activation volume against temperature (strain dependence shown as 1, 3, and 5% contours). Strain-stress relaxation 0 to 2%: ○ ingot A, ● ingot B; 2 to 4%: △ ingot A, ▲ ingot B; 4 to 6%: □ ingot A, ■ ingot B. Strain-rate cycling 0 to 2%: ×, 2 to 4%: ⋇, 4 to 6%: ✳

Fig. 4. Typical strain-rate cycling responses

in σ_μ are anomalous in that positive temperature dependences are observed. The short-range and long-range obstacles determining σ^* in modes 1 and 2 and σ_μ over the complete temperature range, respectively, must be changing in either type, number, or magnitude in a most usual way with temperature.

Due to the complexity of the situation, and the current state of development of activation analysis, the authors consider that attributing specific dislocation-obstacle mechanisms to each mode of deformation would be speculative. Nevertheless, combining the data from Fig. 2 clearly demonstrates that the flow stress peak in γ' is predominately the consequence of an anomalous increase in σ_μ (Fig. 1). In additio the difference in properties of ingots A and B indicates that the mechanisms respons for the positive temperature dependence of σ_μ are very sensitive to relatively small changes in composition and/or dislocation substructure.

References

/1/ R.D. RAWLINGS and A.E. STATON-BEVAN, J. Mater. Sci. 10, 505 (1975).

/2/ S. TAKEUCHI and E. KURAMOTO, Acta metall. 21, 415 (1975).

/3/ K. AOKI and O. IZUMI, J. Japan. Inst. Met. 39, 1282 (1975).

/4/ A.E. STATON-BEVAN and R.D. RAWLINGS, Phil. Mag. 32, 787 (1975).

/5/ J.R. NICHOLLS, Ph.D. Thesis, London University, 1974.

/6/ P. FELTHAM, Acta metall. 17, 1305 (1969).

/7/ J.C.M. LI, Canad. J. Phys. 15, 493 (1967).

/8/ P.H. THORTON, R.G. DAVIES, and T.L. JOHNSTON, Metallurg. Trans. 1, 107.

/9/ P. RANDALL, Ph.D. Thesis, London University, 1971.

/10/ C.Y. CHENG, Scripta metall. 4, 623 (1970).

(Received July 4, 1978)

phys. stat. sol. (a) 49, K91 (1978)
Subject classification: 1.3 and 5; 18.2; 21.1.1

Chemistry Division, Bhabha Atomic Research Centre, Bombay[1)]

Mössbauer and X-Ray Studies of $Fe_{1.67}Ge$

By

P. RAJ, S.K. KULSHRESHTHA, and J.V. YAKHMI

Introduction Results of several investigations using X-rays, polarised neutron diffraction, magnetisation, and Mössbauer technique of the β-phase ferromagnetic hexagonal compounds of the Fe-Ge system show an unusual scatter in the parameters such as the cation distribution, the magnetic moments, and the hyperfine field values. The crystal structure of this phase is reported to be $B8_2$ (Ni_2In type) with the space group $P6_3/mmc$ /1/, or having some distortion with the space group $P\bar{3}m1$ /2/. The values of the magnetic moments assigned to Fe atoms at different crystallographic sites vary widely /3/ and some authors /3, 4/ have even reported a non-colinear magnetic structure for the β-phase. The reported Mössbauer spectra in the magnetically ordered state have been resolved into two sextets /5, 6/ with larger hyperfine fields assigned either to 2(a) Fe atoms /5/ or to 2(d) Fe atoms /6/, as well as into three sextets /3, 7 to 9/ and even a larger number of sextets /10/. Further, in order to explain the origin of three sextets, some authors /7, 8/ have assumed the partial occupancy of 2(c) sites by Fe atoms, with vacancies both at 2(d) and 2(c) sites; whereas others /9/ believe that the vacancies exist at 2(a) sites only. We present our Mössbauer work and X-ray intensity analysis on $Fe_{1.67}Ge$ and conclude that it is the inadequate accounting of anisotropic hyperfine fields and next near-neighbour interactions which is responsible for the unusual scatter in the reported Mössbauer results.

Experimental results and discussion The sample used in this study was the same as utilised earlier for Differential Scanning Calorimetric (DSC) measurements /11/ which showed that the compound undergoes only a usual second-order transition at the magnetic ordering temperature, so that a large discontinuity in the Mössbauer parameters around T_c can be ruled out. X-ray measurements /12/ above and below T_c also did not show any discontinuity in the lattice parameters.

1) Bombay-400085, India.

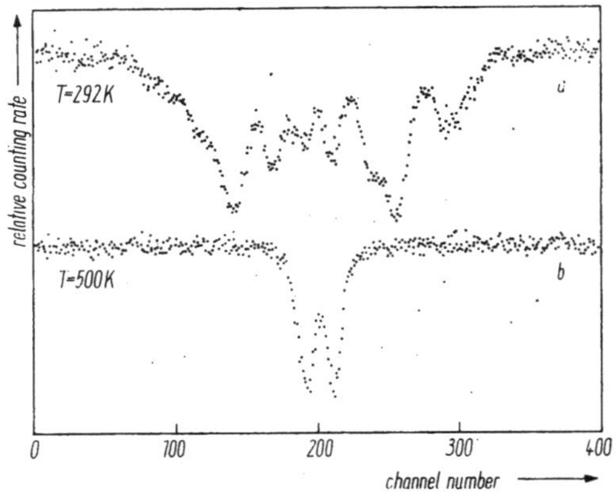

Fig. 1. Mössbauer spectrum of $Fe_{1.67}Ge$ at 292 K (a) and 500 K (b)

Two representative Mössbauer spectra of $Fe_{1.67}Ge$ with respect to ^{57}Co in Pd matrix, at 292 and 500 K are shown in Fig. 1a and b, respectively. The paramagnetic spectrum at 500 K appears to be a doublet with FWHM ≈ 0.4 mm/s and a separation of about 0.5 mm/s. If this doublet were to be resolved into two, three, or a larger number of component doublets, with comparable intensities, no appreciable differences in their IS and QS values can be expected. Our Mössbauer spectrum in the magnetic state (Fig. 1a) is similar to the spectra reported by previous authors /8 to 10/ and from its shape the possibility of resolving into two component spectra can be completely ruled out. Our attempts to fit this spectrum into three sextets with IS and QS values compatible with the paramagnetic range did not succeed. We would like to point out that wherever the IS and QS values of the component spectra in the ferromagnetic state have been explicitly reported /8 to 10/, they are widely different from their paramagnetic values. This implies that either there is a definite structural change occurring at T_c (which is ruled out by DSC /11/ and X-ray studies /12/) or that the logic of fitting the ferromagnetic spectrum into three sextets is erroneous. We have experimentally measured the X-ray line intensities and have also calculated them assuming the atomic distribution proposed by (i) Yasukochi et al. /1/ and (ii) Forsyth and Brown /2/. As seen in Table 1, the agreement between the observed and calculated line intensities is quite good for the $P6_3/mmc$ space group with the atomic distribution as suggested by Yasukochi et al. /1/. However, the line intensities calculated for the $P\bar{3}m1$ space group, with atomic distribution as given by Forsyth and Brown /2/, do not agree satisfactorily with the observed ones, particularly for one of the two most intense lines, (102). Therefore, the magnetic moment values, which are derived on the basis of the latter structural framework /2/ and which also forms the basis of assigning the lower hyperfine field to 2(a) sites /7, 8/, cannot be taken as correct.

Table 1.

X-ray intensity comparison for the two space groups using FeK_{α} radiation

reflection (hkl)	observed intensity	calculated intensity ($P6_3/mmc$)	calculated intensity ($P\bar{3}m1$)
101	18	15	15
002	8	6	7
102	100	93	52
110	96	100	100
202	35	40	50

Further, in order to ascertain whether the lattice vacancies are associated with 2(d) sites /1, 4/ or with 2(a) sites /9/, we carried out the computation of the relative intensities of X-ray peaks by assuming various atomic distributions in the $P6_3/mmc$ space group. It is found that for a fixed 2(a) site occupancy, on shifting part of the 2(d) Fe atoms to 2(c) sites, the intensity of the (101) reflection alone reduces such that it becomes comparable to the (002) reflection. This is contrary to the observed behaviour (Table 1). Thus the assumption of 2(c) sites being occupied by Fe atoms /3, 7, 8/ can be ruled out. Further, by computing the intensity of the (110) reflection, the existence of lattice vacancies at the 2(a) site alone /9/ is also ruled out. It is thus concluded that Fe atoms exist only at two crystallographic sites, 2(a) and 2(d), with vacancies at 2(d) sites only. Therefore, the basis of resolving the composite spectra into three sextets is not justified. The complex shape of Mössbauer spectra in the ordered state can, however, be understood in the following manner.

From the point symmetry it is clear that if there were no vacancies at 2(d) sites, the principal efg axis for both 2(a) as well as 2(d) sites would lie along the c-axis, and with the magnetisation in the basal plane only two sextets should be observed. However, due to the random vacancies at 2(d) sites the orientation of principal efg axes with respect to magnetisation direction not only depends on their number but also on their relative positions. This gives rise to a large number of combinations of asymmetry parameter η and the angles θ and φ. Even the magnitude of hyperfine field depends on the orientation of magnetisation with respect to the principal efg axes. This arises due to anisotropic hyperfine interactions as observed in the case of cubic Laves phase RFe_2 compounds /13/ and hexagonal $(F_{1-x}Mn_x)Sn$ system /14, 15/, wherein crystallographically equivalent Fe atoms experience different hyperfine fields ($\Delta H \approx$ 20 to 30 kOe). We believe that a complete neglect of anisotropic hyperfine fields and the next near-neighbour interactions by earlier authors, led to the unusual scatter in the reported Mössbauer parameters.

Acknowledgement

The authors are grateful to V.S. Jakkal for his help in computing the X-ray line intensities.

References

/1/ K. YASUKOCHI, K. KANEMATSU, and T. OHOYAMA, J. Phys. Soc. Japan 16, 429 (1961).
K. KANEMATSU, K. YASUKOCHI, and T. OHOYAMA, J. Phys. Soc. Japan 18, 1429 (1963).

/2/ J.B. FORSYTH and P.J. BROWN, Proc. Internat. Conf. Magnetism, Nottingham 1964 (p. 524).

/3/ J.M. DANIELS, F.E. MOORE, and S.K. PANDA, Canad. J. Phys. 53, 2428 (1975).

/4/ E. ADELSON and A.E. AUSTIN, J. Phys. Chem. Solids 26, 1795 (1965).
A.E. AUSTIN, J. appl. Phys. 40, 1381 (1969).

/5/ V.G. BHIDE and S.K. DATE, Solid State Commun. 5, 435 (1967).

/6/ E. GERMAGNOLI, C. LAMBORIZIO, S. MORA, and I. ORTALLI, Nuovo Cimento 42 B, 314 (1966).

/7/ H. YAMAMOTO, J. Phys. Soc. Japan 20, 2166 (1965).

/8/ S.C. BHARGAVA and P.K. IYENGAR, phys. stat. sol. (a) 38, 601 (1976).

/9/ I. ORTALLI and C. LAMBORIZIO, Nuovo Cimento, Lettere 12, 147 (1975).

/10/ G.A. FATSEAS, J.L. DORMAN, and L. BROSSARD, J. Physique 32, C 785 (1971).

/11/ J.V. YAKHMI, Ph.D. Thesis, Bombay University, 1975.

/12/ G. ALBANESE and M. MUSCI, phys. stat. sol. 30, K37 (1968).

/13/ G.K. WERTHEIM, V. JACCARINO, and J.H. WARNICK, Phys. Rev. 135, A151 (1964).

/14/ L. HAGGSTROM, T. ERICSSON, R. WAPPLING, and K. CHANDRA, Phys. Scripta 11, 47 (1975).

/15/ S.K. KULSHRESHTHA and P. RAJ, Proc. Internat. Conf. Mössbauer Spectroscopy, Bucharest 1977, in the press.

(Received August 1, 1978)

phys. stat. sol. (a) 49, K97 (1978)

Subject classification: 14.1; 21

Zentralinstitut für Festkörperphysik und Werkstofforschung der Akademie der Wissenschaften der DDR, Institutsteil für magnetische Werkstoffe, Jena (a), and Laboratoire Pierre Weiss, Université Louis Pasteur, Strasbourg (b)

On the Electrical Resistivity of PdH_x

By

W. ANDRÄ (a), H. DANAN (b), R. MATTHEIS (a), and U. ZEIMER (a)

As found by different authors /1 to 4/ the resistivity of hydrogenated palladium in the nonsuperconducting state depends in a characteristic manner on the hydrogen content x as plotted in Fig. 2. We try to explain some features of this curve by means of simple models using the following assumptions:

(i) In the region $0 \leqq x \leqq x_\beta$ the material is composed of two phases: the α-phase which contains a negligible amount of hydrogen x = 0 (at least for T $\leqq$ 300 K) /5/ and the β-phase with $x = x_\beta$. The growing phase consists of small cubes (II in Fig. 1) distributed in form of a simple cubic lattice. The resistivity of the total body is calculated as the resistivity of one unit cell according to the scheme of Fig. 1 where the contribution R_f of the phase boundaries (thickness d, resistivity ϱ_f) are taken into account.

(ii) In the region $x_\beta \leqq x \leqq 1$ we have only the β-phase and the resistivity ϱ obeys a law similar to that used by Burger et al. /1/ $\varrho \sim x(1 - x)$ due to scattering of electrons by protons and vacancies in the hydrogen sublattice /6/. However, this simple relation is only valid for a single phase metal. Therefore, the model is to be applied only to the β-phase as a solid solution of the two components $PdH_{x\beta}$ and H. As a consequence the limits $x = x_\beta$ and x = 1 are to be introduced instead of x = 0 and x = 1. Moreover, the difference $\Delta\varrho$ of the resistivities at $x = x_\beta$ and x = 1 may be taken into consideration. This leads to

$$\varrho(x) = \varrho_\beta + c(x - x_\beta)(1 - x) + \Delta\varrho(x - x_\beta)(1 - x_\beta)^{-1}, \tag{1}$$

where ϱ_β means $\varrho(x = x_\beta)$. The constant c may be determined by adjusting the maximum of $\varrho(x)$ to the experimental result.

(iii) The lattice constant of the β-phase at $x = x_\beta$ is larger than that of the α-phase /5, 7/ and increases linearly with x for $x > x_\beta$ resulting in a volume expansion.

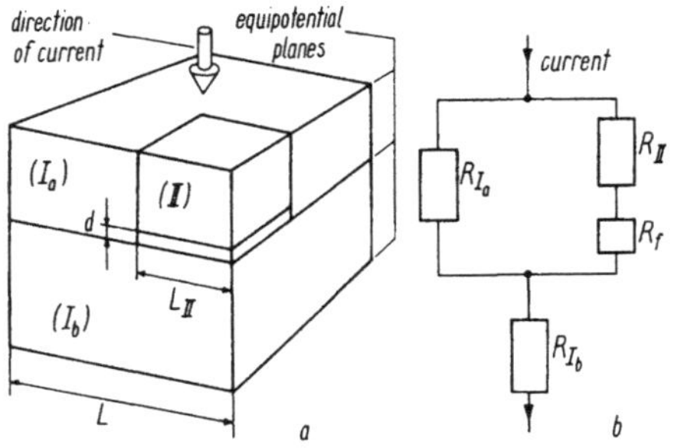

Fig. 1. Geometrical model of a) the two phase body and b) corresponding circuit diagram. For increasing x (hydrogenation) the volume II corresponds to the β-phase, for decreasing x (dehydrogenation) the volume II corresponds to the α-phase

We approximate the volume expansion in the whole concentration range by $V(x) = V_\alpha(1 + \eta x)$, where V_α means the volume before hydrogenation. η can be calculated by $\eta = (a_1/a_o)^3 - 1$, where a_o and a_1 are the lattice constants for x = 0 and x = 1, respectively. With the experimental values of a_o and a_1 we get $\eta = 0.18$. For the case of thin PdH_x films on a substrate this expansion results in an increase of the film thickness D larger than the corresponding thickness increase of a free film. The difference is due to the constraining action of the substrate and the finite compressibility of PdH_x. The thickness D(x) which can differ from D_α (thickness before hydrogenation) up to 18% can be calculated by /4/

$$D(x) = D_\alpha[1 + x\eta(1 + \mu)(3 - 3\mu)^{-1}] , \qquad (2)$$

where μ means Poisson's ratio.

In order to calculate $\varrho(x)$ for $0 \leq x \leq x_\beta$ we denote the relative volume portions of both phases by v_α and v_β with $v_\alpha + v_\beta = 1$. Because of the lattice expansion mentioned above the relative volume v_β of the β-phase is not simply proportional to x:

$$v_\beta = x(\eta + 1)(x_\beta + \eta x)^{-1} . \qquad (3)$$

According to Fig. 1 the resistivity e.g. during dehydrogenation is

$$\frac{\varrho(x)}{\varrho_\beta} = 1 - \frac{(P - 1 - Q)v_\alpha + Q\,v_\alpha^{4/3}}{(P - 1 - Q)v_\alpha^{2/3} + Q(v_\alpha - v_\alpha^{1/3}) + 1 + Q} \qquad (4)$$

with $P = \varrho_\beta/\varrho_\alpha$ and $Q = \varrho_f d_o/\varrho_\alpha L$. The resistance contribution of the phase boundary is introduced by $R_f = \varrho_f d/L_{II}^2$ neglecting the influence of boundaries parallel to the current direction (i.e. $\varrho_f \gg \varrho_\alpha, \varrho_\beta$). d as a function of L_{II} is approximated by $d = d_o L_{II}(L - L_{II})/L^2$.

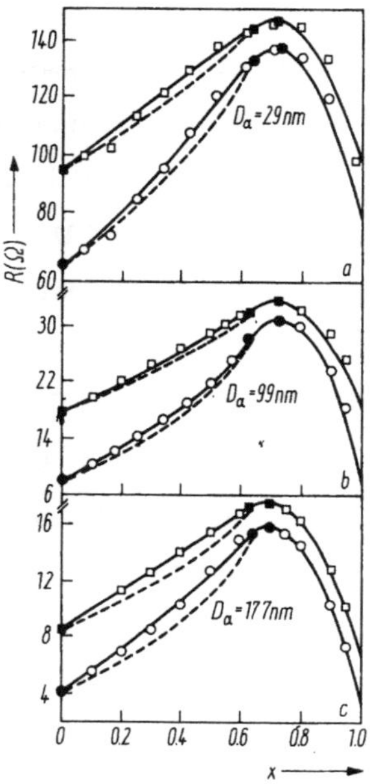

Fig. 2. Curves calculated by means of (1) to (5) and experimental points for T = 77 K (○) and T = 200 K (□) for dehydrogenation with $\mu = 0.4$ and $x_\beta = 0.635$. The curves are adjusted with the values of the black points. The broken curves are calculated with Q = 0 (no phase boundary contribution), the solid curves correspond to a) Q = 0.4 (77 K) and Q = 0.2 (200 K), b) Q = 0.3 (77 K) and Q = 0.3 (200 K), c) Q = 1.4 (77 K) and Q = 0.6 (200 K)

For hydrogenation the indices α and β are to be interchanged resulting in lower resistivity values in agreement with the experimental results /1, 4, 8/.

Since we try to compare the results with experimental values of the resistance R(x) of thin films measured parallel to the film plane we have to take into account the thickness variation (2) by using the relation

$$\frac{R(x)\,D(x)}{R_i\,D_i} = \frac{\varrho(x)}{\varrho_i} \tag{5}$$

where i stands for α and β, respectively.

The comparison of experimental results with the calculated curves (Fig. 2) shows:

(i) For $x > x_\beta$ there is satisfying agreement especially for the location of the maximum R values at x = 0.7 to 0.75 as reported by different authors /1, 2, 4/.

(ii) For $x < x_\beta$ the best agreement is obtained by taking into account the contribution of the phase boundary. Using the Q values given in Fig. 2, the dimension of our samples, and the measured values of R_α we get the following results for $\varrho_f d_o/L$ as a measure for the phase boundary contribution:

Table 1

D_α (nm)	29		99		177	
T (K)	77	200	77	200	77	200
$\frac{\varrho_f d_o}{L}$ (10^{-6} Ωcm)	3.6	2.9	1.3	2.8	5.2	4.7

References

/1/ J.P. BURGER, D.S. MAC LACHLAN, R. MAILFERT, and B. SOUFFACHE, Solid State Commun. 17, 281 (1975).

/2/ E. SCHMIDBAUER, Thesis, München 1967.

/3/ R.J. SMITH and D.A. OTTERSON, J. Phys. Chem. Solids 31, 187 (1970).

/4/ R. MATTHEIS, phys. stat. sol. (a) 45, 219 (1978).

/5/ E. WICKE and G.N. NERNST, Ber. Bunsenges. phys. Chem. 68, 224 (1964).

/6/ N.F. MOTT and G.H. JONES, The Theory of the Properties of Metals and Alloys Clarendon Press, Oxford 1936.

/7/ J.E. SCHIRBER and B. MOROSIN, Phys. Rev. B 12, 117 (1975).

/8/ D.P. SMITH, Hydrogen in Metals, The University of Chicago Press, Chicago (Illinois) 1948.

(Received August 1, 1978)

Pre-printed Titles

Papers to be published

in "physica status solidi (a) - applied research"

Vol. 49, No. 2

Original Papers

H.O.K. KIRCHNER, H. MUGHRABI, F. PRINZ, and H. SCHMID, Wien and Stuttgart

Dislocation Loops in α-Iron

K. BÄRNER and H. BERG, Göttingen

Elastic Anomalies at the Second-Order Phase Transition in MnAs

M. WILKENS and H. FÖLL, Stuttgart

The Black-White Vector $\vec{l}$ of Small Dislocation Loops on Transmission Electron Microscope Images

W. AUCH, W. STEUDLE, J.U. VON SCHÜTZ, and H.C. WOLF, Stuttgart

The Phase Transition in the 1:1 CT-Complex Anthracene/TCNB at 206 K

H.-J. KÖHLER, K. BLÜTHNER, S. LINKE, P. WEBER, and G. ALBRECHT, Jena

Some Properties of Miniature Nb-NbO_x-Pb Tunnel Junctions

W. ZAHN and F. DETTMANN, Jena

Resonance Effects in Quantum Interference Devices with Tunnel Junctions

V.N. BRUDNYI, D.L. BUDNITSKII, M.A. KRIVOV, R.V. MASAGUTOVA, V.D. PROCHUKHAN, and YU.V. RUD, Tomsk and Leningrad

The Electrical and Optical Properties of 2.0 MeV Electron-Irradiated $ZnGeP_2$

K. MAEDA and S. TAKEUCHI, Tokyo

Computer Simulation of Deformation in Two-Dimensional Amorphous Structure

N. RIHON, Liege
Energy Distribution Study of Field-Emitted Electrons from (100) and (111) Nickel Planes

I. TOMÁŠ, Prague
Interaction of Bubble Domains with a Concentric Hard Overlay

B. RABIN, H. TABATABAI, and P. SIFFERT, Strasbourg
Capacitance of Cadmium Telluride Schottky Diodes

D. JÄGER, E. KÄHLER, and R. HEIDEMANN, Münster
Negative Conductance in SCLC Diodes with Traps

C. CLEMEN, X.I. SALDANA, P. MUNZ, and E. BUCHER, Konstanz
Photovoltaic Properties of Some Semiconducting Layer Structures

R. RODLOFF and H. NEUHÄUSER, Braunschweig
Mean Velocity of Edge and Screw Dislocation Groups in Neutron-Irradiated Copper Single Crystals

E.M. SMOKOTIN, G.F. GUSYATSKII, L.M. PROTOPOPOVA, and A.M. KAPITONOV Krasnoyarsk
The Effect of Inhomogeneous Stress on the FMR in Ferrites

J. BAARS, D. BASSETT, and M. SCHULZ, Freiburg
Metal-Semiconductor Barrier Studies of PbTe

E. JOHNSON, J. FERRER, and L.T. CHADDERTON, Copenhagen
Radiolytic Radiation Damage of Sodalite

D. SHAW and E. WATSON, Hull
The Shift of Optical Absorption Edge in CdTe with Temperature and Doping Level

G. SARRABAYROUSE and J. BUXO, Toulouse
The Reversely Biased MIS Tunneling Schottky Barrier. Application to Photodetection

H.J. BLYTHE and L.H. DUNLEAVY, Sheffield
Low-Temperature Magnetic Relaxation in Neutron-Irradiated α-Iron

Papers to be published

in "physica status solidi (b) - basic research"

Vol. 89, No. 2

Original Papers

F.G. GÄRTNER and E. MOLLWO, Erlangen-Nürnberg
IR Absorption of OH and OD Centres and OH/OD, OD/OD, and OH/OD Complexes in Cu-Doped ZnO Single Crystals (I)

F.R. KESSLER and W. OLSOWSKI, Braunschweig
Exciton Contribution to Oscillatory Magnetophotoconductivity in Ge

H. HUMBERG and H. MERZ, Münster
Polarization Dependence of the Photoelectron Energy Distribution Curves of Cesium

M.V. EREMIN, V.N. KALINENKOV, and YU.V. RAKITIN, Kazan and Moscow
Two-Centre Exchange Interactions between Orbitally Degenerate Ions (II)

G.F. KVENTSEL and J. KATRIEL, Haifa
An Exactly Soluble Model of Singlet-Ground-State Ferromagnetism

V.M. KENKRE, Bombay
Model for Trapping Rates for Sensitized Fluorescence in Molecular Crystals

W. STÖSSEL and H.J. WAGNER, Karlsruhe
Comparison of Additional Boundary Conditions Based on Thermoreflectance Spectra of the $A_{n=1}$-Exciton in CdS Crystals at Oblique Incidence

R. VON BALTZ, Karlsruhe
Theory of the Anomalous Bulk Photovoltaic Effect in Ferroelectrics

S.W. KOCH, H. HAUG, G. SCHMIEDER, W. BOHNERT, and C. KLINGSHIRN, Frankfurt (Main) and Karlsruhe
Stimulated Intrinsic Recombination Processes in II-VI Compounds

G. SEIFERT, E. MROSAN, and H. MÜLLER, Dresden and Jena
Electronic Structure of 4d and 5d Transition Metal Clusters

J.A. GAJ, J. GINTER, and R.R. GAŁĄZKA, Warsaw
Exchange Interaction of Manganese $3d^5$ States with Band Electrons in $Cd_{1-x}Mn_xTe$

W.A. THEINER and H.P. GESERICH, Karlsruhe
Anderson Transition in $FeSi_2$ Films

R. SCHWABE, F. THUSELT, H. WEINERT, R. BINDEMANN, and K. UNGER, Leip
Luminescence from Electron-Hole Drops in Undoped GaP

S. ASANO, N. YAMASHITA, and Y. NAKAO, Okayama

Luminescence of the Pb^{2+}-Ion Dimer Center in CaS and CaSe Phosphors

H. FRIEDRICH, YU.V. OZEROV, R. STREHLOW, and A.N. GEORGOBIANI, Berlin and Moscow

On the Electron-Phonon and Exciton-Phonon Interaction in ZnTe

H. SATO, T. BABAUCHI, and K. YONEMITSU, Tokyo

Hall Coefficient of Dilute Aluminium Alloys (II)

B. PERTZSCH and U. RÖSSLER, Regensburg and Stuttgart

Stress Dependence of Shallow Acceptor States in Semiconductors with Zincblende Symmetry

S. EWERT, G. HEILAND, H. LÜTH, U. ROLL, and W. SANDER, Aachen

Molecular Vibrations in a Polymethine Dye Layer Studied by Inelastic Electron Tunneling Spectroscopy

A. MILCHEV, Leipzig

Frequency- and Wave-Vector-Dependent Dielectric Function of a Model Semiconductor

A.D. MOKRUSHIN and R.KH. SABIROV, Moscow

Positron Thermalization as Quantum-Statistical Process

K. ELLMER and J. STURM, Berlin

Energy-Angle Distributions of ^{14}N and ^{20}Ne Ions after Passage through Thin Carbon Foils

H.-D. MAIER, D. MÜLLER, and J. PETERSSON, Saarbrücken

Temperature and Frequency Dependence of the Dielectric Constant and Proton NMR near the Structural Phase Transition of Squaric Acid

E. WIMMER, P. WEINBERGER, A. KOSAKOV, and G. LEONHARDT, Wien and Leipzig

On the Electronic Structure of As_2O_3, Sb_2O_3, and As_2S_3

J. POŹNIAK, Wrocław

Dipolar Relaxation Induced by an AC Field in Various Crystallographic Directions of Me^{2+}-Doped Alkali Halide Crystals (I)

32914

Manuscripts and letters for physica status solidi (b) — basic research and physica status solidi (a) — applied research

should be sent to

DDR-102 Berlin, Neue Schönhauser Str. 20

or

Internationaler Buchversand GmbH, 1 Berlin 19, Eschenallee 21.

The object of "physica status solidi (b)" and "physica status solidi (a)" is to publish rapidly important and topical results concerning fundamental research in the field of solid state physics and aspects of application and experimental techniques of fundamental solid state research, respectively. The issues of "physica status solidi (b)" and "physica status solidi (a)" will be published on the 1st and the 16th day of each month, respectively. Either series will contain the following categories:

1. Review articles written by arrangement with the editors. They will be remunerated.
2. Original papers selected from those which have not been submitted for publication elsewhere, after being reviewed. (The latest date for acceptance of contributions is the 8th and the 23rd day of each month for "physica status solidi (b)" and "physica status solidi (a)", respectively, 50 days before the delivery of the journal.)
3. Short notes published in the off-set part of the journal. They are, if necessary after being reviewed, selected by the editorial staff. (The latest date for acceptance of manuscripts for this part is the 12th and the 27th day of each month for "physica status solidi (b)" and "physica status solidi (a)", respectively, 16 days before the delivery of the journal.)
4. Pre-printed titles of papers to be published in the next issue of both series.

Editorial Information

Manuscripts should be submitted preferentially in English. Besides, papers in German, French, and Russian are admitted. The final decision on the incorporation of the papers in the series (b) or (a) is taken by the editorial staff after consultation of the referee. Authors are asked to observe the following rules:

1. Manuscripts have to be written as concisely as possible. As most, 20 and 4 pages (28 lines per page; abstracts, tables, references, figures, and captions of figures included, where 3 figures are counted as about one page) are admitted for original papers and short notes, respectively. All manuscripts should be type-written on one side only. Two copies are required. Badly arranged manuscripts are returned.
2. The author's name, initials, and institution should be given.
3. One abstract of about ten lines in English and one in German, French, or Russian are required for original papers.
4. Because of lack of space conclusions (if any) must not have the character of a summary (i. e. substantially a repetition of the contents of the abstract).
5. Formulae should not be written too small and with pencil. Separate lines for formulae are desirable. Indistinct writing will lead to a return of the manuscript.
6. Drawings should be enclosed on separate sheets of paper, not exceeding the maximum and minimum sizes of 20 × 30 and 10 × 15 cm², respectively. Photographs should be submitted, on white paper, only for review articles and original papers. Duplicates are required.
7. Captions for the figures should be submitted on separate sheets of paper, and special instructions with regard to the size of illustrations should be given on these.
8. References should be numbered in order of appearance and should be given, on a separate sheet, in the form used in physica status solidi.
9. Units, symbols, etc. should be used in accordance with the recommendations given by the IUPAP.
10. Each paper should contain a proposal for the subject classification according to the classification scheme appearing on the second and third cover of each issue.

The authors will be sent galley proofs of review articles and original papers. Proof reading of short notes and pre-printed titles will be done by the editorial staff.

For each original paper and short note a total of 75 reprints is available free of charge. The number of reprints free of charge per review article is 25. Requests for additional reprints (against payment) should be given together with the submitted mansucript. **The deadline for orders of additional reprints is the 15th and the 30th of each month** for "physica status solidi (b)" and "physica status solidi (a)", respectively, two weeks before the delivery of the journal.